… , when … philosophical, … ual and practical frameworks of modern science arose, and it sees off relativism in the process. While dealing wonderfully in broad sweeps, it offers a wealth of entertaining details' Richard Joyner, *Times Higher Education*, Books of the Year

'Tremendously good' Steven Poole, *New Statesman*

'Fascinating and original . . . Wootton is a marvellous writer with an enviously encyclopaedic knowledge, and he has exciting things to say . . . a stimulating, well-informed and imaginative account' Patricia Fara, *Literary Review*

'David Wootton lucidly describes the individuals, the experiments and the controversies that marked this intellectually turbulent and transformative era. He offers perceptive and novel insights into the nature of science, what distinguishes it from other modes of thought, and why it has been so successful. This fascinating and scholarly book should receive a wide readership' Martin Rees, Astronomer Royal

'Clear, fascinating, ambitious . . . Wootton tells his tales well and portrays characters vividly. He writes with wit, and his book is full of surprises' Robert P. Crease, *Wall Street Journal*

'This is a superb book, at once cogent, revisionist and profound. It offers the most novel and significant account of the Scientific Revolution to appear for many years. Wootton's style is clear and trenchant, and he has a real gift for giving lucid explanations of complex concepts and issues. At its best it is simply rather brilliant' Michael Hunter, Professor of History at Birkbeck, University of London

'Wootton is a dazzling explicator of difficult ideas whose relish for his material is evident' Matthew Price, *Boston Globe*

'Extremely well researched and documented . . . This is bound to become a basic reference in the future' Adhemar Bultheel, the European Mathematical Society

ABOUT THE AUTHOR

David Wootto … ersity of York. His p … *dicine* (2006) and *Ga* … British Academy (200 … nedict Lectures at Bo … ysical Society (2016) …

Title page of Francis Bacon, *Novum organum* (1620), which shows a ship sailing in through the Pillars of Hercules (identified with the strait between Gibraltar and North Africa – the opening from the Mediterranean to the Atlantic) after exploring an unknown world.

DAVID WOOTTON

The Invention of Science

A New History of the Scientific Revolution

PENGUIN BOOKS

PENGUIN BOOKS

UK | USA | Canada | Ireland | Australia
India | New Zealand | South Africa

Penguin Books is part of the Penguin Random House group of companies
whose addresses can be found at global.penguinrandomhouse.com.

First published by Allen Lane 2015
Published in Penguin Books 2016
001

Set in 9/12 pt Sabon LT Std
Typeset by Jouve (UK), Milton Keynes
Printed in Great Britain by Clays Ltd, St Ives plc

A CIP catalogue record for this book is available from the British Library

ISBN: 978-0-141-04083-7

For Alison

Hanc ego de caelo ducentem sidera vidi
(I have seen her draw down the stars from the sky)
– Tibullus, *Elegies*, I.ii

'Eureka!'

– Archimedes (287–212 BCE)

Archimedes in his bath, a woodcut by Peter Flötner (1490–1546) from the first German translation of Vitruvius (*Vitruvius Teutsch*), published by Johannes Petreius in Nuremberg in 1548. Hiero's crown is on the right in the foreground.

Contents

CONCLUSION
The Invention of Science

List of Illustrations

ILLUSTRATIONS IN THE TEXT

PLATE SECTION

INTRODUCTION

> This is the Age wherein (me-thinks) Philosophy comes in with a Spring-tide; and the Peripateticks may as well hope to stop the Current of the Tide, or (with *Xerxes*) to fetter the Ocean, as hinder the overflowing of free Philosophy: Me-thinks, I see how all the old Rubbish must be thrown away, and the rotten Buildings be overthrown, and carried away with so powerful an Inundation. These are the days that must lay a new Foundation of a more magnificent Philosophy, never to be overthrown: that will Empirically and Sensibly canvass the *Phaenomena* of Nature, deducing the Causes of things from such Originals in Nature, as we observe are producible by Art, and the infallible demonstration of Mechanicks: and certainly, this is the way, and no other, to build a true and permanent Philosophy . . .
>
> – Henry Power, *Experimental Philosophy* (1664)

Modern science was invented between 1572, when Tycho Brahe saw a nova, or new star, and 1704, when Newton published his *Opticks*, which demonstrated that white light is made up of light of all the colours of the rainbow, that you can split it into its component colours with a prism, and that colour inheres in light, not in objects.[1] There were systems of knowledge we call 'sciences' before 1572, but the only one which functioned remotely like a modern science, in that it had sophisticated theories based on a substantial body of evidence and could make reliable predictions, was astronomy, and it was astronomy that was transformed in the years after 1572 into the first true science. What made astronomy in the years after 1572 a science? It had a research programme, a community of experts, and it was prepared to question every long-established certainty (that there can be no change in the heavens, that all movement in the heavens is circular, that the heavens consist of crystalline spheres) in the light of new evidence. Where astronomy led, other new sciences followed.

To establish this claim it is necessary to look not only at what happened between 1572 and 1704 but also to look backwards, at the world before 1572, and forwards, at the world after 1704; it is also necessary to address some methodological debates. Chapters 6 to 12, which deal with the core period 1572 to 1704, constitute the main body of this book; Chapters 3, 4 and 5 look primarily at the world before 1572, and Chapters 13 and 14 at the world both somewhat before and somewhat after 1704. Chapters 2, 15, 16 and 17 deal with historiography, methodology and philosophy.

The two chapters of the Introduction lay the foundations for everything that follows. The first chapter briefly suggests what the book is about. The second explains where the idea of 'the Scientific Revolution' comes from, why some think there was no such thing, and why it is a sound category for historical analysis.

1
Modern Minds

> Bacon, of course, had a more modern mind than Shakespeare: Bacon had a sense of history; he felt that his era, the seventeenth century, was the beginning of a scientific age, and he wanted the veneration of the texts of Aristotle to be replaced by a direct investigation of nature.
>
> – Jorge Luis Borges, 'The Enigma of Shakespeare' (1964)[1]

§ 1

The world we live in is much younger than you might expect. There have been tool-making 'humans' on Earth[i] for around 2 million years. Our species, *Homo sapiens*, appeared 200,000 years ago, and pottery dates back to around 25,000 years ago. But the most important transformation in human history before the invention of science, the Neolithic Revolution, took place comparatively recently, between 12,000 and 7,000 years ago.[2] It was then that animals were domesticated, agriculture began, and stone tools began to be replaced by metal ones. There have been roughly 600 generations since human beings first ceased to be hunter-gatherers. The first sailing vessel dates back to 7,000 years or so ago, and so does the origin of writing. Those who accept Darwin's theory of evolution can have no patience with a Biblical chronology which places the creation of the world 6,000 years ago, but what we may term historical humankind (humans who have left written records behind them), as opposed to archaeological humankind (humans who

i I use 'Earth' for the modern, Copernican conception of the Earth as a rotating terraqueous globe, which is one of the planets; 'earth' for the pre-Copernican conception of the world we inhabit, being made up of the element earth, which is stationary at the centre of the universe.

have left only artefacts behind them), has existed only for about that length of time, some 300 generations. Add the word 'great' in front of 'grandparent' 300 times: it will fill just over half a page of print. This is the true length of human history; before that there were two million years of prehistory.

Gertrude Stein (1874–1946) said of Oakland, California, that there was 'no there there' – it was all new, a place without history.[3] She preferred Paris. She was wrong about Oakland: human beings have lived there for 20,000 years or so. But she was also right: the living there was so easy that there was no need to develop agriculture, let alone writing. Domesticated plants, horses, metal tools (including guns) and writing arrived only with the Spanish after 1535. (California is exceptional – elsewhere in the Americas the domestication of maize goes back 10,000 years, as far as any other plant anywhere in the world, and writing goes back 3,000 years).

So the world we live in is almost brand new – older in some places than others but, in comparison to the 2 million years of tool-making history, box-fresh. After the Neolithic Revolution the rate of change slowed almost to a crawl. During the next 6,500 years there were remarkable technological advances – the invention of the water-wheel and the windmill, for example – but until 400 years ago technological change was slow, and it was frequently reversed. The Romans were amazed by stories of what Archimedes (287–212 BCE) had been able to do; and fifteenth-century Italian architects explored the ruined buildings of ancient Rome convinced that they were studying a far more advanced civilization than their own. No one imagined a day when the history of humanity could be conceived as a history of progress, yet barely three centuries later, in the middle of the eighteenth century, progress had come to seem so inevitable that it was read backwards into the whole of previous history.[4] Something extraordinary had happened in the meantime. What exactly was it that enabled seventeenth- and eighteenth-century science to make progress in a way that previous systems of knowledge could not? What is it that we now have that the Romans and their Renaissance admirers did not?[ii]

ii Daryn Lehoux, in a thought-provoking book, asks: 'Are there differences between ancient and modern science? Of course there are. Are those differences fundamental? Did things change suddenly? Can we pinpoint some radically new way of doing things that emerged at some discrete point in history when we got something we call modern science? I think not.' (Lehoux, *What Did the Romans Know?* (2012), 15.) Lehoux thus makes the opposite case to the one made here.

When William Shakespeare (1564–1616) wrote *Julius Caesar* (1599) he made the small error of referring to a clock striking – there were no mechanical clocks in ancient Rome.[5] In *Coriolanus* (1608) there is a reference to the points of the compass – but the Romans did not have the nautical compass.[6] These errors reflect the fact that when Shakespeare and his contemporaries read Roman authors they encountered constant reminders that the Romans were pagans, not Christians, but few reminders of any technological gap between Rome and the Renaissance. The Romans did not have the printing press, but they had plenty of books, and slaves to copy them. They did not have gunpowder, but they had artillery in the form of the ballista. They did not have mechanical clocks, but they had sundials and water clocks. They did not have large sailing vessels that could sail into the wind, but in Shakespeare's day warfare in the Mediterranean was still conducted by galleys (rowed boats). And, of course, in many practical ways, the Romans were much more advanced than the Elizabethans – better roads, central heating, proper baths. Shakespeare, perfectly sensibly, imagined ancient Rome as just like contemporary London but with sunshine and togas.[7] He and his contemporaries had no reason to believe in progress. 'For Shakespeare,' says Jorge Luis Borges (1899–1986), 'all characters, whether they are Danish, like Hamlet, Scottish, like Macbeth, Greek, Roman, or Italian, all the characters in all the many works are treated as if they were Shakespeare's contemporaries. Shakespeare felt the variety of men, but not the variety of historical eras. History did not exist for him.'[8] Borges' notion of history is a modern one; Shakespeare knew plenty of history, but (unlike his contemporary Francis Bacon, who had grasped what a Scientific Revolution might accomplish) he had no notion of irreversible historical change.

We might think that gunpowder, the printing press and the discovery of America in 1492 should have obliged the Renaissance to acquire a sense of the past as lost and gone for ever, but the educated only slowly became aware of the irreversible consequences that flowed from these crucial innovations. It was only with hindsight that they came to symbolize a new era; and it was the Scientific Revolution itself which was chiefly responsible for the Enlightenment's conviction that progress had become unstoppable. By the middle of the eighteenth century Shakespeare's sense of time had been replaced by our own. This book stops there, not because that is when the Revolution ends, but because by that time it had become clear that an unstoppable process of transformation had begun. The triumph of Newtonianism marks the end of the beginning.

§ 2

In order to grasp the scale of this Revolution, let us take for a moment a typical well-educated European in 1600 – we will take someone from England, but it would make no significant difference if it were someone from any other European country as, in 1600, they all share the same intellectual culture. He believes in witchcraft and has perhaps read the *Daemonologie* (1597) by James VI of Scotland, the future James I of England, which paints an alarming and credulous picture of the threat posed by the devil's agents.[iii] He believes witches can summon up storms that sink ships at sea – James had almost lost his life in such a storm. He believes in werewolves, although there happen not to be any in England – he knows they are to be found in Belgium (Jean Bodin, the great sixteenth-century French philosopher, was the accepted authority on such matters). He believes Circe really did turn Odysseus's crew into pigs. He believes mice are spontaneously generated in piles of straw. He believes in contemporary magicians: he has heard of John Dee, and perhaps of Agrippa of Nettesheim (1486–1535), whose black dog, Monsieur, was thought to have been a demon in disguise. If he lives in London he may know people who have consulted the medical practitioner and astrologer Simon Forman, who uses magic to help them recover stolen goods.[9] He has seen a unicorn's horn, but not a unicorn.

He believes that a murdered body will bleed in the presence of the murderer. He believes that there is an ointment which, if rubbed on a dagger which has caused a wound, will cure the wound. He believes that the shape, colour and texture of a plant can be a clue to how it will work as a medicine because God designed nature to be interpreted by mankind. He believes that it is possible to turn base metal into gold,

iii Since the typical well-educated European was male, I use masculine pronouns when writing about the early modern period; I do not do this when writing about our own intellectual life. Similarly, I use 'mankind' when describing early modern views; 'humankind' when expressing my own views. Women were denied membership in all early modern learned societies, but there were a number of significant female scientists, particularly astronomers (Schiebinger, *The Mind Has No Sex?* (1989), 79–101) and alchemists (Ray, *Daughters of Alchemy* (2015)). It has been claimed that Maria Cunitz's *Urania propitia* (1650), a volume of astronomical tables, is 'the earliest surviving scientific work by a woman on the highest technical level of its age' (Swerdlow, '*Urania propitia*' (2012), 81); the book included a foreword by her husband assuring readers that this really was a woman's work, implausible though this must seem. See also below, 28n, 226, 234n, 474 and 569.

although he doubts that anyone knows how to do it. He believes that nature abhors a vacuum. He believes the rainbow is a sign from God and that comets portend evil. He believes that dreams predict the future, if we know how to interpret them. He believes, of course, that the earth stands still and the sun and stars turn around the earth once every twenty-four hours – he has heard mention of Copernicus, but he does not imagine that he intended his sun-centred model of the cosmos to be taken literally. He believes in astrology, but as he does not know the exact time of his own birth he thinks that even the most expert astrologer would be able to tell him little that he could not find in books. He believes that Aristotle (fourth century BCE) is the greatest philosopher who has ever lived, and that Pliny (first century CE), Galen and Ptolemy (both second century CE) are the best authorities on natural history, medicine and astronomy. He knows that there are Jesuit missionaries in the country who are said to be performing miracles, but he suspects they are frauds. He owns a couple of dozen books.

Within a few years change was in the air. In 1611 John Donne, referring to Galileo's discoveries with his telescope made the previous year, declared that 'new philosophy calls all in doubt'. 'New philosophy' was a catchphrase of William Gilbert, who had published the first major work of experimental science for 600 years in 1600;[iv] for Donne, the 'new philosophy' was the new science of Gilbert and Galileo.[10] His lines bring together many of the key elements which made up the new science of the day: the search for new worlds in the firmament, the destruction of the Aristotelian distinction between the heavens and the earth, Lucretian atomism:

> And new Philosophy cals all in doubt,
> The Element of fire is quite put out;
> The Sunne is lost, and th'earth, and no mans wit
> Can well direct him, where to looke for it.
> And freely men confesse, that this world's spent,
> When in the Planets, and the Firmament
> They seeke so many new; they see that this
> Is crumbled out againe to his Atomis.
> 'Tis all in pieces, all cohaerence gone;
> All just supply, and all Relation:

iv The first since Ibn al-Haytham's *Book of Optics* (1011–21). For discussion of Gilbert, see below, pp. 61, 157–8, 304, 315 and 328–9.

Prince, Subject, Father, Sonne, are things forgot,
For every man alone thinkes he hath got
To be a Phoenix, and that then can bee
None of that kinde, of which he is, but hee.

Donne went on to mention the voyages of discovery and the new commerce that followed from them, the compass that made those voyages possible and, inseparable from the compass, magnetism, which was the subject of Gilbert's experiments.

How did Donne know about the new philosophy? How did he know that it involved Lucretian atomism?[v] Galileo had never mentioned atomism in print, although some who knew him claimed that, in private, he made clear his commitment to it; Gilbert had discussed atomism only to reject it. How did Donne know that the new philosophers were seeking new worlds, not only by thinking of the planets as worlds but also by looking for worlds elsewhere in the firmament?

In all likelihood Donne had met Galileo in Venice or Padua in 1605 or 1606.[vi] In Venice he had stayed with the English ambassador Sir Henry Wotton, who was busy trying to obtain the release of a Scotsman, a friend of Galileo, who had been imprisoned for having sex with a nun (a crime that was supposed to carry the death sentence). Perhaps Donne met and talked with Galileo, or with Galileo's English-speaking students; he certainly seems to have met Galileo's close friend Paolo Sarpi.[11] In England he may well have met Thomas Harriot, a great mathematician who was evidently attracted to atomism,[vii] and Gilbert too.[12] As well as, or instead of, Galileo's *Sidereus nuncius*, or *Starry Messenger* (1610), he may have read Kepler's *Conversation with Galileo's Starry*

v Lucretius (*c*.99–*c*.55 BCE) claimed that the universe has no design but is the result of the random interaction of unalterable and indivisible atoms, and that the present universe will eventually be destroyed and replaced – it is just one in an unending sequence of randomly generated universes. Lucretius's poem *On the Nature of Things* was lost during the Middle Ages; it was rediscovered in 1417 and first published in 1473 and there was no complete English translation in print until 1682. Lucretius was a follower of Epicurus (341–270 BCE). We use the word 'Epicurean' to mean someone who seeks physical pleasure but, in the Renaissance, Epicureans were materialists and atheists and consequently unable to acknowledge any good other than physical pleasure.

vi Galileo was living in Padua but frequently visited Venice; equally, Donne, when in Venice, would surely have visited Padua, where there was a significant English and Scottish community.

vii Harriot independently discovered what we now call Galileo's law of fall, and also what we now call Snell's law of refraction, but he never published. See also below, pp. 32, 91, 212, 215, 218, 220–1 and 302.

Messenger (1610), which contained lots of radical ideas about other worlds that Galileo had carefully avoided discussing.

There is another answer. Donne owned a copy of Nicholas Hill's *Epicurean* (which is to say Lucretian) *Philosophy* (1601).[13] That copy – now in the library of the Middle Temple, one of the Inns of Court in London – had previously been owned by his friend and Shakespeare's, Ben Jonson. It had originally been purchased by a fellow of Christ's College, Cambridge – its binding bears the college badge.[14] Its first owner had planned to study it with care, perhaps to write a refutation or a commentary, for it was bound with alternate blank pages on which notes could be made. The pages remained blank. Was it given to Jonson, or did he borrow it and keep it? Was it given to Donne in turn, or did he borrow it and fail to return it? We do not know. We know only that no one took Hill seriously. His book, it was said, was 'full of mighty words and no great matter'. It was 'humorous [i.e. whimsical] and obscure'.[15] The early references to him (in, for example, a satirical verse by Jonson) have more to do with farting than philosophy.[16] At some point before 1610 Donne composed a catalogue of a courtier's library; this was an extended joke, listing imaginary, ridiculous books, such as a learned tome by Girolamo Cardano, *On the Nothingness of a Fart*.[viii] The first entry is a book by Nicholas Hill on the sexing of atoms: how can one tell male from female? Are there hermaphrodite atoms?[ix]

Donne would have learnt from Hill about the possibility of life on other planets, and of planets circling other stars; he would also have learnt that these strange ideas derived from Giordano Bruno.[17] If he read Galileo's *Starry Messenger*, with its account of the moon as having mountains and valleys, Donne would surely have responded exactly as the great German astronomer Johannes Kepler did that spring when he read one of the first copies to arrive in Germany – he saw a remarkable vindication of Bruno's perverse theory that there might be life elsewhere in the universe. If Donne read Kepler's *Conversation* he would have found the link with Bruno spelled out.[18] Jokes about farts were now beside the point. The gathering recognition was too late for Bruno,

viii Brown, '*Hac ex consilio meo via progredieris*' (2008). The Elizabethans took farting very seriously: the Earl of Oxford allowed a fart to escape him when bowing to Queen Elizabeth; mortified, he went abroad for seven years, only to be greeted on his return by the Queen with the words: 'My Lord, I had forgot the fart.' (Trevor-Roper, 'Nicholas Hill, the English Atomist' (1987), 9.)

ix Having discussed with my neighbour in the country the difficulty of sexing her ducklings, I now know, as Donne surely did, that sexing can be far from straightforward.

who had been burnt alive by the Roman Inquisition in 1600; it was probably too late for Hill too, who, according to a later report, committed suicide in 1610, eating rat poison and dying blaspheming and cursing. He was in exile in Rotterdam: he had been caught up in a treasonous plot to prevent James VI of Scotland from succeeding Elizabeth I to the throne of England in 1603 and had fled abroad.[19] Then the death of his son, Lawrence, to whom he was devoted, made further living seem pointless. In 1601 he had chosen to dedicate his only publication not to some great man (there was rather a shortage of great men who wished him well) but to his infant son: 'At my age, I owe him something serious, since he, at his tender age, has delighted me with a thousand pretty tricks.' Hill may not have lived to know it, but suddenly in 1610 Epicurean philosophy had become 'something serious'. A revolution was beginning, and Donne, who only a few years before had mocked the new ideas, who had read Gilbert, Galileo and Hill and perhaps knew Harriot, was one of the first to understand that the world would never be the same again. So by 1611 the revolution was well under way, and Donne, unlike Shakespeare and most educated contemporaries, was fully aware of it.

But now let us jump far ahead. Let us take an educated Englishman a century and a quarter later, in 1733, the year of the publication of Voltaire's *Letters Concerning the English Nation* (better known under the title they bore a year later when they appeared in French, *Lettres philosophiques*), the book which announced to a European audience some of the accomplishments of the new, and by now peculiarly English, science. The message of Voltaire's book was that England had a distinctive scientific culture: what was true of an educated Englishman in 1733 would not be true of a Frenchman, an Italian, a German or even a Dutchman. Our Englishman has looked through a telescope and a microscope; he owns a pendulum clock and a stick barometer – and he knows there is a vacuum at the end of the tube. He does not know anyone (or at least not anyone educated and reasonably sophisticated) who believes in witches, werewolves, magic, alchemy or astrology; he thinks the *Odyssey* is fiction, not fact. He is confident that the unicorn is a mythical beast. He does not believe that the shape or colour of a plant has any significance for an understanding of its medical use. He believes that no creature large enough to be seen by the naked eye is generated spontaneously – not even a fly. He does not believe in the weapon salve or that murdered bodies bleed in the presence of the murderer.

Like all educated people in Protestant countries, he believes that the

Earth goes round the sun. He knows that the rainbow is produced by refracted light and that comets have no significance for our lives on earth. He believes the future cannot be predicted. He knows that the heart is a pump. He has seen a steam engine at work. He believes that science is going to transform the world and that the moderns have outstripped the ancients in every possible respect. He has trouble believing in any miracles, even the ones in the Bible. He thinks that Locke is the greatest philosopher who has ever lived and Newton the greatest scientist. (He is encouraged to think this by the *Letters Concerning the English Nation*.) He owns a couple of hundred – perhaps even a couple of thousand – books.

Take, for example, the vast library (a modern catalogue runs to four volumes) of Jonathan Swift, the author of *Gulliver's Travels* (1726). It contained all the obvious works of great literature and of history, but it also contained Newton, the *Philosophical Transactions* of the Royal Society for the Advancement of Natural Knowledge (the second scientific journal, the *Journal des sçavans*, began publication two months earlier), and Fontenelle's *Entretiens sur la pluralité des mondes* (1686). Indeed, Swift, for all his antagonism towards contemporary science (to which we will return in Chapter 14), was sufficiently familiar with Kepler's three laws of planetary motion to use them to calculate the orbits of imaginary moons around the planet Mars; his hostility was grounded in extensive scientific reading.[x][20] His world was one in which the culture of the elite was much more sharply distinguished from the culture of the masses than it had been in the past but also one in which science was not yet too specialized to be part of the culture of every educated person. Even in 1801 we can still catch Coleridge resolving that 'before my thirtieth year I will thoroughly understand the whole of Newton's works.'[21]

Between 1600 and 1733 (or so – the process was more advanced in England than elsewhere) the intellectual world of the educated elite changed more rapidly than at any time in previous history, and perhaps than at any time before the twentieth century. Magic was replaced by science, myth by fact, the philosophy and science of ancient Greece by something that is still recognizably our philosophy and our science, with the result that my account of an imaginary person in 1600 is

x Swift thought scientific research was a waste of time because it never led to any practical applications, a view forcefully expressed in Part III of *Gulliver's Travels* in his account of the airborne island of Laputa.

automatically couched in terms of 'belief', while I speak of such a person in 1733 in terms of 'knowledge'. The transition was of course still incomplete. Chemistry barely existed. Bleeding, purges and emetics were still used to cure disease. Swallows were still thought to hibernate at the bottom of ponds.[xi] But the changes of the next hundred years were to be far less remarkable than the changes of the previous hundred years. The only name we have for this great transformation is 'the Scientific Revolution'.

§ 3

On the evening of 11 November 1572, soon after sunset, a young Danish nobleman called Tycho Brahe was looking at the night sky. Almost directly above his head he noticed a star brighter than any other, a star that ought not to have been there. Afraid his eyes were playing some sort of trick on him, he pointed out the star to other people and established that they too could see it. Yet no such object could exist: Brahe knew his way around the heavens, and it was a fundamental principle of Aristotelian philosophy that there could be no change in them. So if this was a new object it must be located not in the heavens but in the upper atmosphere – it could not be a star at all. If it *was* a star then it must be a miracle, some sort of mysterious divine sign whose meaning urgently needed to be deciphered. (Brahe was a Protestant, and Protestants maintained that miracles had long ceased, so this argument was unlikely to persuade him.)

In all history, as far as Brahe knew, only one person, Hipparchus of Nicaea (190–120 BCE), had ever claimed to have seen a new star; at least, Pliny (23–79 CE) had attributed such a claim to Hipparchus, but Pliny was notoriously unreliable, so it was easy to assume that either Hipparchus or Pliny had made some sort of elementary mistake.[xii] Now

xi Towards the end of the century the great naturalist Gilbert White was still in two minds on the vexed question of migration versus hibernation: White, *Natural History* (1789), 28, 36, 64–5, 102, 138–9, 165, 167, 188. For a summary of a book which White cites (144), Carl D. Ekmarck's *Migrationes avium* (1757), see Griffiths, 'Select Dissertations from the *Amoenitates academicae*' (1781): Ekmarck held that some birds migrate but that swallows overwinter in ponds. His views are commonly attributed to Linnaeus, who examined his dissertation.

xii Brahe did not count the star of Bethlehem as a true star, for the Gospel of Matthew describes it as moving in the heavens. There had been an even brighter supernova in 1006, but there was no mention of it in the books known to him.

Brahe set about proving that the impossible had in fact occurred by showing, using elementary trigonometry, that the new star could not be in the upper atmosphere but must be in the heavens.[xiii] Soon it became brighter than Venus, and was briefly visible even by daylight, and then it slowly faded away over the course of sixteen months. It left behind a flurry of books in which Brahe and his colleagues debated its location and significance.[22] Also left behind was a research programme: Brahe's claims had caught the attention of the king of Denmark, who supplied Brahe with an island, Hven, and what Brahe later described as a ton of gold to fund the building of an observatory for astronomical research. As a result of his sighting of the new star Brahe was convinced that, if the structure of the universe was to be understood, much more accurate measurements must be made.[23] He designed new instruments, capable of an exquisite precision. When he realized that his observatory shook slightly in the wind, making his measurements imperfect, he moved his instruments into underground bunkers. Over the course of the next fifteen years (1576–91) Brahe's researches at Hven turned astronomy into the first modern science.[24] The nova of 1572 was not the cause of the Scientific Revolution, any more than the bullet which killed Archduke Franz Ferdinand on 28 June 1914 was the cause of the First World War. Nevertheless, the nova marks, quite precisely, the beginning of the Revolution, as the death of the archduke marks the beginning of the war. For the Aristotelian philosophy of nature could not be adapted to incorporate this peculiar anomaly; if there could be such a thing as a new star, then the whole system was founded on false premises.

Brahe had no idea what he was starting as he fretted over the new star that is now named after him – 'Tycho's nova' – and which can still be located in the constellation Cassiopeia, although only with a radio telescope. But since 1572 the world has been caught up in a vast Scientific Revolution that has transformed the nature of knowledge and the capacities of humankind. Without it there would have been no Industrial Revolution and none of the modern technologies on which we depend; human life would be drastically poorer and shorter and most of us would live lives of unremitting toil. How long it will last, and what

xiii Thomas Kuhn thought that, but for Copernicus, Brahe would not have been able to grasp that the new star was in the heavens (Kuhn, *Structure* (1970), 116), although Copernicus had nothing to say about supralunary change, and Brahe was no Copernican. Kuhn's claim is at odds with his broader argument that scientists can identify anomalies, but it is significant that Brahe lived in a culture in which long-established certainties (in religion, for example) were being questioned and overthrown.

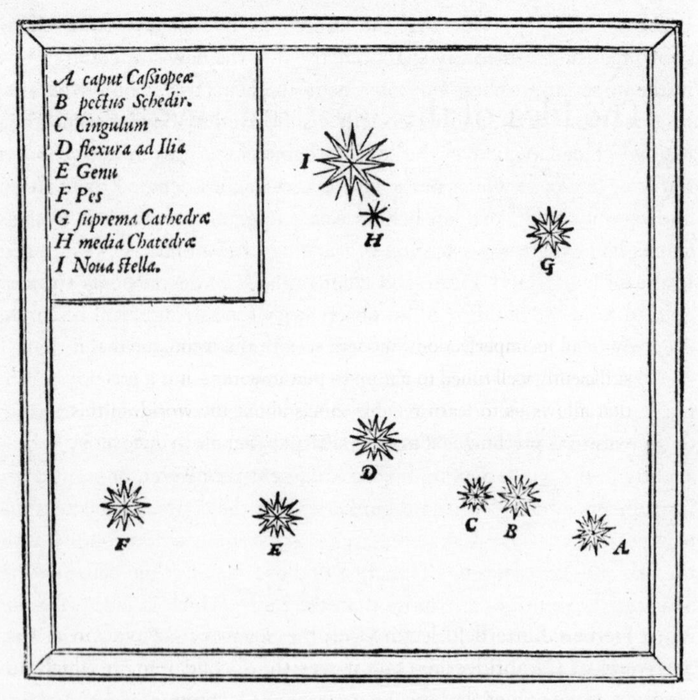

Star map of the constellation Cassiopeia, showing the position of the supernova of 1572 (the topmost star, labelled I); from Tycho Brahe's *The New Star* (1573).

its consequences will be, it is far too soon to say; it may end with nuclear war, or ecological catastrophe, or (though this seems much less likely) with happiness, peace and prosperity. Yet although we can now see that it is the greatest event in human history since the Neolithic Revolution, there is no general agreement on what the Scientific Revolution is, why it happened – or even whether there was such a thing. In this respect the Scientific Revolution is quite unlike, for example, the First World War, where there is general agreement on what it was and a fair amount of agreement on why it happened. An ongoing revolution is a nuisance for historians: they prefer to write about revolutions that happened in the past – when, in reality, this one is still continuing all around us. As we shall see, much of the disagreement on this subject is the result of elementary misconceptions and misunderstandings; once they are cleared out of the way it will become apparent that there really is such a thing as the Scientific Revolution.

2

The Idea of the Scientific Revolution

> With all its imperfections, modern science is a technique that is sufficiently well tuned to nature so that it works – it is a practice that allows us to learn reliable things about the world. In this sense it is a technique that was waiting for people to discover it.
>
> – Steven Weinberg, *To Explain the World* (2015)[1]

§ 1

When Herbert Butterfield lectured on the Scientific Revolution at the University of Cambridge in 1948 it was the second year in which an historian at the university had given a series of lectures on the history of science: he had been preceded the year before by the Regius Professor of History, G. N. Clark, an expert on all things seventeenth century, and the medieval historian M. M. Postan had lectured immediately before Butterfield. It was in Cambridge that Isaac Newton (1643–1727) had written his *Philosophiæ naturalis principia mathematica*, or *Mathematical Principles of Natural Philosophy* (1687), and here that Ernest Rutherford (1871–1937) had split the atomic nucleus for the first time, in 1932. Here, the historians were acknowledging, they were under a particular obligation to study the history of science. They were also keen to insist that the history of science be done by historians, not by scientists.[i 2]

The historians and the scientists at Cambridge shared a common education: Latin was a compulsory entrance requirement. They met over lunch and dinner in their colleges, but they lived in separate mental

i Cambridge, England, was behind Cambridge, Massachusetts: at Harvard, George Sarton gave his first lecture course in the history of science in 1917 and became professor of the history of science in 1940.

worlds. Butterfield began the book based on his lectures, *The Origins of Modern Science* (1949), by expressing the hope that the history of science might serve as a long-needed bridge between the arts and the sciences. He hoped in vain. In 1959 (the year in which Latin was finally dropped as an entrance requirement) C. P. Snow, a Cambridge chemist and a successful novelist, delivered a lecture complaining that Cambridge dons from the sciences and the arts had now more or less stopped speaking to each other.[ii] It was entitled 'The Two Cultures and the Scientific Revolution' – the revolution being Rutherford's revolution, which had led to the creation of the atomic bomb.[3]

In adopting the term 'the Scientific Revolution' a decade before Snow, Butterfield was (it is always said) following the example of Alexandre Koyré (1892–1964).[4] Publishing in French in 1935, Koyré (a German-educated Russian Jew who had been imprisoned in Tsarist Russia at the age of fifteen for revolutionary activity, had fought for France in the First World War, would join the Free French in the Second, and would later become a leading figure in American history of science) distinguished the Scientific Revolution of the seventeenth century, which ran from Galileo to Newton, from 'the revolution of the last ten years'; Heisenberg's classic paper on quantum mechanics had been published exactly ten years before.[iii] For Koyré and Butterfield it was physics, the physics first of Newton and then of Albert Einstein (1879–1955), which symbolized modern science. Now we might give equal prominence to biology, but they were writing before the discovery of the structure of DNA by James Watson and Francis Crick in 1953. As Butterfield was

ii In the years since Snow's lecture the two-cultures problem has deepened; history of science, far from serving as a bridge between the arts and the sciences, nowadays offers the scientists a picture of themselves that most of them cannot recognize. It has become part of the problem, not part of the solution.

iii Koyré, *Études Galiléennes* (1966), 12 (where *révolution* is used as being equivalent to Gaston Bachelard's term *mutation*). Heisenberg, '*Über quantentheoretische Umdeutung kinematischer und mechanischer Beziehungen*' (1925) is the paper which founded modern quantum mechanics; it led to the publication of Schrödinger's equation (which describes how the quantum state of a physical system changes with time) in January 1926 and the formulation of Heisenberg's uncertainty principle (the more precisely the position of some particle is determined, the less precisely its momentum can be known, and vice versa) in 1927. The date on the first edition of Koyré's *Études Galiléennes* is 1939 (although the actual publication was in April 1940 – Costabel, '*Sur l'origine de la science classique*' (1947), 208 – and 1940 was the date Koyré himself sometimes used); consequently, nearly all commentators date Koyré's use of the term 'scientific revolution' to 1939. However, the first essay had already appeared in print in 1935: Murdoch, 'Pierre Duhem and the History of Late-Medieval Science' (1991), 274. So 'the last ten years' means since 1925.

giving his lectures the medical revolution represented by the first modern wonder drug, penicillin, was only just getting under way, and even in 1959 C. P. Snow still thought the important new science was being done by physicists, not biologists.

So at first there was not one Scientific Revolution but two, one exemplified by Newton's classical physics, the other by Rutherford's nuclear physics. It was only very slowly that the first won out over the second in the claim to the definite article.[5] The idea that there is such a thing as 'the Scientific Revolution' and that it took place in the seventeenth century is thus a fairly recent one. As far as historians of science are concerned, it was Butterfield who popularized the term, which occurs over and over again in the course of *The Origins of Modern Science*; but the first time he introduces it he refers to it, awkwardly, as 'the so-called "Scientific Revolution", popularly associated with the sixteenth and seventeenth centuries.' 'So-called' is apologetic; even stranger is his insistence that the term is already in popular use.[6] Where did Butterfield find the term, other than in Koyré (whose work would have been totally unknown to his audience), used specifically about the sixteenth and seventeenth centuries? The phrase 'the scientific revolution of the seventeenth century' originates, it seems, with the American philosopher and educational reformer John Dewey, the founder of pragmatism, in 1915,[iv] but it is unlikely that Butterfield ever read Dewey. Butterfield's source, surely, is Harold J. Laski's *The Rise of European Liberalism* (1936), an immensely successful book which had just been reissued in 1947.[7] Laski was a prominent politician and the leading socialist intellectual of the day; he was enough of a Marxist to have a taste for the word 'revolution'. It was his usage, then, not Koyré's, that Butterfield adopted with some discomfort, believing that it would already be familiar to many of his listeners and readers.

Thus in this respect the Scientific Revolution is not like the American or French revolutions, which were called revolutions as they happened; it is a construction of intellectuals looking back from the twentieth century. The term is modelled on the term 'Industrial Revolution', which

iv Dewey was attacking Marxism: 'our strictly scientific economic interpreters will have it that economic forces present an inevitable evolution, of which state and church, art and literature, science and philosophy are by-products. It is useless to suggest that while modern industry has given an immense stimulus to scientific inquiry, yet nevertheless the industrial revolution of the eighteenth century comes after the scientific revolution of the seventeenth. The dogma forbids any connection.' Dewey, *German Philosophy and Politics* (1915), 6. The phrase runs through Dewey's later writings.

was already commonplace towards the end of the nineteenth century (and which originates, it seems, in 1848, with Horace Greeley – now famous for supposedly saying, 'Go West, young man!'),[8] but which is also an after-the-fact construction.[v] And this of course means that some will always want to claim that we would be better off without such constructions – although it is worth remembering that historians constantly (and often unthinkingly) use them: 'medieval', for example, or the Thirty Years War (terms that could, necessarily, be introduced only after the fact); or, for any period before the Renaissance, 'the state', or, for any before the mid-eighteenth century, 'class' in the sense of social class.

Like the term 'Industrial Revolution', the idea of a scientific revolution brings with it problems of multiplication (how many scientific revolutions?) and periodization (Butterfield took as his period 1300 to 1800, so that he could discuss both the origins and consequences of the revolution of the seventeenth century). As time has gone on the idea that there is something that can sensibly be called *the* Scientific Revolution has come increasingly under attack. Some have argued for continuity – that modern science derives from medieval science, or indeed from Aristotle.[vi] Others, beginning with Thomas Kuhn, who published a book on *The Copernican Revolution* in 1957, followed by *The Structure of Scientific Revolutions* (1962), have sought to multiply revolutions: the Darwinian revolution, the Quantum revolution, the DNA revolution, and so on.[9] Others have claimed that the real Scientific Revolution came in the nineteenth century, in the marriage of science and technology.[10] All these different revolutions have their utility in understanding the past, but they should not distract our attention from the main event: the invention of science.

It should be apparent that the word 'revolution' is being used in very different senses in some of the examples above, and it helps to distinguish three of them, exemplified by the French Revolution, the Industrial Revolution and the Copernican revolution. The French Revolution had a beginning and an ending; it was an enormous upheaval which, in one way or another, affected everyone alive in France at the time; when it began, nobody foresaw how it would end. The Industrial Revolution is rather different: it is rather difficult to say when it began and when it

v See the previous note, and, for example, Butterfield, *The Origins of Modern Science* (1950), 197–8: 'Indeed the scientific, the industrial and the agrarian revolutions form such a system of complex and interrelated changes, that in the lack of a microscopic examination we have to heap them all together as aspects of a general movement . . .'
vi See 'A Note on Greek and Medieval "Science"' (pp. 573–5).

ended (conventionally, it runs from about 1760 to sometime between 1820 and 1840), and it affected some places and some people much more rapidly and extensively than it affected others, but everyone would agree that it began in England and depended on the steam engine and the factory system. Finally, the Copernican revolution is a conceptual mutation or transformation, which made the sun, not the earth, the centre of the universe, and placed the Earth in movement around the sun instead of the sun around the earth. For the first hundred years after the publication of Copernicus's book *On the Revolutions of the Heavenly Spheres* in 1543 only a fairly limited number of specialists were familiar with the details of his arguments, which were only generally accepted in the second half of the seventeenth century.

A failure to distinguish these senses, and to ask which of them the first users of the term 'the Scientific Revolution' had in mind, has caused a tremendous amount of confusion. The source of this confusion is simple: from its first appearances, the term 'the Scientific Revolution' was being used in two quite different ways. For Dewey, Laski and Butterfield, the Scientific Revolution was a lengthy, complex, transformative process, to be compared with the Reformation (which Laski called a theological revolution) or the Industrial Revolution. For Koyré, it was, following on Gaston Bachelard's concept of an 'epistemological break', identified with a single intellectual mutation: the replacement of the Aristotelian idea of place (in which there was always an up and a down, a left and a right) by a geometrical idea of space, a substitution which made possible, he argued, the invention of the idea of inertia, which was the foundation of modern physics.[11] Koyré had a vast influence in America, and his Bachelardian conception of an intellectual mutation was adopted by Thomas Kuhn in *The Structure of Scientific Revolutions*. Laski and Butterfield had a comparable influence in England on works such as Rupert Hall's *The Scientific Revolution* (1954), which denied any connection between the Scientific and Industrial revolutions, and J. D. Bernal's *Science in History*, whose second volume, *The Scientific and Industrial Revolutions* (1965), insisted on the closeness of the connection.

There is a fundamental difference between these two conceptions of the Scientific Revolution. Copernicus, Galileo, Newton, Darwin, Heisenberg and others who have been responsible for particular intellectual reconfigurations, mutations or transformations in science had a very good grasp of what they were doing as they did it. They knew that if their ideas were adopted the consequences would be momentous. It is thus easy to think of scientific revolutions as deliberate acts, conducted by people who achieve what they set out to achieve. Butterfield's Scientific Revolution was not that sort of revolution. Comparisons between the Scientific Revolution and political revolutions are not entirely misleading, for they were both transformative of the lives of all they touched; they both had identifiable beginnings and ends; they both involved struggles for influence and status (in the Scientific Revolution between the Aristotelian philosophers and the mathematicians who favoured the new science). Above all, both political revolutions and the Scientific Revolution have had unintended, not intended, outcomes. Marat aspired to liberty; the outcome was Napoleon. Lenin, when he published *State and Revolution* just two months before the October 1917 revolution, genuinely believed that a communist revolution would lead to the rapid withering away of the state. Even in the American Revolution, which came closest to realizing the ideals which first inspired it, there is a vast gap between Thomas Paine's *Common Sense* (1776), which envisaged a democratic system in which a majority could do more or less whatever they chose, and the complex checks and balances of the American Constitution as analysed in *The Federalist* (1788), which were designed to keep radicals like Paine trussed and tied. In the Scientific Revolution, Bacon and Descartes were amongst those with plans for thoroughgoing intellectual change, but their plans were castles in the air, and neither of them imagined what Newton would achieve. The fact that the outcome of the Scientific Revolution *as a whole* was not foreseen or sought by any of the participants does not make it any the less a revolution – but it does mean it was not a neat epistemological break of the sort described by Koyré.[vii] So, too, when first Thomas Newcomen (1711) and then James Watt (1769) invented powerful new steam engines, neither foresaw that the age of steam would see the construction of a great railway system girdling the Earth – the first public steam railway did not open until 1825. It is this sort of revolution, a revolution of unintended consequences and unforeseen outcomes, that Butterfield intended to evoke by the term 'the Scientific Revolution'.

If we define the term 'revolution' narrowly as an abrupt transformation that affects everybody at the same time, there is no Scientific Revolution – and no Neolithic Revolution, or Military Revolution (following the invention of gunpowder), or Industrial Revolution (following

vii The classic study of the unintended consequences of revolutionary upheaval is de Tocqueville, *The Old Regime and the Revolution* (1856).

the invention of the steam engine) either. But we need to acknowledge the existence of extended, patchy revolutions if we want to turn aside from politics and understand large-scale economic, social, intellectual and technological change. Who, for example, would object to the term 'the digital revolution' on the grounds that it is not a singular and discrete event, localized in time and space?

There is a certain irony in Butterfield's adoption of the retrospective term 'Scientific Revolution', and an even greater one in his choice of *The Origins of Modern Science* for his title. In 1931 he had published *The Whig Interpretation of History*, which attacked historians who wrote as if English history led naturally and inevitably to the triumph of liberal values.[13] Historians, Butterfield argued, must learn to see the past as if the future were unknown, as it was to people at the time. They must think their way into a world in which the values we now hold, the institutions we now admire, were not even imagined, let alone approved. It was not the historians' job to praise those people in the past whose values and opinions they agreed with and criticize those with whom they disagreed; only God had the right to sit in judgement.[viii] Butterfield's attack on the liberal tradition of historical writing in England was salutary, although he soon grasped that the sort of history he was advocating would be unable to make sense of the past, since without hindsight it would be impossible to establish the significance of events; history would become like the Battle of Borodino as experienced by its participants – at least according to Tolstoy in *War and Peace* – and both the readers and the historians themselves would stumble about, unable to make sense of events. Tolstoy, of course, as an omniscient narrator, also provides a running commentary, establishing what it was that the combatants were all, willy-nilly, conspiring to bring about. But later historians have naturally turned the phrase 'Whig history' back against Butterfield himself, accusing him of taking for granted the superiority of modern science over all that went before. The very idea of a book about 'origins' seems to them contrary to the principles he established in *The Whig Interpretation of History*.[ix14] Indeed it is; but the fault lies with Butterfield's early principles, not his later practice, for we really do need

viii It should be obvious that he was not right about this: no one, I trust, would want to read an account of slavery written by someone incapable of passing judgement.

ix Butterfield had been unequivocal: 'The consequences of his [the whig historian's] fundamental misconception are never more apparent than in the whig historian's quest for origins'; 'History is not the study of origins; rather it is the analysis of all the mediations by which the past was turned into our present.' Butterfield, *The Whig Interpretation of History* to understand the origins of modern science if we are to understand our own world.

§ 2

For the most part, scholars in recent years have been reluctant to adopt the term 'the Scientific Revolution', and many have explicitly rejected it. The opening sentence of Steven Shapin's *The Scientific Revolution* (1996) is often quoted: 'There is no such thing as the Scientific Revolution,' he wrote, 'and this is a book about it.'[15] The main source of their discomfort (once one has cleared away confusions over the meaning of the word 'revolution') points to a feature of the study of history that Butterfield simply took for granted and saw no need to discuss: that language is 'the principal working tool' of the historian.[16] The whole of Butterfield's *Whig Interpretation of History* is a critique of anachronistic thinking in history, but Butterfield never discusses a fundamental source of anachronism: the language in which we write about the past is not the language of the people we are writing about.[x] When Butterfield's arguments were restated by Adrian Wilson and T. G. Ashplant in 1988, the central feature of the historian's enterprise had become the fact that texts which survive from the past are written in what amounts to a foreign language.[xi] Suddenly it seemed that there was a hitherto unacknowledged problem with the word 'revolution' and indeed with the word 'science', too, for these are our words, not theirs.[xii]

(1931), 42–3, 47. On the evolution of his views, Sewell, 'The "Herbert Butterfield Problem" and Its Resolution' (2003).

x In *The Origins of Modern Science* there are faint glimmerings of an interest in language; e.g. in a discussion of the origins of the Enlightenment, 'Whereas "reason" had once been a thing that required to be disciplined by a long and intensive training, the very meaning of the word began to change now any man could say that he had it, especially if his mind was unspoiled by education and tradition. "Reason", in fact, came to signify much more what we today should call common sense.' (170)

xi Wilson & Ashplant, 'Whig History' (1988). A crucial source of this perception was Skinner, 'Meaning and Understanding in the History of Ideas' (1969). (Skinner's argument, as originally stated, derived from Wittgenstein, though this is less clear in his revised version of 2002: Wootton, 'The Hard Look Back' (2003)). It only began to impact the history of science belatedly, with Shapin & Schaffer, *Leviathan and the Air-pump* (1985) and Cunningham, 'Getting the Game Right' (1988).

xii The latest revision of the *Oxford English Dictionary* online (March 2014) gives the first use of 'scientific', meaning 'concerned or involved with (esp. natural) science . . . treating of science; having science as its subject', as occurring in 1675, with no further usage recorded until 1757; for 'science' in its modern sense ('the intellectual and practical activity

The word 'science' comes from the Latin *scientia*, which means 'knowledge'. One view to take, a view that derives both from Butterfield's rejection of Whig history and from Wittgenstein (to whom we will turn later in this chapter), is that truth or knowledge is what people think it is.[xiii] On this view astrology was once a science, and so of course was theology. In medieval universities the core curriculum consisted of the seven liberal 'arts' and 'sciences': grammar, rhetoric and logic; mathematics, geometry, music and astronomy (including astrology).[17] They are often now referred to as the seven liberal arts, but each one was originally called both an art (a practical skill) and a science (a theoretical system); astrology, for example, was the applied skill, astronomy the theoretical system.[xiv] These arts and sciences provided students with the foundations for the later study of philosophy and theology, or of medicine or law. These, too, were called sciences – but philosophy and theology were purely conceptual explorations that lacked an accompanying applied skill. They had practical implications and applications, of course – theology was applied in the art of preaching; and both ethics and politics, as studied by philosophers, had practical implications – but there were no university courses in applied theology or philosophy. They were not arts, and it would have been incomprehensible to claim then, as we do now, that philosophy belongs with the arts, not the sciences.[xv]

Moreover, these sciences were organized into a hierarchy: the theologians felt entitled to order the philosophers to demonstrate the rationality of belief in an immortal soul (despite the fact that Aristotle had not been of this view: philosophical arguments against the immortality of the soul were condemned by the theologians of Paris in 1270); the philosophers felt entitled to order the mathematicians to prove that all motion in the heavens is circular, because only circular movement

---

encompassing those branches of study that relate to the phenomena of the physical universe and their laws'), it gives a first usage in 1779 (which would make earlier usages of 'scientific' in the modern sense rather puzzling, but, as we shall see, there is a much earlier usage of 'science' in the required sense).

xiii Thus Shapin states: 'For historians, cultural anthropologists, and sociologists of knowledge, the treatment of truth as accepted belief counts as a maxim of method, and rightly so' (Shapin, *A Social History of Truth* (1994), 4). See 'Notes on Relativism and Relativists', No. 1 (pp. 580–1).

xiv Apart from the liberal arts, the skills of the educated, there were plenty of arts that involved manual labour – mechanical arts – such as goldsmithing or masonry.

xv When did theology stop being a science? Perhaps with Temple, *Miscellanea: The Third Part* (1701), 261.

can be uniform, permanent and unchanging, and to demonstrate that the earth is the centre of all these heavenly circles.[xvi] A basic description of the Scientific Revolution is to say that it represented a successful rebellion by the mathematicians against the authority of the philosophers, and of both against the authority of the theologians.[18] A late example of this rebellion is apparent in Newton's title *Mathematical Principles of Natural Philosophy* – a title which is a deliberate act of defiance.[xvii] An early example is provided by Leonardo da Vinci (d.1519), who in his posthumous *Treatise on Painting*[xviii] wrote: 'No human investigation can be termed true science if it is not capable of mathematical demonstration. If you say that the sciences which begin and end in the mind are true, that is not to be conceded, but is denied for many reasons, and chiefly the fact that the test of experience[xix] is absent from these exercises of the mind, and without it nothing can be certain.' In saying this, Leonardo, who was an engineer as well as an artist, was rejecting the whole of Aristotelian natural philosophy (which is what he means by 'the sciences which begin and end in the mind') and confining true sciences to those forms of knowledge which were simultaneously mathematical and grounded in experience; arithmetic, geometry, perspective, astronomy (including cartography) and music are the ones he mentions. He realized that the mathematical sciences were often dismissed as 'mechanical' (that is, tainted by a close relationship to manual labour), but he insisted that they alone were capable of producing true knowledge. Later readers of Leonardo could not believe he had meant what he said, but he surely did.[19] And, as a consequence of this rebellion of the mathematicians, philosophy in modern times has been demoted from pure science to mere art.

A key part of philosophy, as that discipline was inherited from

xvi Thus Gioseffo Zarlino described the science of music as being 'subordinated' (*sottoposto*) to philosophy (Zarlino, *Dimostrationi harmoniche* (1571), 9). When the Jesuit astronomers in Rome accepted in 1611 that Venus orbits the sun they did so 'to the scandal of the philosophers', who were not accustomed to such insubordination (Lattis, *Between Copernicus and Galileo* (1994), 193).

xvii Before Newton there was Kepler: Kepler's *New Astronomy, Based upon Causes, or Celestial Physics* (1609) deliberately conflates the worlds of the mathematician (who deals with astronomy) and natural philosopher (who deals with physics and with causation in nature).

xviii First published in 1651; the text was constructed around 1540 from Leonardo's notes by his pupil Francesco Melzi, and long circulated in manuscript.

xix Leonardo signs himself with a flourish: '*Leonardo Vinci disscepolo della sperientia*' ('Leonardo, disciple of experience') (Nicholl, *Leonardo da Vinci* (2004), 7).

Aristotle and taught in the universities, was the study of nature – 'nature' coming from the Latin word *natura*, for which the Greek equivalent is *physis*. For Aristotelians, the study of nature was about understanding the world, not changing it, so there was no art (or technology) associated with the science of nature. And because nature was the embodiment of reason it was, in principle, possible to deduce how things had to be. For Aristotle, the ideal science consisted of a chain of logical deductions from incontestable premises.[xx]

When an alternative to Aristotelian natural philosophy, an alternative calling itself at first the 'new philosophy' (a term we have seen John Donne taking up in 1611), developed in the course of the seventeenth century, there was an obvious need to find a vocabulary to describe the new knowledge.[xxi] The word that we use in modern English, 'science', was too vague: as we have seen, there were already lots of sciences. One option – the one most frequently adopted – was to continue to use the terms of Latin origin: 'natural philosophy' and 'natural philosopher'.[xxii] Since these were terms associated with higher status and bigger salaries, it was inevitable that the new philosophers tried to lay claim to them:[20] Galileo, for example, who had been a professor of mathematics, became in 1610 philosopher to the Grand Duke of Tuscany.[xxiii] (Hobbes held that Galileo was the greatest philosopher of all time.)[21] For some, the only real philosophy was natural philosophy: thus Robert Hooke, one of the first people to be paid to carry out experiments, states baldly, 'The Business of Philosophy is to find out a perfect Knowledge of the Nature and Properties of Bodies', and to work out how to put this knowledge to use. This was what he called 'true science'.[22] This usage of the terms 'philosophy' and 'philosopher' survived much longer than one might think. In 1889 Robert Henry Thurston published *The Development of the Philosophy of the Steam Engine*: by 'philosophy' he meant 'science'.

But the term 'natural philosophy' was unsatisfactory because it implied the new philosophy was rather like the old, that it had no

xx For further discussion of Aristotle, see below, pp. 68–74.

xxi The first book claiming to present a 'new philosophy' would seem to be Francesco Patrizi's anti-Aristotelian *Nova de universis philosophia*, or *A New Universal Philosophy*, of 1591.

xxii Of Latin origin in that *philosophia* and *philosophus* were naturalized in classical Latin, although their origins are Greek.

xxiii *Filosofo e matematico primario del ser*[mo] *Gran Duca di Toscana* is the title Galileo uses: Galileo was the duke's only philosopher, and the first amongst his mathematicians.

practical application. There was another option, which was to use an existing phrase that avoided the term 'philosophy' – 'natural science' – and this usage was common in the seventeenth century.[xxiv] (It was only in the nineteenth century that 'science' came to be generally used as a shorthand for 'natural science'.) An even more general term was available: 'natural knowledge'. The student of nature needed a name, hence a new word appeared in the late sixteenth century, 'naturalist' – only later did 'naturalist' come to refer specifically to someone who studied living creatures (as late as 1755 Dr Johnson in his *Dictionary* defined a naturalist as 'a person well versed in natural philosophy'). An alternative to 'naturalist' was 'natural historian', a term derived from Pliny's *Naturalis historia* (78 CE): but Pliny's reputation fell as a consequence of the new science, and unsophisticated natural histories were soon being replaced by more elaborate programmes of observation.

If Latin offered no perfect solution, what about Greek? The obvious solutions were 'physic(s)' (or 'physiology') and 'physician' (or 'physiologist').[xxv] Both sets of terms, like their Greek originals, included the whole study of nature, animate and inanimate – thus Boyle's *Physiological Essays* of 1661 are about natural science as a whole. But both had already been claimed by the doctors (medicine was for a long time the only 'art' based on a science of nature), which was a considerable inconvenience. Nevertheless, English intellectuals in the second half of the seventeenth century used 'physicks' to mean 'knowledge of Nature'

xxiv There are 245 occurrences in Early English Books Online (henceforth EEBO) if one's search includes variant forms and variant spellings; 29 further occurrences of 'sciences natural' and 8 of 'science of nature'. For Galileo's use of the term, see below, p. 511; and for Milliet de Chales', above, n. 18. An alternative term was 'physical science' (25 occurrences). In French, see, for example, Dupleix, *La Physique, ou science naturelle* (1603); the earliest example of *science naturelle* in the singular that I can find is 1586, in the plural 1537. In Italian, Zarlino defines the study of material entities as *scienza naturale*, also called *fisica* (Zarlino, *Dimostrationi harmoniche* (1571), 9), music being a mixed science, partly physical and partly mathematical. 'There was no science in early modern society,' says Adrian Johns (Johns, 'Identity, Practice and Trust' (1999), 1125 – I quote the abstract; see also Johns, *The Nature of the Book* (1998), 6 n. 4, and 42–3: 'There is a sense in which the history of early modern science no longer exists'). Nowhere does he acknowledge that there was such a thing as 'natural science'; in the text he considers only natural philosophy and mathematics, not 'physiology', 'physics', etc. In claiming that *scientia* refers only to 'certain, demonstrative knowledge' he shows a fundamental misunderstanding of the meaning of the term in the seventeenth century, for, besides music, geography and anatomy, to take just two examples, were sciences. The same confusions (which may be labelled 'the Cambridge fallacy') are to be found in Cunningham, 'Getting the Game Right' (1988), and in Henry, *The Scientific Revolution* (2008), 4–5.

xxv There was also a third word, now entirely obsolete, 'physiologer'.

or 'natural philosophy' (as opposed to 'physick', meaning medicine). For the Presbyterian minister Richard Baxter, '[T]rue Physicks is the Knowledge of the knowable works of God,' and for John Harris, who was giving public lectures on the new science from 1698, 'Physiology, Physicks, or Natural Philosophy, is the Science of Natural Bodies,'[23] although he acknowledges that some also use the term 'physiology' to refer to 'a Part of Physick that teaches the Constitution of the Body'. Harris was here still using 'physiology' in a sense which remained commonplace until late in the eighteenth century – it is the original sense of the word, preceding its use to refer to the study of human biology. Someone who studied natural philosophy was a 'physiologist'. It was not until the nineteenth century that 'physiology' was definitively ceded to the doctors, while the natural scientists redefined 'physics' to exclude 'biology' (a word invented in 1799) and, alongside the word 'physics', a new one was introduced, 'physicist'.[24]

A further solution was to invent a term which reflected the way in which the new knowledge crossed over between the traditional disciplines of natural philosophy (which included what we now call physics) and mathematics (which included mechanics and astronomy). Hence the use of terms such as 'physico-mathematical' and 'physico-mechanical', as in 'physico-mechanical experiments', and even the peculiar hybrids 'mechanical philosophy' and 'mathematical philosophy'.[xxvi]

Thus we are not dealing with a transformation reflected in a single pair of terms – 'natural philosophy', which, in the nineteenth century, became 'science'.[25] Instead, there is a complicated network of terms, and a change in the meaning of one term results in adjustments in the meaning of all the others.[26] The most striking innovation of the nineteenth century, as far as the language of science is concerned, was the

xxvi Of these, by far the commonest were 'mechanical philosophy' (62 hits on EEBO) and 'physico-mechanical' (122); 'experimental philosophy' (352 hits – with 24 further hits for 'experimental natural philosophy') was commoner still. 'Mathematical philosopher' is to be found in Benedetti, *Consideratione* (1579), 15, contrasted to 'natural philosopher'; later (49) he puns on the different meanings of *naturale*: the mathematical philosophers are the only ones to take seriously, as the natural philosophers are completely 'natural' (in the sense of simple-minded). The term 'natural philosophy' is not unproblematic for the period before 1650. Gilbert uses the term *philosophia naturalis* only once in *De magnete*, to refer to the old way of thinking (Gilbert, *De magnete* (1600), 116); and in Galileo's *Dialogo* the term appears three times, always to refer to Aristotelian philosophy. As far as Marin Mersenne, a theologian, philosopher and mathematician himself, was concerned, Galileo was not a philosopher but a 'mathematician and engineer' (Garber, 'On the Frontlines of the Scientific Revolution' (2004), 151–2, 156–9). It is only from the 1640s that natural philosophy becomes a crucial category, largely owing to the influence of Descartes.

introduction of the word 'scientist'. But the fact that there were no persons called 'scientists' before 1833, when William Whewell coined the term, does not mean that there was no word for someone who was an expert in natural science – they were called 'naturalists', or 'physiologists', or 'physicians'; in Italian they were *scienziati*, in French *savants*, in German *Naturforscher*, and in English *virtuosi*.[27] Robert Boyle's *The Christian Virtuoso* (1690) is about someone who is 'addicted to Experimental Philosophy'.[28] As terms such as *virtuosi* came to seem old-fashioned they were replaced by the phrase 'men of science', which in the sixteenth and seventeenth centuries was used to refer to all those who had a liberal or philosophical education ('men of a science, not a trade'), but which in the course of the eighteenth century began to be more narrowly used to refer to the people we call 'scientists'.[xxvii]

The word 'scientist' was very slow to become established for the straightforward reason that it was (like our word 'television') an illegitimate hybrid of Latin and Greek. The geologist Adam Sedgwick (d.1873) scribbled in the margin of his copy of a book by Whewell, '[B]etter die of this want than bestialize our tongue by such barbarisms.'[29] As late as 1894 Thomas Huxley ('Darwin's bulldog') insisted that no one with any respect for the English language would use the word, which he found 'about as pleasing a word as "Electrocution"' (a Greek–Latin, rather than a Latin–Greek, hybrid) – and even at that time he was not alone.[xxviii] We can usefully contrast 'scientist' in this respect with the

xxvii For the old usage, Woodward, *Dr Friend's Epistle to Dr Mead* (1719); and for the new, Jurin, *A Letter to the Right Reverend the Bishop of Cloyne* (1744), 18: '[Y]ou must pardon me, if I assure your Lordship, that however glibly it may pass off with the Ladies, the *Ipse dixit* of your Lordship will not, with Men of Science, stand against Experience and well-founded Knowledge. WATER, my Lord, give me Leave to assume so much the Chemist to inform you, can no more draw Oil from Tar, than Fire can Salt from Glass.'

xxviii Ross, '"Scientist": The Story of a Word' (1962), 78. Whewell understood that opposition to the word was based on its etymology: '[S]ome ingenious gentleman [Whewell himself, at a meeting of the British Association for the Advancement of Science] proposed that, by analogy with *artist*, they might form *scientist*, and added that there could be no scruple in making free with this termination, when we have such words as *sciolist*, *economist*, and *atheist* – but this was not generally palatable' (Whewell, 'On the Connexion of the Physical Sciences' (1834), 59). Whewell's motive for raising the issue in print in 1834 may partly have been that 'scientist', unlike 'man of science', is a gender-neutral term – he was reviewing a book by the science writer Mary Somerville. He returned to the subject a few years later in the context of a general discussion of the language of science, in which he asserted, 'The combination of different languages in the derivation of words, though to be avoided in general, is in some cases admissible,' and went on to argue (against the general view) that terminations in *-ist* 'are applied to words of all origins . . . Hence we may make such words when they are wanted. As we cannot use *physician* for a cultivator of physics, I have called uncontroversial word 'microscopist' (1831), a word properly formed because it was made entirely out of Greek materials.[30] If we look at other European languages, only Portuguese has followed English in creating a linguistic hybrid: *cientista*. The claim that '[t]he word "scientist" was not coined until 1833 because only then did people realize it was needed' is thus mistaken: there had long been a perceived need for a word to do the job.[31] The problem was that finding a suitable word – one that did not already have a different usage and was properly constructed – was a genuine obstacle, so that only when the need became absolutely pressing was the obstacle overcome, and only then by breaking what was regarded as one of the basic rules of word formation. Fundamentally, though, the word 'scientist' was merely a new and useful word for a type of person who had long been in existence.[32]

The word 'scientific' falls between classical 'science' and nineteenth-century 'scientist'. *Scientificus* (from *scientia* and *facere*, knowledge-making) is not a term in classical Latin; it was invented by Boethius early in the sixth century. In English, apart from a couple of occurrences in a text of 1589, 'scientific' does not appear until 1637, after which date it becomes increasingly common. It has three main meanings: it can refer to a certain type of expertise ('scientific', as opposed to 'mechanical'; the learning of a scholar or gentleman, as opposed to that of a tradesman); to a demonstrative method (that is, by Aristotelian syllogisms); but, in a third sense (as in 'the scientifick measuring of Triangles', 1645, in a work on surveying), it refers to the new sciences of the Scientific Revolution. In French the word *scientifique* was introduced earlier,

---

him a *physicist*. We need very much a name to describe a cultivator of science in general. I should incline to call him a *Scientist*. Thus we might say that as an Artist is a Musician, Painter, or Poet, a Scientist is a Mathematician, Physicist, or Naturalist.' (Whewell, *The Philosophy of the Inductive Sciences* (1840), cvi, cxiii; 'artist' looks like a Latin–Greek hybrid but is actually, like 'dentist', an import from French.) But, despite Whewell, the word 'scientist' does not appear in Galton, *English Men of Science* (1874), a study of 190 Fellows of the Royal Society. According to a Google ngram, 'scientist' + 'scientists' only becomes more frequent than 'man of science' + 'men of science' in 1882. This was also the year in which the word 'scientist' was first used in the annual presidential address of the British Association for the Advancement of Science; but the great biologist (and classics scholar) D'Arcy Wentworth Thompson was still avoiding it in the 1920s. As one would expect, the word caught on more rapidly in North America than in Britain, where scientists continued to have a classical education. Ross, '"Scientist": The Story of a Word' (1962); Secord, *Visions of Science* (2014), 105 (who is quite wrong to claim that Whewell meant the term to be 'a putdown'; Whewell uses the word 'sciolist' as an example not because he wants to suggest that science is not an honourable vocation but simply because it is, exceptionally, a Latin–Greek hybrid of the sort his opponents reject as unacceptable); Barton, 'Men of science' (2003), 80–90 and n. 33.

in the fourteenth century, in the sense of knowledge-making; in the seventeenth century it was used to refer to the abstract and speculative sciences, and it only begins to be used as the equivalent of the English word 'scientist' – *un scientifique* – in 1895, around the same time as the English word began to be widely used.[33]

In each European language, of course, the pattern was slightly different. In seventeenth-century French we find the equivalent terms to the English 'physician' (*physicien*) and 'naturalist' (*naturaliste*). In French, *physicien* had never been a term used for doctors, so the word was conveniently available to mean a natural scientist, and then to evolve to become the French equivalent for 'physicist'.[xxix] In Italy, by contrast, the link between *fisico* and medicine was already strong in the sixteenth century, and the new philosophers rarely called themselves *fisici*;[34] but then Italian already had to hand a word, *scienziato* (man of knowledge), lacking in English, and still lacking in French (*scientiste* is nearly always used pejoratively to mean someone who makes a fetish of being scientific).

To claim, as is often done, that there was no science until there were 'scientists' is therefore simply to betray an ignorance of the evolution of the language for knowledge of nature, and for the knowers of nature, between the seventeenth and the nineteenth centuries.[35] Those who hesitate to use the words 'science' and 'scientist' for the seventeenth century, convinced that they are anachronistic, do not understand that all history involves translation from one language into another, and that 'science' is simply an abbreviation for a perfectly commonplace seventeenth-century term, 'natural science', just as 'scientist' is simply a substitute for 'naturalist', 'physician', 'physiologist' and 'virtuoso'. The first formal meeting of the group that would become the Royal Society discussed forming an association to promote 'Physico-Mathematicall-Experimentall Learning': they were making perfectly clear that their enterprise was not natural philosophy as traditionally understood but the new type of knowledge that had resulted from the mathematicians invading the territory of the philosophers.[36]

It has also been claimed that there were no scientists in the seventeenth century because there was no professional role for a scientist to occupy. 'There were no scientists in Stuart England,' we are told, 'and all

xxix So, too, in French you find 'physic' in the singular, not the plural, to mean natural science: e.g. Daneau, *Physique françoise, comprenant . . . le discours des choses naturelles, tant célestes que terrestres, selon que les philosophes les ont descrites* (1581).

the men we have grouped together under that heading were in their varying degrees dilettantes.'[37] By the same argument, Hobbes, Descartes and Locke were not philosophers, in that no one paid them to write philosophy; the only proper philosophers in the seventeenth century, it would follow, were scholastic philosophers, employed by universities and Jesuit colleges. Some of the new scientists were, like the new philosophers, in this sense indeed amateurs, not professionals: Robert Boyle, after whom Boyle's law is named, was independently wealthy and a profession would have been beneath his dignity as the son of an earl. John Wilkins, who wrote extensively on scientific questions, was a clergyman and eventually a bishop, but when the Royal Society was founded in 1662 he had already been Warden of Merton College, Oxford, and Master of Trinity College, Cambridge (appointed by the regime of Oliver Cromwell), although his university career had been wrecked by the Restoration, and he was then forced to fall back on ecclesiastical preferment.[xxx] Charles Darwin, too, of course, was an amateur not a professional scientist.[xxxi]

However, it would be quite wrong to think of the new science as primarily an amateur – that is, unpaid – activity. In this respect it is unlike the new philosophy of Hobbes, Descartes and Locke: they belonged to no profession, but the new scientists were for the most part practising science as part of their paid employment. Giovanni Battista Benedetti (1530–90, mathematician and philosopher to the Duke of Savoy),[xxxii] Kepler (mathematician to the Holy Roman Emperor) and Galileo (for eighteen years a professor of mathematics) were not dilettantes or amateurs: they were professional mathematicians, engaging with problems that were part of the university curriculum, even if their solutions to those problems were quite unlike those taught in the universities. Tycho Brahe, as we have seen, received state funding. Mathematical instrument making and cartography were both commercial undertakings (both were conducted by Gerardus Mercator (1522–99), for example).

Nor was there a shortage of such people in Stuart England. Robert

xxx It was a condition of nearly all Oxford and Cambridge fellowships that the holder was ordained, so nearly every academic in England was a clergyman.

xxxi Darwin, as far as I can tell, never described himself as a 'scientist'; but then as late as 1892 it was still possible to claim that 'naturalist' was the right general term for a student of natural science (*OED* s.v. naturalist).

xxxii For the claim to be the duke's philosopher, see the title page to the *Consideratione* of 1579; for the claim to be his mathematician, the title page of the *De temporum emendatione opinio* of 1578.

Hooke (d.1703), Denis Papin (d.1712), and Francis Hauksbee (d.1713) were all paid by the Royal Society to perform experiments, although only Hooke received a regular salary.[xxxiii] Christopher Wren, a founding member of the Royal Society and now best remembered as an architect, was Savilian Professor of Astronomy at the University of Oxford, holding a chair founded in 1619, having previously held the chair of astronomy at Gresham College in London (founded 1597); astronomy was universally recognized as a branch of mathematics, and architecture required mathematical skills. Isaac Newton was Lucasian Professor of Mathematics at Cambridge, holding a chair founded in 1663. In so far as there was a professional role occupied by the new scientists, it was that of mathematician, and there were plenty of people making a profession out of mathematics outside the two universities: Thomas Digges (1546–1595), for example, who played an important role in the largest engineering project of the Elizabethan era, the rebuilding of Dover harbour, and also sought to turn England into an elective monarchy, or Thomas Harriot (d.1621), whose skills as an astronomer, navigator, cartographer and military engineer led to his being hired to form part of Raleigh's expedition to Roanoke (1585).[38] Thus there were plenty of mathematicians who saw the new philosophy as falling within their area of professional expertise.[39] And, naturally, the crucial topics for the new science corresponded neatly with the professional preoccupations of seventeenth-century mathematicians: astronomy/astrology, navigation, cartography, surveying, architecture, ballistics and hydraulics.[40]

It would be perfectly sensible to avoid the words 'science' and 'scientist' when talking about the seventeenth century if the introduction of these words marked a real moment of change, but 'science' is simply an abbreviation for 'natural science', while 'scientist' marks not a change in the nature of science, or even a new social role for scientists, but a change in the course of the nineteenth century in the cultural significance of classical learning – a change which has become incomprehensible to those historians of science who have not received even the rudiments of a classical education.

xxxiii They were followed by John Theophilus Desaguliers, who performed the role of curator of experiments from 1716 to 1743.

§ 3

Although Copernicus, Galileo and Newton were well aware that their ideas were momentous, and we can legitimately describe their work as revolutionary, they never explicitly said to themselves, 'I am making a revolution.' The word 'revolution' was rarely used even in Newton's lifetime to refer to large-scale transformations, and almost never before the Glorious Revolution of 1688, the year after the publication of his *Principia*; even then it was at first confined to political revolutions.[xxxiv][41] Butterfield was right to stress that the historian must aspire to understand the world from the point of view of those alive at the time,[xxxv] but, as we have seen, just understanding the world from their point of view can never be enough. The historian has to mediate between the past and the present, finding a language which will convey to present readers the beliefs and convictions of people who thought quite differently. All history therefore involves translation from the source language – here that of seventeenth-century mathematicians, philosophers and poets – into the target language – here, that of the early twenty-first century.[42] Thus the historian properly translates 'natural science' into 'science' and 'physiologer' into 'scientist'.

But perhaps there is more than an issue of translation here? In Newton's language, it might be claimed, there is not only no single word or phrase equivalent to our word 'revolution', but the very concept is lacking. Newton's culture, it could be argued, was inherently conservative and traditionalist; Newton could not have formulated the idea of a revolution even if he had wanted to. In Chapter 3 we will see that, although it may be a helpful generalization to describe Renaissance and seventeenth-century culture, in many respects, as backward looking, there are important exceptions, and it was the exceptions that made modern science possible. For the moment, though, let us just note that there *is* a word which, for Protestants at least, had many of the

xxxiv An early example of the word in an extended usage is Daniel Defoe, *Robinson Crusoe* (1719): 'The revolution in trade brought about a revolution in the nature of things.' But this is from the eighteenth century, not the seventeenth.

xxxv 'Real historical understanding is not achieved by the subordination of the past to the present, but rather by our making the past our present and attempting to see life with the eyes of another century than our own . . . The study of the past with one eye, so to speak, upon the present is the source of all sins and sophistries in history, starting with the simplest of them, the anachronism.' Butterfield, *The Whig Interpretation of History* (1931), 16, 31–2.

connotations of 'revolution', and that word is 'reformation'. In the space of a few decades, between 1517 and 1555, Luther and Calvin had transformed the doctrines, rituals and social role of Christianity; they had made a revolution, one that gave rise to one hundred and fifty years of religious warfare. And so, before the Scientific Revolution was a revolution, it was a reformation. 'The main Design,' wrote Hooke in 1665, of his own efforts and those of the Royal Society, was 'a *reformation* in Philosophy'.[43] Thomas Sprat, writing a history of the Royal Society in 1667, repeatedly compared the reformation in natural philosophy with the earlier reformation in religion.[xxxvi][44]

Sprat went on to acknowledge that there were some hardliners who were so hostile to all aspects of ancient learning that they wanted to abolish Oxford and Cambridge. He compared these zealots to the men who had set out to abolish episcopacy in England but had ended up executing the king and establishing the Commonwealth:

> I confess there have not bin wanting some forward *Assertors* of *new Philosophy*, who have not us'd any kind of *Moderation* towards them [the universities]: But have presently concluded, that nothing can be well-done in *new Discoveries*, unless all the *Ancient Arts* be first rejected, and their Nurseries abolish'd. But the rashness of these mens proceedings, has rather prejudic'd, than advanc'd, what they make shew to promote. They have come as furiously to the purging of *Philosophy*, as our *Modern Zealots* did to the *reformation* of *Religion*. And the one Party is as justly to be condem'd, as the other. Nothing will suffice either of them, but an utter *Destruction*, *Root* and *Branch*,[xxxvii] of whatever has the face of *Antiquity*.[45]

Thus Sprat acknowledged that some of the advocates of the new science reminded him of the regicides (the monarchy, as much as the episcopacy, had 'the face of *Antiquity*') – which is about as close as he could possibly come to calling them revolutionaries. Sprat was publishing seven years after the restoration of the monarchy and in support of a society founded under royal patronage. He needed to distance himself

xxxvi Later, Peter Shaw wrote of the 'th[o]rough Reformation in Philosophy' that had transformed natural science and medicine (Shaw, *A Treatise of Incurable Diseases* (1723), 3), and Richard Davies, writing in 1740, said that it was about the year 1707, long after the publication of Newton's *Principia*, that 'the Learned began to be sensible how much the Author [i.e. Newton] had done towards a Reformation in Philosophy.' (Davies, *Memoirs of Saunderson* (1741), v).

xxxvii This is a reference to the Root and Branch Bill of 1641, which sought to abolish episcopacy, and led directly to the Civil War.

from any link between radicalism in science and radicalism in politics; it is all the more remarkable that he was willing to press home this comparison between some of the proponents of the new philosophy and the men who had, only a few years before, turned the world upside down.

Naturally, in 1790, Antoine Lavoisier, caught up in the midst of the French Revolution, declared that he was making a revolution in chemistry. Lavoisier, unlike Sprat, speaks our language because he was living through a revolution which transformed the language of politics, shaping the language we still speak. Many French intellectuals were already discussing the possibility of a political revolution in the years preceding 1789, and after 1776 the American Revolution presented them with a model.[46] In France, the word preceded the deed, though not by much.[xxxviii] In the seventeenth century Galileo and Newton knew nothing of this language.[xxxix] But they and their contemporaries were perfectly clear that they were trying to carry out a radical, systematic change: the fact that they did not have the word 'revolution' does not mean that they were obliged to think of knowledge as something stable and unchanging. 'As to our work,' wrote an anonymous member of the Royal Society in 1674, 'we are all well agreed, or should be so, that it is not to whiten the walls of an old house, but to build a new one.'[47] Tearing down the old and starting again from scratch is what revolutions are all about.

xxxviii Remarkably, Lavoisier had written about a revolution in chemistry even before 1789, indeed before 1776: '*L'importance de l'objet m'a engagé a reprendre tout ce travail,*' he wrote in his laboratory notebook in 1772 or 1773, '*qui m'a paru fait pour occasionner une révolution en physique et en chimie.*'

xxxix Is there a term, other than 'reformation', that could be substituted for 'revolution'? (Laslett once appealed for a new label for the Scientific Revolution: Laslett, 'Commentary' (1963).) In 1620 Francis Bacon called for a Great Instauration – 'instauration' here means 'founding', and the term is suitably vague. Bacon hoped a new practical, useful, technological science would come into existence and, eventually (although not as quickly as he had hoped), it did. In the 1660s the Royal Society looked back to Bacon as the first to enunciate the principles of the new science. So anachronism could apparently be avoided by talking about the Great Instauration (as in Webster, *The Great Instauration* (1975)), but it is not clear that this would make any real difference in what is actually meant; and, in any case, Bacon's phrase was not taken up by the members of the Royal Society ('Lord Bacon's Design for the Instauration of Arts and Sciences' is referred to only once in the *Philosophical Transactions*, on 25 March 1677).

§ 4

Just as over-scrupulous historians refuse to use the words 'revolution', 'science' and 'scientist' when writing about the seventeenth century, they baulk at using Butterfield's other word, 'modern', because it, too, seems to them inherently anachronistic. Yet Renaissance books on warfare often included the word 'modern' in their titles to show that they acknowledged the revolutionary consequences of gunpowder.[48] In the Renaissance, modern music was understood to be quite different from ancient music because it was polyphonic rather than monodic – Galileo's father, Vincenzo, wrote a *Dialogue on Ancient and Modern Music*.[49] Modern maps showed the Americas.[xl]

The first history to be written in terms of progress is Vasari's history of Renaissance art, *The Lives of the Artists* (1550).[50] It was quickly followed by Francesco Barozzi's 1560 translation of Proclus's commentary on the first book of Euclid, which presented the history of mathematics in terms of a series of inventions or discoveries. Indeed, the mathematicians (who often spent time with the artists, because they taught them the geometry of perspective)[xli] were already eager to claim that they, too, were making progress, and had begun to publish books with the word 'new' in the title, creating a fashion which spread from mathematics to the experimental sciences: *New Theories of the Planets* (Peuerbach, written 1454, published 1472); *The New Science* (Tartaglia, 1537); *The New Philosophy* (Gilbert, d.1603 – this is the subtitle, or perhaps the proper title, of the posthumously published *Of Our Sublunar World*; the layout of the title page is ambiguous); *The New Astronomy* (Kepler, 1609); *Two New Sciences* (Galileo, 1638); *New Experiments Touching the Void* (Pascal, 1647); *New Anatomical Experiments* (Pecquet, 1651); *New Experiments Physico-mechanical* (Boyle, 1660). The list goes on and on.[51] As the great pioneer of the idea of progress, Bacon wrote *The New Organon* and *The New Atlantis*, and his book on *The Wisdom of the Ancients* (1609) implied a sharp contrast between the ancients and the moderns.

Given all this emphasis on newness, why did scientists not use the

xl '[T]hen studie well these moderne Maps: and with your eie you shall beholde, not onely the whole world at one view, but also every particular place contained therein.' Blundeville, *A Briefe Description of Universal Mappes* (1589), C4r.

xli See Chapter 6.

word 'modern' in the titles to their books? The answer is straightforward. In both Islam and Christianity 'modern philosophy' meant post-pagan philosophy.[52] For William Gilbert, the founder of a new science of magnetism, for example, Thomas Aquinas (1225–74) was a modern philosopher.[53] Consequently, he had no interest in describing his own natural philosophy as 'modern', preferring to call it 'new'. In philosophy, unlike warfare and music, the word 'modern' was unavailable because it had already been put to a different use. The same was true in architecture, where, in the fifteenth century, 'modern architecture' meant Gothic architecture.[54] In science this began to change only in the course of the debate on the ancients and moderns at the end of the seventeenth century. Jonathan Swift still counts Aquinas among the moderns in *The Battle of the Books* (1720), but in doing so he is being deliberately old-fashioned.[55] René Rapin, who was one of the first to set the ancients against the moderns, had redefined the concept of modern philosophy by calling Galileo 'the founder of Modern Philosophy' in 1676 – a judgement particularly surprising coming from a Jesuit, given Galileo's condemnation by the Roman Inquisition in 1633 – but this usage had not caught on in English.[56] Nevertheless, with a definite article, 'the modern Philosophy', or 'the *modern way* of *Philosophy*', could be used, if a little awkwardly, to refer to contemporary science: Boyle was the first to use it thus, in 1666.[57] The phrase 'modern science' was first used by Gideon Harvey in 1699, in the course of an indiscriminate attack on both the old and the new philosophies.[58] The old philosophy is, by the end of the seventeenth century, scholasticism; modern science is the science of Descartes and Newton.

Just as the word 'modern' was slow to establish itself in a scientific context, so too it was only towards the end of the seventeenth century that the word 'progress' and those with similar meanings became commonplace. The proper title of the Royal Society, founded in 1660, is The Royal Society of London for the Improving of Natural Knowledge. 'Improving' implies progress, so it is not surprising that the fuller title of Thomas Sprat's *History of the Royal Society* was *The History of the Institution, Design and Progress of the Royal Society of London for the Advancement of Experimental Philosophy* – 'experimental philosophy' being, of course, yet another term for what we now call 'science', and 'progress' being used here somewhat ambiguously between its old meaning (a journey, and so a process of change) and its new meaning (a process of improvement); 'advancement' is another progress-related word. A year later Joseph Glanvill published *Plus ultra: or the Progress and Advancement of Knowledge since the Days of Aristotle*. By the end of the century progress was taken for granted, as in the title of Daniel Le Clerc's *The History of Physick, or, An Account of the Rise and Progress of the Art, and the Several Discoveries Therein from Age to Age* (1699).[59] Before the word 'progress' became fashionable, Robert Boyle twice used as his epigraph a passage from Galen: 'We must be bold and go hunting for the truth; even if we do not come right up to it, at least we will get closer to it than we are now.'[60] Boyle is using the hunt as a metaphor for progress. It is this triumph of the idea of progress, along with the redefinition of the word 'modern', which marks the end of the first phase of the long Scientific Revolution through which we are still living.[61]

In any case, there were alternatives to the language of progress which served exactly the same purpose – the languages of invention and discovery. In 1598 Brahe insisted the new geoheliocentric system of the cosmos was his own *invention* – he was laying claim to having invented a theory in exactly the same way as he claimed to have invented the astronomical sextant. Others had tried to steal the credit for the geoheliocentric system from him but, properly, it belonged to him alone.[62] Galileo, when in 1610 he announced to the world what he had seen through his telescope, was compared to his fellow Florentine Amerigo Vespucci, to Christopher Columbus and to Ferdinand Magellan.[63] In discovering the moons of Jupiter, Galileo had discovered new worlds, just as the navigators had done. Thereafter every scientist dreamed of making comparable discoveries. Here is the first professional scientist, Robert Hooke (1635–1703), writing in a hurry. Plenty of people, in all ages, he says, have enquired 'into the nature and causes of things':

> But their endeavours, having been only single and scarce ever united, improved, or regulated by art, have ended only in some small inconsiderable product hardly worth naming. But though mankind have been thinking these six thousand years, and should be so six hundred thousand more, yet they are and would be much whereabouts they were at first, wholly unfit and unable to conquer the difficulties of natural knowledge. But this new-found world must be conquered by a Cortesian army, well-disciplined and regulated, though their number be but small.[64]

The Royal Society was to be this 'Cortesian army, well-disciplined and regulated, though their number be but small'. Hooke's imagery is misleading – and he was misled by it. He was opposed not by the Aztecs but by Aristotelian philosophers. He did not need to conquer nature in order to understand it. His army did not need to be disciplined and regulated; competition (as we will see in Chapter 3) provided the only discipline it needed. But he was right about the fundamentals. He had chosen the Cortesian army as his image because he wanted to conjure up in the mind the most convulsive, irreversible transformation recorded in history; because he wanted to discover new worlds; and because he wanted his discoveries to benefit his own society, just as the conquest of the New World had enriched Cortes's Spain. Hooke's key terms were not 'science', or 'revolution', or 'progress', but it is a reasonable translation of his own terms ('natural knowledge', 'newfound world', 'Cortesian army') into our language to say that he was dreaming of what we call the Scientific Revolution.

He was not alone: 'The Aristotelian Philosophy is inept for New discoveries,' wrote Joseph Glanvill in 1661. 'There is an *America* of secrets, and [an] unknown *Peru* of Nature,' yet to be found:

> And I doubt not but posterity will find many things, that are now but *Rumors*, into *practical Realities*. It may be some Ages hence, a voyage to the *Southern* unknown *Tracts*, yea possibly the *Moon*, will not be more strange then one to *America*. To them, that come after us, it may be as ordinary to buy a *pair* of *wings* to fly into remotest *Regions*; as now a pair of *Boots* to ride a *Journey*. And to conferr at the distance of the *Indies* by *Sympathetick* conveyances, may be as usual to future times, as to us in a *litterary* correspondence ... Now those, that judge by the narrowness of former *Principles*, will smile at these *Paradoxical expectations*: But questionless those great Inventions, that have in these later Ages altered the face of all things; in their naked proposals, and meer suppositions, were to former times as *ridiculous*. To have talk'd of a *new Earth* [the New World of the Americas] to have been discovered, had been a *Romance* to *Antiquity:* And to sayl without sight of *Stars* or shoars by the guidance of a *Mineral* [the compass], a *story* more absurd, then the flight of *Daedalus*.[65]

And of course Glanvill was right: we do fly and 'conferr' at a distance; we have been not only to Australia but also to the Moon.

Thomas Hobbes, writing in 1655, thought that there was no astronomy worth the name before Copernicus, no physics before Galileo, no physiology before William Harvey. 'But since these, astronomy and natural philosophy in general have, for so little time, been extraordinarily advanced ... Natural Philosophy is therefore but young.'[66] But it was Henry Power (one of the first Englishmen to experiment with the microscope and the barometer) who, in 1664, gave most eloquent expression to the idea that knowledge was being transformed and that the new knowledge was quite unlike the old:

> And this is the Age wherein all mens Souls are in a kind of fermentation, and the spirit of Wisdom and Learning begins to mount and free it self from those drossie and terrene Impediments wherewith it hath been so long clogg'd, and from the insipid phlegm and *Caput Mortuum* of useless Notions, in which it has endured so violent and long a fixation.
>
> This is the Age wherein (me-thinks) Philosophy comes in with a Spring-tide; and the Peripateticks may as well hope to stop the Current of the Tide, or (with *Xerxes*) to fetter the Ocean, as hinder the overflowing of free Philosophy: Me-thinks, I see how all the old Rubbish must be thrown away, and the rotten Buildings be overthrown, and carried away with so powerful an Inundation. These are the days that must lay a new Foundation of a more magnificent Philosophy, never to be overthrown: that will Empirically and Sensibly canvass the *Phaenomena* of Nature, deducing the Causes of things from such Originals in Nature, as we observe are producible by Art, and the infallible demonstration of Mechanicks: and certainly, this is the way, and no other, to build a true and permanent Philosophy ...[67]

In 1666 the mathematician and cryptographer John Wallis (who introduced the symbol ∞ for infinity) wrote, more circumspectly, 'For, since that *Galilæo*, and (after him) *Torricellio*, and others, have applied *Mechanick* Principles to the salving of *Philosophical* Difficulties; Natural Philosophy is well known to have been rendered more intelligible, and to have made a much greater progress in less than an hundred years, than before for many ages.'[68]

Hooke, Glanvill, Hobbes, Power and Wallis were participants in this transformation; but their understanding of what was taking place was shared by well-informed bystanders. In 1666 Bishop Samuel Parker hailed the recent triumph of 'the Mechanical and Experimental Philosophie' over the philosophies of Aristotle and Plato, and claimed that:

> we may rationally expect a greater Improvement of Natural Philosophie from the *Royal Society*, (if they pursue their design) then it has had in all former ages; for they having discarded all particular *Hypotheses*, and wholly addicted themselves to exact Experiments and Observations, they may not only furnish the World with a compleat *History of Nature*, (which is the most useful part of *Physiologie* [natural science]) but also laye firm and solid foundations to erect *Hypotheses* upon.[69]

Parker (with good reason) thought the great improvement in knowledge was just about to happen, now that the right method of enquiry had been established. Only two years later the poet John Dryden (also with good reason) took the view that it was already well under way:

> Is it not evident, in these last hundred years (when the Study of Philosophy has been the business of all the *Virtuosi* in *Christendome*) that almost a new Nature has been reveal'd to us? that more errours of the [Aristotelian] School have been detected, more useful Experiments in Philosophy have been made, more Noble Secrets in Opticks, Medicine, Anatomy, Astronomy, discover'd, than in all those credulous and doting Ages from *Aristotle* to us? so true it is that nothing spreads more fast than Science, when rightly and generally cultivated.[70]

Dryden's chronology is right: 'these last hundred years' takes us back, almost exactly, to the nova of 1572. His vocabulary is exemplary: he uses *Virtuosi* to mean scientists and 'Science' to mean, well, science.[xlii] He sees that the new science relies on new standards of evidence. He acknowledges the possibility of relativism (how many new natures might there be?) while insisting that the new science is not just some sort of local fashion but an irreversible transformation in our knowledge of nature.[71]

§ 5

One could go on accumulating evidence of this sort for the validity of the idea of the Scientific Revolution, but plenty of scholars would remain unpersuaded and unpersuadable. The anxiety which now troubles historians when they read the words 'scientific', 'revolution', 'modern' and (worst of all) 'progress' in studies of seventeenth-century natural science is not just a fear of anachronistic language; it is a symptom of a much larger intellectual crisis which has expressed itself in a general retreat from grand narratives of every sort.[xliii] The problem, it is claimed, with grand narratives is that they privilege one perspective over another; the alternative is a relativism which holds that all perspectives are equally valid.

xlii This is perhaps the first use of 'science' without qualification to mean 'natural science': the *OED*'s failure to recognize the sense in which the word is used here presumably results from not reading it in context.

xliii The term 'grand narrative' originates with Lyotard, *La Condition postmoderne* (1979).

The most influential arguments in favour of relativism flow from the philosophy of Ludwig Wittgenstein (1889–1951).[xliv] Wittgenstein taught in Cambridge on and off from 1929 to 1947 – he left the year before Butterfield lectured on the Scientific Revolution – but it would never have occurred to Butterfield that he needed to consult Wittgenstein, or indeed any other philosopher, to learn how to think about science. It was not until the late 1950s, following the publication of the *Philosophical Investigations* in 1953, that arguments drawn from Wittgenstein began to transform the history and philosophy of science; their influence can already be seen, for example, in Thomas Kuhn's *The Structure of Scientific Revolutions*.[72] Thereafter it became common to claim that Wittgenstein had shown that rationality was entirely culturally relative: our science may be different from that of the ancient Romans, but we have no grounds for claiming that it is better, for their world was utterly unlike ours. There is no common standard by which the two can be compared. Truth, according to Wittgenstein's doctrine that meaning is use,[73] is what we choose to make it; it requires a social consensus but not any correspondence between what we say and how the world is.[74]

This first wave of relativism was later supplemented by other, profoundly different intellectual traditions: the linguistic philosophy of J. L. Austin, the post-structuralism of Michel Foucault, the postmodernism of Jacques Derrida and the pragmatism of Richard Rorty. The phrase 'the linguistic turn' is often used to refer to all these different traditions, because they share a common sense of how, as Wittgenstein put it, 'the limits of my language mean the limits of my world.'[xlv] As we shall see in a moment, much of the argument about the Scientific Revolution stems from the ramifications of this view.

Within history of science, one post-Wittgensteinian tradition has been particularly important: it is often called Science and Technology

xliv In the history of science literature it is generally taken for granted that Wittgenstein was a relativist. This view seems to me wrong, but I have kept that argument away from my main text; see the note 'Wittgenstein: No Relativist' (pp. 577–80). In the main text, both here and in Chapter 15, I outline what I call a Wittgensteinian position, which can indeed claim to be grounded in Wittgenstein's texts but which is not, I would argue, Wittgenstein's.

xlv Rorty (ed.), *The Linguistic Turn* (1967); Wittgenstein, *Tractatus Logico-philosophicus* (1933), 5.6. Williams, 'Wittgenstein and Idealism' (1973), argues that Wittgenstein was discussing the limits of language *in general*, not the limits of any particular language or any particular speaker (and each speaker, of course, may have access to more than one language). Wittgenstein was surely doing both, and he deliberately moves back and forth between the first person singular and the first person plural to convey both views.

Studies.[75] This movement originated with Barry Barnes and David Bloor at the Science Studies Unit of the University of Edinburgh (founded in 1964), both of whom were deeply influenced by Wittgenstein (Bloor, for example, wrote *Wittgenstein: A Social Theory of Knowledge* (1983)). Barnes and Bloor proposed what they called 'the strong programme'. What makes the strong programme strong is the conviction that the content of science, and not just the ways in which science is organized, or the values and aspirations of scientists, can be explained sociologically. Its essence lies in the principle of symmetry: the same sorts of explanation must, according to this principle, be given for all types of knowledge claims, whether they are successful or not.[xlvi] Thus if I meet someone who claims the earth is flat, I will seek a psychological and/or a sociological explanation for their peculiar belief; when I meet someone who claims that the Earth is a sphere floating through space and orbiting the sun, I must look for exactly the same sorts of explanation for this belief too. The strong programme insists that it is illegitimate to say that the explanation for the second belief is that it is right, or even that people believe it because they have good evidence for it. It thus systematically excludes from consideration the very feature which makes scientific arguments distinctive: their appeal to superior evidence. No follower of Wittgenstein can accept the notion of 'evidence' uncritically – indeed, some would claim they cannot accept it at all. Bertrand Russell first met Wittgenstein in 1911. In the brief obituary he wrote forty years later a memory from their early encounters imposed itself upon him:

> Quite at first I was in doubt as to whether he was a man of genius or a crank, but I very soon decided in favour of the former alternative. Some of his early views made the decision difficult. He maintained, for example, at one time that all existential propositions are meaningless. This was in a lecture room, and I invited him to consider the proposition: 'There is no hippopotamus in this room at present.' When he refused to believe this, I looked under all the desks without finding one; but he remained unconvinced.[76]

No one should be surprised if histories and philosophies of science which start from Wittgenstein seem to miss the very thing that science is all about.[xlvii]

xlvi Also known as the equivalence postulate. See 'Notes on Relativism and Relativists', No. 2 (pp. 581–2).

xlvii Exactly why Wittgenstein held this position in 1911 puzzles Wittgenstein scholars: McDonald, 'Russell, Wittgenstein and the Problem of the Rhinoceros' (1993) (Russell's

Barnes and Bloor are sociologists, and it is perfectly understandable that they insist that they and their fellow sociologists ought to stick to sociological explanations. But they go well beyond that. A relativistic view, that science is not a way of coming to grips with reality, is not the conclusion these scholars draw from their research; it is the assumption (following their interpretation of Wittgenstein) they build into it. In order to justify this, proponents of this position insist that evidence is never discovered: it is always 'constructed' within a particular social community. To say one body of evidence is superior to another is to adopt the viewpoint of one community and reject that of another. The success of a scientific research programme thus depends not on its ability to generate new knowledge but on its ability to mobilize the support of a community. As Wittgenstein puts it, 'At the end of reasons comes *persuasion*. (Think of what happens when missionaries convert natives.)'[77]

These scholars present science as being about rhetoric, persuasion and authority because the symmetry principle obliges them to assume that that is all it can be about. In doing so they go directly against the views of the early scientists themselves. Thus an influential article is entitled '*Totius in verba*: Rhetoric and Authority in the Early Royal Society', though the motto of the Royal Society was *nullius in verba* ('take no man's word for it'; or alternatively, perhaps, 'words don't count') – the claim of the Society's founders being that they were escaping from forms of knowledge based on rhetoric and authority.[xlviii] A form of history which presents itself as acutely sensitive to the language of people in the past proceeds by dismissing out of hand what they said, over and over again, about themselves. Anachronism, driven in disgrace out of the back door, re-enters in triumph through the front.

It may be hard to believe, but proponents of the strong programme have acquired a dominant position within the history of science. The most

memory had played a trick on him: from his contemporary correspondence it is clear that it was a rhinoceros, not a hippopotamus, that was not in the room).

xlviii Dear, '*Totius in verba*' (1985). What does Dear mean by *totius in verba*? He never says. The correct translation of *nullius in verba* is 'take no man's word for it' because it is a quotation from Horace and this is the meaning in its original context (Sutton, '*Nullius in verba*' (1994)); the tag from Horace had already been put to use in Carpenter, *Philosophia libera* (1622), 1st sig. 8v (the text differs from the 1621 edition); but *nullius* can mean *nihil*, and so 'words don't count' is a possible translation. However, *totius in verba* cannot mean 'language [or rhetoric] is everything' (which is, surely, the meaning Dear intends) but must mean 'on the word of the whole' – *totius* and *nullius* are not antonyms in all their usages. I return to *nullius in verba* below, pp. 282 and 294. For Galileo's rejection of the view that success in science can be based on rhetorical prowess, see below, p. 511.

striking example of this approach in action is Steven Shapin and Simon Schaffer's *Leviathan and the Air-pump* (1985), generally acknowledged as the most influential work in the discipline since Thomas Kuhn's *Structure of Scientific Revolutions*.[xlix] The new history of science offered, in Steven Shapin's phrase, a social history of truth.[l] Scientific method, it was now argued, kept changing, so that there was no such thing as *the* scientific method: a famous book by Paul Feyerabend was entitled *Against Method*,[li] its catchphrase 'Anything goes'; it was followed by *Farewell to Reason*.[78] Some philosophers and nearly all anthropologists agreed: standards of rationality were, they insisted, local and highly variable.[79]

But we must reject the Wittgensteinian notion that truth is simply consensus, a notion incompatible with an understanding of one of the fundamental things science does, which is to show that a consensus view must be abandoned when it is at odds with the evidence.[lii] The classic text here is Galileo's 'Letter to Christina of Lorraine' (1615) in defence of Copernicanism. He starts by saying that there are certain matters on which the philosophers have all been agreed, but that he has discovered with his telescope facts that are entirely at odds with their beliefs; consequently, they need to change their views.[80] What seemed to be true can no longer be regarded as true. What Galileo is engaged in here (may even be said to be inventing) is what Shapin and Schaffer call 'the empiricist language-game', according to which facts are 'discovered rather than invented'.[81] That's true; the Wittgensteinian move then consists, in the view of his followers, in maintaining that there are no grounds for thinking that this game has greater validity than any other, and Galileo then becomes no more reasonable than the philosophers he is

xlix See 'Notes on Relativism and Relativists', No. 3 (pp. 582–3).

l See 'Notes on Relativism and Relativists', No. 4 (pp. 583–5).

li *Against Method* did not appear as a book until 1975, but it started life as a conference paper given in 1966 (Feyerabend, 'Against Method' (1970)). The first hardback edition of the book carries on its dust jacket not the usual potted biography of the author but instead his horoscope: Feyerabend was certainly consistent (and playful) in his relativism. For his defence of astrology, see Feyerabend, *Science in a Free Society* (1978), 91–6.

lii Followers of Wittgenstein maintain that a belief system can never be refuted by new evidence; followers of Popper maintain that refutation is straightforward; and followers of Kuhn maintain that new evidence can throw a belief system into crisis and eventually bring about a revolutionary transition to a new consensus. Both the Kuhnian and Popperian positions are, in principle, compatible with an understanding of what science does; the Wittgensteinian position, as presented by his followers, is inherently anti-scientific. I return to this below, in Chapter 15.

opposing.[liii] And at this point Wittgensteinian history of science places itself directly at odds with Galileo's own account of what he is actually doing, and *history of science* is in a direct conflict with *science*.[liv]

When Shapin and Schaffer refer to 'the empiricist language-game' as if it were just one amongst any number of equally valid language-games, they assume that there is no reality outside the language-games of Galileo and his opponents because it is the language-games themselves which define what is to count as real; they assume that 'the limits of my language mean the limits of my world.'[lv] This cannot be true in any absolute sense; Galileo's telescope transformed the world of astronomers before they had any new words for what they could now see – before they even had the word 'telescope'. In writing about his discoveries Galileo was not obliged to write in a way that others found puzzling or incomprehensible: it was what he said that caused consternation, not how he said it. Yet, although the philosophers understood him perfectly well, some of them continued to insist that what he and other astronomers claimed to see couldn't possibly be there. Galileo's world and their world had different limits, although they understood each other perfectly. The limits were not set by their language but by their priorities, by their sense of what was negotiable and what was not.[lvi]

The telescope may seem a special case. Of course our world changes

liii 'Isn't the question this: "What if you had to change your opinion even on these most fundamental things?" And to that the answer seems to me to be: "You don't have to change. That is just what their being 'fundamental' is"' (Wittgenstein, *On Certainty* (1969), §512).

liv Wittgenstein writes: 'Suppose we met people who did not regard that [the propositions of physics] as a telling reason. Now, how do we imagine this? Instead of the physicist, they consult an oracle. (And for that we consider them primitive.) Is it wrong for them to consult an oracle and be guided by it? – If we call this "wrong" aren't we using our language-game as a base from which to *combat* theirs?' (Wittgenstein, *On Certainty* (1969), §609). In this case, instead of consulting Galileo (or Boyle, whose language deliberately echoes Galileo), Shapin & Schaffer consult Wittgenstein and use his language-game as a base from which to combat science.

lv Actually, proponents of the strong programme treat the empiricist language-game as if it were one of a number of equally false language-games, for the only valid game, in their eyes, is the Wittgensteinian meta-game. Everyone is limited by language, except for those who write about how everyone is limited by language. But there is no need to linger over this fatal flaw.

lvi Kuhn argued that there are limits to communication between people who inhabit different intellectual worlds, but there is general agreement that he overstated this argument: Galileo and his critics had difficulties in agreeing but not in communicating; they played by different rules, but they could make sense of their opponents' moves. Kuhn's views are outlined in Sankey, 'Kuhn's Changing Concept of Incommensurability' (1993), and criticized in Sankey, 'Taxonomic Incommensurability' (1998); see also Hacking, 'Was There Ever a Radical Mistranslation?' (1981).

when we introduce a new technology, or go somewhere we have never been before. But every day we experience things for which we have no words, and in such circumstances we are often lost for words, or find ourselves saying that something is 'beyond words'. Only later do we sometimes find the words (love, grief, jealousy, despair) for what we have been feeling all along. 'It did not occur to him,' writes Tolstoy of Andrei, 'that he was in love with Miss Rostov.' And the whole, wonderful point of some experiences – music, sex, laughter – is that there are not and never will be any adequate words to describe them. This does not mean they do not exist.

But even though it is by no means always the case that 'the limits of my language are the limits of my world', we must acknowledge that our language often determines the limits of what we can argue about and understand with precision. Clouds were only named early in the nineteenth century – 'cirrus' and 'nimbus' may sound old because they are Latin, but the Romans had no names for the different types of cloud.[82] Of course, long before there was a language for clouds, people experienced them more or less as we do: one only has to look at seventeenth-century Dutch seascapes to see all the different types of cloud accurately represented, even though the painters had no names for them. Robert Hooke evidently saw clouds perfectly clearly when he asked, 'What is the reason of the various Figure of the Clouds, undulated, hairy, crisped, coyled, confus'd, and the like?'[83] But he is well aware that describing clouds is at the limit of his linguistic capacities. The naming of clouds was a great event in the history of meteorology, after which much more serious discussion and understanding were possible.

When we are studying ideas, linguistic change is the key to finding out what people understood that their predecessors did not. A decade before Galileo's telescopic discoveries, William Gilbert, the first great experimental scientist of the new age, had acknowledged: 'Sometimes therefore we use new and unusual words, not that by means of foolish veils of vocabularies we should cover over the facts [*rebus*] with shades and mists (as Alchemists are wont to do) but that hidden things which have no name, never having been hitherto perceived, may be plainly and correctly enunciated.'[lvii] [84] His book begins with a glossary to help the

lvii This is the nub of the matter. Wittgenstein's whole undertaking, as interpreted by the sociologists, is directed at disputing the idea that there can be perception independent of enunciation. Thus he says, 'the idea of "agreement with reality" does not have any clear application' (*On Certainty*, §215). Science, of course, seeks to show that it does, just as Russell sought to show that there was no hippopotamus in the room.

reader make sense of these new words. Then a few months after Galileo discovered what we call the moons of Jupiter (Galileo does not call them moons, but first stars and then planets), Johannes Kepler invented a new word for these new objects: they were 'satellites'.[lviii] Thus historians who take language seriously need to search out the emergence of new languages, which must represent transformations in what people can think and how they can conceptualize their world.[lix]

It is important here to distinguish this claim from the argument with which this chapter started. The historian has always to learn the language which people in the past used, and must always be alert to changes in that language; that does not mean they need always use that language when writing about the past. Kepler's word 'satellite' acknowledges that Galileo had discovered a new type of entity, but it is perfectly sensible for us to say that what Galileo had discovered were the moons of Jupiter (a terminology used neither by Galileo nor Kepler – the earliest usage I can find is 1665 – and thus, strictly speaking, anachronistic), particularly as, for us, stars (Galileo's term) are fixed, and satellites (Kepler's term) are usually human-made objects launched into space.

Recent history of science, for all its talk of languages and discourses, has not been nearly attentive enough to the emergence in the seventeenth century of a new language for doing natural science, a language discussed in Part Three of this book. Indeed, so invisible has this new language been that the very same scholars who refuse to use the word 'scientist' for anybody before the second half of the nineteenth century happily talk about 'facts', 'hypotheses' and 'theories' as if these were transcultural concepts. This book seeks to remedy this peculiar lapse.[lx] We can state one of its core premises quite simply: a revolution in ideas requires a revolution in language. It is thus simple to test the claim that there was a Scientific Revolution in the seventeenth century by looking for the revolution in language that must have accompanied it. The

lviii *Narratio de observatis Jovis satellitibus*, dated 11 September 1610 but published in 1611 (a modern edition in Kepler, *Dissertatio cum nuncio sidereo* (1993)). In classical Latin *satellitium* means an escort or guard.

lix 'When language-games change, then there is a change in concepts, and with the concepts the meanings of words change.' Wittgenstein, *On Certainty* (1969), §65.

lx It is remarkable that, so long after the 'linguistic turn', the basic history of some key words/concepts which make the scientific enterprise possible remains to be written. This present book can thus be seen, in part, as an extension of Bruno Snell's account of the primitive origins of science: Snell, 'The Origin of Scientific Thought' (1953) (first published in 1929) and Snell, 'The Forging of a Language for Science in Ancient Greece' (1960).

revolution in language is indeed the best evidence that there really was a revolution in science.

There are some features of linguistic change that are worth bearing in mind as we go on. Obviously (as we have already seen in the cases of 'arts' and 'sciences'), over time the meanings of words change. But often words do not simply change their meaning, rather they acquire new meanings, meanings which are sometimes apparently unrelated to their original meaning. We have seen how the word 'revolution' is now used in so many different senses that one source of confusion as to whether there is or is not a Scientific Revolution is a failure to distinguish these senses one from another. When I visit the local branch of my bank, I don't think of this vast business as a tree, but 'branch' here is a dead metaphor. So too is the word 'volume' when used in the context of measurement: first in French, and then much later in English, 'volume' began to be used to refer not to a book but to the space occupied by a three-dimensional object. If I measure the volume of a sphere, the language I am using is a dead metaphor.

When we write of 'laws of nature' the word 'laws', too, is being used in a metaphorical sense. What are laws of nature? To understand the sort of contexts in which the phrase is used it may help to explore its origins; such an exploration may, in the end, help us understand why there is no good answer to the question 'What are laws of nature?' other than an account of how we use the phrase (in this case, as Wittgenstein said, meaning is indeed use). Thus in the United Kingdom we have an unwritten constitution. What is an unwritten constitution? Any decent answer will be full of puzzles and paradoxes but will need to include an account of how the idea that states have constitutions originates with Bolingbroke in 1735, and that the idea of an *unwritten* constitution distinguishes the United Kingdom from the United States and France, the first countries to have written constitutions. Just as the idea of an unwritten constitution contains apparently irresolvable puzzles once written constitutions become the norm (How do we know what the unwritten constitution is? Whence comes its authority?), so crucial concepts that we use in discussing science ('discovery', or 'laws of nature') are inherently problematic, at least for us: the only way to understand them is to recover their history.[85] My argument is that during the seventeenth century the idea of natural science underwent a fundamental revision, and by the end of the century the idea that had taken shape was basically the one that we still have. I don't claim that idea was consistent or coherent; I claim that it was successful, that it provided a template for the discovery of new knowledge and new technologies.[lxi]

§ 6

Much of this chapter has been concerned with the language of science, as much of the book will be, but the book's argument is also and equally about what Leonardo called 'the test of experience'. The first generation of historians and philosophers to study the Scientific Revolution downplayed the importance of new evidence and new experiments, insisting that what really counted was what Butterfield called 'a transposition in the mind of the scientist himself'. The foundations of modern science were, the philosopher Edwin Burtt had insisted back in 1924, metaphysical.[86] According to Koyré, '[I]t is thought, pure unadulterated thought, and not experience or sense-perception . . . that gives the basis for the "new science" of Galileo Galilei.'[87] Thus what Koyré took to be the key concept that made possible the new science, the concept of inertia, was, he held, constructed by Galileo thinking about everyday experience, by a mere thought experiment. This, it seems to me, is to mistake effect for cause, to get the whole story of the new science upside down and back to front.[lxii] The Scientific Revolution is precisely about new experiences and new sense-perceptions. It should be obvious that if all that was required for the Scientific Revolution was new *thinking*, then it would be impossible to explain why it did not take place long before the seventeenth century.[lxiii]

lxi 'For the purpose of understanding and facilitating scientific practice,' writes Hasok Chang, 'I would like to suggest a fundamental re-orientation in our conception of knowledge, to think of it in terms of *ability* rather than *belief*.' (Chang, *Is Water H2O?* (2012), 215; and on 'success', 227–33). I return to this in the final chapter.

lxii And, of course, to misunderstand Galileo: see, for example, Galileo's essay on the tides (Galilei, *Le opere* (1890), Vol. 5, 371–95), in which experience is described as a reliable guide – '*sensate esperienze (scorte sicure nel vero filosofare)*' (378); Stabile, '*Il concetto di esperienza in Galilei*' (2002); Galilei, *Le opere* (1890), Vol. 10, 118 (Galileo to Altobelli), Vol. 18, 249 (Galileo to Liceti) & 69 (Baliani to Galileo). Galileo's father, Vincenzo, had already stressed over and over again the primacy of experience: Palisca, '*Vincenzo Galileo*' (2000).

lxiii If *thinking* were enough to engender the new science it would have begun not with Galileo but with the fourteenth-century philosopher Nicholas Oresme. At the most, one might argue that the recovery of certain classical texts (Archimedes, Lucretius, Plato) was an essential precondition for the new thinking: but this process was complete by the middle of the fifteenth century.

But for thirty years now a second generation of historians and philosophers of science has been attacking the claim that the Scientific Revolution vastly improved humankind's ability to understand nature; adopting a relativist perspective, they have been unwilling to acknowledge that Newton was superior to Aristotle or Oresme, even if only in the sense that his theories made possible better predictions and new types of intervention. Their arguments have convinced almost all anthropologists, nearly every professional historian and many philosophers. But they are wrong. Thanks to the Scientific Revolution, we have a much more reliable type of knowledge than ancient and medieval philosophers ever had, and we call it science. For the first generation, the new science was all in the mind; for the second, it was simply a language-game. These two debates, about thinking and knowing, interlock because both generations have downplayed the idea that the new science was grounded in a new type of engagement with sensory reality. Both thus missed its essential characteristic: that it systematically employed the test of experience.

For the new scientists of the second half of the seventeenth century were in a quite different position from their classical, Arab and medieval predecessors. They had the printing press (a fifteenth-century invention whose impact grew through the seventeenth century), which created new types of intellectual community and transformed access to information; they had a family of instruments (telescopes, microscopes, barometers), all made from glass, that acted as agents of change; they had a new preoccupation with the test of experience, which had now given rise to the experimental method; they had a new, critical attitude to established authority; and they had a new language, the language which we now speak, which made it much easier to think new thoughts. Mutually supporting and interlocking, these diverse elements made possible the Scientific Revolution.

§ 7

In 1748 Denis Diderot, the great Enlightenment philosopher, published, anonymously, an erotic novel entitled *The Indiscreet Jewels* (the word 'jewel' here is a euphemism for the vagina). It was, as he and his publisher must confidently have expected, promptly banned and immediately successful. Chapter 29 is subtitled 'Perhaps the best, and the least read, of this story' – least read because, exceptionally, there is no sex in it. It describes how the protagonist (the sultan Mangogul, a flattering representation of Louis XV) has a dream in which he flies on the back of a hippogriff to a vast edifice suspended in the clouds. There, a great crowd of misshapen individuals are gathered around an old man in a pulpit made of cobwebs; he says nothing, but blows bubbles. All are naked, except for a few bits and pieces of cloth here and there on their bodies – these are fragments of the robe of Socrates, and we discover that we are in the temple of philosophy. Suddenly:

> I saw in the distance a child who was walking towards us with slow but sure steps. He had a small head, a thin body, weak arms, and short legs, but his limbs grew larger and larger as he advanced. Throughout his successive spurts of growth, he appeared to me in many different guises; I saw him point a long telescope skyward, assess the rate of a falling body with the help of a pendulum, ascertain the weight of the air with a tube full of mercury, and break up light with a prism in his hand. He became a huge colossus; his head touched the skies, his feet were lost in the abyss, and his arms stretched from one pole to the other. With his right hand he shook a torch whose rays spread out in all directions lighting the depths of the waters and penetrating into the bowels of the earth.

The colossus strikes the temple, which collapses, and Mangogul stirs.[88]

'What,' Mangogul asks just before he awakes, 'is this gigantic figure?' The answer may seem obvious: Diderot is writing about the transformation in knowledge that we now call the Scientific Revolution. As we shall see, Galileo had pointed his telescope skyward, Mersenne (following Galileo's example) had accurately measured the rate of falling bodies, Pascal had weighed the air, Newton had broken up light with a prism. The new science destroyed the old one taught by the philosophers. But Diderot's name for this newly born colossus is not 'Science', as we might expect. The word 'Science' in French was not and is not nearly specific enough to refer to the new sciences of Galileo and Newton, for, as we have seen, there were and are all sorts of sciences, including now the social sciences. Even 'natural science' would not do, for the philosophers had always claimed to be experts in natural science, and so 'natural science' would no more serve to distinguish the new science from the old than would 'natural philosophy'. Instead, Plato, who has conveniently turned up to explain what is going on, says, 'Recognize Experience, for it is she.'[lxiv] And yet surely there is nothing new about

lxiv '*Reconnoissez l'Expérience, me répondit-il; c'est elle-même*' (Diderot, *Les Bijoux indiscrets* (1748), Vol. 1, 352 [of 370]).

experience? Isn't experience something that all human beings have in common? How then can 'Experience' be the right name for the new sciences?

In answering this question I will keep returning to the problem to which Diderot alerts us when he names his colossus 'Experience' – the difficulty of finding an adequate language with which to describe the new science, a problem which is not just ours when we seek to understand it but rather was a crucial difficulty for those who invented it, and also for those, like Diderot, who wrote in praise of it. Indeed, I shall argue that the new science would not have been possible without the construction of a new language with which to think, a language necessarily cobbled together out of available words and phrases. That language was pioneered in English, where, for example, 'experience' and 'experiment' began to diverge in meaning during the seventeenth century. (Diderot, who began his career by translating books from English into French, was well acquainted with this new language.) Diderot's *expérience* can thus be translated into English, not quite exactly, as 'experiment' (a word which still does not exist in French), and it is immediately apparent that 'experiment' might be a helpful word in describing the new science, a more helpful word, perhaps, than 'experience', though we have already seen that Leonardo identified experience as the key to reliable knowledge. We can mark the beginning of this process of language construction precisely: it starts with a new word, which inaugurated a wider transformation in the role experience had to play, the word 'discovery', which is a word with equivalents in every European language.

In the pages which follow we will see how experience, in the form of observations and experiments directed at making discoveries, came to be something new in the seventeenth century; how this new enterprise of discovery made possible the invention of science; and how this new science began to transform the world, a process which resulted in the modern technologies on which our lives depend. They tell the story of the birth of science, of its infancy, and of its extraordinary transformation into the great colossus under whose shadow we all now live. But Diderot's peculiar chapter also serves as a warning: with its dream framing, its monsters and its allegories, its linguistic slipperiness, it conveys a sense of difficulty. What would a history of experience – of this new sort of experience – be like?

It might seem far easier for us to answer that question than it was for Diderot, for he was still caught up in the triumph of Newtonianism (which came later in France than in England), while we have all the advantages of hindsight. But Diderot had one great advantage over us: graduating from the Sorbonne in 1732, he had been educated in the world of Aristotelian philosophy. He knew how shocking the destruction of that world had been, for he had experienced it at first hand. From a bird's-eye view – the historian's view – the Scientific Revolution is a long, slow process, beginning with Tycho Brahe and ending with Newton. But for the individuals caught up in it – for Galileo, Hooke, Boyle and their colleagues – it represents a series of sudden, urgent transformations. In 1735 Diderot, educated in the old ways, still planned to become a Catholic priest; by 1748, only a little more than a decade later, he was an atheist and a materialist, already at work on the great *Encyclopaedia*, the first volume of which appeared in 1751. The destruction of the temple of philosophy was not, for him, an historical event; it was a personal experience, the moment when he had awakened from a nightmare.

PART ONE

The Heavens and the Earth

> What indeed is more beautiful than heaven, which of course contains all things of beauty?
>
> – Nicholas Copernicus, *On the Revolutions* (1543)[1]

The two chapters of Part One address three intellectual revolutions which transformed the way in which the cosmos was conceived. The first argues that before Columbus discovered America in 1492 there was no clear-cut and well-established idea of discovery; the idea of discovery is, as will become apparent, a precondition for the invention of science. The second shows that the discovery of America disproved a central claim about the world which had been generally accepted before 1492, that there could be no antipodean land masses, for South America lay halfway around the globe from parts of the Old World. An immediate consequence, therefore, which is the subject of Chapter 4, was a radical transformation in the understanding of how the Earth is constructed: the emergence of the concept of the terraqueous globe. This was a crucial precondition for the astronomical revolution which followed. The chapter goes on to reappraise what Thomas Kuhn called the Copernican revolution. As we will see, the Copernican revolution was delayed until the seventeenth century: very few sixteenth-century astronomers accepted Copernicus's claim that the Earth revolves around the sun instead of standing still at the centre of the cosmos. The real revolution in astronomy came with Tycho Brahe's nova, with the abandonment of belief in the crystalline spheres, and with the invention of the telescope. The key date is not 1543 but 1611.

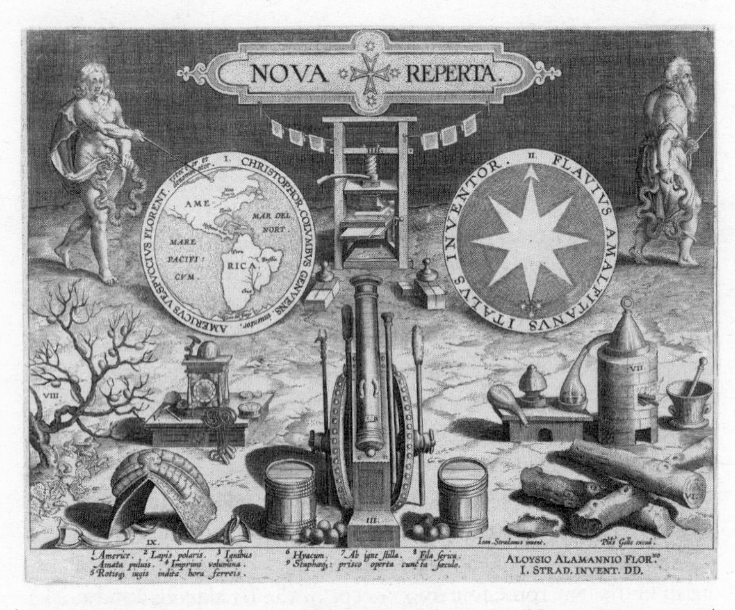

The title page of Johannes Stradanus's *New Discoveries* (*Nova reperta*, *c.*1591) summarizes the knowledge that marks off the modern world from the ancient. Pride of place is given to the discovery of America and the invention of the compass, with, between them, the printing press. Also present are gunpowder, the clock, silk weaving, distillation and the saddle with stirrups.

3
Inventing Discovery

Discovery is what science is all about
– N. R. Hanson, 'An Anatomy of Discovery', 1967[1]

§ 1

On the night of 11/12 October 1492 Christopher Columbus discovered America. Either Columbus, on the *Santa Maria*, who claimed to have seen a light shining in the dark some hours before, or the lookout on the *Pinta*, who actually saw land by the light of the moon, was the first European since the Vikings to see the New World.[2] They thought that the land they were approaching was part of Asia – indeed, throughout his lifetime (he died in 1506) Columbus refused to recognize that the Americas were a continent. The first cartographer to show the Americas as a vast land mass (not yet quite a continent) was Martin Waldseemüller in 1507.[3]

Columbus discovered America, an unknown world, when he was trying to find a new route to a known world, China. Having discovered a new land, he had no word to describe what he had done. Columbus, who had no formal education, acquired several languages – Italian, Portuguese, Castilian, Latin – to supplement the Genoese dialect of his childhood, but only Portuguese had a word (*discobrir*) for 'discovery', and it had acquired it very recently, only since Columbus had failed in his first attempt, in 1485, to get backing from the king of Portugal for his expedition.

The idea of discovery is contemporaneous with Columbus's plans for his successful expedition, but Columbus could not appeal to it because he wrote the accounts of his voyage not in Portuguese but in Spanish and Latin. The nearest classical Latin verbs are *invenio* (find out), *reperio* (obtain) and *exploro* (explore), with the resulting nouns *inventum*, *repertum* and *exploratum*. *Invenio* is used by Columbus to announce his discovery of the New World; *reperio* by Johannes Stradanus for the title of his book of engravings illustrating new discoveries (*c.*1591); and *exploro* by Galileo to announce his discovery of the moons of Jupiter (1610).[4] In modern translations these words are often represented by the word 'discovery', but this obscures the fact that in 1492 'discovery' was not an established concept. More than a hundred years later Galileo still needed, when writing in Latin, to use convoluted phrases such as 'unknown to all astronomers before me' to convey it.[i5]

In all the major European languages the same metaphorical use of a word meaning 'uncover' was soon adopted to describe the voyages of discovery. The lead was taken by the Portuguese, who had been the first to engage in journeys of exploration, attempting, in a series of expeditions beginning in 1421, to find a sea route to the spice islands of India by sailing along the coast of Africa (in the process proving, contrary to the established doctrines of the universities, that the equatorial regions are not too hot for human survival): in that language the word *descobrir* was already in use by 1484 to mean 'explore' (probably as a translation of the Latin *patefacere*, to lay open). In 1486, however, Fernão Dulmo proposed a quite new type of enterprise, a voyage westwards across the ocean into the unknown to find (*descobrirse ou acharse*, discover or find) new lands (this was two years after Columbus had proposed sailing westwards to reach China).[6] The voyage probably never took place, but it would have been one of discovery rather than exploration. Dulmo discovered nothing; but his concept of discovery was soon to take on a life of its own.[ii]

i Compare Giordano da Pisa's account (written in Italian in 1306) of the invention of spectacles, as described in a sermon he had heard: 'It is not yet twenty years since the technique of making spectacles, which enable one to see well, was found [*si trovó*]. It is one of the best techniques and one of the most useful that the world knows, and it is just a short time since it was found: a new technique that never existed previously [*arte novella che mai non fu*]. And the preacher said: "I saw the person who first found it and used it, and had a conversation with him."' Evidently, Giordano has no word for 'invention' or 'discovery', so he relies on a paraphrase: 'a new technique that never existed previously'. So, too, Filarete (died *c.*1469, writing also in Italian) on Brunelleschi's invention of perspective painting: 'Pippo di ser Brunelleschi found out [*inventò*] how to make perspective images, which before him no one had known how to do . . . Although the ancients were clever and cunning, they were not acquainted with perspective.' *Inventare* alone did not in itself adequately convey the idea of finding out something which had never previously been known; Filarete was well aware that his readers would assume all discovery was really rediscovery, so he had in addition to spell out his opposition to that view.

ii A test case for the claim that the concept of discovery is new in 1486 is provided by the fourteenth-century documents (Verlinden, 'Lanzarotto Malocello' (1958)) which discuss the first Portuguese voyages to the Canaries. '*Predictarum insularum fuerunt prius nostri*

The new word began to spread across Europe with the publication in 1504 of the second of two letters written (or supposedly written) by Amerigo Vespucci, in which he described his journeys to the New World in the service of the king of Portugal. This 'Letter to Piero Soderini', written and first published in Italian, had appeared in a dozen editions by 1516. In Italian it used *discoperio* nine times as an import from Portuguese; the Latin translation (based on a lost French intermediary), published in 1507, used *discooperio* twice.[7] This was the first use of this word in the modern sense of 'discover': *discooperio* exists in late Latin (it occurs in the Vulgate), but only to mean 'uncover'. Because it did not exist in classical Latin *discooperio* never established itself as a respectable term; in any case, discovery was such a new concept that at first it required explication. Vespucci helpfully explained that he was writing about the finding of a new world 'of which our forefathers made absolutely no mention'.[iii]

The new word spread almost as fast as news of the New World. Fernão Lopes de Castanheda published *História do descobrimento e conquista da Índia* (i.e. the New World) in 1551, and this was quickly translated into French, Italian and Spanish, and later into German and English, and played a key role in consolidating the new usage. From its

---

regnicole inventores' ['the first finders of these islands came from our kingdom'], 1188; '*avendo délie nos as yllas que trobou e nos gaanou que som no mar do Cabo Nom*' ['having received from him the islands that he found and won for us'], 1197 – here *inventores* and *trobou* seem to be words that imply discovery; but *querentes ad eas insulas, quas vulgo repertas dicimus* ['setting out for those islands which in common speech we call "found"'], 1191 – it turns out that the language of 'finding' is only a popular figure of speech (because, of course, you cannot 'find' an inhabited island, since it can hardly be said to be 'lost'). Moreover, the educated knew there was no such thing as discovery – and, indeed, the Canaries were known to the Romans.

iii Waldseemüller, *The Cosmographiæ introductio* (1907), 88 (translation corrected: see xliv). There is a helpful discussion in Brotton, *A History of the World in Twelve Maps* (2012), 155–6, but Brotton goes on (166–7) to quote a mistranslation of Waldseemüller (from Hessler, *The Naming of America* (2008)), giving the impression that Waldseemüller thought Ptolemy had *some* knowledge of America, and thus implying that even Waldseemüller lacked a fully developed concept of discovery. For the Latin text and a reliable translation, see Waldseemüller, *The Cosmographiæ introductio* (1907), xxviii, 68. See also, for example, Grynaeus, *Novus orbis regionum ac insularum veteribus incognitarum* (1532). Contrast Columbus's insistence that the mainland he had found in his first two voyages was 'well known to the ancients and not unknown, as the envious and ignorant would have it' (quoted from Washburn, 'The Meaning of "Discovery"' (1962), 12). As late as 1535 Oviedo continued to defend the claim that the existence of the New World had simply been forgotten: Bataillon, '*L'Idée de la découverte de l'Amérique*' (1953), 44; O'Gorman, *The Invention of America* (1961), 16. What was new, to their mind, was not the New World but the transoceanic voyage.

Table of book production

Here are some figures, in thousands, for the production of individual copies of printed books: of necessity, they are merely sophisticated estimates. The Printing Revolution was a very large-scale but at the same time very drawn-out process which neatly coincides with the Scientific Revolution (see below, p. 95). In 1500 it was only just beginning to pick up speed:

1450–1500	1500–1550	1550–1600	1600–1650	1650–1700	1700–1750
12,589	79,017	138,427	200,906	331,035	355,073

(From Buringh & van Zanden, 'Charting the "Rise of the West"' (2009), 418.)

first appearances in the titles of books we can see how rapidly the word became established: Dutch – 1524 (but then not again until 1652); Portuguese – 1551; Italian – 1552; French – 1553; Spanish – 1554; English – 1563; German – 1613.

If discovery was new with Vespucci, then surely invention was not? In the sixteenth and seventeenth centuries gunpowder, printing and the compass were the three inventions of modernity most frequently cited to prove the superiority of the moderns to the ancients. All three pre-date Columbus, but I can find no example of them being cited in this way before 1492.[8] It was the discovery of America which demonstrated the importance of the compass; in due course printing and gunpowder might have seemed of revolutionary importance, but as it happens they were only recognized as revolutionary in the post-Columbus age. And there are good reasons for this: the first battle whose outcome was determined by gunpowder is often said to be Cerignola in 1503, and printing had very little impact before 1500.

We are so used to the varied meanings of the word 'discovery' that it is easy to assume it always meant roughly what it means today. 'I've just discovered I'm entitled to a tax refund,' we say. But 'discover' in this sense comes after talk about Columbus discovering the New World; it's the voyages of discovery which give rise to this loose use of 'discover' to mean 'find out', and this loose usage was encouraged by the practice of translating *invenio* as 'discover'. The core meaning of 'discovery', after 1492, is not just an uncovering or a finding out: someone who announces a discovery is, like Columbus, claiming to have got there first, and to have opened the way for all those who will follow. 'We have found the secret of life,' Francis Crick announced to all and sundry in the Eagle pub in Cambridge on 13 February 1953, the day he and James Watson worked out the structure of DNA.[9] Discoveries are moments in an historical process that is intended to be irreversible. The concept of discovery brings with it a new sense of time as linear rather than cyclical. If the discovery of America was a happy accident, it gave rise to another even more remarkable accident – the discovery of discovery.[iv][10]

I say 'more remarkable', for it is discovery itself which has transformed our world, in a way that simply locating a new land mass could never do.[v] Before discovery history was assumed to repeat itself and tradition to provide a reliable guide to the future; and the greatest achievements of civilization were believed to lie not in the present or the future but in the past, in ancient Greece and classical Rome. It is easy to say that our world has been made by science or by technology, but scientific and technological progress depend on a pre-existing assumption, the assumption that there are discoveries to be made.[vi] The new attitude was summed up by Louis Le Roy (or Regius, 1510–77) in 1575.[11] Le Roy, who was a professor of Greek and had translated Aristotle's *Politics*, was the first fully to grasp the character of the new age (I quote the English translation of 1594):

iv In stressing 'discovery' I am providing an explanation for the new cultural values of late-Renaissance Europe. Alternatively, one can stress 'curiosity' as a distinctly European value, but then one has to find an explanation of where the new approval for curiosity (which had traditionally been regarded as a vice) came from; curiosity only begins to be approved of in the late seventeenth century (an early example is Hobbes, *Humane Nature* (1650), 112, where curiosity is defined as 'appetite of knowledge'), and so approval for curiosity would seem to be a consequence, not a cause of, the Scientific Revolution. 'Discovery', I believe, is also a helpful category in distinguishing late-Renaissance Western civilization from other civilizations: for a critique of the claim that there were Chinese voyages of discovery, see Finlay, 'China, the West and World History' (2000).

v Invention, of course, is important as well as discovery. But the crucial modern inventions depend on prior scientific discoveries: in the case of the steam engine, Boyle's law. The designers of the first steam engines did not know about latent heat, but they understood air pressure; it was this that made it possible for them to grasp that the steam engine could be more than a mere toy, as it had been for Hero of Alexandria, and could harness immense power.

vi 'The fifteenth-century Portuguese voyages of exploration had revealed small groups of new islands and expanded European knowledge of already familiar continents, but the revelation that the world included whole new continents undreamed of by the ancients opened up a fissure in time as well as space. Compared to earlier writing on travel, the works from the decades after 1492 demonstrate a heightened sense of novelty and possibility – of just how new and different things were able to be.' (Daston & Park, *Wonders and the Order of Nature* (1998), 147.) See also, for example, Humboldt, *Examen critique* (1836), Vol. 1, viii–x.

> [T]here remayne more thinges to be sought out, then are alreadie invented, and founde. And let us not be so simple, as to attribute so much unto the Auncients, that wee beleeve that they have knowen all, and said all; without leaving anything to be said, by those that should come after them ... Let us not thinke that nature hath given them all her good gifts, that she might be barren in time to come: ... How many [secrets of nature] have bin first knowen and found out in this age? I say, new lands, new seas, new formes of men, maners, lawes, and customes; new diseases, and new remedies; new waies of the Heaven, and of the Ocean, never before found out; and new starres seen? yea, and how many remaine to be knowen by our posteritie? That which is now hidden, with time will come to light; and our successours will wonder that wee were ignorant of them.[12]

It is this assumption that there are new discoveries to be made which has transformed the world, for it has made modern science and technology possible.[13] (The idea that there are 'formes of men, maners, lawes, and customes' also represents the birth of the idea of a comparative study of societies, cultures or civilizations.)[14]

Le Roy's text helps us to distinguish events, words and concepts. There were geographical discoveries before 1486 (when Dulmo changed the meaning of the word *descobrir*), such as that of the Azores, which happened sometime around 1351 – but no one thought of them as such; no one bothered to record the event, for the simple reason that no one was very interested. The Azores were later rediscovered around 1427, but the event still seemed unimportant, and no reliable account survives. The governing assumption was that there was no such thing as new knowledge: just as when I pick up a coin that has been dropped in the street I know that it has belonged to someone else before me, so the first Renaissance sailors to reach the Azores will have assumed that others had been there before them. In the case of the Azores this was mistaken, while it was correct in the case of Madeira, discovered – or rather rediscovered – around the same time, for it was known to Pliny and Plutarch. But no one thought Columbus's discovery of (as he thought) a new route to Asia was insignificant; there were disputes over whether America was or was not a previously unknown land, but no one claimed that any Greek or Roman sailor had made the voyage westward before Columbus. (There was an obvious explanation: the Greeks and the Romans did not have the compass and so were reluctant to sail out of sight of land.) Thus Columbus knew he was making a discovery, of a new route if not of a new land; the discoverers of the Azores did not.

Although there were already ways of saying that something had been found for the first time and had never been found before (and, indeed, people continued to rely on such phrases to convey 'discovery' when writing in Latin), it was very uncommon before 1492 for anyone to want to say anything of the sort, because the governing assumption was that there was 'nothing new under the sun' (Ecclesiastes 1:9). The introduction of a new meaning for *descrobrir* implied a radical shift in perspective and a transformation in how people understood their own actions. There were, one can properly say, no voyages of discovery before 1486, only voyages of exploration. Discovery was a new type of enterprise which came into existence along with the word.

A central concern of the history of ideas, of which the history of science forms part, has to be linguistic change. Usually, linguistic change is a crucial marker of a modification in the way in which people think – it both facilitates that change and makes it easier for us to recognize it. Occasionally, focussing on a linguistic alteration can mislead us into thinking that something important has happened when it hasn't, or that something happened at a particular moment when it actually happened earlier. There is no simple rule: one has to examine each case on its merits.[vii] Take the word 'boredom'. Did people suffer from boredom before the word was introduced in 1829?[15] Surely they did: they had the noun 'ennui' (1732), the noun 'bore' (1766) and the verb 'to bore' (1768). Shakespeare had the word 'tediosity'. 'Boredom' is a new word, not a new concept, and certainly not a new experience (although it may have been a much more frequent experience in the age of Dickens than in the age of Shakespeare, and, where *ennui* was thought of as peculiarly French, boredom was certainly British). Other cases are a little more complicated. The word 'nostalgia' was coined (in Latin) in 1688 as a translation of the German *Heimweh* (homesickness). It first appears in English in 1729, well before 'homesick' and 'homesickness'. The French had had '*la maladie du pays*' since at least 1695. Was homesickness new? I rather doubt it, even if there was no word for it; what was new was the idea that it was a potentially fatal illness requiring medical intervention.[16] It's the absence of a simple rule, combined with the fact

vii Quentin Skinner offers 'originality' as an example of a concept which undoubtedly precedes the word: Skinner, *Visions of Politics* (2002), Vol. 1, 159.

that much linguistic change lies in giving new meanings to old words, that explains why some of the most important intellectual events have become invisible: we tend to assume that discovery, like boredom, has always been there, even if there are more discoveries at some times than at others; we assume the words are new, but not the concepts that lie behind them. This is true of boredom, but in the case of discovery it is a mistake.

Some activities are language dependent. You cannot play chess without knowing the rules, so you cannot play chess without having some sort of language in which you can express, for example, the concept of checkmate. It doesn't matter exactly what that language is: a rook is exactly the same piece if you call it a castle, just as a Frisbee was the same thing when it was called a Pluto Platter. In the absence of the word 'rook' you could use some sort of phrase, such as 'the piece which starts out in one of the four corners', just as you can call a Frisbee a flying disc, but you would find it pretty awkward to keep using phrases like this and would soon feel the need for a specialized word. Individual words and longer phrases can do the same job, but individual words usually do it better. And the introduction of a new word, or a new meaning for an old word, often marks the point at which a concept enters general use and begins to do real work.

Since you cannot play chess without knowing that you are doing so, no matter by what name you call the game, playing chess is what has been called 'an actor's concept', or 'an actor's judgement': you have to have the concept in order to perform the action.[17] It is often difficult to work out quite where to draw the line when dealing with actor's concepts. You can surely experience *Schadenfreude*, malicious enjoyment of the misfortunes of others, without having a word for it, so *Schadenfreude* was not new when the word entered the English language late in the nineteenth century; but having the word made it much easier to recognize, describe and discuss. It led people to a new understanding of human motivation: the word and the concept go together. So, too, people were surely embarrassed by awkward social encounters before the word 'embarrass', which originally meant to hinder or encumber, acquired a new meaning in the nineteenth century, but they were much more aware of embarrassment after they had a word for it. Only then did children begin to find their parents embarrassing. *Schadenfreude* and 'embarrassment' are not quite actor's concepts, in that you can experience them without having a word for what it is that you are experiencing, but the words are intellectual tools which enable us to discuss emotional states it would be difficult to talk about without them, and indeed having the words makes it much easier for us to experience and identify the emotional states in a pure and unambiguous form.

So although there were discoveries and inventions before 1486, the invention and dissemination of a word for 'discovery' marks a decisive moment, because it makes discovery an actor's concept: you can set out to make discoveries, knowing that is what you are doing. Le Roy attacks the idea that everything worth saying has already been said and that all that is left for us to do is to expound and summarize the works of our predecessors, urging his readers to make new discoveries: 'Admonishing the Learned, to adde that by their owne Inuentions, which is wanting in the Sciences; doing that for Posteritie, which Antiquitie hath done for us; to the end, that Learning be not lost, but from day to day may receive some increase.'[18]

It is worth pausing for a moment over Le Roy's vocabulary: he uses *inventer* and *l'invention* frequently; he writes of how 'many wonderful things [such as the printing press, the compass and gunpowder] unknown to antiquity have been newly found'; but he also uses the word *decouvremens*, immediately translatable as 'discoveries': '*decouvremens de terres neuves incogneuës à l'antiquité*'; '*Des navigations & decouvremens de païs*'; truth, he says, has not been '*entierement decouverte*'.[19] In his usage the meaning of the word has not yet moved far from its original reference to voyages of discovery. Did he need the word to formulate his argument? Perhaps not. What he did need was the example of Columbus, who was for him, as by then for everyone else, the proof that human history was not simply a history of repetitions and vicissitudes; that it could become – that it was in the process of becoming – a history of progress.

§ 2

To claim that discovery was new in 1492, when Columbus discovered America (or in 1486, when Dulmo talked of making discoveries; or in 1504, when Vespucci disseminated the new word across Europe) may seem plain wrong. After all, the learned humanist Polydore Vergil published in 1499 a book which has been recently translated under the title *On Discovery* (*De inventoribus rerum*, or *On the Inventors*), and which looks at first sight as if it is a history of discovery through the ages.[20]

Vergil's book was enormously successful, appearing in over one hundred editions.[21] The question Vergil asks himself over and over again is: 'Who invented . . . ?'; and he runs through a seemingly endless chain of topics, such as language, music, metallurgy, geometry. In nearly every case he finds a range of different answers to his question in his sources but, in brief, his argument is that the Romans and the Greeks attribute most inventions to the Egyptians, from whom they were acquired by the Greeks and the Romans; while the Jews and the Christians insist that the learning of the Egyptians came from the Jews, above all from Moses. (Had Vergil consulted Islamic authorities, he would have found agreement that all learning came from the Jews, but with Enoch rather than Moses identified as the key figure.)[22]

There are several striking features about Vergil's display of erudition. He is interested in first founders, rather than in the long-term development of a discipline. So he has virtually nothing to say about progress.[viii] When it comes to philosophy and the sciences, he identifies no significant contributions from Muslims (Avicenna (980–1037) is the only Muslim mentioned, and the Arabs are not even given credit for Arabic numerals), or Christians: nearly everything that matters happened long ago. He does, it is true, mention a small number of modern inventions – stirrups, the compass, clocks, gunpowder, printing – but he has almost nothing to say about new observations, new explanations or new proofs. Aristotle qualifies as an innovator only because he had the first library, Plato because he said God made the world, Aesculapius because he invented tooth-pulling, Archimedes because he was the first to make a mechanical model of the cosmos. Hippocrates of Chios is included not because he wrote the first geometry textbook but because he engaged in trade. Euclid is not mentioned, Ptolemy only as a geographer not as an astronomer, and Herophilus (the ancient anatomist) only for comparing the rhythms of the pulse to musical measures. If we use the word 'discovery' to mean something different from 'inventions' (and, of course, Vergil had only one word, *inventiones*, to cover both meanings), there are only two discoveries in Vergil: Anaxagoras's explanation for eclipses and Parmenides' realization that the Evening Star and the Morning Star are the same star. (We cannot really extend the category of discovery to include, for example, the claim that the blood of a dove, wood pigeon

viii Le Roy's book is, in this and other respects, effectively a reply to Vergil; his key move is to steer clear of the Bible as a source.

or swallow is the best cure for a black eye, although some cultural relativists would argue we should.)

These discoveries are included purely by accident, for Vergil's model is a long chapter in Pliny's *Natural History* (*c*.78 CE) entitled 'On the first inventors of diverse things' which lists numerous inventions (the plough, the alphabet), including some 'sciences' (astrology and medicine), and some technologies (the crossbow) but not a single specific discovery. Pythagoras's theorem (which is only vaguely hinted at by Vergil in discussing the architect's square), Archimedes' principle, the anatomical discoveries of Erasistratus – one could compile a long list of all that is missing from both Pliny and Vergil and could have been included if either of them had been interested in discovery as opposed to first founding, or invention, or innovation. There is a simple test of the claim that there is no discovery in Vergil: in the three early-modern-English translations the word 'discover' appears in a relevant sense only once: 'Orestus, son to Dencalion, discovered the vine about Mount Aetna in Sicily'(1686).[23] Needless to say, Vergil makes no mention of the contemporary voyages of discovery, although he went on revising his text until 1553.

So in ancient Rome, whose texts Vergil knew exceptionally well, and in the Renaissance prior to 1492, there was no concept of discovery.[ix] However, the ancient Greeks did have the concept (they used words related to *eureka*: *heuriskein*, *eurisis*; words which can mean 'invention' or 'discovery'), and developed a literary genre on invention: heurematography.[x] Eudemus (*c*.370–300 BCE) wrote histories of arithmetic, geometry and astronomy. They do not survive except in so far as they were quoted in later works; the history of geometry was an important source for Proclus (412–85), whose commentary on Book I of Euclid was first printed (on the basis of a defective manuscript) in the original Greek in 1533, and then in a far superior Latin translation in 1560. Proclus credits Pythagoras, for example, with discovering the theorem we now call Pythagoras's theorem, and Menelaus the theorem that is

ix The nearest exception in ancient Rome is the introduction to Book IX of Vitruvius's *De architectura*, where, in the course of praising great authors, Vitruvius describes the discovery (as we would call it) of Pythagoras's theorem and of Archimedes' principle. For later readers this was the paradigmatic account of discovery; Vergil, who certainly read Vitruvius, makes no reference to it.

x Consequently, they also had a notion of progress: Dodds, *The Ancient Concept of Progress* (1973).

the mathematical foundation for Ptolemaic astronomy. Had Vergil had an opportunity to read Proclus, some of this might have made its way into his text, but it is unlikely that he would have absorbed the concept of discovery. Much of Greek culture had been assimilated by the Romans, but they had found the concept of discovery indigestible, and it is unlikely that Vergil, trained to think like a Roman, would have responded differently.[xi]

§ 3

Vergil was one of the leading humanist intellectuals of the sixteenth century, by which date a humanist education (that is to say, an education in writing Latin like a classical Roman) was widely accepted as the best way to introduce young men to the world of learning, for it provided skills that were easily transferable to politics and business. But in the universities, as opposed to the schoolrooms, humanist scholarship was not the central concern. From the end of the eleventh century until the middle of the eighteenth century there was a fundamental continuity in the teaching of universities across Europe: philosophy was the core subject in the curriculum, and the philosophy taught was that of Aristotle.[xii] Aristotle's natural philosophy was to be found in four texts: his *Physics*, *On the Heavens*, *On Generation and Corruption* and *Meteorology*, and what we think of as scientific subjects were primarily addressed through commentaries on these texts.[24]

Aristotle believed that knowledge, including natural philosophy, should be fundamentally deductive in character. Just as geometry starts from undisputed premises (a straight line is the shortest distance between two points) to reach surprising conclusions (the square on the hypotenuse is equal to the sum of the squares of the other two sides),

xi The Romans, characteristically, lacked a term for 'innovation': the meaning given by Lewis and Short's dictionary for classical Latin *instauratio* is 'a renewing, renewal, repetition'; the first meaning given for classical *innovo* is 'renew', and for post-classical *innovatio*, 'renewal'. Such meanings presume a cyclical view of history. Thus Marcus Aurelius writes: 'Always remember . . . that all things from everlasting are of the same kind, and are in rotation; and it matters nothing whether it be for a hundred years or for two hundred or for an infinite time that a man shall behold the same spectacle.' (Aurelius, *The Meditations* (1968), Vol. 1, 31.)

xii The Dutch are the only significant exception: in the United Provinces the universities were teaching Cartesian philosophy by the late seventeenth century.

so natural philosophy should start from undisputed premises (the heavens never change) and draw conclusions from them (the only form of movement that can carry on indefinitely without changing is circular movement, so all movement in the heavens must be circular). Ideally, it should be possible to formulate every scientific argument in syllogistic terms, a syllogism being, for example:

All men are mortal.
Socrates is a man.
So Socrates is mortal.

Aristotle explained natural processes in terms of four causes: formal, final, material and efficient. Thus if I construct a table, the formal cause is the design I have in mind; the final cause is my desire to have somewhere to eat my meals; the material cause is various pieces of wood; and the efficient causes are a saw and a hammer. Aristotle thought about the natural world in exactly the same way: that is to say, he saw it as the product of rational, purposive activity. Natural entities seek to realize their ideal form: they are goal oriented (Aristotle's natural philosophy is teleological, the Greek word *telos* meaning 'goal' or 'end'). Thus a tadpole has the form of a juvenile frog, and its goal, its final cause, is to become an adult frog. Somewhat surprisingly, the same principles apply to inanimate matter, as we shall see.

Aristotle held that the universe is constructed out of five elements. The heavens are made out of aether, or quintessence, which is translucent and unchanging, neither hot nor cold, dry nor damp. The heavens stretch outwards from the Earth, which is at the centre of the universe, as a series of material spheres carrying the moon, the sun and the planets, and then, above them all, is the starry firmament. The universe is thus spherical and finite; moreover, it is oriented: it has a top, a bottom, a left and a right. Aristotle never thinks in terms of space in the abstract (as geometers already did), but always in terms of place. He denied the very possibility of an empty space, a vacuum. Empty space was a contradiction in terms.

The sublunary world, the world this side of the moon, is the world of generation and corruption – the rest of the universe has existed unchanging from all eternity. In our world there are four primary qualities (hot and cold, wet and dry), and pairs of qualities belong to each of the four elements (earth, water, air and fire), earth, for example, being cold and dry. These elements naturally arrange themselves in concentric spheres outwards from the centre of the universe. Thus all earth seeks to fall towards the centre of the universe, and all fire seeks to rise to the boundary of the moon's sphere; but water and air sometimes seek to go downwards, and sometimes upwards: Aristotle has no notion of a general principle of gravity.

A tadpole is potentially a frog, and as it grows it develops from potentiality to actuality. The element earth is potentially at the centre of the universe, and as it falls towards that centre it realizes its potential. All water is potentially part of the ocean which surrounds the earth: in a river it flows downhill in order to realize its potential. Water has weight when you take it out of its proper place: try lifting a bucket of water out of a pond. But when it is in its proper place it becomes weightless: when you swim in the ocean you cannot feel the weight of water on you. Aristotle thus does not think of the natural movement of the elements as movement through space; he sees it in teleological terms as the realization of potential. It is essentially a qualitative, not a quantitative, process.[xiii]

Aristotle does occasionally mention quantities. Thus he says if you have two heavy objects, the heavier one will fall faster than the lighter one, and if the heavy one is twice as heavy it will fall twice as fast. But he isn't interested enough in the quantities involved to think this through. Does he mean that if you have a one-kilo bag of sugar and a two-kilo bag of sugar, the two-kilo bag will fall twice as fast as the one-kilo bag? Or does he mean only that if you have a cube made out of a heavy material, say mahogany, and another of the same size made out of a lighter material, say pine, that if one is twice as heavy as the other, it will fall twice as fast? The two claims are very different, but Aristotle never distinguishes between them, nor tests his claim that heavy objects fall faster than light ones, for he takes it to be self-evidently true.

Aristotle distinguished sharply between philosophy (which provided causal explanations) and mathematics (which merely identified patterns). Philosophy tells us that the universe consists of concentric

xiii In the fourteenth century there was a move to think of many qualities (such as hot, cold, or green) as if they were quantities, and to imagine, for argument's sake, being able to measure quantities (such as the acceleration of falling bodies) that were assumed to be unmeasurable: it is often argued that this was a precursor to the Scientific Revolution, but see Murdoch, 'Philosophy and the Enterprise of Science in the Later Middle Ages' (1974) for a cautionary note.

spheres; the actual patterns made by the planets as they move through the heavens is a subject for astronomy, which is a sub-branch of mathematics. Astronomy and the other mathematical disciplines (geography, music, optics, mechanics) took their foundational principles from philosophy but elaborated these principles through mathematical reasoning applied to experience. Aristotle thus distinguished sharply between physics (which is part of philosophy and is deductive, teleological and concerned with causation) and astronomy (which is part of mathematics, and merely descriptive and analytic).

Aristotle was remarkable for his explorations of natural phenomena, studying, for example, the development of the chicken embryo within the egg. But as he was taken up in the universities of medieval and Renaissance Europe his works became a textbook of acquired knowledge, not a project to provoke further enquiry. The very possibility of new knowledge came to be doubted, and it was assumed that all that needed to be known was to be found in Aristotle and the rich tradition of commentary upon his texts. The Aristotle of the universities was thus not the real Aristotle but one adapted to provide an educational programme within a world where the most important discipline was taken to be theology. Just as theology was conducted in the form of a commentary upon the Bible and the Church Fathers, so philosophy (and within philosophy, natural philosophy, the study of the universe) was conducted in the form of a commentary upon Aristotle and his commentators. The study of philosophy was thus seen as a preparation for the study of theology because both disciplines were concerned with the explication of authoritative texts.[xiv]

What did this mean in practice? Aristotle took the view that harder substances are denser and heavier than softer substances; it followed that ice is heavier than water. Why does it float? Because of its shape: flat objects are unable to penetrate the water and remain on the surface. Hence a sheet of ice floats on the surface of a pond. Aristotelian philosophers were still happily teaching this doctrine in the seventeenth century, despite the fact that there were two obvious difficulties. It was incompatible with the teaching of Archimedes, who was available in Latin from the twelfth century, and argued that objects float only if they

xiv It is important to grasp that Aristotle's views were approved both because they were held to be rational and because they were held to be authoritative: when Aristotle's authority was destroyed, with it went the very idea of authority in natural philosophy. See, for example, the awkward squirming of Piccolomini as he nerved himself to dispute with Aristotle: Piccolomini, *Della grandezza della terra et dell'acqua* (1558), 1r–2v.

are lighter than the water they displace. The mathematicians followed Archimedes; the philosophers followed Aristotle. Moreover, ice was easily available in much of Europe: in Florence, for example, it was brought down from the Apennines throughout the summer in order to keep fish fresh. The most elementary experiment will show that ice floats no matter what its shape. The philosophers, confident that Aristotle was always right, saw no need to test his claims.[25]

This indifference to what we would call the facts is exemplified by Alessandro Achillini (1463–1512), a superstar philosopher, the pride of the University of Bologna.[26] He was a follower of the Muslim commentator Averroes (1126–98), who studiously avoided introducing religious categories into the interpretation of Aristotle and so implicitly denied the creation of the world and the immortality of the soul. Achillini's brilliance and the transgressive character of his thought were summarized in a popular saying: 'It is either the devil or Achillini.'[27] In 1505 he published a book on Aristotle's theory of elements, *De elementis*, in which he discussed a question that had long been debated by the philosophers: whether the region of the equator would be too hot for human habitation. He quoted Aristotle, Avicenna and Peter of Abano (1257–1316), and concluded, 'However, that at the equator figs grow the year round, or that the air there is most temperate, or that the animals living there have temperate constitutions, or that the terrestrial paradise is there – these are things which natural experience does not reveal to us.'[28] As far as Achillini was concerned, the question of whether figs grow at the equator was as unanswerable as the question of where the Garden of Eden was located, and neither was a question for a philosopher.

As it happens, the Portuguese, in their search for an oceanic route to the spice islands, which involved sailing south along the coast of Africa, had reached the equator in 1474/75 and the Cape of Good Hope in 1488. In 1505 there were already maps showing the new discoveries. The very next year John of Glogau, a professor at Cracow, pointed out (in a mathematical rather than a philosophical work) that the island of Taprobane (Sri Lanka) was very near the equator, and was populous and prosperous.[29] Experience had ceased to be something unchanging, identical with what was known to Aristotle, but Achillini was professionally unprepared for this development, even though he also lectured on anatomy, the most empirical of all the university disciplines.

By 1505 then the relationship between experience and philosophy needed to be rethought, but Achillini was incapable of grasping the problem.[30] By contrast, Cardinal Gasparo Contarini's book on the elements, published posthumously in 1548, explained that Aristotle, Avicenna and Averroes had all denied that the equator was habitable: 'This question, which for many years was disputed between the greatest philosophers, experience has solved in our times. For from this new navigation of the Spaniards and especially of the Portuguese it has been discovered that there is habitation under the equinoctial circle and between the tropics, and that innumerable people dwell in these regions . . .'[31]

For Contarini, experience had taken on a new kind of authority. He died in 1542, the year before the publication of Copernicus's *On the Revolutions* and Vesalius's *On the Fabric of the Human Body*. It was not yet apparent that, once experience was accepted as the ultimate authority, it was only a matter of time before there emerged a new philosophy which would bring the temple of established knowledge crashing down. By 1572 it would be.

§ 4

Before Columbus the primary objective of Renaissance intellectuals was to recover the lost culture of the past, not to establish new knowledge of their own. Until Columbus demonstrated that classical geography was hopelessly misconceived, the assumption was that the arguments of the ancients needed to be interpreted, not challenged.[32] But even after Columbus the old attitudes lingered. In 1514 Giovanni Manardi expressed impatience with those who continued to doubt whether human beings could withstand the equatorial heat: 'If anyone prefers the testimony of Aristotle and Averroes to that of men who have been there,' he protested, 'there is no way of arguing with them other than that by which Aristotle himself disputed with those who denied that fire was really hot, namely for such a one to navigate with astrolabe and abacus to seek out the matter for himself.'[33] Sometime between 1534 and 1549 Jean Taisnier, a musician and mathematician, remarked that Aristotle was sometimes mistaken; he was challenged by a representative of the pope to produce a convincing example of Aristotle being wrong. His opponents felt sure he would be unable to do so. His response was a lecture attacking Aristotle's account of falling bodies, the weakest point in his physics.[34]

It is difficult for us to grasp the extent to which this continued to be the case long into the seventeenth century.[xv] Galileo tells the story of the professor who refused to accept that the nerves were connected to the brain rather than the heart because this was at odds with Aristotle's explicit statement – and stood his ground even when he was shown the pathways of the nerves in a dissected cadaver.[35] There is the famous example of the philosopher Cremonini, who, despite being a close friend of Galileo, refused to look through his telescope. Cremonini went on to publish a lengthy book on the heavens in which no mention was made of Galileo's discoveries, for the simple reason that they were irrelevant to the task of reconstructing Aristotle's thinking.[36] In 1668 Joseph Glanvill, a leading proponent of the new science, found himself arguing against someone who dismissed all the discoveries made with telescopes and microscopes on the grounds that such instruments were '*all deceitful and fallacious*. Which Answer minds me of the *good Woman*, who when her Husband urged in an occasion of difference, *I saw it, and shall I not believe my own Eyes?* Replied briskly, **Will you believe your own Eyes, before your own dear Wife?**[xvi] And it seems *this Gentleman* thinks it unreasonable we should *believe ours*, before his *own dear Aristotle*.'[37] Even the great seventeenth-century anatomist William Harvey, who discovered the circulation of the blood, referred to Aristotle approvingly as 'the great Dictator of Philosophy'; to Walter Charleton, a founding member of the Royal Society and an opponent of scholasticism, he was quite simply 'the Despot of the Schools'.[38]

§ 5

Thus religion, Latin literature and Aristotelian philosophy all concurred: there was no such thing as new knowledge. What looked like new knowledge was, consequently, simply old knowledge which had been mislaid, and history was assumed to go round in circles. On a large

xv Edmund O'Meara wrote in his *Pathologia hæreditaria generalis* (Dublin, 1619, 62–4), 'I marvel indeed with what arrogance anyone presumes to fight against experience, the discoverer of all science and knowledge, unless it should suffice for a reason that many are both ashamed and irked to admit anything new that contradicts their firm opinions, from which they cannot bear to retreat so much as an hair's breadth, lest they seem not to have been right before; you may see many revering Hippocrates, Galen, and Aristotle so fatuously, not to say idolatrously, that they think whatever they did not say must not be said, whatever they did not know must not be known.' (Translation from Lower, *Richard Lower's 'Vindicatio'* (1983), 201–2.)

xvi Italics and bold in the original.

scale, the whole universe was held (at least if you put revealed truth to one side and listened to the astrologers) to repeat itself. 'Everything that has been in the past will be in the future,' wrote Francesco Guicciardini in his *Maxims* (left to his family at his death in 1540, and first published in 1857).[39] As Montaigne put it in 1580, 'the beliefs, judgements and opinions of men . . . have their cycles, seasons, births and deaths, every bit as much as cabbages do.'[40] He could quote the best authorities: 'Aristotle says that all human opinions have existed in the past and will do so in the future an infinite number of other times: Plato, that they are to be renewed and come back into being after 36,000 years' (an alarming thought, given that the Biblical chronology implied that the world was only six thousand years old; Cicero's figure of 12,954 years was not much better). Giulio Cesare Vanini wrote (in 1616; he was executed for atheism three years later): 'Again will Achilles go to Troy, rites and religions be reborn, human history repeat itself. Nothing exists today that did not exist long ago; what has been, shall be.' On a smaller scale the history of each society was supposed to involve an endless cycle of constitutional forms (*anacyclosis*), from democracy to tyranny and back again, and it was a small step to assume that cultures, too, recurred, along with constitutions.[41]

For Platonists, there could be no such thing as genuinely new knowledge, for Plato insisted that the soul already knew the truth, so that whatever seemed new was in fact really reminiscence (*anamnesis*). In the *Meno* Socrates persuaded an uneducated slave boy that he already knew that the square on the hypotenuse is equal to the sum of the squares of the other two sides. And of course it is true that a discovery often involves a recognition of the significance of something already known. When Archimedes cried 'Eureka!' and ran naked through the streets of Syracuse we say that he had discovered what we call Archimedes' principle. We could equally say that Archimedes had recognized the implications of something he already knew – that he displaced water when he got into the bath. Recognition and reminiscence imply that our present and future experiences are just like our past experiences; discovery implies that we can experience something that no one has ever experienced before. The idea of discovery is inextricably tied up with ideas of exploration, progress, originality, authenticity and novelty. It is a characteristic product of the late Renaissance.

The Platonists' doctrines of recurrence and reminiscence were not the real problem, however; both were endorsed by Proclus, who still wrote, as the Greeks did, in terms of discovery. The real obstacle, apart from unquestioning belief in Aristotle, was the even more unquestioning belief in the Bible. Where the Greeks and the Romans believed that human beings had started out as little better than animals and had slowly acquired the skills required for civilization, the Bible insisted that Adam had known the names of everything; Cain and Abel had practised arable and pastoral agriculture; the sons of Cain had invented metallurgy and music; Noah had built an ark and made wine; his near-descendants had set out to build the Tower of Babel. The suggestion that the various skills required for civilization had had to be invented over a long a period of time or that there were significant types of knowledge of which Abraham, Moses and Solomon had no conception was simply unacceptable. The Greeks, the early Church Fathers pointed out, acknowledged their debt to the Egyptians, and it was easy to see that the Egyptians had acquired their learning from the Jews. 'Stop calling your imitations inventions!' Tatian (*c.*120–80) cried in exasperation, rejecting wholesale the claim that the Egyptians and Greeks had made discoveries unknown to the Jews.[42]

Christianity not only imposed an abbreviated chronology; the liturgy was constructed around an endless cycle, the annual recurrence of the life of Christ. 'Annually, the Church rejoices because Christ has been born again in Bethlehem; as the winter draws near an end he enters Jerusalem, is betrayed, crucified, and the long Lenten sadness ended at last, rises from death on Easter morning.' At the same time the sacrament of the Mass affirms 'the perpetual contemporaneity of The Passion' and celebrates the 'marriage of time present with time past'.[43]

The idea of discovery simply could not take hold in a culture so preoccupied with Biblical chronology and liturgical repetition on the one hand, and secular ideas of rebirth, recurrence and reinterpretation on the other. Francis Bacon complained in 1620 that the world had been bewitched, so inexplicable was the reverence for antiquity. Thomas Browne protested in 1646 against the general assumption that the further back in time one went, the nearer one approached to the truth. (He surely had in mind Bacon's insistence that the reverse is the case, that *veritas filia temporis*, 'truth is the daughter of time.')[44] Symptomatic of this backward orientation of orthodox culture is the title of one of the most important books to describe the new discoveries of Columbus and Vespucci: *Paesi novamenti retrovati* (Vicenza, 1507; *Lands Recently Rediscovered*). In the German translation that followed a year later this became *Newe unbekanthe Landte* (*New Unknown Lands*).[45] This revision marks the first, local triumph of the new.

Of course, it is natural for us to think that there was much that was 'new' before 1492. But what looks new to us generally did not look so new (or at least not quite so incontrovertibly new) to contemporaries. An interesting test case is provided by the revolutionary developments in art that took place in Florence early in the fifteenth century. Leon Battista Alberti, returning there in 1434 after long years of exile (Alberti had, in his own view and in the view of his fellow Florentines, been born in exile, in 1404, and had spent the greater part of his adult life in Bologna and Rome), was astonished at what he saw. The new dome of Florence's cathedral, designed by Brunelleschi, 'vast enough to cover the entire Tuscan population with its shadow', towered over the city, and a group of brilliant artists – Brunelleschi himself, Donatello, Masaccio, Ghiberti, Luca della Robbia – were producing work which seemed unlike anything that had gone before. 'I used both to marvel and to regret that so many excellent and divine arts and sciences, which ... were possessed in great abundance by the talented men of antiquity, have now disappeared and are almost entirely lost,' he wrote in 1436. But now, looking at the achievements of the Florentine artists, he thought 'our fame should be all the greater if without preceptors and without any model to imitate we invent [*troviamo*] arts and sciences hitherto unheard of and unseen.'[46] Brunelleschi's dome was '[s]urely a feat of engineering, if I am not mistaken, that people did not believe possible these days and was probably equally unknown and unimaginable among the ancients'. Faced with achievements that seemed to have no classical parallel, Alberti nevertheless felt obliged to express himself with great caution: 'surely', 'if I am not mistaken', 'probably'.[xvii] Notably, Alberti singles out here Brunelleschi's dome, not the art of perspective painting, which was his real subject: he and his successors were quite unclear as to whether the latter technique was brand new or simply a rediscovery of the techniques used by the ancient Greeks and Romans for painting stage sets, as described by Vitruvius. Alberti himself claimed (characteristically) in 1435 that the technique of perspective

xvii Compare Pierre Guiffart's defence of Pascal's vacuum experiments in 1647: 'Although the experiences of M. Pascal look new to us, they have some appearance of having been formerly practiced, and several ancients having taken from them the basis for maintaining that there could be a void in nature ...' How else, he wonders, could Epicurus and Lucretius have been confident of the existence of a vacuum? Guiffart is old-fashioned; Pascal does not argue like this, and even Guiffart has to admit at last that the experiments may in fact be unprecedented. (Quoted from Dear, *Discipline and Experience* (1995), 191; French text at Pascal, *Oeuvres* (1923), 9.)

was 'probably' unknown to the ancients; Filarete in 1461 insisted it was completely unknown to them; but Sebastiano Serlio in 1537 took exactly the opposite view, saying bluntly 'perspective is what Vitruvius calls *scenographia*.'[47]

In such circumstances the conviction that there was no new knowledge to be had might bow and crack, but it would not quite shatter. To get a sense of its resilience one need only think of Machiavelli, almost a hundred years later, who opens the *Discourses on Livy* (*c.*1517) with a reference to the (relatively recent) discovery of the New World and a promise that he, too, has something new to offer, only to turn sharply and insist that in politics, as in law and medicine, all that is required is a faithful adherence to the examples left by the ancients, so that it turns out that what he has to offer is not a voyage into the unknown but a commentary on Livy. Unsurprisingly, Machiavelli thought it perfectly obvious that, despite the invention of gunpowder, Roman military tactics remained the model that all should follow: the whole purpose of his *Art of War* (1519) was to write for those who, like himself, were *delle antiche azioni amatori* (lovers of the old ways of doing things).[48]

Naturally, Copernicus, half a century after the discovery of America, was careful to mention Philolaus the Pythagorean (*c.*470–385 BCE) as an important precursor in proposing a moving Earth.[49] Copernicus's disciple Rheticus, in the first published account of the Copernican theory, held back any reference to heliocentrism for as long as he possibly could, for fear of alienating his readers.[50] The text of Thomas Digges's *Prognostication* of 1576 emphasized the absolute novelty and originality of the Copernican system; but the illustration which accompanied the text made no mention of Copernicus, claiming to represent 'the Cælestiall Orbes according to the most auncient doctrine of the Pythagoreans', and in later reprints this phrasing was picked up in the table of contents and the chapter headline.[51] Even Galileo, in his *Dialogue Concerning the Two Chief World Systems* (1632), repeatedly coupled Copernicus's name with that of Aristarchus of Samoa (*c.*310–230 BCE), to whom he (mistakenly) attributed the invention of heliocentrism.[52] What was new was not yet admirable, and thus it presented itself, as best it could, within the carapace of the ancient. Few were prepared, as Le Roy was, to embrace novelty wholeheartedly.

Within a backward-looking culture the crucial distinction was not between old knowledge and new knowledge but between what was generally known and what was known only by a privileged few who had obtained access to secret wisdom.[53] Knowledge, it was assumed, was never really lost. It either went underground, becoming esoteric or occult, or it was mislaid and would eventually turn up after centuries of lying neglected in some monastery library. As Chaucer had written in the fourteenth century:

> ... out of olde feldes, as men seyth,
> Cometh al this new corn from yer to yere,
> And out of old bokes, in good feyth,
> Cometh al this newe science that men lere.[54]

The discovery of America was crucial in legitimizing innovation because within forty years no one disputed that it really was an unprecedented event, and one that could not be ignored.[55] It was also a public event, the beginning of a process whereby new knowledge, in opposition to the old culture of secrecy, established its legitimacy within a public arena. The celebration of innovation had begun even before 1492, however. In 1483 Diogo Cão erected a marble pillar surmounted by a cross at the mouth of what we now call the Congo River to mark the furthest limit of exploration southwards. This was the first of what became a series of pillars, each designed to demarcate the boundaries of the known world, and thus to replace the Pillars of Hercules (the Strait of Gibraltar), which had done so for the ancient world. Then, after Columbus, the Spanish joined in. In 1516 Charles (the future Charles V King of Spain and Holy Roman Emperor) adopted as his device the Pillars of Hercules with the motto *plus ultra*, 'further beyond', a motto later adopted by Bacon. (There is no satisfactory translation of *plus ultra* because it is ungrammatical Latin.)[56] João de Barros was able to claim in 1555 that Hercules' pillars, 'which he set up at our very doorstep, as it were, ... have been effaced from human memory and thrust into silence and oblivion'.[57] One of Galileo's opponents, Lodovico delle Colombe, complained in 1610/11 that Galileo behaved like someone setting sail on the ocean, heading out past the Pillars of Hercules and crying, '*Plus ultra!*', when of course he should have recognized that establishing Aristotle's opinion was the point at which enquiry should stop.[58] Poor Lodovico, he does not seem to have realized that the discovery of America had made ridiculous the claim that one should never venture into the unknown. Still, in June 1633, during Galileo's trial, his friend Benedetto Castelli wrote to him, remarking that the Catholic Church seemed to want to establish new columns of Hercules bearing the slogan *non plus ultra*.[59]

But it took more than a century to make innovation – outside geography and cartography – respectable, and then it became respectable only among the mathematicians and anatomists, not among the philosophers and the theologians. In 1553 Giovanni Battista Benedetti published *The Resolution of All Geometrical Problems of Euclid and Others with a Single Setting of the Compass*, the title page of which announces boldly that this is a 'discovery' ('*per Joannem Baptistam de Benedictis Inventa*'); he was following Tartaglia, who claimed to have invented his *New Science* (1537). But Tartaglia and Benedetti were exceptional in boasting of their achievements. A better marker of the new culture of discovery is the publication in 1581 of Robert Norman's *The Newe Attractive*. Norman announced on his title page that he was the discoverer of 'a newe . . . secret and subtill propertie', the dip of the compass needle. Although he knew neither Greek nor Latin (though he did know Dutch), he knew enough about discovery to compare himself to Archimedes and Pythagoras, as described by Vitruvius. He had joined the ranks of those 'overcome with the incredible delight conceived of their own devises and inventions'.[60] When Francesco Barozzi's *Cosmographia* was translated into Italian in 1607 its title page declared that it contained new discoveries ('*alcune cose di nuovo dall'autore ritrovate*'); there had been no mention of these on the title page of the original edition of 1585. By 1608 it was possible to complain that 'nowadays the discoverers of new things are virtually deified'. One precondition for this, of course, was that, like Tartaglia, like Benedetti, like Norman, and like Barozzi's translator on his behalf, they made no secret of their discoveries.[61]

Two decades later a pupil of Galileo, newly appointed as a professor of mathematics at Pisa, complained that 'of all the millions of things there are to discover [*cose trovabili*], I don't discover a single one,' and as a consequence he lived in 'endless torment'.[62] Since time began there have been impatient young men, anxious that they would fail to live up to their own expectations; but Niccolò Aggiunti may well have been the first to worry that he would never make a significant discovery. In Galileo's circle, too, all that counted was discovery.

What was remarkable about the knowledge produced by the voyages of discovery was not just that it was indisputably new, nor that it was public. Geography had been transformed, not by philosophers teaching in universities, not by learned scholars reading books in their studies, not by mathematicians scribbling new theorems on their slates; nor had it been deduced from generally recognized truths (as recommended by Aristotle) or found in the pages of ancient manuscripts. It had been found instead by half-educated seamen prepared to stand on the deck of a ship in all weathers. 'The simple sailors of today,' wrote Jacques Cartier in 1545, 'have learned the opposite of the opinion of the philosophers by true experience.'[63] Robert Norman described himself as an 'unlearned Mechanician'. The new knowledge thus represented the triumph of experience over theory and learning, and was celebrated as such. 'Ignorant Columbus,' wrote Marin Mersenne in 1625, 'discovered the New World; yet Lactantius, learned theologian, and Xenophanes, wise philosopher, had denied it.'[64] As Joseph Glanvill put it in 1661:

> We believe the *verticity* of the *Needle* [i.e. that the compass points north], without a Certificate from the *dayes* of *old*: And confine not our selves to the sole conduct of the *Stars*, for fear of being wiser then our Fathers. Had *Authority* prevail'd here, the Earths *fourth part* [America] had to us been *none*, and *Hercules* his Pillars had still been the worlds *Non ultra: Seneca*'s Prophesie [that one could sail west to reach India] had yet been an unfulfill'd Prediction, and one moiety of our *Globes*, an empty *Hemisphear*.[65]

What is important here is not, despite what Diderot says, the idea that experience is the best way to acquire knowledge. The saying *experientia magistra rerum*, 'experience is a great teacher', was familiar in the Middle Ages: you don't learn to ride a horse or shoot an arrow by reading books.[66] What is important is, rather, the idea that experience isn't simply useful because it can teach you things that other people already know: experience can actually teach you that what other people know is wrong. It is experience in this sense – experience as the path to discovery – that was scarcely recognized before the discovery of America.

Of course, the geographical discoveries themselves were only the beginning. From the New World came a flood of new plants (tomatoes, potatoes, tobacco) and new animals (anteaters, opossums, turkeys). This provoked a long process of trying to document and describe the previously unknown flora and fauna of the New World; but also, by reaction, a shocked recognition that there were all sorts of European plants and animals that had never been properly observed and recorded. Once the process of discovery had begun, it turned out that it was possible to make discoveries anywhere, if only one knew how to look. The Old World itself was viewed through new eyes.[67]

There was a second consequence of describing the new. For classical and Renaissance authors every well-known animal or plant came with a complex chain of associations and meanings. Lions were regal and courageous; peacocks were proud; ants were industrious; foxes were cunning. Descriptions moved easily from the physical to the symbolic and were incomplete without a range of references to poets and philosophers. New plants and animals, whether in the New World or the Old, had no such chain of associations, no penumbra of cultural meanings. What does an anteater stand for? Or an opossum? Natural history thus slowly detached itself from the wider world of learning and began to form an enclave on its own.[68]

§ 6

The noun 'discovery' first appears in its new sense in English in 1554, the verb 'to discover' in 1553, while the phrase 'voyage of discovery' was being used by 1574.[69] Already by 1559 it was possible, in the first English application for a patent, made by the Italian engineer Jacobus Acontius, to talk of discovering, not a new continent, but a new type of machine:

> Nothing is more honest than that those who by searching have found out things useful to the public should have some fruit of their rights and labours, as meanwhile they abandon all other modes of gain, are at much expense in experiments, and often sustain much loss, as has happened to me. I have discovered most useful things, new kinds of wheel machines, and of furnaces for dyers and brewers, which when known will be used without my consent, except there be a penalty, and I, poor with expenses and labour, shall have no returns. Therefore I beg a prohibition against using any wheel machines, either for grinding or bruising or any furnaces like mine, without my consent.[70]

His suit was eventually granted with the statement: '[I]t is right that inventors should be rewarded and protected against others making profit out of their discoveries.'[xviii] This may seem like an extraordinary shift in meaning, for it is easy to see how you might 'uncover' something that is already there, much harder to see how you could uncover something that has never previously existed; but it must have been facilitated by the range of meanings present in the Latin word *invenio*, which

xviii Note that Acontius uses 'invention' and 'discovery' interchangeably. We normally distinguish between inventions and discoveries, but the distinction was slow to establish itself and is even now clearer in our use of 'invention' than of 'discovery'. We cannot say, as John Ray did in 1691, that the salivary ducts are 'of late invention'; but we can, perhaps, still write, as H. W. Haggard did in 1929, of 'the discovery of the obstetrical forceps'. (Both examples from the *OED*, 'invention' and 'discovery'.)

covered both finding and inventing. In 1605 this new idea of discovery was generalized by Francis Bacon in *Of the Proficiency and Advancement of Learning*. Indeed, Bacon claimed that he had discovered how to make discoveries:

> And like as the West Indies [i.e. the Americas as a whole][xix] had never been discovered, if the use of the Mariners' Needle [the compass], had not been first discovered; though the one be vast Regions, and the other a small Motion. So it cannot be found strange, if Sciences be no further discovered, if the Art itself of Invention and Discovery, hath been passed over.[71]

Bacon's claim to have invented the art (i.e. technique) of discovery depended on a series of intellectual moves. First, he rejected all existing knowledge as being unfit for making discoveries and useless for transforming the world. The scholastic philosophy taught in the universities, based on Aristotle, was, Bacon insisted, caught up in a series of futile arguments which could never generate new knowledge of the sort he was looking for. Indeed, he rejected the idea of a knowledge grounded in certainty, in proof. Aristotelian philosophy had been based on the idea that one ought to be able to deduce sciences from generally accepted first principles, so that all science would be comparable to geometry. Instead of proof, Bacon introduced the concept of interpretation; where before scholars had written about interpreting books, now Bacon introduced the idea of 'the interpretation of nature'.[72]

What made an interpretation right was not its formal structure but its usefulness, the fact that it made possible prediction and control. Bacon pointed out that the discoveries which were transforming his world – the compass, the printing press, gunpowder, the New World – had been generated in a haphazard fashion. No one knew what might happen if a systematic search for new knowledge was undertaken. Thus Bacon rejected the distinction, so deeply rooted in his society, between theory and practice. In a society where a sharp line was drawn between gentlemen, who had soft hands, and craftsmen and labourers, whose hands were hard, Bacon insisted that effective knowledge would require cooperation between gentlemen and craftsmen, between book-learning and workshop experience.

Bacon's central claim was thus that knowledge (at least, knowledge of the sort that he was advocating) was power: if you understood

xix cf. Blundeville (1594): 'America, which we now call the West Indies' (*OED* s.v. 'West Indies').

something, you acquired the capacity to control and reproduce nature's effects.[xx] Far from the products of human expertise being necessarily inferior to the products of nature, human beings were in principle capable of doing far more than nature ever did, of doing things 'of a kind that before their invention the least suspicion of them would scarcely have crossed anyone's mind, but a man would simply have dismissed them as impossible'.[73] Where the goal of Greek philosophy had been contemplative understanding, that of Baconian philosophy was a new technology. Bacon's ambitions for this new technology were remarkable: it was to be a form of 'magic'; that is, it was to do things which seemed impossible to those unacquainted with it (as guns seemed a form of magic to Native Americans).[74]

With this – the discovery of discovery – went a new commitment to what Bacon, when writing in English, called 'advancement', 'progression' or 'proficiency' (using the word in its original sense of 'moving forward'), and his translators from 1670 called 'improvement' or, quite simply, 'progress'. If the discovery of America began in 1492, so too did the discovery of progress. Bacon was the first person to try to systematize the idea of a knowledge that would make constant progress.[75] In his lifetime he published three books outlining the new philosophy – *The Advancement of Learning* (1605 and, in an expanded Latin version, 1623), *The Wisdom of the Ancients* (1609), *The New Organon* (1620, the first part of a projected but unfinished larger work, *The Great Instauration*) – followed, posthumously, by *The New Atlantis* and *Sylva sylvarum* in 1626. Despite its Latin title, *Sylva sylvarum* is written in English. In Latin *silva* is a wood but also the collection of raw materials needed for a building. So *Sylva sylvarum* is literally 'the wood of woods', but, effectively, *The Lumber Yard*. *Organon* is the Greek word for a tool (Galileo calls his telescope an organon),[76] so *The New Organon* provides the tools, the mental equipment, and the *Sylva sylvarum* provides the raw materials for Bacon's enterprise.[77]

The books were read, but they had little influence, and the demand for them was modest: it took, for example, twenty-five years for *The New Organon* to appear in a second edition. Bacon had no disciples in England until the 1640s. (He had more influence in France, where a number of his works appeared in translation.)[78] The reason for this is

xx Bacon is often quoted as saying 'Knowledge is power'; in fact, the closest he comes to this is 'Human knowledge and power come to the same thing': Weeks, 'Francis Bacon and the Art–Nature Distinction' (2007), 123.

very simple: Bacon made no scientific discoveries himself. His claims for his new science were entirely speculative. It was only in the second half of the seventeenth century that he was rescued from relative obscurity and hailed as the prophet of a new age.

§ 7

While Bacon was writing about discovery, others were making discoveries. Slowly and awkwardly, in the course of the sixteenth century, there had come into existence a grammar of scientific discovery: discoveries occur at a specific moment (even if their significance comes to be apparent only over time); they are claimed by a single individual who announces them to the world (even if many people are involved); they are recorded in new names; and they represent irreversible change. No one devised this grammar, and no one wrote down its rules, but they came to be generally understood for the simple reason that they were based on the paradigm case of geographical discovery.[xxi] An early example of someone evidently confident in his understanding of how the rules worked is the anatomist Gabriele Falloppio. He recounted that when he first went to teach at the University of Pisa (1548) he told his students that he had identified a third bone in the ear (other than the hammer and anvil bones) which the great anatomist Andreas Vesalius had not noticed – it is, after all, the smallest bone in the human body. One of his students advised him that Giovanni Filippo Ingrassia, who was teaching in Naples, had already discovered this bone and named it *stapes*, or 'the stirrup'. (Ingrassia had made the discovery in 1546, but his work was published only posthumously in 1603.) When Falloppio published in 1561 he acknowledged Ingrassia's priority and adopted the name he had proposed for the new bone. His admirable behaviour did not go unnoticed: it was a textbook example for Caspar Bartholin in 1611.[79] Falloppio knew the rules and was determined to play by them, for he wanted his own discovery to be duly acknowledged. Ingrassia could keep the stirrup bone; Falloppio had discovered the clitoris.[80] It might be thought that it wouldn't be very hard to discover the clitoris; but the standard view, inherited from Galen, was that men and women had exactly the same sexual parts, although they were folded differently; consequently, anatomists gave them the same names – the ovaries

xxi On 'paradigm', see 'Notes on Relativism and Relativists', No. 5 (pp. 585–6).

(as we now call them), for example, were simply the female testicles. The discovery of the clitoris was therefore another remarkable triumph of experience over theory, since it had no male equivalent but was unique to the female anatomy.[81]

The anatomists were thus pioneers in carefully recording who claimed to have discovered what: Bartholin's 1611 textbook begins its account of the clitoris by recording the competing claims of Falloppio, whose claim he favours, and Realdo Colombo, Falloppio's colleague and rival at the University of Padua (although he suspected, as we might, that it had already been known to the ancients).[82] As a former student of medicine, and as a professor at Padua, where many of the new anatomical discoveries were made, from 1592 Galileo was certainly familiar with this new culture of priority claims: Falloppio's best student, Girolamo Fabrizi d'Acquapendente, who had discovered the valves in the veins, was both his physician and a personal friend.

When Galileo pointed his telescope towards Jupiter on the night of 7 January 1610, he noticed what he took to be some fixed stars near the planet. Over the next night or two the relative positions of the stars and Jupiter changed oddly. At first, Galileo assumed that Jupiter must be moving aberrantly and the stars must be fixed. On the night of 15 January he suddenly grasped that he was seeing moons orbiting Jupiter. He knew he had made a discovery, and at once he knew what to do. He stopped writing his observation notes in Italian and started writing them in Latin – he was planning to publish.[83] The moons of Jupiter were discovered by one person at one moment of time, and from the beginning – not retrospectively – Galileo knew not only exactly what it was that he had discovered but that he had made a discovery.

Because Galileo rushed into print, his claim to priority was undisputed. He later claimed that it was in 1610 that he first observed spots on the sun, but he was slow to publish, and in 1612 both he and his Jesuit opponent Christoph Scheiner were advancing competing priority claims.[84] They disagreed about how to explain what they had seen, but at least they agreed that the drawings they both published were illustrations of the same phenomena. Matters are not always so straightforward. The classic problem case is the discovery of oxygen. In 1772 Carl Wilhelm Scheele discovered something he called 'fire air', while, independently, in 1774 Joseph Priestley discovered something he called 'dephlogisticated air' (phlogiston being a substance supposedly given off in burning – a reverse oxygen). In 1777 Antoine Lavoisier published a new theory of combustion which clarified the role of the new gas, which he named

Johannes Hevelius with one of his telescopes (from *Selenographia*, 1647, an elaborate map of the moon). Hevelius, who lived in Danzig, Poland, built one enormous telescope 150 feet long. He also published an important star atlas. (No drawing or engraving records Galileo's telescopes, and the two of his manufacture which survive are not as powerful as the one he was using in 1610/11. Thus we do not know what his own astronomical telescopes looked like.)

'oxygen', meaning (from the Greek) 'acid-producer', because he mistakenly thought it to be an essential component of all acids. (The nature of acids was not clarified until Sir Humphry Davy's work in 1812.) Even Lavoisier did not understand oxygen properly: discovery is often an extended process, and one that can be identified only retrospectively.[85] In the case of oxygen it can be said to have begun in 1772 and ended only in 1812.

There are some who claim not just that some discoveries are difficult to pinpoint but that all discovery claims are essentially fictitious. They assert that discovery claims are always made well after the event, and that in reality (if there is a reality here at all) there is never one discoverer but always several, and that it is never possible to say exactly when a discovery is made.[86] When did Columbus discover what we now call

America? Never, for he never realized that he had not arrived in the Indies.[87] Who did discover America? Waldseemüller, perhaps, sitting at his desk, for he was the first to grasp fully what Columbus and Vespucci had done.

The straightforward example of the discovery of the moons of Jupiter shows that these claims are superficially plausible, but mistaken. One mistake is the argument that discovery claims are necessarily retrospective because 'discovery' is an 'achievement word', like checkmate in chess.[88] Passing a driving test is an achievement of the sort they have in mind: you can only be sure you have done it when you get to the end of the test. But any competent chess player can plan a mate several moves ahead; and they know that they have won the game not *after* they have moved their piece but as soon as they see the move they need to make. Galileo's discovery of the moons of Jupiter was not like checkmate, or winning a race: he did not plan for it or see it coming. Nor was it like an ace at tennis: you only know you have served one after your opponent has failed to make a return. It was like singing in tune: he knew he was doing it at the same time as he did it. Some achievements are necessarily retrospective (like winning the Nobel Prize, or discovering America), some are simultaneous (like singing in tune) and others can be prospective (like checkmate). Scientific discoveries come in all three forms. As we have seen, the discovery of oxygen was retrospective. The classic example of simultaneous discovery is Archimedes' cry of 'Eureka!' He knew he had found the answer at the very moment he saw the level of the water in his bath rise – which is why he was still naked and dripping wet as he ran through the streets shouting out the good news. So, too, with the discovery of the moons of Jupiter: Galileo had 'a Eureka moment'.[xxii]

The really remarkable cases are those of prospective discovery, since they straightforwardly refute the claim that all discoveries are necessarily retrospective constructions. Thus in 1705 Halley, having noticed that a particularly bright comet reappeared roughly every seventy-five years, predicted that the comet now known as Halley's Comet would return in 1758. The comet reappeared just in time, on Christmas Day 1758, but in any case Halley had modified his prediction in 1717 to 'about the end

xxii Gasparo Aselli actually cried 'Eureka!' when he discovered the lacteal vessels of the lymphatic system by accident during the vivisection of a dog in 1622: Bertoloni Meli, 'The Collaboration between Anatomists and Mathematicians in the Mid-seventeenth Century' (2008), 670.

of the year 1758, or the beginning of the next'.[89] When did Halley make his discovery? In 1705, of course, when he noticed the pattern of regular reappearance; though his improved prediction of 1717 is worthy of note. Surely he didn't make his discovery in 1758, when he was long dead. The discovery was *confirmed* in 1758 (and the comet was consequently named Halley's Comet in 1759), but it was *made* in 1705; we are not reading something retrospectively into Halley's statements when we say that he predicted the return of the comet. Similarly, Wilhelm Friedrich Bessel predicted the existence of Neptune on the basis of anomalies in the orbit of Uranus. The search for the new planet began long before it was finally observed in 1846.[90]

Wittgenstein explained that there are some terms we have and use quite reliably which we cannot adequately define. Take the term 'game'. What is it that soccer, darts, chess, backgammon and Scrabble have in common? In some games you keep score, but not in chess (except in matches). In some games there are only two sides, but not in all; indeed, some games – solitaire, keepie-uppie – can be played by one person on their own. Games have, Wittgenstein said, a 'family resemblance', but this does not mean that the term – or the difference between a game and a sport – can be adequately defined.[91]

So, too, as the category of discovery has developed over time it has come to include many radically different types of event. Some discoveries are observations – sunspots, for example. Others, such as those of gravity and natural selection, are commonly called theories. Some are technologies, for instance the steam engine. The idea of discovery is no more coherent or defensible than the idea of a game; which means that the idea presents philosophers and historians with all sorts of difficulties, but it does not mean that we should stop using it. In this respect, indeed, it is typical of the key concepts that make up modern science. But in the case of discovery we have a straightforward paradigm case – the case from which the whole language derives. This is Columbus's discovery of America. Who discovered America? Both Columbus and the lookout on the *Pinta*. What did they discover? Land. When did they discover it? On the night of 11/12 October 1492.

Both Columbus and the lookout, Rodrigo de Triana, had a claim to have made the discovery. A great sociologist, Robert Merton (1910–2003), became preoccupied with the idea that there are nearly always several people who can lay claim to a discovery, and that where there are not this is because one person has so successfully publicized his own claim (as Galileo did with his discovery of the moons of Jupiter) that other claims are forestalled.[92] Merton was a great communicator. We owe to him indispensable phrases which encapsulate powerful arguments, such as 'unintended consequence' and 'self-fulfilling prophecy'; one of his phrases, 'role model', has moved out of the university into daily speech. Like all great communicators, he loved language: he wrote a whole book about the word 'serendipity', another about the phrase 'standing on the shoulders of giants', and he co-edited a volume of social science quotations.[93] Yet he complained that no matter how hard he tried he could not win support for the idea of multiple discovery (itself an idea, he pointed out, that had been discovered many times).

We cannot, somehow, give up the idea that discovery, like a race, is a game in which one person wins and everyone else loses. The sociologist's view is that every race ends with a winner, so that winning is utterly predictable. If the person in the lead trips and falls the outcome is not that no one wins but that someone else wins. In each race there are multiple potential winners. But the participant's view is that winning is an unpredictable achievement, a personal triumph. We insist on thinking about science from a participant's viewpoint, not a sociologist's (or a bookmaker's). Merton was right, I think, to find this puzzling, as we have become used to thinking about profit and loss in business from both the participant's point of view (the chief executive with a strategic vision) and from that of the economy as a whole (bulls and bears, booms and busts). Similarly, in medicine we have become used to moving back and forth between case histories and epidemiological arguments. I do not know when I will die, but there are tables which tell me what my life expectancy is and insurers will insure me on the basis of those tables. Somehow we are mesmerized by the idea of the individual role in discovery, just as we are mesmerized by the idea of winning – and of course this obsession serves a function, as it drives competition and encourages striving.

For Merton discoveries are not singular events (like winning), but multiple ones (like crossing the finishing line). Jost Bürgi discovered logarithms around 1588 but did not publish until after John Napier (1614). Harriot (in 1602), Snell (in 1621) and Descartes (in 1637) independently discovered the sine law of refraction, though Descartes was the first to publish. Galileo (1604), Harriot (*c.*1606) and Beeckman (1619) independently discovered the law of fall; only Galileo published.[94] Boyle (1662) and Mariotte (1676) independently discovered Boyle's law. Darwin and Wallace independently discovered evolution (and published jointly in 1858). The most striking multiples are those when several people claim to make a discovery at almost exactly the same time. Thus Hans Lipperhey, Zacharias Janssen and Jacob Metius all claimed to have discovered the telescope at around the same time in 1608. You might think that those who think that the idea of discovery is a fiction would welcome such cases, but no: as far as they are concerned, multiple discoveries are fictions, too. One far-fetched strategy they have used to undermine such cases is to maintain that in every case where claims are advanced for several different people having made one discovery, they have each in fact discovered different things. That is, Priestley and Lavoisier did not both discover oxygen; they made very different discoveries.[95] It is obvious, though, that Lipperhey, Janssen and Metius all discovered (or claimed to have discovered) exactly the same thing.

But let us go back to our original example, that of sunspots (putting aside the case of the telescope, where it may be suspected that there was one true discoverer, and the others were trying to steal his idea). Between 1610 and 1612 four different people discovered sunspots, each independently of the others: Galileo, Scheiner, Harriot (who did not publish) and Johannes Fabricius. It is just possible that Galileo stole the idea from Scheiner, or Scheiner from Galileo, but the other two were certainly independent, both of each other and of the first two. There really can be multiple, simultaneous discoveries of the same thing. If one wants to say that the four were making different discoveries because they interpreted what they saw somewhat differently, then one also has to say that when Copernicus saw Venus rise in the morning sky he was looking at a different planet from the one seen by every other astronomer since Ptolemy, for he was looking at a Venus that orbited the sun while they were looking at a Venus that orbited the earth.[96] Nevertheless, they could all agree on the coordinates of the planet they were looking at, and no one has ever claimed that Copernicus discovered Venus. (One might, on the other hand, argue that the first person to realize that the Morning Star and the Evening Star were the same object – Thales or Parmenides, according to the ancient historians – *had* discovered Venus.)[97]

§ 8

As we have seen, Bacon, who built a philosophy of science around the idea of discovery, took Columbus as his model; five years later Galileo was being hailed as the Columbus of astronomy: *quasi novello Colombo* (the diminutive is affectionate).[98] With discovery came the competition to be first. Columbus was determined to insist that he had seen land first because Ferdinand and Isabella of Spain had promised a lifetime pension to the first person to sight land. He offered de Triana the second prize of a silk jacket. Galileo rushed his book on his telescopic discoveries through the press because he was afraid someone else would beat him to it – and he was right to be fearful, as Harriot was already making astronomical observations with a telescope. In particular, he wanted to get his book out in time for copies to be shipped to Frankfurt ahead of the spring book fair.[99] Galileo was racing against unknown, imaginary competitors from the moment he realized Jupiter had moons. (He knew nothing about Harriot, but he knew that telescopes were becoming commonplace, and that soon everyone would be using them to look at the heavens.)[xxiii]

Since we live in a society constructed around competition, we have a tendency to take competitive behaviour for granted as a universal aspect of social life. But we should be cautious about doing so. The noun 'competition' first appears in English in 1579, the verb 'compete' in 1620. In late-sixteenth-century French *concurrence* still means 'coming together' and not yet 'competing'; in early-seventeenth-century Italian *concorrente* is only beginning to take on its modern meaning. Nor was there an obvious synonym, at least in English: 'rival' (noun, 1577; verb, 1607) and 'rivalry' (1598), are roughly contemporary with 'competition' and reflect the need for a new language for competitive behaviour, a behaviour which is as much the result as the cause of the new culture of discovery.[100]

Different individuals responded in different ways to the new pervasive spirit of competition. In the case of the great mathematician Roberval the result was a pathological conviction that other people were stealing his ideas. As Hobbes remarked of his friend, 'Roberval has this peculiarity: whenever people publish any remarkable theorem they have discovered, he immediately announces, in papers which he distributes, that he discovered it first.'[101] Newton waited nearly thirty years before publishing a full account of his discovery of calculus; no one

xxiii Strictly speaking, Galileo knew Harriot existed, as he had read William Gilbert's *On the Magnet*, in which Harriot is mentioned in passing, and described as 'most scholarly'. Harriot managed to make a telescope capable of resolving the moons of Jupiter very soon after he read Galileo's *Starry Messenger*: he probably read it in July and he was certainly observing the moons in October – he could not observe them earlier because Jupiter was too close to the sun (Roche, 'Harriot, Galileo and Jupiter's Satellites' (1982)).

could have seemed less interested in claiming priority. By the time he published it, in 1693, he was well behind Leibniz, who had published his somewhat different version in 1684. Yet in the years after 1704 a bitter dispute broke out between them as to whether Leibniz had seen a manuscript of Newton's and stolen his ideas. Newton's friends had to pressure him to publish his great work, the *Principia* (1687), in which he explained gravity. Two years later Leibniz published an alternative theory. This gave rise to a further dispute as to whether Leibniz had arrived at his theories, as he claimed, independently, or had cobbled them together on the basis of a reading of the *Principia*. The first charge against Leibniz was mistaken and unfair, but Newton pursued it ruthlessly, even writing himself what was supposed to be an impartial judgement by the Royal Society on the rights and wrongs of the dispute. The second accusation was, recent scholarship has shown, well founded. In this respect, Leibniz was indeed a plagiarist. With no good reason and with every good reason, Newton became caught up in what was to prove the most bitter and protracted of all priority disputes, complaining furiously that he had been 'robbed of his discoveries'.[102]

If Newton, who had for so long seemed indifferent to such matters, could not resist entering into battle on his own behalf, it was because nothing less was expected of him by his friends and disciples. He was immersed in a culture that was obsessed with claims to priority (Newton himself was the target of accusations of plagiarism from Hooke, who insisted he had given him the inverse-square law, a gift which Newton refused to acknowledge.)[103] There was more to the culture of the new science (as we shall see) than competition alone, but competition was at its heart; indeed, there would have been no science without it.

The existence of competition among scientists is in itself evidence that the idea of discovery is present; and where there is no competition, there is no concept of discovery. The claim that discovery was to all intents and purposes new is a strong claim, but it is easy to test it again (for we have already tested it once by looking for discoveries in Vergil's *On Discovery*).[104] When was the first priority dispute? By this I mean not a priority dispute that has been constructed later by historians (Who first discovered America, Columbus or the Vikings?), but one that led to conflict at the time. Well before the dispute over who had discovered sunspots (1612 onwards) there was a bitter dispute in the years after 1588 between Tycho Brahe and Nicolaus Reimers Baer, known as Ursus (the Bear), over who had discovered the geoheliocentric cosmology (Brahe published shortly before Ursus, but Ursus claimed independent development – the hypothesis, he argued, was not really new – both of which Brahe hotly denied).[xxiv][105] The two also lodged rival claims to have invented the mathematical technique of prosthaphaeresis, which was important for making lengthy calculations before the invention of logarithms (logarithms are another multiple discovery, since they were independently discovered by John Napier in 1614 and Joost Bürgi in 1620).[106] But Brahe and Ursus did not invent the priority dispute; rather, they cared about priority because mathematicians had been taking it seriously since at least 1520.[107]

In 1520 Scipione del Ferro discovered a method to solve cubic equations. Del Ferro taught the method to one of his students, but it was independently discovered by Niccolò Fontana, known as Tartaglia (a nickname meaning 'stammerer'). Tartaglia beat del Ferro's student in a public 'duel' in which they competed to show off their mathematical prowess (and to recruit students; in the city states of Renaissance Italy mathematical education was crucial to commercial success, but the pool of potential students was limited, resulting in fierce competition for them between mathematicians). The mathematician and philosopher Girolamo Cardano persuaded Tartaglia to pass the secret on to him by misleading him into expecting a significant financial reward, but Cardano was sworn to secrecy, and Tartaglia coded the secret into a poem so that he could later demonstrate his own priority. Cardano sometime afterwards discovered that del Ferro had known the secret before Tartaglia, so held himself to be released from his oath and published the technique in 1545 – which led to a bitter dispute between Cardano and Tartaglia and a further 'duel' between a pupil of Cardano and Tartaglia (which Cardano's pupil won).[108]

This little episode demonstrates clearly what the preconditions are for priority disputes. First, there must be a close-knit community of experts who share criteria by which to identify what constitutes success (this is apparent in the 'duels'). Second, this community of experts must have a shared base of knowledge which enables them to establish whether a result is not only right but also new. Third, there must be ways of establishing priority – Tartaglia's coded poem is a device to demonstrate that he already has the solution even though he is keeping it secret. (In 1610, using a similar method, Galileo published anagrams to prove that he had discovered the phases of Venus and the strange shape of Saturn, even though he had not yet announced these discover-

xxiv On Brahe's cosmology, see below, pp. 193–4.

ies; Robert Hooke first announced what we now call Hooke's law of elasticity by publishing an anagram in 1660; and Huygens, in the discovery of Saturn's moon (now called Titan) and the ring of Saturn, similarly relied on anagrams to protect his claims to priority.)[109] Finally, there must be a mechanism for publicizing one's knowledge – Cardano, for example, publishes a book. In normal circumstances it is *publication* which creates, in the first place, a community of experts and an established body of knowledge (these are, as it were, two sides of one coin), and then makes possible an indisputable claim to priority.

It is not impossible to imagine priority disputes without the printing press, but in fact there are no priority disputes that we know of which predate printing.[xxv] If we go back, for example, to ancient Rome, where Galen engaged in public disputes with other doctors rather like the duels between mathematicians in Renaissance Italy, we find plenty of competition between self-proclaimed experts; what is missing is agreement about what constitutes expertise, and so on how to identify a winner.[110] Galen's extraordinary logorrhoea – his surviving works amount to 3 million words, and represent perhaps a third of what he wrote – is the consequence of his obsessive and futile effort to overcome this insuperable obstacle. Ironically, in the universities of medieval Europe every doctor came to accept that Galen was the embodiment of medical learning. In Rome there had been competition between several medical schools (empirics, methodists, rationalists) without any clear-cut winner; in the medieval university there was one winner and no competition;[xxvi] in the Renaissance the printing press created for the first time the conditions for real competition, that is, for both conflict and victory.

In anatomy the process began rather later than in mathematics. In 1543 Andreas Vesalius published *The Fabric of the Human Body*, in which he identified scores of errors in Galen's works. He was competing with Galen, but there was not yet a community of anatomists in competition with each other, and Vesalius was not in the business of claiming priority for his discoveries. Rather, he established a base line which enabled others to make priority claims. (As we have seen, both Ingrassia

xxv The printing revolution is a theme that runs through this book: we will return to it in Chapters 5–8 and 17; and see above, p. 60.

xxvi There was plenty of disputation in the medieval universities, but that is quite different from real intellectual competition: participants were expected to be able to argue either side of a case, so disputation was a test of rhetorical skill but not a competition to get to the truth.

and Falloppio were able to claim a discovery, the stirrup bone, on the basis of having seen something Vesalius had failed to mention.)

One of Merton's fundamental claims about science is that scientific knowledge is public knowledge – knowledge that has been made available for others to question, test and dispute.[111] Privately held knowledge is not really scientific knowledge at all because it has not survived the test of peer review. So there can be no science until there is a reliable way of publicizing knowledge. And discoveries that are never made public, or made public only long after they have been made, are not really discoveries at all.[xxvii] Priority disputes are an infallible indicator that knowledge has become public, progressive and discovery oriented. Thus their first appearance in a discipline marks a crucial moment in the history of that discipline, the beginning of what, in retrospect, we may call 'modernity'. We have seen that they appear first in mathematics, and that by 1561 Falloppio was engaged in a priority dispute with Colombo over who had discovered the clitoris.[112] Since Colombo was recently dead, and Falloppio died in 1562, the dispute had to be continued by Falloppio's student Leone Carcano. A century later, in the years after 1653, there was a bitter dispute between Thomas Bartholin and Olof Rudbeck over the discovery of the human lymphatic system.[113] Where there are vituperative priority disputes, ways of resolving them must be sought. Brahe embarked on a court case against Ursus (who died before the matter came to trial), but it was obvious the courts lacked the required expertise.[114] So the dispute between Reinier de Graaf and Jan Swammerdam over the discovery of the egg within the ovary, which began in 1672, was referred to the Royal Society in London for adjudication.[115] The Society awarded priority to neither of the contestants, but to Niels Steno.

Of equal importance to priority disputes is the naming of discoveries. Scientific discoverers have often claimed the right to name their discoveries, in imitation of the discoverers of new lands – Ingrassia named the stirrup bone, Galileo named the moons of Jupiter the Medicean planets,

xxvii To distinguish so sharply between public and private knowledge may seem to diminish the achievements of those who make discoveries but never publish them. But, as we shall see (e.g. below, pp. 198, 303–4, 331, 340, 348, 358 and 424), science exists only where there is a community of scientists; it is from competition within this community that progress results. Another way of making this argument is in terms of Popper's three worlds. Popper distinguished between the world of physical objects, the world of mental processes and the third world, the world of 'problems, conjectures, theories, arguments, journals and books'. Science belongs to this third world. (Popper, *Objective Knowledge* (1972), 107. See also his earlier statements, Popper, *The Logic of Scientific Discovery* (1959), 44–7.)

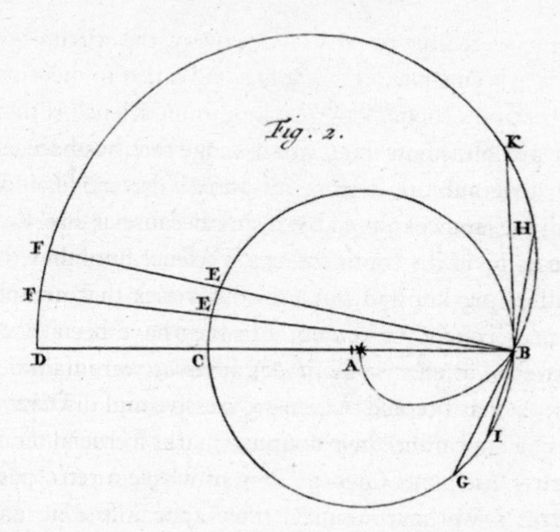

The mathematical curve called 'Mr P[ascal]'s Snail', taken from Roberval's *Mathematical Works* (1731).

and Lavoisier named oxygen. Often, discoveries are named after their inventors; from 1597 it became standard to distinguish three systems of the cosmos, those of Ptolemy, Copernicus and Brahe.[116] Étienne Pascal, father to Blaise, discovered a striking mathematical curve in 1637: in 1650 it was named Pascal's Snail by his friend Gilles de Roberval – or rather, perhaps out of respect for Étienne Pascal's modesty (for he was still alive), Mr P's Snail.[117]

Such names are themselves priority claims, made on behalf of the original discoverers by their admirers, and the implicit analogy is with the naming of America.[xxviii] This explains why there are no parts of the body named after Hippocrates or Galen, no stars named after Ptolemy, no creatures named after Aristotle or Pliny. The naming game is inseparable from the discovery game; it could not exist before the voyages of discovery. Indeed, for the naming game to get going scientists have to make priority claims that can be taken forward on their behalf. Even Vesalius, the first great anatomist of the Renaissance, was not in the business of claiming priority, which is why, for all his originality, there are no parts of the body named after him.

xxviii Such claims, of course, are often advanced for reasons of national pride: see, already, Wallis, 'An Essay of Dr John Wallis' (1666), 266; and Anon., 'An Advertisement Concerning the Invention of the Transfusion of Bloud' (1666), 490.

§ 9

There was an immediate realization that there was something new about the geographical discoveries in Waldseemüller's decision in 1507 to call the lands explored by Vespucci 'America'; very quickly this became the name of the continent as a whole.[118] Eponymy (the naming of things after people) had not previously been a common practice. It was not, of course, unknown: Christianity, after all, was named after Christ. And, on the same model, heresies were named after their originators: Donatism and Arianism, for example. There were also some cities named after their founders: Alexandria after Alexander, Caesarea after Augustus Caesar, Constantinople after Constantine.[xxix] An important set of astronomical tables, the Alfonsine Tables, were named after the man who had sponsored them, Alphonso X, king of Castile (1221–84).[119]

As the Portuguese navigators explored the coast of Africa they charted and named it, often borrowing names from the local peoples, or using the names of saints. At last, in 1488 Bartolomeu Dias reached the southernmost tip of the continent, which he named the Cape of Good Hope. Beyond the Cape, the furthest point that Dias reached he named the Rio do Infante, the River of the Crown Prince, after Prince Henry, now called the Navigator.[120] Columbus named the islands he had discovered San Salvador, Santa María de la Concepción, Fernandina, Isabella, Juana and Hispaniola; and the first Spanish city he named La Navidad; all these names refer to Christian doctrine or the Spanish royal family. The only feature of the New World named after a commoner before 1507 seems to have been the Rio de Fonsoa, named after a sponsor of the 1499 expedition.[121] Eponymy received an enormous boost from the practice of naming new lands after their discoverers' patrons (the Philippines after Philip II of Spain; Virginia after Elizabeth I, the Virgin Queen; Carolina after Charles I), but these were still nearly always kings and queens (an exception is Van Diemen's

xxix Galileo mentions these on the opening page of *The Starry Messenger* (1610). On a map of 1339 the island we now call Lanzarote is labelled Insula de Lanzarotus Marocelus; Lancelotto Malocello had laid claim to the island *c.*1336 (Verlinden, 'Lanzarotto Malocello' (1958)). Labelling an island 'Lanzarote's Island' is not quite the same thing as naming it Lanzarote, and it is unclear when the transition from the one to the other took place – it was certainly after 1385. Liechtenstein, like Lanzarote, is named after its owner, but this happened only in 1719.

Land, named after the Governor General of the Dutch East Indies in 1642, but only much later renamed after Abel Tasman, who discovered it).

Like the idea of discovery itself, eponymy was soon carried over from geography into science. Just how new this was is indicated by Galileo's attempt in 1610, when he was naming the newly discovered moons of Jupiter 'the Medicean stars', to find a precedent for naming a star after a person. The only example he could find was an attempt by Augustus to name a comet after Julius Caesar (a futile attempt, because, of course, the comet, now known as Halley's, quickly disappeared).[122] Augustus, naturally, was claiming that Caesar was not a person but a god, for the planets were all named after gods (and the principle continued to be respected in the naming of newly discovered planets: Uranus, Neptune, Pluto).[xxx][123] In Latin the days of the week were named after the planets (including the sun and the moon, which were planets in the Ptolemaic system); in the Germanic languages some of them were renamed after pagan gods. Amerigo Vespucci, on the other hand, was no divinity, no emperor and no king. Eponymy had been brought down to earth with a bump.

In geography the discovery game and the naming game went hand in hand; but in science the second was slower to get going than the first. This is not obvious to us because in due course classical discoveries came to be linked to the names of their discoverers. 'The principle of Archimedes' (that an object will float if the weight of liquid displaced is equal to the weight of the object) does not seem to have been named as such until 1697.[124] A 1721 etymological and technical dictionary contains only two examples of eponymy, aside from the three astronomical systems of Ptolemy, Brahe and Copernicus (or Pythagoras): the Fallopian tubes and a nerve called the *accessorius Willisii*, discovered by Thomas Willis (1621–75).[125]

So when did eponymy in science begin? As we have seen, eponymy in geography is very rare before the naming of America, and America remained an exception in being named after a commoner. Cicero had used the adjectives *Pythagoreus*, *Socraticus*, *Platonicus*, *Aristotelius* and

xxx There are twelve elements named after people, and plenty of stars, comets and asteroids – but no planets or planetary moons. Herschel wanted to name Uranus after King George III; and Le Verrier wanted to name Uranus after Herschel and Neptune after himself (Hoskin, 'The Discovery of Uranus' (1995), 175; Morando, 'The Golden Age of Celestial Mechanics' (1995), 218, 220). But the classical convention of naming the planets (and now planetary moons) after the gods has survived, at least for our solar system.

Epicureus, so it is natural that we find early on adjectives for other philosophers – *Ippocratisa* (*c.*1305), *Thomista* (1359), *Okkamista* (1436) and *Scotista* (1489) – although many of these words were very slow to enter the vernacular; apart from *Epicureus* (which appears in the Wycliffite Bible of 1382), I cannot find any of them – not even 'Platonist' – in English before 1531 (when 'Scotist' appears).[126]

What seems to us the obvious process of labelling ideas and discoveries by attaching to them the names of their authors (at the moment I suffer from no fewer than three medical conditions named after their discoverers) was not commonplace before the invention of discovery.[xxxi][127] 'Algorithm', after the Latin form of the name of the Persian mathematician al-Khwārizmī (780–850), goes back at least to the early thirteenth century, but that seems to be an outlier.[128] 'Menelaus's Theorem', named after Menelaus of Alexandria (70–140), which is the mathematical foundation of Ptolemaic astronomy, had been explicitly attributed to Menelaus by Proclus in the fifth century. In 1560, in the margin of his translation of Proclus, Francesco Barozzi named it Menelaus's theorem (*Demonstratio Menelai Alexandrini*), though it was known by the Arabs and the medieval commentators as the Transversal Figure.[129] In the index, but not in the text or marginal notes, Pythagoras's theorem was named as such (it had previously been known as Dulcarnon, from the Arabic for 'two-horned', after the appearance of the diagram that accompanied it). Indeed, the index demonstrated a systematic determination to link ideas with their original authors wherever possible, and in the text and the index Barozzi even carefully labels one comment 'the *scholium* of Francesco Barozzi'. Since every idea now had to have an author, where an author could not be found their absence was noted – Barozzi's *scholium* was a reply to 'the *scholium* of an unknown author', found in an old manuscript.[130] This was new: Vitruvius, first published in 1486, had described Plato's method for doubling the area of a square, the invention of Pythagoras's set-square (both practical applications of Pythagoras's theorem), and the discovery of

xxxi A simple example of how eponymy has come to seem normal to us, but was not normal before the discovery of America, is provided by the religious orders whom we call the Dominicans and the Franciscans. They were founded in 1216 and 1221; but the Order of Preachers and the Order of Friars Minor, as they are properly called, were informally renamed after their founders, Dominic and Francis, much later: *Dominicanus* dates only to 1509 (1534 in English), *Franciscanus* to 1515 (again, 1534 in English). *Dominicanus* and *Franciscanus* from Latham (ed.), *Dictionary of Medieval Latin from British Sources* (1975); 'Dominican' and 'Franciscan': Erasmus, *Ye Dyaloge Called Funus* (1534). (The *OED* gives 1632 for 'Dominican'.)

Archimedes' principle, but the indexes to the various editions of Vitruvius show that names only very slowly came to be associated with ideas. The German translation of 1548 is the first to provide an extensive range of name entries, but even it, though it has entries for Archimedes and Pythagoras, has no entry for Archimedes' principle or Pythagoras's set-square.[131]

In 1567 the great Protestant logician and mathematician Petrus Ramus referred to 'the laws of Ptolemy' and 'the laws of Euclid'.[132] But Ramus was looking far back into the past. Indeed, one can formulate a general law (Wootton's law, of course, since our subject is eponymy): where a scientific discovery took place before 1560 and is named after its discoverer (or supposed discoverer), the naming of that discovery took place long after the event. Thus, to take an example at random, Leonardo Pisano, known as Fibonacci, is the supposed discoverer of the Fibonacci number series. He wrote in 1202, but it was named after him in the 1870s.[133]

If 1560 marks the effective beginning of eponymy in science, the practice only began to be widespread and employed to refer to contemporary discoveries after 1648, the year in which the standard vacuum experiment (involving a long glass tube closed at one end and a bath of mercury) came to be known as Torricelli's experiment.[xxxii] (The experiment had first been performed in 1643, but it was not generally known at first that Evangelista Torricelli was its inventor; in 1650, as we have seen, Roberval named a mathematical curve after Étienne Pascal.) In 1651 Pascal reacted with horror to the suggestion that he had tried to pass Torricelli's experiment off as his own: everyone understood, he insisted, that this would be the scholarly equivalent of theft.[134] What seemed obvious to Pascal, that one could 'own' an idea or an experiment, would have puzzled anyone before 1492.[xxxiii] Indeed 'plagiary' becomes a word in English only in 1598, 'plagiarism' in 1621, 'plagiarize' in 1660, 'plagiarist' in 1674.[135] In 1646 Thomas Browne collected numerous examples from Greek and Roman authors of texts being

xxxii I have searched through Bartholin, Bartholin, and others, *Institutiones anatomicae* (1641), looking for a body part named after a person, but without success; discoverers are scrupulously recorded, but their discoveries are not yet named after them.

xxxiii This idea long precedes the legal claim to authorial copyright, which did not exist in British law until 1710, and was established later elsewhere. Printers could claim monopoly rights to print a text within a particular jurisdiction; authors, as such, had no rights at all at law: Kastan, *Shakespeare and the Book* (2001), 23–6.

copied wholesale and reissued under another author's name.[xxxiv] 'The practise of transcription in our dayes was no monster in theirs: Plagiarie had not its nativitie with printing,' he concluded, 'but began in times when thefts were difficult' because books were in short supply.[136] What was new was not the practice of copying others but the idea that this was something to be ashamed of. It did not occur to Browne that the notion of intellectual property owed as much to Columbus as to the printing press.

From around the middle of the seventeenth century there is a flood of adjectives that appear in English for scientific experiments, theories or discoveries, all based on the names of scientists. In 1647 Robert Boyle referred to 'the Ptolemeans, the Tychonians, the Copernicans'.[137] This is followed by 'Galenic' (1654), 'Helmontian' (1657),[138] 'Torricellian' (1660), 'Fallopian' (1662),[139] 'Pascalian' (1664), 'Baconist' (1671),[140] 'Euclidean' (1672), 'Boylean' (1674) and 'Newtonian' (1676).[141] In the early eighteenth century scientific laws begin for the first time to be named after their discoverers. (The idea of a scientific law was itself new, which is why even now there are no laws named after ancient or medieval mathematicians and philosophers; unlike Ramus, we don't speak of the laws of Euclid and Ptolemy, for Ramus meant by 'law' a mathematical definition, not a regularity in nature.) Thus we have Boyle's law (1708),[142] Newton's law (1713)[143] and Kepler's law (1733).[144] The mapping of the moon, beginning with van Langren in 1645, provided a crucial precedent for eponymic naming, helping to transfer it from geography to astronomy. The early selenographers had so many features to name that they found themselves honouring ancients as well as moderns, opponents as well as allies. Giovanni Battista Riccioli, a Jesuit supporter of Brahe, named a crater after Copernicus. This does not prove, as some have imagined, that he was a secret Copernican, only that there were plenty of craters to go round.

xxxiv In the Middle Ages the word *auctor* primarily means 'authority': 'No "modern" writer could decently be called an *auctor* in a period in which men saw themselves as dwarfs standing on the shoulders of giants, i.e. the "ancients"' (Minnis, *Medieval Theory of Authorship* (1988), 12). Even in the seventeenth century, Shakespeare was referred to as an 'author' only after he was dead: Kastan, *Shakespeare and the Book* (2001), 69–71. Relevant here, of course, is Foucault's famous discussion of the author function: '*Qu'est-ce qu'un auteur?*' (1969), in Foucault, *Dits et écrits* (2001), Vol. 1, 817–49.

§ 10

Discovery is not in itself a scientific idea but rather an idea that is foundational for science: we might call it a metascientific idea. It is difficult to imagine how one could have a form of science (in the sense in which we now use the term) which did not claim to have made progress and did not present that progress in terms of specific acquisitions of new knowledge. The metaphor of uncovering, the paradigm case of voyages of discovery, the insistence that there is one discoverer and one moment of discovery, the practice of eponymy, and other more recent ways of marking discovery, such as the award of the Nobel Prize (1895) or the Fields Medal (1936) – these are surely aspects of a local culture, but *any* scientific culture would need an alternative set of concepts that fulfilled the same function of marking, and inciting, change. As we have seen, Hellenistic science, the science of Archimedes, provides an interesting test case. It had many of the characteristics of what we call 'science' (indeed, the first modern scientists were simply trying to imitate their Greek predecessors) and it had, in heurematography, a rudimentary understanding of science as discovery.[145] Nevertheless, no ancient Greek struck a medal with the word *Eureka* on it and started awarding it to successful scientists as we award the Fields Medal to successful mathematicians. By contrast, Galileo's *Starry Messenger* (1610) opens with Galileo asserting (a little indirectly, for modesty's sake) his own claim to immortal fame, a claim that no statue or medal, he tells us, could adequately recognize.[146] There were, as yet, no prizes or medals for scientific achievement; but in Galileo's imagination such rewards already existed. Francis Bacon, in his *New Atlantis* (1627), imagined a gallery containing the statues of the great inventors (such as Gutenberg) and discoverers (such as Columbus).[147] In 1654 Walter Charleton called for 'a Colossus of Gold' to be erected in Galileo's honour.[148] The Nobel Prize is merely a new version of Charleton's colossus.

Discovery started as a local concept, symbolized at first by the erecting of new 'pillars of Hercules' by Portuguese explorers advancing along the coasts of Africa. With it came the word *descubrimento*, meaning first 'exploration' and then 'discovery'; and then this word, in its vernacular equivalents, spread across Europe. Is this a local story, or a cross-cultural story? Discovery began as a concept confined to a particular activity (sailing towards Asia), and a particular culture (fifteenth-century Portugal), but it soon became a concept available throughout Western Europe. It was the essential precondition for the new era of intellectual revolution, for it is a necessary concept which any society that saw itself as improving knowledge would have to develop. The wide dissemination of words meaning 'discovery' across Europe in the sixteenth and seventeenth centuries reflects, in the first place, the diffusion of a new type of cartographic knowledge, a form of knowledge that may have been local at first but rapidly became cross-cultural (just as the Portuguese ocean-going ship, the carrack, was quickly imitated across Europe). It is worth noting that the new geographical discoveries were swiftly accepted right across Europe – you did not need to be Spanish to believe that Columbus had discovered a new continent. But it also reflects, in the second place, the spread of a new culture, a culture oriented towards progress. Once the idea of discovery had established itself, it expanded out of geography into other disciplines. This, too, is a form of cross-cultural transmission.

For a considerable time – some centuries – the new scientific knowledge was confined within the boundaries of Europe and the ships and colonies of Europeans abroad. The whole of Europe proved capable – some regions more so than others, of course – of abandoning its old theories and adopting new ones, of rejecting the idea of knowledge as fundamentally complete and embracing the idea of knowledge as a work in progress. The new knowledge did not spread in the same rapid, confident way outside Europe.[149] There are various explanations for this but, crucially, European culture allowed considerable space for competition and diversity. European societies were everywhere fragmented and divided, with many local jurisdictions (such as self-governing cities and universities), each state in competition with every other state, and everywhere, religious authority in tension with secular authority. And, of course, Europe had inherited Greek culture as well as Latin: the new science could claim to be continuing a respectable intellectual programme, to be in the tradition of Pythagoras, Euclid, Archimedes and even, in certain respects, Aristotle.

Thus the category 'discovery' proved to be capable of disseminating across the various local cultures of Renaissance Europe, but it did not fare well elsewhere. Other cultures (and to some degree Catholic cultures in Europe after the condemnation of Copernicus) were not prepared to accept radical intellectual change. My argument is that a discovery concept of some sort is a crucial precondition for systematic innovation in the knowledge of nature; there is a logic to innovation, and if knowledge is to be geared towards innovation it has to respect that logic. But the idea of discovery does not bring with it cultural uniformity; rather, it encourages diversity. It is compatible with all sorts of different forms of new knowledge, with the geocentrism of Riccioli as well as the heliocentrism of Copernicus, with Descartes' denial of a vacuum and Pascal's acceptance of one, with Newton's view of uniform space and time and Einstein's theory of relativity; it does not lead of necessity to any particular type of science. Moreover, the social practice which we label 'discovery' can be confused, contradictory and paradoxical: it really is not always obvious who made a discovery or when they made it. So, on the one hand, discovery is more than a local practice, it is a precondition for science; on the other, it relies on contingent and local ways of determining what counts as a discovery and what does not. The existence of the idea of discovery is a necessary precondition for science, but its exact form is variable and flexible; where it meets resistance, as it did in the Ottoman empire and in China, then the scientific enterprise itself cannot take root.[xxxv]

With the emergence of the idea of discovery and the consequent development both of priority disputes and of the determination to link every discovery to a named discoverer, something that is recognizable as modern science begins to appear for the first time. And with the new science came a new kind of history.[xxxvi] Here, for example, is the second paragraph of the entry for 'magnet' from a technical lexicon of 1708:

> Sturmius in his *Epistola Invitatoria dat. Altdorf* 1682 observes that the attractive quality of the magnet hath been taken notice of beyond all history. But that it was our countryman Roger Bacon who first discovered the *verticity* of it, or its property of pointing towards the pole, and this about 400 years since. The Italians first discovered that it could communicate this virtue to steel or iron. The various declination of the needle under different meridians was first discovered by Sebastian Cabbott, and its inclination to the nearer pole by our countryman Robert Norman.[xxxvii]

xxxv I take it that Joseph Needham's great study of *Science and Civilization in China* demonstrated that Chinese technology was superior to European through the Middle Ages, but not that China had an intellectual enterprise corresponding to the European conception of natural science.

xxxvi Although here a classical precedent must be acknowledged in the work of Eudemus of Rhodes (*c.*370–300 BCE): Zhmud, *The Origin of the History of Science* (2006).

xxxvii 'Declination' here means variation east and west, while 'inclination' means variation up or down from the horizontal. I use 'variation' as a general term for both, and when they are to be distinguished I call the first 'variation' and the second 'dip'. I avoid 'declination', which seems to me too easily confused with 'dip'.

> The variation of the declination, so that 'tis not always the same in one and the same place, he observes, was taken notice of but a few years before, by Hevelius, Auzout, Petit, Volckamer, and others.[150]

Such histories are not just histories of foundings, they are also histories of progress.

Thus we can sum up the argument so far very straightforwardly. The discovery of America in 1492 created a new enterprise that intellectuals could engage in: the discovery of new knowledge. This enterprise required that certain social and technical preconditions be met: the existence of reliable methods of communication, a common body of expert knowledge and an acknowledged group of experts able to adjudicate disputes. First cartographers, then mathematicians, then anatomists, and then astronomers began to play the game, which was inherently competitive and immediately gave rise to priority disputes and, more slowly, to eponymic naming. Inseparable from the idea of discovery were the ideas of progress and intellectual property. In 1605 Bacon claimed to have identified the basic method for making discoveries and ensuring progress, and in 1610 Galileo's *Starry Messenger* confirmed the idea that there was a new philosophy of nature which had an unprecedented capacity to make discoveries.

Of course, the discovery game had its antecedents and its precedents. The best example is the patent. In 1416 the Venetian government granted a patent, to last fifty years, to Franciscus Petri, the inventor of a new fulling machine. In 1421 the great engineer and architect Brunelleschi was given a three-year patent by the city of Florence on a new design for a barge for carrying marble. In 1474 the republic of Venice formalized its own patent system by requiring that those wanting to claim a monopoly first register their new inventions with the state. (This became the model for the first English grant of a patent, to Acontius in 1565.)[151] Before Columbus discovered America, there was already a reward announced, should he be successful. But patents do not last for ever, and they give you a claim only within a particular jurisdiction. Columbus's reward was to last only a lifetime, and since he never expected to discover an unknown land (instead of a new route to a known land) it never occurred to him to claim naming privileges. There is, on the other hand, no limit of time or space on a discovery – it represents a new form of immortality. And, in any case, the social and technical prerequisites for the discovery game were only just coming into existence in 1492, for it was the printing press (invented *c.*1450) which carried news of the discoveries, first of Columbus, then of Cardano, Tycho Brahe, Galileo and all the rest. It was the printing press that established a common knowledge base against which these discoveries could be measured.[152]

What was not yet clear in 1610 was how best to conduct this new enterprise. Bacon thought he had the answers, but he was wrong. In fact, he had very bad judgement when it came to good science, dismissing the work of Copernicus and Gilbert out of hand. But Bacon was not alone in having bad judgement (in Chapter 4 we will be exploring some of the mistakes made by the early scientists). Often these mistakes were obvious. The great Galileo dedicated much of his life to proving the movement of the Earth by claiming it was the only possible cause of the tides. It was his determination to present this argument as decisive that led to his condemnation by the Inquisition. His argument did not explain the facts: if he was right, high tide should have been at the same time every day, and there should have been only one each day. The only contemporary it convinced was Giovanni Battista Baliani, who, in order to make Galileo's theory (more or less) work, had to put the Earth in orbit around the moon! And yet Galileo was absolutely sure his argument was conclusive.[153]

In the course of the first century after the publication of Vesalius's anatomy and Copernicus's cosmology (both appeared in 1543) a set of values was slowly devised for how best to conduct the intellectual activity that we now call science: originality, priority, publication, and what we might call being bomb-proof: in other words, the ability to withstand hostile criticism, particularly criticism directed at matters of fact, came to be regarded as the preconditions of success. The result was a quite new type of intellectual culture: innovative, combative, competitive, but at the same time obsessed with accuracy. There are no *a priori* grounds for thinking that this is a good way to conduct intellectual life. It is simply a practical and effective one if your goal is the acquisition of new knowledge.

It was apparent right from the start that discovery, priority and originality were ambiguous and, at the limit, incoherent; and that these values conflicted with the obligation to check and check again before publication. Take discovery, which was regarded as the highest form of originality. Was it de Triana, Columbus, Vespucci or Waldeseemüller who discovered America? The credit went to Columbus, because it was his expedition which got there first, even though he never knew where he was: the importance of the discovery outweighed his failure to understand exactly what he had done. Galileo understood this when he rushed *The Starry Messenger* through the press – but the same Galileo held back his discovery of the law of the acceleration of falling bodies for more than thirty years, determined not to publish before either victory was assured or death was at hand. (Harriot and Beeckman also discovered the law of fall; both died without publishing.) Copernicus, similarly, had delayed and delayed the publication of *On the Revolutions*. There was a constant tension between the aspiration to be first and the fear of not being believed, of being regarded as an eccentric and a fool.

Despite all the resulting conflicts and contradictions, which remain with us still, it was the idea of discovery that made the new science, and the new set of intellectual values that underpinned it, possible. When you think about it, this is a simple and obvious truth – but it is one that historians of science, who want to maintain that every culture has its own science and that they are all equally valid, have failed to grasp. The discovery enterprise is no more universal than cricket or baseball or soccer; it is peculiar to the post-Columbus world, and it can only survive within a society which fosters competition. It is the only enterprise that produces, in Pierre Bourdieu's phrase, 'trans-historical truths'.

And, of course, the triumph of the discovery enterprise was not complete until well into the eighteenth century. The old ideas had too much authority, particularly because they were grounded in the Biblical narrative, simply to disappear without trace. Most striking here is the case of Newton, who, after he had made and published his great discoveries in the *Principia*, began to suspect that they were not new but merely rediscoveries. Surely Moses had known all this already? He planned a second edition, in which he would demonstrate that everything that was thought to be new in his book was really old. As Fatio de Duillier, who was working as his assistant, wrote in 1692: 'Mr Newton believes to have discovered good evidence [*avoir decouvert assez clairement*] that the Ancients, such as Pythagoras, Plato, etc., had all the demonstrations which he gives regarding the true System of the World, which are based on Gravity . . .'[154] Newton acquired a mass of material intended to establish this peculiar thesis. But three warnings need to be entered at this point: first, when Newton wrote the *Principia* he did not yet have this theory and he had not sought to advance his new physics by reading ancient sources; second, Newton himself was aware of resistance to his theory, which as a result was largely withheld from the second edition when it finally did appear in 1713; and third, as far as Newton's contemporaries were concerned, his discoveries were entirely new. Newton's theory that the ancients understood gravity was a private eccentricity, a useful defence (we may suspect) against the pride that might be engendered by the realization that he was the greatest scientist of all time; only one or two of his closest friends were prepared to take it seriously. The old conviction that there was no new knowledge had momentarily resurfaced, only to sink almost without trace beneath the tide whose very existence it denied.

4
Planet Earth

an utterly insignificant little blue-green planet.

– Douglas Adams, *The Hitchhiker's Guide to the Galaxy* (1979)[1]

§ 1

The voyages of discovery brought about an astonishing transformation in geographical knowledge from 1460 onwards. Where the known world in the first half of the fifteenth century was more or less identical to the world known to an educated Roman at the time of Christ, by the beginning of the sixteenth century it was clear that there were extensive inhabited territories that had been unknown to the Greeks and Romans. Where the conventional view had been that lands close to the equator must be uninhabitable, this had turned out to be nonsense. This expansion of the known world was carefully recorded by cartographers, and it brought about the first great triumph of experience over philosophical theory.

The subject of this chapter, though, is not the voyages of discovery as such. In the wake of Columbus's discovery of America a silent revolution occurred, the invention of what we now call 'the terraqueous globe'. This revolution took place in the space of a few years and encountered no (or almost no) resistance. It is of profound importance, but it is completely invisible in the standard historical literature. Thomas Kuhn once wrote:

> A historian reading an out-of-date scientific text characteristically encounters passages that make no sense . . . It has been standard to ignore such passages or to dismiss them as the products of error, ignorance, or superstition, and that response is occasionally appropriate. More often, however,

> sympathetic contemplation of the troublesome passages results in a different diagnosis. The apparent textual anomalies are artifacts, products of misreading.[2]

My subject is a whole library of texts that at first sight make no sense. For fifty years now historians of science, inspired by Kuhn, have sought out such texts in order to demonstrate their expertise, their capacity to make sense of the apparently nonsensical, but these particular texts have been almost completely ignored. Why? Because they point to something that isn't supposed to exist: a silent revolution. According to Kuhn, revolution always brings with it disputation and conflict;[3] since there was virtually no disputation, it is all too easy to assume that there can have been no revolution. It is this very anomaly, on the other hand, which makes these texts the perfect place to embark on a new, post-Kuhnian history of science.

What shape is 'the earth'? The answer to this question must seem obvious. Surely everyone knew that the earth is round? In the nineteenth century it was claimed, in all seriousness, that Columbus's contemporaries thought the world was flat and expected him to sail over the edge.[4] This story is balderdash. But the fact that everyone (or at least every properly educated person) thought that you could in principle sail around the world (and in 1519–22 Magellan did just that) does not mean that they thought it was round. Columbus, strangely, thought that the old world, known to Ptolemy, was half of a perfect sphere, but the new world, he believed, was shaped like the top half of a pear, or like a breast; he had the impression he was sailing uphill as he left the Azores behind him.[5] The stalk, or nipple, of this other hemisphere was the location of the terrestrial paradise.[6] 'The earth' (or rather the agglomerate of earth and water) bulged.

This view, that the agglomerate of earth and water was not a perfect sphere, was universally accepted in the later Middle Ages, and the new cosmography required its refutation.[i][7] According to Aristotle, the universe is divided between a supralunary zone, where nothing changes and movement is always in circles, and a sublunary zone. In the sublunary zone the four elements – earth, water, air and fire – which form the basis of all our daily experience of matter are to be found. These

i Early modern cartographers describe themselves as cosmographers because they mapped both the heavens and the Earth and regularly produced paired globes; the word 'cosmography' originates in ancient Greek and is thus the traditional term, while the word 'cosmology' is relatively modern: it does not pre-date the second half of the sixteenth century.

The concentric spheres which make up the universe, from Jodocus Trutfetter, *A Textbook of Natural Philosophy* (*Summa in tota[m] physicen*), 1514. Within the sublunary zone there are four distinct spheres: earth, water, air and fire; outside them are the planets, including the sun and the moon. The zodiac of the fixed stars is the outermost visible sphere, with three invisible spheres beyond.

elements naturally arrange themselves in concentric circles around a common centre: earth surrounded by water, water surrounded by air, and air surrounded by fire. This arrangement, however, is not perfect, for dry land emerges from the water, and on the land all four elements interact. It is this interaction of the elements which makes living creatures possible, and without it the universe would be sterile.[8]

This account presented Muslim and Christian philosophers with a problem which had not worried their pagan predecessors: how did it come about that the four elements did not form perfect concentric circles?[9] They seized on this partly because it enabled them to introduce into philosophy a creator God unknown to Aristotle and Ptolemy. According to Genesis, God had gathered the waters together on the third day of creation in order to make dry land. So a simple answer was that the existence of dry land was a miracle. Since the waters of the oceans were higher than the land (higher, it was regularly said, than the highest mountains; otherwise, you would not find water springing from the ground near mountain peaks),[ii] it was easy to conclude that the oceans were held back from flooding the land, as they had in Noah's Flood, by divine Providence. The philosophers found such an answer unsatisfactory, even though something similar was to be found in Pliny's *Natural History*,[10] and sought a natural explanation. If the initial separation required divine intervention, how should one characterize the relationship between earth and water since the Flood?

The problem was a simple one, and the range of possible answers was limited. In the course of a 250-year period, all the possibilities were fully explored.[11]

1. The waters have been displaced from their original position, and their sphere now has a centre other than the centre of the universe. This view implies that ships sail uphill as they sail out on the ocean (we still acknowledge this traditional view when we use the term 'the high sea' or 'the high seas'). It was held by Sacrobosco (*c.*1195–*c.*1256), who wrote the standard textbook on astronomy used in medieval and Renaissance universities, and after him by Brunetto Latini (1220–94), Ristoro d'Arezzo (writing in 1282), Paul of Burgos (1351–1435) and Prosdocimo di Beldomandi (d.1428). In 1320 Dante took it to be the standard view (though his text, the *Quaestio de aqua et terra*, was unknown until it was first published in 1508).
2. The earth (as distinct from the sphere of water) is no longer a sphere; rather, as the result of the growth of a boss, or tumour,

ii It was understood that water evaporates from the oceans and becomes rain, and that the rain feeds rivers which flow to the ocean. But it was maintained that rainfall alone could not explain the size of the rivers or the existence of springs emerging from underground; springs were fed directly, it was argued, from the ocean. This view survived into the eighteenth century, and is refuted, for example, by Vallisneri, '*Lezione accademica intorno all'origine delle fontane*' (1715), who explained how the underground movement of water is affected by different rock formations.

it has acquired an elongated, irregular shape, so that its centre of gravity (the point around which it would hang without moving if suspended) corresponds to the centre of the universe but its geometrical centre doesn't. The boss is what makes dry land possible. This was the view of Giles of Rome (1243–1316), who calculated that the earth's diameter must have been stretched to almost twice its original length, and of Dante. The problem with this view was that it meant abandoning the idea that the universe was created out of nested spheres – a high price to pay, and one that few were prepared to contemplate.

3. If it could be argued that the earth might not be a true sphere then, equally, the waters might not be. Some suggested that the waters are not truly spherical but rather oval in shape, with the result that the oceans are deeper at the poles; this was held by Francesco di Manfredonia (d.*c.*1490) to be a partial explanation for the appearance of dry land. The weakness of this argument, as Francesco must have realized, was that if the waters were ovoid then there should be a belt of dry land at the equator and nowhere else; consequently, this argument was insufficient on its own, as he was obliged to recognize.
4. The earth is still a sphere, but it is no longer at the centre of the universe. This was the view of Robertus Anglicus (1271), but it was bound to have few supporters, as it ran against a core principle of Aristotelian philosophy: that the proper place for the earth was the centre of the universe. This difficulty, however, simply provoked the philosophers to think harder. Suppose, they argued, the earth is a sphere but its composition is not homogeneous: the action of the sun has made dry land less dense than it originally was, shifting the centre of gravity of the whole mass. Thus the centre of gravity of the earth still coincides with the centre of the universe, but its geometrical centre does not. Water, on the other hand, remains symmetrically arranged about the centre of the universe. This was the view of the fourteenth-century Parisian philosophers, of John of Jandun (1286–1328), Jean Buridan (*c.*1300–*c.*1358), Nicholas Bonet (d.1360), Nicholas Oresme (*c.*1320–82) and Albert of Saxony (*c.* 1320–90).[12] It preserved the system of nested spheres, and it had the great advantage of making water

> always flow downhill (which it does not in the first option above). As a modification of this view, one could argue that the centre of the universe corresponds to the centre of gravity of the aggregate of the two spheres of earth and water. This was the view of Pierre d'Ailly (1351–1420), despite the fact that he had read Ptolemy's *Geography*, copies of which began to circulate in the Latin West around 1400. By 1475, in one or another variation, this was the standard belief.

These four views standardly took it for granted that the sphere of water was considerably larger than the sphere of earth. The conventional idea from 1200 to 1500 (mistakenly attributed to Aristotle) was that it was ten times larger, as it was held that each element exists in the same quantity, but a quantity of water occupies ten times the volume of the same quantity of earth while air occupies ten times the volume of water, and fire occupies ten times the volume of air.[13] The relative size of the spheres and the extent of their displacement with regard to each other determines the size of the zone of dry land. It was commonplace to regard this as approximately one quarter of the earth/water globe, but it could extend to up to half of the earth/water globe. The first view assumed that the known world was all that there was to be known; the second implied that there was more land as yet undiscovered. This was usually held to lie in the southern hemisphere, and was sometimes thought to be inhabited.

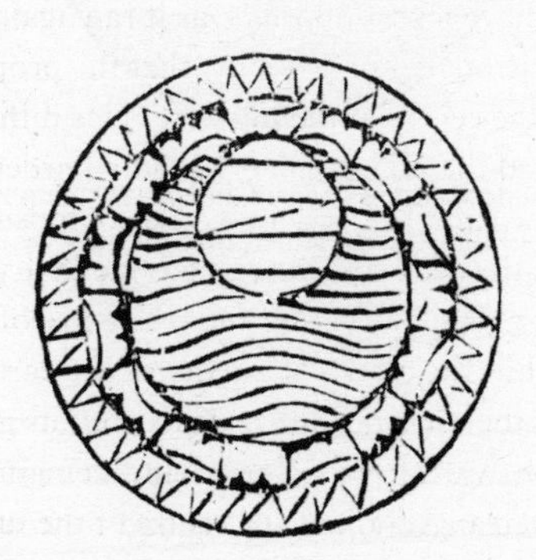

The spheres of earth, water, air and fire from Sacrobosco's *Sphere* (*Sphaera mundi: Joannis de Sacro Busto sphæricum opusculum*), Venice, 1501. The earth floats like an apple in a bucket. The orientation is not north/south; rather, Jerusalem, the centre of the known world, is at the top.

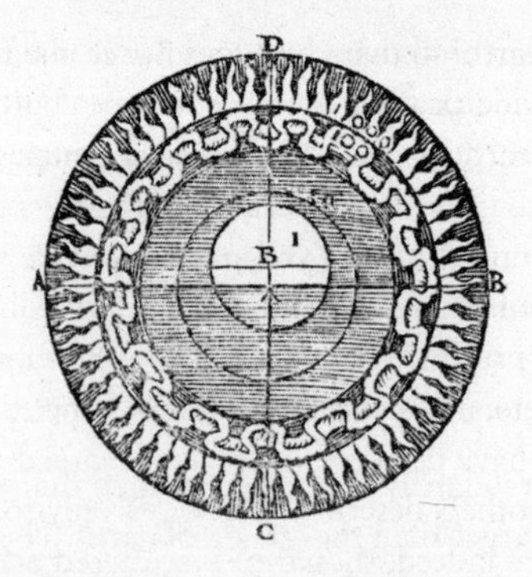

The distinct centres of the spheres of water (centred on A) and earth (centred on B), from Sacrobosco's *Sphere* (*Sphera volgare novamente tradotta*), Venice, 1537. The two are marked as having relative volumes of 10:1, although, as Copernicus shows, were this to be the case, the sphere of land would not overlap with the centre of the sphere of water, which is here taken to be the centre of the universe.

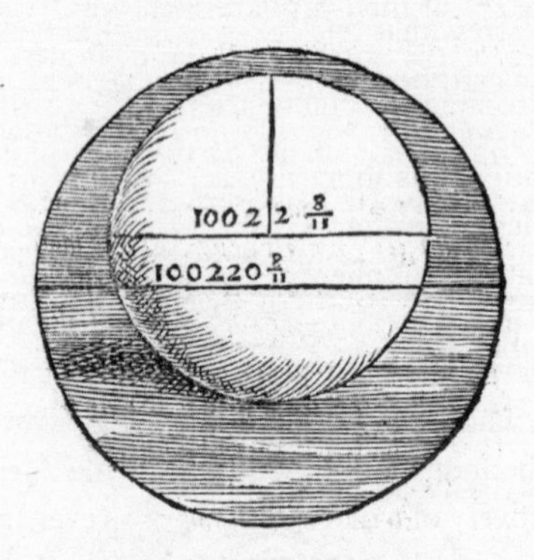

The relative and absolute volumes of earth and water, again from *Sphera volgare*, 1537. Copernicus would have complained that the two spheres were not in fact drawn to the same scale.

It was generally acknowledged that there was a limited range of possible causes of a change in the relationship between earth and water. Either God had acted directly, heaping up and concentrating the waters in order to clear space for dry land, or the sun had acted on the earth to dry it out, or the stars had acted to pull either the waters or the earth out of position.

But, lastly, we come to the fifth view: there are no separate spheres of earth and water, there is less water than earth, and the oceans lie in the concavities of the earth, so that earth and water make up a single aggregate sphere. This, which is the modern conception (although of course we no longer think of 'earth' as one of the four elements), was held by Robert Grosseteste (*c.*1175–1253), Andalò di Negro (1260–1334), Themo Judaei (mid-fourteenth century) and Marsilius of Inghen (1340–96). Of these four, the opinions of Robert Grosseteste and Marsilius of Inghen were made available in print to Renaissance readers (though Marsilius was read by philosophers, not astronomers); but knowledge of the existence of this notion will have been reasonably widespread throughout the fifteenth century, for others described it, though only to reject it. It implies that land could be – indeed, should be – scattered across the whole surface of the Earth, a view endorsed by Roger Bacon (1214–94), probably under the influence of Grosseteste, and by the author of *The Travels of Sir John Mandeville* (*c.*1360).[14] Of all the views, this is the only one straightforwardly compatible with the existence of antipodes (that is, bodies of land directly opposite each other on the globe).

It is essential to stress that this last belief found no support in the fifteenth century. For astronomers and geographers in 1475 (the year in which Ptolemy's *Geography* was first printed, though the first Latin manuscript translation was in 1406), the basic choice was between an account of the sphere of *water* as displaced from the centre of the universe and an account of the sphere of the element *earth* as displaced from the centre of the universe (but still overlapping with it). To support Columbus's voyage you did not have to think that these theories were wrong; you simply had to agree that going west might nevertheless be a quicker route to the Indies than circumnavigating Africa or going overland. After the discovery of a new continent, however, the outmoded view of Grosseteste once again became respectable among the philosophers.

Thus in 1475 there was general agreement that the centre of the two spheres of earth and of water were no longer identical, and indeed there was now a puzzle about three other centres: Where was the geometrical centre of the universe? Did it correspond to the centre of one of the spheres and, if so, which one? And if the earth was not homogeneous, where was its centre of gravity? Finally, where was the centre of gravity of the conjoined spheres of earth and water? Where the Aristotelian universe had one centre, there were now potentially five different ways of defining the centre of the universe.

§ 2

Late-medieval and Renaissance students learnt their astronomy by studying the *Sphaera*, or *Sphere* (*c*.1220) of Johannes de Sacrobosco, who taught in Paris but may have been English (in which case his name was presumably originally John of Holywood).[15] His textbook was first printed in 1472 and went through more than two hundred editions.[16] In addition there were numerous commentators who sought to explicate the text, beginning with Michael Scot (*c*.1230), including Giambattista Capuano da Manfredonia (*c*.1475),[iii] and culminating with Christoph Clavius (1570), the leading Jesuit astronomer of the late sixteenth century. The *Sphere* was still the standard text from which Galileo lectured when he was a professor at the University of Padua (1592–1610); the last edition for students, in 1633, conveniently marks the demise of Ptolemaic astronomy as a living tradition. In line with the notion that the globe was made up of two non-concentric spheres, one of earth and one of water, and following the example of Ptolemy's *Almagest* (which had been available in the Latin West from the twelfth century), Sacrobosco proved separately that the surface of the earth was curved (he showed how this could be made apparent to someone travelling either north–south or east–west), and that the surface of the water was curved. (This was evident because a lookout on top of a ship's mast could see further than someone standing on the deck.) Modern commentators assume that Sacrobosco had proved that the Earth is round;[17] he had done nothing of the sort, and the medieval commentators did not claim that he had, for neither he nor they believed that the two spheres shared a common centre.

It should now be apparent that when medieval philosophers talked of 'the earth' they normally meant the sphere of the element earth which, where it showed above the ocean, constituted dry land; this sphere floated in an ocean of ocean, itself a larger sphere. The term 'the earth' was, however, inherently ambiguous. We find John of Wallingford (d.1258), for example, distinguishing in the space of two sentences between a) the earth, meaning dry land; b) the earth, meaning the element earth, whose centre is the centre of the universe; and c) the whole globe, i.e. the agglomeration of earth and water.[18] The third usage (which looked back to Cicero's *Dream of Scipio*) was distinctly

iii Confusingly, Giambattista at the beginning of his career was called Francesco.

unphilosophical for anyone who accepted the dominant two-spheres theory, so unphilosophical that it is difficult to find examples of *terra* being used in this sense in the later Middle Ages or early Renaissance, except by Latinizing humanists such as Petrarch.[19] To all intents and purposes, the notion that the earth/water assemblage was to be thought of as a single globe or sphere disappeared around 1400. Even before 1400 it had never been the dominant view. The earth/water agglomeration was no longer round.

All these late-medieval discussions took place within the context of a geographical knowledge which corresponded to that of the ancients. No one believed that the earth was flat (it consisted of a portion of a sphere), but the habitable earth could be represented fairly accurately on a flat surface. This habitable earth had a centre, which was generally taken to be Jerusalem. However, there was another centre: measuring from west to east, from the Fortunate Isles (the Canaries) to the Pillars of Hercules (which marked the limits beyond which it was impossible to travel), there existed a notional location on the equator called Arim, or Arin, believed to be 10 degrees east of Baghdad. For the Arabs, and for astronomers relying on Arabic sources, Arim represented the degree zero of longitude and latitude.[20] It was universally accepted that dry land was confined to one hemisphere, the rest being covered by ocean. Of the dry land, the furthest northern and southern parts were uninhabitable because they were too cold or too hot, and so the habitable portion of the earth represented approximately one half of the whole of the dry land, one sixth of the surface of the whole agglomeration of earth and water.

As Dante pointed out in 1320, there was an obvious problem here, for the arguments of the philosophers and the maps of the geographers did not match up. If the philosophers were right and the habitable earth was a sphere floating on the surface of a larger globe of water, then a map should show the habitable earth as a circle. In fact, maps showed it as shaped like a cloak spread out on the ground; but the known world was referred to as the *orbis terrarum*, the circle of lands, as if it had the required form. Dante, unlike the philosophers, took his geography seriously, but no philosopher could have found his abandonment of the fundamental principle that the universe was made out of spheres entirely satisfactory.

Where the Aristotelian, idealized scheme of concentric spheres was symmetrical on every axis, the medieval elaborations (with the exception of the fifth) were each symmetrical around one axis only. Moreover, this axis was not the north–south axis of the poles but an axis through Jerusalem and the geometrical centre of the universe. Had late-medieval philosophers tried to imagine (which, of course, very few of them did) an Earth spinning in space on a north–south axis, then many of them would have been sure that the centre of gravity of the Earth (of either the sphere of earth or the sphere of water) was not on the north–south axis; such a spinning globe would have a natural tendency to wobble. The exception was the Parisian philosophers, for whom the centre of gravity of both the earth and the water remained coincident with the centre of the universe. Entirely logically, the only medieval philosopher of significance to take seriously the theory of the diurnal rotation of the Earth was a Parisian, Nicholas of Oresme (1320–82). Crucially, Oresme, unlike other philosophers who accepted (as he did) that there were two spheres of earth and water with separate geometrical centres, did not accept that the sphere of water was in itself bigger than the sphere of earth. He claims that if the two spheres had the same centre, water would inevitably cover the whole surface of the earth – except perhaps for a few mountain tops. And he describes the sphere of water as like a cloak or hood covering the earth. The result is that he has, as the illustrations accompanying his *Livre du ciel et du monde* (1377) show, a conception of the Earth as in effect a single globe, capable of rotating on its axis (but, since it is wrapped in the sphere of water, incapable of having antipodes).[iv] As it happens, Oresme's text was never published, and cannot have circulated widely because it was written in French.[21]

Thus the two-spheres theory of the world was shared by nearly all philosophers, astronomers and cartographers (despite the difficulties it was known to present) until the late fifteenth century, and the rediscovery of Ptolemy's *Geography* was integrated into it without too much difficulty.[22] The Portuguese explorers reached the equator in 1474/5 (it is not difficult to tell when you have reached the equator: the Pole Star disappears from view), discovering a new heaven and new stars, but

iv See plate 3. It is important to forget about Australia and New Zealand (largely unexplored until late in the eighteenth century) when thinking about antipodes in this context. Two places are antipodes to each other if they are directly opposite each other on the globe. The two-spheres theory as generally propounded makes antipodes impossible, as it confines all dry land within one hemisphere. Oresme, exceptionally, held that the distance from Africa to India going west was probably less than the distance going east, so he evidently held that there were areas of dry land across more than 180 degrees of the joint sphere of earth and water near the equator, and thus that there were, as a limit case, true antipodes there; but he maintained there could be be no antipodes for higher latitudes as at least half the sphere of the earth must be covered in water.

they found no uninhabitable zone: this required some minor rethinking, but little more.[23] It was true that Ptolemy in the *Geography* (unlike the *Almagest*) treated earth and water as a single sphere, and this was obviously bound to be of interest. After the translation of Ptolemy's *Geography* there is a record of a terrestrial globe being made in 1443 'according to Ptolemy's description'.[24] Columbus read Ptolemy and was convinced that earth and water formed one sphere; he produced a small globe to illustrate his planned voyage. At the same time he chose to reject Ptolemy's account of the extent of the habitable world, preferring that of Marinus of Tyre (*c.*100–150), who had claimed that it extended more than halfway around the globe – a view difficult to reconcile with the two-spheres theory. But there was as yet no general crisis for the two-spheres theory: the geographers summoned by Ferdinand and Isabella to advise on Columbus's plans had no hesitation in dismissing them out of hand.[25]

That crisis began with Columbus's landfall in 1492. In 1493 Peter Martyr described Columbus as returning from 'the Western Antipodes'. In a notarial certificate drawn up by Valentim Fernandes in 1503 Pedro Álvares Cabral's discovery of Brazil in 1500 is described as the discovery of 'the land of the Antipodes'.[26] (He was right: Brazil is antipodal to the eastern extremity of the world known to the ancients.) But the decisive event was the publication in 1503 of the first letter written (or supposedly written) by Vespucci, entitled *Mundus novus*, which went through twenty-nine editions in the space of four years.[27] (It was Vespucci's second letter which introduced the word 'discovery' to a European audience; his first had already destroyed the medieval cosmography.) Vespucci's claim was that he had encountered a vast new landmass which formed no part of the previously known world – he had found a New World. Moreover, it was clear that this land mass, although it was only one quarter of the way around the globe from his starting point, was halfway around the globe from other parts of the known world. And Vespucci had sailed 50 degrees south of the equator: this was not just the equatorial antipodes that some exponents of the two-spheres theory had envisaged. Antipodes had become a reality, and there was no longer any way of fitting the Earth's land masses into one hemisphere.

Thus what was disturbing about these antipodes was not that they implied that some people were 'upside down' compared to other people – you had to be fairly unsophisticated to have difficulties with this idea – but that the two-spheres theory could accommodate

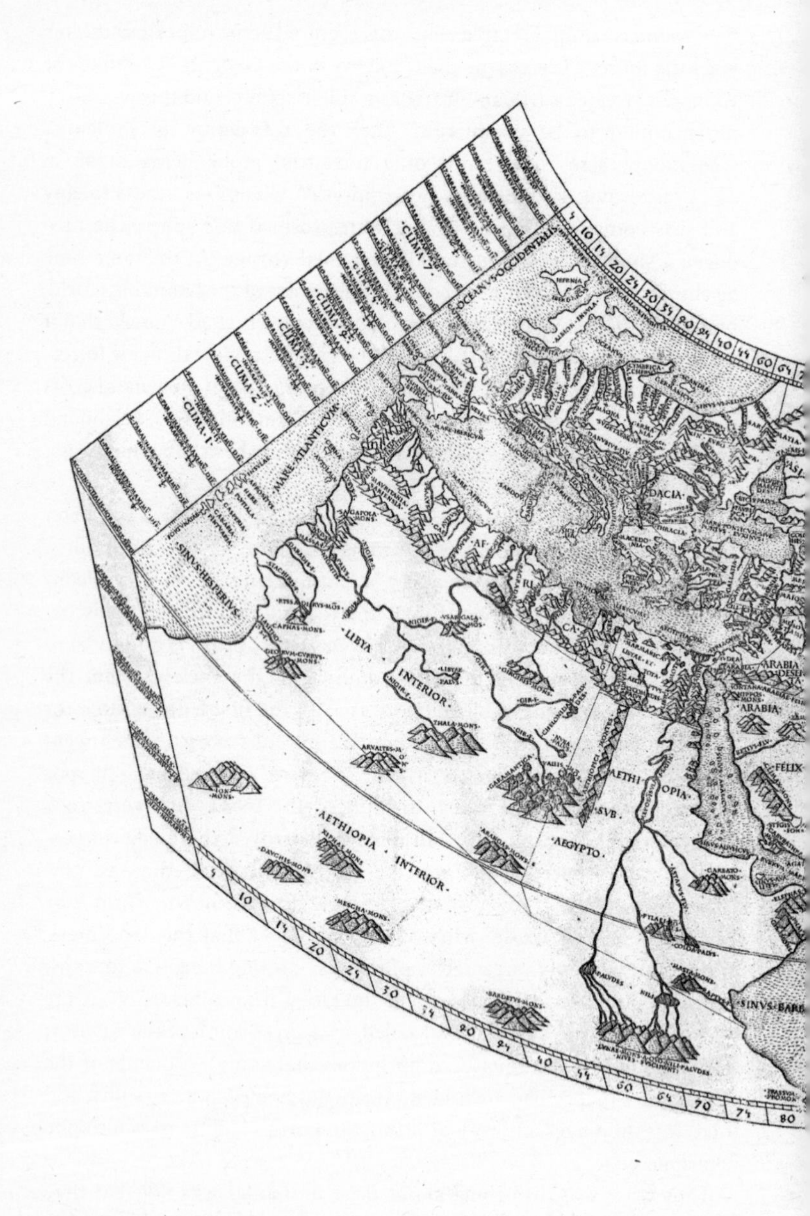

OCEANVS OCCIDENTALIS
MARE ATLANTICVM
SINVS HESPERIVS
CLIMA 7
CLIMA 2
CLIMA 1
DACIA
LIBYA
INTERIOR
AETHIOPIA
INTERIOR
AETHI OPIA
SVB
AEGYPTO
ARABIA
FELIX
SINVS BARB
5
10
15
20
25
30
40
60
65
70
75
80

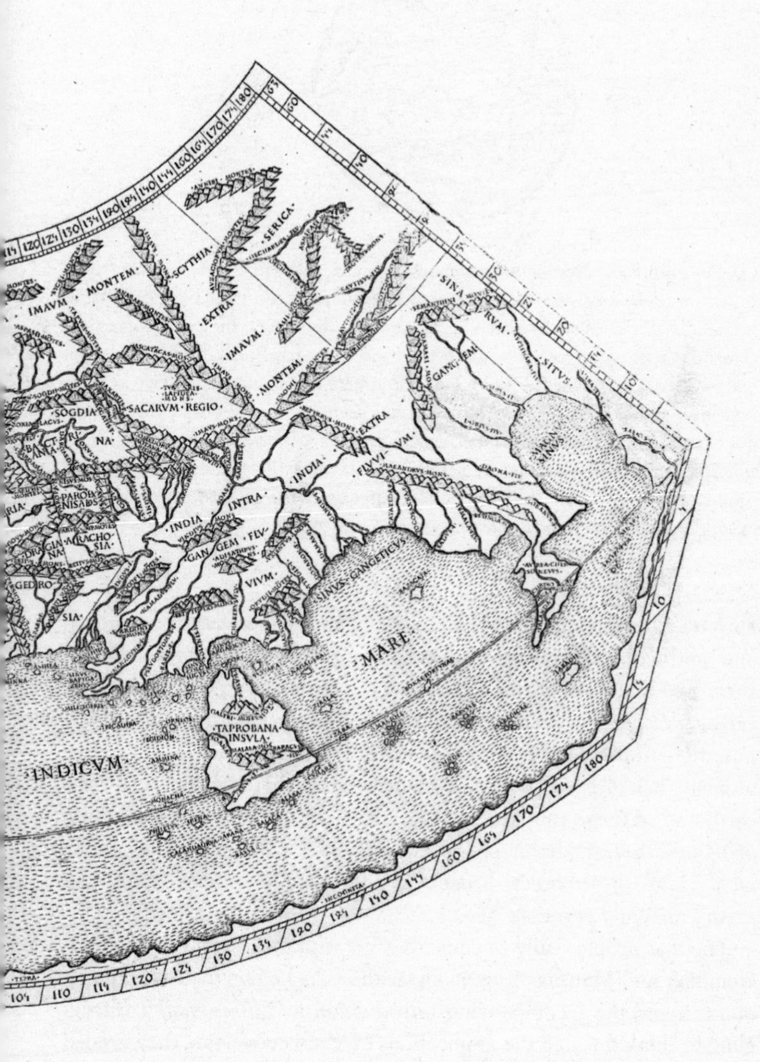

The map of the world from Ptolemy's *Geography*, printed in Rome in 1490. The same plates had previously been used in two earlier editions (Bologna, 1477; Rome, 1478), and they are consequently the earliest printed illustrations for the *Geography*.

Clavius's representation, in his commentary on Sacrobosco (1570; but taken here from the revised edition of 1581) of the standard account of the relationship between water and earth, which he rejects. The dots mark the two geometrical centres, that of the sphere of water (below) and that of the sphere of earth (above). Since discussion of whether there was one sphere of earth/water or two spheres was inseparable from discussion of whether there were antipodes (which could not exist on the two-spheres model, except perhaps in a brief band where the two spheres met if they were of similar size), Clavius's illustration also includes (non-existent) antipodes, which are underwater. Since the antipodes are known to exist, this traditional model must be wrong.

antipodes only as a limit case, along the boundary between the northern and southern hemispheres, and only then if the sphere of water was shrunk so that its diameter was almost the same as that of the sphere of earth.[28] Vespucci's claim required a major reconsideration of the supposed relationship between the elements of water and earth. Up to this moment it had been possible to believe both that the spheres of the earth and of the ocean were round, and that the zone of dry land (the *orbis terrarum*, the habitable world) had, as the Bible put it, four corners.[29] Now these corners became, in John Donne's phrase, 'the round earth's imagin'd corners'.[30]

The first people really to come to grips with this were Martin Waldseemüller and Matthias Ringmann, as they worked on their world map of 1507 and the accompanying *Introduction to Cosmography*.[v] Struggling to think through the implications of Vespucci's claim, they needed a way of referring to what we call the Earth, or the world – the single globe formed of land and sea. They called it *omnem terrae ambitum*, the

v See plate 6.

whole circumference of the Earth, of which, they explained, Ptolemy knew only a quarter.

Other early world maps present themselves as illustrations of the *orbis terrarum*. In classical Latin, from which the phrase derives, an *orbis* is usually a flat disc, but sometimes it is an orb or globe. When Cicero writes of the *orbis* he sometimes means the habitable dry land, a disk rising above the waves, and sometimes the whole globe of land and ocean. This ambiguity was carried through into the Renaissance. Thus Ortelius's 1570 atlas was entitled *Theatrum orbis terrarum*, the theatre of the sphere of lands. The frontispiece makes clear that the *orbis* is a globe, but the plural *terrae* implies a collection of maps of different countries. Mercator, exceptionally, used the phrase *orbis terrae* – in 1569 the word *terra* is beginning to mean Earth, or world (as in Planet Earth); one word has been substituted for Waldseemüller and Ringmann's clumsy phrase. By 1606 Ortelius's *Theatrum* could be translated into English as *The Theatre of the Whole World*. Only later, in 1629, was a satisfactory technical term invented to identify unambiguously this new entity: it was called 'the terraqueous globe'.[31]

We can trace in detail the progress of this new concept in the years after the publication of Waldseemüller's and Ringmann's *Introduction to Cosmography* in 1507. The first sign of change is to be found in a physics textbook published in Erfurt in 1514. The author, Jodocus Trutfetter, presents the one-sphere theory first, although he then goes on to explain the view that the sea is higher than the land; he notes that the most recent cosmographers have claimed that there are inhabited antipodes at the eastern and western extremes of the world, although he balances this by explaining that Augustine had denied the possibility of antipodes. If the text is cautious, the accompanying illustration is not: it shows only three sublunar spheres, of earth, air and fire. Evidently, earth and water now make one sphere.[vi][32]

In 1515 Joachim Vadianus, a man of many talents (he was the Poet Laureate of the Habsburg empire), published in Vienna a little pamphlet,

vi If we turn from print to manuscript sources, there is a clear statement of a new theory in a text written between 1505 and 1508 by Duarte Pacheco Pereira (Morison, *Portuguese Voyages to America* (1940), 132–5): 'It follows, therefore, that the earth contains water and that the sea does not surround the earth, as Homer and other authors affirmed, but rather that the earth in its greatness surrounds and contains all the waters in its concavity and centre; moreover, experience, which is the mother of knowledge, removes all doubt and misapprehension.' This theory seems to lie somewhere between what became the new standard theory and Bodin's theory, which we will come to shortly.

The first sophisticated illustration of the earth and water as making a single sphere where the two elements interlock: from Joannes de Sacro Bosco, *Opusculum de sphaera* (1518), edited by Tanstetter. There are now three sublunary spheres, not four.

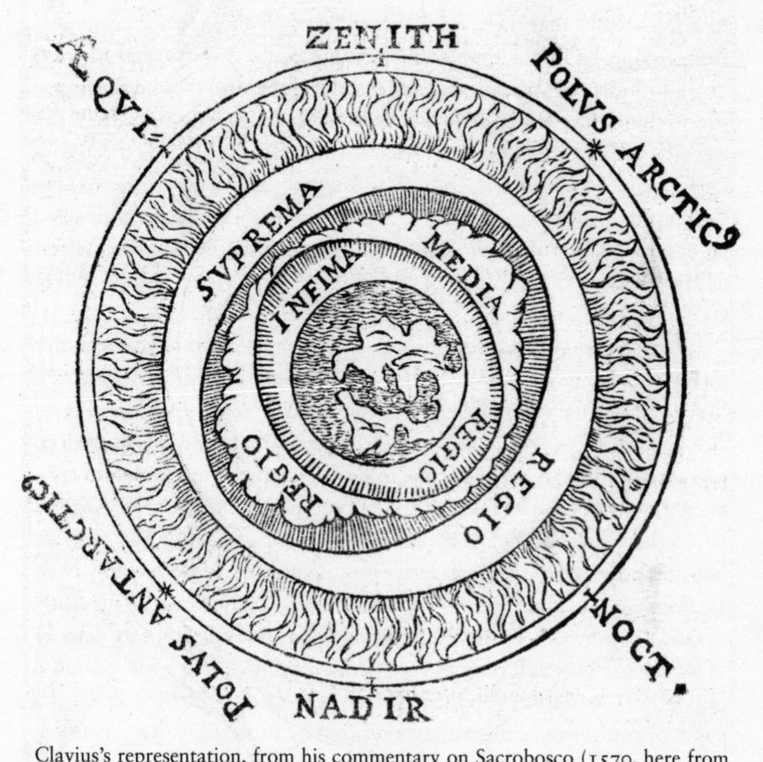

Clavius's representation, from his commentary on Sacrobosco (1570, here from the 1581 edition) of the relationship between earth, water, air and fire. Earth and water make one sphere, surrounded by three levels of the atmosphere (the weather being generated in the middle level) – only the outermost of these levels is a perfect sphere, beyond which is the sphere of fire.

Habes lector, or *Dear Reader* (reprinted half a dozen times), in which he suggested, in the light of the discovery of America, that, contrary to the standard interpretation of Aristotle, habitable land was scattered almost randomly across the surface of the globe, and that earth and water were so intermingled as to form one sphere.[33] The geometrical centre of the globe and its centre of gravity were, he asserted, one and the same. As for Augustine's fear that to admit the existence of antipodes would be to acknowledge that there were human beings who were not descended from Adam, he had a simple answer: one could travel overland from Spain to India, almost halfway around the globe, and there was no reason to think that any inhabited land was set at a vast distance from the rest (the implication being that America was close to Asia). Three years later, again in Vienna, George Tannstetter (also known as George Collimitius), who was in close collaboration with Vadianus, published an edition of Sacrobosco's *Sphere* which contains the first illustration of the 'modern' conception of the globe as made up of interlocking land and sea.[34]

In 1531 Jacob Ziegler published in Basle an elaborate commentary on Book II of Pliny's *Natural History*. In it he interpreted Pliny's account of how the waters are higher than the earth in terms of the medieval two-spheres theory, only to conclude, bluntly, that modern discoveries had shown this view of the globe to be fallacious, as land was not confined to only one hemisphere of the globe.[35] In the same year as Ziegler's book, there appeared at Wittenberg an edition of Sacrobosco with an introduction by the leading Lutheran theologian and educator, Melanchthon.[36] Melanchthon's introduction praised astronomy as the study of God's handiwork, but it also went on to provide an elaborate defence of astrology. This edition was repeatedly reprinted, and widely pirated (in Catholic countries the introduction was often printed without the name of the author, since all texts written by Protestants were banned; in earlier copies Melanchthon's name is often blotted out on the title page). A crucial new illustration showing the earth/water globe was copied from an edition of the *Sphere* produced by Peter Apian in 1526, and, through the influence of the Wittenberg edition, it became the new standard; it was even copied in the much-reprinted commentary on Sacrobosco produced by Christoph Clavius, the first edition of which appeared in 1570.[37]

In 1538 the Wittenberg presses produced a new, elaborate version of the Melanchthon edition which included 'volvelles', paper instruments or illustrations with circular moving parts.[38] In this edition (which also

Peter Apian's new illustration to show that the earth is round, later copied by Melanchthon and Clavius, from Sacrobosco, *Sphaera . . . per Petrum Apianum . . . recognita ac emendata* (1526).

went on to be frequently reprinted and copied) the conventional chapter headings into which Sacrobosco's text had been divided were revised. Where earlier editions had had a chapter proving the earth was a sphere, and another proving that the waters were a sphere, this new edition presented a whole section as being about water and earth making up one globe. The text itself had not been changed (as it was, for example, in an edition for use in schools which appeared in Leiden in 1639), but the new heading, *Terram cum aqua globum constituere*, transformed its meaning.[39] From 1538 the new understanding of earth and water as making up a single sphere became an orthodoxy among both Protestant and Catholic astronomers.

In 1475 the two-spheres theory of the world was universally held by philosophers and astronomers; by 1550 every expert had abandoned it.[40] That did not mean, however, that certain aspects of the old theory could not be preserved within the new. One might think that the adoption of the theory of the terraqueous globe automatically meant acknowledging that the seas are lower than the dry land, but the contrary view seemed to be clearly established by scripture and by innumerable respectable authorities. So the Jesuit Mario Bettini (1582–1657) argued that when God had turned the separate spheres of earth and water into one sphere by opening up cavities in the earth to absorb the bulk of the water, it had been necessary to compensate for the fact that (since water is, by definition, lighter than earth) the centre of gravity of the new terraqueous globe was in danger of not coinciding with the centre of the universe; consequently, the waters bulged outwards so that their weight was equal to that of the earth they had displaced. Gaspar Schott (1608–66, also a Jesuit) accepted this argument as the explanation for the origin of most rivers. Their headwaters, he thought (as this illustration seeks to show), lie below the highest point of the sea (high sea level: F), but above the shoreline (low sea level: BC). It was, he held, an open question whether there were rivers that originated at a point above high sea level (E). Thus the doctrine that the seas are higher than the land survived well into the second half of the seventeenth century.[vii][41] Obviously, the notion that the height of a mountain could be measured from sea level could establish itself only after this view had been abandoned. Still, this was not the old two-spheres theory, and it was now axiomatic that earth and water had a single centre, which was both the geometrical and the gravitational centre of the globe. I can find only two people who, after the publication of Waldseemüller's map, sought to defend the old theory against its attackers: the new reality was incompatible with the old theories.

One morning in August 1578 a debate broke out at the breakfast table of the Duke of Savoy, Emanuele Filiberto, as to why rivers run to the sea. An Averroist philosopher who was present, Antonio Berga, insisted that, as the sea was higher than the land, it could not simply be because water naturally flows downhill. Berga went on to appeal to the old orthodoxies: the sphere of water is ten times greater than the sphere of earth, the two spheres do not have the same geometrical centre, and the oceans are higher than the land. Berga's views were disputed by Giovanni Battista Benedetti, who was officially the duke's mathematician and philosopher, and, since the honour of both men was now at stake, the dispute rumbled on after the meal was over. Benedetti told Berga to read Piccolomini, and he put some of his own arguments on paper for the duke to read; Berga published a refutation of Piccolomini, and implicitly of Benedetti; and Benedetti responded, cruelly mocking Berga (who showed his lack of expertise by confusing the Antarctic and the Arctic) and calling him 'half Huguenot' in his philosophy (this was tit-for-tat, as Berga had dismissed the new theories as philosophical heresies).[42] Berga, it must be stressed, made no attempt to claim that he had support among contemporary philosophers for his antiquated views: if

vii Kepler had argued in 1618 that the belief that the seas are higher than the land is the consequence of a visual illusion: Kepler, *Epitome astronomiae Copernicanae* (1635), 26–7 (a view echoed in Froidmont, *Meteorologicorum libri sex* (1627)).

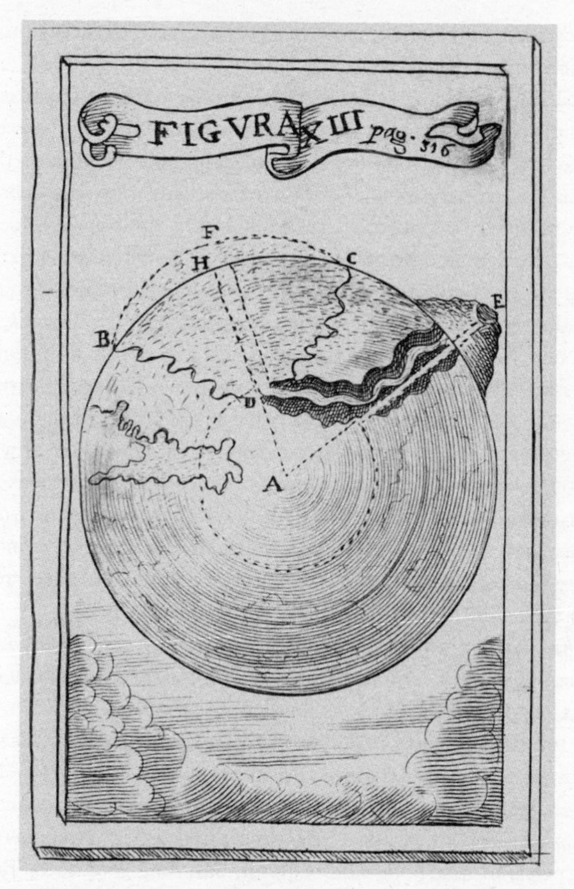

Schott's illustration from *Anatomia physico-hydrostatica fontium ac fluminum* (1663) to show how the surface of the ocean curves upwards and how water from the ocean travels underground through fissures in the earth to emerge as springs and rivers. The fact that the ocean is higher than the land explains why water can spring out of the ground above the level of the shore, although Schott acknowledges that the relative heights of the tops of mountains and the ocean have not been established.

there were others who thought as he did they were too sensible to entrust their arguments to print. For to preserve the old orthodoxy it would have been necessary to insist that the world's land masses were confined to one hemisphere.[43] Berga side-stepped this issue and, as far as I can tell, only one person was so foolish as to explicitly present this argument.[viii]

Still, one would have expected there to be a range of alternative theories proposed to account for the new evidence. One could, for example, argue that far from there being one earthly sphere floating in the ocean it was now apparent that there were two. This view was expressed by those (echoed by Copernicus) who described the New World as *altera orbis terrarum*, another sphere (or circle) of land masses. It was put forward in all seriousness in 1535 by Oviedo (Gonzalo Fernández de Oviedo y Valdés), writing the official Spanish history of the discovery of the New World.[44] But for Copernicus this was merely a turn of phrase, for it was evident that you could not have two spheres of earth and at the same time place the element earth at the centre of the universe. A universe in which there were two earths within one sphere of water was no longer an Aristotelian universe. *Altera orbis terrarum* was a catchy phrase which could not be turned into a viable theory. So the two-spheres theory was abandoned, even while efforts were made by some conservative thinkers to preserve the traditional claim that the seas are higher than the land.

One author, however, was not so easily defeated. In his *Universae naturae theatrum* of 1596, Jean Bodin argued that the new continents were simply vast plates floating on a bottomless ocean. He held that the element earth is heavier than the element water, but that (following Aristotelian orthodoxy) heavier objects can, if they have the correct shape, float on lighter objects. The floating continents would displace their own weight in the water (according to Archimedes' principle) but, in a striking non sequitur, only one seventh of their bulk would be below the waves. To make matters worse, Bodin clung to the traditional belief that the ocean bulges up above the land, higher than the highest mountain tops, although it was hardly compatible with his account of the continents as floating high above the waves. Bodin was sure that one could have floating land masses; he believed there were

viii Agostino Michele, *Trattato della grandezza dell'acqva et della terra* (1583), 13, takes this view. Michele is essentially an autodidact, and so not to be taken too seriously. It is possible that he was misled by the fact that our antipodes see some of the same stars that we do (because in the course of a night we see more than half the sphere of stars); the only places that see none of the same stars as their antipodes are the North and South poles. He was certainly misled by the fact that Vespucci had made clear that he had not been to the antipodes of western Europe: it does not follow that he had not been to the antipodes of somewhere in the Old World. For the conclusive nature of the argument from geography, see Benedetti, *Consideratione* (1579), 14.

reliable reports of islands that sneakily changed their position during the night – but the big continents, he thought, remained in one place. Thus Bodin proposed, not a terraqueous globe but an aquaterreous one in which (as one annotator summed up his thesis in the margin of the text) *terram aquis supernatare*, the earth floats on the surface of the waters.[45]

Bodin's motives for this strange argument are complex. In the first place, he was clear that land was not confined to one hemisphere, so the old two-spheres theory would not do. Secondly, he had read in Copernicus a demonstration that if the earth were one tenth the size of the waters it would be entirely immersed if any part of it overlapped with the centre of the sphere of water. So he decided that the only solution, if one wanted to retain the right ratio between water and land, was to break up the land and scatter it across the surface of the waters. In doing so, he completely abandoned two principles which had been fundamental to Aristotle: that the element earth is a sphere, and that the element earth is at the centre of the universe. Yet he came closer, he believed, to the Old Testament account of creation.

So peculiar was Bodin's theory that Gaspar Schott, writing two generations later, simply could not understand it.[46] He interpreted Bodin, quite wrongly, as advocating a very large sphere of earth floating in a sphere of water, thus retaining the core principles of a traditional Aristotelian argument. He drew an elaborate diagram to explicate what he took to be Bodin's theory, although his drawing is quite unlike Bodin's own. Schott's complete incomprehension suggests that it would have been difficult for Bodin to persuade other scholars that his views made sense. Anyone who closely examined them would have been forced to conclude that his account of how heavier-than-water bodies might float was riddled with inconsistencies because Archimedes and Aristotle were simply incompatible, and it is very difficult to see how a stable theory could have been generated which was based on Bodin's concept of floating continents.

What are we to make, then, of the peculiar story of the almost silent demise of the two-spheres theory? There had been good evidence against it long before Vespucci reached the New World. Giles of Rome and Dante had pointed out that if the theory was correct the land emerging from the waters should have a circular shape, and it did not. Dante, entirely sensibly, said that one should establish whether something was the case (*an sit*) before determining why it was the case (*propter quid*);

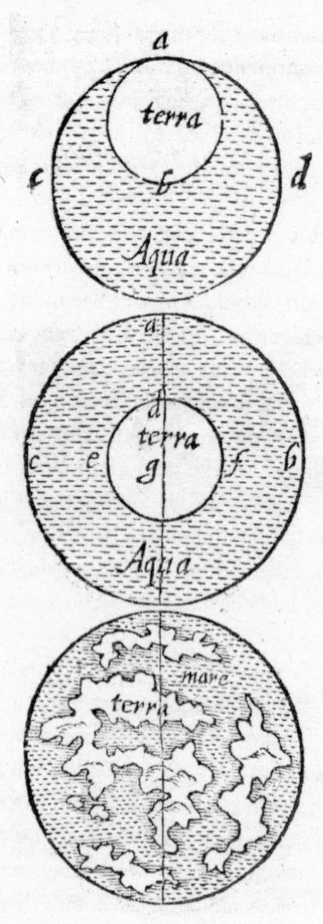

Jean Bodin's illustration to show his new theory of the relationship between earth and water, from the *Universae naturae theatrum* (1596). The middle image shows the standard late-medieval conception of a sphere of earth one-tenth the size of the sphere of water; the top image shows that such a sphere of earth will not overlap with the centre of the universe; and the bottom image shows Bodin's own conception of a series of flat plates of earth floating on the oceans.

in his view, the evidence falsified the two-spheres theory, even if that theory was an elegant reinterpretation of Aristotle.[ix] Moreover, the

ix Compare Dante's contemporary, Levi ben Gerson, who held that there was observational evidence incompatible with Ptolemy's theory of epicycles: 'No argument can nullify the reality that is perceived by the senses; for true opinion must follow reality, but reality need

Schott's version of Bodin's new theory of the relationship between land and water, from his *Anatomia physico-hydrostatica* (1663).

early, embattled proponents of what would later be known as the terraqueous-globe theory, Andalò di Negro and Themo Judaei, had pointed to the circular shape of the shadow of the earth during eclipses of the moon (a phenomenon already known to Aristotle) as proof that there was only one terraqueous sphere, not two overlapping spheres. Water, they insisted, was not simply transparent: a sphere of water

---

not conform to opinion.' (Goldstein, 'Theory and Observation' (1972), 47). Such claims could justify minority views; prior to 1492 they were never conclusive in settling an intellectual debate.

would cast a shadow, and no such shadow was to be seen.[47] Copernicus recycled this argument in *On the Revolutions* (1543).

In the fourteenth century, evidence, good evidence, against the two-spheres theory had been presented, and it had been brushed aside. In the early sixteenth century the voyages of Vespucci provided further evidence against that theory, and it was decisive. Was the quality of the evidence different? It was. There are two important features to Vespucci's voyages (for all that modern scholars debate how many voyages he made, and whether he wrote the accounts of his voyages that were published in his name). First, there was no disputing the importance of the discoveries in the New World, for the simple reason that they had become matters of state, the concern of kings. How could scholars ignore what governments took seriously? Second, and even more importantly, these discoveries were *new*. When Andalò di Negro invoked the shadow of the Earth as seen in eclipses of the moon, or Dante invoked the shape of dry land in the known world, they were appealing to information that had long been available. It was easy to assume that these arguments had already been taken into account, somehow, somewhere, by the advocates of the two-spheres theory, for in a manuscript culture no one can hope to have every relevant text to hand. But it was evident that Vespucci's information was quite simply unprecedented: it needed to be addressed here and now.

The invention of discovery, acting in combination with the printing press, transformed the balance between evidence and theory, tilting it away from the reinterpretation of old arguments and towards the acquisition and interpretation of new evidence. As far as the two-spheres theory was concerned, the voyages of Vespucci were deadly. The new facts were killer facts. As it happens, this is the first occasion since the establishment of universities in the thirteenth century on which a philosophical theory was destroyed by a fact.[x] Astonishing as it may seem, there is no previous occasion on which new empirical evidence determined the outcome of a long-standing debate between philosophers. Aristotle, for example, had argued that the nerves are all connected to the heart; Galen had shown they were connected to the brain; but Aristotelian philosophers, both ancient and medieval, had continued to follow Aristotle's teaching, as if Galen did not

x At more or less the same time, as we have seen (above, p. 72), it was generally acknowedged that the claim that the tropics were uninhabitable had been refuted by experience.

exist.[xi] In 1507 the relationship between theory and evidence changed, and changed for ever.

§ 3

In 1543 Copernicus published *On the Revolutions*, in which he argued that, far from standing still at the centre of the universe, the Earth orbits the sun once a year and turns every twenty-four hours on its axis.[48] Copernicus was a canon of the Cathedral of Warmia in Polish Prussia and had studied extensively in Italy (astronomy at Bologna and medicine at Padua). He begins his great work by running through a set of conventional arguments drawn from Sacrobosco: the heavens are spherical; the earth is spherical; the waters are spherical. In the last sentence of Book 1, Chapter 2, Copernicus rejects the argument (taken from Pliny and the Bible) that the waters are higher than the land. Then in Chapter 3 he stresses the importance of the discovery of America: earth and water make one globe in which the centre of gravity and the geometrical centre coincide. The waters cannot be, as many medieval philosophers had claimed, ten times as extensive as the earth, for if they were, and the earth is round and rises above the surface of the water, then simple geometry shows that no part of the earth will coincide with the centre of the universe. There really are antipodes and antichthones; 'Indeed geometrical reasoning about the location of America compels us to believe that it is diametrically opposite the Ganges district of India' (a calculation rather different from that made by Vadianus, who had placed India and Africa as each other's antipodes). Thus Copernicus argued for a spherical Earth – appealing to the evidence of the shape of the Earth's shadow cast on the moon during eclipses to confirm that the Earth was to all intents and purposes a perfect sphere, the occasional mountain and valley notwithstanding – a crucial first step towards arguing that the Earth rotates on a north–south axis.

By 1543 the broad outline of Copernicus's argument for the Earth as a single globe was conventional. But we know that Copernicus had first formulated his views by 1514, for at that date at least one copy of his preliminary sketch, the *Commentariolus*, or *Little Commentary*, was in

xi Similarly, Ptolemy had demonstrated that no homocentric planetary system could account for the observed phenomena, but philosophers were still trying to produce such a system well into the sixteenth century.

existence.[49] He gives us two accounts of the development of his thinking, one at the beginning of the *Little Commentary* and the other at the beginning of *On the Revolutions*. From them we learn that he had long been dissatisfied with conventional astronomical theories, that he had engaged in a systematic programme of reading in an attempt to identify alternatives, that the idea that the Earth moved had at first seemed to him absurd but that he had persisted with it, determined to see if it could provide the basis for a new account of the movements of the heavens.

Those few commentators who have grasped that Copernicus's doctrine that the earth and the seas form one sphere was relatively new have quite correctly concluded that there was a fundamental obstacle which Copernicus had to overcome before he could envisage a rotating Earth: he had to envisage the Earth as spherical (to push at the limits of possibility, as symmetrical on a north–south axis, or, as an absolute minimum, having its centre of gravity on its north–south axis).[50] Edward Rosen has argued that the geographical information in Book I, Chapter 3, of *On the Revolutions* (such as the claim that America is the antipodes of the Ganges) is based on Waldseemüller's map of 1507, the book that accompanied it, and another map by John Ruysch published in the same year.[51] If so, it seems Copernicus came to view the Earth as a spherical globe some time between 1507 and 1543. But when exactly?

Here we have no useful source of information other than the *Little Commentary*. It begins with a number of axioms. The second is '*centrum terrae non esse centrum mundi, sed tantum gravitatis et orbis Lunaris*' ('the centre of the earth is not the centre of the universe [because the sun, not the Earth, is at the centre of the universe], but only the centre of gravity and of the lunar sphere'). As we saw in Chapter 3, the late-medieval view was that the earth overlapped with the centre of the universe but that there were at least three relevant centres of gravity: the centre of the earth, towards which solid objects fell; the centre of the sphere of water, towards which water descended; and the centre of gravity (that is to say, the point of balance or equilibrium) of the two spheres. One of these three centres was held to be the centre of the universe. *Centrum terrae esse centrum gravitatis* simply cuts through this debate in the fewest possible words, rejecting the arguments of the Parisian School and demonstrating that Copernicus already subscribed by 1514 to the argument that Vadianus was to be the first to publish (in 1515), and which Copernicus was to repeat in *On the Revolutions*: the geometrical and gravitational centre of the Earth are one and the same.

Secondly, Copernicus describes the rotation of the Earth as follows: '*Alius telluris motus est quotidianae revolutionis et hic sibi maxime proprius in polis suis secundum ordinem signorum hoc est ad orientem labilis, per quem totus mundus praecipiti voragine circumagi videtur, sic quidem terra cum circumfluis aqua et vicino aere volvitur.*' Rosen translates this as: 'The second motion, which is peculiar to the earth, is the daily rotation on the poles in the order of the signs, this is, from west to east. On account of this rotation the entire universe appears to revolve with enormous speed. Thus does the earth rotate together with circumjacent waters and encircling atmosphere.'

We need to be a little more precise: *terra cum circumfluis aqua et vicino aere volvitur* means 'the earth rotates, together with the water and the neighbouring air which flow around it.'[52] On the traditional view (explicitly rejected by Copernicus in *On the Revolutions*), the earth floats like an apple in a larger sphere of water.[53] But here the water is compared to the neighbouring air – both lie on the surface of the land and flow around it and across it. Prefigured here, then, is the claim later made in *On the Revolutions*, that 'finally, I think it is clear that land and water together press upon a single center of gravity; that the earth has no other center of magnitude; in that, since earth is heavier, its gaps are filled with water; and that consequently there is little water in comparison with land, even though more water perhaps appears on the surface.'

So if we look carefully at the text of the *Little Commentary* we find, in telegraphic form, what will become the argument of *On the Revolutions*.[54] Three conclusions follow from this. First, the *Little Commentary* cannot have been written before 1507. There is independent evidence to support this view, for in 1508 Lawrence Corvinus wrote a poem in which he implies that Copernicus believed at that time that the sun moved in the heavens; in other words, he had not yet adopted heliocentrism, even though he had already formulated 'wonderful [new] principles'.[55] Second, Copernicus was one of the first since the fourteenth century to reject the two-spheres, several-centres theory of the earth, which helps to explain the emphasis he places upon this argument in *On the Revolutions*, despite the fact that by 1543 he was knocking on an open door. Indeed, other Copernicans must have found Copernicus's emphasis on this point hard to understand, so quickly had it become uncontentious. Thomas Digges, when he translated the key parts of Book I into English, dropped the discussion of the roundness of the earth altogether, for he simply took it for granted that the Earth is a 'ball of earth and water'.[56]

With this chronology in mind, we can now address an important question: was Copernicus's adoption of the terraqueous-globe theory the key event which led to his switch from geocentrism to heliocentrism? It has been suggested that Copernicus originally considered a geoheliocentric theory, that is to say a theory in which the sun goes round the earth and the planets go round the sun – the theory later advocated by Tycho Brahe.[57] I doubt this, because Copernicus seems to have assumed that the correct theory must already have been formulated: he needed to read until he found it. He was not looking for a brand-new theory; he did not yet have a conception of knowledge as progressive. Still, if Copernicus did consider geoheliocentrism, it seems clear he quickly abandoned it, presumably when he recognized that such a theory was incompatible with belief in physical spheres carrying the planets, as the orbit of Mars around the sun would intersect with that of the sun around the earth. As soon as he turned to consider a more radical theory, heliocentrism (more radical in that it involved a moving Earth, but more conservative in that it was compatible with belief in physical spheres, and in that it had already been formulated by ancient philosophers), he will have realized that he had to determine the shape of the earth/water aggregate, because his Earth had to be capable of rotating on its axis and flying through space.

Sacrobosco's theory, that the waters had been displaced from the centre of the earth, would have had to be dismissed out of hand, for how could these waters turn evenly around the centre of the earth if that was not their centre? The Parisian view, that the centre of gravity of the earth corresponded to the centre of the sphere of water, will have seemed at first sight like a viable option. But Copernicus was a competent mathematician. He would quickly have realized, as he pointed out in *On the Revolutions*, that if, as was generally assumed, the sphere of water was ten times as large as the sphere of land, then the sphere of land would not overlap at all with the centre of the sphere of water, so the centres of gravity of the earth and of the water could not be made to coincide. Even if he shrank the sphere of water considerably, it would be difficult to get the centre of gravity of the sphere of earth to correspond with the centre of the sphere of water, unless one assumed that dry land was radically different from elementary earth – and a large part of the sphere of earth would have to be made up of theoretically 'dry' land, even though it was below the level of the waters. Pierre d'Ailly, and after him Gregor Reisch (1496), had tried to overcome this difficulty by treating earth and water as a single aggregate when identifying a centre of gravity which could coincide with the centre of the universe: the result was

REVOLVTIONVM, LIB. I. 2

set octaua, diameter eius nō posset esse maior, quàm quæ ex centro ad circumferentiam aquarum: tantū abest, ut etiā decies maior sit aqua. Quòd etiam nihil intersit inter centrum grauitatis terræ, & centrum magnitudinis eius: hinc accipi potest, quòd conuexitas terræ ab oceano expaciata, non continuo semper intumescit abscessu, alioq arceret quàm maxime aquas marinas, nec aliquo modo sineret interna maria, tamq uastos sinus irrumpere. Rursum à littore oceani non cessaret aucta semper profunditas abyssi, qua propter nec insula, nec scopulus, nec terrenum quidpiam occurreret nauigantibus longius progressis. Iam uero constat inter Ægyptium mare Arabicumq sinum uix quindecim superesse stadia in medio ferè orbis terrarum. Et uicissim Ptolemæus in sua Cosmographia ad medium usq circulum terram habitabilem extendit, relicta insuper incognita terra, ubi recētiores Cathagyam & amplissimas regiones, usq ad LX. longitudinis gradus adiecerunt: ut iam maiori longitudine terra habitetur, quàm sit reliquum oceani. Magis id erit clarum, si addantur insulæ ætate nostra sub Hispaniarum Lusitaniæq Principibus repertæ, & præsertim America ab inuentore denominata nauium præfecto, quam ob incompertam eius adhuc magnitudinem, alterū orbem terrarum putant, præter multas alias insulas antea incognitas, quo minus etiā miremur Antipodes siue Antichthones esse. Ipsam enim Americam Geometrica ratio ex illius situ Indiæ Gangeticæ è diametro oppositam credi cogit. Ex his demum omnibus puto manifestum, terrā simul & aquā uni centro grauitatis inniti, nec esse aliud magnitudinis terræ, quæ cū sit grauior, dehiscētes eius partes aqua expleri, & idcirco modicam esse cōparatione terræ aquam, etsi superficietenus plus forsitan aquæ appareat. Talem quippe figurā habere terram cum circumfluentibus aquis necesse est, qualem umbra ipsius ostendit: absoluti enim circuli circumferentijs Lunā deficiētem efficit. Non igitur plana est terra, ut Empedocles & Anaximenes opinati sunt: neq Tympanoides, ut Leucippus: neq Scaphoides, ut Heraclitus: nec alio modo caua, ut Democritus. Neq rursus Cylindroides ut Anaximāder: neq ex inferna parte infinita radicitus crassitudinę submissa, ut Xenophanes, sed rotūditate absoluta, ut Philosophi sentiūt. a ij

A copy of the first edition of Copernicus (from Lehigh University), with a contemporary annotation. The reader is working out the logic of Copernicus's claim that the traditional account of the relationship between earth and water is internally contradictory, as the volume of water cannot be ten times the volume of earth if the sphere of earth is to overlap with the centre of the sphere of water – which it must do if the earth is still to be at the centre of the universe, despite no longer having its own centre coincide with it. Exactly the same point caught the attention of Bodin in the *Theatrum*. (I am indebted to Noel Malcolm for painstakingly transcribing this nearly illegible annotation.)

a theory which claimed that for some purposes 'the Earth' was to be thought of as consisting of two spheres; for other purposes it was to be thought of as consisting of one sphere.[58] Either way, there could be antipodes, but only along the margin between the two spheres.

Copernicus tells us that he engaged in a systematic programme of reading as he struggled with the formulation of his new astronomy.[59] Michael Shank has suggested that in the course of this reading Copernicus obtained a copy of the compendium of astronomical texts published by the Giunta press in Venice in 1508. There he would have found

Grosseteste's brief exposition of the one-sphere theory. But he would also have found a commentary on Sacrobosco by Giambattista Capuano (first published in 1499) which is the only pre-Copernican work to discuss how one might formulate an astronomical theory based on the concept of a moving earth.[60] Crucially, Capuano discusses not only the familiar idea (expounded by Oresme) that the earth rather than the heavens might rotate daily, but also the possibility that it might move through the heavens on an annual path comparable to the path normally assigned to the sun. If this text indeed fell into Copernicus's hands (and Copernicus had been studying in Padua between 1501 and 1503, when Capuano was lecturing on astronomy, so he may have already heard his lectures, or read an earlier printed edition), then we can be sure that he read it with great care. Capuano formulated a series of objections to a moving earth which were to become classical – for example, if you throw an object straight upwards in a moving boat it will fall behind the boat.[61] If the earth rotated, he argued, we would all simply be drowned, as every day our bit of earth would turn under the waves – as it must, according to the two-spheres theory. If one argued that earth, water and air all rotated together, so that they were all stationary with regard to each other, then why are there always violent winds blowing at the tops of mountains? Capuano believed these winds were caused by the movement of the spheres being transmitted to the upper atmosphere. Copernicus's careful formulation in the *Little Commentary*, that the earth rotates together with the *neighbouring* air, seems almost designed to leave scope for a failure of the upper atmosphere to rotate along with the earth, thus providing an alternative explanation for the winds on mountain peaks. Reading Capuano would have left Copernicus in no doubt that he needed an account of what sort of body the Earth is, along with an account of what happens when objects fall on a moving Earth. (Copernicus's account explains that falling objects move with the moving Earth, but he does not extend this to claim that a falling object on a moving ship would move with the ship.)

If we imagine that Copernicus had reached this point in his thinking soon after 1508, then the geographical discoveries of Amerigo Vespucci and the maps and commentaries of Waldseemüller and Ringmann will have been crucial for him in developing his heliocentric theory, for they provided a definitive solution to the problem of the Earth's shape. It is evident from the text of *On the Revolutions* that the concept of the terraqueous globe was of fundamental importance to him; this was surely the last building block in the construction of the new theory.[62] Without

Vespucci there would have been no Copernicanism, for Copernicanism required a modern theory of the Earth.

Can we test the claim that Copernicanism required a modern theory of the Earth? At first, it would seem impossible: all we have to go on is two texts by Copernicus. But there are three other early presentations of the claim that the Earth moves: Copernicus's disciple Rheticus's *First Narration* (1540), the first appearance in print of an account of Copernicus's theories; Celio Calcagnini's short treatise arguing that the Earth rotates on its axis (pre-1541, and thus pre-Copernicus); and a text by Rheticus (1542/3) dealing with Biblical arguments against the motion of the Earth. Although these are all too late for there to be any need to demonstrate at length that the Earth is one sphere, we could expect to find the modern theory of the Earth clearly referred to in each of them when they discuss the movement of the Earth – and we do. Each of them thinks it necessary to stress that the Earth is a perfectly round ball, globe or sphere.[63]

§ 4

What are the implications of claiming that the Earth is a planet? Copernicus did not discuss the question; but his successors were bound to. In the summer of 1583 a strange little Italian gave a series of lectures in Oxford.[64] We know him as Giordano Bruno, but he liked to invent long names and titles for himself, names, it was said, longer than his body. The opening words of this letter of his provoked laughter:

> Philotheus Jordanus Brunus Nolanus, doctor of a more sophisticated theology, professor of a more pure and innocent wisdom, known to the best academies of Europe, a proven and honoured philosopher, a stranger only among barbarians and knaves, the awakener of sleeping spirits, the tamer of presumptuous and stubborn ignorance, who professes a general love of humanity in all his actions, who prefers as company neither Briton nor Italian, male nor female, bishop nor king, robe nor armour, friar nor layman, but only those whose conversation is more peaceable, more civil, more faithful, and more valuable, who respects not the anointed head, the signed forehead, the washed hands, or the circumcised penis, but rather the spirit and culture of mind (which can be read in the face of a real person); whom the propagators of stupidity and the small-time hypocrites detest, whom the sober and studious love, and whom the most noble

minds acclaim, to the most excellent and illustrious vice-chancellor of the University of Oxford, many greetings.[65]

When he walked to the lectern he rolled up his sleeves, as if he were a juggler about to perform a trick. As he spoke he bobbed and dipped like a dabchick or little grebe. He lectured, as all academics did, in Latin, but he spoke Latin with a Neapolitan pronunciation; the dons of Oxford (who found their own English pronunciation of Latin civilized and sophisticated) laughed at him for saying *chentrum*, *chirculus* and *circumferenchia* (which, as it happens, is now the approved pronunciation). But mostly they took exception to his Copernicanism. Twenty years later, George Abbott, who would eventually become Archbishop of Canterbury, remembered it as if it were yesterday: 'he undertooke among very many other matters to set on foote the opinion of Copernicus, that the earth did goe round, and the heavens did stand still; wheras in truth it was his owne head which rather did run round, & his braines did not stand stil.'[66]

It was forty years since Copernicus had published *On the Revolutions*. His new astronomy had certain evident advantages over the established astronomy of Ptolemy. According to Plato and Aristotle, all movement in the heavens should be circular and unchanging and, as we have seen, in the Renaissance there were still philosophers (such as Girolamo Fracastoro (1477–1553), the first to think seriously about contagious diseases) trying to construct a simple model of the universe which consisted of spheres nested around a common centre. But, try as they might, the philosophers could not get such models to fit what actually happens in the heavens. What Ptolemy had managed to achieve was a system that accurately predicted movements in the heavens. The Ptolemaic system, like those of Plato and Aristotle, claimed that the moon, the sun and all the planets circled around the earth. But in order to predict accurately the movement of these heavenly bodies it employed a complex system of deferents (circles), epicycles (circles on circles), eccentrics (circles rotating around a displaced centre) and equants. The equant was a device for speeding up and slowing down the movement of a body in the heavens by measuring its movement not from the centre of a circle but from another point. By this means the movement could be described (or misdescribed) as constant; it was thus a method of cheating on the fundamental principle insisted on by the philosophers that heavenly movement should be circular and unchanging. (For strict Aristotelians, even the epicycle was a cheat, as they wanted all the circular movements to have a common centre.)

Copernicus proposed to abolish the equant, and to eliminate an epicycle for each planet further from the sun than the Earth by showing how the movement of the Earth created an apparent movement in the sky equivalent to an epicycle. Copernicus also claimed that his system was preferable because it specified more tightly the characteristics of the system as a whole. Ptolemaic philosophers had never been sure, for example, whether Venus or the sun was closer to the earth (the right answer, in our terms, being that sometimes it is one, and sometimes the other, but this was an unacceptable answer within the Ptolemaic system), while Copernicus's system placed the heavenly bodies in a fixed order.[67]

It used to be thought that Copernicus initiated an intellectual revolution – indeed Thomas Kuhn called his first book *The Copernican Revolution* (1957). But in this Kuhn was mistaken. Throughout Europe astronomers took a keen interest in what Copernicus had to say, but, with only a very few exceptions, they took it for granted that his account of a moving Earth was simply wrong. If the earth moved, we would be aware of it; you would feel the wind in your face. If you dropped an object from a tall tower, it would fall towards the west. If you fired a cannon to the west, the ball would go further than if you fired it to the east. Since none of these things happened, all the leading astronomers – Erasmus Reinhold (1511–53), Michael Maestlin (1550–1631), Tycho Brahe (1546–1601), Christoph Clavius (1538–1612) and Giovanni Magini (1555–1617) – were confident that Copernicus was wrong. Still, they were fascinated by the simplicity of his techniques for calculation, and thrilled at the idea that it might be possible to junk the equant. In an extraordinary labour of love, every surviving copy of the first (1543) and the second (1566) editions of *On the Revolutions* has now been studied to identify the marginal comments written by its first readers, with the result that we can tell very reliably what they liked and what they disliked, what they found credible and what they found incredible.[68] They liked Copernicanism as a mathematical device; they had no time for it as scientific truth. They read it as the prefatory letter (now known to have been written by Osiander, and added without Copernicus's permission) encouraged them to read it, as a purely hypothetical construction.

In 1583 there were, as far as we know, only three competent astronomers in the whole of Europe who accepted Copernicus's claim that the Earth travelled around the sun: in Germany, Christoph Rothmann (who did not publish, and eventually abandoned Copernicanism); in Italy, Giovanni Benedetti (who published a few sentences on the question in

1585); and, in England, Thomas Digges (who had published in support of Copernicanism in 1576).[xii] So it must have simply astonished the dons of Oxford to hear this peculiar Italian, as he dipped and dodged, chucked and chirred, defending Copernicanism as the literal truth.

We do not know how far Bruno got in his exposition of Copernicanism. He was stopped after he had given three lectures; he was accused of merely reciting passages from the Renaissance Platonist philosopher Ficino (who had written in praise of the sun), while giving the impression that the words were his own. This is quite possible – Bruno does similar things in his published texts and, as we have seen, the concept of plagiarism was a novel one.[xiii] But we know what Bruno wanted to say because, after he was driven out of Oxford, he took refuge with the French ambassador in London, and there he set about writing a series of works, of which the most famous is *The Ash Wednesday Supper*, in defence of his position.[69] In the course of eighteen months Bruno published six books in London, all of them written in Italian.[xiv] Before and after his time in England, Bruno published only in Latin (with the solitary exception of a play, *Il candelaio*, published in Paris in 1582), so his choice of Italian, when his books must have mainly been sold to Englishmen (though some will have been carried to the great book fair in Frankfurt), seems odd. But Italian was the language of Dante and of Petrarch. Educated Englishmen could read it; by using it, Bruno signalled that he was addressing himself to poets and courtiers, not to professors of mathematics or philosophy.

The English were hostile to foreigners and to Catholics. If you were too obviously foreign, as Bruno was, you risked being beaten up in the street. Bruno hardly dared venture outdoors. In the dialogues he wrote he describes himself as mixing with the elite of English society, but he later claimed this was fiction not fact.[70] Still, his books must have sold, or his printer would have stopped printing them. Bruno himself was penniless, and astonished to see that the dons of Oxford wore great,

xii Twenty-five years later, all of these being dead, the numbers were similar: in 1608 we can count Kepler, Galileo, Harriot and Stevin. Remarkably, prior to the nova of 1572, Copernicus had only one firm supporter: Rheticus.

xiii Henry Savile's lectures on astronomy, delivered in Oxford in the early 1570s, contained 'long passages . . . copied verbatim from Ramus' (Goulding, 'Henry Savile and the Tychonic World-system' (1995), 153).

xiv *La Cena de le Ceneri* (1584); *De la causa, principio, et uno* (1584); *De l'infinito universo et mondi* (1584); *Spaccio de la Bestia Trionfante* (1584); *Cabala del cavallo Pegaseo–Asino Cillenico* (1585); *De gli heroici furori* (1585).

jewelled rings on their fingers – we can be sure there were none on his – so he cannot have been providing his printer with a subsidy.

These books mark a true revolution. Copernicus had described a spherical universe with the sun at its centre. He had acknowledged that it might be possible to conceive of an infinite universe, but he had refused to pursue that line of thought, saying, 'Let us therefore leave the question whether the universe is finite or infinite to be discussed by the natural philosophers' (Copernicus himself being a mathematician, not a philosopher).[71] Bruno seized on Copernicanism to argue for an infinite and eternal universe. The stars, he said, were suns, and the sun a star: here he was following not Copernicus but Aristarchus of Samos (310–230 BCE). Thus there could be other inhabited planets in the universe; even the sun and the stars might be inhabited, for they could not be equally hot all over, and there might be creatures, quite different from ourselves, who thrive on heat. Moreover, there was nothing to show that the other planets were different from the Earth. Bruno argued that the moon and the planets could be presumed to have continents and oceans, and that they shone, not by their own light (as was generally assumed; even the moon was assumed to be translucent at least), but solely by reflected light.[72] Thus, seen from the moon, the Earth would look like a gigantic moon; seen from even further away, it would be a bright star in the sky. The Earth, Bruno thought, would shine brightly because the seas would reflect more light than the land. (Here, as Galileo later showed, he was wrong – which is why when astronomers, after the discovery of the telescope, began to make maps of the moon they named the dark patches, not the light patches, seas.) Thus Bruno imagined an infinite universe, with numberless stars and planets, all possibly inhabited by extraterrestrial life forms.[73] Since Bruno did not believe that Christ was the saviour of mankind (he was a sort of pantheist), he did not have to worry about how the Christian drama of sin and salvation was played out in this infinity of worlds.

Bruno was not the first to imagine an infinite universe with extraterrestrial life. Nicholas of Cusa, in his *On Learned Ignorance* (1440), had argued that only an infinite universe was appropriate for an infinite God. Nicholas thought the earth was a heavenly body which from a distance would shine like a star, an idea which caught Montaigne's attention.[74] But Nicholas assumed that the earth and the sun were similar bodies. A habitable world was, Nicholas thought, hidden behind the shining visible surface of the sun; as for the earth, it, like the sun, was surrounded by a fiery mantle which was invisible to us, and which you would see only if you viewed the earth from outer space. Thus Nicholas made the earth into a heavenly body, but simultaneously he made the sun into a terrestrial one.[xv] Bruno, by contrast, was the first to distinguish stars and planets as we do now, making the sun a star and the planets, including the Earth, dark bodies shining by reflected light.

Bruno tried to resolve the standard arguments against Copernicanism by adopting the principles of the relativity of location and of movement; in his universe (unlike in those of Aristotle and Ptolemy) there was no up or down, no centre or periphery, no left and right and no way of telling if one was moving or stationary except by comparison with other objects.[xvi] Oresme and Copernicus had adopted the principle of the relativity of movement when considering two bodies, the sun and the earth – the movement of the sun that we perceive can equally be caused by the sun moving or the Earth turning – but they had not extended the argument to the more complicated circumstances considered by Bruno. Thus, Bruno argued, you can be in the cabin of a ship sailing across a calm sea and be quite incapable of telling whether you are moving or stationary; and if you throw something straight up in the air, it falls back into your hand, it doesn't drift backwards towards the stern of the ship as the ship moves on.[75] And Copernicus's universe had a centre; he could not imagine (or at least could not acknowledge the possibility of) a universe in which location was purely relative. Bruno also made some radical and ill-judged alterations to the Copernican system, designed in part to eliminate basic objections to it (such as that Mars and Venus should greatly change in size if they are sometimes quite near and sometimes very far from the Earth).[76]

In 1585 Bruno's host, the French ambassador, was withdrawn from England, and Bruno had no choice but to leave with him. He wandered around Europe (carrying with him his copy of Copernicus, which is

xv This view was still held by Isaac Beeckman, who has a good claim to be the founder of mechanical philosophy, in the years after 1616: Berkel, *Isaac Beeckman* (2013), 98–9.

xvi For extensive discussion of left and right, up and down in the universe, see Oresme, *Le Livre du ciel et du monde* (1968), 315–55, originally written in 1377. Oresme held that one of the major advantages of having the earth rather than the heavens rotate would be (since anticlockwise rotation is the correct direction of rotation, in that it involves the right hand passing above the left) that 'up' would become north rather than, as it must be if the heavens rotate around the earth, south. This would place us in the more noble 'upper' hemisphere. This did not feature in later arguments in favour of Copernicanism (although it was still important to Calcagnini; Calcagnini, *Opera aliquot* (1544), 391), presumably because cartographers had by the mid-sixteenth century already abandoned the Aristotelian and Arabic south-is-up orientation for the modern north-is-up orientation.

now in the Biblioteca Casanatense in Rome), and in 1592 he was arrested in Venice and handed over to the Roman Inquisition. After eight years of solitary confinement in the dark, and after prolonged torture, he was burnt alive in one of the main squares of Rome, the Campo de' Fiori, on 17 February 1600. He had refused to recant his heresies, including his belief in other inhabited worlds.[xvii] His books were banned throughout Catholic Europe.

Bruno is important to our story not because he was brave (though he was), or brilliant (though he was), but because he was, on occasion, right. Bruno's revisions to, and misunderstandings of, Copernicus were misconceived. The infinite and eternal universe theory has been replaced, in the course of the last fifty years, by the Big Bang theory (so recent that it was named only in 1949).[77] But we now know that the sun is a star, that other stars have planets, and there is every reason to think that there is life elsewhere in the universe. We are not at the centre of the universe: rather, the Earth is just another planet. Bruno would find himself more at home in our universe than would Cardinal Bellarmine, the man who played the key role in his trial, as he played the key role in the Catholic Church's condemnation of Copernicanism in 1616. On crucial points Bruno was right before anyone else: he was the first to say in print that the preface to *On the Revolutions* was not by Copernicus, and he was the first modern to insist that the planets shine by reflected light.[xviii]

§ 5

It is worth comparing Bruno to Thomas Digges. In 1576, a few years before Bruno's Oxford lecture, Digges had published a sixth edition of his father Leonard's perpetual almanac, *A Prognostication Everlasting*. (The book had first been published in 1555, and it went through thirteen editions that we know of, the last in 1619.)[78] The primary purpose

xvii Belief in a heliocentric universe was not yet condemned by the Church: it was forbidden only in 1616 and remained so until 1758, when the Index omitted the general ban on books teaching heliocentrism; Copernicus himself continued to be banned until 1822. As it happens, we do not know exactly what the charges against Bruno were, as the file recording his trial was taken to Paris (along with the file of Galileo's trial and other Inquisition documents) when Napoleon conquered Rome. After Napoleon's defeat the papacy reclaimed the files, but many disappeared on the return journey, probably sold for recycling to cover expenses.

xviii Before Bruno, this view had been maintained by Al-Battani (858–929) and Witelo (*c.* 1230–*c.*1290): Horrocks, *Venus Seen on the Sun* (2012), 73.

of the *Prognostication* was to enable its readers to predict the weather by a combination of astrology (the locations of the planets) and meteorology (phenomena in the atmosphere, such as rainbows and clouds). But the *Prognostication* also allowed you to determine when to let blood, purge (induce diarrhoea) and bathe (modern readers will find it odd to see bathing listed as if it were a medical intervention; Digges, father and son, recommend that one should not bathe when the moon is in Taurus, Virgo or Capricorn: these are earth signs, and hence at odds with water); how to tell the time from the rising of a star or from the moon; how to calculate sunrise, sunset, high tide and low tide, and the length of the day for any date. It was an eminently practical work – it provided a compass rose, for example, which you could copy on an enlarged scale, and a plan for a device for locating the planets in the heavens, which you could use as a blueprint, or (by adding a plumb line and a magnetic compass) you could turn the book itself into a paper instrument. Leonard also supplied some information that had no practical purpose: he showed the relative sizes of the sun, the planets, the earth and the moon, he explained how an eclipse of the moon could occur, and he gave the dimensions of the heavens: from the earth (which he, of course, assumed was at the centre of the universe) to the sphere of the fixed stars was, he said, 358,463 miles – and a half. To this successful work Thomas now added a translation (with a few revisions and additions of his own) of what he took to be the key sections of Book I of Copernicus's *On the Revolutions*.

Few copies of the *Prognostication*, in any of its editions, survive. It was a cheap publication, aimed at minor gentlemen and farmers, the sort of thing that is used for lighting fires once it becomes obviously outdated. If most almanacs were designed to last only for a year, even a perpetual almanac would soon become grubby and dog-eared. By the 1640s, if it survived that long, the print and layout of most copies would have looked hopelessly old-fashioned: the first eight editions were printed throughout in blackletter; then there were three in which the bulk of the book was printed in a humanist typeface but the translation of Copernicus remained in blackletter, perhaps to signify its intellectually serious content; the whole text was given a modern appearance only in 1605. As maritime compasses became cheaper and more widely available, instructions on how to make your own would have become increasingly irrelevant. By the eighteenth century astrology itself was generally regarded as outdated. Pages of tables and designs for instruments were probably often torn out for ready reference, leaving mutilated copies. Most copies will have been thrown away long before it occurred to anyone that the book was worth preserving simply because it was old and rare. Nobody published a proper study of the 1576 edition until 1934.[79]

And then, overnight, this edition became an object not only of great rarity (there are plenty of rare, ephemeral pamphlets) but also of great value. Every auctioneer, every librarian, was on the lookout for it. For it was now recognized that Thomas Digges had not only included in it the first substantial defence of Copernicanism by an Englishman, or in English,[80] he had also included an illustration of the cosmos which showed the stars arranged not in a sphere but stretching out to the limits of the page and beyond – the first illustration of an apparently infinite universe. This illustration spreads over two pages, and it appears to have been added as an afterthought as the book went through the press. Binders were never quite sure what to do with it – to incorporate it as a fold-out page, or as a two-page spread. Often it must have been damaged, torn, or either left as a loose page or omitted altogether. Of the first edition, only seven copies are known to survive, and not one has come on the market since the importance of the volume was recognized. The very wealthiest collectors have had to make do with copies of the later editions.

The 1576 edition of the *Prognostication* is a little puzzle in which we find the whole problem of the early modern history of science in miniature. It represents an intellectual breakthrough: Digges was the first competent astronomer explicitly to propose an infinite universe. (Nicholas of Cusa had argued that an omnipotent God would surely make an infinite universe, but this was a philosophical, not an astronomical argument.)[81] Moreover, Digges was not an insignificant figure in the new astronomy. In 1573 he had published a study of the nova which had appeared the previous year.[82] And yet at the very same time he was happily engaged in using the new astronomy to predict the weather and to decide when doctors should bleed their patients. He published his new, Copernican account of the cosmos alongside his father's old, Ptolemaic account. He knew the Copernican system could work only if the cosmos was much bigger than the Ptolemaics had imagined, but he did not correct his father's figures for the dimensions of the universe. His father had provided an illustration of the Ptolemaic cosmos in which the outermost sphere was labelled 'Here the learned do appoint the habitacle of God and the Elect.' Thomas's illustration, modelled on his father's, also mingles astronomy and theology: its outermost zone (now an infinite space rather than a sphere) is also labelled 'the habitacle of the Elect'. How can the old and the new, the past and the future, rational science and superstition, live side by side here without any sign of discomfort? This question requires several answers.

The first answer is that Copernicus himself was less of a revolutionary than is commonly supposed. In all his published work, Copernicus made no mention of astrology – but nor is there anything to suggest that he would have disputed the standard view, that astronomy existed to make astrology possible.[83] Copernicus's universe is different from Ptolemy's in that the sun, not the Earth, lies at (or rather, to be exact, very close to) its centre. But it is in other respects exactly like Ptolemy's: it is made up of a series of spheres, nested one within the other. It is finite in size.[xix] All movement within it (outside the immediate vicinity of the Earth) is determined by the fundamental principle that heavenly movement is circular and therefore unchanging. Ptolemy, Copernicus thought, had betrayed this principle not (as strict Aristotelians thought) by adding epicycles to deferents in order to explain why the planets sometimes appear to move backwards in the sky, but by introducing the equant in order to speed them up and slow them down. Copernicus achieved the same effect by different means.

Historians of astronomy trade insults on the question of whether there are equants in Copernicus or not; the answer is that there are no equants, but there are mechanisms designed to replicate equants.[84] Historians of Arabic astronomy point out that the mechanisms used by Copernicus had already been invented by the Arabs, and argue that

xix Copernicus certainly must have expected his audience to believe in the spheres and the finite universe (the two issues are connected, since a universe made of spheres is necessarily finite), and he was read by contemporaries as believing in them himself. But did he? He explicitly sidestepped the question of whether the universe was infinite, and his disciple Rheticus crossed out the words *orbium coelestium* ('Heavenly Spheres') on the title page of presentation copies (Gingerich, *An Annotated Census* (2002), xvi, 32, 135, 153, 209; information not available to Rosen, in his note on this subject: Copernicus, *On the Revolutions* (1978), 333–4). Rosen thinks Copernicus believed in material spheres because he uses the words *sphaera* and *orbis*; but Kepler uses these words in the *Epitome astronomiae Copernicanae* (the first three books are titled *De doctrina sphaerica*) and he certainly did not believe in material spheres: he merely used the conventional vocabulary when he had to in order to make himself understood. Barker, 'Copernicus, the Orbs and the Equant' (1990), points out that the Copernican spheres are not properly 'nested' as there are gaps between them, but otherwise takes a conventional line. In order to have a Copernicanism with spheres there needs to be an account of how the Earth/moon system is attached to a sphere, which is noticeably lacking. I suspect Copernicus intended to leave scope for doubt about his views on both issues.

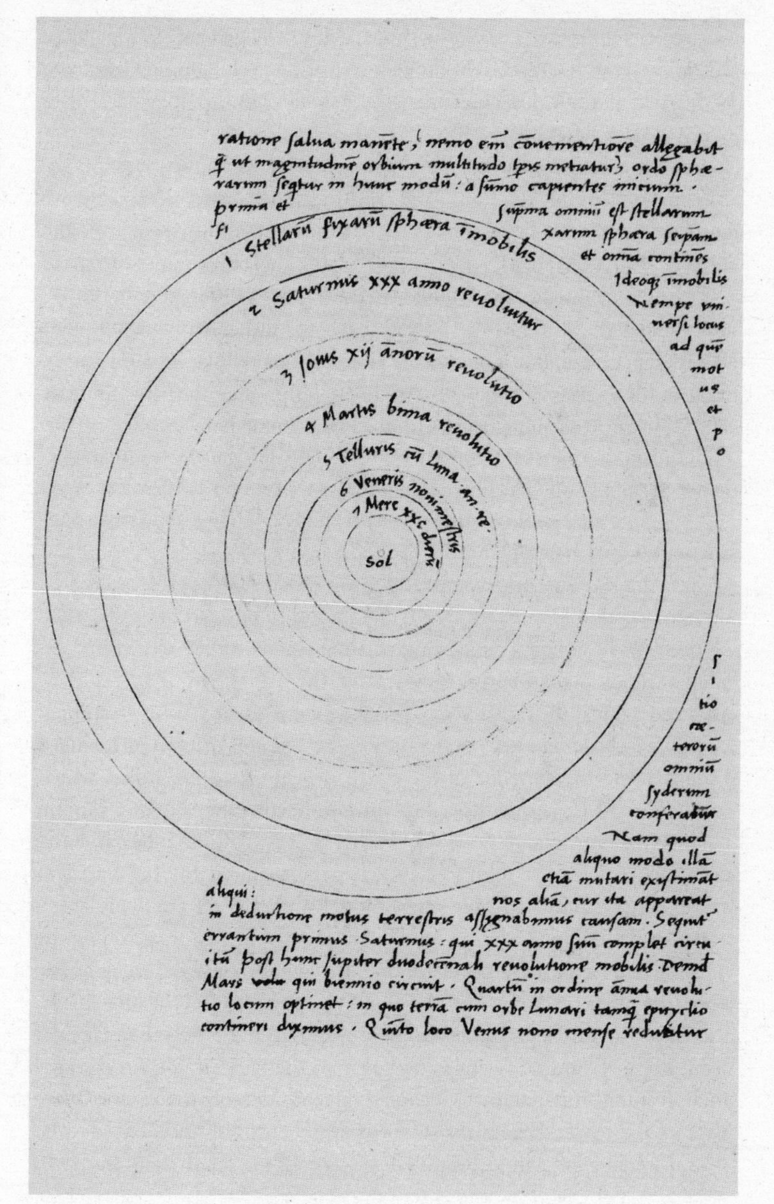

Copernicus's own diagram showing the heliocentric cosmos from the original manuscript of *On the Revolutions* (1543). The moon is not drawn but is mentioned in the inscription. The sphere of fixed stars is the outer ring.

Copernicus borrowed them without acknowledgement, rather than inventing them from scratch, though no one has yet identified a printed book or a manuscript describing the key mechanism to which he is likely to have had access.[xx] [85]

For the first two generations of astronomers reading Copernicus the crucial point about his book was not that it advocated heliocentrism but that it took the principle of circular movement more seriously and applied it more systematically than Ptolemy had. One consequence of Copernicus's mathematical model was that it was easier to make calculations using his system than Ptolemy's, and many astronomers proceeded to publish Copernican tables of planetary locations even though they thought that Copernicanism was not a plausible description of how the cosmos is organized. (Just as everyone happily uses the underground map for London, even though it distorts the distances between the stations; its great advantage is that it makes it easy to work out which route to follow and where to change, while a spatially accurate map would be much harder to read.)

But Digges was not a conventional reader of Copernicus, for he understood that Copernicus really did intend to be taken literally when he described the Earth as moving and the sun as stationary. In his version of Book I of *On the Revolutions* the arguments that could be adduced against the movement of the Earth are given a more prominent position. According to the perfectly standard figures offered by Leonard Digges, the circumference of the Earth measures 21,600 miles, which means that, if Copernicus is right and the Earth rotates once a day on its axis, this movement alone requires us to travel at 900 miles an hour, quite apart from the additional movement required for the Earth to travel in a vast circle around the sun once a year. If we are flying along at 900 miles an hour, it was argued (and remember that those doing the arguing had never travelled faster than they would on a galloping horse, about 30 miles an hour), then we ought to be able to feel the movement; the wind should be rushing through our hair. Birds, when they take off from trees, should be swept away towards the west. If you drop something from the top of a tower it should fall considerably to the west of the base of the tower. Digges insists these arguments are mistaken (and he thus may have influenced Bruno's discussion of the relativity of motion).

xx And it may be noted that Oresme had grasped the crucial principle, that a combination of circular movements could produce the appearance of rectilinear movement, apparently independently of any Arabic source. (Kren, 'The Rolling Device' (1971).)

If you climb to the top of the mast of a moving ship, Digges argues, and lower a plumb line, the plumb line will descend vertically to the bottom of the mast; it will not stream out backwards until the plumb ends up in the water behind the ship. This is a slightly different (and less convincing) experiment to the one later imagined (or performed) by Galileo, where you drop an object from the top of a mast, but it serves the same purpose of establishing that the concept of verticality is relative: a plumb line or a falling body on a moving ship will make a line vertical to the deck of the moving ship, not vertical to a fixed point on the surface of the Earth. Galileo also demonstrated that if you throw an object straight up in the air when on a moving ship, it falls not far behind you but straight back in your hand: this was a direct refutation of an argument from Giambattisa Capuano, who may be the source of all these moving-ship experiments, some real, and some merely thought experiments. Thus Digges had not simply translated Copernicus but strengthened his argument at its most vulnerable point.[86]

After the discovery of his illustration of the cosmos, Digges was given the credit for being the first to portray the stars as arranged not in a sphere but scattered over the outer margins of the page until they disappeared, and he certainly thought the stars went on for ever. But Digges's universe has a centre, so it is not really infinite, for an infinite universe could have no centre. He thinks each star is larger than the whole solar system; they have to be a truly astonishing distance away or there would be some measurable change in their relative positions as the Earth moves on its vast orbit around the sun, and so they must be absolutely gigantic if they are to remain visible.[87] It follows that Digges does not think of the sun as a star, or of the stars as suns. Moreover, his universe is shaped by his theology. The space occupied by the stars is heaven, the habitation of God, the angels and the elect. The solar system is the zone of sin and damnation. This sinful world is, Digges tells us, a dark star – 'this litle darcke starre wherein we live'.[88]

In fact, Digges's image of the universe – its boundless extent, its identification of the stars with heaven and of the Earth with hell (hence, perhaps, Mephistopheles' famous line in Marlowe's *Doctor Faustus* (1592): 'Why this is hell, nor am I out of it'), its description of the Earth as a dark star – comes from a popular poem commonly read by English schoolboys at the time, *The Zodiake of Life* (Latin, 1536) by Marcello Palingenio Stellato.[89] Digges knew the eleventh book of the poem off by heart '& takes mutch delight to repeate it often'.[90] What Digges had done was put the sun rather than the earth at the centre of Stellato's universe.

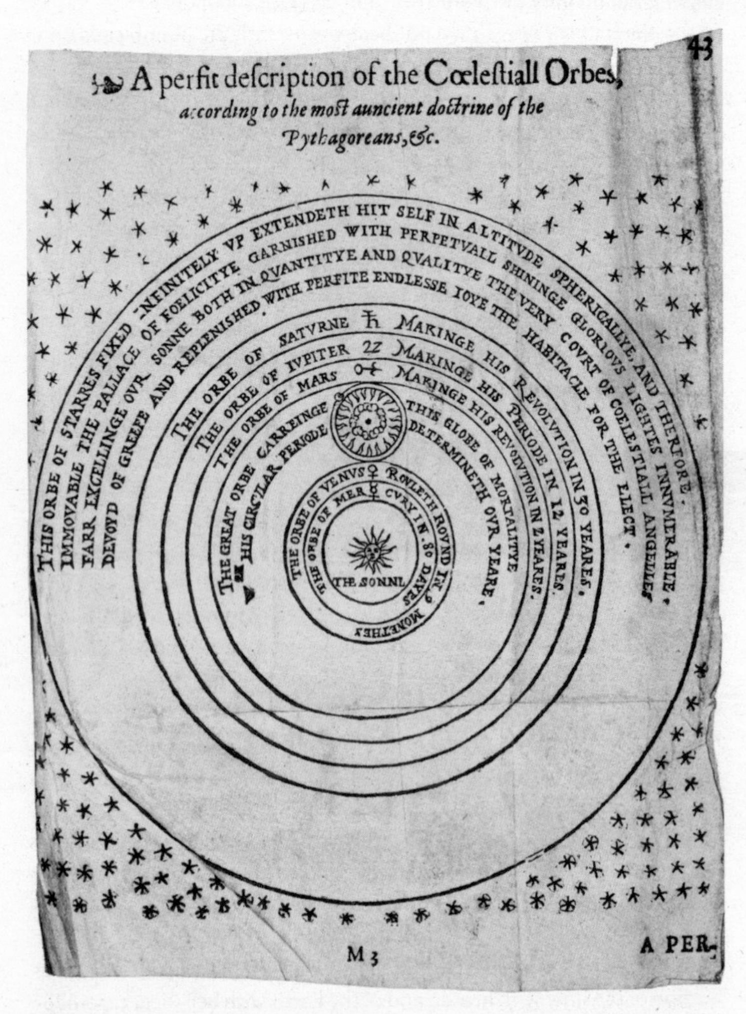

Digges's image of the Copernican cosmos, with the stars extending outwards to the edge of the page, symbolizing a universe without bounds (from the *Prognostication* – this is from the Linda Hall Library copy of the 1596 edition, but the illustration first appears in 1576).

Stellato had been posthumously condemned by the Inquisition for denying the divinity of Christ (heretical works he had written had been found among his papers after his death), and his body had been dug up and burnt, but Protestant Europe knew nothing of his rejection of Christianity (although there are plenty of indications of it in the *Zodiake*), and his anticlericalism and his determinism made it possible to read him as, if not actually a Protestant, then at least as sympathetic to Protestant views.[91] The fact that the *Zodiake* had been placed on the Index only confirmed this reading. For his English publishers, and presumably for Digges, he was 'the most Christian poet' (1561), 'the godly and zealous poet' (1565), 'the excellent and Christian poet' (1576), although Bruno, astutely, read him as a kindred spirit. It never occurred to Digges that the Earth might shine like a star, or that the planets are other Earths. The sun and the Earth are unique, and the universe has a centre.

Stellato and Digges were not the only ones to think of the Earth as a dark star.[92] In 1585 Giovanni Battista Benedetti published a collection of essays in which he dealt, among much else, with issues in contemporary cosmology. Like Digges, Benedetti was a realist Copernican. But he was more radical than Digges. Noticing that the path of the moon is in effect an epicycle around the path of the Earth, and that the planets appear to travel through epicycles, Benedetti proposed a remarkable hypothesis: what we think of as planets, he suggested, are merely the shining moons circling dark planets. These hidden planets (they are 'cloaked', to adopt the terminology of *Star Trek*) are Earth-like and presumably carry life. Benedetti's proposal was based on the assumption that the moon and the Earth are made of quite different types of substance, the moon being much more reflective than the Earth, although less reflective in the darker patches, where much of the sun's light is absorbed rather than being reflected. Benedetti held that the universe is spherical but that it is surrounded by unbounded empty space.[93]

Digges and Benedetti had not read Bruno, so they had not encountered his theory that, from a distance, the Earth would be indistinguishable from a star. A great proto-scientist, William Gilbert (1544–1603), the founder of modern studies of magnetism and electricity, however, had read Bruno, and adopted his arguments wholesale. Gilbert copied from Digges his illustration of an unbounded universe. But Gilbert understood that, seen from the moon, the Earth would shine like a vast moon; and that, seen from further away, it would shine like a star (here, he directly argued against Benedetti). The moon, he thought, had continents and oceans, just like the Earth. Like Bruno, he thought the oceans would be brighter than the land. He saw no reason why the other planets should not be just like the Earth.[94]

Gilbert drew, before the invention of the telescope, the first map of the moon, and as a result discovered its libration, the fact that it appears to turn slightly, up and down and from side to side, as it faces the Earth. This confirmed his conviction that the planets float freely in space. Moreover, Gilbert was the first to break completely with the notion that movement in the heavens must be circular: his planets trace complicated paths through the void; such a path could explain why the moon appears to wobble in the sky. Gilbert's *On the Universe* was never finished (he died in 1603, but the section on cosmology appears to date from the early 1590s), and it was not published until 1651. Bacon read it in manuscript, but had no time for it: Gilbert's preoccupation with magnetism seemed to him an irrational obsession, and as a result he had 'built a ship out of a shell'.[95]

§ 6

Digges, Bruno, Benedetti and Gilbert were members of the tiny group of realist Copernicans. They were bold pioneers of the new philosophy. Yet it would be wrong to think that they shared any common understanding of what natural science is or how it should be conducted. Digges was a proper mathematician. He taught surveying, navigation, cartography and military engineering. He experimented with mirrors and lenses; some think he had a secret telescope. He tried to measure the distance from the Earth of the supernova of 1572 and established that it was in the heavens – thus refuting that central claim of Aristotelian philosophy, that there was never any change in the heavens. (Digges thought it was a miraculous event and provided advice to the English government on what it might portend.)[96]

Benedetti was a comparable figure to Digges: he was an adviser on mathematical and engineering questions to Duke Emanuele Filiberto of Turin, and he published on perspective, the construction of sundials (which itself involves questions of perspective, since one must project the path of the sun on to a flat surface), calendar reform, the physics of falling bodies and the question of earth and water. But his cosmological arguments are purely speculative and philosophical.

Gilbert was a physician (he was briefly the personal physician to first

Elizabeth I and then James I) who chose to embark on a programme of experimental enquiry into the workings of magnets, and he evidently had close links to the experts who made compasses and taught navigation. His study of the moon's libration shows that he was looking for new observations with which to solve cosmological issues.

There is an old-fashioned way of writing the history of early modern science in which Copernicus, Digges, Benedetti and Gilbert are presented as scientists, although none of them used the word of themselves. The assumption is that they were engaged in an activity which is continuous with modern science; indeed, they were all Copernicans, and the publication of *On the Revolutions* is often (wrongly) taken to mark the beginning of modern science. Not Bruno, however, despite his Copernicanism. Bruno read Copernicus, lectured and wrote about him; often he was right where Copernicus was wrong. But he had no interest in measurement or experiment; he thought Copernicus was excessively preoccupied with mathematical problems. Copernicus, Digges and Benedetti called themselves mathematicians; Bruno and Gilbert called themselves philosophers. Copernicus and Digges wrote books on astronomy; Benedetti on *physica* (natural science); Gilbert on *physiologia* (the study of nature). None of them was a scientist, because science, as we understand the term, did not yet exist. Newton, however, was a scientist – who can doubt it? Sometime between the 1600s and the 1680s, science was invented.

PART TWO

Seeing is Believing

> They are deceiv'd that acquiesce to things which they have heard, and believe not what they have seen.
>
> – Thomas Bartholin, *The Anatomical History* (1653)[1]

Part Two begins in the early fifteenth century, and then follows issues related to sight through into the eighteenth century. Our starting point in Chapter 5 is the invention of perspective painting, which involved the application of geometrical principles to pictorial representation. These same principles led astronomers to take a new interest in measuring distances in order to establish exactly where certain objects – new stars – were in the heavens. Such activities established a new confidence in the power of mathematics to come to grips with nature, and this chapter follows this process through to Galileo. The second chapter, Chapter 6, looks at the impact of telescopes and microscopes on people's sense of scale: human beings suddenly came to seem insignificant in the vast spaces which the telescope opened up, while the microscope exposed a world in which complexity seemed to reach right down to the smallest imaginable creatures, so that it became commonplace to imagine that fleas might have fleas, and so on, *ad infinitum*.

5
The Mathematization of the World

> Philosophy is written in this very great book which always lies open before our eyes (I mean the universe), but one cannot understand it unless one first learns to understand the language and recognize the characters in which it is written. It is written in mathematical language and the characters are triangles, circles and other geometrical figures; without these means it is humanly impossible to understand a word of it; without these there is only clueless scrabbling around in a dark labyrinth.
>
> – Galileo, *The Assayer* (1623)[1]

§ 1

Double-entry bookkeeping goes back at least to the thirteenth century. The principle of double entry is simple: every transaction is entered twice, as a credit and as a debit. So if I buy a bar of gold worth £500, I credit £500 to my current account, and I debit £500 to my list of assets. If I borrow £500, then £500 is a debit to my current account, and a credit to my list of liabilities. In the Renaissance the standard system involved three books. First, a 'waste book', in which you recorded everything exactly as it happened, in as much detail as possible: you could refer to this in the event of any future dispute or confusion. Then a register in which you turned your record into a list of transactions. And then the account book proper, with debits and credits on facing pages. If you checked the account book against the register, and the debits against the credits, then you could be confident that the books were accurate; and every time you balanced the books you could establish whether you were making money or losing money. Accounting thus became the basis for rational investment choices and made it possible to decide how to divide up the profits of a partnership.[2]

Teaching bookkeeping was one of the main ways by which Italian mathematicians earned a living: this is what you learned at the *scuola d'abbaco*, the school for those going into business, where arithmetic and accountancy were taught. Double entry, like any mathematical technique, depends on abstraction. Bookkeeping turns everything into a notional cash value, even though you do not actually know if you will ever sell it or what you will get for it if you do. When two partners divide up the profits of a business between them they assign a notional book value to the stock in hand.

There would seem to be no connection between bookkeeping and science. But Galileo, who had probably taught bookkeeping himself when he was scrabbling a living together in the years between 1585, when he ceased to be a university student, and 1589, when he obtained his first university appointment, thought there was. When people complained to Galileo that his law of fall did not correspond to the real world, because falling objects do not accelerate continuously since they are retarded by air resistance, he replied that there was no contradiction between the world of theory and the real world, because:

> [w]hat happens in the concrete . . . happens in the same way in the abstract. It would be novel indeed if computations and ratios made in abstract numbers should not thereafter correspond to concrete gold and silver coins and merchandise . . . Just as the bookkeeper who wants his calculations to deal with sugar, silk, and wool must discount the boxes, bales, and other packings, so the mathematical scientist, when he wants to recognize in the concrete the effects which he has proved in the abstract, must deduct the material hindrances, and if he is able to do so, I assure you that things are in no less agreement than arithmetical computations. The errors, then, lie not in the abstractness or concreteness, not in geometry or physics, but in a calculator who does not know how to make a true accounting.[3]

Double-entry bookkeeping thus represents an attempt to make the real world, the world of bolts of silk, bales of wool and bags of sugar, mathematically legible. The process of abstraction it teaches is an essential precondition for the new science.

§ 2

Another source of income for mathematicians in Galileo's day was teaching the geometrical principles of perspective representation.[4] Galileo's own mathematics teacher, Ostilio Ricci, taught perspective to painters. Perspective painting was a more recent invention than double-entry bookkeeping. It began sometime between 1401 and 1413 when Filippo Brunelleschi produced a most peculiar work of art.[5] The object itself no longer survives; we last hear of it in 1494, when it is listed among the effects of Lorenzo the Magnificent, the Medici ruler of Florence, on his death.[6] Our only half-decent description of it was written in the 1480s, by Antonio Manetti, who was twenty-three years old when Brunelleschi died.[7] Manetti's account is puzzling and unsatisfactory, but it is all we have. There have been endless attempts to reconstruct exactly what Brunelleschi did, and why, for contemporaries were clear that this little object represented the birth of perspective painting.[8] Every attempt at reconstruction runs into difficulties and we have not a single word from Brunelleschi to help us. But we must do the best we can.

The object was a painting on a wooden panel about twelve inches square. It showed the Baptistry in Florence, an octagonal building, and something of the buildings on either side. The top part of the painting, where the sky would have been, was covered in burnished silver. (Brunelleschi had trained as a goldsmith, so producing a flat, silvered surface would have been easy for him.) In this painting, in the centre near the bottom, Brunelleschi made a hole, and the viewer was invited to look through the back of the painting. If they stood in exactly the right place, the place where Brunelleschi intended them to stand when he made the painting, and held up a mirror in front of them while looking through the painting from behind, they would see the image of the painting overlapping with the real Baptistry, and by raising and lowering the mirror they could convince themselves that the painting looked just like the real thing. Because they were looking at both the painting and the real world through one eye, the painting would look more nearly three-dimensional, and the real world more two-dimensional, so the two would become more like each other.[9] In the burnished silver of the painting the sky was reflected, so that clouds (if there were any) could be seen; reversed on the silver, the clouds would have been reversed again in the mirror so that they, too, would have corresponded to reality. It seems fair to say that Brunelleschi's image aspires to exemplify what philosophers call a correspondence theory of truth, in which a statement or representation is true if it corresponds to external reality.[10]

It is clear that this strange peep-show set-up ensured that the viewer looked at both the painting and the Baptistry with one eye – geometrical perspective depends on a single point of view. But why use a mirror?[11] Why not look at the painting directly through a small hole in a board? Evidently, once Brunelleschi had silvered the upper part of his painting, he needed to place it where it could reflect the sky, and then the mirror became necessary if it was to reflect the sky above the Baptistry as it was at that particular moment and at the same time overlap with the real Baptistry. What is not clear is whether this was his original motive or if this was a feature of his peep-show of which he then decided to take advantage.

I want to stress the strangeness of this procedure. If you lowered not the mirror but the painting, what you would see is yourself. Even looking at the painting in the mirror, what you would see when you looked straight at it is the pupil of your own eye, and you would learn that there is a point in the picture which corresponds to or reflects the eye of the painter. This would later be called the centric point, and it is the spot where the vanishing point is placed in a vanishing-point construction. The viewer, in playing their part in the peep-show performance, was constantly reminded of their own role: at one moment they made reality appear and disappear; at another they made themselves the object of their own inspection. Brunelleschi's ingenious construction serves a double function: it demonstrates that art can successfully imitate nature, so that the two are almost indistinguishable; and it demonstrates that even when art is at its most objective (or rather, especially when art is at its most objective), we make it and we find ourselves in it. It is an exercise, simultaneously, in a new objectivity and a new subjectivity.

After he had produced this image Brunelleschi produced a second one, which we also learn about from Manetti, of the Florentine town hall and the square surrounding it. This time he cut away the board at the skyline, so that a viewer would see the real sky (a neater solution, in many respects, than burnished silver). This time there was no mirror. This object, too, was obviously site specific: you would stand where Brunelleschi stood when he painted it; when raised up the painting would perfectly mask and exactly reproduce the real buildings; when lowered you would see the real buildings. Back and forth you could go, confirming the exact correspondence between reality and image, making and unmaking your own world.

It is evident that both paintings avoided the obvious method of demonstrating depth in a two-dimensional image, which is to show orthogonals, parallel lines running at right angles to the picture plane and converging on a vanishing point. The most straightforward example is a tiled floor.[i] Instead, both pictures must have used two-point perspective, where lines which are neither parallel to the picture plane nor at right angles to it converge on distance points to the left and right of the picture plane itself. If Brunelleschi wanted to experiment with depth of field, why not use a vanishing-point perspective, which would have been straightforward, and indeed familiar to him? Ambrogio Lorenzetti's *Annunciation* of 1344, for example, uses a tiled floor and converging parallels to create a sense of depth of field.[ii] Lorenzetti had not mastered all the complexities of perspective construction – see how the back of Mary's chair is higher than the front, and how the angel's left foot is no further back than his right knee. But he did know how to make a tiled floor recede into the distance. If Brunelleschi was simply trying to create an impression of depth, he could have simply shown an interior with a tiled floor.

So what was Brunelleschi up to? A standard view (which can appeal to Vasari's *Lives of the Artists* (1550) for support, though Vasari was writing long after the event) is that Brunelleschi was illustrating the geometrical principles of perspective drawing, codified by Alberti more than twenty years later in 1435, in *On Painting*, a work that founded a long tradition of writing texts on geometrical perspective.[12] We can reasonably assume that Brunelleschi had a fairly sophisticated grasp of geometry. We know he had only a limited education; his father had ensured that he learnt some Latin, probably with a view to his following in his own footsteps as a notary, but Brunelleschi had decided to apprentice as a goldsmith. Then he had turned from jewellery to architecture (he is most famous for having designed the dome of Florence's cathedral in 1418, a work based on classical models and quite unlike any medieval construction). However, if Brunelleschi had mastered the geometry of perspective as early as 1413 it becomes a little difficult to explain why there are no surviving paintings which embody these principles before 1425. Indeed, it used to be thought that Brunelleschi had produced his demonstration images around 1425 simply because scholars wanted to envisage these images as immediately provoking new art and new theories. Recent documentary evidence, however, strongly implies (as indeed does Manetti's text) that Brunelleschi's images were produced

i Some illustrations of Brunelleschi's images impose such a chequerboard pattern on the piazzas in the foreground in order to make it more obvious that these are representations of three dimensions – but these patterns correspond to nothing in the real world, and so to nothing in his images.

ii See plate 11.

earlier. This obliges us to reconsider just exactly what it was that he had accomplished.[13]

It has been argued that both Brunelleschi and Alberti were applying to painting the theories of medieval optics, which ultimately derived from an eleventh-century Arabic author, Ibn al-Haytham, known in the West as Alhazen, whose works were available in both Latin and Italian translations. These works on optics were about 'perspective', a term which meant 'the science of sight'. Alhazen had shown how light travels in straight lines, so that vision depends on a cone of straight lines extending outward from the eye to the object. Depth of field is thus not experienced directly; it is a result of binocular vision and of our capacity to interpret the way in which things closer to us look bigger and things further from us look smaller, so that, to judge distance, we need a reference point – an object whose distance or size is known to us. It is easy to understand why Alhazen had concerned himself only with how we see, not with how we might represent the world in a painting: representational art was forbidden in the Islam of his day. But it is harder to understand why his medieval successors had not developed his theories in order to show how they might be put to work by artists.[14]

One claim is that even if the university experts did not discuss painting explicitly, the painters learnt about their theories. Giotto (1266–1337) did his most important work in Franciscan churches, and as it happens the libraries of the friaries attached to those churches contained the crucial texts on perspective. The friars who commissioned his work had, as followers of St Francis, a love of the natural world, and thus a desire for a new realism in art. They would have wanted him to produce a sense of depth because they knew from studying the theory of vision that we make sense of the world by turning a two-dimensional sensation (light rays falling on the eye) into a three-dimensional experience. Giotto's art, which included using *trompe l'oeil* to create the illusion of non-existent columns, was, it is suggested, the result of a dialogue with his employers.[15] This is very likely, but there is an important caveat: the medieval theory of vision provided the elements of a theory of what we now call perspective (what came to be called in the Renaissance 'artificial perspective'), but it did not provide a systematic account of how to create the illusion of three-dimensionality. If it had, Giotto would have completed the perspectival revolution, Brunelleschi's images would have been unnecessary, and Alberti would have had nothing new to say. It may have seemed to a contemporary that 'there is nothing which Giotto could not have portrayed in such a manner so as to deceive the sense of sight,'[16] but we may doubt whether Giotto aspired to create images which corresponded throughout to visual reality. Is the angel flying through the wall of the *Annunciation of St Anne* intended to be an accurate representation of what Mary saw?[iii] The question is surely misplaced. The reality Giotto wants to convey is not just visual, while in Brunelleschi's test pieces the whole and sole point is geometrical precision.

We do know that Brunelleschi, in his search for new architectural forms, had made a study of the surviving classical buildings in ancient Rome, a study which involved taking measurements and drawing plans and elevations. He would thus have been familiar with the basic principle that objects appear smaller as they recede into the distance, a principle analysed by Euclid, and familiar in the Middle Ages.[17] This principle makes it possible to work out how big an object is if you know how far away it is and measure the angle, as seen from where you are, between its top and its bottom. Brunelleschi would have had plenty of practice in applying this knowledge when he set out to measure the height of surviving classical buildings in Rome in 1402–4.[18] But there was nothing new about this principle, and the resulting knowledge would have been most obviously useful in drawing standard elevations, rather than perspective images, so it is hard to see why it would suddenly produce a new type of artistic representation.

So we seem to have various elements that might contribute to an answer to the question of what made possible the invention of perspective painting – the application of geometry, medieval optics, the surveying of old buildings – but they do not seem quite sufficient.[19] The crucial missing element, I believe, is provided by the Florentine artist known as Filarete ('lover of excellence'), who wrote a work on architecture completed around 1461 – this is indeed our earliest source.[20] Filarete was twenty-three years older than Manetti, and thus perhaps had a better understanding of Brunelleschi's world. Filarete was convinced that Brunelleschi had come up with his new method of perspective representation (which he does not describe in any detail) as a result of studying mirrors. The mirror is indeed the obvious source of a correspondence theory of art (and truth). It not only provides the appearance of three dimensions on a two-dimensional surface, it also makes it easy to answer the question 'How big does the Baptistry look from here?' Trying to answer that question by measuring angles and distances would be

iii See plate 10.

much more complicated than simply holding up a mirror. The mirror works as a scaling device; it is able to do this because it reflects the cone of rays coming from an object as they pass through a plane. This alerts us to a feature of Brunelleschi's peep-show I have not previously mentioned: he stood, according to Manetti, inside the porch of the cathedral. His view would therefore have been framed by the porch; indeed, his picture may have simply reproduced the view within the frame, as if he were looking through a window.

Some have concluded from Filarete's comments that Brunelleschi's board was burnished all over – that he painted on a mirror. But Manetti, who had held the board in his hand, would surely have noticed this. Much more likely is that he had his board and a mirror side by side on an easel. This explains the peculiarly small size of Brunelleschi's first image: good-quality mirrors were very rare and very expensive at the beginning of the fifteenth century (the revolution associated with Venetian mirrors occurred a century later), and glass mirrors were always small.[21] Working from a mirror, of course, would produce a reversed image, hence Brunelleschi's interest in looking at his painting reflected in a mirror – and, happily, he had one to hand. It is true that the Baptistry is a symmetrical building, which means that a reversed image would have been very like a normal view; but Manetti tells us that the square on either side of the Baptistry could be seen, and even a symmetrical building will have markers on it (shadows and moss, for example) which are not symmetrical. Working with a mirror would also have involved Brunelleschi in an endless struggle: he would have wanted to see the Baptistry reflected without distortion in the mirror, but if he placed himself directly in front of the mirror what he would have seen is himself (which is why it is easy to use mirrors for self-portraits). The peculiar feature of his peep-show construction, that the viewer is looking at himself as well as at the painting, simply recapitulates this earlier tension.

It would have been at the point when he looked at his painting in a mirror in order to see it the right way round that Brunelleschi realized he could lay burnished silver on the board to reflect the sky. And it is also at this point that he would have made an unfortunate discovery: looking at his image in a mirror will have had the effect of halving its height. A painting made to correspond exactly in size to the Baptistry when seen from the porch of the cathedral would have ended up one quarter the size because the effect of the mirror would have been to double the apparent distance between the viewer and the Baptistry.[22] Of course, Brunelleschi could have foreseen this problem and scaled up his image to allow for it – but we know he did not, because we know he wanted the viewer to stand where the painting had been made, in the porch, and it is easy to show that, standing in that position, an image one foot square would have corresponded to the apparent size of the Baptistry. In order to scale up to allow for the second reflection, Brunelleschi's board would have needed to have an area of four square feet, not one.

So what had Brunelleschi learnt from his peep-show, apart from the difficulties of working with mirrors? He was demonstrating in this first image that perspective drawing involves establishing a picture plane through which the image is viewed. He took this new understanding forward into his second image, of the town hall. Perhaps this time he worked from an image created in two mirrors (a procedure Filarete recommends). Perhaps he began by looking through translucent parchment and inking an outline directly on to it. Alberti was the first to devise (*cuius ego usum nunc primum adinveni*; 'the use of which I found out for the first time just recently', with *primum adinveni* often translated as 'discover') a method of looking through a grid and using the lines of the grid as reference points – or at least he claimed to have invented this method in the Latin text of *On Painting* (1435); the claim does not appear in the Italian version.[23] When Alberti says that he does not understand how anyone could ever achieve even moderate success in perspective representation without using his method, one begins to suspect that Brunelleschi may have beaten him to it, and the revision to his text may be taken as confirmation that he had belatedly discovered as much.[24] This method later became well known and was illustrated, for example, by Leonardo, Dürer and Vignole (see plate 16).

If this reconstruction is right – that Brunelleschi started by representing what he saw in a mirror – then he was learning to understand that perspective drawing involves establishing a picture plane through which the scene is viewed, and the task of the artist is to construct an image that corresponds to the image as it would appear on a piece of glass placed on that plane. It was this principle that Alberti invoked when he compared a painting to a window that you look through to see the scene beyond, and which led Dürer to claim that the word 'perspective' comes from the Latin *perspicere* in the sense of 'to see through', when in fact it comes from its sense of 'to see clearly'.[25] What Brunelleschi had discovered was not the vanishing-point or the distance-point construction; he had not engaged in elaborate measurements or sophisticated geometrical constructions, even if he had the competence to perform them. He had learnt to think of the painted surface as a piece of glass that you look through. He had also learnt something of enormous importance: that for a perspective construction to work, the artist and the viewer must have their eye located in the same place, and to this place there corresponds a point in the picture directly opposite the artist's eye. A perspective painting appears to be a totally objective representation of reality, yet it depends on having a viewer prepared to look at it in the right way, and when they do this, the viewer can in effect locate themselves in relation to the picture. Brunelleschi's paintings did not have vanishing points, but they did have situated viewers.

§ 3

Roughly two decades separate Brunelleschi's first studies and Masaccio's famous painting of the Trinity (*c.*1425), the first large-scale painting which fully masters the technique of perspective representation.[iv] Masaccio's painting shows Christ on the cross in front of a chapel with a barrel vault – but of course the chapel does not exist; it is entirely a painted chapel. Here is the difference between Brunelleschi's studies and Masaccio's painting: Brunelleschi was representing reality; Masaccio is representing an imaginary space. You can use the various picture-plane techniques to paint reality; but if you want to paint an imaginary world you have to work out how to construct that world so that it appears convincing and aesthetically satisfying.[26] You have to decide where you want the vanishing point and/or the distance points to be. You have to sketch out a grid of converging lines. You have to apply the principles of geometry. And we know that this is precisely what Masaccio did: we can see the lines he scored in the plaster on which he painted.[27] We know that Brunelleschi discussed perspective with Masaccio,[28] and that Alberti was soon to write a textbook on geometrical perspective.

Thus it would seem to be Masaccio who was responsible for the next stage in the development of perspective painting, and it was of course the crucial stage, as most Renaissance art was religious art, and religious art is almost never a direct representation of presently existing reality. Of course painters had models. Masaccio's patrons, who had paid for his painting, appear on either side, kneeling. Masaccio may well have gone to look at an actual barrel-vaulted chapel and copied actual columns. But to fit these elements on to this wall he had to do

iv See plate 12.

sketches, draw converging lines, work out scales and foreshortening. He had to construct a theoretical space which became a painted space.

Perspective painting thus involves the application of theory to particular circumstances. It provides an abstract account of lines in space, travelling from the object, through the picture plane, to the eye, and an account of how these lines appear on the picture plane itself. It trains the eye to think in terms of geometrical shapes. A straightforward example of this is provided by Father Niceron's *Curious Perspective* of 1652.[29] Niceron explains how to produce anamorphic shapes: shapes like the skull in Holbein's *Ambassadors*, which takes form as a skull only if you view it at a sharp angle to the picture's surface. But first he has to train the reader in understanding and representing shapes.

Take his demonstration of how to draw a chair. First, he shows how to draw a simple rectangular box. Then he adds a back and feet to it. The result looks like a Bauhaus chair for the simple reason that it is a chair made from the simplest of geometrical forms. It looks nothing like a seventeenth-century chair because no seventeenth-century chair would have been entirely devoid of curves and decoration – just look at the curly ribbon which serves as a label to get a sense of the period's aesthetic. It is an abstract or theoretical chair, a geometer's chair, not a real chair. Learning to look like this involved learning to isolate mathematical shapes within more complex objects.

Naturally, as soon as they became familiar with the geometrical techniques of perspective representation, artists became fascinated by mathematical shapes and the difficulties of drawing them. Leonardo himself provided the illustrations to Luca Pacioli's *On Divine Proportion* (1509). The two were evidently good friends; they were both employed by Ludovico Sforza, Duke of Milan, and they fled Milan together in 1499 when the city fell to the French, ending up in Florence, where for a while they roomed together. In a portrait of Pacioli we see two such shapes: one, a dodecahedron (a regular solid with twelve faces), is sitting on top of a book written by Pacioli; the other, a rhombicuboctahedron (a symmetrical solid with twenty-six faces), is made of glass sheets and has been half filled with water.[v] It hangs from the finest thread in empty space, a decorative object, as interesting for the way in which it catches the light as for its geometrical form.[30]

Pacioli is caught explaining a problem from Euclid to a pupil: Euclid is open on his desk, he is drawing the figure required to understand the

v See plate 18.

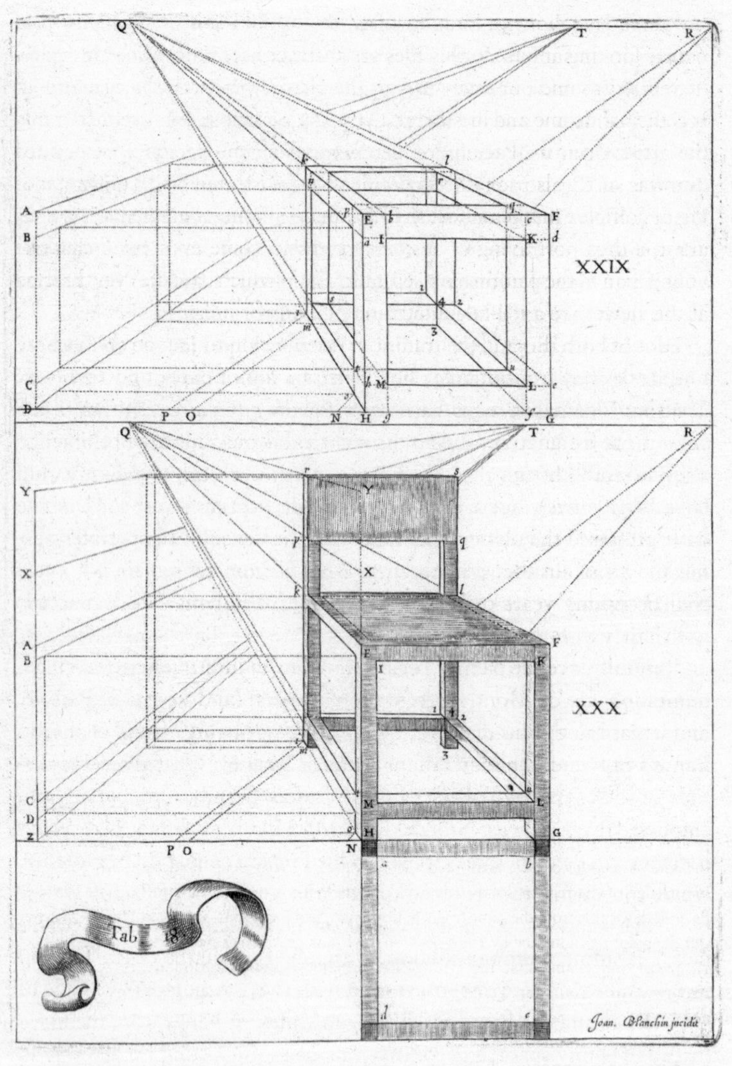

From Niceron's *Curious Perspective* (1652): a chair reduced to a problem in geometrical construction.

problem on a slate, and scattered across the table he has his mathematical drawing instruments and the little tubular case that holds them. Unlike his pupil, Pacioli is not looking at us (he is thinking deep thoughts); but we are looking at him, for his eyes are at the centric point, directly opposite the artist's and our own (as is emphasized by the stylus he is holding). It is the handsome and aristocratic young man whose eyes are directed at the artist, or at us. Pacioli is a mathematician; the person who painted him was surely also a mathematician, as demonstrated by his understanding of complex mathematical shapes.[vi] In portraying a mathematician the artist is thus portraying a version of himself: some even think that the young man in the painting is a self-portrait, in which case the eyes directed at the viewer are a tell-tale indication of a mirror image.[vii]

I doubt both this and the traditional attribution to Jacopo de' Barbari. On the desk, in front of the young man, is a slip of paper on which a fly is sitting. The paper reads 'Iaco. Bar. Vigennis. P. 1495'. This has been taken to be a signature, and so the painter has been identified as Jacopo de' Barbari, although this looks nothing like his work and he was not twenty (*vigennis*), but a good deal older, in 1495.[viii] No one seems to have proposed the obvious explanation, that the slip of paper identifies not the artist but the young man ('P.' for *pictum* not *pincit*), who may well be twenty years old. There are plenty of Italians called Giacomo with last names beginning 'Bar' (Bardi, Barozzi, Bartolini, Bartolozzi, and so on). Since the painting originally bore an inscription dedicating it to Guidobaldo da Montefeltro, Duke of Urbino (and a pupil of Pacioli), and it was hung in his dressing room, we can probably assume that Iaco. Bar. was a friend of his, and that it is the prince's eyes that meet with his.

vi It is presumably for this reason that Pacioli's first biographer, Bernardino Baldi, writing in the late sixteenth century, attributed the painting to Piero della Francesca, whose expert knowledge of the regular solids was well known. Piero was, according to Baldi, a friend of Pacioli; they came from the same town, Sansepolcro, and Piero may have been Pacioli's teacher. But the painting cannot be by Piero because Piero was dead when the painting was made.

vii Jacob Soll claims that the young man is Guidobaldo da Montefeltro himself and remarks. 'An accountant would never again be painted in a relation of superiority to a nobleman.' This is on the face of it implausible; and in any case we have a good portrait of Guidobaldo, attributed to Raphael, and this is not he. (Soll, *The Reckoning* (2014), 50; quotation from caption to illustration.)

viii There is no direct evidence regarding de' Barbari's date of birth, but he was described as old and ill in 1512; his earliest securely dated work is 1500. It used to be assumed he was born between 1440 and 1450; it is now argued that he was born in the 1470s, but the argument is circular in that it largely depends on accepting the Pacioli portrait as being by him, while acknowledging that it does not look like his work. (Gilbert, 'When Did a Man in the Renaissance Grow Old?' (1967); Levenson, 'Jacopo de' Barbari' (2008).)

Why record, even in abbreviated fashion, the young man's name? Again the obvious explanation would seem to be that the painting is some kind of memorial – perhaps he is dead, perhaps he has gone away.

Thus the painting belongs to the court life of Urbino. It was in Guidobaldo's library that Polydore Vergil wrote *On Discovery*. Working in this superb room, which not only contained an abundance of books but was embellished with gold and silver, gave Vergil such a distorted view of the world that he claimed that in his day any scholar, no matter how needy, could lay their hands on any book they wanted.[31] Guidobaldo's court was later made famous by Castiglione, who set his *Courtier* (1528) there, notionally locating the imaginary discussions he recorded in 1507. Guidobaldo himself never appears in Castiglione's book: he lies sick in bed while his wife Elisabetta takes charge.

The portrait of Pacioli illustrates the way in which, once perspective had been discovered, mathematics and art went hand in hand. Piero della Francesca wrote a number of mathematics texts (two survive: *The Abacus Treatise* and *The Short Book on the Five Regular Solids*) which deal with practical problems such as how to work out how much grain there is in a conical heap, or how much wine there is in a barrel, as well as a book on perspective, *On Perspective for Painting*.[32] Such problems turn real objects – heaps of grain, barrels of wine – into abstract shapes so that mathematical principles can be applied to them. Pacioli's publications reproduce material from Piero's books wholesale. Pacioli was not only a friend of Leonardo but also of Alberti, with whom he stayed for some months as a young man. He was not himself an artist, but *On Divine Proportion* discusses the golden section, the principles of architecture and the design of typefaces. Pacioli is now known primarily for the fat book on which the dodecahedron sits: *A Compendium of Arithmetic, Geometry, Proportions and Proportionality* (1494). This was a textbook of applied mathematics, and included within it was the first published account of double-entry bookkeeping – double entry was not new, but printing was, so Pacioli was taking advantage of an obvious opportunity.[33]

§ 4

Perspective painting generally involves a peculiar form of abstraction: the construction of a vanishing point. It is worth noting that the term itself is relatively modern: in English it dates to 1715. Alberti calls it the centric point (*il punto del centro*), and in many early texts it is simply referred to as the horizon.[34] But Alberti is perfectly clear that the image in a one-point-perspective painting extends 'to an almost infinite distance'.[35] This, to a Renaissance intellectual, is a deeply puzzling concept. Aristotle's universe is finite and spherical; moreover, it is not surrounded by infinite space and there is no such thing as empty space. Indeed, Aristotle does not really have a concept of space as distinct from the objects which fill it. So, for Aristotle, all space is finite, all space is place, and the idea of an infinite extension is conceptually contradictory, just like the idea of a vacuum.[36]

This is not true, of course, in Euclidean geometry, where parallel lines can be extended indefinitely without meeting (nor, it may be added, in the optics of Alhazen). But what you can see when you look across an infinite distance is precisely nothing. It helps, then, if you want to work with a vanishing point, to have a concept of nothing. Euclid lacked the number zero, which was introduced into Western Europe in the early thirteenth century with what we call the Arabic numerals (actually, it is the only one of the ten numerals that is Arabic; the others are Indian). Arabic numerals made possible the paper-based accounting of double-entry bookkeeping. Zeros are wonderfully useful, even if deeply mysterious; perhaps only a culture which had the number zero could have made sense of the idea that a vanishing point could both be a point of non-seeing and the key to the interpretation of a painting.[37]

As a consequence of the vanishing point, artists found themselves living simultaneously in two incommensurable worlds. On the one hand, they knew the universe to be finite. On the other, perspective geometry required them to think of it as infinite. A good example of this is provided by Cesare Cesariano's commentary on Vitruvius (1521). Cesariano provides perfectly conventional illustrations of the Aristotelian universe as a series of bounded spheres. But when he introduces the idea of measuring distances he imagines measuring the distance to the sun and the planets and onwards and outwards for ever: he explicitly states that the lines from the spectator through points T and M (picture below) extend to infinity. Perspective thus introduced an anomalous concept of infinity into a finite universe.[38]

Handling this presented problems for artists. In early perspective paintings the vanishing point is often hidden by a seemingly casually placed foot or a bit of drapery. In religious images, the lurking presence of the infinite could be put to good use. Thus the vanishing point in Masaccio's *Trinity* is just above the top of the tomb, in apparently featureless space.

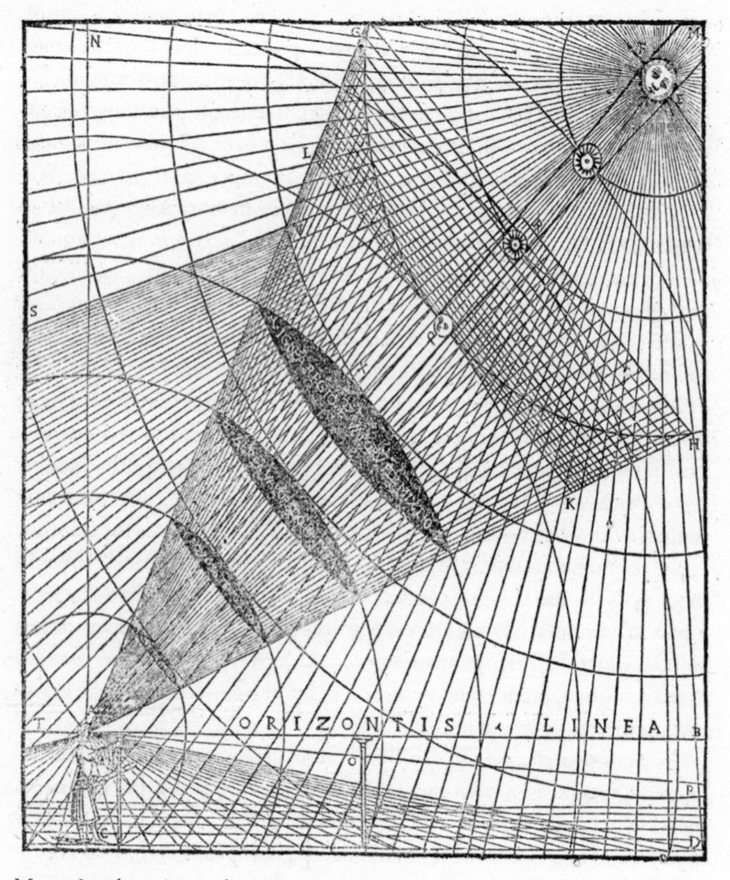

Measuring the universe, from Vitruvius, *De architectura* (1521), with commentary by Cesare Cesariano.

But the painting originally had an altar standing in front of it, and the vanishing point would have lain immediately behind the host when the priest elevated it at the dramatic high point of the Mass, the moment when transubstantiation occurs. This is the point to which the viewer's eyes are drawn. (So successful was Masaccio's painting in providing a setting for the host that it was soon being copied in the design of tabernacles – wooden boxes constructed to hold the consecrated host.) Similarly, in Masaccio's *Tribute Money* the vanishing point lies behind Christ's head.[39]

One particular subject which encouraged artists to explore the vanishing point was the Annunciation. Mary's womb was compared to a closed garden ('A garden enclosed is my sister, my spouse; a garden enclosed, a fountain sealed up,' it says in the Song of Songs), so a closed door leading to a garden was often placed at the vanishing point.[40] But the incarnation of Christ restores to human beings the possibility of salvation, reopening the gates of Eden, which had been closed against Adam and Eve, and opening to believers the gates of paradise. So an open door leading to a garden could symbolize salvation. And, of course, God is infinite, so the Annunciation represents an encounter between the finite human and the infinite divine: in Piero della Francesa's *Annunciation* the vanishing point seems to be used simply to evoke the presence of infinity, and the swirling patterns of marble become a symbolic representation of a God who cannot be seen, or comprehended.[ix]

In secular paintings, however, the vanishing point had to be kept under control, for the human world is finite and limited. Thus in a painting of an ideal city, dating to 1480–4 and attributed to Fra Carnevale, the two lines of buildings on either side of a piazza point into the far distance, but that space is blocked by a temple, in which a half-open door suggests that one could explore further, but only within an enclosed space.[x] If there is infinity to be found here, it is within a closed religious space. In Uccello's *Hunt by Night* there is an alarming multiplication of vanishing points, all of which lead into the darkness. One has a strong sense of how easy it would be to become lost, or for the stag to escape; the painting is a play on the idea of disappearance, for the viewer's sight disappears into the dark rather than into an infinite distance.

§ 5

By the middle of the fifteenth century artists were experimenting with the idea of infinite, abstract, undifferentiated space. They knew this idea was problematic and anomalous, but they also knew that without it there could be no perspective representation. Art had escaped, or partially escaped, from Aristotle, and it had done so under the guidance of geometry and optics. But perspective also encouraged a new way of looking at the world in three dimensions and of recording what one had seen. This made it possible to see things no one had seen before and to do things no one had done before.

ix See plates 13 and 14.
x See plate 17.

Before perspective drawing, if you wanted to design a piece of machinery, you had to make it, or make a model of it. There was no substitute for working with three-dimensional materials. But once engineers had acquired the capacity to draw three dimensions they could design with a pen or a pencil (the pencil was invented around 1560) in their hand. Leonardo (1452–1519) designed plenty of machines that were never built, many (such as flying machines) that never could be built. Plate 15 shows his design for a ratchet winch. The winch itself is shown in the drawing to the left; on the right, the winch is shown taken apart (or 'exploded') to demonstrate its assembly. Each wheel is attached to a ratchet system. If you pull on the lever attached to the right of the winch assembly, one wheel grips and turns the axle, which lifts the weight. If you push, the other wheel grips, but it is geared so that the axle still turns in the same direction and the weight continues to rise. Since you can exert more force pulling and pushing on a lever than you can turning a crank, this is more efficient than a crank mechanism would be at lifting weights. Leonardo's drawing is clear enough for a model of the machine to have been built and demonstrated to be functional. It is only a step from a drawing such as this to a modern blueprint. Leonardo's sketch is already implicitly drawn to scale, with a detail of the ratchet mechanism shown at a higher level of magnification.[41]

Of course, making a real machine from a drawing is not straightforward. What tools would you require to construct Leonardo's winch? If you were lifting a heavy weight and pulling hard on the lever, considerable force would be placed on the pegs that drive the mechanism. Out of what sort of wood would you need to make them? Books of drawings produced in the early modern period are mainly intended to advertise an engineer's skills, not to provide you with the information you would need to do the job yourself. Even the elaborate plates in the great *Encyclopaedia* (1751–72) of Diderot and d'Alembert seem to be there to help you know what can be done, not to teach you how to do it. Nevertheless, there are examples of the successful transfer of designs through the medium of print. In 1602 Tycho Brahe published his *Instruments for the Restoration of Astronomy*, which provided elaborate illustrations of the new instruments he had devised for conducting astronomical observations. In Peking, in the 1670s, Ferdinand Verbiest, a Jesuit astronomer, was able to build instruments based on his designs without ever having seen Brahe's originals.[42]

Apart from being an artist, an architect and an engineer (all professions requiring skills which interlock in their use of geometry and

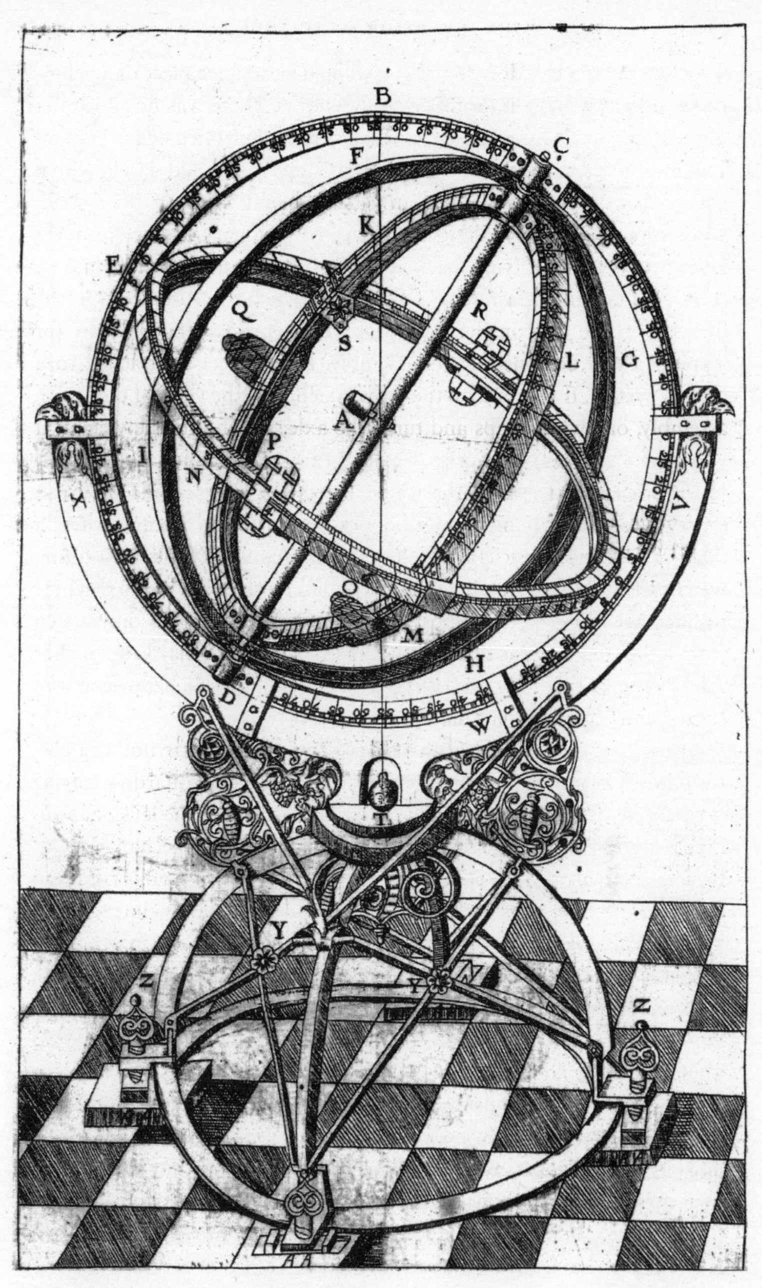

Brahe's design for an equatorial armillary sphere, from his *Astronomiae instauratae mechanica* (1598).

The imperial observatory in Peking, from Ferdinand Verbiest, *Xinzhi Yixiangtu* ('pictures of newly made instruments'), which was produced from 1668 to 1674, showing instruments built by the Jesuit missionary on the basis of Brahe's designs.

perspective), Leonardo did careful anatomical studies based on the dissection of animals and human beings. But, although he seems to have had plans to publish, he never did. The revolution in anatomy came with the publication of Andreas Vesalius's *On the Fabric of the Human Body* (1543). Vesalius (who was teaching at the university of Padua) employed artists from Titian's workshops in Venice to produce illustrations to the highest possible standard. The illustrations were keyed by letters to textual labels. Leonardo, in his drawing of the winch, was already using letters as labels, and the practice of course has its origins in geometrical diagrams, but Vesalius was the first person to make systematic use of it in anatomy. Vesalius could thus show the reader what he had seen in the body. The engraved plates produced in Venice were then carried over the Alps to Basle, as Vesalius did not trust the Venetian printers to produce work of sufficiently high quality.

The whole point of Vesalius's *Fabric* is to insist that the evidence of one's senses must take priority over Galen's text. Medieval anatomists had frequently lectured by reading Galen aloud and commenting on his text, while assistants opened up the body: the body was intended to illustrate what Galen said, not to correct him when he was wrong. But, even when medieval anatomists had performed their own dissections, what they found (or thought they had found) was what Galen had told them to find. Mondino de Liuzzi (1270–1326), for example, the author of the first medieval textbook on how to perform a dissection, had plenty of hands-on experience, but he still found at the base of the human brain the *rete mirabile* (miraculous network) of blood vessels that Galen claimed was there, despite the fact that it isn't there at all – it is only present in ungulates. Leonardo carried out dissections, but he still thought he found a channel linking the male penis to the spinal cord and so to the brain: down it he believed flowed material which became part of the ejaculate and was essential for generation. The first anatomist regularly to disagree with Galen on the basis of direct experience was Jacopo Berengario da Carpi, whose *Anatomy* was published in 1535, only a few years before Vesalius's *Fabric*.[43] Only in a culture where the authority of the great classical authors such as Ptolemy and Galen had begun to be undermined could a project like Vesalius's *Fabric* be undertaken. In this respect the coincidence in date between the great works of Copernicus and Vesalius points to an underlying correspondence: both lived in a world where respect for antiquity had been fatally weakened, at least among the intellectually adventurous, by the new culture of innovation.

Galen's text had never been accompanied by illustrations – Galen explicitly said that illustrations were worthless – because in a manuscript culture complex illustrations degrade rapidly each time they are copied.[xi][44] Thus it was often far from clear exactly what Galen was

xi This is why no copy of Ptolemy's *Geography* survives accompanied by the maps that Ptolemy describes, and no copy of Vitruvius's great book on architecture (produced while Augustus was emperor, 27BC–14AD) survives accompanied by diagrams. The diagrams that describing. With Vesalius, on the other hand, it is easy to see what he is talking about. Vesalius claimed to identify scores of errors in Galen and thus undermined his authority, just as Columbus's discoveries had undermined the authority of Ptolemy. But even more important for later anatomists was the fact that, where anatomical details did not appear in Vesalius's illustrations, or were incorrectly shown, one could confidently say that he had made a mistake. Sophisticated printed illustrations based on perspective drawings thus turned anatomy into a progressive science, where each generation of anatomists was able to identify mistakes and oversights in the work of their predecessors. In anatomy, discovery does not begin with Vesalius: rather, he provides the base line that allows others to claim to have made discoveries.

The techniques employed by Vesalius in anatomy were at the same time also being employed in botany, where authors faced a similar difficulty to that faced by Vesalius himself: should they portray actual specimens, with all their flaws and faults, and thus accurately reflect reality, or should they provide idealized images of the perfect specimen, as Vesalius had done with his musclemen? Should they show the plant at a moment in time, or show both flower and fruit in a single illustration? Just as Vesalius's images had made possible the reliable identification of parts of the body and progress in anatomical knowledge, so the new illustrated botanies made possible reliable knowledge of the different species and progress in their naming and identification. But progress involves discrimination: Conrad Gesner, the first compiler of natural historical information in the age of print (*Historiae animalium*, 1551–8) often supplies illustrations which he labels as false, and even Vesalius at one point illustrates a mistaken claim of Galen's. The convention that seems basic to us – that illustrations represent reality – was not immediately obvious.[45]

Thus by 1543 two revolutions had come together to make possible a new type of science. On the one hand, there was perspective painting, grounded in geometrical abstraction; on the other, the printing of engraved plates, supplemented by text produced on a printing press. Perspective painting goes back to 1425; engraved prints to at least 1428; the printing press to 1450. The fall of Constantinople, one consequence of which was a flood of Greek manuscripts and Greek-speaking scholars entering the Latin-speaking West from the East (and thus

originally accompanied Vitruvius's text were in any case evidently few in number and extremely rudimentary. The first illustrated edition appeared in 1511.

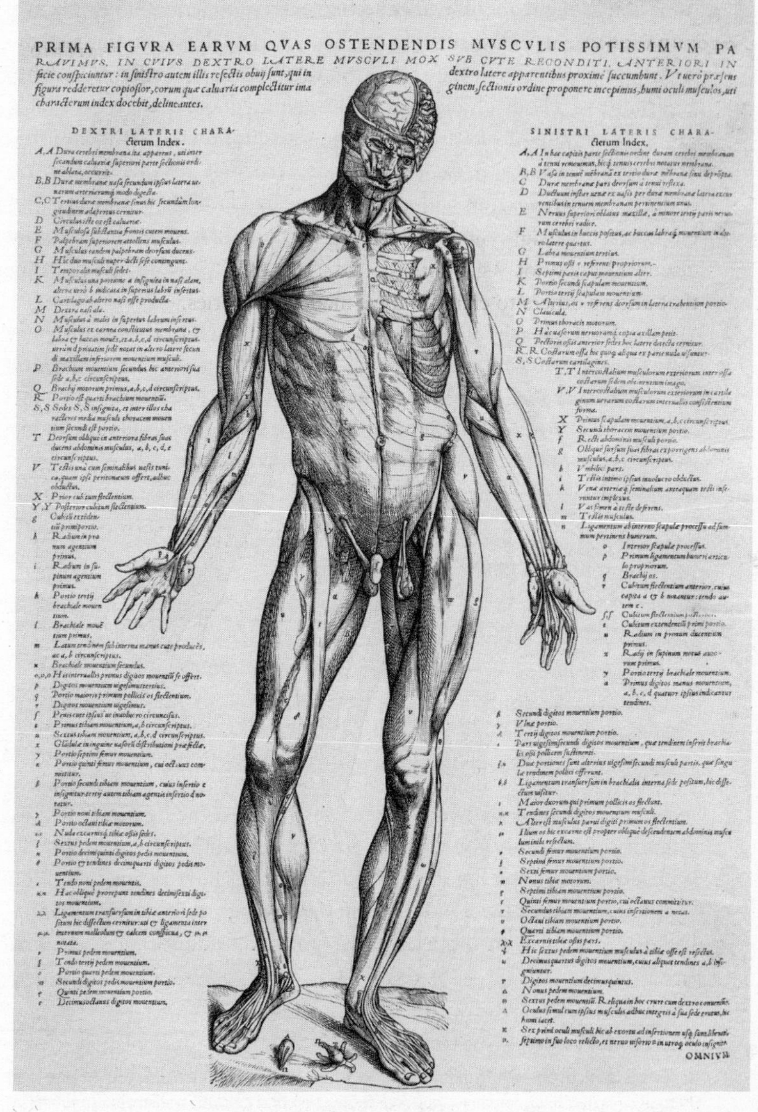

First illustration of the muscles of the body, from Vesalius, *On the Fabric of the Human Body* (1543).

improved knowledge of the Greek originals of Galen's texts) occurred in 1453.[xii] Why then did it take a further century to complete the transformation brought about by the mechanical reproduction of perspective images? There are two answers to this question. First, the immediate priority of publishers in the years after the invention of printing was to publish the vast body of religious, philosophical and literary texts which had been inherited from the past – first the Latin texts, and then, for a more limited audience, the Greek texts. The first reliable edition of Galen, on which Vesalius had worked, appeared in Basle in 1538; it was there that Vesalius insisted that his *Fabric* should be printed. Second, a long cultural revolution had to take place, in which book learning came to seem of lesser importance than direct experience. That revolution, as I have argued, began with Columbus.

Alongside the great works of Copernicus and Vesalius we may place *On the History of Plants* by Leonhart Fuchs, which appeared the year before (1542), and which contained 512 exact images of plants. In his preface Fuchs writes:

> Though the pictures have been prepared with great effort and sweat we do not know whether in the future they will be damned as useless and of no importance and whether someone will cite the most insipid authority of Galen to the effect that no one who wants to describe plants would try to make pictures of them. But why take up more time? Who in his right mind would condemn pictures which can communicate information much more clearly than the words of even the most eloquent men? Those things that are presented to the eyes and depicted on panels or paper become fixed more firmly in the mind than those that are described in bare words.[46]

Fuchs's words represent two distinct revolutions: the demoting of the authority of antiquity (Galen is 'the most insipid authority'; it is difficult to imagine how shocking these words must once have seemed) and the recognition of the power of images in the new age of mechanical reproduction.[47] These were both essential preconditions for the Scientific Revolution.

xii This is sometimes said to be the date at which the Renaissance properly begins. The alternative earlier date, for those who like to imagine that cultural transformations can be confined within precise dates, is Petrarch's rediscovery of Cicero's letters to Atticus in 1345 – symbolizing the rediscovery of the cultural legacy of ancient Rome, while the fall of Constantinople serves as a marker for the rediscovery of the cultural legacy of ancient Greece.

§ 6

In 1464 a German astronomer, Johannes Müller, known as Regiomontanus (Regiomontanus being a Latin version of the place he came from, Königsberg), gave a lecture at the University of Padua.[48] Regiomontanus had recently completed an exposition and commentary on Ptolemy's astronomy, begun by his mentor Georg Peuerbach. This was to become the standard textbook in advanced astronomy for the whole of the sixteenth century, and in it Peuerbach and Regiomontanus did not hesitate to criticize Ptolemy for his errors. In 1464 Regiomontanus was writing a pathbreaking guide to plane and spherical trigonometry (*On All Sorts of Triangles*), which laid out the mathematical foundations of astronomical calculations. He had learnt Greek in Vienna in order to read Ptolemy in the original, and in Italy he had been able to read in Greek Archimedes (who had been translated into Latin in the Middle Ages but was not yet available in print) and Diophantus (who was not yet available in Latin; Diophantus (*c.*210–*c.*290) was the originator of algebra).

Regiomontanus was one of the first to benefit from the supply of ancient Greek texts that reached Italy after the fall of Constantinople. At the time of his Paduan lecture, less than a decade after the publication of the Gutenberg Bible, the printing revolution was only just beginning to get under way: Euclid, for example, was first printed in Latin only in 1482, in Greek in 1533, in Italian in 1543 and in English in 1570. Regiomontanus's lecture thus marks a key moment in the reacquisition of Greek mathematics, and points towards the ambitious programme for the publication of mathematical texts that Regiomontanus developed, though he died before it could be carried out.

Regiomontanus spoke in praise of the mathematical sciences, and he praised them by denigrating the Aristotelian philosophy taught in the universities. Even Aristotle, he said, if he came back to life, would not be able to make sense of what was said by his modern disciples. 'This [i.e. that the texts are incomprehensible] no one unless mad has dared to assert of our [mathematical] sciences, since neither age nor the customs of men can take anything away from them. The theorems of Euclid have the same certainty today as a thousand years ago. The discoveries of Archimedes will instill no less admiration in men to come after a thousand centuries than the delight instilled by our own reading.'[49] Regiomontanus's praise of the mathematical sciences did not, however, imply uncritical admiration for contemporary mathematicians. Only the year before he had written, 'I cannot [but] wonder at the indolence of the typical astronomers of our age, who, just like credulous women, receive as something divine and immutable whatever they come upon in books ... for they believe in writers [such as Ptolemy] and make no effort to find the truth.'[50] This theme, that one should turn from the study of books to the study of the real world, was to be repeated over and over again by the advocates of the new sciences as they set themselves up in opposition to the old philosophy. It was, for example, one of Galileo's favourite rhetorical tropes: the suggestion was still as radical in the 1620s as it had been in the 1460s, for the hold over university education of the traditional curriculum had not been diminished. Galileo also shared Regiomontanus's conviction that Euclid and Archimedes ('the divine Archimedes', as he called him) provided the only models for reliable knowledge.[51]

In 1471 Regiomontanus worked out a procedure for measuring the parallax of heavenly bodies and so their distance from the Earth.[52] His procedure presumed use of a cross-staff, an instrument invented by Rabbi Levi ben Gerson (1328).[53] The cross-staff is a very simple instrument, a calibrated shaft along which a bar slides. You sight along the shaft and move the bar back and forth until you have lined up its ends with two points, and the angle can then be read off from the scale on the shaft. You can use a cross-staff, for example, to measure the angle between the horizon and the sun at midday. If you know the date and have the right tables, you can then read off your latitude (this, of course, involves squinting at the sun; the backstaff was invented in 1594 to enable you to take this measurement without having to peer at the sun). Alternatively, at night, you could measure latitude directly by measuring the angle between the horizon and the Pole Star. The cross-staff is merely one of a series of instruments, such as the quadrant and the sextant, designed for measuring angles by taking sightings. Before it was invented, the astrolabe (copied in medieval Europe from Islamic models) had provided a sighting device, and also a method of measuring the height of the sun from its shadow. With this device you could establish your latitude if you knew the time of day but, rather more important for most users, you could tell the time of day if you knew your latitude and the date. Specialist forms of all these instruments were developed for surveying, for astronomy and for navigation, but the basic principle that angles could be used to determine distances or times was the same for all of them.[xiii][54]

xiii A rather different instrument was the nocturnal: it enabled you to tell the time at night, provided you knew the date, from the position of the stars in the constellation of the Great

INTRODVCTIO

GEOGRAPHICA PETRI APIANI IN DOCTISSIMAS VERneri Annotationes, cōtinens plenum intellectum & iudicium omnis operationis, quæ per sinus & chordas in Geographia confici potest, adiuncto Radio astronomico cum quadrante nouo Meteoroscopii loco longe vtilissimo.

HVIC ACCEDIT Translatio noua primi libri Geographiæ CL. Ptolemæi, Translationi adiuncta sunt argumenta & paraphrases singulorum capitum: libellus quoq; de quatuor terrarum orbis in plano figurationib. Authore Vernero.

LOCVS etiam pulcherrimus desumptus ex fine septimi libri eiusdem Geographiæ Claudii Ptolemæi de plana terrarum orbis descriptione iam olim & à veterib. instituta Geographis, vnà cum opusculo Amirucii Constantinopolitani de his, quæ Geographiæ debent adesse.

ADIVNCTA est & epistola IOANNIS de Regio monte ad Reuerendissimum patrem & Dominum D. Bessarionem Cardinalem Nicenum, atq; patriarcham Constantinopolitanum, de compositione & vsu cuiusdam Meteoroscopii armillaris, Cui recens iam opera PETRI APIANI accessit Torquetum instrumentum pulcherrimū sanè & vtilissimum.

Cum Gratia & Priuilegio Imperiali.

Cross-staff being used for surveying and astronomy – from the title page of Petrus Apianus, *Introductio geographica*, 1533.

In surveying, if you knew how far away a building was it was now easy to calculate its height. Suppose you wanted to scale the walls of a fortress which was on the other side of a river. You could take two measurements in a straight line with the building and, from the distance between the measurements and the difference between the angles as measured with a cross-staff, you could calculate the height of the walls and make ladders of the right length. The basic principles involved had been described by Euclid and were well understood in the Middle Ages. They are exactly the same principles as are involved in perspective painting. But, where perspective painting takes a three-dimensional world and turns it into a two-dimensional surface, Regiomontanus was now trying to take a two-dimensional image – the night sky – and turn it into a three-dimensional world. To do so you have, in effect, to move from monocular vision to binocular vision.

The principle of parallax enables you to do this. It is a variation of the basic principle that if you know one angle and one side of an equilateral or right-angled triangle, then you can determine the other angles and sides. It thus requires not one measurement but two. Hold up your finger in front of you, close your left eye, and note where your finger appears to be against the background. Then switch eyes. Immediately, your finger will jump to the right. If you know the distance between your eyes and measure the angle that corresponds to the apparent shift in your finger's position, then you can calculate how far away your finger is – although, of course, no one would bother. In this case the distance between your eyes is a significant proportion of the distance between your eyes and your finger; if you were trying to measure the distance to an object that was very far away, then you would have to set up two observation spots that were far apart – or at least so it would seem.

Regiomontanus grasped that an astronomer does not have to travel in order to get two observation points that are, in effect, far apart.[55] If the heavens rotate around the centre of the universe, and if that centre is at or near the centre of the Earth, then the observation point of the astronomer, who is on the surface of the Earth, changes in its relationship to the heavens as they move simply because the astronomer is not looking at the heavens from the centre of the universe but from a point distant from that centre.

Bear, which rotate around the Pole Star. Here, the angle being measured is not between the observer and two distant objects, but rather the stars are being read as if they were the hands of a clock.

Imagine you are standing at the dead centre of a merry-go-round on which the horses are arranged in three concentric circles. At the centre there is a stationary, round platform around which the circles of horses, each taking the same time to complete a circuit, revolve. As you look outwards and the horses turn around you, the relative position of the horses will remain the same – a horse which is in line with another horse at one moment will still be in line with it a quarter of a revolution later. But if you take a few steps in any direction, until you reach the edge of the stationary platform, then the relative position of the horses will appear to change all the time. Moreover, if you know the size of the stationary platform and the distance to the outer ring of horses, then you can use changes in the relative position of horses in the other two rings to work out how far away they are. Regiomontanus thus saw that you could measure the parallax of heavenly bodies by taking two observations from the same place but at different times, rather than by taking two observations from different places but at the same time.

According to Aristotle, comets exist in the upper atmosphere. They must, because comets come into existence and disappear, while the heavens continue the same for ever. Comets must therefore be sublunary, not supralunary: below the moon, not above it. Aristotle's hypothesis was that they represent some sort of exhalation from the earth catching fire. As far as we know, no one had actually tried to measure the parallax of a comet before 1471; it had simply been assumed that the Aristotelian theory was obviously correct.

Although Regiomontanus worked out how to make such a measurement in 1471, the full account of his procedure was not published until 1531. Unfortunately, in 1548 a text, apparently by Regiomontanus, was published which claimed to measure the parallax of the comet which had appeared in 1472, and to confirm that it was close to the Earth because the parallax was a whopping 6 degrees – placing it much closer than the moon, which has a diurnal parallax of about 1 degree. Some cunning detective work has shown that this text is not by Regiomontanus: it must have been found among his papers when he died, and was presumably in his handwriting, so that it was published as being by him, but it does not employ his methods and it had in fact already been published during Regiomontanus's lifetime by someone else, an anonymous physician of Zurich (tentatively identified as Eberhard Schleusinger). We now know this, but no one in the sixteenth century realized it, and it has caused a great deal of confusion in the historical literature.[56] Sixteenth-century astronomers accepted in good faith the apparently solid evidence that Regiomontanus had confirmed the traditional account of the distance of comets from the Earth; we now know that there is no reason to think that Regiomontanus had actually applied the system of measurement he had described in 1471; in order to apply it, one would, in any case, first have to work out how to handle the fact that comets are moving, not stationary, objects. In 1532, however, Johannes Vögelin measured the parallax of the comet that appeared in that year and claimed to confirm the erroneous result of pseudo-Regiomontanus.

Then in 1572 Brahe's nova appeared in the sky. For a time it was the brightest object in the heavens, other than the sun and the moon, brighter even than Venus. Such events occur only once in a thousand years or so. And, unlike a comet, the new star stood still, which made it much easier to measure its parallax. All over Europe astronomers were obsessed with it – and since they now knew Regiomontanus's real technique for measuring parallax, they naturally tried to apply it. Some found a measurable parallax, but others insisted that there was no parallax to measure. Accurately measuring parallax was far from easy, particularly as it required a more exact measure of time than any sixteenth-century clock could provide, but showing that there was no measurable parallax was much more straightforward. All one had to do was hold up a taut thread as a sighting device and find two stars that were exactly in line with the nova but north and/or south of it; if the same stars were exactly in line with the nova later the same night, then there was no parallax to measure. This simple technique was employed by Michael Maestlin, Kepler's teacher.[57] And if there was no parallax, then the comet must be a vast distance away, far further than the moon, whose parallax was quite easy to measure; it must be a supralunary, not a sublunary body.

How to explain the appearance of a new star in the heavens? Since there could be no natural explanation, assuming the star was indeed in the heavens, the event was clearly a miracle, a sign sent by God. The finest astronomers and astrologers – Thomas Digges in England, Francesco Maurolico in Italy, Tadeàš Hàjek in Prague – racked their brains in an attempt to work out what this might portend and hastened to publish their conflicting conclusions.[58]

The new star of 1572 was followed by the comet of 1577, and here again parallax measurement placed the comet beyond the moon. Where a nova could possibly be regarded as a miracle, a comet was too commonplace an occurrence to be handled in this way, so if comets were supralunary phenomena, Aristotle was wrong.[59] Brahe also worked on a further problem that could be solved by measuring parallax: a crucial difference between the Ptolemaic system on the one hand and the Copernican and Tychonic systems on the other was that under these modern systems Mars must at times approach much closer to the Earth than under the Ptolemaic system. Brahe at first thought that he had obtained a reliable figure for the parallax of Mars which proved the Ptolemaic system was mistaken, although he later realized that there were problems with it. Regiomontanus's procedure for measuring parallax ideally involved comparing the apparent position of a celestial object soon after dark with its apparent position not long before dawn, thus maximizing the parallax to be measured. Neither the nova of 1572 nor the comet of 1577 set in the night sky as viewed from northern Europe, so the ideal procedure was inapplicable; in the case of Mars, there was no choice but to make measurements when the planet was nearly in line with the sun, and thus it never rose high above the horizon at night. In measuring the location of an object near the horizon, Brahe had to allow for the refraction caused by the greater thickness of atmosphere through which its rays had passed, and eventually he found that he had miscalculated this allowance, thus vitiating what he had hoped would be a key argument against the Ptolemaic system. His long series of measurements of the location of Mars, however, was to prove invaluable to Kepler when it came to calculating Mars's 'orbit' (as he called it; he invented the term as used in astronomy) on Copernican assumptions, and demonstrating that it was best understood as an ellipse.[60]

In 1588 Brahe published *Concerning the Recent Phenomena of the Aetherial World*, Book II (Book I, on the nova of 1572, was published posthumously in 1602), a definitive study of the comet of 1577, in which he reviewed the extensive literature it had provoked and argued that those observations which found no parallax were the only reliable ones, and that Aristotle was therefore wrong when he claimed that comets were sublunary phenomena.[61] But he went further: in place of the Ptolemaic and Copernican systems, he proposed his own geoheliocentric system, which was geometrically equivalent to Copernicanism but had a moving sun and a stationary Earth. Since his calculations implied that comets moved through the crystalline spheres of the planets, and since his geoheliocentric system required Mars to cut through the sphere of the sun, Brahe abandoned the whole theory of solid spheres, and argued that the sun, moon and planets floated freely in the heavens, like fish in the sea. Brahe's caution about committing himself to this, the dissolution of the celestial spheres, is probably what had caused the delay in publication.[xiv] It is now generally regarded as a much more important marker of the beginning of modern astronomy than the publication of Copernicus's *On the Revolutions*.[62]

§ 7

This story is a fine example of two fundamental features of the Scientific Revolution. First, path dependency. Once Regiomontanus's true system for measuring parallax had been published, astronomers embarked on a path that could only lead, sooner or later, to decisive evidence being produced that was at odds with central claims made by Aristotle and Ptolemy (although Regiomontanus would have been astonished to learn this). The fact that there was a long delay does not mean that Regiomontanus's contribution was not decisive; it means only, first, that there was a delay in the publication of his work and, second, that the nova of 1572 simplified and clarified the issues, producing a classic revolutionary crisis. Certain features of the Ptolemaic system (such as geocentrism) could survive this shock, as Brahe's geoheliocentric system demonstrated, but key features common to both the Ptolemaic and Copernican systems (unchanging heavens, solid spheres) could not. By 1650 this was universally acknowledged; indeed, no competent astronomer defended the Ptolemaic system as understood (for example) by Regiomontanus after Galileo's discovery of the phases of Venus had been corroborated in 1611.[63]

This claim – that new observations were fatal to old theories – is at odds with much recent philosophy of science, which insists that observations and theories are both malleable, and that, consequently, there are always ways in which the phenomena can be saved. A standard approach is to distinguish between data (raw observations, made, for example, with a thermometer in boiling water) and phenomena (interpretations of the data, for example, that the boiling point of water at sea level is 100 degrees Celsius). Theories, it is said, explain phenomena, not data, and it is always possible to open up a gap between data and phenomena as well as between phenomena and theories.[64] But in the case of

xiv A crucial role was evidently played by Christoph Rothmann's *Discourse on the Comet* (1585), which directly attacked the doctrine of the spheres: Granada, Mosley & others, *Christoph Rothmann's Discourse* (2014).

the geometrical sciences of the seventeenth century the gaps between data and phenomena, and between phenomena and theories, are intended to be virtually non-existent.

In the case of Brahe's observations of the nova and the 1577 comet the data was an absence of diurnal parallax, the phenomenon that needed to be explained was that these new bodies were in the supralunary and not the sublunary world, and the immediate theoretical conclusion that resulted was that there was change in the heavens. What tied data, phenomenon and theory together was a geometrical argument (that if there was no observable parallax, the new bodies must be much further away than the moon) which was unbreakable, providing the initial observations were reliable. This was not true in all cases where diurnal parallax was observed; as we have seen, refraction might make it possible to open up a gap between data and phenomena, and even if Brahe's measurements of Mars's parallax had been correct, they would not have helped decide between his cosmology and that of Copernicus. But in the cases of the nova of 1572 and the comet of 1577 the data necessitated the phenomenon, and the phenomenon falsified the established theory.

Obviously, if Brahe was to confirm the claim that his argument was unbreakable, he had to provide an explanation for the fact that not everyone had produced observations which showed a complete absence of identifiable parallax. In *Recent Phenomena II*, consequently, Brahe went carefully through the observations of those whose results differed from his but (conveniently) corresponded to the results that traditional astronomy would have predicted, and identified their errors: one astronomer had measured the distance between the comet and a star but had then confused that star with another when repeating the measurement; another had added when he should have subtracted; a third had made two measurements an hour apart when he should have made them as close together as possible; and a fourth had confused two different systems of heavenly coordinates. Thus Brahe identifies elementary mistakes which neatly explain why their results are different from his; the observations are not, he wants to insist, subjective or personal, but objective and reliable, and once they are granted the rest follows of necessity.

Of course, the mere existence of a diversity of results made it difficult to persuade everyone that Brahe's arguments were conclusive. Galileo was still going over the measurements of parallax for the nova of 1572 in his *Dialogue Concerning the Two Chief World Systems* of 1632. There he argued that you could not just pick the measurement that

Brahe's observatory: the curved scale is a quadrant for measuring elevations that is built into the wall; inside it is a *trompe l'oeil* section of Brahe's observatory, with a giant figure of Brahe himself. The image comes from the 1598 printing of his *Astronomiae instauratae mechanica*. The painting above the quadrant was done in 1587 by Hans Knieper, Hans van Steenwinckel the Elder and Tobias Gemperle, who were responsible, respectively, for the painting of the landscape at the top, the three pairs of arches representing the three areas of Uraniborg and the portrait of Brahe.

suited your purposes best (as opponents of Brahe continued to do); that the accuracy of instruments would vary and so there would never be uniformity in observations; that outlying results were almost certain to be mistaken; and that results were likely to cluster around the correct measurement. Thus it might be impossible to decide if any particular one of the thirteen measurements he surveyed was correct, but one could identify a range within which the correct measurement almost certainly lay and be confident that all of the outlying measurements were mistaken.[65] Galileo was here making the distinction (to use the terminology of Bogen and Woodward) between data and phenomena, and using the distinction to develop the first theory of observational error.

Arguments over the location of novas and comets in the heavens thus continued even after 1610, after which date the traditional Ptolemaic system was abandoned by all competent participants. Within a year or two of Galileo's telescopic discoveries no one disputed that the moon had mountains, Jupiter had moons, Venus had phases and the sun had spots, and thus Galileo's observations were conclusive in a way that Brahe's measurements of diurnal parallax should have been, but weren't.[xv]

The second fundamental feature of the Scientific Revolution is the impact of the printing press. By the early sixteenth century the printing revolution was well under way. We have seen the impact the publication of Vesalius's *Fabric* had on anatomy. It was only printing that ensured that a significant number of astronomers had access to Regiomontanus's text on parallax after 1531. Printing made it possible for Brahe to

xv There is a wrinkle to this story. Galileo's first venture into astronomy was to insist that parallax measurements proved that the nova of 1604 was in the heavens, and this was still an argument of fundamental importance to him in 1632. But in 1618 (shortly after the Catholic Church had condemned Copernicanism) he rejected the claim that parallax measurements proved that comets were in the heavens: they might, he argued, simply be a reflection or refraction of sunlight, like a rainbow, in which case they would have no measurable parallax. The argument was dreadfully weak: it offered no explanation of why comets did not seem to move with the observer, as rainbows do, nor of why they were visible from all parts of the Earth and at all hours of the night. It amounted to a defence of Aristotle against the new astronomy. If Galileo really took it seriously (there is no indication that anyone else, apart from his pupil Castelli, did) then Brahe's attempt to tie indissolubly together data, phenomena and theory was misconceived. But it is unlikely that Galileo believed what he was saying. He was engaged in a polemic with the Jesuits, who had abandoned Ptolemy and adopted Brahe; so Galileo was happy to make any argument, however wild, which diminished Brahe's authority and might bring his readers to share his own fundamental belief (which, in the circumstances of the time, he dared not state) that there could be no coherent astronomical theory aside from Copernicanism. If he did take it seriously, then (as with his argument that the tides prove that Copernicus is correct: the Earth is moving) contemporaries were right to ignore him, and we should, too. Wootton, *Galileo* (2010), 157–70.

survey a wide range of publications (there were over a hundred on the comet of 1577, though many were merely astrological prognostications) and demonstrate that the four best observers had produced results compatible with his own.[66] It also ensured that Brahe's new system was quickly known throughout Europe, so that his arguments could be tested against the nova of 1604 and the comets of 1618. Printing created a community of astronomers working on common problems with common methods and reaching agreed solutions. This community had not existed in 1471 (which is another reason why it took so long for Regiomontanus's method of measurement to have an impact). When did it come into existence? Kepler, reasoning from astrology, dated the moment of transition to 1563: the great planetary conjunction of that year had transformed the world of learning; certainly it had led to a flood of astrological publications.[xvi] My preferred date would be 1564, the very next year, which saw the first published catalogue of the Frankfurt book fair. The Frankfurt catalogues were circulated throughout Europe, setting up for the first time a truly international trade in books.[67]

Before 1572 astronomers measured the positions of the sun, the moon and the planets (the sun and the moon were, technically, planets, too, according to the Ptolemaic system) in the heavens in order to predict their future movements. They had inherited some rough calculations of the size and distance of the sun, the moon and the stars, but distances did not really matter: all they sought to do was predict the angles that defined an object's position in the heavens at any particular time, and to do so by manipulating the Ptolemaic armament of deferents, epicycles and equants, which together amounted to what was called an hypothesis – a term which meant a mathematical model which produced reliable predictions. But with Tycho Brahe, measurements of distance suddenly became critical. Where before it had always been possible to 'save the phenomena', that is, adjust the hypothesis to fit the phenomena (if necessary by adopting two incompatible hypotheses, one to predict movement east and west, and the other movement north and south), Brahe's observations were simply incompatible with established

xvi Kepler: 'Every year, especially since 1563, the number of writings published in every field is greater than all those produced in the past thousand years. Through them there has today been created a new theology and a new jurisprudence; the Paracelsians have created medicine anew and the Copernicans have created astronomy anew. I really believe that at last the world is alive, indeed seething, and that the stimuli of these remarkable conjunctions did not act in vain.' *De stella nova* (1608), quoted from Jardine, *The Birth of History and Philosophy of Science* (1984), 277–8.

theories, whether Ptolemaic or Copernican (for Copernicus, it was assumed, had continued to believe in solid spheres carrying the planets through the heavens).[68] By 1588 astronomy had become concerned with the organization of the heavens in three dimensions, not just in two.

§ 8

Historians of science have often (and rightly) suggested that the key to the Scientific Revolution is 'the mathematization of nature'.[xvii][69] Aristotle and Ptolemy had assumed that the heavens were mathematically legible, and indeed Ptolemy had devised techniques for reading them. One aspect of the Scientific Revolution consists in the extension of mathematical theories to include sublunary phenomena. Where Aristotelian physics was preoccupied with qualities – the four elements (earth, air, fire, water) embody the four qualities (hot and cold, dry and wet) – the new physics was preoccupied with movements and quantities that could be measured, and it quickly led to attempts to measure the speed of falling bodies, the speed of sound and the weight of air. Where Aristotle had assumed that each element behaved differently, the new physics assumed that all heavy objects could be thought of as the same. Where Aristotelian physics had depended on all five of the senses, the new physics relied only on the sense of sight. With Galileo's discovery of the parabolic path of projectiles (1592) and the law of fall (1604) the sublunary world began to be mathematically legible, and Newton went on to show that the same physical principles were at work in the heavens and on Earth. But long before this the Aristotelian demarcation between superlunary and sublunary had been shattered by Brahe. From 1572 onwards Aristotelian philosophy was facing a crisis from which it could not escape without sacrificing fundamental claims that it had long regarded as unquestionable.

According to Aristotle, the sublunary elements were naturally at rest, while the supralunary spheres rotated in endless circles. Even before he discovered the law of fall, Galileo had been questioning the distinction between the two worlds. In the early manuscript *On Motion* (pre-1592) he suggested that if you were to slide a stone across a perfectly smooth surface it would continue for ever. He was thinking of circular

xvii I prefer the phrase 'the mathematization of the world' rather than 'of nature' because 'nature' seems to me to imply that biology is involved, rather than just physics.

movement – the stone would circumnavigate the earth – but he was also questioning the idea that rest was more natural than movement, and insisting on the legitimacy of theoretical abstraction, for of course perfectly smooth surfaces exist only in the mind.[70] His first discovery of a mathematical principle underlying sublunary movement was the identification of the parabola as the path of a projectile such as a cannon ball – of its path, that is, in a theoretical world in which there is no air resistance and in which cannon balls do not spin as they fly. After Galileo's death practical tests showed that the path of a real cannon ball is very different from Galileo's theoretical model; his student Torricelli was not in the least put out, any more than he would have been to be told that there are no perfectly smooth surfaces.[71] Galileo, Descartes and Newton constructed a new universe in which matter was inert and its behaviour (at least in theory) mathematically predictable, and in which movement and location were relative rather than absolute.

The traditional historiographical emphasis on the new physics assumes that the mathematization of the world began in the seventeenth century. But perspective painting provided the first glimpse of this new universe. It may be no coincidence that Galileo learnt his mathematics from Ostilio Ricci, who also taught perspective to artists, or that Brahe had a remarkable *trompe l'oeil* painting on the wall of his observatory (above, p. 196). The mathematization of the sublunary world begins not with Galileo but with Alberti, not in the seventeenth century but in the fifteenth. Alberti's *On Painting* starts with a simple exposition of the principles of geometry, in which he defines points, lines and surfaces and continues with a basic exposition of optics, which was traditionally regarded as a branch of mathematics. He also wrote a more sophisticated textbook on geometry for the use of artists, *Elements of Painting*. From perspective painting, the new mathematical learning spread to cartography. Waldseemüller's introduction to his world map (1507) begins with elementary geometry for cartographers: they need to understand circles and axes in order to make sense of longitude and latitude, of poles and antipodes. There was nothing particularly new in this: Cicero had thought geography was a branch of geometry.[72] Sebastiano Serlio's popularization of Vitruvius (1537) begins with a book expounding the elementary principles of geometry, starting with points and lines, right angles and triangles. But the systematic application of geometry to real-world questions – to architecture, to optics, to cartography, to astronomy, to physics (for Galileo claimed to be able to demonstrate some of his laws of fall through geometrical arguments) – was incompatible, in anything but the short term, with the survival of Aristotelianism.

With geometry came abstraction. This is beautifully illustrated by a diagram Peter Apian drew for his *Cosmographic Book* (1524). It shows how the measurement of longitude and latitude depends upon reference to an imaginary grid. To simplify things, Apian treats this grid as if it were a flat surface, not a sphere. And he presents it in perspective, with two parallel lines converging towards a vanishing point. It is, in fact, just like the grid used by artists to establish the picture plane, and its representation requires the same techniques as the representation of a tiled floor. The great art historian Erwin Panofsky claimed that the tiled floors in perspective paintings were the first abstract system of coordinates; he was wrong, because Ptolemy had already invented longitude and latitude as a system of coordinates, but he was right to think that perspective painting implies an abstract system of coordinates.[73]

In the bottom-left-hand corner of Apian's diagram he has drawn some mountains, suggestive of a real place, perhaps even a reference to the Alps. But these merely serve to conceal the fact that cartography transforms place into space. This seems wrong, because we use maps to get from one place to another: surely maps are about places? But maps function by substituting symbols (here, pins stuck into an imaginary board) for real places and by locating places in abstract space. In Apian's diagram there is nothing to suggest that Venice is a port but Vienna is not; that Erfurt and Nuremberg are Protestant while Munich and Prague are Catholic; that these cities differ in size or belong to states. Real cities have been replaced by coordinates; real places by theoretical spaces.

Geometry also acquired new importance as a result of the invention of gunpowder: fortifications had now to be built to resist cannon balls, which fly (as seen from a bird's-eye view) in straight lines. In order to provide raking and flanking fire along every wall, forts needed to be designed on the page, with carefully measured angles and distances. Bastions (called by the French the *trace italienne* and by the American colonists the 'star fort'), were being built not just in Europe but in Asia and the New World – wherever cannons were fired – from the late fifteenth century on, so every commanding officer required some knowledge of geometry. The new science of fortification was taught by mathematicians, Galileo among them.[74] In Shakespeare's *Othello*, Iago is furious to find Michael Cassio, 'a great arithmetician', whose knowledge of warfare is all book-learning, has been promoted ahead of him.[75]

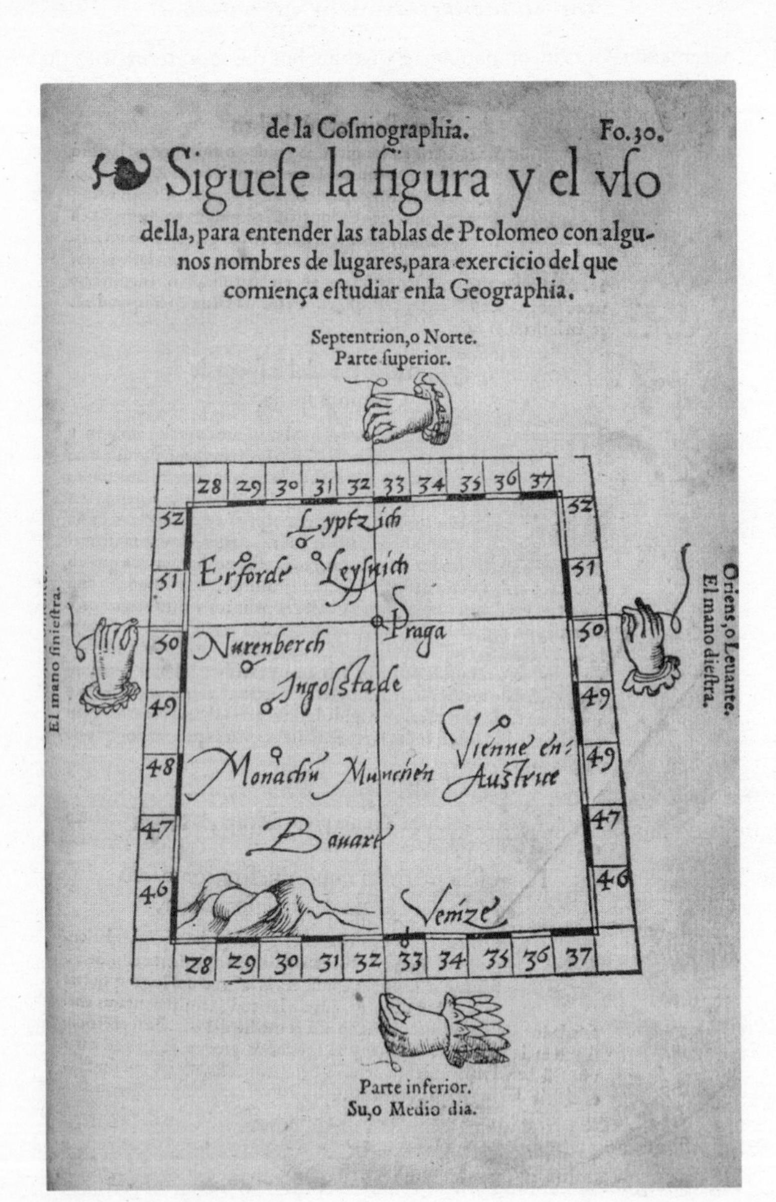

Peter Apian's diagram illustrating longitude and latitude, from his *Cosmographicus liber* (1524).

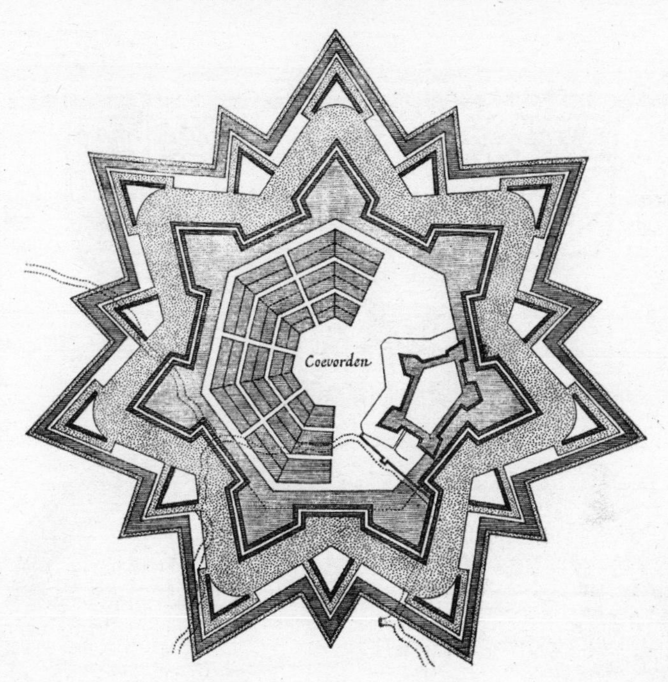

Fortification plan of Coevorden in the Netherlands, as laid out in the early seventeenth century by Maurice of Nassau, Prince of Orange. Simon Stevin advised Maurice of Nassau on the design of fortifications, and Descartes served in his army.

Alberti had written, 'Mathematicians measure the shapes and forms of things in the mind alone and divorced entirely from matter.'[76] But this divorce between mathematics and matter soon turned into a marriage. In Galileo's famous words, which provide the epigraph to this chapter, the book of the universe is written in geometrical figures. The claim went back to Pythagoras and Plato, but Renaissance Platonists had been interested in number mysticism rather than real mathematics. It is first boldly restated by Tartaglia in his *New Science* (1537), the new science being ballistics. The frontispiece to that book shows Euclid controlling the gate which leads not only to knowledge of ballistics but to all philosophy.[77]

Tartaglia went on to publish the first translation of Euclid into a modern vernacular language (Italian, 1543) and to devise new instruments and techniques for surveying (*Diverse Problems and Inventions*,

Frontispiece to Niccolò Tartaglia's *New Science* (1537). Euclid controls the gate to the castle of knowledge, where a mortar and a cannon are being fired to illustrate the path of projectiles. Entry to the inner redoubt requires one to pass through the mathematical sciences, with Tartaglia himself standing among them; within is Philosophy, accompanied by Aristotle and Plato.

1546), techniques that could be used to calculate how far away a target was. In 1622, for example, a fleet of Dutch ships tried to seize the Portuguese colony of Maçao. A Jesuit mathematician did the geometry calculations to determine the distance to a stockpile of gunpowder the Dutch had brought ashore and the angle of elevation at which the cannon should be set. A direct hit turned the tide of the battle and ensured that Maçao remained a Portuguese colony.[78] Thus if we ask how did the Scientific Revolution become mathematized, the answer is through perspective painting, cartography (and the related sciences of navigation and surveying) and ballistics. It was these subjects which gave mathematicians like Tartaglia, Brahe and Galileo (and, as we have seen, Leonardo, too) the confidence that it was they, not the philosophers, who knew how to make sense of the world. Painting, cartography and ballistics do not strike us as cutting-edge sciences but, once, they were.

The different mathematical sciences were interlocked: anyone with the skills to excel in one had the skills to excel in others. Alberti was an architect, a painter and a mathematician; Piero della Francesca was a mathematician and a painter; Pacioli was a mathematician and an architect; Leonardo was a painter and a military engineer; Digges published on surveying and astronomy; the greatest cartographers (Mercator, the Cassinis) were also important astronomers, just as the greatest astronomers (Brahe, Halley) were also cartographers. These were not independent sciences but a family group which held a set of geometrical techniques and measuring instruments in common. According to the standard translation of *On the Revolutions*, Copernicus wrote, 'Astronomy is written for astronomers'; but he didn't. He wrote, '*mathemata mathematicis scribuntur*' ('mathematics is written for mathematicians'): Copernicus expected every mathematician to be able to follow his arguments. He, like virtually all the others, worked in more than one field: he published on monetary reform as well as astronomy. As for Kepler, he published not only on optics but also on the mathematical analysis of the volumes of wine barrels (a subject directly related to a problem in his astronomy, that of calculating the area of an ellipse) and on how best to stack cannon balls.[xviii]

xviii The first scientists nearly all contributed to several disciplines (the only major exceptions are a few doctors, such as William Harvey). Gilbert, the physician and experimental scientist, had been a mathematical examiner when a fellow of St John's College, Cambridge, and his fundamental goal was to prove the truth of Copernicanism. Galileo contributed to physics and astronomy, while teaching fortification and optics. Stevin published on algebra, engineering, astronomy, navigation and accountancy. Huygens worked on the mathematics

Moreover, it was not only the painters who were concerned with how to represent three dimensions in two: this was also a central problem for the cartographers, who had to project the globe on to a flat surface (some even think that one of Ptolemy's solutions to this problem influenced Brunelleschi),[79] and for the sundial makers (always mathematicians; sometimes – Regiomontanus, Benedetti – of the first rank), who had to work out how the sun's movement through three dimensions would be projected on to a flat dial. One set of images captures this overlapping of interests better than any other. Albrecht Dürer made two trips to Italy (1494–5; 1505–7) to learn the latest artistic techniques. He published on geometry as applied to painting and architecture (*Four Books on Measurement*, 1525). In 1515 he published, in association with the astronomer and cartographer Johann Stabius, a pair of celestial charts showing the northern and southern hemispheres: these were the first printed star charts, and the first, printed or not printed, to show the heavens with a clearly marked system of coordinates. They were accompanied by the first perspective drawing of the entire Earth as a sphere. Here, geometry, perspective painting and cartography merge into one.

§ 9

The belief that the tools of the mathematician (specifically, the tools provided by geometry) were the correct ones to use to understand the world made possible all sorts of new representations. But did it significantly alter society's control over the natural world, or the control of

---

of the pendulum and the design of watches, but he also discovered the rings of Saturn. Mercator and the Cassinis (beginning with Giovanni Domenico Cassini) were both cartographers and astronomers. Boyle published on physics and chemistry. Newton practised alchemy as well as publishing on physics and optics. Daniel Bernoulli published on astronomy as well as probability. Even the relatively uneducated Leeuwenhoek, the first great microscopist, who had no Latin, qualified as a surveyor. Often, we have lost sight of their range of interests: Brahe and Halley are well-known as astronomers, but who remembers that they were also cartographers? Copernicus is now remembered only as an astronomer, but he was an expert on monetary reform who published a treatise, *On the Theory of Coinage* (1526), in which he formulated what we now call Gresham's Law, that bad money drives out good. The fundamentally interdisciplinary character of the new science continued to be true at least until Leonhard Euler (1707–83), who reconceptualized ballistics and planetary orbits, disagreed with Newton on optics and wrote at length about music. As they moved from discipline to discipline, the new scientists carried with them a set of assumptions about how to construct new knowledge. It is these assumptions that are the very core of the Scientific Revolution.

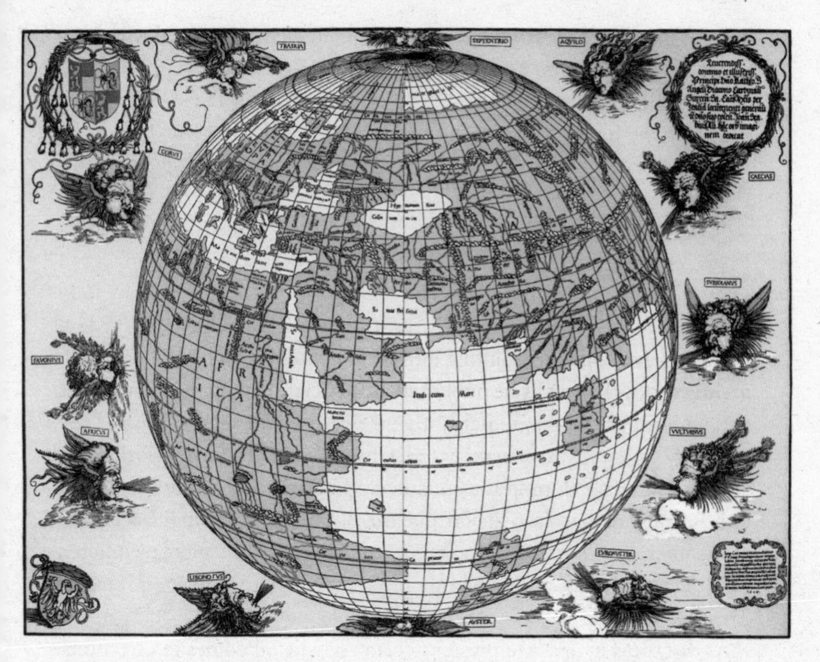

Dürer's World Map of 1515. It is part of a set including maps of the northern and southern skies. Dürer's map illustrates how rapidly the notion of the Earth as a globe established itself after the publication of Waldseemüller's map in 1507; it also demonstrates Dürer's complete mastery of perspective representation.

one social group over another? Vesalius's purpose was not only to enhance understanding but also to improve surgery. However, without anaesthetics, antibiotics or reliable methods of controlling blood loss through tourniquets and sutures (let alone transfusions), surgery continued to be painful, risky and often fatal. The knowledge acquired in dissection had few, if any, practical applications.[80]

It was very different, of course, with the sciences of cartography and navigation, ballistics and fortification. But it is important to distinguish between the first pair and the second: one deals with space and place; the other with percussive force. Once sailors began to sail for any length of time out of sight of land, they needed new tools (compasses, and devices such as the ship's astrolabe – a specialized version of the astrolabe for use at sea – or the backstaff for establishing true north by the sun and the stars), new charts and maps, and new supplies (hardtack or ship's biscuit). There is a tendency in the modern literature to think of maps as technologies for domination, as reflections of imperial culture.[81] This seems to me mistaken, even though John Donne compared mapping the heavens to owning them:

> For of Meridians, and Parallels,
> Man hath weaved out a net, and this net throwne
> Upon the Heavens, and now they are his owne.
> Loth to goe up the hill, or labour thus
> To goe to heaven, we make heaven come to us.[82]

European world maps placed Europe at the centre, but when the Chinese were shown a world map by Matteo Ricci and complained that China should be at the centre, he promptly produced a new map which did just that.[xix] Mercator's projection (1599), when used to produce a world map, shrinks countries close to the equator and makes the northern countries appear much larger than they really are, but this is an entirely accidental consequence of constructing a projection which enables a course plotted on a chart to be used directly for navigational purposes; the Mercator projection, which shows a three-dimensional globe on a flat surface, distorts distances in order to preserve accuracy when it comes to directions. These maps were originally intended as tools for sailors, not assertions of European supremacy; they look like ideological statements only to people who do not use them for navigation.

Moreover, until the eighteenth century, cartographers were mainly interested in producing maps for the purposes of navigation. What generals wanted were not accurate maps but sketch maps that showed the roads along which troops and supplies should move.[83] Such maps focused on roads, passes and fords, ignoring everything to left or right of the main route. They did not show abstract space (which is what the open ocean is) but real place. Commanders wanted plans of fortifications, and bird's-eye views (of a sort pioneered by Leonardo) which enabled them to identify where guns could be brought forward without being exposed to enemy fire, or where an advance could be blocked or ambushed. Thus, visualizing the application of power on land required quite different techniques from at sea, and for a long time cartography

xix For a wonderfully confused discussion of this issue, see Mignolo, *The Darker Side of the Renaissance* (2010), 219–26, which attacks Ricci for behaving 'as if geometry were the warranty of a non-ethnic and neutral ordering of the shape of the earth'. This complaint makes sense only if one assumes (as Mignolo does) that there is no such thing as objective knowledge, that all knowledges are ethnic and partisan.

enhanced maritime power, not terrestrial power (which is why the Dutch, who were almost exclusively dependent on maritime power, were so preoccupied with cartography).

This brings us back to the harsh truth, that, although, in themselves, maps, compasses, backstaffs and hardtack may be neutral devices, by making it possible for ships to sail the oceans they enabled Europeans to bring gunpowder technology (whether in the form of cannon fired from floating fortresses or landing parties armed with muskets) to bear upon societies which had no comparable means of defending themselves.[84] Cartography came as a technological package, along with gunpowder weaponry; and gunpowder weaponry really is about power, and nothing else. So while cartography and navigational instruments taken in isolation are neutral devices, in practice they are part of a package of technologies that ensured Western global dominance for five hundred years.

§ 10

My argument so far is that the seventeenth-century mathematization of the world was long in preparation. Perspective painting, ballistics and fortification, cartography and navigation prepared the ground for Galileo, Descartes and Newton. The new metaphysics of the seventeenth century, which treated space as abstract and infinite, and location and movement as relative, was grounded in the new mathematical sciences of the fifteenth and sixteenth centuries, and if we want to trace the beginnings of the Scientific Revolution we will need to go back to the fourteenth and fifteenth centuries, to double-entry bookkeeping, to Alberti and Regiomontanus. The Scientific Revolution was, first and foremost, a revolt by the mathematicians against the authority of the philosophers. The philosophers controlled the university curriculum (as a university teacher, Galileo never taught anything but Ptolemaic astronomy), but the mathematicians had the patronage of princes and merchants, of soldiers and sailors.[85] They won that patronage because they offered new applications of mathematics to the world. That involved the invention of many new instruments for improved measurements on Earth and in the heavens – cross-staffs, sextants, quadrants – and it was driven by a new obsession with accuracy. Accuracy and certainty: these were the watchwords of the new science.

Regiomontanus may have been one of the first, but he was not the last, to see in the mathematical sciences a new type of reliable knowledge. In 1630 Thomas Hobbes, who had received a conventional humanistic and scholastic education at Oxford, came across a copy of Euclid's *Elements* 'in a gentleman's library' in Geneva. It was open at Proposition 47 of Book I (which we now call Pythagoras's theorem). 'From that moment on he was in love with geometry.'[86] He soon aspired to construct a new science of morality and politics on geometrical principles. What Hobbes had realized is that nothing is more certain than the truths of mathematics. Two plus two always equals four; the square on the hypotenuse is always equal to the sum of the squares of the other two sides. These are universal truths: to understand them is to adopt them.[xx] For two centuries or so, from Regiomontanus (d.1476) to Hobbes (d.1679), Euclid and Archimedes provided the crucial examples of how to construct a new sort of knowledge, the sole defence-works against the sort of doubt so eloquently expressed by Sextus Empiricus and Montaigne.[87] But if the Revolution begun by the mathematicians was to be successful it needed to identify other ways of establishing and communicating universal truths. It is to these that we now turn.

xx See 'Notes on Relativism and Relativists', No. 6 (pp.586–7).

6
Gulliver's Worlds

> But the most hateful sight of all, was the lice crawling on their clothes. I could see distinctly the limbs of these vermin with my naked eye, much better than those of a European louse through a microscope, and their snouts with which they rooted like swine. They were the first I had ever beheld, and I should have been curious enough to dissect one of them, if I had had proper instruments, which I unluckily left behind me in the ship, although, indeed, the sight was so nauseous, that it perfectly turned my stomach.
>
> – Jonathan Swift, 'A Voyage to Brobdingnag',
> *Gulliver's Travels* (1726)

§ 1

One day towards the beginning of 1610 Johannes Kepler was walking across a bridge in Prague when a few snowflakes settled upon his coat.[1] He was feeling guilty, because he had failed to give his friend Mathias Wacker a New Year's present. He had given him *nichts*, nothing. On his coat the snowflakes melted and turned into nothing. Watching them, Kepler evidently grasped two things more or less simultaneously. Each snowflake was unique, but they were all alike in that they were all six-cornered. This got Kepler thinking about two-dimensional six-cornered shapes and how they form a lattice: the cells of a honeycomb, or the seeds of a pomegranate. And about how the only shapes that one can use to tile a floor, if all the tiles are the same, are triangles, squares and hexagons. And about the patterns you can make if you pile cannon balls. Kepler thought he could work out the most space-saving way of piling spheres: his claim has become known as the Kepler conjecture (that the best arrangements are ones in which the centres of the spheres in each layer are above the centres of the spaces between the balls in the layer below) and was finally proved true for any regular lattice in 1831, and for any possible arrangement of spheres in 1998. For Kepler, this was applied mathematics: Thomas Harriot had been asked by Sir Walter Raleigh in 1591 how cannon balls should be piled on the decks of ships in order to get as many on board as possible, and Harriot had passed the problem on to Kepler.

Kepler was the first person we know of to imagine that snowflakes might be worth close inspection, and the little pamphlet he wrote about them (*On the Six-cornered Snowflake*, 1611) is now hailed as the founding text of crystallography. But he wrote it because he had also thought of a pun that he could not resist making. The Latin for snowflake is *nix*, almost the same as the German for nothing. If you give someone a snowflake, you are giving them nothing, for soon it will melt; he could give his friend a little book about snowflakes and it would be both something and nothing. He would no longer have to feel embarrassed about having given him nothing; now he could take pride in it.

Like Galileo, Kepler believed that the book of nature is written in the language of geometry. In his first major work, *The Cosmographic Mystery* (1596), he had argued that the spacing between the planets in the Copernican system was the spacing you would get if the five Platonic solids had been nested within each other in a certain order (working from the inside out: octahedron, icosahedron, dodecahedron, tetrahedron, cube). If God was a mathematician (and who could doubt it), then one must expect to find a mathematical logic in the most unexpected places, for example in the organization of the solar system or in a snowflake.

Kepler was thus conceptually prepared to find a mathematical order in the snowflake. But he was surprised to find himself looking for it there, of all places, and to find the same order at work in the great and the little. Indeed, he found himself considering the possibility that diamonds and snowflakes are formed by the same shaping agency, which could be neither cold nor vapour, but must be the Earth itself:

> But I am getting carried away foolishly, and in attempting to give a gift of almost Nothing, I almost made Nothing of it all. For from this almost Nothing I have very nearly recreated the entire universe, which contains everything! And having before shied away from discussing the tiny soul of the most diminutive animal [the chigger], am I now to present the soul of that thrice greatest animal, the orb of the earth, in a tiny atom of snow?[2]

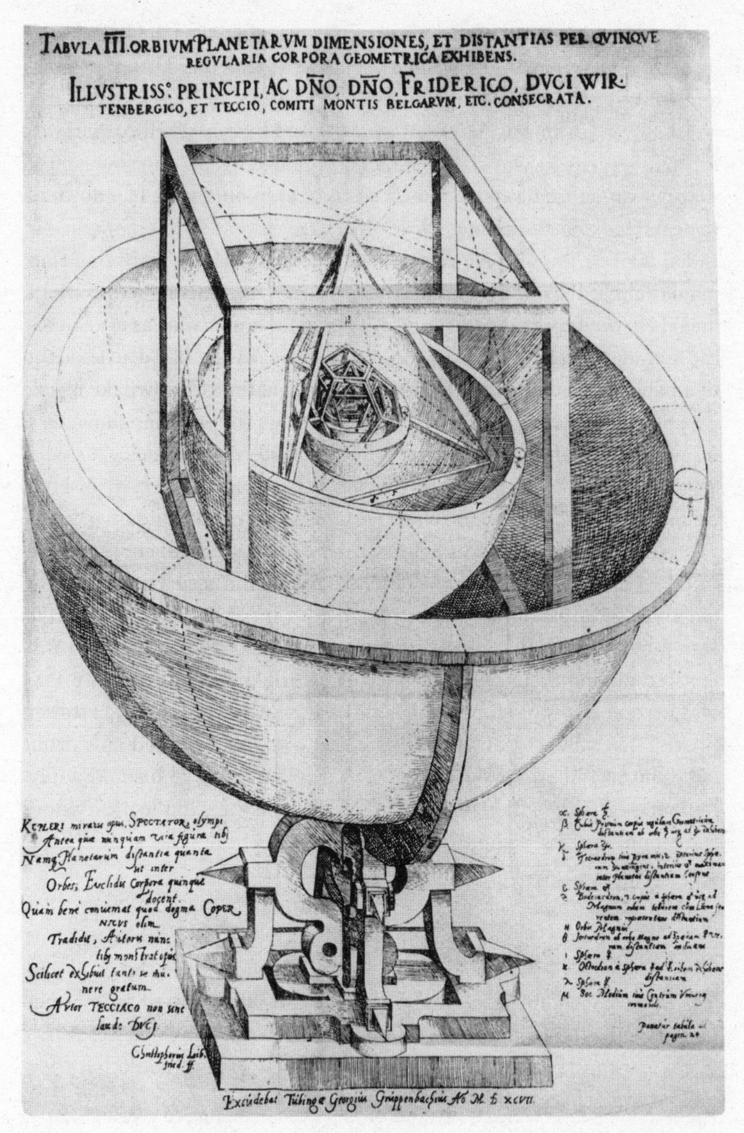

Kepler's representation of the five Platonic solids (cube, dodecahedron, icosahedron, octahedron and tetrahedron) nested within each other, from the *Mysterium cosmographicum* (1596). Kepler argued that the size of the planetary orbs corresponded to the size of an orb which would just fit within each of the solids if the solids were nested inside each other in the right order. He saw this as proof of God's pleasure in mathematical symmetry showing in his design of the universe.

Kepler enjoys his joke about nothing. He even imagines a local doctor dissecting the chigger, the smallest creature visible to the human eye – and thus of course impossible to dissect.[3]

A couple of months later, on 15 March, Kepler's world was transformed. His friend Wacker came racing round in his carriage, so excited that he was shouting out his news without even bothering to get out of the carriage and go indoors. Word had arrived that in Venice someone called Galileo, using some sort of new instrument, had discovered four planets circling a distant star. Bruno was right – the universe was infinite and there were other earths; and Kepler, who had always insisted that the sun and the Earth were unique, was plain wrong. Kepler describes them shouting at each other and laughing, Wacker delighting in his triumph and Kepler laughing off being in the wrong, and laughing, too, with delight at the thought of such an extraordinary discovery.[4]

Galileo's book (dedicated to Cosimo II de' Medici, the ruler of Florence; Galileo would soon move from Venice to Florence) had been published on 13 March; on 8 April a copy reached Prague in the diplomatic post and was presented by the Florentine ambassador to the emperor, who passed it straight on to Kepler.[5] It turned out that the rumour Wacker had picked up was wrong.[i] In fact, Galileo had discovered moons circling Jupiter, not planets circling a distant star. Bruno was not necessarily right after all, although the new discovery certainly proved that Copernicus had been entitled to claim that the Earth could be a planet and at the same time have a moon going around it, which had seemed deeply implausible to the defenders of Ptolemy (for whom the moon was one of the planets) and of Brahe.

§ 2

The story of Galileo's discoveries is, it seems, straightforward. In 1608 the telescope was invented in the Netherlands. It was a chance discovery made, perhaps, by Hans Lippershey, a spectacle maker (two other spectacle makers disputed Lippeshey's priority claim). In 1609 Galileo, who

i In the seventeenth century it took about three weeks for news to get from northern Italy to Prague, so Wacker's news must have come from someone who had learnt about the book as it went through the press, although Galileo had certainly tried to keep his discovery under wraps for as long as possible – Wacker's source was presumably in Florence, where reports of Galileo's discovery began to spread when he asked for permission to dedicate his book to Cosimo de' Medici and to name the new stars after him.

had never seen a telescope, worked out how to make one.[6] It had an obvious application in warfare, both on land and sea, and so he persuaded the Venetian government to reward him for his invention. They were somewhat irritated to discover within a few days that telescopes were becoming widely available and that Galileo had taken them for a ride. Galileo's first telescope had a usable magnification of 8x; by the beginning of 1610 he had managed to produce one that had a magnification of 30x and he had begun to explore the heavens.[7]

There is a standard phrase that is used over and over again in the literature: 'Galileo turned his telescope to the heavens.' Of course he did, in the autumn of 1609. Harriot did the same thing in England four months before Galileo (his first telescope had a magnification of 6x).[8] The puzzle lies in the enormous effort that Galileo put into improving his telescope, grinding on his own equipment two hundred lenses in order to end up with ten telescopes with a magnification of 20x or better. For what is strange about these ten telescopes is that they were too good for their obvious, military, use. Their field of view was tiny – Galileo could see only part of the moon at a time. Held with two hands, they shook and wobbled so anything you were looking at kept slipping out of the field of view: some sort of tripod or mount was essential.

How do we know that Galileo's telescope was too good for naval and military use? If you are looking for ships at sea, the curvature of the globe means that the limit of how far you can see is determined by the horizon. From a height of 24 feet the horizon is only 6 miles away: the maximum distance from which a lookout on one galley could see another is about 12 miles. The practical range of cannon fire was about a mile, so in a battle on land that was the crucial range for improving vision. In 1636, towards the end of his life, Galileo entered into negotiations with the Dutch. He had a cherished scheme for working out the longitude of a ship by using the moons of Jupiter as a clock (a reliable sea-going chronometer was not invented until 1761). At that time there was not a single telescope in the whole of the Netherlands capable of 20x magnification – and yet the Dutch had plenty of fine telescopes perfectly adequate for military and naval use.[9] If telescopes with a magnification of 20x had had a practical application, they would have had them.[ii] It is clear then that Galileo had turned his telescope

ii Between 1622 and 1635 Isaac Beeckman, in the Netherlands, struggled to produce a telescope comparable in quality to the ones Galileo had in 1610. It seems he failed. Berkel, *Isaac Beeckman* (2013), 68–9.

into an instrument that was good for only one purpose – looking at the heavens. He had turned it into a scientific instrument. Others, including Harriot, raced to catch up with him.

It is important here to distinguish between the impact of the telescope and that of the microscope. The two are basically the same thing, so as soon as Galileo had a telescope he could use it to study flies, for example. He later devised a better, table-top instrument and studied how flies could climb up glass. But the first publication to represent what could be seen through a microscope, a single broadsheet entitled 'The Apiarium' (about bees, in honour of Pope Urban VIII, the symbol of whose family, the Barberini, was the bee) did not appear until 1625, and the first major publication was Hooke's *Micrographia* of 1665.[10] The telescope, on the other hand, transformed astronomy almost overnight, while the microscope was slow to be adopted (and, towards the end of the century, quick to be abandoned).[11] The reason for this is simple: there was an established body of astronomical theory, and what was seen with the telescope was at odds with it. Astronomers could scarcely dispute the relevance of the telescope to their studies. But the microscope brought into vision a world previously unknown; it was hard to establish how the new information it produced related to established knowledge. The telescope addressed directly issues that were already under discussion; the microscope opened up new lines of enquiry whose relevance to current concerns was not obvious. That the telescope flourished and the microscope languished is one of the signs that the Scientific Revolution can properly be understood as a revolution – that is, a revolt against a previous order. Both telescope and microscope produced new knowledge, but in the seventeenth century only the telescope directly endangered the existing order.

In 1609, however, it was far from obvious that the telescope was going to transform astronomy: if it had been, there would have been large numbers of astronomers trying to make high-powered telescopes (as there were as soon as Galileo published his discoveries). Why did Galileo take it seriously as a scientific instrument? It would seem evident that he thought there was something that he would be able to see if his telescope was powerful enough. What? There is only one possible answer to this question: he was looking for mountains on the moon. Orthodox teaching was that the moon, being a heavenly body, was a perfectly smooth, round sphere. The variations in its colouring, however they were to be explained, were certainly not due to any surface irregularity. But Galileo was familiar with Plutarch, who had claimed that the moon had a landscape of mountains and valleys.[12] Kepler was so taken with this idea that in 1609, as part of his exchanges with Wacker, he had begun to write a story – the first work of science fiction – about a voyage to the moon (it was eventually published posthumously, in 1634).[13] From the moon, he argued, one would have the illusion that the moon was stationary and that the Earth floated through the sky. Kepler was not alone in imagining a moon with features like the earth. In 1604 someone close to Galileo (perhaps Galileo himself) had published an anonymous tract in Florence which claimed that there were mountains on the moon:

> There are also on the moon mountains of gigantic size, just as on earth; or rather, much greater, since they are [even] sensible to us. For from these, and from nothing else, there arise in the moon scabby little darknesses, because greatly curved mountains (as Perspectivists teach) cannot receive and reflect the light of the sun as does the rest of the moon, flat and smooth.[14]

When Galileo pointed his improved telescope at the moon in 1609 he was able to pick out something much more striking and unambiguous than 'scabby little darknesses' (which were presumably what we now call craters). He was able to show that along the terminator – the boundary between the light and the dark parts of the moon – which would be a smooth, unbroken line if the moon was a perfect sphere, one could see dark marks where there should be light, and light patches where there should be dark. These, he argued, were shadows and highlights such as you would find on a mountain range as it caught the rising sun. He had confirmed Plutarch's theory and, whether he liked it or not, he had reopened the question of the existence of other habitable worlds.[15] As John Donne put it in 1624 (with perhaps a backward look to Nicholas of Cusa, or Bruno):[iii]

> Men that inhere upon *Nature* only, are so far from thinking, that there is anything *singular* in this world, as that they will scarce thinke, that the world it selfe is *singular*, but that every *Planet*, and every *Starre*, is another *World* like this. They find reason to conceive, not only a *pluralitie* in every *Species* in the world, but a *pluralitie of worlds*.[16]

In *The Starry Messenger* (1610) Galileo acknowledged no debts, except to Copernicus; Plutarch, Nicholas of Cusa, Bruno and della Porta do not rate a mention, which seemed unfair to Kepler (who

iii See Chapter 4 above.

evidently thought some of his own contributions to the field were, for that matter, neither irrelevant nor insignificant).[17] Telescopic astronomy was presented as a brand-new beginning – which indeed it was.

It so happens that Harriot had already seen exactly what Galileo had seen. We have a sketch he drew on 26 July 1609. Looking at it, it is perfectly clear that the terminator is irregular, but this irregularity is what information scientists call 'noise': it makes no sense and conveys no information. We have another sketch by Harriot, dated 17 July 1610.[18] This difference this time is that Harriot had now read Galileo's *Starry Messenger*, which had been published in the spring. Now what he saw was exactly what Galileo had seen. Indeed, it seems clear that what he was doing was comparing Galileo's illustration with what he could see through his telescope, for both Galileo's illustration and Harriot's feature a large circular boss. In fact, there is no such prominent object on the moon, and scholars have suggested that Galileo deliberately enlarged a crater to enable the viewer to, as it were, zoom in on a tell-tale feature.[19] Harriot, looking at the moon, saw the irregular terminator, the highlights and shadows, the mountain ranges and valleys that Galileo had described – and he also convinced himself that he saw Galileo's imaginary crater. Once Galileo had described what he had seen, once he had trained viewers in how to look, it was almost impossible to dispute that the moon had mountains and valleys; but Galileo could only understand what he was looking at because he had a better telescope than Harriot, and because he (unlike Harriot) was accustomed to looking at perspective paintings. The anonymous author of 1604 had been quite right to insist that the theory of perspective would provide the key to interpreting the image of the moon.

Having observed the moon, Galileo turned his telescope to Jupiter and discovered that Jupiter had moons. According to conventional Ptolemaic astronomy, all heavenly bodies revolved around the Earth; a difficulty with Copernicanism was that it not only put the Earth in motion and the sun at the centre of the universe, but it required the moon to revolve around the Earth at the same time as the Earth revolved around the sun. Jupiter's moons made this arrangement rather less implausible than it had seemed. Galileo now rushed to publish his discoveries, which transformed astronomy in the space of a few months – the time it took for others to acquire telescopes with which they could corroborate his findings.

But there is something more to this story than at first meets the eye. Galileo had not only made a remarkable discovery; using his telescope, he had seen something where, before, there was apparently nothing at

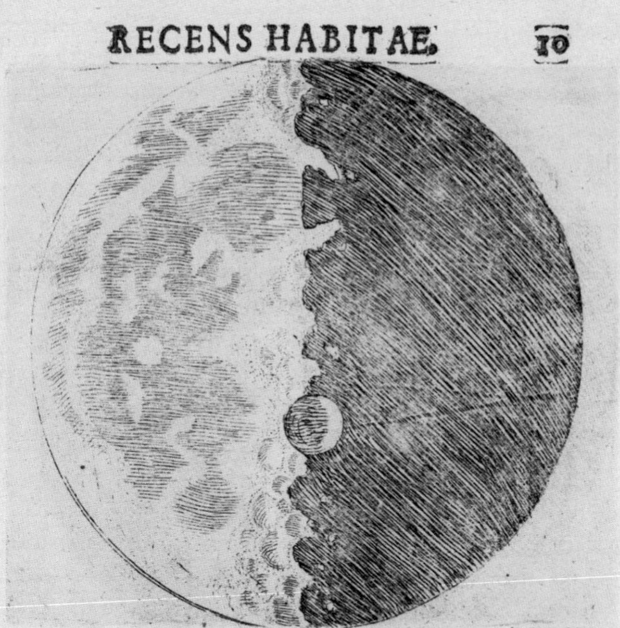

RECENS HABITAE. 10

Hæc eadem macula ante ſecundam quadraturam nigrioribus quibuſdam terminis circumuallata conſpicitur; qui tanquam altiſſima montium iuga ex parte Soli auerſa obſcuriores apparent, quà verò Solem reſpiciunt lucidiores extant; cuius oppoſitum in cauitatibus accidit, quarum pars Soli auerſa ſplendens apparet, obſcura verò, ac vmbroſa, quæ ex parte Solis ſita eſt. Imminuta deinde luminoſa ſuperficie, cum primum tota fermè dicta macula tenebris eſt obducta, clariora mõtium dorſa eminenter tenebras ſcandunt. Hanc duplicem apparentiam ſequentes figuræ commoſtrant.

One of Galileo's illustrations of the moon, from *The Starry Messenger* (1610). Galileo's purpose is to show how the terminator (the line between the light and the dark sides of the moon) is not smooth but jagged, proof that the moon is not a perfect sphere. On either side of the terminator one can see shadows (on the light side) and highlights (on the dark side), just as when the sun rises or sets over a mountain range the peaks are illuminated before the valleys.

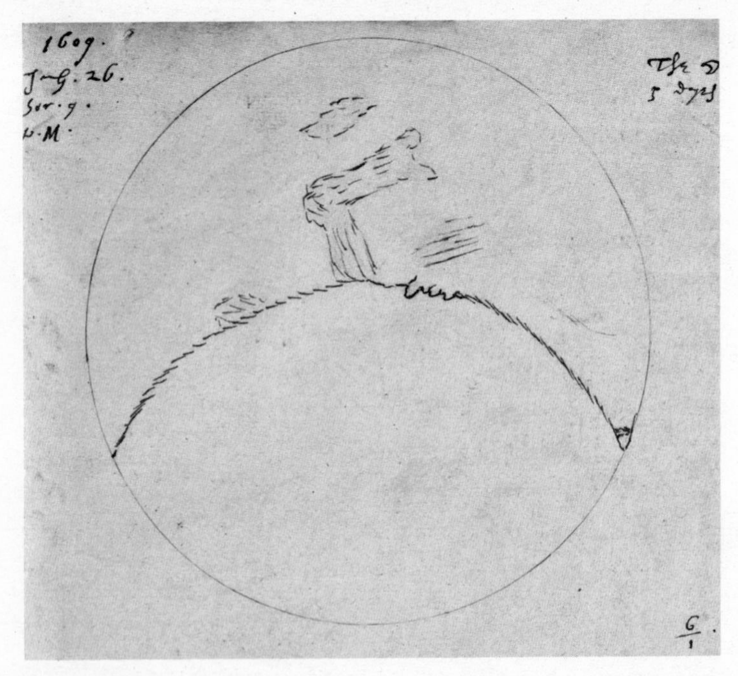

Harriot's first drawing of the moon as seen through his telescope, before he had read Galileo: Harriot has not grasped that the pattern of light and dark can be interpreted as showing that the moon has mountains and valleys, so that it is for him meaningless.

all to see. In the winter of 1609–10 he had transformed what seemed to be nothing into something. The idea that out of almost nothing you could re-create the entire universe was plainly ridiculous, yet that was what Galileo was now doing. The idea that one might dissect a chigger was also ridiculous in 1610 – but, thanks to the microscope, that too would soon be possible.

Nobody, not even Galileo, was better prepared for this new world where nothings became somethings than Kepler. He rapidly wrote a letter to Galileo (which was soon published in Prague, in Florence and in Frankfurt as the *Dissertatio cum Nuncio sidereo*) praising Galileo's discoveries, even though others suspected Galileo of telling lies, and Kepler had not yet confirmed the discoveries with his own eyes. Perhaps, he said, if, as Galileo claimed, there were mountains on the moon, Bruno had been partly right – maybe the moon was inhabited, and life was not confined to the Earth. Kepler tried making his own telescope, but it

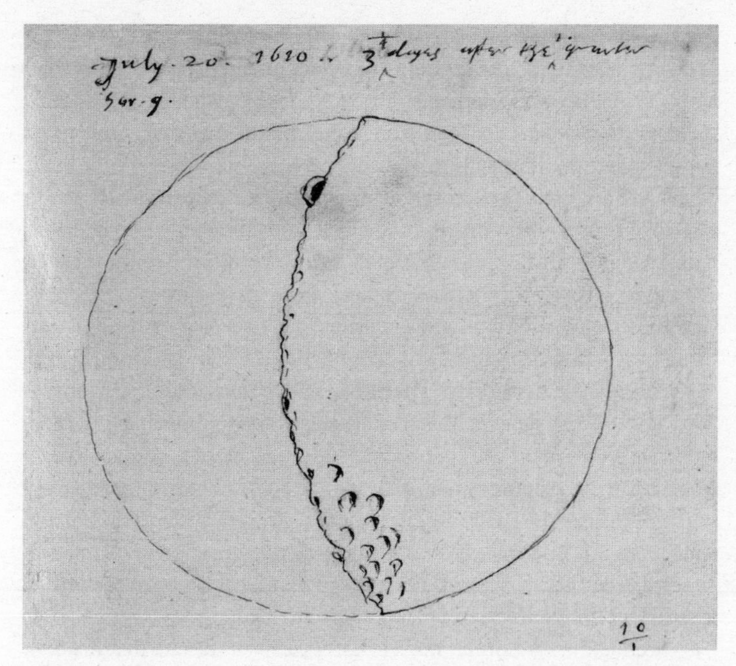

Harriot's drawing of the moon after he had read Galileo's *Starry Messenger*: under the influence of Galileo, Harriot draws a large circular object which appears in Galileo's illustration but which is not actually to be seen on the moon. One suggestion is that Galileo deliberately magnified a typical crater in order to bring out its structure as shown by the shadows and highlights to be seen on it; Harriot may have been doing the same thing, or he may have been genuinely persuaded that the structure was there, since with a good telescope he would have been able to see only a part of the moon's disc at a time.

wasn't good enough to see the moons of Jupiter. On 5 September he managed to get hold of a telescope Galileo had sent to the elector of Cologne, and finally he saw for himself. Kepler had described his snowflakes as like little stars; now, everywhere he turned his telescope he found them, thick enough to make a snowstorm.

§ 3

It is easy to assume that the discoveries reported in *The Starry Messenger* are the most important that Galileo made with the telescope. This is not the case. It would seem that soon after its publication Galileo first observed sunspots, which could be regarded as definitive proof that there is change in the heavens, but at first he did not know what to make of them: it was not until April 1611 that he started to draw other people's attention to them.

In October 1611 Galileo, who had now moved to Florence, began to observe Venus through his telescope. His motivation was simple: Venus was a problem for both the Ptolemaic and Copernican systems because, according to both theories, its distance from the Earth varied greatly. According to the Ptolemaic system, it travelled on a large epicycle, which brought it sometimes closer to and sometimes further from the Earth. According to the Copernican system, as both Venus and the Earth travelled around the sun, the distance between them must alter radically: sometimes they must be on opposite sides of the sun, and sometimes Venus must come between the Earth and the Sun and be, relatively speaking, very close to the Earth. Yet, although Venus is sometimes brighter and sometimes less bright in the sky, it was difficult to see the variation that either theory must predict. Galileo had a further motive for looking at Venus. He had argued that the moon was an opaque body, shining solely by reflecting the light of the sun. He had explained the fact that the dark side of the moon seems sometimes to shine by its own ghostly light by claiming that the moon was being illuminated by light reflected from the Earth; that just as we on Earth see moonshine, so on the moon there is earthshine, and it is much brighter there than moonshine is here. If Venus was, similarly, an opaque body, it would, like the moon, have phases. So Galileo wanted to see if Venus had phases.

He must have realized from the beginning that if Venus *did* have phases, the nature of them would establish whether Ptolemaic astronomy was well founded or not. Ptolemaic astronomers were unable to agree on whether Venus was closer to the Earth than the sun. If Venus was closer to the Earth than the sun, its phases would range from crescent to half and never pass the half-illuminated point. However, if Venus was further from the Earth than the sun, its size would vary considerably over time, but it would nearly always be a full circle, and never be much less than that.[20]

Before 1611 the competition between the three alternative accounts of the cosmos – the Ptolemaic, the Copernican and the Tychonic – represents a genuine case of under-determination. In the Ptolemaic, or geocentric, system, which had been in existence for many centuries, the stars, sun, planets and moon all circle the Earth, but the planets and sun also move on other circles (epicycles). In the Copernican, or heliocentric, system, effectively new in 1543, the planets (of which the Earth is now one) circle the sun, but the moon circles the Earth. In the Tychonic, or geoheliocentric, system, invented as an alternative to Copernicanism in 1588, the planets circle the sun, and the sun and moon circle the Earth. These three systems are, when fully articulated, geometrically equivalent, which is to say that, although they combine circles in different ways, they produce identical predictions of the apparent locations of bodies in the heavens when viewed with the naked eye from the Earth.[iv] A Ptolemaic combination of a circle and an epicycle to predict the movement of a planet produces exactly the same result as the Copernican combination of the orbit of the planet with the orbit of the Earth, and that produces exactly the same effect as Brahe's combination of the orbit of the sun with the orbit of the Earth (just as taking one step forward and then two steps to the left is equivalent to taking two steps to the left and then one step forward) – which is why it was impossible to choose between them on the basis solely of information relating to the position of the planets in the sky.[v]

There was a widespread view that it ought to be possible to construct a fourth system which would better meet the requirements of Aristotelian philosophy: a homocentric system in which all the circles shared a common centre, ideally the Earth. Despite the efforts of major intellectual figures, such as Regiomontanus (1436–76), Alessandro Achillini and Girolamo Fracastoro (1478–1553), no one managed to construct a successful version of this system: it could not be made (as we would say) to fit the facts.[vi][21] (Even the Copernican system did not achieve homocentrism, as the moon circled the Earth, rather than the sun.)

iv Transits of Mercury (which occur every seven years or so) and of Venus (which occur in pairs, but more than a century apart), for example, can be seen only with a (low-power) telescope: Kepler thought he had seen a transit of Mercury in 1607 with a camera obscura, but he was wrong (Van Helden, *Measuring the Universe* (1985), 96–9). The first transit of Mercury was seen by Gassendi in 1631; of Venus by Horrocks in 1639. Horrocks's work led to a radical reassessment of the scale of the solar system: Van Helden, *Measuring the Universe* (1985), 95–117; Horrocks, *Venus Seen on the Sun* (2012). Proper naked-eye observations of novae and comets could produce results incompatible with celestial-sphere astronomy, but these results were irrelevant to planetary predictions.

v Their equivalence is demonstrated in Swerdlow, 'An Essay on Thomas Kuhn's First Scientific Revolution' (2004), 106–11.

vi As Copernicus put it in the prefatory letter to Pope Paul III in *De revolutionibus*: 'those who have trusted homocentrics have been able to establish nothing certain that corresponds to the phenomena' (translation from Barker, 'Copernicus and the Critics of Ptolemy' (1999), 345).

After Galileo discovered the phases of Venus in 1610 and thus proved that Venus orbited the sun, the Ptolemaic system ceased to be viable, although it was still possible to argue that some planets (Mercury, Venus, Mars) orbited the sun, and others (Saturn, Jupiter) orbited the Earth; this was the conclusion of Riccioli's *New Almagest* of 1651. Now there were only two (or two and a half) surviving systems, and intelligent and well-informed people had difficulty choosing between them for another half-century or so. So between 1610 and 1710 (say) cosmological theories were under-determined, in that there were at least two systems for which a strong case could be made, but not undetermined, in that everyone agreed that the Ptolemaic and homocentric systems were clearly not viable.

Galileo began to observe Venus in June 1610, as soon as it distanced itself enough from the sun to be visible. At first there was nothing interesting to see, as Venus was a full circle in his telescope; it was evidently on the far side of the sun. But at the beginning of October it became apparent that Venus was changing shape: slowly, it was moving towards being a half-circle. Day by day, Galileo watched this change carefully. On 11 December he sent Kepler a cipher which, when decoded, said, 'The mother of love [i.e. Venus] imitates the shapes of Cynthia [the moon].'[22] By this point Galileo knew both that Venus had phases (which meant that it was an opaque body shining by reflected light) and that the range of phases it displayed were incompatible with Ptolemaic astronomy, which required Venus always to be *either* further from the Earth than the sun *or* closer to the Earth than the sun. He waited a while longer until he was absolutely sure, and then on 30 December he wrote to his pupil Castelli (who had asked him in a letter Galileo will have received on 11 December – a letter which evidently provoked him to register his discovery with Kepler – whether Venus might not have phases) and to the leading mathematician in Rome, Christoph Clavius, announcing his discovery. On 1 January 1611 he wrote to Kepler, deciphering his earlier message, and Kepler went on to publish his correspondence with Galileo in his *Dioptrice* (1611).[23]

Clavius and Kepler will have had no difficulty confirming immediately that Venus had phases: all they had to do was point a decent telescope in the right direction. But a Venus with phases is perfectly compatible with Ptolemaic astronomy; what isn't is a Venus whose phases go from crescent to full: such a Venus must be in orbit around the sun. You do not need to observe the whole sequence of phases. All you need to do is either see Venus move from nearly full to half (as Galileo had in December), or see her move from crescent to nearly half.

When Galileo announced his discovery Venus was moving towards the sun: conjunction occurred on 1 March. There was nothing interesting to be seen, because all the phases that occurred between 1 January and 1 March would be repeated in reverse order as Venus emerged from the conjunction. On 5 March Galileo announced his intention to leave for Rome; on the nineteenth he was still impatiently waiting for a litter to carry him and complaining that he had a deadline to meet.[vii] Within a day or two he had left; Galileo was thus in Rome as the Jesuit astronomers turned their telescopes on Venus and watched it move towards a half-circle. It was probably during March that Clavius made revisions to a new edition of his *Sphere*: he records with care Galileo's discoveries to date (he makes no mention of sunspots, which Galileo had not yet drawn to his attention); he mentions the phases of Venus, and he says that astronomers are going to have to revise their theories in the light of these new findings.[24] What he does not say is as important as what he does: he does not say that Venus orbits the sun. Similarly, in April, Cardinal Bellarmine asked the Jesuit astronomers whether Galileo's discoveries had been confirmed. They said they had (though they reported that Clavius thought it might be possible to regard the moon's mountains as internal, not external, structures), and included the phases of Venus; they make no mention of Venus orbiting the sun.[25]

On 18 May, however, the Jesuit astronomers threw a party for Galileo. Odo van Maelcote delivered a lecture in which he announced that, although they had not yet seen Venus as a full circle (this would not happen for another few months, as Venus approached the sun and passed behind it in December 1611), they had seen enough to be sure that Venus did not revolve around the Earth. The philosophers in the audience were scandalized by this claim; Galileo was naturally thrilled to have been vindicated and feted. Clavius by this time was very ill, and we do not know what he made of this new evidence.[26]

It is essential to understand that what Maelcote announced was a killer fact: the Ptolemaic model in which all the planets (including the sun and the moon) revolve around the Earth had been proved to be

vii Galileo says he is hoping to celebrate Easter in Rome, but he acknowledges that his primary reason for going is 'to close the mouths of my detractors once and for all'. It was the emergence of Venus, not the Church calendar, which was the source of his sense of urgency. (Galilei, *Le Opere* (1890), Vol. 11, 67, 71.)

wrong. It was evident that Venus travelled around the sun (and this would become clearer and clearer as the months advanced towards the next conjunction) and, presumably, Mercury did too. After 18 May the Ptolemaic system, which had survived for more than 1,400 years, was fatally wounded. The choice was now between Copernicanism (all the planets, including the Earth, going around the sun); Brahe's system (all the planets going around the sun, and the sun going around the Earth, which remains stationary at the centre of the universe); or a compromise between Brahe and Ptolemy, in which the inner planets go round the sun and the outer planets go round the Earth. No competent astronomer defended the traditional Ptolemaic system once they had heard that Venus had a full set of phases; you had to be an ill-informed philosopher to do so. Moreover, it was generally acknowledged that the Tychonic system was incompatible with belief in solid heavenly spheres. Now, anyone who wanted to believe in solid spheres would have to imagine the sun going round the Earth, an epicycle on a deferent, and then Mercury and Venus going round the sun, of necessity cutting through the sphere of the sun. It is not surprising that this was regarded as further evidence against solid spheres (which Clavius had defended up until the last).[27]

According to contemporary history and philosophy of science, there are no such things as killer facts. We have seen already that the two-spheres theory could not survive the discovery of America; now we find that traditional Ptolemaic astronomy could not survive the discovery of the phases of Venus. Thus in August 1611 the anti-Copernican mathematician Margherita Sarrocchi described the phases of Venus as a 'geometrical demonstration that Venus goes round the sun'. The Jesuit astronomer Christoph Grienberger wrote to Galileo from Rome on 5 February 1612 confirming that the annual changes of Venus, 'just like the monthly changes of the Moon, very clearly demonstrate that it goes around the Sun'.[28] Galileo, in his first letter to Mark Welser about sunspots, which was written on 4 May 1612 (and published in 1613), says of the phases of Venus: 'These ... will not leave room for anyone to be in any doubt ... that its revolution is about the Sun.'[29] On 25 July 1612 Galileo's opponent on the issue of sunspots, the Jesuit astronomer Christoph Scheiner, wrote to Welser, describing the phases of Venus as an 'ineluctable argument': 'Venus goes around the Sun: the prudent man will scarcely dare to doubt it in the future.'[30] And Galileo writes in his third letter on sunspots, dated 1 December 1612, that the phases of Venus 'serve as a single, solid, and strong

The frontispiece to Giovanni Battista Riccioli's *New Almagest* (1651). Hanging in the balance being held by Astraea, the goddess of justice, are the competing world systems of Tycho Brahe and Copernicus; Riccioli is one of the last major astronomers to insist on the superiority of the Tychonic system. Discarded on the floor is the Ptolemaic system, which became indefensible when Galileo discovered the phases of Venus, and Ptolemy himself lies slumped in the background. In Riccioli's version of the Tychonic system Jupiter and Saturn orbit around the Earth, not the sun.

argument to establish its revolution around the Sun, such that no room whatsoever remains for doubt'.[31] No one was so foolish as to dispute these claims.[viii]

It is easy to show that conventional Ptolemaic astronomy was thriving until 1610 and went into crisis immediately afterwards: one only has to look at publications of the standard textbook, Sacrobosco's *Sphere*, and of the more advanced textbook, Peuerbach's *Theoricae novae planetarum*. Included in the figures for Sacrobosco are, for example, editions of Clavius's *Commentary*, which went through fifteen editions between 1570 and 1611, with a solitary final edition in 1618. (By comparison, there are only two editions of Kepler's *Epitome of Copernican Astronomy*, the first part of which was first published in 1618.) Clavius was published in Rome, Venice, Cologne, Lyons and Saint-Gervais. No textbook emerged capable of replacing Sacrobosco, Peuerbach and Clavius for the simple reason that no new consensus was established on the question of how the universe was organized until the eventual triumph of Newtonianism well into the eighteenth century – by which point the vernacular languages had replaced Latin, so no textbook could hope to have the international presence that they had had.

viii Ariew, 'The Phases of Venus before 1610' (1987), argues that Ptolemaic astronomy could have been adapted to allow for the phases of Venus by making the sun the centre of Venus's epicycle, which would in effect have made Venus – and by implication Mercury – moons of the sun. This would certainly have presented difficulties for the theory of the celestial spheres, and was regarded by contemporaries as being closer to the Tychonic than the Ptolemaic system. He also argues (Ariew, 'The Initial Response to Galileo's Lunar Observations' (2001)) that the scholastic account of the lunar surface could survive Galileo's telescopic discoveries; but here he does not adequately distinguish between blotches (seas) on the moon's surface and shifting patterns of light and dark (which Galileo interpreted as shadows and highlights). In March, Clavius, who was no slouch, could offer no solution to the problems presented by Galileo's new discoveries (see Clavius, *Opera mathematica* (1611), Vol. 3, 75; translated at Lattis, *Between Copernicus and Galileo* (1994), 198); this is perhaps the source of John Wilkins's statement that 'Tis reported of *Clavius*, that when lying upon his Death-bed, he heard the first Newes of those Discoveries which were made by *Gallilaeus* his Glasse, he brake forth into these words: *Videre Astronomos, quo pacto constituendi sunt orbes Coelestes, ut haec Phaenomena salvari possint:* That it did behoove Astronomers, to consider of some other *Hypothesis*, beside that of *Ptolomy*, whereby they might salve all those new appearances.' Wilkins, *A Discourse* (1640) II:21), but it is worth bearing in mind that the real crisis for the Ptolemaic system came not in March but in May, so that by the end of his life, (he died in February 1612) Clavius may indeed have been willing to recognize that the Ptolemaic system was indefensible–so at least Margherita, Sarrocchi was in August 1611.

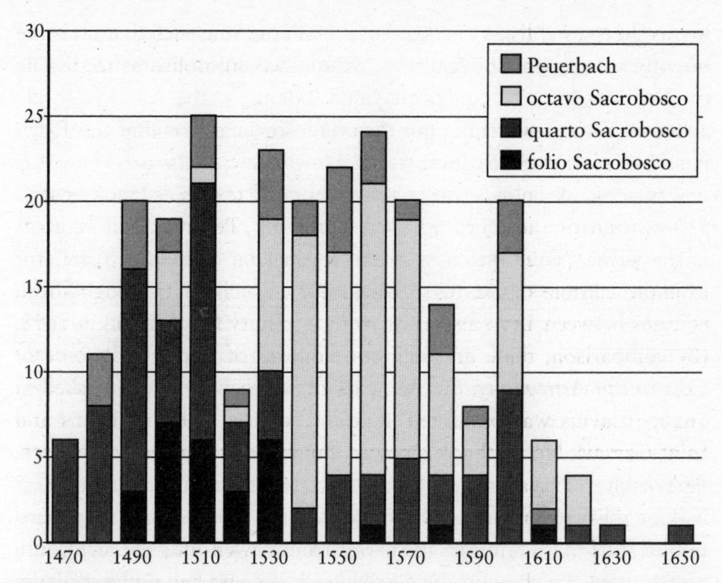

This chart shows the number of distinct editions of Sacrobosco's *Sphere* (folio, quarto and octavo editions are shown separately) and of Georg von Peuerbach's *Theoricae novae planetarum* – the two standard textbooks, one elementary and one advanced, for the study of astronomy in Renaissance universities. The figures at the base of columns are the first year of the decade, so 1470 refers to the decade 1470–1479. It is apparent that the publication of Copernicus (1543) had no effect on sales of these books, but there does seem to have been a dip following the comet of 1577 and the publication of Tycho's new system in 1588. However, demand was fully re-established in the decade 1600–1609, and not just for the new more complex commentaries such as that of Clavius, published in fat quartos, but also for cheap octavo editions. Demand collapsed, however, immediately after Galileo's telescopic discoveries. From this evidence it would appear that it was the telescope that killed off Ptolemaic astronomy. Publications of Sacrobosco are taken from the list in Jürgen Hamel, *Studien zur 'Sphaera' des Johannes de Sacrobosco* (2014) and of Peuerbach from Worldcat. I am indebted to Owen Gingerich (as always) for discussing this chart with me, and for suggesting breaking down the publications of Sacrobosco by format.

§ 4

Thus by 1611 not only was it generally accepted that the moon was a body rather like the Earth, in that it had mountains, but also, Venus had become an opaque body like the Earth and the moon. It followed that, if the Earth was inhabited, so other heavenly bodies might be; and if

Venus shone brightly in the sky above the Earth, so the Earth must shine brightly in the sky above Venus. Scholastic philosophers had fertile imaginations and had often imagined looking at the Earth from far away, even from the stars; but they had not imagined that the Earth would shine like the brightest stars.

The telescope itself provides a form of space travel; as Hooke put it, a 'transmigration into heaven, even whil'st we remain here upon earth in the flesh'.[32] Everyone now began to imagine looking at the Earth from deep space. Milton would imagine the Earth as a 'pendent World, in bigness as a star', and Pascal would go further, imagining what it would be like to look at the Earth from deepest space only to lose sight of it: 'an imperceptible point on the vast bosom of nature'. This became a new commonplace. For Locke, the Earth is not a point but a spot: 'our little spot of Earth', 'this spot of the Universe'.[33] The idea that the Earth was tiny compared to the universe, or that one might imagine looking at it from far away, was not new; what was new was the expansion of scale that came with the new astronomy, so that the Earth could be simultaneously held to be a bright star if seen from another planet and invisible if seen from deep space, and the hold that this idea of seeing the Earth from a vast distance had over the imaginations of the educated.[ix]

Galileo's telescope made two ideas that had previously seemed abstract and theoretical suddenly seem plausible and perfectly realistic: there might indeed be other inhabited worlds, and space might indeed be infinite. A whole literature was soon dedicated to these notions.[34] As

ix The claim that the earth is a dot compared to the universe had already been made by Pliny, *Natural History*, II:68. It became a standard topic of discussion: Giambattista Capuano da Manfredonia discussed it at length in the fifteenth century (Gaurico, Prosdocimus & others, *Spherae tractatus* (1531), 78rv). In 1505 Alessandro Achillini had asked if the earth would look like a tiny dot (*punctum*) if seen from space. The answer was that it would simply be invisible, lost against the blackness of the sky: Achillini, *De elementis* (1505), 85r; see also Barozzi, *Cosmographia* (1585), 32. Piccolomini, *De la sfera del mondo* (1540), 10v–11r, maintains that the earth would be *almost* invisible if seen from the stars. (For Achillini, the earth is the sphere of the element earth: the sphere of water would be invisible because transparent; otherwise, it would be visible during eclipses of the moon.) Copernicus, who realized his universe needed to be much larger than Ptolemy's, had insisted that the Earth was merely a *punctum* compared to the cosmos, but his *punctum* is a mathematical point, which is by definition infinitesimal: Copernicus, *On the Revolutions* (1978), 13. See also Benedetti, *Consideratione* (1579), 29; here, Benedetti writes as a Ptolemaic astronomer, though his thinking may be influenced by his commitment to Copernicanism. Bodin was under the delusion that Copernicanism rescued the Earth from being a mere point: Bodin, *Universæ naturæ theatrum* (1596), 581 = Bodin, *Le Théâtre de la nature universelle* (1597), 838.

early as 1612–13, John Webster in *The Duchess of Malfi* refers to Galileo's telescope as making visible 'another spacious world i' th' moon'.[35] In England, Francis Godwin's fictional *The Man in the Moone* appeared posthumously in 1638 – it had been written some time after 1628 – and was translated into French and German. Its account of a voyage to the moon marks the beginning of science fiction in English.[36] Godwin was a bishop and a crackpot; he seems to have believed that he had invented the radio.[37] John Wilkins, also a bishop and later a founder of the Royal Society, published his non-fiction *The Discovery of a World in the Moon* the same year (in which he argues that it may one day be possible to travel to the moon and suggests that the moon may be inhabited) and *A Discourse Concerning a New World and Another Planet* in 1640 (the first part is a reprint of the *Discovery* and the second an account of how our world is now, thanks to Copernicus, known to be a planet).[38] But by far the most important of such works was Cyrano de Bergerac's posthumous *The States and Empires of the Moon* (1657), an account of a journey to the moon, followed shortly by *The States and Empires of the Sun*.[39] Cyrano was later turned into a fictional character by Edmond Rostand, but the real Cyrano has little (apart from a large nose) in common with the fictional character. A lover of men, not women, and an atheist, he used the device of space travel to criticize everything he disliked about the real world. His work, inevitably, had to be toned down for publication and did not appear unbowdlerized until 1921. Even so, his moon book went through at least nineteen editions in French and two in English before the end of the century.

Fiction provided a useful disguise for dangerous ideas such as the atheism and materialism of Cyrano. But as the century went on it became less necessary to adopt such a disguise. Pierre Borel published *A New Treatise Proving a Multiplicity of Worlds: That the Planets are Regions Inhabited and the Earth a Star* (French, 1657; English, 1658), the first work since Bruno to argue that the planets were inhabited worlds. Borel believed that visitors from outer space had already arrived; not little green men, but birds of paradise. No one has ever found their nests, he claims, so it is evident that they visit us from another planet.[40] Influenced by Borel, John Flamsteed, the future first Astronomer Royal, concluded that all the stars were accompanied by 'systemes of planets, like our earth inhabitable & fild with creatures, perhaps more obedient to the lawes of their maker, then its [our Earth's] inhabitants'.[41] Borel was followed by two other works of popularizing science. Fontenelle's *Conversations on the Plurality of Worlds* sought to promulgate the

Frontispiece to Francis Godwin's posthumous and anonymous *The Man in the Moone* (1638), arguably the first work of science fiction. The hero flies to the moon in a vehicle powered by gansas (swans).

The frontispiece to John Wilkins, *A Discourse Concerning a New World* (1640; reprinted 1684): Copernicus and Galileo discuss the Copernican system, which is illustrated behind them; like Digges, Wilkins assumes the stars are spread out through an unbounded space. Copernicus presents his ideas hypothetically; Galileo says he has confirmed them with his telescope; and Kepler whispers in his ear, saying, 'If only you could confirm it by flying there!'

cosmology of Descartes: it appeared in at least twenty-five French editions between 1686 and Fontenelle's death in 1757; in the same period there were ten editions of two English translations.[x][42] It was followed by Christiaan Huygens' *Kosmotheoros* (1698), yet another posthumous work, which appeared in Latin, French and English.[43]

By 1700 every educated person was familiar with the idea that the universe might be infinite and that there were probably other inhabited worlds. Indeed, the idea had become entirely respectable, so that we find it being given forceful expression in Richard Bentley's Boyle Lectures against atheism (1692):

> [W]ho will deny, but that there are great multitudes of lucid Stars even beyond the reach of the best Telescopes; and that every visible Star may have opake Planets revolve about them, which we cannot discover? Now if they were not created for Our sakes; it is certain and evident, that they were not made for their own. For Matter hath no life nor perception, is not conscious of its own existence, nor capable of happiness, nor gives the Sacrifice of Praise and Worship to the Author of its Being. It remains therefore, that all Bodies were formed for the sake of Intelligent Minds: and as the Earth was principally designed for the Being and Service and Contemplation of Men; why may not all other Planets be created for the like Uses, each for their own Inhabitants which have Life and Understanding? If any man will indulge himself in this Speculation, he need not quarrel with revealed Religion upon such an account. The Holy Scriptures do not forbid him to suppose as great a Multitude of Systems and as much inhabited, as he pleases.[44]

The result was a quite new sense of the insignificance of human beings.[xi] 'Man is the measure of all things,' Protagoras (*c*.490–420 BCE) had said, and, once, this was literally true. The foot as a unit of measurement is based, naturally, on the foot. An ell (Italian, *braccio*; French, *aulne*) is the length of the forearm. A mile is a thousand Roman paces. Galen defines hot or cold in the patient in simple terms: a hot patient is one

x Fontenelle entertained himself with the idea that in the Milky Way stars and planets were crowded so close together that birds could fly from planet to planet; in a more serious moment he argued that a creature from the moon would drown in our atmosphere, making interplanetary travel feasible only between planets with similar atmospheres (Rawson, 'Discovering the Final Frontier' (2015); Fontenelle, *Entretiens sur la pluralité des mondes* (1955), 98–9, 134).

xi An exception in this respect is Cavendish, *The Description of a New World* (1666): she manages to write about extraterrestrial life without describing space travel, and without invoking a disorientating sense of the universe's vast expanse.

who is hotter than the physician's hand. In Galen's view the hand of a healthy person was designed to be the proper measure of hot and cold, damp and dry, soft and hard. As late as 1701 Newton wanted to take blood heat as one of the two fixed points for a temperature scale (the lower point being freezing water); it was one of three fixed points in Daniel Gabriel Fahrenheit's scheme, still widely used, of 1720, while a few years later John Fowler thought the upper fixed point should be the hottest water that can be endured by a hand held still.[45] Time was measured against a day divided into twenty-four hours, but in ordinary life short periods of time were measured subjectively, in Ave Marias or Paternosters: the time it took to say the Hail Mary or the Lord's Prayer. Only where weight was concerned was man not the measure. Man ceased to be the measure of all else only with the adoption of the metric system in France in 1799.[46] The basic unit of measurement (from which volumes and weights were derived) became the metre, originally defined as one ten millionth of the distance from the equator to the North Pole. The metric system merely completed a process that had begun with the invention of the telescope, which definitively destroyed the idea that the universe was made on the same scale as man.

§ 5

According to orthodox Christian thinking (at least until Pascal), the universe had been made to provide a home for humankind. The sun was there to provide light and heat by day, the moon and stars light by night. There was a perfect correspondence between the macrocosm (the universe as a whole) and the microcosm (the little world of the human body). The two were made for each other. The Fall had partly disrupted this perfect arrangement, forcing human beings to labour to survive; but the original architecture of the universe was still visible for all to see. Platonism, with its account of the universe's creation by a divine craftsman, the Demiurge, could be invoked to support this view – indeed, the idea of microcosm and macrocosm derived from neo-Platonism; but even Aristotelian philosophy, which held the universe to be eternal, assumed that human beings have all the faculties required to understand the universe.

For human-centredness had not been confined to measurement. Magnifying glasses were known to the ancient Greeks, and spectacles were in use from the thirteenth century. But lenses were used to correct imperfect vision, not to see things which could not be seen by someone with good eyesight. Again, the assumption was that God had given us eyes that were, when healthy, quite good enough for our purposes.[xii] Moreover, human beings were made in the image of God: a view hardly compatible with the notion that their senses were defective.

In the half-century or so between 1610 and 1665 this delightful picture of the universe as a home for humankind, an extension of Eden, was fatally undermined, and with it the notion that man is the proper measure of all things. This transformation has three distinct but interlocking components: first, humanity was displaced from the centre of the universe, which implied the possibility of intelligent life elsewhere; second, the correspondence between microcosm and macrocosm was shattered, so that the universe was no longer made to fit around us; and third, size became relative and scale became arbitrary – stars became snowflakes and snowflakes became stars.[47] This great transformation has escaped proper attention because it does not have a label, and it does not have a label primarily because it is three transformations rolled into one.

In fact, all three had a single cause: the telescope, whose impact was the same for anyone who looked through it. Here, for example, is one of the first uses of the word 'telescope' in English, in a religious tract from the English Civil War:

> This sober honest *Mercury* [i.e. news-sheet] coming to my hands, I thought it no great *Error* if I gave it that entertainment which I sometimes give even the *Phrantick* Bedlam *Pamphlets*: I must confesse it was to me a kind of *Eye-salve*, for I looked formerly at the wrong end of the *Perspective* [i.e. telescope], and the transgressions of our Welsh *Itinerants*, palliated with the name of *Saints*, seemed but small *Atoms* in a large Sun-shine. This *Book* is a new *Telescope*; it discovers what we could not see before; and the *Spots* in this Spiritual *Moon*, are *Mountains*.[48]

The telescope and the microscope do exactly the same thing: they turn atoms into mountains, and, if you look through the other end, mountains into atoms. This is the Scaling Revolution, as we might call it, by which, as William Blake put it in 'Auguries of Innocence', you can 'see a world in a grain of sand' or, alternatively, see a world as a grain of sand.

xii As late as 1689 the philosopher and doctor John Locke (who was, in general, perfectly open to new ideas) rejected the microscope as an instrument that might be of use to medicine on the grounds that such a use implied that God had not adequately equipped us to care for our own health, a view incompatible with proper respect for the deity. Locke, *An Essay* (1690), 140–1.

The classic reflection on this revolution is Voltaire's story *Micromégas* (1752; the name combines the Greek words for 'small' and 'large)', which describes the visit to Earth of a 20,000-foot-tall giant from one of the planets of Sirius, accompanied by an inhabitant of Saturn one third his size. For them, human beings are barely visible to the naked eye.[49]

The Scaling Revolution was not entirely without precedent. Lucretian atomism presented a picture of a universe which is frequently dissolved and remade, where the processes of nature are interactions between atoms which are invisible to us, and where sensations such as smell and taste are dismissed as subjective interpretations caused by the shape and movement of atoms. It was his familiarity with atomism which made it possible for Bacon, exceptionally and presciently, to dismiss human sensory organs as inherently defective and often misleading.[50] But if atomism suggested the existence of an invisible world of micro-mechanisms, it did not imply the existence of an invisible world of micro-organisms. That world was discovered by the Dutchman Antonie van Leeuwenhoek when, in 1676, he was the first to see living creatures invisible to the naked eye. Van Leeuwenhoek's discovery was met with initial scepticism: Hooke, in England, could see nothing comparable through his own microscope: but then he was using a compound microscope, not the tiny glass bead (a simple microscope) with which Leeuwenhoek achieved astonishing levels of magnification. It took four years for Leeuwenhoek's discovery to be confirmed. Galileo's discovery of the moons of Jupiter had been confirmed within a few months.

The first microscopists thought there was no limit to what they might see. Henry Power, who published just before Hooke, but whose book had little impact because there were only three illustrations, and those of poor quality, thought that eventually the microscope might reveal 'the Magnetical Effluviums of the Loadstone, the Solar Atoms of light (or *globuli aetherii* of the renowned Des-Cartes), the springy particles of air ...'[51] Hooke may actually have hoped to see the physical basis of memory, the 'continued Chain of Ideas coyled in the Repository of the Brain'.[52] Instead, the microscope hit a limit with Leeuwenhoek's single-celled organisms (1676). Hooke had shown that the louse was every bit as complicated a creature as a lizard. Leeuwenhoek dissected them, exploring their genitalia and discovering their sperm. Such experiences created the assumption that the very smallest creatures were as complex as the largest, and had the same sorts of organs. Far from recognizing that protozoa were different in character from larger organisms, Leeuwenhoek's work seemed to imply that they were the same. Size appeared to be irrelevant.

This assumption was crucially important when it came to trying to understand reproduction. The general view was that all life came from an egg (or at least all life visible to the naked eye; microscopic life was thought to generate spontaneously), even though no one had actually seen a mammalian egg. A contemporary of Leeuwenhoek's, Jan Swammerdam, showed that butterflies, which had been regarded as new creatures born out of the pupa, were already present within the caterpillar: their organs could be identified by dissection. Marcello Malpighi showed that the parts of the full-grown tree were present in the seed.[53] This led to the doctrine of preformationism: the adult existed fully formed within the egg. It followed logically that the egg already included the eggs of the next generation, that preformation implied pre-existence, and indeed that Eve had contained within her all future human beings to the end of time, each one fully formed within an egg within an egg within an egg, and so on. Thus Pascal's dream of worlds within worlds became a serious theory that every human individual was already present in Eve's ovaries (to which one might want to add all the humans who never happened to be born – the children nuns might have had if they had married, for example).[54]

Ovism, as this was called, seems to us the most fantastical of theories. It had obvious defects: it could not, for example, explain the inheritance of characteristics from the father; in 1752 Maupertuis showed that polydactylism could be inherited in the male as well as the female line. Preformationism assumed that new life was never created, yet in 1741 Abraham Trembley showed that you could cut a polyp into a dozen pieces and it would turn into a dozen polyps. Above all, ovism seems to us utterly impossible: how could every human being that ever existed or ever will exist be contained, fully formed, within Eve's ovaries? Yet this was not seen as a serious problem at all. The idea that there might be worlds within worlds had become entirely respectable. Only when cell theory established itself in the 1830s was preformationism abandoned. Only at this point did it become clear that the Scaling Revolution had its limits, that the idea of worlds within worlds was fantastical, not real.

Jonathan Swift knew all about Leeuwenhoek's discoveries when he wrote in 1733:

> So, Nat'ralists observe, a Flea
> Hath smaller Fleas that on him prey,

And these have smaller Fleas to bite 'em;
And so proceed *ad infinitum*.[xiii]

Yet long before Leeuwenhoek such creatures had existed in the imagination of those who had grasped the full implications of the Scaling Revolution. Cyrano writes of them, and Pascal (d.1662), who as far as we know never looked through a microscope, and certainly never saw Hooke's famous image of a flea (published in 1665), imagined someone inspecting a scabies mite:

> Let a mite be given him, with its minute body and parts incomparably more minute, limbs with their joints, veins in the limbs, blood in the veins, humours in the blood, drops in the humours, vapours in the drops. Dividing these last things again, let him exhaust his powers of conception, and let the last object at which he can arrive be now that of our discourse. Perhaps he will think that here is the smallest point in nature. I will let him see therein a new abyss. I will paint for him not only the visible universe, but all that he can conceive of nature's immensity in the womb of this abridged atom. Let him see therein an infinity of universes, each of which has its firmament, its planets, its earth, in the same proportion as in the visible world; in each earth animals, and in the last mites, in which he will find again all that the first had, finding still in these others the same thing without end and without cessation.[55]

Borges summarizes Pascal thus: 'There is no atom in space which does not contain universes; no universe that is not also an atom. It is logical to think (although he does not say it) that he saw himself multiplied in them, endlessly.'[56]

But which then, in all these endless universes, nested within one another, would be the real Pascal? The answer is that we could not possibly tell. This is quite different from the world of Rabelais. *Pantagruel* (1532) and *Gargantua* (1534) play with size-shifting: a whole army, for example, lives within a giant's mouth. But these are pre-telescopic texts and there are always clues as to who is normally sized and who is miniaturized or gigantized. Giants eat, drink and defecate against the backdrop of a normally sized word. In *Gulliver's Travels*, on the other hand, Swift creates a (more modest) version of Pascal's world. When

xiii Swift, *On Poetry* (1733), 20. The conceit goes back to Power, *Experimental Philosophy* (1664), 20: 'Fleas and Lice may have other Lice that feed upon them, as they do upon us.'

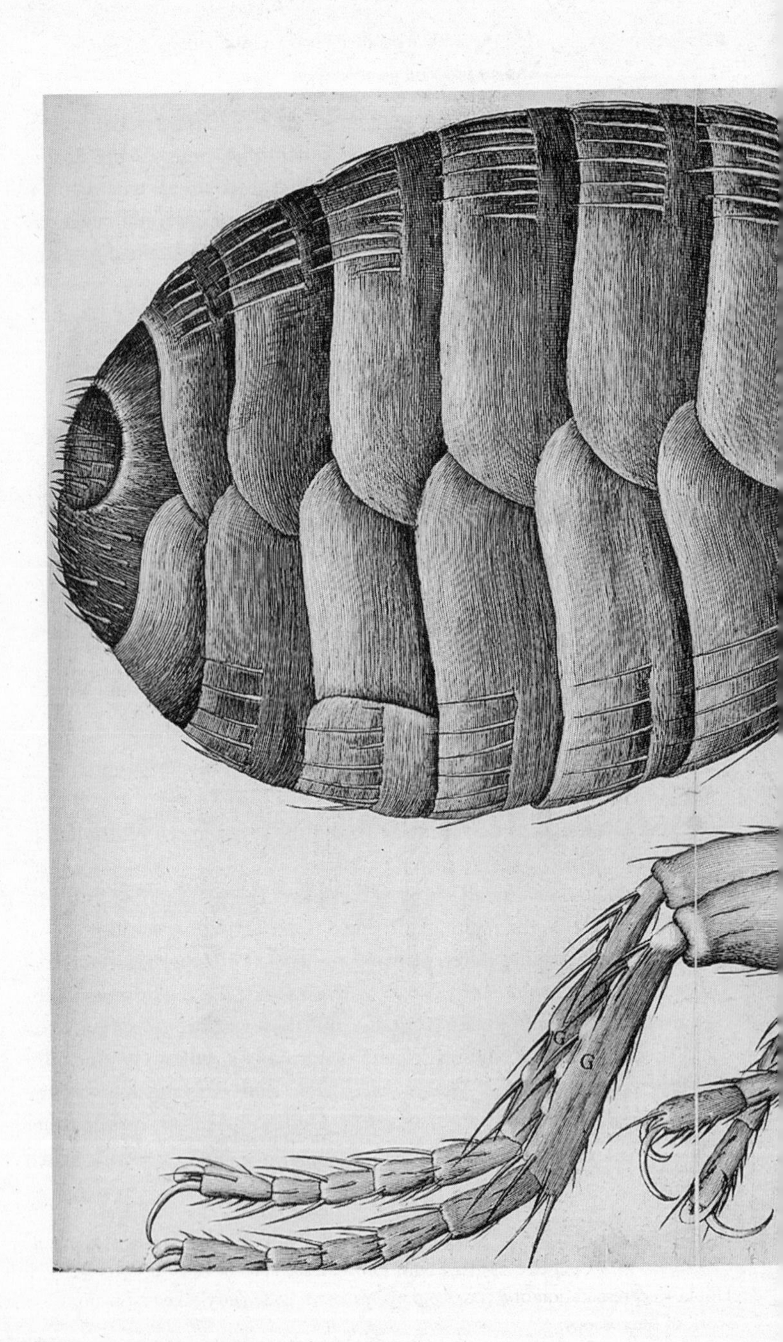
G
G

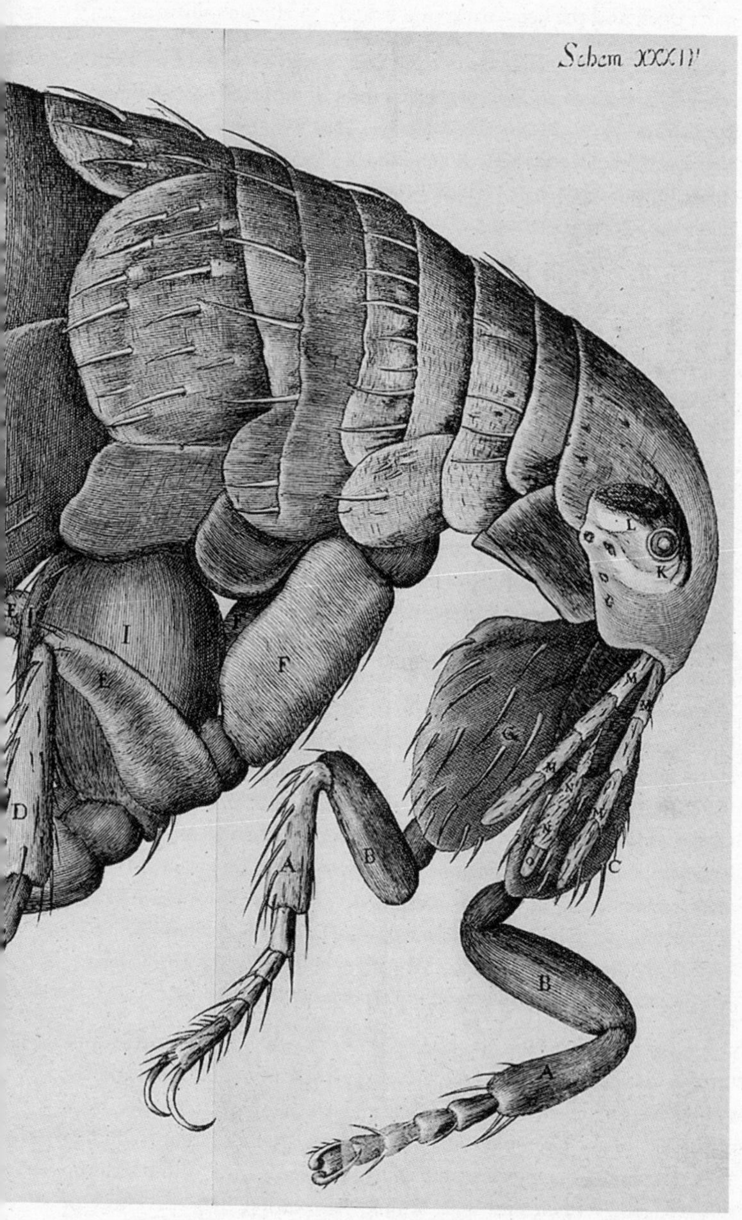

Hooke's representation of a flea, from his *Micrographia* (1665), the first major work of microscopy.

Gulliver finds himself among the Brobdingnagians the wasps are the size of partridges, and the lice correspond exactly to Hooke's illustration:

> I could distinctly see the Limbs of these Vermin with my naked Eye, much better than those of an *European* Louse through a Microscope; and their Snouts with which they rooted like Swine. They were the first I had ever beheld, and I should have been curious enough to dissect one of them, if I had proper Instruments (which I unluckily left behind me in the Ship) although indeed the Sight was so nauseous, that it perfectly turned my Stomach.[57]

Is it Gulliver or the Brobdingnagians who are the wrong size? The Brobdingnagians, we would say, but that is only because we know that Gulliver's size is our size. Swift had read Cyrano, and *Gulliver* is a cunning variation on the by then conventional themes of science fiction, one in which islands are substituted for planets.

The central message that readers were bound to take away from such texts is that human beings have a mistaken sense of their own importance. Cyrano was absolutely explicit, attacking

> the unsupportable Pride of Mankind, who perswade themselves, that Nature hath only been made for them; as if it were likely that the Sun, a vast Body, Four hundred and thirty four times bigger than the Earth, had only been kindled to ripen their Medlars, and plumpen their Cabbages. For my part, I am so far from complying with their Insolence, that I believe the Planets are Worlds about the Sun, and that the fixed Stars are also Suns, which have Planets about them, that's to say, Worlds, which because of their smallness, and that their borrowed light cannot reach us, are not discernible by Men in this World: For in good earnest, how can it be imagined, that such spacious Globes are no more but vast Desarts; and that ours, because we live in it, hath been framed for the habitation of a dozen of proud Dandyprats? How, must it be said, because the Sun measures our Days and Years, that it hath only been made, to keep us from running our Heads against the Walls? No, no, if that visible Deity shine upon Man, it's by accident, as the King's Flamboy by accident lightens a Porter that walks along the Street . . .[58]

So even before the microscope had been put to serious use, the telescope created a vertiginous sense of the infinite vastness of the universe and the insignificance of human beings when viewed, in the mind's eye, from outer space. In the Lucretian universe the gods are indifferent to human beings, and human beings are an accidental consequence of the random jostling together of atoms. The Scaling Revolution had the effect of forcing even those who continued to believe in a divine architect to recognize the coherence of this view. Even Kepler and Pascal, who wanted to think of themselves as inhabiting a universe made by God for man's salvation, found that they had no choice but to recognize that the universe was so vast, and the tiniest creatures within it were so exquisitely detailed, that it was either infinite, or might as well be. 'The eternal silence of these infinite spaces frightens me,' wrote Pascal.[59] Like it or not, even those who insisted that Bruno was wrong when he described an infinite universe were forced to imagine what it would be like if he were right.

Moreover, by expanding our range of vision, the telescope and microscope made it easier to recognize the limitations of our sensory apparatus when deprived of artificial aids. Pascal's friend Roberval suggested that human beings perceive light, but they simply lack the senses that they would need to discover what light is;[60] when he goes to the moon Cyrano is told:

> [T]here are a Million of things, perhaps, in the Universe, that would require a Million of different Organs in you, to understand them. For instance, I by my Senses know the cause of the Sympathy, that is betwixt the Loadstone and the Pole, of the ebbing and flowing of the Sea, and what becomes of the Animal after Death; you cannot reach these high Conceptions but by Faith, because they are Secrets above the power of your Intellects; no more than a Blind-man can judge of the beauties of a Land-skip, the Colours of a Picture, or the streaks of a Rain-bow . . .[61]

Locke agreed: other creatures on other planets may have senses that we lack, but we cannot even begin to imagine what they are like:

> He that will not set himself proudly at the top of all things; but will consider the Immensity of this Fabrick, and the great variety, that is to be found in this little and inconsiderable part of it, which he has to do with, may be apt to think, that in other Mansions of it, there may be other, and different intelligent Beings, of whose Faculties, he has as little Knowledge or Apprehension, as a worm shut up in one drawer of a Cabinet, hath of the Senses or Understanding of a Man.[62]

What is the poor worm doing shut up in a drawer? Presumably he is a woodworm, not an earthworm, and Locke's furniture was crawling with them.[xiv]

xiv Michael Hunter (personal communication) suggests the worm is a specimen in a curiosity cabinet; but it needs, or so it seems to me, to be a living creature, not a specimen, for only a living creature can have knowledge and apprehension.

§ 6

It might be thought that Copernicus was responsible for the destruction of the correspondence between microcosm and macrocosm. But this would be a mistake. There was only one major scale shift in Copernicus's universe: the stars were required to be at a vast distance from the solar system, given that there was no measurable change in their relationship to each other in the sky while the Earth orbited the sun in the course of a year, and so consequently they must be very big if they were not to be invisible. But the sun and the planets remained the same size, and Copernicus still continued (it seems) to believe that the universe consisted of nested spheres. Copernicus's universe was no longer Earth-centred, but it was still Earth-friendly, and there was no reason to think it was not the product of benevolent design. There was nothing in his argument which might imply that the Earth was just another planet, or that the universe had not been created for the benefit of human beings. The universe still had a centre, and the sun and the Earth were still unique objects.

The key change occurred in 1608 with the invention of the telescope and the microscope. Instruments are prostheses for thinking, and act as agents of change. Before 1608 the standard scientific instruments – cross-staffs, astrolabes, and so forth – were all designed to make naked-eye measurements of degrees of a circle. Even the vast sextants and quadrants built by Tycho Brahe were simply enlarged sighting devices. These instruments were no different in principle from those used by Ptolemy, and although, by making parallax investigations of comets and novae, they could be used to undermine the traditional belief in the translucent spheres that supported the planets (as had still been accepted by Copernicus), they reinforced the assumption that human beings were the perfect observers of the cosmos, and the cosmos itself was designed to support human life.[xv]

xv Right up to his death in 1687 the great Polish astronomer Johannes Hevelius insisted on making measurements of stars and planets only with naked-eye instruments: others claimed that the invention of the telescopic sight and the eyepiece micrometer made much greater accuracy possible, but Hevelius was not convinced (rightly, it turns out, because his instruments were designed to allow the human eye to function with maximum acuity, so that he could distinguish down to five seconds of arc, while Hooke, the great advocate of telescopic sights, claimed to have demonstrated experimentally that the human eye was incapable of distinguishing below thirty seconds of arc). Hevelius was, however, happy to use telescopes

These were not the only specialist instruments: alchemists had a specialized equipment of stills, crucibles and retorts, but these were simply a variety of containers to which heat could be applied (alchemy was frequently defined as trial by fire). They provided no new information about humanity's place in the universe. The printing press not only transformed the dissemination of knowledge but also, by making exact visual information widely available, brought about a revision in the traditional conception of what knowledge is.

After 1608 a new range of instruments made the invisible visible. The thermometer (*c.*1611) and the barometer (1643) made it possible to see temperature and air pressure, the first of which had previously been a subjective sensation, while the second is, under normal circumstances, something of which human beings are completely unaware. The barometer and Boyle's air pump (1660) made it possible to see what happened when living creatures or flames were subjected to a vacuum. We might add to these Newton's prisms, which visually demonstrated the fact that white light is made of light of different colours for the first time (1672). So by the end of the century there was quite an array of new instruments, but none of the others had an impact comparable to that of the telescope: originally intended to serve as a simple tool in warfare and navigation, it transformed not only astronomy but also how human beings envisaged their own significance.[63]

§ 7

In these last two chapters we have been looking at the ways in which intellectual change has knock-on consequences. The discovery of America killed off the two-spheres theory of the Earth. Copernicanism led to the idea that the planets shine by reflected light, which was confirmed by the discovery of the phases of Venus; and this killed off the Ptolemaic system. There was nothing arbitrary or contingent about these changes; they were as inevitable as the discovery of America once Columbus had set sail. These were intellectual transformations of fundamental importance, yet historians of science barely discuss them. They have become dark stars themselves – effectively invisible.

Why? Since Kuhn's *Structure* history of science has focused on

---

for other purposes, such as mapping the moon, and eventually built a massive 150-footer. Buchwald & Feingold, *Newton and the Origin of Civilization* (2013), 44–52.

controversy between scientists,[64] the assumption has been that every major new theory is contentious, and that there is nothing inevitable about the process by which one theory supplants another. This approach has been extraordinarily illuminating. But, in shining a light on controversy, it has left in the shadows all those changes which took place almost silently and were inevitable – indeed, could be seen to be inevitable at the time. Nobody (or, rather, only a few confused and ill-informed individuals) sprang to the defence of the two-spheres system after 1511. Nobody defended the Ptolemaic account of Venus after 1611. By 1624, eleven years after he had made public his discovery that Venus had a full set of phases, Galileo could take it for granted that no competent person would defend the Ptolemaic system.[xvi] It is easy to find evidence to support the claim that it was the telescope that killed off the Ptolemaic system, despite Thomas Kuhn's claim that Copernicanism was in the ascendant well before 1610 and the telescope made little difference.[xvii] As we have seen, editions of Sacrobosco's *Sphere*, the elementary textbook for the Ptolemaic system, and of Peuerbach's *Theoricae*, the more advanced textbook, dropped off sharply after 1610. The evidence is clear: Ptolemaic astronomy was unaffected by Copernicus; it went into crisis with the new star of 1572, but by the end of the sixteenth century it had fully recovered. The telescope, on the other hand, brought about its immediate and irreversible collapse.

Sometimes there are real, live, enduring controversies in science. In the seventeenth century such conflicts took place between those who believed in the possibility of a vacuum and those who did not, between those who believed in the possibility of a moving Earth and those (after

xvi Kuhn's discussion of the phases of Venus (Kuhn, *The Copernican Revolution* (1957), 222–4) is deeply unsatisfactory. He takes for granted a central issue that was resolved only by the discovery of the phases: that planets shine by reflected light. He describes the phases as providing 'strong evidence that Venus moves in a sun-centred orbit' when one might rather say that they provide conclusive evidence. And he concludes, 'None of the controversies discussed above, except perhaps the last [i.e. the phases of Venus], provides direct evidence for the main tenets of Copernicus' theory' – a pretty substantial exception, and one that throws in question his central claim that evidence is never decisive in settling scientific disputes. (Wootton, *Galileo* (2010), 178–9.) In *Structure*, relying on his mistaken argument that Copernicanism had triumphed before the invention of the telescope (it is worth pointing out, in response, that in 1632 Scheiner claimed that the hybrid Ptolemaic/Tychonic model was now universally accepted), Kuhn presents the phases of Venus as being influential, but influential 'particularly among non-astronomers' (Kuhn, *Structure* (1970), 155).

xvii Kuhn, *The Copernican Revolution* (1957), 220: 'Coming when it did, Galileo's astronomical work contributed primarily to a mopping-up operation, conducted after the victory was clearly in sight.'

1613, supporters of Brahe rather than Ptolemy) who did not. Sometimes the outcome really does teeter and hang in the balance. But, at other times, vast, well-constructed, apparently robust intellectual edifices are swept away with barely a murmur because, to paraphrase Vadianus, experience really can be demonstrative. If you concentrate on controversy, then it begins to look as if progress in science is arbitrary and unpredictable. If you assume that there is no major change without controversy, then your central assumption is never tested. The relativist thesis appears to be confirmed because evidence that would challenge it is never even considered. The picture changes radically if you look more broadly at intellectual change; then the demise of the two-spheres theory and the dark-star theory emerge as striking examples of intellectual change that took place without any controversy at all. Yet these were not minor theories: one was held by the best philosophers of the later Middle Ages; the other by the cleverest Copernicans of the late sixteenth century. The importance of an intellectual change simply cannot be measured by the amount of controversy it generates.

PART THREE

Making Knowledge

No theory of knowledge should attempt to explain why we are successful in our attempts to explain things . . . there are many worlds, possible and actual worlds, in which a search for knowledge and for regularities would fail.

– Karl Popper, *Objective Knowledge* (1972)[1]

Part Three contains the central chapters of this book. All of them are concerned with the development of a new language for thinking about, talking about and writing about science. In each chapter questions of language are intertwined with direct engagements with nature on the one hand, and with broader conceptual and philosophical questions on the other. The argument is simple: the language we use when thinking about scientific questions is almost entirely a construction of the seventeenth century. This language reflected the revolution that science was undergoing, but it also made that revolution possible.

7
Facts

Facts alone are what are wanted in life.

– Thomas Gradgrind in Dickens, *Hard Times* (1854)

The fact can only have a linguistic existence, as a term in a discourse, and yet it is exactly as if this existence were merely the 'copy', purely and simply, of another existence situated in the extra-structural domain of the 'real'.

– Roland Barthes, 'The Discourse of History' (1967)[1]

. . . the so-called facts proved never to be mere facts, independent of existing belief and theory.

– Thomas Kuhn, *The Trouble with the Historical Philosophy of Science* (1992)[2]

§ 1

We have seen that Renaissance science went beyond Greek science. Archimedes cried 'Eureka!', but the Renaissance invented discovery, priority disputes and eponymy. Vitruvius described something like perspective painting, but the Renaissance invented a new combination of subjectivity and objectivity, the situated viewer and the vanishing point. Cicero thought cartography was a branch of geometry, but the Renaissance developed a whole range of new mathematical disciplines and demonstrated their power to make sense of the world. Above all, the Renaissance acknowledged the existence of killer facts: facts which required the abandonment of well-established theories. Clearly, there were some fundamental changes before 1608, but in many respects

Renaissance science was essentially an extension of classical science. Regiomontanus and Galileo saw themselves as disciples of Archimedes. They would have been puzzled by the claim that they had something he did not (accurate though that claim would be for Galileo, and perhaps for Regiomontanus, too). In 1621 Kepler published the second part of his *Epitome of Copernican Astronomy*. He described it as 'a supplement to Aristotle's *On the Heavens*' because he assumed it would be part of a programme of education still based on Aristotle.[3] By 1700 this sense of continuity had been destroyed: the moderns knew that they were different from the ancients. If one thing marks out that difference, it is 'the fact'.

We take facts so much for granted that there have been few attempts to write their history, and none of them satisfactory.[4] Yet our culture is as dependent on facts as it is on gasoline. It is almost impossible to imagine doing without facts, and yet there was a time when facts did not exist. What did the map of knowledge look like before the invention of the fact? On the one hand there was truth, on the other opinion; on the one hand there was knowledge, on the other experience; on the one hand there was proof, on the other persuasion. Opinion, experience and persuasion were necessarily unreliable and unsatisfactory; knowledge had to be built on firmer foundations. The story of the fact is a story in which the lowest and most unreliable form of knowledge was magically transformed into the highest and most reliable.

What we are concerned with in this chapter is what the *OED* lists as 'fact' sense 8a: 'A thing that has really occurred or is actually the case' – although dictionaries don't distinguish clearly enough between an agency idea of a fact (something that has occurred because someone has done it) and an impersonal idea of a fact (something that has occurred in the course of nature).[5] How did you refer to this something before facts were invented? In Greek there was the phenomenon, but the phenomena were malleable – they could be 'saved' or 'salved'; while facts are stubborn. In Latin there was the thing: *res*.[6] The Romans said '*res ipsa loquitur*'; we say, 'the facts speak for themselves.'[7] Wittgenstein wrote in the *Tractatus* that 'The world is the totality of facts, not of things.' There is no translation for this in classical Latin or Elizabethan English.[i]

i The best one could do in Latin is 'The world is the totality of true judgements (*sententiae*) not of things' – but this is a quite different claim, as judgements are made while facts are not, and *sententiae* would include value judgements, such as 'murder is wrong.'

In English, before the fact, there were particulars.[ii] Phenomena are too subjective: they are appearances, not realities; things and particulars are too much in the real world: none of them corresponds to that peculiar blend of reality and thought which is the fact. This peculiar blend is what Barthes was referring to when he described facts as linguistic yet claiming to be copies of the real.

What is a fact? Naturally, philosophers disagree. My subject is what the philosophers call Humean facts. According to Hume, 'All the Objects of human Reason or Enquiry may naturally be divided into two Kinds, *viz. Relations of Ideas* and *Matters of Fact*. Of the first kind are the sciences of *Geometry*, *Algebra*, and *Arithmetic* ... [which are] discoverable by the mere operation of thought ... Matters of fact, which are the second object of human reason, are not ascertained in the same manner; nor is our evidence of their truth, however great, of a like nature with the foregoing.'[8] Relations of ideas deal with matters that are definitionally or necessarily true, such as 2 + 2 = 4 or all bachelors are unmarried. Matters of fact deal with matters that are contingently true, that is to say matters that could be otherwise (such as that the Earth has only one moon, or that I was born in January). Relations of ideas are purely logical; our knowledge of contingent matters, of matters of fact, depends on evidence: testimony, experience, documents.

In ordinary language, in our culture, we often ignore this distinction between relations of ideas and matters of fact, so I might say that it is a fact that the square on the hypotenuse is equal to the sum of the squares of the other two sides. Indeed, because all facts are by definition true we tend to think all truths are facts. But, for Hume, and for anyone writing about facts in the seventeenth century, the relationship is not reciprocal; Pythagoras's theorem is not a fact but a deduction (unless I have been measuring the squares on one particular triangle). It should immediately be apparent that in a culture in which experience was taking on a new significance, the term 'fact' was invaluable because it identified that type of knowledge which is grounded in experience. And it should also be apparent that the distinction between the two types of knowledge had a quite different significance in a world where there were two conflicting approaches to knowledge – Aristotelianism, on the one hand,

ii '... an Indigested Heap of Particulars'; 'we desire that Men should learne and perceive, how severe a Thing the true *Inquisition* of *Nature* is; And should accustome themselves, by the light of Particulars, to enlarge their Mindes, to the Amplitude of the World; And not reduce the World to the Narrownesse of their Mindes.' Bacon, *Sylva sylvarum* (1627), A1r, 74.

relying primarily on relations of ideas; and experimental science on the other, relying primarily on matters of fact – and intellectuals were under pressure to choose between them. Hume, in distinguishing relations of ideas from matters of fact, is recapitulating the fundamental intellectual conflict which gave rise to the Scientific Revolution.

What is a fact? It is a sort of trump card in an intellectual game. If you are playing Rock, Paper, Scissors, you can never be sure who will win. Intellectual life was a bit like that when the fact was invented – some thought reason should win; some authority (particularly where questions of faith were concerned); and still others wanted to rely on experience or experiment. But when facts entered the game everything changed because there is no arguing with the facts: they always win. Facts are a linguistic device which ensures that experience always trumps authority and reason. As Hume acknowledged, 'there is no reasoning . . . against matter of fact.'[9] The quotations chosen by the *OED* to illustrate the meaning of the word tell their own story: 'Facts are stubborn things' (1749); 'Facts are more powerful than arguments' (1782); 'One fact destroys this fiction' (1836).

We get an insight into the world before the fact by reading a book which tries to talk about facts without having the word or indeed the concept: Thomas Browne's *Pseudodoxia epidemica*, or *Vulgar Errors*, of 1646. Browne's aspiration is to rid the world of false beliefs (such as the notion that elephants have no knees, or the claim that beavers when fleeing from the hunt bite off their own testicles), but in doing so he compares himself to David taking on Goliath: '[W]ee are often constrained to stand alone against the strength of opinion; and to meet the Goliah and Gyant of Authority, with contemptible pibbles [pebbles], and feeble arguments, drawne from the scrip and slender stocke of our selves.'[10] Browne lived in a world in which what we would call facts seemed impotent when faced with authority. He thought reason, opinion and authority should all give ground before experience, but he lacked the language in which to express this simple idea. He couldn't say, as Hume did and we do, that there's no arguing with the facts.

We take facts so much for granted that it comes as a shock to learn that they are a modern invention. There is no word in classical Greek or Latin for a fact, and no way of translating the sentences above from the *OED* into these languages. The Greeks wrote of *to hoti*, 'that which is', and scholastic philosophers asked *an sit*, 'whether it is'. But there is plenty of scope for arguing with 'is' statements, and one could hardly describe them as stubborn or powerful. Of course, words and things are not always identical: one could have the idea of a fact, or a procedure for establishing facts, without the word 'fact'; indeed, I will shortly argue that there is an important distinction between fact-establishing and the language of the fact. I have already said that Vespucci's voyages produced killer facts – but there is no word corresponding to 'fact' in Vespucci's texts, either in the original languages (*The New World* was published in Latin in 1503; *The Letter* in Italian in 1505), or in the numerous early translations.

In Latin the word that is most frequently translated as 'fact' by modern translators is *res* (thing). But things and facts are not the same. A thing exists without words, but a fact is a statement, a term in a discourse. Things are not true, but facts are. Things and facts are not the same. Nevertheless, we treat facts as if they are equivalent to things, and dictionary definitions of 'fact' slide between defining facts as things and defining them as true beliefs. Thus, according to the *American College Dictionary*, a fact is 'something that actually exists; reality' and also, or alternatively, 'a truth known by actual experience or observation; something known to be true'.[11] Our understanding of facts is thus Janus-faced: at one moment we regard them as things, reality itself; at the next they are true beliefs, statements about reality. The result is that the grammar of the fact is profoundly problematic. In so far as facts are real, they are not true or false; in so far as they are statements, they are. It would be a mistake to think that one can resolve this contradiction: the whole point of the fact is that it inhabits two worlds and claims the best of both. It is precisely this quality that makes facts the raw material of science, for science, too, is a peculiar amalgam of the real and the cultural. Facts and science are made for each other.[12]

Facts are not just true or false; they can be confirmed by an appeal to evidence. The statement 'I believe in God' is either true or false, but only I can know for sure which, because it refers to a purely internal state of mind; it is inherently subjective. If I practise certain religious observances, then there are grounds for thinking the statement is true, but it is difficult to see how one could ever prove it. There are people who carry on practising religious observances although their faith has (temporarily, they hope) deserted them. But I can prove that I have been baptized or married: these are documented facts. They are objective states of affairs.

A contemporary philosopher has distinguished between three sorts of facts: brute facts, language-dependent facts and institutional facts. Let us look at some of his examples:

1. 'Mt Everest has snow and ice near its summit.' This is objectively true or false, and does not depend on my language or my subjective experience (although I need an appropriate vocabulary, of course, to communicate this truth to someone else). It is a brute fact.
2. Today is Thursday 6 June 2013. This is true, but it depends on a convention for numbering years and numbering and naming months and days. It is a language-dependent fact.
3. This is a ten-pound note. This is only true because this piece of paper has been issued by the Bank of England and is of the approved form. It is an institutional fact. Much of social reality consists of institutional facts: property, for example, or marriage.[13]

These categories imply that we find Everest to be covered in snow; we make today Thursday; and the bank decrees that this is legal tender. So some facts are found, some made and some decreed. Nothing could be more straightforward – except we never talk about finding, making or decreeing facts; instead, we 'establish facts'.[iii] I have been able to trace the phrase back to 1725 and, of course, its great advantage is that 'establish' shares the ambivalence of the concept of a fact: we may establish that such and such is the case, establish a base camp and establish a business – it applies to words, deeds and things.

That's not the only problem with this classification. European knowledge of Mount Everest depends on a long history of discovery, exploration, surveying and cartography. In 1855 a British trigonometric survey of India measured a mountain labelled Peak XV as being 29,002 feet high ('high' here means higher than sea level, although two hundred years earlier it was possible to believe that the oceans were higher than the highest mountains). In 1865 this mountain was officially

iii According to Bruno Latour, the great French philosopher of science Gaston Bachelard held that '*un fait est fait*' – a fact is an artefact, or facts are made (Latour, 'The Force and the Reason of Experiment' (1990), 63). But I can't find the phrase in Bachelard. So, too, Steven Shapin attributes to Ludwik Fleck the view that facts are made or invented (Shapin, 'A View of Scientific Thought' (1980)), but the language is not Fleck's. Latour and Shapin want to convey the idea that facts are made, but without taking responsibility for it. Pierre Bourdieu has complained about those who adopt 'a typical strategy, that of advancing a very radical position (of the type: the scientific fact is a construction or – *slippage* – a fabrication, and therefore an artefact, a fiction) before beating a retreat, in the face of criticism, back to banalities, that is, to the more ordinary face of ambiguous notions like "construction", etc.' (Bourdieu, *Science of Science* (2004), 26–7).

named by the Royal Geographical Society. So when I say that there is snow at the top of Mount Everest I am relying on a shared knowledge that there is a place we have agreed to call Mount Everest, that that place is a very high mountain, and that therefore it is not at all surprising that there is snow and ice near its summit. This 'brute' fact has become a brute fact for us; but that is because the process by which the mountain was discovered, measured, named and made famous has become invisible to us. In fact, Mount Everest is itself a language-dependent and institutionally defined entity. Making statements about Mount Everest shareable involves more than just finding a snow-covered mountain; it involves creating a shared language.[iv] 'Peak XV has snow and ice near its summit' would have been an equally true statement, but one that would only ever have made sense to a small group of surveyors and cartographers; to anyone else it would have been meaningless.

Or take today's date. This isn't just a linguistic convention. It is an institutional fact because contracts depend on the interpretation of dates. In Britain and the British Empire the Gregorian calendar, the calendar we use now, was introduced by law in 1752. The day after Wednesday 2 September 1752 was Thursday 14 September. Over most of Continental Europe, Wednesday 2 September (British-style) was already 13 September. At the same time as the date was changed, the start of the year was moved from 25 March to 1 January, so 1 January to 24 March 1752 never existed. Dates are not just named, they are also, like money, decreed.

The social and technological process by which we establish facts becomes invisible to us because we naturalize it. Language-dependent and institutional facts come to seem like brute facts to us: this is true for social institutions, like money, but even more so for claims about the natural world which are, in truth, theory dependent: we have naturalized the idea that the heights of mountains should be measured from sea level, an idea that would have made no sense in the Middle Ages. A couple more examples should make this clear. I know my date of birth: my parents told me, it is recorded on my birth certificate, my driving

iv Postmodernists often assume that because facts are expressed in language they are inherently contestable. Thus Jonathan Goldberg declares his opposition to those who 'suppose . . . that there are certain unquestionable facts; rather than allowing that what counts as factual is itself a discursive formation' (Goldberg, 'Speculations: *Macbeth* and Source' (1987), 244). It should be obvious that facts can both be unquestionable and only have meaning within certain discursive conventions: e.g. Washington is the capital of the United States of America, or 1 centimetre equals approximately 0.39 inches.

licence, my passport, and in all sorts of official records. It is a true objective fact, and if I had a stroke and forgot my date of birth I could establish it without any difficulty. However, I don't know Shakespeare's date of birth. He probably never knew his date of birth. The only official record tells us the date on which he was baptized.

You may think that of course Shakespeare must have known his own date of birth, even if we do not. You would be wrong. In 1608 Galileo was in correspondence with Christina of Lorraine, the wife of Ferdinand I, Grand Duke of Tuscany. Christina wanted Ferdinand's horoscope to be cast, but she was not sure when he had been born and offered two alternative dates more than a year apart: 19 July 1548 and 30 July 1549.[14] Galileo had to cast two horoscopes, work out which seemed better to fit the life of Ferdinand so far, thus decide his date of birth and so predict his future. Here was a great prince (originally a younger son, it is true, and so not expected to inherit) about whose year, let alone day, of birth there was genuine doubt. We all know the day on which we were born not because there is anything natural or even normal about such knowledge but simply because we live in a world in which such knowledge has been institutionalized.

When Marin Mersenne, a friar and mathematician in Paris, read Galileo's *Dialogue Concerning the Two Chief World Systems* (1632), he came across measurements of the relative speed of falling bodies expressed in *braccia*, arm's lengths or ells, the standard Italian unit of measurement.[v] But how long was Galileo's *braccio*? Mersenne wrote to him, asking, but never got a reply. A few years later when he was in Rome, he sought out a shop which sold measuring sticks and acquired a Florentine *braccio*. He then checked Galileo's measurements and decided they were wrong.[15] But did Galileo make his measurements in Florence, or earlier in his life in Venice? The Venetian *braccio* was longer than the Florentine, which would have made Galileo's measurements much more nearly accurate. In all probability Galileo did not worry about producing absolutely precise measurements, simply because he knew that in Rome, in Venice, in Florence and in Paris different units of measurement were used: precision was pointless when units of measurement were local. Indeed, in Florence and Venice two different *braccia* were used for different purposes. Thus it is true to say that Galileo made measurements relating to falling bodies; but he did not manage to turn those measurements into facts, as far as Mersenne was concerned,

v See above, p. 234.

because the measurements were language dependent, and behind the linguistic differences lay institutional decrees: the length of a Florentine *braccio* was determined by the Florentine state to ensure that merchants did not cheat their customers. What looks like a brute fact – the distance a body falls in a certain period of time – turns out to be in part dependent on language and institutions. Mersenne wanted to assess Galileo's claims by getting the facts straight; this turned out to be very far from straightforward because establishing facts depends upon instruments, even instruments as simple as measuring sticks, which have to be standardized.[16]

We live in societies that mass-produce facts: packages are marked with weights, road signs tell you distances and, in some countries, the populations of the towns you are passing through. We not only mass-produce them, we distribute them as efficiently as we distribute the post: my utility statements, for example, tell me how much electricity I have used; my bank statements tell me how much money I have to spend. Before the Scientific Revolution facts were few and far between: they were handmade, bespoke rather than mass produced, they were poorly distributed, they were often unreliable. Nobody, for example, knew what the population of Great Britain was until the first census in 1801; the first serious attempt at an estimate was made by Gregory King in 1696; before him, John Graunt had estimated the population of London in 1662. Before that numbers were hopelessly unreliable, and no one bothered to produce population figures for whole countries. In 1752 David Hume published an essay, 'Of the Populousness of Ancient Nations', pointing out that the figures we find in classical texts are meaningless.[17] Thus, according to Diodorus Siculus, writing in the first century BCE, the city of Sybaris in 510 BCE could field an army of 300,000 free men; add in women, children, old men and slaves, and Sybaris was apparently much, much bigger than London at the time Hume was writing (a total population of approximately 700,000, according to modern estimates). So was Agrigentum, which, according to Diogenes Laertius, in the third century had a population of 800,000. Yet these were only minor cities in their day, whereas London was the greatest commercial capital the world had ever seen. Hume's essay marks an intellectual shift because he expects numbers to be accurate; no one before 1650 or so complained that Diodorus Siculus's or Diogenes Laertius's numbers were untrustworthy, because they expected nothing else and their own numbers were equally unreliable.

It was not just science that made this new world, it was also the state, which was busy taxing citizens, borrowing money and putting armies in the field. The stock market required figures for profits and loss, capital and turnover. But states had been doing all these things for thousands of years without getting the numbers right. Merchants had been making and losing money since time immemorial. The idea that accurate figures could make a fundamental difference started with double-entry book-keeping in the thirteenth century; it then spread to the sciences, and outwards from both accountancy and science to government.

In 1662, for example, John Graunt published the numbers dying in London, the cause of death and his estimate of their age at death. From these he produced the first calculations of life expectancy for different age groups, and so the first reliable figures which could provide a basis for pricing life insurance. He lived in a new world of statistical accuracy. It was from the scientists, from men like William Petty, a first-generation fellow of the Royal Society, who surveyed Ireland, that Gregory King, a government administrator, effectively an accountant, acquired the conceptual tools that enabled him to calculate (very approximately) what we would call the Gross National Product of Britain and France in 1696 in order to work out which had the greater resources for winning the war they were fighting. (King's enterprise involved calculating not just the number of human beings and their taxable income but also the populations of cows, sheep and rabbits.)[18] We have something the Greeks and Romans did not, which is reliable facts and accurate statistics, and, in so far as they relate to more than the affairs of a particular business enterprise, these date back to the Scientific Revolution of the seventeenth century.

In stressing that facts are 'established', and that you have to learn how to establish them, I do not want to imply that they are subjective or culturally relative. Everest was just as covered in snow and just as tall before it was named in 1865 as it was after it was named, but finding and sharing facts about Everest required a naming process, a measuring process, a mapping process. Everest was there before 1865, but there were no facts about Everest before 1865. The facts about Everest were established, and this involved a triple process of finding, making and decreeing.

§ 2

Let us get down to business, then, with a particular example of fact-establishing (and let's for the moment set aside the anachronism implicit in using the word 'fact' when describing the activities of people who did

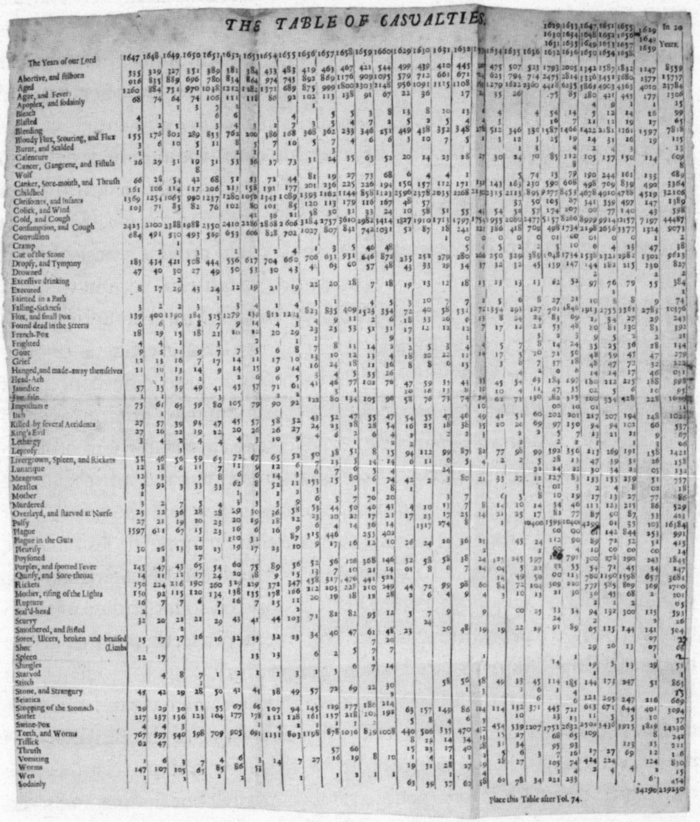

THE TABLE OF CASUALTIES.

The Years of our Lord	1647	1648	1649
Abortive, and Stilborn	335	329	327
Aged	916	835	889
Ague, and Fever	1260	884	751
Apoplex, and Sodainly	68	74	64
Bleach			1
Blasted	4	1	
Bleeding	3	2	5
Bloody Flux, Scouring, and Flux	155	176	802
Burnt, and Scalded	3	6	10
Calenture	1		
Cancer, Gangrene, and Fistula	26	29	31
Wolf			
Canker, Sore-mouth, and Thrush	66	28	54
Childbed	161	106	114
Chrisomes, and Infants	1369	1254	1065
Colick, and Wind	103	71	85
Cold, and Cough			
Consumption, and Cough	2423	2200	2388
Convulsion	684	491	530
Cramp			1
Cut of the Stone		2	1
Dropsy, and Tympany	185	434	421
Drowned	47	40	30
Excessive drinking			2
Executed	8	17	29
Fainted in a Bath			
Falling-Sickness	3	2	2
Flox, and small Pox	139	400	1190
Found dead in the Streets	6	6	9
French-Pox	18	29	15
Frighted	4	4	1
Gout	9	5	12
Grief	12	13	16
Hanged, and made-away themselves	11	10	13
Head-Ach		1	11
Jaundice	57	35	39
Jaw-fain	1	1	
Impostume	75	61	65
Itch		1	
Killed by several Accidents	27	57	39
King's Evil	27	26	22
Lethargy	3	4	2
Leprosy			1
Livergrown, Spleen, and Rickets	53	46	56
Lunatique	12	18	6
Meagrom	12	13	
Measles	5	92	3
Mother	2		
Murdered	3	2	7
Overlayd, and starved at Nurse	25	22	36
Palsy	27	21	19
Plague	3597	611	67
Plague in the Guts			
Pleurisy	30	26	13
Poysoned		3	
Purples, and spotted Fever	145	47	43
Quinsy, and Sore-throat	14	11	12
Rickets	150	224	216
Mother, rising of the Lights	150	92	115
Rupture	16	7	7
Scal'd-head	2		
Scurvy	32	20	21
Smothered, and stifled			2
Sores, Ulcers, broken and bruised (Limbs	15	17	17
Shot			
Spleen	12	17	
Shingles			
Starved		4	8
Stitch			
Stone, and Strangury	45	42	29
Sciatica			
Stopping of the Stomach	29	29	30
Surfet	217	137	136
Swine-Pox	4	4	3
Teeth, and Worms	767	597	540
Tissick	62	47	
Thrush			
Vomiting	1	6	3
Worms	147	107	105
Wen	1		1
Sodainly			

Place this Table after Fol. 74.

Graunt's table of mortality from his *Natural and Political Observations* (1662). Graunt compiled statistics for the number born and dying each year, and the causes of death, from the annual bills of mortality that were published in London. He used these to calculate life expectancies for each age group, and to estimate the population of London, which he concluded was 460,000 – not, as had been claimed, 7 million or so.

not yet have the word). On the night of 19 February 1604, in Prague, Johannes Kepler was out measuring the position of Mars in the sky with a metal instrument called a quadrant.[19] The type of measurement he was trying to make was perfectly familiar to astronomers: such measurements had been made since Ptolemy. But in Kepler's view Ptolemy's measurements were not accurate enough, nor were any of those made since, except for Tycho Brahe's. On this particular night it was bitterly cold with a biting wind. Kepler found that if he removed his gloves his hands were soon too numb to manage his instrument; if he kept them on he could barely make the fine adjustments necessary. The wind was too strong to keep a candle alight, so he had to read his measurements and write them down by the light of a glowing coal. The results, he felt sure, were unsatisfactory – he was out, he thought, by 10 minutes of a degree (a minute being one sixtieth of a degree). On a modern school protractor you cannot distinguish ten minutes of a degree, and only one astronomer before Kepler would have thought such a measurement unsatisfactory. Ptolemy and Copernicus had regarded 10 minutes as precisely the acceptable margin of error. But Kepler had worked with Tycho Brahe, who had devised new instruments capable of measuring with astonishing accuracy to a single minute.

Kepler was worried about such tiny numbers because he had a quite different understanding of astronomy from anyone before him. The goal of previous astronomers had been to construct mathematical models that would successfully predict the location of the planets in the heavens. They had all shared the assumption that such models must involve various combinations of circular movement, because the philosophers had instructed them that all movements in the heavens was required to be circular. The problem for Kepler was that circles, eccentrics and epicycles were geometrical constructions; there was no evidence that any such gearing existed in the heavens. Moreover, his predecessors had been quite happy to use two distinct models for each planet: one to calculate its movement from east to west, and the other its movement from north to south.

Kepler knew there were no crystalline spheres in the heavens, and so he understood that the planets were moving through empty space on what he called an 'orbit'. Kepler replaced orbs with orbits because his ambition was to replace geometry with physics. ('Orbit' used in this sense was a marker of Kepler's key innovation; previously, an orbit was the track left by a wheel in the ground, or the socket into which the eye is set. An orbit is physical, while an orb is a geometrical abstraction.)[20]

In order to understand the movements of the planets, Kepler thought about ferrymen trying to row across a fast-flowing river. If you were steering a planet through space, he wondered (for Kepler was prepared to imagine intelligences guiding the planets), how would you locate yourself, and how would you keep on course? An eccentric, which involved a perfect circle around an unmarked point in featureless space, seemed to him an impossibility. Kepler was convinced one had to think of forces flowing through space – his inspiration was provided by Gilbert's recently published study of the magnet – and ask oneself how a celestial helmsman would take his bearings.[21] Consequently, he insisted on using one single mathematical model to account for a planet's movement across the heavens. When he tried to employ this method for Mars using Brahe's combination of circles, he could get satisfactory results for the longitudes (2 minutes of error), but then the latitudes failed. When he readjusted the geometry to get the latitudes right, the error in the longitudes climbed to a value that would once have been dismissed as insignificant but which Kepler now thought intolerable: a full 8 minutes.[22]

The truth is that, if Kepler had been determined to find a system of circles that would fit, he could have done so, as he later recognized. But instead he began to play around with other mathematical models, and he discovered that he could produce results to a satisfactory standard of accuracy (better than 2 minutes) if he modelled the orbit as an ellipse with the sun as its focus. Previous astronomers would have rejected this solution because it did not involve circular movement. Kepler, however, was delighted with it, because he could imagine that there was some sort of physical force involved which caused the planet to swing through space, speeding up as it approached the sun – the source of the force – and slowing down as it moved away from it: and he was right, of course, for that force is gravity.[23]

Kepler did not have the word 'fact' (he wrote of phenomena, observations, effects, experiments, of *to hoti*), but he certainly had the idea and knew that facts were what he was after. He chose to place on the title page of his *New Star* (1606) the image of a hen pecking around in a farmyard, with the motto *grana dat e fimo scrutans* ('hunting about in the dung, she finds grain'). He presented himself not as a great philosopher but as someone prepared to grub around for facts. And because he had to make his facts credible he was obliged to adopt many of the rhetorical techniques that were, according to the literature, invented much later: the apparently prolix recounting of irrelevant details (the glowing

JOANNIS KEPPLERI
Sac. Cæs. Majest. Mathematici
DE
STELLA NOVA
IN PEDE SERPENTARII, ET
QUI SUB EJUS EXORTUM DE
NOVO INIIT,
TRIGONO IGNEO.
LIBELLUS ASTRONOMICIS, PHYSICIS, META-
physicis, Meteorologicis & Astrologicis Disputationibus,
ἐνδόξοις & παραδόξοις plenus.
ACCESSERUNT
I. DE STELLA INCOGNITA CYGNI:
Narratio Astronomica.
II. DE JESV CHRISTI SERVATORIS VERO
Anno Natalitio, consideratio novissimæ sententiæ LAV-
RENTII SVSLYGÆ Poloni, quatuor annos in usitata
Epocha desiderantis.
Cum Privilegio S. C. Majest. ad annos XV.

PRAGAE
Typis PAULI SESSII, impensis AUTHORIS.
ANNO M. DCVI.

Title page of Kepler's *New Star* of 1606.

coal by which he read his instruments on the night of 19 February 1604); the determination to report failures (he presents what he calls his war on Mars as an almost endless series of defeats) with the same care as successes; the insistence on involving the reader as if they were really present.[24] In the *New Star* he even introduces us to his wife, as though we were visiting them at home, explaining that he had found it difficult to refute the arguments of the Epicureans, who thought the universe was the product of chance. But his wife is a more redoubtable adversary than he is:

> Yesterday, when I had grown tired of writing and my mind was full of dust motes from thinking about atoms, she called me to dinner and served me a salad. Whereupon I said to her, 'If one were to throw into the air the pewter plates, lettuce leaves, grains of salt, drops of oil, vinegar and water and the glorious eggs, and all these things were to remain there for eternity, then would one day this salad just fall together by chance?' My beauty replied, 'But not in this presentation, nor in this order.'[25]

The purpose of such irrelevant details – the pewter plates, the glorious eggs – is to create what Roland Barthes called 'the reality effect'.[26] We can trust Kepler, we are to understand, because he tells us what really happened. In the nineteenth century this sort of narrative became the ideal of the historian (*wie es eigentlich gewesen ist*, 'just as it actually happened', as Ranke put it), but in the seventeenth century it was not the historians but the scientists who aspired to realism as a literary style. (There are exceptions, however, Newton and Descartes being the most striking.) Since the new science had yet to establish its claim to authority, that claim had to be asserted by appealing to reality. In a world without peer review, texts, in order to convey trustworthiness, reliability and accuracy, had to employ literary devices. In the case of Kepler's *New Astronomy* (1609) the quest for realism took what seems to the modern reader a most peculiar form: instead of outlining his new astronomy, Kepler presents an historical narrative of his quest for a new astronomy, with all his false turns and mistakes carefully recorded. In order to make facts, Kepler not only had to freeze his fingers on February nights, he also had to devise literary forms that would convince the reader that he had gone to extreme lengths to get his facts (and his theories) right; even the title page declares that his new astronomy was 'worked out at Prague in a tenacious study lasting many years'.[27]

Of course, not everyone found these strategies helpful: Galileo complained that he found Kepler unreadable. For him, drama rather than historical narrative was the way of constructing the appearance of reality. Galileo's *Dialogue Concerning the Two Chief World Systems* has as its frontispiece the image of Aristotle, Ptolemy and Copernicus in front of the sort of curtain that would be raised at the start of a theatrical performance. Galileo, by presenting a dialogue in which he himself never appears on stage, was able (in principle, at least) to avoid taking responsibility for any of the arguments proposed. But he also wanted to give the reader the sense of being present at a real argument, one from which Copernicanism emerged as indisputably victorious. Unfortunately, these two objectives were directly at odds with each other, and Galileo's success at the second undermined his rather half-hearted efforts at the first. There is a paradox here. Galileo's *Dialogue*, although apparently set in a real place (Venice), is transparently fictional: one of his characters, Simplicio, was imaginary, and the two others were dead (Salviati in 1614; Sagredo in 1620). But the purpose of the fiction is to create a sense of reality which will convince the reader that the information in the dialogue is perfectly genuine. The facts are true, even if the protagonists are fictions.

I started with Kepler on the night of 19 February 1604 because every fact has a local history: Kepler had an interest in telling that history in order to convince his reader that his measurements were accurate, and the historian has an interest in telling it in order to catch facts in the process of being both established and narrated. One reason that Kepler had to go to such lengths was that he could not simply baldly state the facts, because there was no tradition of taking facts at face value. The key term in philosophy, even in Kepler's and Galileo's time, was 'phenomena'. As far as Aristotle was concerned, phenomena included everything which was generally accepted to be the case.[28] So if people generally believe that mice are spontaneously generated in straw, then the task of the philosopher is to explain why this is so, not to question whether it is so.[29] Moreover, phenomena were malleable. Ptolemy had based his astronomy on measurements, just as Kepler did. But Ptolemy and his followers had assumed that measurements could be regarded as approximations: in practice, there were bound to be minor discrepancies between theoretical predictions and actual measurements. Furthermore, they were under no obligation to be consistent. It was perfectly acceptable to use one hypothesis (or model) to account for a planet's movement along the plane of the ecliptic, and another, conflicting model to account for its deviation above and below this plane;[30] thus saving, or salving, the phenomena.[31] Kepler, by contrast, was looking for a perfect match. He may have been making

The frontispiece to Galileo's *Dialogue* (1632). Aristotle (*left*), shown as a feeble old man; Ptolemy (*centre*), wearing a turban because he comes from Egypt; and Copernicus, wearing the clothes of a Polish priest, stand on the shore of the Florentine port of Livorno debating questions of physics and astronomy. But Copernicus looks nothing like the image of Copernicus that appears in other sources, where he is always portrayed as young and clean-shaven. Indeed, the Latin translation by Bernegger soon corrected this 'error', providing a more accurate representation of Copernicus. It seems Galileo has decided to present himself in the role of Copernicus. Over the heads of the three philosophers hangs the curtain which rises at the beginning of a theatrical performance – a device that was used by Galileo's engraver, Stefano della Bella, for the frontispieces of plays. Thus Galileo implies that the arguments presented in the book are not to be regarded as true, for Copernicanism had been condemned by the Church.

measurements very like the ones Ptolemy had made, using instruments very like the ones Ptolemy had used, but in his enterprise the measurements (what we call the facts) had a new status, a new authority.

§ 3

Facts are not only established, they are 'disestablished'. Of course, we do not say that: facts are by definition true, so when they are discovered to be false they simply cease to be, like Tinkerbell, who can live only if the children believe in her. Facts are based in experience and they are refuted by experience. The ancient Greeks and Romans believed that if you rubbed a magnet with garlic it ceased to work.[32] Plutarch, Ptolemy and all sorts of other authors believed this implicitly. It was, for them, in the words of Daryn Lehoux, an example of 'unproblematic facticity'.[33] You could get the magnet to start working again by smearing it with goat's blood. Sophisticated thinkers (what Thomas Browne in 1646 called 'grave and worthy Writers') went on believing this well into the seventeenth century.[34] In 1589 Giambattista della Porta (a Neapolitan aristocrat whose *Natural Magick* was one of the great best-sellers in the century between 1560 and 1660) protested:

> But when I tried all these things, I found them to be false: for not onely breathing and belching upon the Loadstone after eating of Garlick, did not stop its vertues: but when it was all anointed over with the juice of Garlick, it did perform its office as well as if it had never been touched with it.[35]

This appeal to experience, however, was not new. Plutarch had claimed 'palpable experience' of the disempowering effect of garlic upon the magnet.[36] Something in the nature of experience had, it would seem, changed between Plutarch and della Porta.

But there was nothing new (we are told) in della Porta's approach to evidence: he was every bit as credulous about many matters as all those who had believed in the power of garlic. For example, he believed in spontaneous generation: not only that there were barnacles that could hatch into geese (hence the barnacle goose; even Kepler believed geese came from barnacles) but that putrefying sage generated a bird 'like a blackbird'. And he believed that bears love honey because in their quest for it they are stung, and the bee stings on a bear's mouth draw down the thick humour that normally clouds its vision, thus bears love honey because it improves their eyesight. If della Porta happens to hold the same view as we do on the subject of garlic and magnets, it is not because he is better at handling evidence than Plutarch or Ptolemy. All he has done is isolate the question of garlic and magnets from the larger context in which it was normally placed, that of sympathy and antipathy. This has made it possible for him to come up with a new answer to the question: 'What happens when garlic meets a magnet?' His answer happens to be our answer.[37]

But the claim that della Porta had isolated the study of magnets from the question of sympathy and antipathy is rather odd, because right at the beginning of his book, when he discusses these powerful forces through which so much can be accomplished, his examples are the standard ones, such as the antipathy between man and the wolf, which ensures that if a man sees a wolf he is rendered speechless. Included, almost inevitably, is the case of garlic and the magnet (I quote from the 1658 English translation):

> Hither belongeth that notable Disagreement that is betwixt Garlick and the Load-stone: for being smeared about with Garlike, it will not draw iron to it; as *Plutark* hath noted, and after him *Ptolomaeus:* the Load-stone hath in it a poisonous vertue, and Garlick is good against poison: but if no man had written of the power of Garlick against the Load-stone, yet we might conjecture it to be so, because it is good against vipers, and mad dogs, and poisonous waters. So likewise those living creatures that are enemies to poisonous things, and swallow them up without danger, may shew us that such poisons will cure the bitings and blows of those creatures.[38]

We should not to be too critical of della Porta's acceptance of the doctrine of sympathy and antipathy. Even Descartes, who would soon set out to reject all received opinions, believed in 1618 (probably on the authority of della Porta) that a drum covered in lamb's skin would fall silent if it sensed the vibrations from one covered in wolf's skin. Not death – not even tanning – could undo the antipathy between lamb and wolf.[39] And yet only a few decades later Walter Charleton poured scorn on those who held this view:

> [I]t hath been affirmed by many of the Ancients, and questioned by very few of the Moderns, *that a Drum bottomed with a Woolfs skin, and headed with a Sheeps, will yeeld scarce any sound at all*; may more, that a *Wolfs skin will in short time prey upon and consume a Sheeps skin*, if they be layed neer together. And against this we need no other Defense than a downright appeal to *Experience*, whether both those Traditions deserve

> not to be listed among Popular Errors; and as well the Promoters, as Authors of them to be exiled the society of Philosophers: these as Traitors to truth by the plotting of manifest falsehoods; those as Ideots, for beleiving [*sic*] and admiring such fopperies, as smell of nothing but the Fable; and lye open to the contradiction of an easy and cheap Experiment.[40]

In the middle years of the seventeenth century experience stopped being something that accorded naturally with the statements of previous authorities and became a caustic solvent of fabulous beliefs.

How, though, are we to explain the peculiar fact that della Porta both believes in and rejects the garlic/magnet antipathy?[vi] A first step is to recognize that *Of Natural Magic* is a book written over a period of more than thirty years. Indeed, it is two books. The first edition, divided into four 'books' or sections, was published in 1558 (and went through sixty editions in five languages over the course of seventy years).[41] The second, consisting of twenty books, was published in 1589. The relationship between the two editions is a complex one. A good deal of material that had been in the 1558 edition disappeared in 1589. The main reason for this is straightforward. Della Porta had been put on trial by the Inquisition in 1577–8 accused of black magic. (His arrest probably dates back to 1574, when he was ordered to close the academy he had founded to enquire into the secrets of nature.)[42] He had continuing problems with Catholic censorship, for a time being banned from publishing.

The second edition of *Of Natural Magic* had been worked through carefully to remove possible causes of offence. Introduced was a careful sentence to bring della Porta's discussion of the soul in line with Christian teaching, and all references to the world soul, the *anima mundi*, were now carefully turned into quotations.[vii] Gone, inevitably, is a chap-

vi One might think that Renaissance authors simply weren't very concerned with being consistent. I don't think this is right; inconsistency is often (but not in this case!) the result of our misinterpreting what they are saying. For a striking example, compare Stephen Orgel's interpretation of Helkiah Crooke's *Microcosmographia* (1615) on the anatomical differences between the sexes, in the course of which Orgel claims that for Crooke 'the scientific truth or falsehood' of an argument is simply not an issue, that he sees 'no need to reconcile the conflicting scientific arguments', with the result that he is quite happy to propound arguments which are mutually incompatible (indeed, in Orgel's view, consistency is a 'post-Enlightenment' virtue), with Janet Adelman's rebuttal: Orgel, *Impersonations* (1996), 21–4; Adelman, 'Making Defect Perfection' (1999), 36–9 and n. 29.

vii 'For God the first cause and beginner of things, as *Macrobius* saith, of his own fruitfulnesse hath created and brought forth a Spirit, the Spirit brought forth a Soul (but the truth of Christianity saith otherwise), the Soul is furnished partly with reason, which it bestows ter in which della Porta described an experiment with an unguent supposedly used by witches to enable them to fly to the sabbat: he had conveniently made the acquaintance of a witch who agreed to provide a demonstration of her powers. Having rubbed herself all over with an ointment (della Porta provided two recipes, one based on the fat of young children, the other on bat's blood), she had simply fallen into a deep sleep, but her body had never left the room in which she was locked, although when she awoke she described flying over seas and mountains. The clear implication was that the sabbat was an hallucination, not a reality.[viii] Gone, too, were various procedures which might be suspected of being magical, including a lengthy discussion of amulets. Or at least anyone looking for them with the first edition in hand would think they were gone. But a recipe for finding out if your wife is faithful or not (put a magnet, engraved with an image of Venus, under her pillow: if she is faithful, she will make amorous advances to you in her sleep; if not, she will kick you out of the bed) is simply moved into the expanded section on magnets, and rendered relatively innocuous by being transformed into a bit of scholarly learning. Also displaced into a relatively safe position, to the last book, a collection of miscellaneous remarks entitled 'Chaos', was a recipe guaranteed to make women tear their clothes off and dance wildly: heat hare's fat over a lamp until it smokes. The recipe is incomplete, since the lamp ought to be inscribed with mysterious characters and there is an incantation to be mumbled. Even so, della Porta was obviously relying on the censors being too careless to pay close attention right to the very end, for these mysterious characters and mumbled incantations would surely have implied black magic to any suspicious reader.

Religious censorship was not the only pressure upon della Porta's text. We know he was engaged in a lengthy search for the secret of how to turn base metal into gold; indeed, briefly in the 1580s, he thought he had found it.[43] Such a secret could not be widely disseminated, as there would be no point in making gold if everyone was at it. In both editions of *Of Natural Magic* he assured the reader that he was not going to promise mountains of gold, but in both he pointed out that he had

---

up Divine things, as heaven and the stars (for therefore are they said to have Divine Spirits) and partly with sensitive and vegetative powers, which it bestows upon frail and transitory things. Thus much *Virgil* well perceiving, calleth this Spirit, The soul of the World . . . But the truth of Christianity holdeth that the Souls do not proceed from the Spirit, but even immediately from God himself.' Della Porta, *Natural Magick* (1658), 7–8.

viii Reginald Scot naturally seized on this story in his *Discovery of Witchcraft* (1584).

sometimes expressed himself obscurely, hiding the truth from the unsophisticated reader, and he provided recipes for enterprises which amounted to making both fake and real silver and gold: increasing the weight of a bar of gold, for example.[44]

Despite acknowledging that most alchemists are con artists, della Porta assures us that he is different and offers only reliable information. Both editions of *Of Natural Magic* open with a promise that della Porta will speak only from personal experience:

> Many men have written what they never saw, nor did they know the Simples that were the Ingredients, but they set them down from other mens traditions, by an inbred and importunate desire to adde something, so Errors are propagated by succession, and at last grow infinite, that not so much as the Prints of the former [i.e. the original ingredients] remain. That not onely the Experiment will be difficult, but a man can hardly reade them without laughter.[45]

The first edition provides two examples of such errors. Cato and Pliny had said that a bottle made of ivy could be used to establish whether wine had been adulterated with water, as the wine would be expelled, leaving the water behind. And Galen had said that it was false to claim that crushed basil spontaneously generated scorpions: della Porta had tested this by putting crushed (crushed, but not torn!) basil outdoors on earthenware tiles, and not only had scorpions been generated, but other scorpions, attracted by the smell of the basil, had come flocking. (Della Porta does not bother to explain how he could distinguish the freshly generated scorpions from those who were simply passing by.)

Della Porta is thus a puzzling case. He has the notion that much information is unreliable and needs to be tested, and yet he seems incapable of performing sensible tests. Part of the problem is that he is incorrigibly dishonest, insisting he has seen and done things he cannot possibly have. This dishonesty has recently been shown to lie at the heart of his discussion of magnetism in the second edition, which is largely plagiarized from an anonymous manuscript written by a Jesuit philosopher, a Venetian nobleman teaching at the Jesuit college in Padua, Leonardo Garzoni (1543–92).[46] The evidence is clear: not only does della Porta echo Garzoni's words, but he misunderstands some of the experiments he claims to have performed, and so misreports them. Ironically, della Porta was soon complaining that his own discussion of magnetism had been plagiarized by William Gilbert.[47]

Only one incomplete copy of Garzoni's text survives. It had formed part of the Paduan library of Giovanni Vincenzo Pinelli, a library in which Galileo read hard-to-find books, and the whole library had been sold on Pinelli's death and loaded on to ships for transport from Venice to Naples. One ship, carrying part of the library, was seized by pirates, who were dismayed to discover the cargo consisted of nothing but old books; they tossed some boxes overboard out of sheer frustration, and, having seized the crew (who, unlike the cargo, could easily be sold), left the ship to drift until it was wrecked. Many of the books and manuscripts that survived the waves were burnt by fishermen as if they were driftwood. Pages were torn out to plug holes in boats, or to stretch across windows (window glass was still a luxury).[48] By the time the owner's representatives arrived to lay claim to what remained, much damage had been done. This particular manuscript escaped the waves and the flames, but evidently fell into the hands of the fishermen, for part of it has disappeared.[49]

But there must once have been other copies, for it is not only a crucial unacknowledged source for della Porta but also an acknowledged one for *The Magnetical Philosophy* of the Jesuit Niccolò Cabeo (1629). William Gilbert, whose *On the Magnet* (1600) is generally held to mark the beginning of modern experimental science, is heavily dependent on Garzoni, although probably only as transmitted by della Porta.[50] Garzoni devised over a hundred experiments, many of which were copied by Gilbert. There is a good case for regarding him, not Gilbert, as the founder of modern experimental science (we'll come back to this problem in the next chapter). How did della Porta get hold of Garzoni's text? It is possible he read it when he was in Venice in 1580/81 – but of course only if it had been written by then, of which we cannot be confident. The great Venetian historian and scientist Paolo Sarpi was in Naples from 1582 to 1585, and della Porta tells his readers that he has learnt most of what he knows about magnetism from him; Sarpi may have supplied him with a copy of Garzoni's text. Della Porta was also a lay brother of the Society of Jesus: this seems to have been part of his efforts to prove his religious orthodoxy after his trial for heresy; so he may have had other ways of gaining access to Jesuit philosophy.

A further problem is presented by della Porta's fulsome thanks to Sarpi. He writes:

> I knew at Venice R. M. *Paulus* the Venetian, that was busied in the same study: he was Provincial of the Order of servants, but now a most worthy

> Advocate, from whom I not onely confess, that I gained something, but I glory in it, because of all the men I ever saw, I never knew any man more learned, or more ingenious, having obtained the whole body of learning; and is not onely the Splendor and Ornament of Venice or Italy, but of the whole world.[51]

This is the only occasion on which della Porta thanks a named individual. Sarpi, at some point, wrote a brief treatise on magnetism, which no longer survives. Perhaps della Porta had read it. But one reason for introducing Sarpi is evidently to provide cover in case della Porta's plagiarism is exposed; by acknowledging a debt to Sarpi della Porta can deny having ever read Garzoni, for he can claim that any similarity between their texts is the result of what he has learnt from Sarpi.

Garzoni is not very interested in garlic and magnets; but his tract begins by stating that there is much nonsense written about magnets, and that reliable knowledge has to be based on experiments. He explains the basic equipment you need: a couple of magnets, some small iron bars, some iron pointers. And he remarks that with this equipment you will find it easy to establish that garlic and diamonds do not disempower magnets; you can do the experiment whenever you like. Della Porta was clearly intrigued, and if he did not actually perform a number of the experiments described by Garzoni which he claimed to have, it seems he did perform these particular experiments. Having insisted repeatedly (twice in the first edition, once in the second) that garlic disempowers magnets, he now reports that 'when I tried all these things, I found them to be false: for not only breathing and belching upon the Loadstone after eating of Garlic, did not stop its vertues: but when it was all anoynted over with the juice of Garlic, it did perform its office as well as if it had never been touched with it.'[52]

Having disproved the supposed capacity of garlic to disempower the magnet (sailors themselves, we are told, had no time for this story, for 'Sea-men would sooner lose their lives, then abstain from eating Onyons and Garlick'), della Porta goes on to show that the popular belief (which he had happily accepted in the first edition) that a diamond also disempowers the lodestone is false:

> I tried this often, and found it false; and that there is no Truth in it. But there are many Smatterers and ignorant Fellows, that would fain reconcile the ancient Writers, and excuse these lyes; not seeing what damage they bring to the Common-wealth of Learning. For the new Writers, building on their ground, thinking them true, add to them, and invent, and draw

> other Experiments from them, that are falser then the Principles they insisted on. *The blinde leads the blinde, and both fall into the pit.* Truth must be searched, loved and professed by all men; nor must any mens authority, old or new, hold us from it.[53]

Indeed, della Porta's experiments led him to adopt the opposite view: that you can use a diamond to make, rather than unmake, a magnet.

False, too, is the claim (again made in the first edition, and a commonplace in the literature from Pliny onwards) that goat's blood re-empowers it:

> Since therefore there is an Antipathy between the Diamond and the Loadstone; and there is as great Antipathy between the Diamond and Goats blood, as there is sympathy between Goats blood and the Loadstone; We are from this Argument proceeded thus far, that when the vertue of the Loadstone is grown dull, either by the presence of the Diamond, or stink of Garlick, if it be washed in Goats blood it will then recover its former force, and be made more strong: but I have tried that all the reports are false. For the Diamond is not so hard as men say it is: for it will yield to steel, and to a moderate fire: nor doth it grow soft in Goats blood, or Camels blood, or Asses blood: and our Jewellers count all these Relations false and ridiculous. Nor is the vertue of the Loadstone, being lost, recovered by Goats blood. I have said so much, to let men see what false Conclusions are drawn from false Principles.[54]

When he is writing about garlic and magnets della Porta reads just like a modern, and yet only a few pages earlier (before the passages stolen from Garzoni) he had been as addlepated as ever: it would surely be possible, he suggests, to communicate with someone at a distance (even with someone held fast in prison) if you are each equipped with a compass with the alphabet engraved around the dial. If one person points the needle of their compass at a letter, the other compass needle will swing and point at the same letter. Yet again, 'false conclusions are drawn from false principles,' though at least he does not claim to have tried the method and proved that it works.[ix]

So it is impossible to turn della Porta into a cautious empiricist concerned to get his facts right, for all his own repeated assertions to the contrary, or to lay claim to him as a modern thinker, for all his insistence that he intends to show 'how exceedingly this later Age hath surpassed

ix For later accounts of the same device, Passannante, *The Lucretian Renaissance* (2011), 1–2; Glanvill, *The Vanity of Dogmatizing* (1661), 202–4.

Antiquity'.[55] And yet this assessment is also obviously wrong: when it comes to magnets and their supposed interactions with garlic, diamonds and goat's blood, della Porta is a modern empiricist, determined to get the facts straight, even if it involves sacrificing a cherished theory. There appear to be two della Portas. One just talks the talk; the other, surprisingly, walks the walk.

There is a simple explanation for this. Della Porta reads like a modern when Garzoni is doing his thinking for him, and like Pliny when he is thinking for himself. So these little nuggets of problematic facticity – the inability of garlic to disempower a magnet or of goat's blood to re-empower it – found their way into della Porta's text. And of course he was delighted to include them. What better proof could there be, after all, of his oft-repeated claim that he relied on experience not authority?

And yet della Porta could not bring himself to rethink his world in the light of this simple discovery. And so garlic and the lodestone still found their traditional place in his crucial chapter on sympathy and antipathy, in which we learn that a wild bull tied to a fig tree becomes tame, that basilisks are frightened by the crowing of cocks, that a well-washed snail cures drunkenness, that the sight of a wolf renders a man speechless, and that garlic disempowers magnets. The refutation of the supposed antipathy between garlic and magnets was like a loose thread on a sweater: pull on it and the whole thing would unravel. So della Porta simply tucked the thread back into place and pretended there was nothing wrong.[56]

It is easy to eliminate an obvious alternative. One could suggest that the new section on magnetism was added at the last minute, and that della Porta simply failed to revise his introduction in the light of his new conclusions. This will not do, for the new section on magnetism contains material that della Porta had moved out of its original position as he sought to make his book inoffensive to the censors; it would seem evident, then, that he wrote or revised the section on magnetism at the same time as he revised the opening chapters. In any event, the whole work must have been carefully reviewed before publication to ensure that it would satisfy both the Congregation of the Index (who were in charge of ecclesiastical censorship) and the Inquisition (who prosecuted heresy). Della Porta must have realized that he was contradicting himself.

So, willy-nilly, a little fragment of problematic facticity was let loose in the world. Anyone with access to a compass and a clove of garlic could perform their own test, which is why the unmaking of the old 'fact' was so important. Much harder to lay one's hands on a wild bull, a basilisk or a wolf. As della Porta's book went through edition after edition, translation after translation, it carried (for those who read far enough; some hardly read beyond the chapter on sympathy and antipathy) a powerful antidote to the old beliefs. Della Porta had paid no more than lip service to the idea that all intellectual authority must be regarded with suspicion and all claims to experience put to the test, but even he had a part to play in rendering the old certainities problematic. Thus Bernardo Cesi in his *Mineralogia* of 1636 reports the old story that garlic disempowers the magnet but is sufficiently impressed by della Porta's vehement denials to be (almost) persuaded by him. He finds it harder to abandon the belief that diamonds disempower the magnet because it is unanimously supported by the most distinguished authors; nevertheless, he faithfully reports della Porta's insistence that he has direct experience to the contrary. In the end, though, Cesi was prepared to carry on, as della Porta was, as if nothing had really happened, as if one could both believe the old stories and disbelieve them at the very same time. After all, had Cesi not said earlier in his book that '[w]e know by daily experiments that the power of the lodestone is weakened by garlic'?[57]

At this point it looks as though we have solved the problem with which Lehoux presents us. Lehoux wants to argue that there is no real difference between Plutarch and della Porta, which is fair enough; he could not have made the same argument if he had compared Plutarch and Garzoni. But there is a further issue we need to explore. Plutarch, Garzoni and della Porta all appeal to experience. But let us look at what Plutarch says. It is that '[w]e have palpable experience of these things.' Cesi writes, 'We know by daily experiments that the power of the lodestone is weakened by garlic.' Arnold de Boate wrote in 1653 that the lodestone 'hath an admirable vertue not onely to draw Iron to it self, but also to make any iron upon which it is rubbed to draw iron also. It is written notwithstanding, that being rubbed with the juyce of Garlick, it loseth that vertue, and cannot then draw iron, as likewise if a Diamond be layed close upon it.' Compare Plutarch's 'We have palpable experience', Cesi's 'We know' and de Boate's 'It is written' with della Porta's 'when I tried all these things' or Garzoni's invitation to us to get our own equipment and make our own tests. Garzoni's and della Porta's facts are based not on a collective knowledge, on a shared understanding, but on direct, personal experience. Lehoux tells us that an anonymous referee, reading his text, 'correctly pointed out that there is logically another possibility [than that Plutarch's 'experience' and della Porta's are the same sort of thing]: that Plutarch may mean something significantly different by "experience" than we do'.[58]

The referee was right: Plutarch's experience was an *indirect* experience, just as Aristotle's phenomena were based on *other people's* experiences; Garzoni's and della Porta's experiences were based on real, personally performed tests.[59] A good example is provided by Pietro Passi, writing in 1614, who denies that diamonds disempower magnets: 'for I have carried out tests here in Venice, in order to clarify the question, in the presence of Padre Don Severo Sernesi . . . and I used twenty diamonds . . .' supplied by a jeweller of the highest repute.[60] Or take Thomas Browne, who dismissed the garlic/magnet antipathy as 'certainly false' in 1646. How did he know? 'For an Iron wire heated red hot and quenched in the juice of Garlick, doth notwithstanding contract a verticity from the Earth, and attracteth the Southern point of the Needle. If also the tooth of a Loadstone be covered or stuck in Garlick, it will notwithstanding attract; and Needles excited and fixed in Garlick until they begin to rust, do yet retain their attractive and polary respects.'[61] Browne does not use the first person singular, but his careful use of detail (the red-hot iron, the rusting needles) suggests direct experience, not conventional presumption. Jacques Rohault in 1671 certainly does use the first person singular: '[T]hese are stories [about magnets and garlic] which are refuted by a thousand experiments that I have performed.'[62]

We can see an early example of the new standard of evidence at work in Anselmus Boëtius de Boodt's study of minerals published in 1609. De Boodt, who came from Bruges, was educated at Padua, and became the personal physician to the Emperor Rudolph II. He accepted that modern scholars must be right when they claim that garlic has no effect on magnets, since the sailors agree with them; as for the capacity of diamonds to disempower a magnet, he reported both the traditional view and the experiments (or supposed experiments) of della Porta, and then cautiously added that he had not conducted a test of his own.[63] He was suspicious of the claim made by Pliny and others that there is a sort of magnet which repels iron instead of attracting it; he had never been able to see this for himself or find a reliable first-hand report of it. He also doubted the often-repeated claim that there is a stone (the pantarbe) which attracts gold as the magnet attracts iron, and that there is another type of magnet which attracts silver: he could find no direct eyewitness testimony in either case.[64] And he rejected the claim that diamonds cannot be smashed with a hammer: in recent times, every diamond tested had proved to be frangible. There could be no need to resort to goat's blood to soften the diamond.[65]

Of course, the new opponents of the claim that garlic destroyed the power of the magnet – della Porta, William Barlowe (1597),[66] Gilbert, Browne – were not immediately victorious. The old views were upheld by Jan Baptist van Helmont (1621), Athanasius Kircher (1631) and Alexander de Vicentinis (1634).[67] The last attempt to give them a serious scientific formulation appears to be Robert Midgeley's *New Treatise of Natural Philosophy* (1687).[68] How was this possible? The best answer is provided by Alexander Ross's reply (1652) to Browne:

> I know what I said but now (*Book* 2. *c.* 3.) of the Garlick in hindring the Load-stones attraction, is contradicted by Doctor *Brown*, and before him by *Baptista Porta*; yet I cannot believe that so many famous Writers who have affirmed this property of the garlick, could be deceived; therefore I think that they had some other kinde of Load-stone, then that which we have now. For *Pliny* and others make divers sorts of them, the best whereof is the Ethiopian. Though then in some Load-stones the attraction is not hindred by garlick, it follows not that it is hindred in none; and perhaps our garlick is not so vigorous, as that of the Ancients in hotter Countries.[69]

In other words, Ross knew perfectly well that he would not be able to confirm the story by testing it, yet he continued to believe it nevertheless. 'Grave and worthy Writers' trumped his own experience and that of his contemporaries.

The correct response to this is to be found in one of Aesop's fables, 'The Braggart'. An athlete boasts that in Rhodes he performed the most astonishing jump and claims he can produce eyewitnesses to testify to it. Which meets with the reply: 'Imagine this is Rhodes. Jump here.' (In Latin, *hic Rhodus, hic saltus*).[x] Thus the alchemist George Starkey insisted that he did not just rely on testimonials but was prepared to be put to the test whenever his critics chose to name a time and place: *hic Rhodus, hic saltus*.[70]

I don't mean to suggest that we moderns are unlike Ross in that we believe only things we have experienced for ourselves. But, like de Boodt, we believe things (at least where science is concerned) only if we are

x A phrase later made famous by Marx in the *Eighteenth Brumaire* in the variant form *Hic Rhodus, hic salta*. (For an account of the derivation of Marx's phrase in a pun by Hegel, http://berlin.wolf.ox.ac.uk/lists/quotations/quotations_by_ib.html, accessed 22 Dec. 2014.)

confident that they can be traced back to a direct experience, or a series of direct experiences, and can survive retesting.[71] If I wanted to persuade you of continental drift, for example, I would point you to the classic papers on paleomagnetism and we could then go and make our measurements in the field. Boyle laid out the rules for the new knowledge in the methodological preface to his *Physiological Essays* of 1661 (revised in 1669). There, he distinguishes, as I have done, between writers who insist on their direct experience of matters of fact, or at least their reliance on identifiable witnesses who have had direct experience, and those who uncritically report established traditions. His policy, he says, is not to quote the second sort (naming Pliny and della Porta as examples):

> [T]he occasions I have had of looking into divers matters of fact deliver'd in their Writings, with a bold and an impartial Curiosity; have made me conclude so many of those Traditions to be either certainly false, or not certainly true, that except what they deliver upon their own particular Knowledge, or with peculiar Circumstances that may recommend them to my belief, I am very shy of building any thing of moment upon foundations that I esteem so unsure.

For, he insists, he does not appeal 'to other Writers as to Judges, but as to Witnesses, nor employing what I have found already publish'd by them barely as Ornaments to imbellish my Writings, and much less as Oracles by their Authority to demonstrate my Opinions, but as Certificates to attest Matters of fact'.

Having chosen to trust only a few, reliable writers, Boyle expressed contempt for the rest, the Plinys and della Portas:

> [W]hen vain Writers, to get themselves a name, have presum'd to obtrude upon the credulous World such things, under the Notion of Experimental Truths, or even great Mysteries, as neither themselves ever took the pains to make tryal of, nor receiv'd from any credible Persons that profess'd themselves to have try'd them; in such cases, I see not how we are oblig'd to treat Writers that took no pains to keep themselves from mistaking or deceiving, nay, that car'd not how they abuse us to win themselves a name, with the same respect that we owe to those, who though they have miss'd of the Truth, believ'd they had found it . . .

Honest mistakes must be clearly distinguished from a failure to take pains. What Boyle was advocating was a disciplined, organized distrust of other authors; this was the logical consequence of trying to find out not what books say but what '[t]hings themselves would incline me to think'.[72]

Boyle never discussed garlic and magnets, but he did address the old belief that it was impossible to crush a diamond unless it had first been softened in goat's blood. Too frugal to experiment with one of his own diamonds, he sought advice from someone with first-hand experience:

> Notwithstanding the (lately mention'd) wonderful Hardness of Diamonds, there is no Truth in the Tradition, as generally as 'tis receiv'd, that represents Diamonds as uncapable of being broken by any External force, unless they be soften'd by being steep'd in the Blood of a Goat. For this odd Assertion, I find to be contradicted by frequent practice of Diamond Cutters: And particularly having enquir'd of one of them, to whom abundance of those Gems are brought to be fitted for the Jeweller and Goldsmith, he assur'd me, That he makes much of his Powder to Polish Diamonds with, only, by beating board Diamonds (as they call them) in a Steel or Iron Morter, and that he has that way made with ease, some hundreds of Carrats of Diamond Dust.[73]

The assertion that goat's blood softens diamonds seemed odd to Boyle, because he had escaped from the old conceptual framework of sympathy and antipathy, according to which there was a natural sympathy between the lodestone and goat's blood and a natural antipathy between the diamond and goat's blood. But all that was required to abolish this conceptual scheme was an insistence on direct as opposed to indirect experience.[xi]

The result of such an approach, which looks to us just like common sense but was revolutionary at the time, was a transformation in the reliability of knowledge.[74] William Wotton, in his *Reflections upon Ancient and Modern Learning* (1694), put it like this:

xi The epistemological sloppiness that, in Western culture, characterizes references to experience before the Scientific Revolution is not a necessary characteristic of pre-scientific societies. The Matses, an Amazonian tribe, are obliged to specify, whenever they use a verb, 'exactly how they come to know about the facts they are reporting . . . There are separate verbal forms depending on whether you are reporting direct experience (you saw someone passing by with your own eyes), something inferred from evidence (you saw footprints on the sand), conjecture (people always pass by at that time of day) or hearsay (your neighbour told you he had seen someone passing by). If a statement is reported with the incorrect evidentiality form, it is considered a lie. So if, for instance, you ask a Matses man how many wives he has, unless he can actually see his wives at that very moment, he would answer in the past tense and would say something like . . . "There were two last time I checked."' (Deutscher, *Through the Language Glass* (2010), 153)

> *Nullius in verba* ['Take no man's word for it', i.e. defer to no authority][xii] is not only the motto of the *ROYAL SOCIETY*, but a received Principle among all the Philosophers of the present Age. And therefore, when once any new Discoveries have been examined, and received, we have more Reason to acquiesce in them than there was formerly ... So that, whatsoever it might be formerly, yet in this Age general Consent ... especially after a long Canvass of the Things consented to, is an almost infallible Sign of Truth.[75]

Here again, we are coming up against one of the conditions of possibility of the new science. I can trace the unmaking of the garlic-disempowers-magnets pseudo-fact to della Porta's direct experience; I can do this because I have a number of key books to hand. I know about Garzoni because his tract has, at long last, made its way into print. In a manuscript culture claims to experience cannot be traced back in this way. Plutarch could not go back beyond the 'we' that seemed to him secure enough; he could not point to any direct experience. Printed books, by improving access to information, make it far easier to establish and refute facts. In the course of a few years della Porta's personal experience came to be shared with the whole of educated Europe. As Wotton put it in 1694, 'Printing has made Learning cheap and easie.'[76] It may at first seem strange, but it is the printing press that made it possible to privilege the eyewitness account over all others, simply by making available a much greater range of accounts from which to choose.[77]

What we are seeing when we read della Porta is a moment of transition, not only between ancient beliefs and modern ones but also between a manuscript culture, in which experience is unspecific, indirect and amorphous (and in which a conman like della Porta can hope to get away with all sorts of wild claims; when he died he left behind a manuscript in which he claimed to have invented the telescope),[78] and a print culture, in which experience is specific, direct, documented and retrievable. In a print culture it becomes possible to apply the peculiarly high standards of a court of law (Roman or common law) to anything and everything. In comparison to the world of print, manuscript culture is

xii Horace, *Epistles* I.i, lines 14–15:

Nullius addictus iurare in verba magistri,
quo me cumque rapit tempestas, deferor hospes.

'I am not bound to swear allegiance to any master; where the storm drives me I turn in for shelter.' See above, p. 44.

one of rumour and gossip. The printing press represents an information revolution, and secure facts are its consequence.

Garzoni's text, with its invitation to test garlic and magnets, existed, until 2005, only in manuscript. However, it seems to have been written with a view to publication. It is this that underlies his decision to declare war on 'false rumour and the opinions of some, based on unreliable and untrustworthy foundations'.[79] Unfortunately, his success in demolishing the supposed antipathy between garlic and magnets remained hidden from history until very recently; it is della Porta who established the new fact within the world of learning, not Garzoni.

Yet the printing press alone does not explain the unique authority now given to eyewitness testimony. In the post-Columbus, post-Galilean world no one could dispute that important discoveries depended solely on the corroboration of eyewitnesses.[80] As we saw in Chapter 3, the very concept of discovery depended on the conviction that there could be new experiences, unlike any that had gone before. Moreover, many discoveries were made by men of low social status, men like Columbus himself, or Cabot, who had discovered the variation of the compass. Thus we suddenly find sailors and jewellers being asked to resolve disputes between philosophers and gentlemen. Bacon had seen clearly that this was the direction the new philosophy must take. But the revolution was long and slow: if Garzoni marks its beginning in the 1570s or 1580s, neither Browne in the 1640s nor Boyle in the 1660s marks its end. For us, to whom the privileged status of eyewitnesses seems obvious, this great revolution has ceased to be visible, and it is almost impossible to conceive of ourselves living in a world – a world that was never real, but always imaginary – in which garlic disempowers lodestones and goat's blood softens diamonds.

§ 4

Kepler had lots of facts, and della Porta had one or two, but neither had the word 'fact' in the modern sense. Where does that word come from? In 1778 Gotthold Lessing wrote a little essay on the German word for 'fact', *Tatsache*: 'The word is still youthful,' he said. 'I remember perfectly the time before anyone used it.'[81] But the word itself, at least in English, French and Italian, is not new. It's source is the Latin verb *facio*, 'I do'. *Factum*, the neuter past participle, means 'that which has been done'. Throughout Europe, wherever the influence of Roman law was felt, the law concerned itself with the *factum* – the deed, or crime. Thus 'the fact of Cain' was the killing of Abel.[82] In Shakespeare's *All's Well that Ends Well*, Helena says:

> Let us assay our plot; which, if it speed,
> Is wicked meaning in a lawful deed
> And lawful meaning in a lawful act,
> Where both not sin, and yet a sinful fact:[83]

The play on words here depends on 'fact' being both a synonym for 'deed' and 'act' but also a word used specifically for unlawful deeds and acts. We still use this (now somewhat archaic) language when we talk of 'an accessory after the fact' – someone who helps a criminal after the crime has been committed.

In England the jury was the judge of fact (Did Joe kill Tom? The jury determined whether Joe did the deed); the judge was the authority on questions of law (Under what circumstances can someone kill someone else in self-defence? Is this document correctly drawn up?) One could appeal against the judge's interpretation of the law and his guidance to the jury, but not against the jury's determination of the fact.[84] It should be stressed that there was nothing natural about this legal conception of the fact; it was a construction of the thirteenth century, when the jury was introduced as a substitute for trial by ordeal.[85] But it meant that the fact had a peculiar status in English law: once determined, it could never be disputed. Hence the peculiar feature of the word 'fact' in its modern usage that a fact (unlike a theory) is always true: facts are infallible because juries determined facts and they were held to be infallible (or at least incorrigible and indisputable, which amounts to the same thing).

Between the Latin *factum* and the modern English 'fact' there was a barrier that had to be crossed: a *factum* requires an agent, a fact does not. The barrier is clear in principle, although there are inevitable ambiguities in practice. Bacon (d.1626), when writing in a posthumously published text about the capacity of the imagination to act on bodies, insists it is wrong 'to mistake the Fact or Effect; And rashly to take that for done, which is not done'.[86] Thus witches often claim to be responsible for events that would have happened anyway. A 'fact' here is still an action or deed (although Dr Johnson's *Dictionary* of 1755, quoting Bacon, says otherwise).[87] When Noah Biggs describes, in 1651, how sunflowers turn to follow the sun he calls it a 'matter of fact' but he also calls it a 'thing done'; he is treating sunflowers as agents, with what he calls instincts.[88] He is stretching but certainly not breaking the convention that old-fashioned facts have agents. The same thing happens a year later when Alexander Ross, previously chaplain to Charles I, discusses an ancient story, one told by Averroes and rejected by Thomas Browne, of a woman who got pregnant from bathing in water in which men had bathed before her; Ross thinks an instinctive attraction between womb and semen may be at work.[89] A second ambiguity occurs, as here, in discussing historical events which don't quite count as actions. History was held to be concerned with facts – with things people had done. In his diary entry of 1 September 1641 John Evelyn, then in the Netherlands, records a visit 'to see the monument of the woman pretended [claimed] to have been a Countess of Holland [and] reported to have had as many children at one birth as there are days in the year. The basins were hung up in which they were baptized together with a large description of the matter of fact in a frame of carved work in the church of Lysdun, a desolate place.' Although births are not exactly actions, they easily slip into the realm of historical facts.[90]

When and where was the language of the fact invented? Only quite recently historians thought there was a straightforward answer to this question. Francis Bacon invented the fact; from Bacon the fact entered the English language and was adopted by the Royal Society. So historians began to write about 'Baconian facts'.[91] English philosophy has always been thought to be peculiarly empiricist; on this account, it seemed that England had created and invented the culture of the fact.[xiii]

Unfortunately, this story just won't do. Crucially, the fact isn't English. Galileo and his correspondents happily discuss facts, but there are Italian usages from much earlier, from the 1570s.[92] According to the established scholarship, the French discovered the new word only in the 1660s[93] – and yet Montaigne uses *faict* to mean fact no less than five times, one of which dates from 1580 (before his journey to Italy), and the rest from 1588 (they occur in three crucial essays, 'On Regret', 'On Experience' and 'On the Lame'). It is worth remarking that in three of these five cases Florio, who was the first to translate Montaigne into English, felt that he could stretch the English word 'fact' to cover the

xiii Let me add as an aside that a word that we think of as naturally paired with the word 'fact', the word 'fiction', enters English in the 1590s; the term 'fiction' predates the modern idea of fact.

meaning of Montaigne's *faict*, but in two he could not.[xiv] Similarly, Montaigne's disciple Charron in *De la sagesse* twice uses *faict* to mean fact – but in neither case does Samson Lennard's English translation of 1608 feel it can use the English word 'fact'.[94] We can be sure that Montaigne and Charron were not alone: Jean Nicot's *Thresor de la langue françoyse, tant ancienne que moderne* of 1606 contains a couple of examples of the noun *fait* being used in the modern sense: *articuler faits nouveaux* can refer to new deeds or new things; 'the facts' can be those things, quite generally, which underpin an argument.[95]

Nor is the word 'fact' Baconian. Bacon never uses the word in its modern meaning in English in print, and he uses *factum* three or perhaps four times in print, but the crucial text, the *Novum organum* of 1620, was not translated into English in time to have any influence.[96] The failure of Bacon (or, for that matter, Florio) to introduce the word 'fact' into everyday English is plain as a pikestaff from Browne's failure to use the word in its impersonal sense despite his familiarity with Montaigne and Bacon, his love of Latinate language and his evident need for a word (other than 'pibbles') to describe his weapons of war. As far as Browne was concerned, the word he needed did not exist.

§ 5

The man with a far stronger claim than Bacon to have introduced the word 'fact' into English is Thomas Hobbes, who discusses facts in the first part of his *Elements of Law, Natural and Politic*, written in 1640 (but

xiv The three occasions on which Florio follows Montaigne are: 'I have no body to blame for my faultes or misfortunes, but my selfe. For in effect I seldome use the advise of other[s] unlesse it be for complements sake, and where I have need of instruction or knowledge of the fact' (Montaigne, *Essayes* (1613), 456); 'True it is, that proofes and reasons grounded upon the fact and experience. [*sic*] I untie not: for indeede they have no end; but often cut them, as *Alexander* did his knotte' (582); 'And were it not, that what I doe for want of memorie, others more often doe the same for lacke of faith, I would ever in a matter of fact rather take the truth from anothers mouth, then from mine owne' (605). But where Montaigne has '*Je vois ordinairement que les hommes, aux faicts qu'on leur propose, s'amusent plus volontiers à en chercher la raison qu'à en chercher la verité*', Florio has 'I ordinarily see, that men, in matters proposed them, doe more willingly ammuze and busie themselves in seeking out the reasons, than in searching out the trueth of them'(578). And where Montaigne has '*joinct qu'à la verité il est un peu rude et quereleux de nier tout sec une proposition de faict*', Florio has 'since truely, it is a rude and quarelous humour, flatly to deny a proposition' (579).

not published until 1650, under the title *Humane*[xv] *Nature*).[xvi] Hobbes had been a secretary to Bacon, but he had also (according to Aubrey) met Galileo, whom he undoubtedly admired; either or both may have been behind his use of the word 'fact'. Hobbes circulated the *Elements* among his friends, and so the word first appears in print in English in its modern sense in a text written by one of Hobbes's friends: that bundle of contradictions – for he was a Protestant and a Catholic, an Aristotelian and an atomist, a faithful royalist and a friend of Cromwell – Sir Kenelm Digby. In his book on the immortality of the soul, published in Paris in 1644, Digby argues that the fantasies women have while having intercourse can affect the appearance of their children; so if you think of your lover as a bear you may have a furry baby. We can establish 'the verity of the fact', he says, without having knowledge of the cause.[97]

Next, in 1649, comes a translation of a book by Jan Baptist van Helmont on the weapon salve (an ointment which cures wounds by being applied not to the body but to the weapon), a book first published in Latin in 1621 and now translated with a lengthy introduction by Digby's friend Walter Charleton. Charleton is well aware that his usage of the word 'fact' is unusual: the first time he introduces it (in the preface), he paraphrases it with a Latin phrase, *de facto*; the second, with a Greek word, *hoti*.[xvii][98]

xv In modern English there are two distinct words, 'human' and 'humane', but in seventeenth-century English there is one word, usually spelt 'humane', with two distinct meanings. 'Humane' here, and in the title of Locke's *Essay*, means 'human'.

xvi '[I]f a Man seeth in present that which he hath seen before, he thinks that that which was antecedent to that which he saw before, is also antecedent to that he presently seeth: As for Example, He that hath seen the Ashes remain after the Fire, and now again seeth ashes, concludeth again there hath been Fire: And this is called again *Conjecture* of the past, or Presumption of the Fact . . . When a Man hath *so often* observed like Antecedents to be followed by like Consequents, that *whensoever* he seeth the Antecedent, he looketh again for the Consequent; or when he seeth the Consequent, maketh account there hath been the like Antecedent; then he calleth both the Antecedent and the Consequent, *Signs* one of another, as Clouds are Signs of Rain to come, and Rain of Clouds past' (Hobbes, *Humane Nature* (1650), 37–8); ' . . . there be *two Kinds* of Knowledge, whereof the *one* is nothing else but *Sense*, or Knowledge *original* . . . and *Remembrance* of the same; the *other* is called *Science* or Knowledge of the *Truth of Propositions*, and how Things are called; and is derived from *Understanding* . . . And of these two Kinds of Knowledge, whereof the former is *Experience of Fact*, and the later of *Evidence of Truth*; as the *former*, if it be great, is called *Prudence*; so the *latter*, if it be much, hath usually been called, both by Ancient and Modern Writers, *Sapience* or Wisdom: and of this *latter, Man* only is capable; of the *former, brute Beasts* also participate' (60–1, 64–5).

xvii 'An example, *de facto*' (Prolegomena to Helmont & Charleton, *A Ternary of Paradoxes* (1649), c2v); 'Wherefore I come directly to the examination of the Hoti, or matter of

After these works by Hobbes's friends came Hobbes's *Humane Nature* (the first part of the *Elements*) in 1650. Here Hobbes distinguishes between two types of knowledge: science, which is, as Hume would later say, about the relationship between ideas; and what he calls prudence, which is about facts. And Hobbes sticks in other respects (apart from his innovatory use of the word 'facts') to an old-fashioned vocabulary: knowledge of facts is derived from testimony and from signs (which we would call evidence; we have seen the English translation of della Porta calling them 'prints', as in paw prints), while knowledge of concepts is accompanied by what he calls evidence (which we would call understanding). Next comes a book written in Paris but published in England, Hobbes's *Leviathan* (1651), a text widely read and deeply influential, but one to which few thinkers wanted to acknowledge their debt because it was generally held to be wickedly atheistical. There, for the first time in English, we are told that there are historical facts (the actions of men and women), the subject of civil history, and natural facts, the subject of natural history.[xviii] And, for Hobbes is nothing if not consistent, the same vocabulary appears in his *Of Libertie and Necessitie* of 1654.[99]

I have been able to find only one unambiguous use of the word 'fact' in print in English before 1658 outside these three authors; this is in a work called *The Modern States-Man* by G. W., published in 1653. Unfortunately, we cannot identify G. W. with any confidence, but the likelihood is that he had been reading Hobbes.[100] The ambiguous usages of Noah Biggs and Alexander Ross, discussed earlier, also follow the publication of Hobbes's *Humane Nature*. So where did the three friends Hobbes, Digby and Charleton get the idea of the fact from? The right answer, I suspect, is from several places. Hobbes, as we have seen, knew both Bacon and Galileo. Digby was writing in France, but he also spoke Italian like a native – though this alone cannot have been a sufficient condition for using the new word, or we would find it in William Harvey and Thomas Browne, both of whom were educated in Padua. All three – Hobbes, Digby and Charleton – had a great deal of reading in

---

Fact.' (d1r). For examples from the translation, accompanied by the original Latin, see endnote 98.

xviii 'The Register of *Knowledge of Fact* is called *History.* Whereof there be two sorts: one called *Naturall History;* which is the History of such Facts, or Effects of Nature, as have no Dependance on Mans *Will;* Such as are the Histories of *Metalls, Plants, Animals, Regions,* and the like. The other, is *Civill History;* which is the History of the Voluntary Actions of men in Common-wealths.' Hobbes, *Leviathan* (1651), 40.

common, reading which surely included Montaigne, Galileo and Bacon. But there is one source which we cannot doubt: Charleton's Latin source, van Helmont, used *factum* to mean 'fact' (although Charleton used 'fact' when speaking in his own voice and not simply when translating van Helmont's Latin).

So far, what we have learnt is this: there were facts in Italian, French and Latin before there were facts in English; and the key role in the introduction of the word 'fact' into the English language was played not by Bacon but by Hobbes. There is a delicious irony in this because Hobbes believed factual knowledge was a truly inferior kind of knowledge, science consisting solely of deductive knowledge. Hobbes's thinking is straightforward: we may define facts as necessarily true, and say that something mistaken for a fact is not a fact at all, but mistakes are frequent; the supposed facts are often not facts. And when we try to draw conclusions from the facts we often go astray because we have misunderstood their significance. Hobbes even sketched out what would later become the classic problems of induction: Hume's, that just because the sun has risen every morning until now, it does not follow that it will rise tomorrow; and Popper's, that just because all the swans you have seen are white, it does not follow that there are no black swans (indeed, there are, in Australia) – in order to show the limitations of arguments from facts.[101] Hobbes was the first serious philosopher of the fact because he understood facts, but he did not trust them.

The next important contribution to the philosophy of the fact came in 1662, with the publication of *The Logic of Port-Royal*. The four final chapters of that work, apparently composed after 1660, and probably written by Antoine Arnauld, are famous for outlining modern probability theory for the first time; they are also the first extended discussion in French of the concept of the fact, for facts are here defined as contingent events, and contingent events are more or less probable. Thus 'it snowed on Christmas Day' is perfectly credible if it is a statement about Sydney, Nova Scotia, but suspect if it is a statement about Sydney, New South Wales. Where did Arnauld get his idea of the fact from? Not from Hobbes, whose key discussions of the subject were at this point available only in English. Arnauld's preoccupation with the fact developed out of the great dispute over whether Jansenism, of which Arnauld was the leading light, was heretical. After 1653 this dispute turned on the question of whether the five Jansenist propositions condemned by the pope as heretical were or were not to be found in Jansen's *Augustinus*. The pope, Arnauld argued, had authority in matters *de jure* but not in matters *de facto*. On the crucial matter of fact, whether the propositions were contained in the book (and it was a matter of fact rather than of act, because Jansen could hardly have intended to place in his work propositions that had yet to be formulated), the pope was simply wrong, and one could still defend the teaching of the *Augustinus* properly interpreted, while accepting the pope's authority to condemn the five propositions. In the course of this dispute, and with the example of Montaigne to hand, Arnauld reinvented the idea of the fact.[xix]

Following Arnauld's example, Blaise Pascal published his own defence of Jansenism in 1657. Written while he was in hiding from the authorities and published under the pseudonym Louis de Montalte, the *Provincial Letters*, which came out at first one by one, on illegal presses, contained frequent usages of the word for 'fact' in its modern sense.[xx] They were also brilliantly funny, and devastating for the Jesuits, against whom they were directed. Quickly, the text was translated into English, and published first in 1657 and then in an expanded edition in 1658.[102] The translation was organized by Henry Hammond, a royalist clergyman, but the book did not go unnoticed among members of the Royal Society: John Evelyn translated a sequel, *Another Part of the Mystery of Jesuitism*, which appeared in 1664.[103] In the *Provincial Letters* the word 'fact', and particularly the phrase 'matter(s) of fact', occurs over and over again, dozens of times. 'Matters of fact', as opposed to matters of law and of faith, becomes an intellectual slogan and a powerful political weapon. Pascal had never used the word 'fact' in its modern sense in his scientific writings, but now (even though it is most unlikely that any of his English readers knew the true identity of Louis de Montalte) he had given it respectability as the indispensable word required in any attack upon received opinion or in any dispute with established authority.

xix For example, Arnauld & Nicole, *Response au P. Annat* (1654), Avant-Propos: 'Or tout le monde demeure d'accord, que les Papes sont point garands de la verité des faits qu'on leur propose, et qu'ils rapportent en suitte dans l'expositif de leurs Rescrits et de leurs Constitutions . . .' In Arnauld, *Première lettre apologétique de Monsieur Arnauld Docteur de Sorbonne; À un évêque* (1656), 12, he denies that the pope and the bishops can determine '*un point purement de fait, & dont les yeux sont juges*'.

xx Pascal, *Les Provinciales* (1657), 48–9 (4th letter): '*Ce n'est pas icy un point de foy, ny mesme de raisonnement. C'est une chose de fait. Nous le voyons, nous le sçavons, nous le sentons.*'

§ 6

This complicates our main story, the dissemination of the idea of the fact in England. So far my argument has been that – in the language of epidemiologists – the index case is the circulation in manuscript of Hobbes's *Elements*, and that the word 'fact' spreads from there, first among the friends of Hobbes, and then a little more generally. But if we look at England in 1658, the year Cromwell dies, facts are now newly established in the language, thanks not to the friends of Hobbes, but to Pascal. It was also in 1658, in a text first published in French but then immediately translated into English, that Sir Kenelm Digby, returning to the question of the weapon salve, gave a clear definition of the new usage:

> In matter of fact, the determination of existence, and truth of a thing, depends upon the report which our senses make us. This businesse is of that nature, for they who have seen the effects, and had experience thereof, and have been carefull to examine all necessary circumstances, and satisfied themselves afterwards, that there is no imposture in the thing, do nothing doubt but that it is real, and true. But they who have not seen such experiences, ought to refer themselves to the Narrations, and authority of such, who have seen such things.[104]

Was Digby here echoing Hobbes or Pascal? We cannot tell.

Originally, the weapon salve is an unguent applied to the weapon which has caused a wound and thus cures the wound. One recipe involves bear grease, boar fat, powdered mummy (as in Egyptian mummy) and moss that has grown on a skull. Della Porta gives this recipe: 'Take of the moss growing upon a dead man his scull, which hath laid unburied, two ounces, as much of the fat of a man, half an ounce of Mummy, and man his blood: of linseed oyl, turpentine, and bole-armenick, an ounce; bray them all together in a mortar, and keep them in a long streight glass.'[105] Van Helmont, it is worth remarking, provoked the fury of the Jesuits, his co-religionists, by suggesting that the skull of a Jesuit would be ideal – he was hostile to the Jesuits because they had little difficulty persuading people to believe in their miracles, while his own scientific facts were met with scepticism. Digby was advocating a much simpler chemical powder which could be dissolved in water and easily carried with one into battle.

Because the weapon salve involved action over a distance, it defied a fundamental principle of Aristotelian physics, that action requires contact. Van Helmont, Charleton and Digby argued that this was no bar to an effective cure; they wanted to redescribe the weapon salve as 'magnetical' because the magnet provides a paradigm case of action over a distance. Their fundamental claim is that, although some argue that such cases are miraculous or demonic, they are in fact perfectly easy to reproduce. As Charleton wrote in 1649, he had no choice but to believe:

> until my Scepticity may be allowed to be so insolent, as to affront the evidence of my own *sense*, and question the verity of some *Relations*, whose Authors are persons of such confessed integrity, that their single Attestations oblige my faith, aequall with the strongest demonstration. Among many other *Experiments*, made by my self, I shall select and relate onely one: and that most ample and pertinent . . .[106]

And he goes on to report a test where the unguent was applied by a sceptical clergyman, so that there could be no suspicion of either cheating or demonic involvement. The weapon salve was thus to be brought within the sphere of experimental science, and strange facts were to be naturalized. There is a deep irony in the fact that the idea of factual knowledge was first promoted not, as some have thought, to help interpret Boyle's vacuum-pump experiments but to convince sceptics of the real efficacy of the weapon salve.[xxi]

By 1654 Charleton, whom we earlier met translating van Helmont, had become one of the insolent sceptics. He announced that he had changed his mind about the weapon salve. He had three objections to it: the theory underpinning it was incoherent (why could the unguent not cure any wound in its vicinity?); the claim that it worked needed to be tested by comparing a group treated with the weapon salve with a control group who had not been in order to establish that it worked better than no treatment at all; and, in any case, its supposed efficacy was, he now suspected, an illusion because the occasions on which it had appeared to succeed were widely reported, while its failures were lost in oblivion:

xxi Contrast Shapin & Schaffer, *Leviathan and the Air-pump* (1985), 24, where 'Boyle and the experimentalists' are taken as originating the preoccupation with matters of fact; Hobbes is seen simply as an opponent of arguments from facts (22), not as someone who had originated the language that Boyle uses. It is a strange feature of the book that it takes as its theoretical foundation Wittgenstein's concept of language-games (15, 22), but never considers the possibility that developing a new language-game might involve changing the meanings of words.

> [M]any of those stories [of its success] may be Fabulous; and were the several Instances or Experiments of their Unsuccessfulness summed up and alledged to the contrary, they would, doubtless, by incomparable excesses overweigh those of their successfulness, and soon counter-incline the minds of men to a suspicion at least of Error, if not of Imposture in their Inventors and Patrons.[107]

The facts which had previously convinced him now seemed mere chance occurrences. The principle at issue was straightforward: natural facts must be replicable and reproducible if they are to count as facts at all. We can see here in miniature how the idea of the fact was inseparable from questions of evidence and probability. And, of course, as replication became the test, historical facts, which had once seemed so solid and reliable, came to seem increasingly fragile and tenuous.

Quite suddenly, in the five years after the publication of Pascal's *Mystery of Jesuitisme* (1657) and Digby's *Late Discourse* (1658) on what he called the powder of sympathy, the word 'fact', in its new sense, was naturalized in the English language. This is the analogous moment in English to the revolution Lessing lived through a hundred years later in German; and the extraordinary success of Digby's *Discourse* – it went through twenty-nine editions – may have had a good deal to do with it. But Pascal's *Provincial Letters* surely had an even greater influence. Before 1658 instances of the use of the word are so few and far between that one can reasonably doubt that it exists in English except in a metaphorical or stretched sense or as a private idiolect. After 1663 facts are everywhere. In Germany the culture of the fact was created in the 1770s; in England and France it happened in the early 1660s.

In England the fact not only became linguistically commonplace; it also became institutionally entrenched, for the Royal Society's official aim was to establish new facts. According to their statutes of 1663:

> In all Reports of Experiments to be brought into the Society, the matter of fact shall be barely stated, without any prefaces, apologies, or rhetorical flourishes; and entered so in the Register-book, by order of the Society. And if any Fellow shall think fit to suggest any conjecture, concerning the causes of the *phaenomena* in such Experiments, the same shall be done apart; and so entered into the Register-book, if the Society shall order the entry thereof.[108]

Here the fundamental distinction between facts and explanations, which goes back to Montaigne and beyond, is restated. When the Royal

Society adopted as its motto *nullius in verba* it was committing itself not to scepticism about facts (experience distilled into words) but about the conjectures concerning the causes of inherently malleable phenomena which had been the central enterprise of scholastic natural philosophy. The society was to defer to the word of no authority but stick to the facts. *Nullius in verba* implied that facts are not words but things caught up in the net of language, like fish in a keep-net. And so when Sprat wrote what was somewhat misleadingly called *The History of the Royal-Society* – a work he began writing in 1663, when the society was three years old, and published in 1667 – facts were given a central role. Matters of fact, he insisted, must always trump authority, no matter how ancient; and facts were the sole concern of the Society: '[T]hey only deal in matters of *Fact*.'[109]

How did the fact enter the mainstream of English-language intellectual life? The first thing to note is that not everyone was quick to adopt it: there are no facts, for example, in Hooke's *Micrographia* or in Newton's *Optics* (both rely on the word 'observations').[110] More surprisingly perhaps, there are no facts in Robert Boyle's *New Experiments* of 1660, his first account of the air-pump experiments, only phenomena. In *Leviathan and the Air-pump* Steven Shapin and Simon Schaffer have argued that the production of facts is at the heart of Boyle's experimental method: the air pump is a machine for creating matters of fact. But this is not the case in the *New Experiments*. Boyle was already familiar with the English word 'fact' in its modern usage; he had used it in 1659 in a prefatory letter to a short work on the preservation of anatomical specimens (and his sister used it in a letter later that year).[111] He went on to use it three times in the *Sceptical Chymist* of 1661, eight times in the *Physiological Essays* of the same year (both texts were written some time before they were published), and it finally appears in the context of his vacuum experiments in his *Defence of his New Experiments* in 1662. This suggests that it took a while for Boyle to think of the word 'fact' as a respectable term to be used in natural philosophy when engaging with scholastics and Cartesians. It wasn't, for example, a word that his great predecessor Pascal had used when writing about his vacuum experiments (which we will come to in the next chapter). The *Sceptical Chymist* and the *Physiological Essays* were works heavily influenced by van Helmont; it took a while for this new terminology to cross over from the topics discussed by the followers of Paracelsus, the iatrochemists, into those discussed by the mathematicians. It looks as though at first Boyle wanted to keep these two sides of his intellectual life, with their different vocabularies, separate. But the word rapidly became fashionable, and by 1662 he could no longer resist it.

§ 7

What made the word 'fact' respectable in philosophical English? The standard argument is that the fact became important in the 1660s because it represented a way of ending (or sidestepping) debates; in a society which had been torn apart by Civil War, natural philosophers were keen to identify a route to agreement, an end to dispute.[112] I am sure this is true; certainly, Joseph Glanvill insists in *The Vanity of Dogmatizing* (1661) that a key merit of the new philosophy is that it will put an end to disputation, although, in France, as we have seen, the word 'fact', far from ending the dispute over Jansenism, was poured as petrol on the flames. Facts can provoke disputes as well as settling them.[113] In any case, my story so far invites a more local history. Hobbes was left out of the Royal Society – there is an extended literature on why[114] – but Digby, Charleton and Boyle, all readers of van Helmont, were among the first members. A simple explanation would be that the word 'fact' acquired its significance as a result of their influence; had the initial membership been very slightly different, scientists might still be discussing 'phenomena', not 'facts', and the fact might have entered English, as it entered German, only in the eighteenth century.

But if Hobbes was excluded from the Society, might the word 'fact' not also have been excluded? Was it not a dangerous word, too closely connected to Hobbes and to dubious stories about sympathetic magic told by Digby – someone whom John Evelyn, another early member, could dismiss as an arrant mountebank?[115] Was it not a word, thanks to Pascal, irretrievably associated with religious polemics of the sort that the members of the Royal Society were determined to avoid? The simple answer to this might be that the members of the Royal Society were familiar with Bacon's use of the word in Latin. There is, however, no sign – not the slightest fragment of evidence – that they had been struck by this, and Sprat went out of his way to criticize Bacon's insufficiently critical approach to questions of evidence.[116] Bacon was not their model.

There is another possible reason why the word 'fact' suddenly became respectable. Late in 1661 Thomas Salusbury, the librarian to the Marquis of Dorchester, published the first volume of his *Mathematical*

Collections and Translations, which contained the first English translations of Galileo's *Dialogue Concerning the Two Chief World Systems*, his *Two New Sciences* and his 'Letter to Christine of Lorraine'. The book is rare, and presumably had few readers, but those few would have found frequent occurrences of the word 'fact' in Salusbury's translations, particularly in the *Letter*. Earlier that year Joseph Glanvill published his *The Vanity of Dogmatizing*, in which he criticized Hobbes but praised Digby, in the process taking over his use of the phrase 'matter of fact'.[117] He also took over from Galileo's *Two New Sciences* Galileo's paradoxes regarding the movement of a wheel, and summarized Galileo's arguments in favour of a moving Earth, urging interested readers to read the *Dialogue* for themselves; since copies of the Latin edition of the *Dialogue* were notoriously difficult to obtain, and the Italian edition was a true collector's rarity, Glanvill was probably aware that an English translation was about to appear, and he may even have been shown Salusbury's translations. It may have been Galileo, in Salusbury's translations, who made Digby's usage respectable.[118]

Like Glanvill, the members of the Royal Society were, naturally, interested in Galileo. Charleton had drawn on his work extensively in his defence of Epicurean natural philosophy, the *Physiologia Epicuro-Gassendo-Charletoniana* of 1654. Boyle was desperate to show that polished marble plates would cease to cohere in a vacuum because Galileo had said as much.[119] Evelyn recorded a suggestion that the Society might take as its coat of arms a representation of a pair of crossed telescopes surmounted by the Medicean planets.[120] John Wilkins, who, along with Henry Oldenburg, originally shared the role of secretary to the Society, was the author of two works arguing that the moon is like the Earth, and the Earth is a planet. It was Wilkins who supervised Sprat's *History*, in which Galileo's discoveries are reported at length.[xxii]

Salusbury's *Mathematical Collections* may thus have been crucial to the success of the word 'fact'; they rescued it from Hobbes and van Helmont, from the weapon salve and the powder of sympathy, from furry babies and virgin births. They also rescued it from Pascal and religious dispute. They made it respectable. The answer to the question 'To whom do we owe the word "fact" in English?' is therefore perhaps to

xxii The Marquis of Dorchester was elected to the Royal Society in 1663, although he played no part in its proceedings – one might think that he was rewarded for employing Salusbury as his librarian. Wilding, 'The Return of Thomas Salusbury's *Life of Galileo* (1664)' (2008), 260.

Montaigne, Galileo, Bacon and van Helmont (although they wrote in French, Italian and Latin); certainly to Hobbes, Digby and Charleton; evidently to Pascal; and finally, perhaps, to Salusbury, as the translator of Galileo. It was this complex and ambiguous inheritance that the Royal Society adopted when it wrote the word into its Statutes. And what was Boyle's role in all this? Like Digby and Charleton, he was a reader of van Helmont, so the word 'fact' came naturally to him. Unlike them, he was not a pioneer in the use of the new word, and he waited until it had become respectable before extending its use to new fields. In this area he was, it would seem, a follower, not a leader.

So the word 'fact' in its modern sense becomes respectable in English only after 1661, while in French it was at first a word particularly associated with Jansenism. But for Digby and Charleton, who happened to be in the right place at the right time, but for the rapid translation of Pascal, and but for Salusbury's translation of Galileo, there might have been no culture of the fact in England for another hundred years – there was no guarantee the English would become obsessed with facts a century before the Germans. So, too, but for the dispute over the *Augustinus*, the French might have remained in the old world of proof and persuasion, deduction and experience, truth and opinion. And without the fact, the new notion that knowledge is grounded in evidence, not authority, might have received only the sort of inconsistent and unreliable endorsement that we have seen it receiving from della Porta.

But the words are one thing, the concepts are another. The word 'fact' tells us very little about the establishing and refuting of facts. In astronomy this is true; but in all the other fields of scientific enquiry the word consolidates a conceptual revolution.[121]

According to the standard principles of Renaissance education, there were fundamentally two types of argument: arguments from reason, and arguments from authority. There were various sorts of arguments collected under the general heading of 'authority': arguments from 'custom, public opinion, antiquity, the testimony of the skilled in their own art, the judgement of the wise, or the many, or the best'.[122] So when, in 1651, Pascal drafted an introduction to his unfinished treatise on the vacuum he started by distinguishing two sources of knowledge: reason and authority. How do we know the names of the kings of France through history? From authority: documentary testimony is classified under authority. Then, suddenly, and out of nowhere, he introduces sense experience as an adjunct of reason (though some writers had classified the senses under authority). So, decisions about the existence of a vacuum should be made not by appeal to authority but on the basis of sense experience and reason. Where does Pascal's own testimony to the outcome of his experiments fit? He does not say. We find exactly the same confusion in Browne. He wants to attack authority, and so he naturally appeals against authority to reason; but as a consequence he feels an obligation to insist that testimony, as a form of authority, is relevant only in very restricted circumstances. The fact that almost all his arguments are in the end arguments from testimony never seems to occur to him.[123]

In Hobbes this traditional schema was revolutionized. As far as he was concerned, there were only two sources of knowledge: reason on the one hand, and on the other sense experience, memory and testimony, all of which established matters of fact. Within this schema there was no scope for custom, public opinion, antiquity or the judgement of the wise, but the place of testimony was clear: it, like memory, represented a surrogate form of immediate sense experience. Hobbes would not have said that we derive our knowledge of the kings of France from authority; he would have said we derive it immediately from testimony and, ultimately, from sense experience.

The word 'fact' symbolized this new status given to testimony. Everybody understood that it was a word imported from the law courts, and with it came an established set of standards for judging the reliability of testimony. These standards were not peculiar to any one system of law but were generally acknowledged throughout Europe. Browne, even while dismissing testimony as relevant only to morality, rhetoric, law and history, and irrelevant to natural philosophy, summed up the basic principle: 'in Law both Civill and Divine, that is only esteemed *legitimum testimonium*, or a legall testimony, which receives comprobation from the mouths of at least two witnesses; and that not onely for prevention of calumny, but assurance against mistake.'[124] His problem with admitting testimony to natural philosophy was that it would then be necessary to accept what he called 'aggregated testimony' – in other words, the indirect experience of people who simply voice what everyone believes to be the case. He could not imagine turning the republic of letters into a vast law court.

Thus, before the invention of the fact an appeal to testimony was seen as an appeal to authority (even Digby, writing in 1658, had thought of eyewitnesses as authorities): witnesses, we might say, were thought of as character witnesses, not as eyewitnesses. After the fact, eyewitness testimony became a form of virtual witnessing, hence Boyle's insistence that he did not appeal 'to other Writers as to Judges, but as to Witnesses'. With testimony distinguished from authority, what was previously authority became, in Glanvill's words, simply 'old and useless luggage'. Sprat was even blunter: getting rid of the tyranny of the ancients simply involved throwing out what he called 'the rubbish'.[125]

After the invention of the fact, testimony could be approached with a form of systematized distrust. In the end, all systems of knowledge require one to put one's trust in someone, something or some procedure.[126] But to emphasize the undoubted role of trust in the new science is to run the risk of missing the large part of the iceberg which is hidden under the water: Boyle claimed to be trustworthy because he had learnt to distrust the della Portas of this world, and he hoped to teach others to read his own work in the same sceptical spirit as that in which he read the work of della Porta. The new science was, compared to what had gone before, based on distrust, not trust.

§ 8

It should be evident that many of the supposed facts we have been looking at – the weapon salve, for example – are rather odd, and always have seemed odd. There are cases of 'problematic facticity' and others of 'unproblematic facticity', and the language of the fact seems to be employed first of all to deal with cases of problematic facticity. Lorraine Daston has distinguished what she calls 'strange facts' from 'plain facts'.[127] She argues that strange facts came first and plain facts came later; first, there were Siamese twins, hermaphrodites, furry babies and virgin births, then there was Boyle's air pump. In England, she claims, plain facts replaced strange facts much earlier than in France; in other words, facts became regularized and routinized.

Let me suggest another story: strange facts are always aspirant plain facts. As Isaac Beeckman (who is best known for having introduced Descartes to the corpuscular philosophy) put it in 1626, referring to Simon Stevin's motto 'Wonder is no wonder':

> In philosophy, one must always proceed from wonder to no wonder, that is, one should continue one's investigation until that which we thought strange no longer seems strange to us; but in theology, one must proceed from no wonder to wonder, that is, one must study the Scriptures until that which does not seem strange to us, does seem strange, and that all is wonderful.[128]

This binary distinction between the natural and the supernatural seems straightforward to us, but it was in fact revolutionary, since it implied the abolition of the realm that had previously been held to lie between the natural and the supernatural, the realm of the preternatural, of ghosts and witches, marvels and monsters.[129]

The difficulty, of course, lay in knowing how and when to distinguish between philosophy and theology. *The Logic of Port-Royal* sketches what could go wrong when it describes people who are too credulous when it comes to miracles. They gulp down (*abreuver*), it says, a strange fact (*ce commencement d'étrangeté*), and when they encounter objections to it they change their story to meet them; the strange fact can survive only if it is turned into a plainer fact, which, in this case, involves moving it further and further from whatever truth there may have been in it to begin with. Almost imperceptibly, the supposedly supernatural is transformed into the natural.[130]

It was with the intention of turning strange facts into plain facts that Charleton and Digby insisted that the weapon salve could be reliably reproduced; it might be strange, but it was no stranger than magnetism. Galileo insisted that the mountains on the moon were just like mountains on Earth; the moons of Jupiter rather like our moon; the phases of Venus just like the phases of our moon; the spots on the sun rather like clouds. At every step he took the strangest facts and made them as plain as possible. Even Boyle's vacuum pump produced 'strange facts' in the eyes of Aristotelians and Cartesians.[xxiii] For them, a vacuum was an impossibility, so all the experiments that appeared to establish the existence of such a thing were strange indeed.

The simplest way of making strange facts plainer was to replicate them. In the *Reports of the Society for Experiments*, the society formed in Florence to carry forward a Galilean programme of experimentation after the great man's death, which took as its motto *provando e riprovando* ('test and test again'), there is a paradigmatic example: faced with a plausible but problematic result, they repeated their experiments using a different methodology. Thus they assured themselves that they had not been taken in by a spurious result.[131] If the strangeness of the fact could not be undone, at least the evidence for it could be reinforced

xxiii Unlike the Aristotelians and the Cartesians, Hobbes had no metaphysical objections to a vacuum, but he could not accept that light could pass through a vacuum, which meant the Torricellian space could not be a vacuum: Malcolm, 'Hobbes and Roberval' (2002), 187–96, corrects Shapin & Schaffer on this question.

so that it was turned into a stubborn fact; this was how, Arnauld argued, we could be confident in the miracles reported by St Augustine, for, strange as they might be, who could doubt his veracity? So, from the beginning, strange facts and plain facts existed in an uncomfortable struggle in which the strange facts were constantly trying to promote themselves into being acknowledged as plain facts, or at the very least as stubborn facts. As Arnauld recognized, the question of where to draw the line between facts that were too strange to be credible and facts that were strange but stubborn was far from straightforward.

A good example is provided by meteorites. Eighteenth-century English and French scientists rejected the ample testimony as to the reality of meteorites, as we reject stories of alien abduction. On 13 September 1768 a large meteorite, weighing seven and a half pounds, fell at Lucé, Pays de la Loire. Numerous people (all of them peasants) saw it fall. Three members of the Royal Academy of Sciences (including the young Lavoisier) were sent to investigate. They concluded that lightning had struck a lump of sandstone on the ground; the idea of rocks falling from outer space was simply ridiculous. There are other similar cases.[132] On 16 June 1794 a large meteorite exploded over Siena. The shower of rocks that fell on the city was seen by large numbers of academics and by English noblemen. Abbot Ambrogio Soldani published a whole illustrated book of testimonies. This was the first meteorite fall to be accepted as being (in some sense) genuine. It helped that there were lots of witnesses, and that they were educated and wealthy. It helped that the testimonies were published. But it also helped that the event could be made to seem less strange. Eighteen hours before the meteorites fell, Vesuvius, 320 kilometres away, had erupted; so it was possible to imagine that the rocks had been fired out of Vesuvius, even though they fell out of the northern sky, not the southern. This was clearly preferable to imagining that they had fallen from outer space.[133] The meteorite that fell on Lucé was too strange; the ones that fell on Siena were not so strange. As Arnauld had argued in *The Logic of Port-Royal*, plain facts beat strange facts every time.

§ 9

In this chapter we have looked at a series of local histories: Kepler's efforts to make accurate meaurements of Mars; the introduction of the word 'fact' into English; the weapon salve. If we look too closely at the details we are in danger of missing the big picture – for facts only became stubborn when experience became public and the printing press played a crucial role in turning private experience into a public resource, undermining established authority as it did so. The first new science grounded in what we would now call facts was the anatomy of Vesalius (1543), which depended on the public space of the anatomy theatre and the public space of the printed book to rebut the previously uncontested authority of Galen. Even Browne (1646) was not drawing his 'pibbles' from the 'scrip and slender stocke' of himself but from his ample library. Thus printed books brought with them a new freedom to contest authority. The epigraph to Rheticus's *First Narration* (1540) is a quotation from the second-century Platonist Alcinous: 'Free in mind must be he who desires to have understanding.' It was echoed by Kepler in his *Conversation with Galileo's Sidereal Messenger* (1610); by Galileo in his *Discourse on Floating Bodies* (1612); and the Elseviers made it the epigraph to the Latin translation of Galileo's *Dialogue Concerning the Two Chief World Systems* (1635).[134] In 1581 the exiled Hungarian bishop Andreas Dudith found himself playing host in Breslau to two astronomers, the Englishman Henry Savile and the Silesian Paul Wittich. 'I do not always grasp their ideas,' he wrote, 'but I marvel at their freedom [*libertas*] in judging the writings of the ancients and the moderns.'[135] In 1608 Thomas Harriot complained to Kepler that he could not yet philosophize freely: at the time he was suspected of atheism, and his two patrons, Sir Walter Raleigh and the Earl of Northumberland, were locked up in the Tower of London, the one convicted and the other suspected of treason.[136] In 1621 Nathanael Carpenter published his *Philosophia libera*, or free philosophy. Pascal in 1651 insisted that scientists should have 'complete liberty'.[137] The epigraph to Salusbury's *Mathematical Collections* (1661) is '*inter nullos magis quam inter PHILOSOPHOS esse debet aequa LIBERTAS*' ('between none more than philosophers ought there to be an equal liberty'). There is something inherently egalitarian and liberating about the new inter-related worlds of the book and of the fact. Indeed, we may say that the new science aspired to the creation of that social sphere which was idealized in the seventeenth century as 'the republic of letters' and which the eighteenth century was to label 'civil society'.[138]

Bruno Latour, in an important essay entitled 'Visualization and Cognition: Drawing Things Together', which originally appeared in 1986, claimed that the printing press made facts 'harder'; before printing, facts were too soft to be reliable.[139] What made the Scientific Revolution,

Latour argues, is not the experimental method, or commercial society – both had been around for centuries – but the printing press, which turned private information into public knowledge, private experience into communal experience. Bruno Latour is a bold – even sometimes a rash – thinker, never afraid to push an argument too far; but in this case I think he does not push his argument far enough. The printing press did not make facts harder, it made them – outside a few narrowly specialized fields, such as astronomy – possible. Latour rightly thinks that books represent a special class of objects. Many objects exist to be exchanged and consumed: sacks of grain, for example, are turned into bread. They are, in Latour's language, mutable mobiles. Gold and silver coins, which one might think of as immutable mobiles, are always being melted down and recycled – they are hard but mutable mobiles; books, on the other hand, are good for nothing other than reading (except, ocasionally, burning). They are the first true immutable mobiles.

This phrase, 'immutable mobiles', sums up neatly the epistemological paradox of the fact: facts can be moved around, transferred from one person to another, without being degraded, or so at least the story goes. In this they are quite unlike testimonies, which degrade as they pass from ear to ear in an endless game of Chinese whispers; eighteenth-century probability theorists actually devised formulae for calculating this rate of degradation. It was argued that such formulae could be used to date the Second Coming: the last trump would blow before the testimony for the resurrection of Christ would degrade to the point that belief in it would cease to be rational.[140] Testimonies degrade, facts do not, and yet both are grounded in the very same sensory experience. Facts are made in the image not of people, who misremember, misquote and misrepresent, but of books, immutable but mobile. The fact, you might say, is an epistemological shadow originally cast by a material reality: the printed book.

Gutenberg's *Bible* was published in 1454/5. But the printing revolution took a long while to get going. The comet of 1577 provoked more than 180 publications discussing its significance; Brahe's book on the subject provided not only his own measurements of the comet's parallax, which placed it firmly in the heavens, but an extended review of the measurements and arguments of others. Thus the printing press took scattered astronomers and astrologers belonging to different cultures and holding varied intellectual commitments and allowed the widespread exchange and comparison of ideas. This new community had a physical manifestation in the catalogues of the Frankfurt book fair, which, as we have seen, began in 1564.[xxiv]

The book fair accelerated the growth of an international trade in books, what the Jacobean poet Samuel Daniel called 'the intertraffique of the mind.'[141] By 1600 William Gilbert could complain that intellectuals were expected to navigate 'so vast an Ocean of Books by which the minds of studious men are troubled and fatigued'.[142] In 1608, for example, Galileo came across a book in a catalogue whose title was *De motu terrae*, *On the Movement of the Earth*, and naturally tried to get hold of a copy; two years later he was still trying to track one down, appealing to Kepler for help. It is not surprising the Venetian booksellers had been unable to help Galileo, as I cannot find *De motu terrae* in the Frankfurt catalogues either; but the book exists, so Galileo must have seen it listed in some other catalogue. Had he obtained a copy, he would have been disappointed, as its subject is earthquakes, not Copernicanism.[143] At the end of his life the same international trade meant that Galileo could find a publisher, Elsevier, for his *Two New Sciences*, the manuscript of which had been smuggled out of Italy, and which was published in Leiden, not in Latin or Dutch but in Italian, just as the illustrated edition (1590) of Thomas Harriot's *Brief and True Report of the New-found Land of Virginia* was published in Frankfurt, with simultaneous editions in English, Latin, French and German.

Markets do not always work for the best; according to Gresham's law (first formulated by Copernicus, as it happens), bad money drives out good:[144] but at the Frankfurt book fair, year by year, slowly but surely, good facts drove out bad. As scholars became aware of this process they began to publish books which were compilations of errors which had once been learned errors but could now be dismissed as nonsense. The doctors led the way, with Laurent Joubert's *Popular Errors* (first printed in French in 1578, it was reprinted ten times in six months and frequently thereafter, and also translated into Italian and Latin); Girolamo Mercurii's *Of the Popular Errors of Italy* (in Italian, 1603, 1645, 1658); and James Primerose's *Errors of the Crowd* (seven editions in Latin from 1638, with translations into English and French). Thomas Browne's *Vulgar Errors* took all error as its subject, though he, too, was a doctor (five English editions from 1646, with translations into French, Dutch, German and Latin). And what was in many ways the founding text of the Enlightenment, Pierre Bayle's vast *Critical Dictionary* (1696,

xxiv Above, p. 198.

with eight French editions in fifty years, plus two translations into English and one into German) was originally intended to be simply a compendium of errors.[145] This struggle against error brought into being the footnote: the mechanism for ensuring that every fact could be traced to an authorizing statement.[146]

Thus the printing press strengthened the hand of the innovators by making it possible for them to pool information and work together. It replaced the professorial lecture, the voice of authority, by a text in whose margin you could scribble your dissent. It replaced the manuscript, read more or less in isolation from other texts, with a book which could be consulted in a library, surrounded by competing authorities. It introduced the index as a ready route to the location of information on specific texts, to make it easier to set one authority against another.[xxv] And, by fostering a constant clash of arguments and ideas (Riccioli against Copernicus; Hobbes against Boyle), it forced each side in an argument to adapt and change. What the printing press did, quite simply, was undermine 'the dishonourable tyranny of that Usurper, Authority' and strengthen evidence.[147] It was the perfect tool for the Scientific Revolution.

The printing press also fostered a sort of intellectual arms race, where new weapons (the astronomical sextant, invented by Brahe; the telescope, improved by Galileo; the pendulum clock, invented by Huygens (1656) – astronomers had long sought an accurate way of measuring time) were constantly being brought up to the front line. It's not surprising that Kepler's *New Astronomy* (1609) is full of military metaphors; indeed, he presents the whole book as a war over the motions of Mars. Riccioli's *New Almagest* (1651) puts vast arrays of evidence and argument to the test, evidence and argument largely generated within Riccioli's lifetime, and assembled from Paris and Prague, Venice and Vienna, from books with nothing in common other than that they had

xxv Eisenstein, *The Printing Press as an Agent of Change* (1979), Vol. 1, 88–107; Ong, *Orality and Literacy* (1982), 121–3; William Wotton was among the first to stress the importance of the index (Wotton, *Reflections upon Ancient and Modern Learning* (1694), 171–2); Wolper, 'The Rhetoric of Gunpowder' (1970), 593, thinks this 'the worst argument' for the importance of printing. Words fail me. It seems to be the index which drives the introduction of foliation and pagination to supplement the signatures used by printers (*pace* Blair, 'Annotating and Indexing Natural Philosophy' (2000), 76; see Smith, 'Printed Foliation' (1988)): the 1521 translation of Vitruvius, for example, boasts on its title page that it contains a sophisticated index of a new sort, but the index is keyed to folio numbers which are not actually printed in the book and which readers would have been obliged to add by hand.

all passed at some point through the Frankfurt fair. Such a book is simply inconceivable within a manuscript culture.

I am presenting here a version of what is called the Eisenstein thesis, first propounded by Elizabeth Eisenstein in *The Printing Press as an Agent of Change* (1979). The Eisenstein thesis has never been popular with historians.[148] Historians like microhistories, not macrohistories. They like to be able to point to specific evidence that clinches an argument: but in the case of the printing revolution we are talking about a long, slow transformation. Quite properly, historians have insisted that manuscript culture ran alongside print culture right through the sixteenth and seventeenth centuries; thus, there survive some sixty manuscript copies of Leonardo's *Treatise on Painting*, all apparently produced between 1570 and 1651 (when it was first printed).[149] Knowledge was often spread as much through the correspondence of figures such as Nicolas-Claude Fabri de Peiresc (1580–1637), an astronomer and collector, Mersenne and Samuel Hartlib (*c.*1600–62), a Baconian reformer who sought to promote useful knowledge, as through the printing press. Even books, once annotated, were valuable for their unique contents: Brahe who was himself at the centre of an extended network of correspondence, tracked down individual copies of *On the Revolutions* because he wanted to read the annotations written in them by their previous owners.[150] But he also had his own printing press, and he was fortunate that, after he died, Kepler saw his unpublished works into print. Indeed, Kepler gave the printing press a prominent place in the frontispiece to the Rudolphine tables, which celebrated the progress of astronomy from the ancient world to the modern era. We can point to such contemporary witnesses, but in the end we are dealing with a question of scale: 5 million manuscripts produced in fifteenth-century Europe; 200 million books produced in sixteenth-century Europe; 500 million in the seventeenth century.[151] Even if the book did not have significant advantages over the manuscript, when it comes to illustrations, for example, the increase in the sheer quantity of available information would have been sufficient to generate a major cultural revolution.

Once the printing press had been invented, the concept of the fact (and with it the extension of the process of establishing reliable facts from astronomy to other disciplines) became inevitable, just as it was inevitable that the telescope would eventually be used to discover the phases of Venus, and that, once maritime compasses became widely available, someone would test the supposed antipathy between garlic and magnets. The question was not whether, but when, where and by whom.

The frontispiece to Kepler's Rudolphine tables (1627). The figures, from left to right, are the astronomers Hipparchus, Copernicus, an anonymous ancient observer, Brahe and Ptolemy, each surrounded by symbols of his work. The pillars in the background are made of wood; those in the foreground of brick and marble, symbolizing the progress of astronomy. Astronomical instruments designed by Tycho Brahe serve as decorations. The figures on the cornice symbolize the mathematical sciences, with Urania, the muse of astronomy, in the centre. Kepler's patron, the Holy Roman emperor Rudolph II, is represented by the eagle. On the base, from left to right, are Kepler in his study, a map of Brahe's island of Hven and a printing press.

How is one to understand this peculiar characteristic of the fact which I have labelled its 'hardness'? The unidentified G. W., who had, I believe, been reading Hobbes but who went far beyond Hobbes in embracing the fact, tried to describe it in 1653. Even the world of contingency, he insisted, was subject to what he called 'a determinate cognoscibility':

> For matters in fact are as certain in being and reality, as demonstrations . . . indeed all such effects as lurk in probable causes, that seem to promise very fairly, may be known also in an answerable, and proportionable manner, by strong, and shrewd conjectures: thus the Physician knows the disease, the Mariner forsees a storm, & the Shepherd provides for the security of his flock.[152]

Matters of fact are as certain as demonstrations (i.e. deductions, or logical proofs); the claim seems to us hard to dispute, since facts are by definition true. This whole passage in G. W., which I have abbreviated here, is almost entirely stitched together out of phrases lifted, without acknowledgement, from Nathaniel Culverwell's posthumous *An Elegant and Learned Discourse of the Light of Nature*, which had appeared only the year before. Nowadays, we would call this plagiarism, but that would be entirely to miss the point. Culverwell, for example, had said, 'Matters of fact are as certain in being and reality, as demonstrations,' but he wrote of matters of fact that were historical and legal events, and were only a small part of the larger class of contingent events. Culverwel was writing about old-fashioned facts (facts = deeds), not modern facts (facts = events), while G. W. made all contingent events matters of fact; unlike Culverwell, he was writing about Humean or, rather, Hobbesian facts. Moreover, Culverwell insisted that, in general, our knowledge of contingent matters is deeply imperfect, being based either on 'meer testimony' (if they are matters of old-fashioned fact) or on 'crackt and broken' empirical generalizations (if they are matters of experience).[153] G. W., by contrast, is happy to entrust himself to 'strong, and shrewd conjectures'.

Although the writers use almost exactly the same words, Culverwell's *Elegant and Learned Discourse* is on one side of the line which divides pre-modern from modern thought, and G. W.'s *Modern States-Man* is (as its title would suggest) on the other. G. W. borrowed from Culverwell precisely because there was no danger that anyone would think he was saying what Culverwell had said. Over the course of the next fifty years or so the fact, which had previously existed in a sort of intellectual limbo where it could have only a ghostly existence as a 'phenomenon', came to be the very foundation of all knowledge. In 1694 William Wotton summarized the new science in a phrase: 'Matter of Fact is the only Thing appealed to.'[154] In 1717 J. T. Desaguliers began his *A Course of Experimental Philosophy* with the words: 'All the knowledge we have of Nature depends upon Facts.'[155] In 1721 Count Marsigli of Bologna visited the Royal Society and reported: '[A]ll speculation unsupported by observation or experiment is utterly rejected. In England all study and teaching is based on fact.'[156] It is easy to read past sentences like these, because we now swim in a sea of facts and think that they are merely a recitation of the obvious. But in early-eighteenth-century Italy, where scholasticism still dominated university teaching, there was nothing self-evident about these new English values, just as the Declaration of Independence's claim that all men are created equal had once been anything but self-evident.

What is the significance of the fact? The postmodernists were not the first to challenge the claim that knowledge of matters of fact is true knowledge. It had already been disputed by Hobbes; it would soon be contested by Hume; in any case, every pre-modern thinker up to and including Culverwell was familiar with the arguments that established the unreliability of empirical knowledge. Still, despite all the arguments, we moderns – and, indeed, we postmoderns – place our faith in facts. Without facts there can be no reliable knowledge. It is not books as physical objects that are required to underwrite the fact; it is sources that don't alter and change from one day to the next, of which books remain the clearest manifestation. If you cite a book (or a photographic reproduction of a book on the Web), there is no need to write 'Accessed on . . .' because the text remains the same no matter when you access it. It is the fixity of its text that makes it an immutable mobile, and it is immutable mobiles that are needed if facts are to endure into the post-print age.

8
Experiments

> Thus the discovery of the barometer transformed physics, just as the discovery of the telescope transformed astronomy . . . The history of science has its own revolutions, just like the history of nations . . . with this significant difference, that revolutions in science . . . successfully achieve what they set out to do.
>
> – Vincenzo Antinori, '*Notizie istoriche*' (1841)[1]

§ 1

On 19 September 1648 Florin Périer, brother-in-law of the French mathematician Blaise Pascal, accompanied by a group of local dignitaries from Clermont-Ferrand, set out to climb the Puy-de-Dôme in the Massif Central.[i2] Below them, in a monastery garden, they had left an inverted tube sitting in a bowl of mercury. The height of the mercury in the tube was just over 26 inches (they measured in *pouces*, or inches, but their inches were very slightly longer than English inches). When they reached the summit, some 3,000 feet higher by their calculation, they mounted another barometer (as we would call the instrument: the word, in both English and French, dates to 1666, preceded in English a year earlier by 'baroscope'). The height of the mercury in the tube was a little more than 3 inches lower at the summit than in the garden of the monastery, and they obtained the same result when they took their barometer apart and reassembled it in several different places on the summit. On the way down they repeated the experiment a couple of times at a point nearer the foot than the summit of the mountain: there the mercury was an inch lower than in the monastery garden. One of

i Before 1654 Pascal was primarily a mathematician. From the autumn of that year he devoted his life to religion and began his *Pensées*, unfinished at his death in 1662.

these repetitions was carried out by a M. Mosnier. The next day they carried out the same experiment at the base and at the top of the tower of the cathedral church in Clermont: the difference was small (about two tenths of an inch) but measurable. Pascal, learning of this last result, carried out similar experiments in tall buildings in Paris and, as quickly as possible, published an account of them. Looking back in 1662, Boyle hailed the experiment at the Puy-de-Dôme as the *experimentum crucis*, the crucial experiment, which had validated a new physics.[3] Indeed, this was the first experiment to be hailed with this phrase, later made famous by Newton when he used it in connection with his prism experiments, which demonstrated that a ray of white light is made up of a spectrum of coloured rays.[4]

This is the first 'proper' experiment, in that it involves a carefully designed procedure, verification (the onlookers are there to ensure this really is a reliable account), repetition and independent replication, followed rapidly by dissemination.[5] The experiment was intended to answer a question: was there some natural resistance to the creation of an apparently empty space at the top of the tube (because, as Aristotle claimed, nature abhorred a vacuum), or was the height of the mercury (and so the size of the empty space) determined solely by the weight of the air? Pascal always claimed to be the inventor of this experiment, but the philosopher René Descartes insisted that he had originally suggested it to Pascal, and their joint friend Marin Mersenne was busy trying to organize the very same experiment when Pascal beat him to it. (Mersenne had trouble getting hold of sufficiently long, robust glass tubes 'hermetically sealed' at one end, although he seems to have gone to the same supplier as Pascal, who had no difficulty – perhaps Pascal was buying all the tubes that could be made.)[ii]

There was general agreement that the experiment showed that the height of the mercury was determined by the weight of the air, but there was no agreement as to whether the space at the top of the tube was properly considered to be a vacuum: Pascal thought it was, but Descartes thought it contained a weightless aether (without which, he thought, light would be unable to cross from one side of the tube to the other) capable of passing through glass; and Pascal's friends Mersenne and

ii Shea, *Designing Experiments* (2003), 107–9. Mercury was produced in large quantities because it was used to extract gold and silver from their ores, and it was widely available because it was used in medicine, particularly to treat syphilis. So getting hold of mercury was no problem; it was the glass tubes that were hard to obtain.

Roberval thought that there was some rarefied air in the space. This story is normally told as though Pascal was right and Mersenne and Roberval wrong, although actually all three were right: the space was in effect a vacuum, but it did contain some air under extremely low pressure.[6] Pascal's interpretation of the experiment was in direct contradiction to Aristotle's claim that nature abhors a vacuum.

If we have something they didn't, then experiment would seem a good candidate. As we saw in the last chapter, it is not always easy to define the threshold for saying that a culture 'has' something, but language usually provides a useful marker. This is less true when we turn to experiments. *Experientia* and *experimentum* ('experience' and 'experiment') are more or less synonymous in classical, medieval and early modern Latin, and all the modern languages that have both words initially reflect the Latin usage.[7] In modern English the distinction is clear: going to the ballet is an experience; the Large Hadron Collider is an experiment. But this distinction emerged slowly and became firmly established only during the course of the eighteenth century. The *OED* gives 1727 as the last date at which 'experiment' as a verb was used to mean 'experience', and 1763 as the last date at which 'experience' as a noun was used to mean 'experiment'.[iii] Insensitive to this change of meaning, scholars frequently translate the word *experimentum* in Latin texts as 'experiment', thus often giving a totally false impression of its meaning, which is commonly 'experience'.

Something close to the modern distinction is, however, to be found in Francis Bacon, who distinguishes between two sorts of experience: knowledge acquired randomly (by 'chance') and knowledge acquired deliberately (by 'experiment').[iv] By this definition, though, going to the ballet is an experiment, while learning that the seats are uncomfortable and the drinks sold in the bar overpriced is a chance experience. Moreover, it is quite wrong to think that Bacon is an advocate of an experimental (in our sense), as opposed to an experiential, science. He does think experiments can supplement experiences and provide crucial information, but he attacks William Gilbert for studying the magnet

iii In line with these dates, the first clear distinction between 'experience' and 'experiment' along modern lines seems to be Christian Wolff in 1732: Schmitt, 'Experience and Experiment' (1969), 80.

iv Bacon, *Novum organum*, Vol. 1, 82: '*Restat experientia mera, quae, si occurat, casus; si quaesita sit, experimentum nominatur*' (Bacon, *Works* (1857), Vol. 1, 189). But Bacon then complicates matters by using the adjective *experimentalis* to cover all knowledge grounded in experience; and he has no verb for 'to experiment' other than *experiri*.

through a narrow experimental programme which concerns itself only with magnets: 'For no one successfully investigates the nature of a thing in the thing itself; the enquiry must be enlarged so as to become more general.'[8] Hobbes, for his part, neatly distinguishes experiment from experience, but not as we do. For him, several experiments amount to experience – experiment is particular; experience general.[9]

At first sight one would think that Henry Power's *Experimental Philosophy* of 1664 is a book about experiments in the modern sense, and indeed it includes numerous experiments involving mercury and glass tubes; but the first section of the book is concerned with 'experiments' made with a microscope. Power is, however, well on the way to our modern usage, because, although he says the book is about 'new experiments, microscopical, mercurial, magnetical', he labels each section of his microscopical reports an 'Observation', and each section of his mercurial reports an 'Experiment'. 'Observation' used in this modern sense (rather than in the sense of a practice, such as a religious observation or observance) was relatively new in English, although it exists in classical Latin (*observatio*): the *OED* gives 1547 as the first usage of 'observation' and 1559 as the first of 'observe' in this new sense. Over time, observation became an adjunct to experiment, both producing reliable facts in place of the unreliable, unspecific 'experience' which underlay so much classical and medieval discussion.[10]

In French and Portuguese the old confusions (as they must seem to an anglophone) still exist. In French there is a verb, *expérimenter*, which corresponds to both 'to experience' and 'to experiment'; there is still no noun that corresponds to the English 'experiment', although you can *faire une expérience*, where *expérience* means 'experiment', and in nineteenth-century French *expériences* in the plural always meant experiments, not experiences. French has also acquired the word *expérimentation*, which is now sometimes used as if it were equivalent to 'experiment'.[v][11] There is also in French (and Portuguese) an adjective, *expérimental*, classically used in the phrase *philosophie expérimentale*. The word *expérimental* was used solely in a religious, usually a mystical, context until the translation of Sprat's *History of the Royal-Society* into French in 1669, when the phrase *philosophie expérimentale* was introduced into the language. It is something of a puzzle that the word 'experiment' did not come with it, despite its respectable Latin antecedents.

v Descartes used the word *expériment* once in a letter (Clarke, *Descartes' Philosophy of Science* (1982), 41, n. 2).

There are other words just as ambiguous as sixteenth-century 'experience'/'experiment'. A striking example is 'demonstration'. In classical Latin you demonstrate something by pointing it out with your finger. But in the Middle Ages the word *demonstratio* was used to refer to a deduction or proof in philosophy or mathematics: thus you can demonstrate or prove that all the angles of a triangle add up to two right angles. In French this remained the meaning of the word until very late: it is only in the fourth edition of the Dictionary of the Académie Française (1762) that the use of the word in contexts where you show someone what you are talking about (a demonstration in anatomy, for example) is recorded. In English the two meanings ('demonstration' as deduction; 'demonstration' as pointing out) exist together from an early date. Thus both Aristotelian philosophers and the new scientists produced demonstrations, but they meant radically different things by the word.

Another striking example is 'proof'. On the one hand, we use the word to refer to proofs, deductions and demonstrations in mathematics, geometry and logic. On the other, we talk about 'the proof of the pudding', alcohol being 40 proof and proving a gun. 'Proof' thus covers both necessary truths and practical tests, and it has the same etymological root as 'probe' and 'probability'. This ambiguity comes from the Latin (*probo*, *probatio*) and is found in all the modern languages derived from Latin (Spanish *probar*, Italian *provare*, German *probieren*, French *prouver* – though in French there is also *éprouver*, to test, so in modern French *prouver* has lost the sense of 'to test'). A proof, at least in mathematics and logic, is an absolute; you either prove something or you do not. On the other hand, evidence (to use our modern English word) is something you can have more of, or less of. In Roman law two witnesses may provide a full proof of guilt; one witness and a confession may do the job; or one witness and circumstantial evidence (the accused's knife, for example, was found in the victim). Renaissance lawyers were trained to talk about half a proof or a whole proof.

In cases where a proof was not complete and there was no alternative way of obtaining evidence, torture was (in countries which followed the principles of Roman law) legally used, from the thirteenth century to the eighteenth, in the hope of acquiring complete evidence. In the case of della Porta, for example, the tribunal of the Inquisition voted to torture him mildly (*leviter*) in view of his poor health; then a week later, luckily for della Porta, they changed their minds. We have no record of della Porta's thoughts and feelings during this intervening week.[12] Perhaps he became so ill at the prospect of torture that he could no longer be tortured (for you had to pass a medical examination and be declared fit before you underwent torture; the Inquisition was scrupulous about such matters). Someone against whom an incomplete proof of guilt existed (someone who had been tortured without confessing – Machiavelli in 1513, for example) was neither guilty nor innocent but could properly be punished for having given grounds for suspicion (this is what happened to della Porta, and to Galileo when he was tried by the Inquisition in 1633; Machiavelli had the good luck to be released under an amnesty). Francis Bacon, when he writes about experiments, uses the phrases 'the inquisition of nature' and 'nature vexed'. Does this mean torturing nature to extract an answer?[13] Bacon had himself seen men suspected of treason racked, though torture was not normally used in English legal proceedings. In a world where legal metaphors were constantly employed when knowledge was being discussed (as we have seen, the word 'fact' is itself a dead legal metaphor) questions of proof always carried with them the possibility of torture as a (metaphorical) mode of proceeding, but in English law 'inquisition' (an inquest is an inquisition) and 'vexation' carry no necessary implication of torture.

William Gilbert, writing *On the Magnet* in Latin, is, as one would expect, alert to the difficulty of using words like 'proof' and 'demonstration' to describe what experiments do. His preferred term is the post-classical word *ostensio*, a display or showing, which he defines as 'a manifest demonstration by means of a body'. In other words, he is not providing a demonstration in the logical or mathematical sense, but he is making some physical reality apparent. He intends, he says, to show you things as if he were pointing to them with his finger. When you read his book you are a 'virtual witness' of his experiments.[14]

§ 2

This chapter began in 1648, with Pascal's Puy-de-Dôme experiment, but Pascal was not the first experimental scientist. Take, for example, the evolution of Galileo's thinking on the question of buoyancy. He started out as an admirer of Archimedes. In an early unpublished text from the 1590s he sought to demonstrate that Archimedes' principle, that a body floats when it displaces its own weight in water, is necessarily true.[15] The text of Archimedes had been available in Latin from the twelfth century, and had been first published in 1544. Early editions of Archimedes come with illustrations which show objects floating in a vast ocean of water, an ocean which stretches right around the globe, and Galileo drew such sketches in his own text.

It is perfectly correct to claim that in a boundless fluid a floating body displaces its own weight in water. But as he revised his text Galileo went on to represent objects floating in containers, such as a tank standing on a table top. When you put a block of wood into a tank, the level of water in the tank rises. Galileo at first thought that the volume of water above the old surface corresponded to the volume of water displaced by the object, and the weight of the water above the old surface corresponded to the total weight of the object, according to Archimedes' principle. As we shall see, this is false. Unlike previous interpreters of Archimedes, Galileo had asked himself what sort of experimental apparatus would serve to illustrate Archimedes' principle; what he hadn't grasped was that this apparatus would serve to show that Archimedes' principle is incomplete.

Twenty years later, in 1612, Galileo found himself in a dispute with Aristotelian philosophers. Heavier-than-water objects, they assured him, float if they have the right shape. Thus a chip of ebony, which is heavier than water, floats if placed on the surface of a bucket of water. Provoked, Galileo embarked on a series of experiments to study floating bodies. Chips of ebony, he found, float if they are dry to start with and are placed gently on the surface of the water – but so do metal needles. If they are already wet all over, they sink. Galileo was exploring the phenomenon we call surface tension.

Galileo also wanted to construct an object which would fully immerse but not sink – an object with the same specific gravity as water. He took some wax, mixed it with iron filings and shaped it into a ball: when he had the mixture right it floated just below the surface of the water. In this case, he wrote, in a draft text, the volume and weight of the water displaced correspond to the volume and weight of the ball, according to Archimedes' principle – except they don't. He was still repeating his old mistake.

At this point, Galileo had a sense that something was wrong. He went back to his old thought experiment and began to study it carefully, this time with the help of real tanks, real blocks of wood and pieces of marble. He tried floating the same block of wood and sinking the same block of marble in three different tanks, and worked out the mathematical formula which determines the extent to which the water in the tank is raised by the introduction of the blocks. He now understood the issue of displacement in terms of volume, and it was an easy step then to grasp it in terms of weight.[16] It was now clear to him that when a block of marble is put into a tank so that only part of it is submerged it does not displace water equivalent to the volume of that part of the block which ends up under water; it displaces water equivalent only to the volume of that part of the block which is under *the original* surface level. Consequently, a block of wood floating in a tank displaces less than its own weight in water. According to Archimedes' principle, if water were to occupy the volume of the block which is under water, that water would weigh the same as the whole block. Archimedes' principle did not apply.

Galileo proceeded to confirm his new theory with a very simple experiment.[17] He took a small rectangular tank and put into it a large block of wood which fitted snugly. He poured in water until he reached the exact moment when the wood began to float. He was trying to show – and succeeded in showing – that the ratio of the depth of the water to the overall height of the block corresponds to the ratio of the weights of equal volumes of wood and water. But he was also showing something very odd, which followed from his new discovery: a very small amount of water could be made to float a very large and heavy object; indeed, the actual water in the tank could weigh less than the block of wood it was lifting – which, according to Archimedes' principle as traditionally understood, was impossible. (You can do this yourself by putting a small amount of water inside a wine cooler and then floating a bottle of wine in it.)

Now, at last, Galileo had a secure grasp of the principle that when you introduce the block of wood into the tank and the level of water in the tank rises, the volume of water displaced corresponds only to the portion of the block below the old, lower water-line, which is much less than the space occupied by the portion of the block which is below the new, higher water-line. The more closely the tank fits around the block the more powerful this effect, because the water is not being displaced sideways by the introduction of the block (as it would be in a boundless fluid), it is being raised upwards. Galileo had established that the relationship between the weight of a floating object and the weight of the water displaced by it in a bounded container is not comparable to that of two weights at either end of a balance but rather to that of two weights at either end of a lever. Archimedes' principle is a limit case, not a universal principle. Without intending to, Galileo had devised an elementary hydraulic press.[vi]

vi Much later, Pascal drew upon Galileo's work when studying pressures in liquids in order to understand how the air supports a column of mercury in a barometer. Like Galileo

Galileo published these results in 1612, and they provoked a brief flurry of debate, but they were unnoticed outside northern Italy. The philosophers weren't convinced and continued as before, and the mathematicians weren't impressed – this wasn't maths as they understood it. Galileo had been an experimental scientist for some time, a decade or so – in fact, since he had read William Gilbert's *On the Magnet*. But this was the first time he had published the results of a series of experiments. Gilbert and Galileo were developing a new type of science, based upon systematic experimentation. But very few people paid any attention.

§ 3

There is nothing new about the idea of testing a theory; it is perfectly easy to show that Ptolemy and Galen had carried out experiments, and the *Optics* of the first great experimental scientist, Ibn al-Haytham (965–*c*.1040), had been translated into Latin by 1230 (at which point Ibn al-Haytham acquired his Western name of Alhazen).[vii][18] It was soon widely available in manuscript and appeared in print in 1572. The puzzle is why Ibn al-Haytham's example was not followed more extensively, for it would be difficult to overestimate the significance of his achievements. Using a rigorously experimental method, he refuted the standard extromission theory of sight (that sight is made possible by rays that go out from the eye) and defended the intromission theory (that sight is made possible by rays that enter the eye from the object); he produced the first full statement of the law of reflection, and also studied refraction; he designed the first true camera obscura; he made enormous advances towards an understanding of the physiology of the eye (although he failed to grasp that an upside-down image is projected through the lens on to the retina at the back of the eye); and he laid the intellectual foundations of the science of artificial perspective. Medieval

---

before him, Pascal understood that the pressure exerted by a fluid is not determined by the total weight of the fluid, but by the amount of weight brought to bear on any given area. For example, take a long thin tube (ten foot long should be plenty) and stick it in the top of a wooden barrel containing a liquid; and then fill the tube with water. It requires only a small amount of water to fill the tube, but because the height is great it exerts an enormous pressure – enough to burst the barrel. This experiment is known as 'Pascal's barrel', although there is no evidence he conducted it himself; he did conduct experiments designed to illustrate the principle involved, but that had already been described by Mersenne.

vii See above, p. 168.

optics was heavily dependent upon his contribution and he was unquestionably the best example of an experimental scientist before Gilbert.[viii]

If Ibn al-Haytham offered plenty of real experiments, medieval philosophy is also full of thought experiments designed to test the implications of theories.[19] What would happen, for example, if you drilled a tunnel down through the centre of the Earth and then dropped an object into the tunnel? Would it stop when it reached the centre, its natural resting place? Or would it shoot past? Would it oscillate back and forth until it came to rest? Obviously, this thought experiment is not one that could be carried out in practice (and no one tried to use a pendulum as a substitute),[ix] but often experiments are described in a way that makes it difficult to tell whether they have actually been performed or not, and this continued to be true in the seventeenth century. Boyle complained that Pascal described experiments (carried out under 20 feet of water) that he could not possibly have performed, and modern historians have made the same complaint against Galileo (although, it has to be said, nearly always mistakenly).[20]

The puzzle is therefore not whether there was any experimental science before the Scientific Revolution, for it is easy to find examples; rather, it is why there was so little, particularly given the example provided by Ibn al-Haytham and the prevalence of thought experiments. It is not difficult to identify a number of relevant factors.

First, experimentation involves manual labour. Although it has been argued that Christianity, particularly the monastic tradition, placed a higher value on it than had the ancient world, there was still considerable resistance within medieval and indeed Renaissance culture to physical work. The first experimenters were happy to use their hands. We are told that Galileo loved making little machines as a child (Newton certainly did),[21] and that Torricelli was very skilled with his hands. Experimentation was a hands-on business.

Second, the dominant position acquired by Aristotelian natural philosophy within the medieval universities resulted in a double inhibition

viii But note Sabra's judgement: 'It remains true, however, that the confirmatory experiments in I. H.'s Optics differ in at least one respect from the discovery experiments of seventeenth-century optics: they do not reveal new properties, such as the diffraction, double refraction or dispersion of light; and . . . they lack measurement.' (Ibn al-Haytham, *The Optics: Books 1–3, on Direct Vision* (1989), Book 2, 18–19.)

ix Oresme understood that a weight hung from a beam (which we would call a pendulum) could mimic the action of such a falling stone; but there is nothing to suggest that he then experimented with pendulums. (Clagett, *The Science of Mechanics in the Middle Ages* (1959), 570.)

on experimentation. First, where Aristotle had discussed a subject at any length, it was assumed that adequate knowledge of it was already available (one reason why optics could develop as an intellectual discipline was that the first important treatment of the subject was by Euclid, not Aristotle); and second, the Aristotelian tradition insisted that the highest form of knowledge was deductive, or syllogistic, knowledge.

Medieval philosophers such as Robert Grosseteste (*c.*1175–1253) worked out a fairly sophisticated account of how one might work from experience to theoretical generalization, and then use theoretical generalizations to deduce the facts (or, rather, phenomena) of experience. But the whole point was that this procedure was only to be employed when there were no obvious first principles from which to work, and that it was (perfectly correctly) seen as being entirely compatible with Aristotle's understanding of scientific knowledge. Thus Grosseteste claimed that we can know that all movement in the heavens is circular from first principles (if the movement was not circular, empty space would be opened up between the heavenly orbs, and this is impossible, as a vacuum is impossible), but we cannot deduce the shape of the Earth from first principles. Consequently, we must fill this gap by relying on experience, and experience provides convincing evidence (for example, eclipses occur earlier in the day at points further to the east and later in the day at points further to the west; the Pole Star sinks towards the horizon as one travels south) that the earth is spherical.[22]

Experience and experiment are thus to be invoked only to fill gaps in a fundamentally deductive system of knowledge, never to question the reliability of deductive knowledge itself; and these gaps were always of limited significance within a curriculum centred upon Aristotle's texts. (Grosseteste's view that the shape of the earth was a purely empirical question was not without consequences, for it opened up the intellectual space for him to adopt a one-sphere theory of the Earth.) Grosseteste's own practice demonstrates a remarkable indifference to experimental procedure; thus he formulated a general principle of refraction, but he simply assumed that it must, like the law of reflection, involve equal angles, and never conducted the elementary tests which would have shown that this assumption was misplaced. He produced a new theory of the rainbow which emphasized the role of refraction, where Aristotle had only mentioned reflection; but there is no evidence that Grosseteste ever conducted experiments to test his theory.[23] In 1953 Alistair Crombie published a book entitled *Robert Grosseteste and the Origins of Experimental Science*. Over the course of his life Crombie slowly retreated from the claims made in that book. By 1994 he was prepared to write:

> It is difficult to tell whether so independent a thinker as Robert Grosseteste saw himself as doing and finding out something novel, beyond his authorities[,] as distinct from discovering their real meaning. It seems unlikely. Roger Bacon [1214–94, a follower of Grosseteste, and often heralded as an exponent of experimental science in the Middle Ages] saw contemporary scientific work as a recovery of ancient forgotten knowledge. Perhaps from this mentality, as well as from uncritical literal copying, came the medieval habit of reporting reported observations and experiments as if they were original.[24]

Third, experimentation involves both a study of the external world and a capacity to generalize. It requires an ability to move back and forth between the concrete and the abstract, the immediate example and a scientific theory, and this movement is conceptually and historically problematic. The Greeks never thought of knowledge (*episteme*) as being knowledge of the external world, because for them reason was always universal and eternal; the mind was as one with what it knew.[25] In the Middle Ages, for example, Grosseteste adopted a neo-Platonist view that true knowledge was based on illumination, and the perfect form of knowledge was that of the angels, who needed no sensory experience of reality to know the divine mind and the universe through it.[26] This had a continuing influence into the early modern period: Descartes tried to recapture a Platonic conception of knowledge as being that which is self-evidently true, and even Galileo sought whenever possible to present his new sciences as mathematical demonstrations, not empirical extrapolations. Within this tradition knowledge is primarily mental, conceptual, theoretical and, in the end, mathematical.

Mathematicians, as members of an intellectual discipline, were thus torn between two types of knowledge: Plato and Euclid appeared to justify a purely abstract, theoretical form of knowledge; while the applied sciences of astronomy, cartography and fortification encouraged an empirical, practical orientation. Among the ancients, Archimedes appeared to have bridged this divide by showing how theory could be put to practical purposes, but the tension between the two approaches still continued in Newton, who sought to present his knowledge, in so far as possible, as pure theory, while insisting that it was based on evidence and had practical applications.

Catholic Christianity, in contrast, was committed to the belief that the truth lies outside us: the crucifixion of Christ and the transubstantiation of the host during Mass are not events in the mind but in the external world. Aristotle was thus interpreted by philosophers as grounding knowledge in sensation, and sensation was reinterpreted as knowledge of a reality external to the perceiver. But (and it is a big but) the truths of religion are not normally amenable to sensory perception; in the Mass, the bread and the wine continue to look like bread and wine. Hence the importance of miracles, when sensory perception confirms divine truth.

The medieval emphasis on external reality opened the way to nominalism, which, in reaction against Platonism and Platonizing interpretations of Aristotle, insisted that only concrete individuals exist and that abstractions are merely mental fictions – but in doing so it left little scope for a movement back from the particular to the general. Things are as they are not because of any sort of natural order or necessity, but because God chose to make them so. The world itself is a sort of miracle, and what happened yesterday need not happen tomorrow.[27]

Experimentation thus required a deeply problematic balancing act between Platonic idealism and a crude empiricism. Experimenters have to insist on the particularity of experience, but they also have to claim that general conclusions can be drawn from specific examples. Underlying experimentation, therefore, there must be a theory of the regularity and economy of nature; the natural world has to be the sort of world that could, in principle, be interpreted through experiments. 'For who doubts,' asked Newton's associate Roger Cotes, 'if gravity be the cause of the descent of a stone in Europe, but that it is also the cause of the same descent in America?'[28] In addition, we have to be equipped to interpret the world; our senses have to pick the features which matter: Diderot doubted whether a blind person could ever come to recognize the universe as orderly, and so as divinely created. When the experimental method is successful in explaining what was previously inexplicable, it thus not only establishes particular scientific theories, it also confirms the validity of the general approach which underpins experimentation. Successful experimentation builds confidence in the experimental method; failure undermines it.

A further problem was that an experiment is an artefact. Aristotelian philosophy drew a sharp distinction between the natural and the artificial: understanding one provided no basis for understanding the other. In some cases this is obvious: a kite won't help me understand how a bird flies, or a steam engine how a muscle works. Natural objects have, for an Aristotelian, their own internal formative principles, while artificial objects are made according to a design imposed from outside. The distinction between nature and artifice went even further: it was assumed that the rules governing the behaviour of artificial objects were different to those that operated in the world of nature, so that a machine might enable one to cheat nature by getting out more work than one put in. Galileo was the first to show that this could never happen.

It is evident that we can often understand what we make better than we can understand what nature produces, and this principle can be extended, for example, to mathematics, where we determine the rules of the enterprise. Thus in 1578 Paolo Sarpi wrote:

> We know for certain both the existence and the cause of those things which we understand fully how to make; of those things which we know by experience alone we know the existence but not the cause. Conjecturing it then we look only for one that is possible, but among many causes which we find possible we cannot be certain which is the true one.[29]

Sarpi gives mathematics and clocks as examples of knowledge where we have certainty because we have made what we know, and astronomy as an example of knowledge where we can come up with a possible right answer (the Copernican system, say) but can never be sure that it is correct. Sarpi never shared his friend Galileo's conviction that Copernicanism was obviously right.

Such an approach implies that the sort of knowledge obtained by experiment need not be a reliable guide to how nature works. The fact that I can make a vacuum in the laboratory, for example, need not mean that a vacuum can ever occur in nature. It is often said that William Harvey demonstrated that the heart is a pump; but in *De motu cordis* he never compared the heart to a pump – pumps are artificial, after all, and hearts are natural. It would be dangerous to rely on such a comparison.[30] In contrast, the 'maker's knowledge' principle implies that if I do make a vacuum in a laboratory then I have a real understanding of what it is that I have made.[31] Confidence in experimental knowledge thus requires the natural/artificial distinction to be undermined and replaced by the conviction that by performing procedures that correspond to natural processes I can have true knowledge of those processes.

The first person to insist as a matter of principle that knowledge of artefacts could count as knowledge of nature was Francis Bacon, who said 'artificial things differ from natural things not in form or essence, but only in the efficient.'[32] Thus, knowledge of an artificial rainbow gives you (as we shall see in a moment) causal understanding of a natural rainbow, even though you have produced the artificial rainbow by different means. In a case like this the experimental method requires you to move smoothly back and forth between nature and artifice. Gilbert claimed that the little spherical magnets that he used were equivalent to the Earth; Pierre Guiffart, who had observed Pascal's first vacuum experiments, said of the Torricellian tubes, 'In them one sees a little miniature of the world,' in that one actually sees the weight of the air.[33] Such claims were not straightforward – Jesuit scientists strenuously opposed Gilbert's claim that the Earth itself was a magnet, and opponents of the vacuum protested that the Torricellian tube was deceptive, for it appeared to contain nothing in the space above the mercury when it surely contained something.

To a certain degree, if the world is orderly and predictable, it is because we have worked to make it so by developing technologies that give us control over nature. If we can model its processes, it is because we have developed our own capacities for making nature-like artefacts. It was therefore inevitable that the advocates of the experimental method in the seventeenth century would come to insist that the universe is like a clock, for clocks are embodiments of the principles of order, regularity and efficiency and, moreover, it is we who have made them. If we think of God as a clockmaker, then we can be confident that he will have made a world amenable to experimental enquiry. In the Middle Ages the heavens had been compared to clockwork; now the same principle of regularity was, it was claimed, to be discovered in the sublunary world.[34]

Finally, there was, of course, in the Middle Ages, no culture of discovery. Even Ibn al-Haytham's discoveries were hard to integrate into a system of knowledge that was backward looking, so the extromission theory of vision continued to be the standard theory simply because it was the one upheld by the authors who had antiquity on their side.

These five factors help explain the limited success of experimental science in a medieval context. Take, for example, Theodoric of Freiberg (*c.*1250–*c.*1310), who carried out the most remarkable experimental work in the whole of the Christian Middle Ages. Theodoric provided the first satisfactory account of the rainbow.[35] This involved direct criticism of Aristotle.[x] Aristotle had said that rainbows are the result of reflection, while Theodoric showed that they were the result of two refractions and

x But Aristotle had pointed out that one could make an artificial rainbow, which is why medieval philosophers felt authorized to conduct experiments in order to understand the rainbow: *Meteorologica*, Book 3, Part 4, 374a35–374b5; Newman, *Promethean Ambitions* (2004), 242. (The whole chapter, 238–89, is an important discussion of medieval experimentation.)

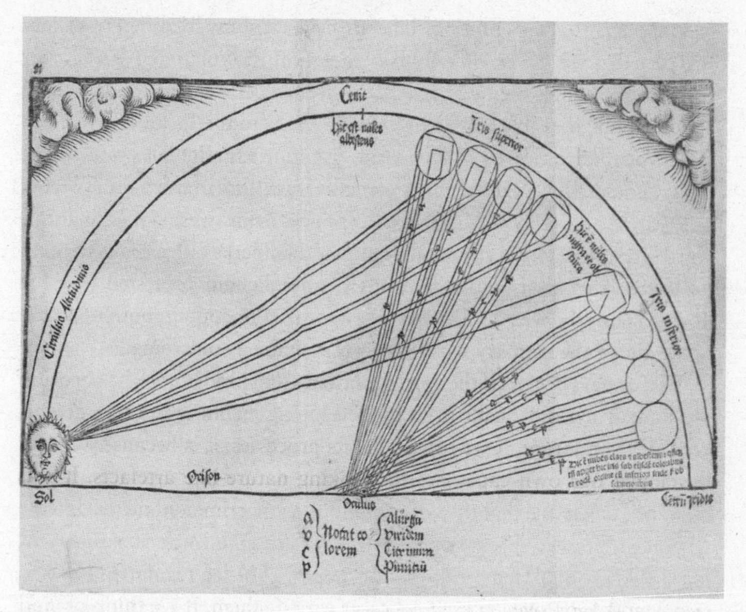

The illustration accompanying the late-thirteenth-century Theodoric of Freiberg's study of the rainbow when it appeared in print in Trutfetter's textbook of 1514. It shows that in forming a rainbow each ray from the sun is twice refracted and twice reflected as it passes through a drop of water before reaching the eye. As it emerges from the drop of water the white light has been split into an array of colours.

two reflections within each drop of water. Aristotle had denied that the colour yellow is really present in the rainbow, and had identified only three colours; Theodoric insisted that yellow was a fourth colour in the rainbow. Theodoric's analysis depended partly on examining rainbow-like images that he encountered in daily life: in the spray thrown up by a turning water-wheel, in the drops of dew on a spider's web. But he also studied what happened when a ray of light entered a glass ball filled with water, on the theory that this would provide a good model of what happened when a ray of light entered a raindrop (he used a urine bottle, which was a standard piece of equipment for any medieval doctor and had a spherical bulb). Around the same time a similar experiment was independently performed within a camera obscura by Kamal al-Din al-Farisi, who was, like Theodoric, drawing on the example presented by Ibn al-Haytham, and who will also have had access to urine bottles.[36]

Only three manuscripts of Theodoric's short tract on the rainbow survive, and we know of only one medieval discussion of his discovery.[37] Regiomontanus, it is true, planned to publish him, but, where other texts Regiomontanus had planned to publish appeared in print in due course, Theodoric's little treatise did not.[38] In 1514 a summary of his argument was presented in a physics textbook intended for students at Erfurt (and an even shorter one appeared in 1517, this time without any illustrations).[39] There is no evidence that these summaries had any influence. Theoderic's work then disappears entirely from view until it was rediscovered in the nineteenth century. When Descartes produced his study of the rainbow he had to start from scratch, despite the fact that he was very largely simply repeating the work both of Theodoric and of al-Farisi.[40] Thus it is important to see that when we hail Theodoric as an important scientist our judgement is essentially anachronistic: he did not seem important to his contemporaries or successors, and his influence is negligible. His work would have been much more likely to have been preserved and copied if it had taken the form of a commentary on Aristotle's *Meteorology* – the most widely read commentary was that of Themo Judaei, which contains no reference to Theodoric – and if it had not depended on elaborate illustrations which were difficult to copy accurately.

A similar point may be made about the work of Ibn al-Haytham. Only one complete manuscript of the original Arabic text of *The Optics* survives (the vast majority of Ibn al-Haytham's work – he wrote two hundred texts – has been lost), and the only Arabic commentary on his optical work that we know of (before the modern day) is that of al-Farisi (1309).[41] Ibn al-Haytham was much more widely discussed in the Latin West than the Muslim East, but even in the West he was treated as a text, not as a handbook of experimental practice. No one, as far as we know, replicated his experiments. Thus in both Arabic and medieval culture experimentation had an uncertain status: it existed, but it was not admired or imitated. It was recognized as a form of knowledge, but only at the margins. In both cultures Ibn al-Haytham was seen as a model to be imitated only when it came to looking for an explanation for the rainbow; for the vast majority of medieval authors learning was something to be found in books and to be tested by abstract reasoning; it was not to be found in things and tested by experimentation.

§ 4

Thus there was nothing unprecedented about experimentation in 1648; there were good precedents, one of which (Ibn al-Haytham's *Optics*) was widely known, if rarely imitated. Rather, the significance and status of experimental knowledge underwent a peculiar transformation in the course of the seventeenth century. It moved from the margins to the centre.[xi] Kant claimed that the experimental method of the seventeenth century (he cites Galileo, Torricelli and the chemist Georg Stahl) represented 'the sudden outcome of an intellectual revolution', the moment when natural science entered on 'the highway of science'; this judgement is sound not if it is understood as claiming that Galileo and Torricelli were the first to conduct experiments but, rather, read as maintaining that previously experiments were seen as no more than a byway.[42] Above all, experimentation began to engage directly with central claims made by Aristotle. At the same time, those who performed experiments ceased to be lonely and isolated individuals; they became members of an experimental network. Why exactly did the significance and status of experimental knowledge change? We need to look more closely to understand what happened.

The first major field for experimental enquiry in the early modern period was the magnet, a subject on which there was virtually no classical commentary (because the compass was unknown in antiquity), which meant that the experimental approach faced fewer obstacles than with any other topic. Moreover, the importance of the compass in navigation meant that the magnet was bound to be a subject of discussion. The first attempt at an experimental study of the magnet was reported in Pierre de Maricourt's *Letter on the Magnet* (1269); Pierre describes the polarity of magnets, shows that like poles repel each other and unlike poles attract each other, and describes how iron can be magnetized. Despite surviving in thirty-nine manuscript copies, there is no evidence that it led to further experimental work until it finally appeared in print in 1558.[43] Like Ibn al-Haytham, like Theodoric, Pierre de Maricourt had no immediate successors.

xi As a result, the word 'experiment' is five times more frequent in English in the 1660s than it had been in the 1640s, and this despite the fact that its use to mean 'experience' was in rapid decline: Pumfrey, Rayson & Mariani, 'Experiments in 17th-century English' (2012), 404.

By 1522 Sebastian Cabot had discovered the variation of the compass: the needle does not point to true north but either somewhat to the east or to the west, and it varies in the extent to which it diverges from true north depending on where you are on the surface of the globe. This discovery presented fundamental difficulties for any account of how the compass worked, but it also raised the exciting possibility that variation might be sufficiently regular for it to be used to measure longitude. Since longitude was the missing piece of knowledge for oceanic navigators, all the early modern studies of the magnet were alert to the possibility that they might be able to fill this gap.

Chronologically speaking, Leonardo Garzoni's treatises (discussed in Chapter 7) are the first major work of modern experimental science, but here chronology is misleading, for in crucial respects they are simply a continuation of the erratic medieval tradition of experimentation. Their conceptual apparatus is Aristotelian, and they seek to address a gap or anomaly in the Aristotelian scheme of knowledge. They respond to the age of exploration and discovery, but only with the intention of preserving and protecting the conceptual apparatus of traditional philosophy. And, just like the medieval experimentalists before him, Garzoni had almost no impact. As far as his colleagues were concerned, his work was of marginal interest unless it could be shown to result in a method for identifying longitude. Only one copy of his manuscript survives, and later Jesuit theorists rediscovered his work only because they needed ammunition to use against William Gilbert. It is true, as we have seen, that Garzoni's treatises were picked up by della Porta, who stole material wholesale from them, and who at least tested his claims about garlic and diamonds, but this was because the subject of magnetism, which involved hidden and inexplicable forces, fell squarely within the territory of natural magic; it is not because della Porta was converted by Garzoni to a new way of thinking, or a new, more reliable experimental practice.

Gilbert's *On the Magnet* takes us into a different world; indeed, he claims to be engaged in a new type of philosophizing. (It is an interesting question whether he would have made this assertion with the same confidence if he had read Garzoni.) For Gilbert, the experimental method is an alternative, not a supplement, to Aristotle. The goal of his philosophy is to make new discoveries, not to patch and mend an existing body of knowledge. De Maricourt and della Porta were crucial sources for Gilbert: we can tell that he repeated their experiments with care, which is how he knew that della Porta was copying from a half-understood source. He also had the advantage that the dip of the compass (that is, its tendency to point downwards from the horizontal by different amounts in different places) had been discovered by Robert Norman in 1581.[44] Gilbert was the first to recognize that the Earth itself is a magnet, and that this is why the compass needle points to the north. Others before him, including Digges, had grasped that the compass needle was not attracted towards a particular location, either in the heavens or within the Earth, but they had not taken the further step of thinking of the whole Earth as a magnet.[45] But this was not the limit of Gilbert's ambition. He sought to show (building on a suggestion by de Maricourt) that a magnet has a natural tendency to rotate on its axis; this, he claimed, provided an explanation for at least one of the three movements attributed to the Earth by Copernicus. Thus Gilbert made magnetic experiments relevant to a well-established branch of natural knowledge, astronomy. But at the same time he failed in his ultimate objective: he could not produce an experiment in which his magnets spontaneously rotated.

Gilbert's Copernicanism guaranteed a hostile response from orthodox Catholic scholars once Copernicanism had been condemned in 1616.[46] Thus Niccolò Cabeo, who was largely reproducing the arguments of Garzoni, continued to insist that there were two separate phenomena, the attraction of iron by magnets and the tendency of magnets to point towards the pole, not, as Gilbert claimed, one underlying phenomenon to which both could be reduced. It also guaranteed a favourable response from Copernicans such as Galileo and Kepler: Galileo said his own method was rather like Gilbert's, and Kepler took Gilbert's account of magnetism as a model for the sort of forces which might drive the planets in orbit around the sun. But, after Gilbert, work on the magnet failed to discover the regularities that make experimental knowledge possible. Not only did variation and dip diverge from place to place, but in 1634 a group of English experimenters claimed that variation fluctuated over time. This depended on their confidence in the reliability of measurements taken decades apart; others were quick to dismiss such findings as the result of defective technique. Nature could not be so capricious. Eventually, however, the naysayers were forced to concede that not only did variation change over time, but so did dip.[47] If successful experimentation depends on nature being economical and regular, then the study of the magnet after Gilbert seemed to undermine the conviction that this was so.

What do de Maricourt, Norman, Garzoni, della Porta and Gilbert have in common? Essentially, nothing. De Maricourt was a mathematical scholar and, it would seem, a soldier. Norman was a sailor who took advice from men of learning. Garzoni was a Jesuit, a Venetian patrician and a scholastic philosopher. Della Porta was a Neapolitan nobleman who had made a profession out of occult learning. Gilbert was an English doctor and the advocate of a new philosophy. Della Porta was preoccupied with sympathy and antipathy, while Garzoni and Gilbert refused to use such categories. This 'nothing' is important because it undermines the standard account of the origins of experimental science. It will not do to claim, as Marxists do, that sixteenth- and seventeenth-century experimentation involved a new collaboration between intellectuals and artisans when already, in 1269, de Maricourt had said that anyone studying the magnet must be 'very diligent in the use of his own hands', so manual dexterity was not new in the sixteenth century; and there is nothing to suggest that Garzoni, who certainly was dextrous, had links to the world of the skilled craftsman.[48] The study of variation and dip evidently depended on collaboration between navigators and intellectuals, but so did the whole science of cartography. It leaves us at the same time with a problem. If Gilbert is an example of a new type of scientist, what makes this new science possible?

The compass allows you to navigate out of sight of land and, naturally, Edward Wright's prefatory letter to *On the Magnet* mentions the circumnavigations of the Earth by English sailors. But in his preface Gilbert presents himself as sailing on a quite different ocean, an ocean of books. And, indeed, he had either bought books in vast quantities or had had access to a remarkable library, for *On the Magnet* begins with the first systematic literature review. Gilbert had read everything that had ever been written on the magnet. No ancient or medieval author (at least not since the great library of Alexandria went up in flames in 48 BCE) could have done this. Gilbert can confidently declare that he has made new discoveries because he knows exactly what was known before. He insists that knowledge comes not from books *only*, but from the study of things; yet the simple truth is that the ocean of books is as important to his researches as are the oceans.

So we might want to claim that it is the book – or rather, in this case, the well-stocked library – that transforms the status of experiment; by crystallizing past knowledge, the library makes new knowledge possible. In the case of anatomy, a single book, Vesalius's *Fabric of the Human Body*, had functioned as a whole library, but each new field required a similar enterprise of assimilating existing knowledge before new discoveries could begin. Printing made facts, as we saw in Chapter

7; and it begins to look for a moment as if it may also have made the new experimental philosophy.

But what is important to Gilbert is not just book learning, or even the performing of experiments. There is a third element to his new science. He thanks

> [s]ome learned men . . . who in the course of long voyages have observed the differences of magnetick variation: the most scholarly Thomas Hariot, Robert Hues, Edward Wright, Abraham Kendall, all Englishmen; Others there are who have invented and produced magnetical instruments, and ready methods of observation, indispensable for sailors and to those travelling afar: as William Borough in his little book on *The Variation of the Compass* or Magneticall Needle, William Barlowe in his *Supply*, Robert Norman in his *Newe Attractive*. And this is that Robert Norman (a skilful seaman and ingenious artificer) who first discovered the declination [i.e. dip] of the magnetick needle.[49]

Gilbert thus acknowledges a little community of experts, many of whom are known to him personally (Harriot, Borough and Norman for example; Edward Wright was a close collaborator). Where all previous experimenters, from Galen to Garzoni, appear to have worked in isolation, we have here for the first time a functioning scientific community, and the quality of Gilbert's work depends in part on his belonging to this community. There is no doubt that the later discovery of the variation of variation depended upon having a close-knit community of experts using the same instruments and techniques who acknowledged the accuracy of each other's measurements over long periods of time.

§ 5

It would be difficult to overestimate the impact of Gilbert's *On the Magnet*, not because everyone was interested in magnets but because for the first time the experimental method had been presented as one capable of taking over from traditional philosophical enquiry and transforming philosophy. Central to Gilbert's enterprise was the claim that you could reproduce his experiments and confirm his results: his book was, in effect, a collection of experimental recipes. In Padua in 1608 Galileo copied Gilbert's technique for arming a magnet by coiling iron wire around it, used it (without acknowledging his debt to Gilbert) to create what he claimed was the strongest magnet in the world and promptly sold his super-magnet for a lot of money to the Grand Duke of Florence.[50] Others, too, were surely copying and testing Gilbert's experiments, although no one else seems to have found a way of making money out of doing so. It is worth noting, then, that there is no record of anyone ever having claimed that Gilbert's experimental results could not be replicated. Replication is a vexed issue in the history of science but, where the history of magnetism is concerned, matters are straightforward: good results can be replicated, and bad ones (which would include some of Garzoni's) cannot.

There can be no replication if there is not some form of publication, or, at least, communication. Two scientists discovered the law governing the acceleration of falling bodies at around the same time – Harriot and Galileo.[51] Harriot kept his results to himself; Galileo finally published in 1632, some decades after he had made his discovery. He claimed that, leaving air resistance aside, heavy objects and light objects would fall at the same speed, so that if one dropped a musket ball and a cannon ball, or a wooden ball and a lead ball, of the same size, simultaneously from a tall building, they would hit the ground at the same moment. Soon all sorts of people were dropping objects off tall buildings and getting rather varying results (it is actually much harder than one might think to drop two object simultaneously, and very difficult to measure how far apart they are when the first one hits the ground). In France, Marin Mersenne went to elaborate lengths in 1633 to replicate Galileo's experiments and conduct accurate measurements. Where orthodox Aristotelianism said that objects fell at a constant speed, and the heavier they were the faster they fell, Galileo maintained that they accelerated as they fell, and that they all did so according to the same principle. His claims provoked a widespread concern with the replication of his experiments precisely because the viability of Aristotelian orthodoxy was now at stake.[52]

In 1638 Galileo also published the claim that if a column of water in a suction pump or in a tube with an air-tight seal exceeded a certain height (32 feet, he said), the column would descend, leaving a vacuum above it. He thought (rightly) that the key issue was the weight of the water: just as a dangling rope would break under its own weight if it was long enough, so at a certain point a column of water would break. What held the column of water together was a natural force, the resistance to the creation of a vacuum, and this force was a major factor in understanding the strength of materials. This was contrary to orthodox Aristotelian philosophy, which said that there could be no such thing as a vacuum.

Galileo had clung to his account despite the fact that a friend, Giovanni Battista Baliani, had suggested an alternative explanation for the fact that suction pumps ceased to work if they were asked to lift water more than 18 *braccia* (roughly 35 feet) – a figure established by experience. The explanation, Baliani said, was that up to a certain point the weight of the water was balanced by the weight of the air pressing down on all of us all the time; above that point the column of water could not rise and, in a pump without leaks, a vacuum would be created. There was no 'resistance' to the creation of a vacuum.[53] Either way, Galileo's and Baliani's claims struck at the very heart of Aristotelian physics. What Gilbert had aspired to do, Galileo had actually done: he had announced discoveries which were entirely at odds with the established philosophy. At first, a guerrilla army picks off enemy outposts and interferes with communications, but then, if it gathers strength, it must eventually move from hit-and-run attacks to a massed engagement with the enemy. Galileo's *Dialogue Concerning the Two Chief Systems* was a full-scale engagement with traditional astronomy; his *Two New Sciences* (1638) represented a full-scale attack on Aristotelian physics. Dispute over Copernicanism was distorted by the intervention of the theologians, but the dispute over physics could take place without such interference. Battle was engaged.

In Rome, sometime after 1638, a group of orthodox philosophers set out to prove that Galileo was wrong about the vacuum. Gasparo Berti built a long lead tube with a window in one end, filled it with water and sealed it at both ends, then upended it in a barrel of water and removed the seal from the bottom. At first, no empty space appeared at the top, but then he realized that he needed to measure the height of the column not from the ground but from the top of the water in the barrel and raised the tube a little higher; at once the column of water descended and an empty space appeared. But was it actually empty? Light travelled through it. A bell was fitted into the top of the tube, and it could be made to ring, so it seemed there was air present. (The vibrations of the bell must have been carried by the strut supporting it, rather than by the air.) The results were inconclusive. They had neither refuted nor confirmed Galileo's claim; they had simply produced an anomaly. And so the philosophers turned their minds to other things; later, no one could even remember the year in which they had performed this experiment, and no one wrote about it at the time.

In Florence, however, Galileo's disciple Torricelli heard in 1643 of Berti's experiments and realized that he could simplify matters by using

This is Schott's representation (some twenty-five years after the event) of the first attempt to create a vacuum at the top of a tall tube filled with liquid, in an attempt to disprove Galileo's claim that a vacuum would appear if the column of water was more than about 11 metres high. The space at the top of the tube has been expanded to include a bell, on the theory that if there was a vacuum no sound would be audible. Gaspare Berti's experiments were inconclusive (a sound was heard from the bell, suggesting there was no vacuum), but provided the inspiration for Torricelli's decision to substitute mercury for water. Berti's experiment was first described in print in Niccolò Zucchi's *Experimenta vulgata* (1648).

a denser liquid. A tube containing mercury would need to be one-fourteenth the height of a tube containing water: if 32 feet of water was the crucial height in order to generate a vacuum, then only a little over 2 feet of mercury would be needed. So he repeated the experiment with mercury, and reproduced the anomalous space. He had reached the same conclusion as Baliani: the space contained a vacuum, and the weight of the mercury was balanced by the weight of the air. We live, he wrote, under an ocean of air. Since the air on some days seemed to be heavier than on others, he reasoned that he ought to be able to measure the changing weight of the air. But his barometer produced puzzling and inconsistent results (it had probably been damp when he introduced the mercury), and he put it aside – and then died before he could learn of the successes of others.[54]

In France, Mersenne received a rather garbled account of Torricelli's experiment and tried unsuccessfully to reproduce it, but he lacked the right sort of glass tube. Shortly afterwards, toward the end of 1644, he travelled to Florence, where he met Torricelli, and then to Rome, where he may have seen Torricelli's experiment; on his return to France he tried unsuccessfully to reproduce the experiment, but his glass tubing was not of high enough quality. In the autumn of 1646 Pierre Petit and his friend Blaise Pascal successfully performed the experiment in Rouen; Petit had heard about it, but neither had seen it done before. Pascal then reinvented (for he had heard nothing of it) the experiment originally performed in Rome, substituting red wine for water to make it easier to see the result. Berti's experiment had been performed in a public place, but there is no evidence that it attracted any attention. Pascal's experiment was different: it was put on as a public show, but still there is no reason to think that Pascal intended to publish any time soon. When others, however, began to debate the significance of what he had done he rushed his own account into print (in 1647) in order to establish his claim to priority. Pascal sent copies of his booklet to all his friends in Paris and to every town in France where he thought there were people who might be interested in reading it – presumably, to the local booksellers, for somewhere between fifteen and thirty copies went to Clermont-Ferrand alone; Mersenne sent copies to Sweden, Poland, Germany, Italy, and indeed all over the place. The status of experimentation was changing; and Pascal and Mersenne made every effort to bring this about.[55]

According to Vincenzo Viviani, the young Galileo in around 1590 had dropped objects off the Tower of Pisa and the whole university had gathered to watch. Viviani was probably right to say that Galileo performed this experiment, and if he did he was not the first: similar experiments had been performed by Giuseppe Moletti in 1576 (but never published) and by Simon Stevin (who published in 1586, but attracted no attention, partly because he wrote in Dutch).[56] But there is absolutely no evidence, other than Viviani's account much later, to suggest that crowds gathered to watch Galileo's early experiments. In assuming they did, Viviani is reading back into Galileo's youth a view about the status of experiments which became established only in the 1630s: Viviani became Galileo's assistant in 1639 at the age of seventeen, and wrote Galileo's Life in 1654. Pascal's experiments of 1646, on the other hand, really did draw crowds.

Up to this point, the story of the vacuum experiments is a story of accidents and near-misses. Berti's experiment had reached no clear conclusions; Baliani and Torricelli had got the theory right, but Torricelli's experiment had not quite worked, and the initial reports that had reached France had not conveyed his theory, nor provided sufficient detail to enable the experiment to be reproduced. From 1646 Torricelli's experiment was widely performed, even though mercury was expensive and it continued to be difficult to get sufficiently sturdy long tubes sealed at one end. In fact, Torricelli's experiment rapidly became famous: the phrase 'famous experiment' is first used in English in 1654 to refer to it, and for an Italian author in 1663 it is *famosissima*.[57]

Once Torricelli's experiment had been accepted as a basic model, it was possible to devise all sorts of variations. Three are of great importance. First, Pascal invented a way of putting a barometer within the anomalous space at the top of the Torricellian tube: when the mercury dropped in the main tube to a height of 27 inches, it dropped to zero (or very near zero) in the second barometer within the Torricellian space. Various revisions and improvements on this experiment were devised by Pascal and others; in its different forms, the 'void in the void' experiment seemed to confirm that there was no (or almost no) air pressure within the Torricellian tube. Second, Pascal devised the Puy-de-Dôme experiment. Third, Roberval devised an experiment where a carp's bladder which had been flattened and sealed tight was placed at the top of the Torricellian tube. When the mercury dropped, the bladder was left behind and swelled up as if it had been pumped full of air. The interpretation of this experiment was far from straightforward, but Roberval argued that if there had been air in the carp's bladder when it seemed there was none, so, too, could there be air in the Torricellian space (although only in the most minute quantities).[58]

In thinking about the expansion or rarefaction of air, Roberval invented the concept of 'the spring of the air', which Boyle went on to make famous in his *New Experiments* of 1660, and which became systematized in Boyle's law (1662).[59] It is true that Roberval did not use the word 'spring', borrowing the Latin word *elater* (which shares a root with 'elastic'; *elater* is the translation for 'spring' in the Latin version of the *New Experiments*) from Mersenne;[60] but Boyle's failure to acknowledge any debt to Roberval for the concept shows that intellectual property in theories was slower to become established than intellectual property in other sorts of discovery, such as the design of an experiment. Already in 1662, however, Boyle was careful to acknowledge that a number of people had contributed to the formulation of Boyle's law.[61] And the concept of intellectual property certainly had been established by 1677, when Oldenburg, writing in the *Philosophical Transactions*, complained that a Latin translation of Boyle's works, produced in Geneva without his permission, failed to record the date when the originals were first published, which could give the false impression that Boyle had stolen from others when they in fact had stolen from him. In the *Second Continuation* Boyle, or rather his publisher on his behalf, returned to the issue: '[F]or though some Writers have with Ingenuity enough cited the Name of our Author in their Works, yet more have done otherwise, transferring not a few of his Experiments, together with the Ratiocinations explaining them, after the manner of Plagiaries into their Books, making no mention of his Name at all.'[62] A few years earlier, the Chief Justice, Matthew Hale, writing anonymously on the Torricellian experiments, had been anxious to insist that he had cited his sources, in order to 'avoid, as much as I can, the imputation of a Plagiary'.[63]

That someone who originates a new idea has a right to be acknowledged seems to us obvious, but this idea was fundamentally new. If we look back to the Parisian philosophers of the fourteenth century, for example, to Oresme, Buridan, John of Saxony and Pierre d'Ailly, we find ourselves in a world where scholars reported each others' arguments but failed to record who originated any particular line of argument, so that historians still cannot write the history of the school of Paris in terms of who influenced whom; being first was not what

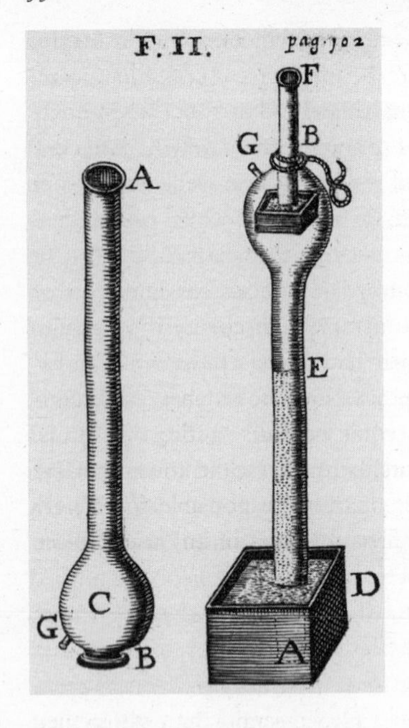

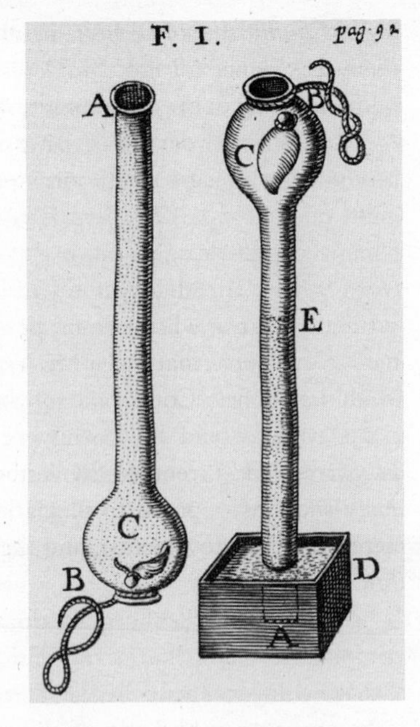

a) Adrien Auzout's void-in-the-void experiment, from Jean Pecquet, *Experimenta nova anatomica* (1651). In this experiment, a barometer inserted into the Torricellian void at the top of a first barometer measures the air pressure there: the mercury in the second tube does not rise, marking the absence of any air pressure in the space; when air is introduced into the space at the top of the first barometer the mercury in that barometer drops down into the tank, while the mercury in the second barometer rises to a height of 27 inches.

b) Gilles de Roberval's carp-bladder experiment. A carp bladder, from which all the air has been squeezed out, and which has been tied off, is introduced into the Torricellian void. It promptly swells up, demonstrating the extraordinary elasticity of the little bit of air left in the bladder. Roberval thought this justified the conclusion that there is always some air, however little, in the Torricellian void.

mattered to fourteenth-century philosophers. This world still existed in 1629 when Niccolò Cabeo published his *Magnetical Philosophy*, which is almost entirely drawn, and indeed much of it taken verbatim, without acknowledgement, from Leonardo Garzoni's unpublished manuscript; it still existed in 1654, when Pascal completed his *Treatise on the Equilibrium of Liquids*: in it he drew heavily on works by Stevin, Benedetti, Galileo, Torricelli, Descartes and Mersenne, but made no mention of any of his predecessors.[64] It still existed in 1660, when Boyle (who had a highly developed sense of propriety and surely had no consciousness of doing wrong) borrowed without acknowledgement from Roberval, but was fast disappearing by 1682, when he complained about the borrowings (still, perhaps, entirely innocent) of others. In 1687 Boyle's friend David Abercromby announced his intention of writing a treatise which would be a history of discovery through the ages – the book that Polydore Vergil had failed to write. Included would be all 'new Contrivances, whether Notions, Engines, or Experiments'. This would be a study of what he calls 'authors', that is, discoverers and inventors: 'By Authors, here are meant, those that are really such [as opposed to plagiaries or those responsible for merely incremental improvements], and the first Inventors of any useful piece of Knowledge'.[65]

From 1646 to 1648 a small group of experimenters (Pascal, Roberval, Auzoult, Petit, Périer, Gassendi, Pecquet) scattered across France were working on vacuum experiments simultaneously. What held them together was their common friendship with Mersenne, with whom they exchanged letters and at whose home they met when they were in Paris. They had a variety of professional commitments, but they regarded themselves as first and foremost mathematicians, and many of them made major contributions to pure mathematics.[66] They competed with each other and collaborated, and (for the most part) trusted each other enough to be confident that their own contributions would be recognized. They freely circulated manuscripts among themselves. Roberval, for example, never published his vacuum experiments, but a letter he wrote describing the early history of the experimental programme in France was published in Poland, a number of his experiments were described in print by an opponent, and his carp's-bladder experiment was published by Pecquet in 1651 in a volume primarily devoted to new anatomical studies (this was translated from Latin into English in 1653). Publication was important within this group, but no more important than private and semi-public correspondence: Mersenne wrote letters to Italy, Poland, Sweden and Holland announcing Pascal's Puy-de-Dôme experiment.[67] It is also significant that Mersenne's friends collaborated without agreeing with each other. There was agreement on the value of experimental enquiry, not on how to interpret the results.

Mersenne died in 1648, and little further progress in research on the vacuum was made in France. But Pascal's and Pecquet's works were read in England (where Henry Power immediately replicated Pecquet's experiments) and Italy, as was Gaspar Schott's *Mechanica* (1657). Schott reported not only Berti's original experiments in Rome but also von Guericke's construction of a vacuum pump.[68] Von Guericke had demonstrated how two hemispheres from which the air had been extracted were held so tightly together by air pressure that teams of horses could not pull them apart. It was Schott's book that inspired Robert Boyle to construct his own air pump in England, and it is perhaps not a coincidence that vacuum experiments were recommenced in Florence in 1657. If one drew up a list of all the people known to have performed barometer experiments between Torricelli in 1643 and the discovery of Boyle's law in 1662, it would reach a hundred names without too much difficulty. These hundred people are the first dispersed community of experimental scientists.[xii]

Experiments produce new knowledge, but if that knowledge does not circulate there is little opportunity for further progress. Torricelli's barometer represents the first piece of experimental apparatus that became standardized and widely available; and endless variations on the experiments that could be performed with it (such as releasing insects into the Torricellian space) were devised. This was the first time experimentation had had an audience (symbolized by the small crowd surrounding Périer on the summit of the Puy-de-Dôme) and it was the first time it became a collaborative and competitive process.

As one would expect, this first successful experimental community changed the way in which scientific communities were constructed and what they were used for. Mersenne's community was an informal group which met and exchanged letters, though he expressed a wish to form a proper college that would function mainly through correspondence. There had been earlier semi-scientific academies: della Porta had formed an academy (which the Inquisition had forced him to close) dedicated to the pursuit of secret knowledge, while both he and Galileo had belonged to the Accademia dei Lincei (the lynx-eyed) founded by Prince

xii 'In the experimental form of life, the practice of producing true knowledge was deemed to be totally dependent upon collective labour. Real experiments had to be performed, seen, and believed to be performed, by a witnessing community.' (Shapin, 'Boyle and Mathematics' (1988), 43.) Gilbert already had a small, local community of experts, but the Torricellian experiment produced an international network of a quite new sort.

Cesi.[xiii] Bacon had imagined a scientific community at work in his utopian *New Atlantis* (1626). Mersenne was not the first, and nor would he be the last, to establish an 'invisible college' through correspondence: his own network grew out of that built up by Peiresc, and similar networks were founded by Hartlib and Oldenburg in England (Oldenburg's network becoming identified with the Royal Society when he became, jointly with Wilkins, its first secretary).[69]

The extraordinary success of the Torricellian network, as we may term the group of people involved in barometer experiments, was a major factor in leading to the establishment of the Accademia del Cimento in Florence (1657), the Académie de Montmor in France (1657), the Royal Society in England (1660) and the Académie Royale in France (1666). The Accademia del Cimento published a single book, but the Royal Society published the first journal, the *Philosophical Transactions* (from 1665), devoted to the new science. In France the *Journal des sçavans* began publishing the same year: it covered a wide range of academic subjects but declared in its first issue that one of its major concerns would be to announce new discoveries.

Thus the informal Torricellian network marks the effective beginning of the institutionalization of science, driven by the conviction that collaboration and exchange would lead to more rapid progress. As one would expect, this was accompanied by a new commitment to the idea of scientific progress. In a draft preface to an unpublished book on the vacuum (*c.*1651), Pascal distinguished between forms of knowledge that were historical in character and depended on the authority of the sources on which they relied (theology was the key example), and forms of knowledge that depended on experience. In the case of the latter, each generation, he claimed, knew more than the one before, so progress was continuous and uninterrupted ('all mankind continually makes progress as the world grows older').[70] Pascal says, indeed, that each generation sees further than the one before. He almost certainly has in mind the famous saying that we are dwarfs standing on the shoulders of giants. The saying originates with Bernard of Chartres in the twelfth century, but is now usually quoted from one of Newton's letters: 'If I have seen further it is by standing on the shoulders of giants.' Newton was engaging in false modesty (and the shoulders he had been standing on most immediately were those of Hooke, who was distinctly short of

xiii *Lynceus* is an adjective meaning sharp-sighted; Accademia dei Lincei is often translated Academy of the Lynxes, but that would be Accademia delle Linci.

Schott's representation of the Magdeburg hemispheres in *Experimenta nova* (1672). In 1654 in Regensburg, and then again in 1656 in Magdeburg, Otto von Guericke evacuated the air from a copper sphere with an air pump. He then attached teams of horses to the sphere, which consisted simply of two hemispheres mated together, but they were unable to draw them apart. This demonstrated the force of air pressure acting on the hemispheres and inspired Boyle's construction of an air pump.

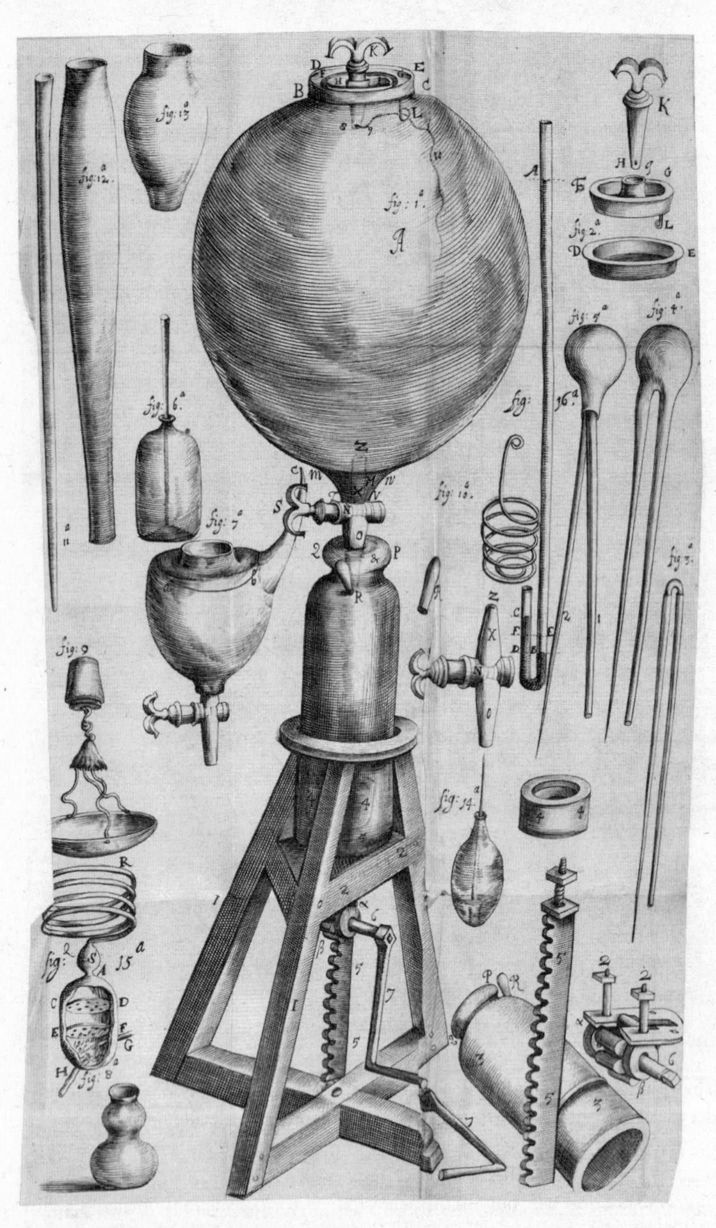

Boyle's first air pump, designed and made by Robert Hooke, from Boyle's *New Experiments Physico-mechanical* (1660).

The frontispiece of the English translation of the experiments of the Accademia del Cimento: Nature is shown turning her back on Aristotle and being introduced by the Accademia to the Royal Society. Published in Italian in 1666, the *Saggi* were presented to the Royal Society (they were published in a luxury edition intended not for sale but for presentation); after considerable delay Richard Waller translated them as *Essayes of Natural Experiments* (1684). The frontispiece was drawn by Waller himself.

stature).[71] Since Pascal's purpose was to undermine respect for antiquity, he had no intention of repeating the claim that the ancients were giants as compared to us; he assumes that each generation is equal to any other in its abilities.

At first sight, Pascal's claim that progress has been continuous seems nonsense: we would accumulate experience from generation to generation only if we had a reliable way of recording and transmitting it. Pascal is, however, assuming a book and not a manuscript culture (we are back with Gilbert's 'ocean of books'); it is only since the invention of printing that knowledge has been effectively recorded and transmitted. Moreover, he recognizes that not every individual makes progress; rather, it is what he calls *l'homme universel*, human beings collectively. It was through his collaboration with other scientists that Pascal came to have the sense of belonging to a collectivity greater than himself. The Torricellian network solved problems more efficiently than any individual could on their own. As history, the preface is bunk, because in 1651 progress was new. But since then it has indeed been uninterrupted and continuous; as an account of modern science, the preface is spot on.

Thus experiments are not new. The first person who rubbed two sticks together to make a fire was conducting an experiment. Galen, Ibn al-Haytham and Theodoric of Freiberg performed experiments. What is new is the scientific community that is interested in experiments. We can see a foreshadowing of this in the group of people which surrounded Gilbert as he conducted experiments on magnets, but for the most part these were fairly uneducated navigators. It takes proper form in the years after the publication of Galileo's *Dialogue Concerning the Two Chief World Systems* in 1632. Even Daryn Lehoux, who thinks the Romans had everything we have, acknowledges one exception:

> There were no ancient universities, no scientific conferences, no journals where investigators published their results. So, too, no *New Scientist*, no science pages in the *New York Times* where the newest work could be reported, compared, commented on. From these modern sources there often emerges an understanding, among professionals and among the scientifically literate public, of something we might call the 'consensus in the field' on many issues.[72]

Kuhn had a particular term for 'the consensus in the field'. He called it 'normal science', as opposed to revolutionary science. The Torricellian barometer was the first experimental apparatus around which a normal science developed. There had been stable, consensus-based sciences before: Ptolemaic astronomy, for example, or Vesalian anatomy. But this is the first time a consensus developed around what the English call an 'experiment'.

In the course of the seventeenth century the Latin words *experientia* and *experimentum* and, with them, the English words 'experience' and 'experiment' began to diverge in meaning. Thus from 1660 on 'experimental philosophy' became a widely used label for a science relying on experiments; no one wrote of an 'experiential philosophy'.[73] The divergence has its roots back in the early thirteenth century, when translators of key Arabic texts, such as Ibn al-Haytham's *Optics*, had chosen the Latin *experimentare* rather than *experiri* to translate the Arabic *i'tibar* and to describe experiments in optics.[74] *Experimentum* was, as a consequence, the word generally used by medieval philosophers to describe artificially constructed experiences. It will have been the obvious choice for Gilbert in his *On the Magnet*. Slowly, 'experiment' became in English a technical term for something scientists do; but not, as we have seen, in French. In Italy, Galileo normally wrote of *esperienza* in Italian, when in Latin he would have written of *experimentum*. *Esperimento* and *sperimentare* were neologisms and, though they were to be found in the dictionary of the Accademia della Crusca (1612), they were not used in any of the classic texts of Tuscan literature. But *esperienza* was too broad a term to identify the procedures of the new science, and after Galileo's death his disciples formed the Accademia del Cimento (*cimento* means 'test' or 'proof' in the practical sense, like the English word 'assay' and the French word *essai*, so this was an academy devoted to experiments). The ultimate success of *esperimento* reflects, just like the phrase *philosophie expérimentale* in French, the influence of the English language and of English science. In English, the phrase the 'experimental method' first appears in 1675.[xiv]

Thus, in the case of the word 'experiment', linguistic change lags behind both theory and practice. If language provides only very limited assistance, how are we to know an experiment when we see one? The

xiv In a review of a book, *Collegium experimentale*, by Johann Christophorus Sturm in the *Philosophical Transactions* (Anon, 'An Accompt of Some Books' (1675), 509): 'The Learned Author of this Piece ... takes notice, that in this one Age [i.e. century], not yet elapsed, there hath been a far greater progress made in Natural Philosophy, than in many Ages before, and that by means of that happy *Experimental* method, embraced and exercised by the *Royal Societies* of *England* and *France* ...' Sturm's book is about *inventa et experimenta physico-mathematica* – physico-mathematical discoveries and experiments.

answer is simple: an experiment is an artificial test designed to answer a question. The Latin term for this, well known to medieval and Renaissance philosophers, is *periculum facere*, to make a test or trial of something.[75] Such a test usually involves controlled conditions, and often requires special equipment.

§ 6

Up to this point, when I have tried to identify what it is that we have that they (Greek, Roman and medieval philosophers) didn't, my response has been to locate a conceptual tool, such as discovery or the fact, or a technical breakthrough, such as measuring the non-parallax of comets, or an instrument, such as the telescope. This chapter has pointed instead to a sociological reality – the scientific network – and more concretely, the small crowd that surrounded Périer on the summit of the Puy-de-Dôme. It would be wrong to overemphasize the difference between conceptual and sociological explanations: discoveries have to be announced, facts accepted, experiments replicated; the concepts are grounded in a sociological reality – the audience (created, above all, by the printing press). The scientific network is another term for that sociological reality: Pascal announced his discoveries to Mersenne's network, and persuaded them that his facts were right by getting them to repeat his experiments. The new concepts and the new social organization are two sides of one coin. If earlier scientists, like Harriot in England, failed to publish, or, like Galileo in Italy, at least as far as his new physics was concerned, were slow to publish, it is partly because they were not confident there was an audience for what they had to say. The success of the Torricellian barometer *created* an audience for the new science.

In stressing that science is a community activity I do not mean to imply (any more than Kuhn did) that science *only* has a social history, or (as relativists have it) that science is whatever scientists agree on. There had been earlier attempts to build communities whose goal was to advance knowledge: sixteenth-century doctors, for example, had formed networks and had published the resulting correspondence.[76] What such communities had never been able to do, however, was to build consensus around the problems to be worked on and the solutions that would be regarded as satisfactory. They never established anything resembling normal science. The key to normal science is replication.

Over and over, in the years after 1647, scientists filled long glass tubes, sealed at one end, with mercury and inverted them in baths of mercury. They provided each other with helpful guidance: don't even breathe in the tube, says Pierre Petit, or you will contaminate the mercury with water; put the apparatus on a blanket, says Henry Power, and have a wooden spoon to hand, so that any mercury you spill is captured and you can scoop it up.[77] They invented endless variations: but nobody who performed a variation had not also performed the basic experiment. And over and over they got the same results.[xv] If the Torricellian barometer had not been easy to replicate, it would never have become the first famous experiment. When the Accademia del Cimento was formed in 1657 its motto was a tag from Dante, *provando e riprovando*, and they tested and tested again. Successful replication (not intellectual coherence or the support of authority) was now the mark of reliable knowledge.

I am here arguing against a powerful tradition in recent historiography of science which insists that replication is always problematic, and that in the end what counts as successful replication is always decided by the intervention of authority.[78] According to these historians, replication is a social artefact, not a natural fact. The classic study is *Leviathan and the Air-pump*, by Steven Shapin and Simon Schaffer.[79] That book, which has been described as the most influential work in the history of science after Kuhn's *Structure of Scientific Revolutions*, presents a number of arguments which have become famous.[80] It argues that Boyle, through his air-pump experiments, was a pioneer in making facts: from the previous chapter we can see that this view is mistaken, unless one focusses narrowly on the use of the word 'fact'. Gilbert, Kepler and Pascal all established facts. The book also contends that Boyle developed a novel technique for winning support by turning readers into virtual witnesses: virtual witnessing is important, but Boyle was hardly a pioneer in this respect either. It maintains that the dispute between Boyle, who claimed (albeit cautiously) to have made a vacuum, and his opponents, who claimed that he had not, was not resolved

xv As Pascal said of the miniaturized Puy-de-Dôme experiment performed in tall buildings: 'All the curious can test it themselves whenever they like' (Pascal, *Oeuvres complètes* (1964), 687). Gassendi did indeed promptly repeat it: Koyré, *Études d'histoire de la pensée scientifique* (1973), 330; and Boyle, for example, carried out a version of the miniaturized experiment on the roof of Westminster Abbey (Boyle, *A Defence* (1662), 51–2 = Boyle, *The Works* (1999), Vol. 3, 52–3). On the ease with which the Torricellian experiment could be repeated, Glanvill, *Plus ultra* (1668), 60–1.

because Boyle had the better arguments but because he had the more powerful social position.

Here it is important to compare the disputes in which Boyle was engaged with those in which Pascal was engaged. Boyle constructed his air pump because the glass globe out of which the air was evacuated provided a better site for experiments than the space at the top of a barometer. Boyle, for example, could place a lit candle in his globe, or a bird; although insects and frogs could be passed through the mercury into the Torricellian space, flames and birds could not. In order to demonstrate that his experimental space was equivalent to that in a barometer, Boyle had to repeat the standard experiments, such as the void in the void and the carp's bladder, and show that he obtained the same (or virtually the same) results. In so far as Boyle's 'vacuum' was indistinguishable from Torricelli's, the arguments in England were fundamentally the same as those which had already taken place in France. Against those who maintained that, since light was able to pass through the Torricellian space, there must be within it some mysterious aether, a substance which appeared to have no weight and was present everywhere, Pascal had replied that, as the nature of light was unknown, it was futile to insist that it required some imaginary substance as its medium for transmission. To claim that there existed a substance with no measurable attributes was, he maintained, to turn physics into a fantastical story, like Don Quixote. Pascal and Boyle both 'won' this argument in so far as they succeeded in casting doubt on the legitimacy of appealing to substances whose existence might be theoretically convenient but could not be demonstrated experimentally.

However, a standard argument against Boyle's experiments with his air pump was that the pump leaked, and that therefore it was incapable of making a vacuum. This was true, and in the void-in-the-void experiment the mercury never fell below half an inch. The Torricellian barometer, by contrast, did not leak, although it was difficult (perhaps impossible) to prevent some air being trapped inside the tube or released by the mercury. Boyle was right, however, to claim that his results were very similar to those previously obtained in experiments with barometers. If Boyle appeared to win in an English context it is because his most forceful opponent, Hobbes, was a more isolated, and thus less dangerous, figure than Descartes; Hobbes was fatally weakened by his reputation for atheism, while Descartes carefully placed his arguments within a wider context which was compatible with Christianity. It is important not to overstate the local and limited successes of Pascal and Boyle: belief in a vacuum, along with Copernicanism, ultimately triumphed only when Newton's theory of universal gravitation (published in 1687) gave an account of how gravitational forces operated across empty space. Pascal and Boyle were then hailed as the discoverers of the vacuum which, it was now held, made up most of the universe. But in the 1660s the situation was quite different: in England, Henry Power, for example, continued to oppose Boyle's claims on the basis of experimental evidence (and a sympathetic attitude to Cartesianism), just as Roberval had opposed Pascal's.[81]

In 1661 Christiaan Huygens built his own air pump and began repeating the standard experiments. He tested the quality of his machine by introducing a water barometer to provide a sensitive measure of how much air (if any) remained in the experimental space. Huygens pumped away, air was removed, but the water level failed to fall. The tube remained full. The water Huygens was using had been purged of air in order to ensure it would not release air into the experiment; Huygens found that he could only get it to behave as expected if he introduced a bubble of air into it. When reports of this reached Boyle, he (quite naturally) dismissed them as nonsensical, but Huygens came to London and showed that the same results could be produced on Boyle's apparatus. (The cause of this puzzling phenomenon is the tensile strength of water, without which trees could not grow above ten metres tall.) In the light of his results, Huygens abandoned his previous belief in a vacuum (despite the fact that the anomaly disappeared when a bubble of air was introduced into the water, he decided that some previously unknown substance was supporting the column of water), while Boyle decided to carry on as if nothing had happened.[82]

What is crucial here is that the two conflicting results did not properly have equal status. The void-in-the-void experiment had been carried out time after time in barometers, with water, wine and mercury, and it had been performed with at least five different configurations of the apparatus, yet no result comparable to that produced by Huygens had ever been seen. Boyle therefore decided to see if he could produce anomalous suspension with mercury (something Huygens had not managed to do) by carefully purging it of air, as this would make it evident that there was something misleading about Huygens' result. As he put it: '[T]he sustentation of tall Cylinders of Mercury in the Engine [i.e. the air pump] seem'd to me to have too little Analogy with all the Experiments yt hath been hitherto made about those of Torricellius.'[83]

But even before performing the experiment with the air pump, Boyle succeeded in producing anomalous suspension of mercury (to a height of 52 inches) in the open air. At this point it was clear that the phenomenon had nothing to do with what was or was not present within a supposed vacuum. Huygens agreed: it took less than two years (a shorter period of time than it would now seem, given the difficulties of travel and communication in the seventeenth century) for him to accept that his anomalous result was irrelevant. One can reasonably say, then, that Huygens had been wrong to claim that his result was in some way 'better' than Boyle's, and wrong to abandon his earlier beliefs in the light of it. Boyle's result was the right result, and Huygens' result was simply a deeply puzzling anomaly; we can say this with confidence now, but it was also apparent to intelligent observers at the time, and to Huygens himself as soon as Boyle demonstrated the anomalous suspension of tall cylinders of mercury, both in and out of the air pump.[84]

The experimental method depends on independent replication, and one claim made by sociologists of science is that truly independent replication never takes place: in order to make a new experiment work, it is claimed, scientists always have to spend time in the company of scientists who have already performed it, picking up unwritten tricks of the trade. But Petit together with Pascal replicated the Torricellian experiment independently, and Valerio Magni in Warsaw either replicated it in 1647 or reinvented it. Others seem to have performed the experiment entirely independently, on the basis only of written descriptions – Henry Power, for example. The simple truth is that replication of the Torricellian experiment was unproblematic; it follows that the sociologists of science are wrong.[xvi]

If we look back over the history of experimentation in this way we can begin to understand the significance of what happens in the seventeenth century, symbolized by the small crowd which accompanied Périer to the top of the Puy-de-Dôme. Why were they there? Périer was surely glad to have witnesses, but they were there because they believed this was an opportunity to watch history being made. Their presence marks the beginning not of discovery itself but of a *culture* of discovery, a culture which was now shared by government-office holders and sophisticated clergymen. (Périer names two clergymen, two government-office holders and a doctor as accompanying him.) Moreover, publication

xvi I discuss below, pp. 521–2, Simon Schaffer's account of Newton's *experimentum crucis*; again, the claim that replication was a social artefact turns out to be false. On replication as requiring a transfer of tacit knowledge which ensures that it is (almost) never truly independent: Collins, 'The TEA Set' (1974); Collins, 'Tacit Knowledge, Trust and the Q of Sapphire' (2001); and Pinch in Labinger & Collins (eds.), *The One Culture?* (2001), 23.

ensured that there was a much larger number of 'virtual onlookers'. As Walter Charleton put it, the little group of French experimenters 'seemed unanimously to catch at the experiment [of Torricelli], as a welcome opportunity to challenge all the Wits of Europe to an aemulous [actuated by a spirit of rivalry] combat for the honour of perspicacity'.[85] And when Boyle coined the phrase *experimentum crucis* to honour the Puy-de-Dôme experiment he was marking the beginning of a new era in which philosophical disputes would be resolved by experimentation.

§ 7

The issue of replication is central to a topic of foremost importance for any understanding of the Scientific Revolution: the demise of alchemy.[86] Boyle and Newton devoted an enormous amount of effort to alchemical researches. Boyle seems to have spent much of his life trying to turn base metal into gold, although our knowledge of his activities is limited because most of the relevant papers were (as best we can tell) destroyed on the instructions of his first biographer, Thomas Birch.[87] Boyle believed he was on the verge of succeeding, so close indeed that he thought it prudent to campaign (successfully) for a change in the law, which condemned to death anyone making gold.[88]

Like many alchemists, Boyle was convinced that the quest for the philosopher's stone (which would turn base metal into gold) involved a spiritual element. He believed he had seen transmutation performed; he evidently thought it likely that the anonymous stranger who had performed it, in his presence and with his assistance, was an angel, no less.[89] He had been singled out for this special revelation. Such beliefs made Boyle the perfect target for the sophisticated con artist. One part of Boyle's correspondence concerning alchemy survives by chance: it was written in French, a language unknown to Henry Miles, Birch's assistant, who was tasked with sorting through Boyle's papers and deciding which ones to throw out. From it, we learn that a Frenchman called Georges Pierre persuaded Boyle that he (Pierre) was the agent of the Patriarch of Antioch, the head of a society of alchemists that had members in Italy, Poland and China. To become a member Boyle had to hand over his own alchemical secrets, but also valuable gifts – telescopes, microscopes, clocks, luxurious fabrics, large sums of money. In return, Pierre reported on the manufacture of a homunculus in a glass vial. Pierre told a good story: one meeting of his secret society had, he assured

Boyle, been disrupted by disgruntled employees who had blown up the castle in which the society was meeting. And Pierre went to great lengths: he planted stories about the Patriarch of Antioch in Dutch and French newspapers on the offchance that Boyle would come across them. In fact, when Pierre was supposed to be in Antioch, he was actually in Bayeux, having a jolly time with his mistress. And his wild stories had already acquired him the nickname in his home town of Caen of 'honest Georges'.[90]

How could Boyle, one of the key figures of the Scientific Revolution, be so totally convinced of the reality of alchemical transmutation? The answer is that alchemy was a self-fulfilling enterprise. Those who practised it were convinced that the philosopher's stone had been successfully produced in the past. As George Starkey, who collaborated closely with Boyle in his alchemical researches, put it: '[T]he wise Philosophers with all their might have sought & found, & left the record of their search in writing, withall so veyling the maine secret that only an immediate hand of god must direct an Artist who by study shal seeke to atteyne the same.'[91] Like Boyle, Starkey believed he had held the stone in his hand, and with it he claimed that he had been able to turn base metals into gold and silver – or at least into a sort of gold and a sort of silver, for the gold had proved unstable, and the silver, though very like silver, weighed too much.[92]

Starkey sought to discover this and other lost secrets by careful study of the alchemical texts, which were written, as he acknowledged, in a deliberately impenetrable language. Since early in the seventeenth century the word 'hermetical' (meaning 'in the tradition of the mythical author Hermes Trismegistus', a supposed contemporary of Moses, to whom numerous works were attributed) had been given a new meaning: those experimenting with chemicals began to refer to containers being 'hermetically closed', in other words, airtight, the word 'hermetical' becoming a pun on the idea of impenetrability.[93] When Starkey failed to achieve the results he hoped for (bankrupting himself and reducing his wife and children to penury in the process) it never occurred to him that the texts were wrong;[94] he was convinced he had simply misinterpreted them or failed to follow their instructions with sufficient precision. Thus alchemists had in principle procedures for verification ('trial by fire'), although verification was constantly deferred. They had no procedure for falsification.

Starkey referred to the adage 'Hear the other side', a fundamental principle of natural justice, as 'hateful'.[95] The other side he refused to hear were those who dismissed alchemy as a deception, a delusion, a fantasy. Yet among scholastic philosophers these were the majority, from Aquinas and Albertus Magnus on. Alchemy had long been mocked by sceptics – by Reginald Scot in the *Discovery of Witchcraft* (1584) and by Ben Jonson in *The Alchemist* (1610) to take but two examples among many. Belief could be sustained only if a peculiar authority was granted to obscure books, or better still ancient manuscripts newly discovered in locked chests. 'Alchemy was inalienably a textual as well as an experimental science,' writes Brian Vickers, and in paintings the alchemist was always shown surrounded by books or manuscripts as well as by the impedimenta of his laboratory.[96] Missing from the paintings, though, is the most important moment of all, the moment when one person was persuaded to put their trust in another. Thus Boyle left at his death 'a kind of Hermetic legacy to the studious disciples of that art' (a legacy which does not survive, and was presumably destroyed). Included were many alchemical recipes he had not tried but which he was sure were efficacious, for they had been '(though not without much difficulty) obtained, by exchange or otherwise, from those that affirm[ed] they knew them to be real, and were themselves competent judges, as being some of them disciples of true adepts, or otherwise admitted to their acquaintance and conversation'.[97] Difficulty was itself the guarantor of authenticity; in the absence of anyone who could reliably be identified as a true adept (i.e. someone capable of producing the philosopher's stone) a mere claim to acquaintance and conversation was sufficient to authenticate an incomprehensible text as carrying hidden meaning. Boyle believed because he wanted to believe.

In the recent literature there has been an extended effort to present alchemy (or, as its historians now prefer to call it, 'chymistry') as the first experimental science; out of alchemy, we are told, came modern chemistry.[98] In showing that many alchemical recipes can indeed be performed in a modern laboratory, these scholars have rendered apparently incomprehensible texts meaningful and reinstated alchemy as a laboratory science. But if this argument is pressed too far it becomes difficult to explain why, in the eighteenth century, modern chemistry established itself not as a continuation of but as a refutation of alchemy. Why did Birch destroy Boyle's alchemical papers, not celebrate them?

There has been little written on the end of alchemy, yet an activity which was respectable in the eyes of Boyle and Newton had become entirely disreputable by the 1720s.[99] John C. Powers has argued that this was the result of a series of 'rhetorical' moves by chemists in the Académie des Sciences, such as Nicolas Lémery (1645–1715), who adopted many of the experimental findings of the alchemists, and also many of their attacks on those who brought their art into discredit by telling tall tales; at the same time they dismissed as ridiculous the quest for the philosopher's stone. The implication is that deep down they were alchemists, they were just not prepared to admit it. Powers does not consider taking the eighteenth-century chemists at their word. The advocates of the new chemistry insisted that they had no time for texts whose meaning was impenetrable. They protested that 'that sect of chemists [i.e. the alchemists] . . . writes so obscurely that in order to understand them one must have the gift of divination.'[100] They were, they insisted, interested only in chemical processes which they could reproduce in their own laboratories and could then have certified by their colleagues. 'Each *memoir*,' produced by the advocates of the new chemistry, writes Powers, 'presented a limited investigation into a specific question or set of questions, and the chemist relied solely on the account of his experiments to persuade his audience [to] accept his conclusions.'[101] Powers describes these as 'purported' experiments, but of course they were real.

What made it possible to consign alchemy to the dustbin of history was a new understanding of what chemists were trying to do. For the alchemists, including Boyle and Newton, the fundamental enterprise was one of *transmuting* one substance into another. But in 1718 Étienne François Geoffroy, the son of a pharmacist and the holder of the chair in chemistry at the Jardin des Plantes in Paris, an institution established for the training of pharmacists, published a 'Table of Different Relations Observed in Chemistry'. Geoffroy's table lists what he calls 'the principal materials with which one usually works in chemistry' (a total of twenty-four), but he left out all sorts of substances that chemists frequently worked with. The principle of selection is revolutionary: the materials he lists combine with each other to form new stable compounds – but each of these compounds can be broken down, if the right chemical procedures are followed, to release their original components. Geoffroy's twenty-four substances thus survive even when they are combined with other substances: they are not transmuted when they enter into such combinations. Geoffroy was a long way from having a modern theory of elements of the sort propounded by Lavoisier towards the end of the century, but he did have a research programme which had escaped entirely from the concept of transmutation. It is thus Geoffroy, not (as is often claimed) Boyle, who marks the beginning of modern chemistry.[102]

Geoffroy's work appeared in a context where chemists were already trying to escape from alchemical thinking. What killed alchemy was not experimentation (Starkey, Boyle and Newton were indefatigable in their pursuit of experimental knowledge), nor the development of learned networks devoted to new knowledge (alchemists were very effective at seeking each other out and worming information out of each other, always on the basis of exchanging one secret for another), nor even Geoffroy's recognition that chemical combination did not imply transmutation. What killed alchemy was the insistence that experiments must be openly reported in publications which presented a clear account of what had happened, and they must then be replicated, preferably before independent witnesses. The alchemists had pursued a secret learning, convinced that only a few were fit to have knowledge of divine secrets and that the social order would collapse if gold ceased to be in short supply. Some parts of that learning could be taken over by the advocates of the new chemistry, but much of it had to be abandoned as incomprehensible and unreproducible. Esoteric knowledge was replaced by a new form of knowledge which depended both on publication and on public or semi-public performance. A closed society was replaced by an open one.[xvii]

If in thinking about alchemy we keep our eyes on individuals like Boyle, we are in danger of missing the role of institutions, both formal, such as the Royal Society and the Académie des Sciences and informal, such as Mersenne's circle. Many of the founding members of the Royal Society – Digby and Oldenburg, for example, as well as Boyle – were preoccupied with alchemy. But alchemical transmutations were never discussed at the meetings of the Royal Society, and only one brief publication by Boyle in the *Transactions* dealt with alchemical matters; it served effectively as an advertisement, announcing his interest in the hope that others would contact him.[103] Everyone (except perhaps Boyle) was clear that the principles on which the Royal Society was based – the free exchange of information, the replication of experiments, the publication of results, the confirmation of 'facts' – were at odds with the principles of the alchemists. Boyle and Newton were both alchemists and participants in the new scientific community, but for the most part they were perfectly clear that the two sides of their lives were separate, just as Pascal was clear that his religious life, which was intense and demanding, was separate from his scientific life. Boyle, it is true, wanted

xvii Karl Popper's approach to the philosophy of science began to be widely known in the English-speaking world with the publication of Popper, *The Open Society and Its Enemies* (1945); his classic 1935 account of scientific progress as being based on falsification not verification was eventually translated from German as Popper, *The Logic of Scientific Discovery* (1959).

to bring alchemy a little into the public eye, if only in order to make it easier for alchemists to identify each other – Newton immediately told him off, recommending 'high silence'. Boyle, he complained, was 'in my opinion too open & too desirous of fame'.[104]

Pascal, as we have seen, maintained that the fundamental difference between science and religion was that in science there were no truths that could not be questioned, while religion depended on accepting certain truths as being beyond question. For the alchemists, the reality of the philosopher's stone was beyond question; within a generation their appeal to authority, to ancient texts and secret manuscripts, seemed hopelessly misplaced. Alchemy was never a science, and there was no room for it to survive among those who had fully accepted the mentality of the new sciences. For they had something the alchemists did not: a critical community prepared to take nothing on trust. Alchemy and chemistry were both experimental disciplines, but the alchemist and the chemist had different forms of life and belonged to different types of community.[xviii] An important consequence follows from this argument: we should not really expect to find reliable science before scientific communities began to take shape in the 1640s. And this seems right. Had Galileo belonged to a functioning scientific community, to take just one example, he would have been firmly discouraged from placing his theory of the tides at the centre of his defence of Copernicanism.[105]

Consequently, we do not have to wait until the publication of Geoffroy's table in 1718 to hear the death knell of alchemy. According to the new historians of chymistry, alchemy and chemistry were a single undifferentiated discipline until the publication of the third edition of Nicolas Lémery's textbook in 1679, when distinctions between them began to be drawn;[106] by the 1720s the two had been effectively separated. Here, though, is the *Plus ultra* of Joseph Glanvill, published in 1668. It contains fulsome praise of Boyle as someone whom pagans would have worshipped as a god, but its approach to alchemy/chymistry clearly foreshadows that of the eighteenth century:

> I confess, Sir, that among the *Aegyptians* and *Arabians*, the *Paracelsians*, and some other *Moderns*, *Chymistry* was very *phantastick*, *unintelligible*, and *delusive*; and the *boasts*, *vanity*, and *canting* of those *Spagyrists*

xviii '"So you are saying that human agreement decides what is true and what is false?" – It is what human beings *say* that is true and false; and they agree in the *language* they use. That is not agreement in opinions but in form of life.' (Wittgenstein, *Philosophical Investigations* (1953), para. 241 (emphasis in original))

> [alchemists], brought *scandal* upon the *Art*, and exposed it to *suspicion* and *contempt*: but its late *Cultivatours*, and particularly the *ROYAL SOCIETY*, have refin'd it from its *dross*, and made it *honest*, *sober*, and *intelligible*, an excellent *Interpreter* to *Philosophy*, and *help* to *common Life*. For *they* have laid aside the *Chrysopoietick* [gold-making], the *delusory Designs* and *vain Transmutations*, the *Rosie-crucian Vapours*, *Magical Charms*, and *superstitious Suggestions*, and form'd it into an *Instrument* to know the *depths* and *efficacies* of *Nature*.[107]

Glanvill would presumably have been shocked to learn that Boyle and Newton did not share his views, but it was he, not they, who had grasped the relationship between alchemy and the new science. The *Lexicon technicum* of 1704 expressed the rapidly emerging consensus:

> ALCHYMIST, is one that studies Alchymy; that is, the Sublimer Part of Chymistry which teaches the Transmutation of Metals and the Philosopher's Stone; according to the Cant of the Adeptists, who amuse the

A family of alchemists at work, an engraving by Philip Galle, after a painting by Pieter Bruegel the Elder, published by Hieronymus Cock, *c.*1558.

> Ignorant and Unthinking with hard Words and Non-sense: For were it not for the Arabick Particle Al, which they will needs have to be of wonderful vertue here, the word would signifie no more than Chymistry. Whose Derivation see under that Word. This Study of Alchymy hath been rightly defined to be, *Ars sine Arte, cuius principium est mentire, medium laborare, & finis mendicare*: That is, an Art without an Art, which begins with Lying, is continued with Toil and Labour, and at last ends in Beggery.[108]

The demise of alchemy provides further evidence, if further evidence were needed, that what marks out modern science is not the conduct of experiments (alchemists conducted plenty of experiments), but the formation of a critical community capable of assessing discoveries and replicating results. Alchemy, as a clandestine enterprise, could never develop a community of the right sort. Popper was right to think that science can flourish only in an open society.[109]

9
Laws

Nature and Nature's Laws lay hid in Night:
GOD said, *Let Newton be!* And all was Light.

– Alexander Pope, epitaph for
Sir Isaac Newton (published 1735)

§ 1

On 10 November 1619 a young French soldier, René Descartes (1596–1650), found himself stranded in Ulm.[1] He was in the service of Duke Maximilian of Bavaria, a Catholic, and the terrible pan-European conflict afterwards known as the Thirty Years War was just beginning. There was the prospect of fighting ahead; but immediately it was winter, and there was, literally, nothing for a soldier to do. Descartes had received a good, but largely conventional, education from the Jesuits, and had spent two years at university studying law, to please his father; but in 1619 he had no reason to think he would ever pursue any other profession than soldiering. But, trapped by winter, Descartes shut himself up in a room heated by a stove and sat and sat, and thought and thought. He reached the conclusion that what was wrong with all existing systems of knowledge was that they had been cobbled together by lots of different people over long periods of time. What was needed was a complete new beginning. One person, starting from scratch, should redesign the whole of philosophy, including natural science.

Excited and exhausted, Descartes fell asleep and dreamed three dreams. In the first he was attacked by phantoms and a great wind, and he had some sort of debilitating weakness on one side of his body. He tried to get into a chapel to pray but was unable to enter. Waking from this nightmare, Descartes prayed and tried to compose his spirits. When he fell asleep again he heard a great clap of thunder and, as it seemed to him, opened his eyes to see the room filled with sparks from the fire; it was not clear to him quite when he became properly awake, but eventually the sparks disappeared and he fell asleep again. In his third dream there was a big book, a collection of poems. Opening it, he found the words *Quid vitae sectabor iter?* 'What road in life shall I follow?' A stranger entered and gave him another poem, beginning with the words *est et non*, 'it is and it is not.' Descartes tried to find this poem in the book, but the book and the stranger disappeared. (In the *Corpus poetarum* of Pierre de la Brosse (1611) these two poems are to be found on facing pages.) Lying half awake, Descartes tried to interpret his dreams. The first two, he thought, were to be understood as showing that he had lived his life badly up to now, the third as mapping out his future: his path must be to commit himself to the philosophical task of establishing what is and what is not.

For the rest of his life Descartes dated his new life as a philosopher from these dreams. He began work on a series of rules for thinking which would help him establish the truth; he sold some property so that he could be self-sufficient and concentrate on his great enterprise. Fourteen years later, and living in the Protestant Netherlands, Descartes was about to publish a series of treatises in natural science when he heard of the condemnation of Galileo and decided that, since his philosophy supported Copernicanism, he dare not publish for fear of condemnation by the Catholic Church (though Descartes would have been in no danger had the Church condemned him, as in the Netherlands he was safely out of the Inquisition's reach). In 1637 he finally published the *Discourse on Method* and three essays on mathematics and natural science. In 1644 he published a summary of his philosophy, *The Principles.*

By the time Descartes published the *Discourse* he had decided that the best way to introduce his philosophy was through applying scepticism absolutely to the limit. How can we know that the world is real? How can we know that we are not dreaming? How can we know that we are not being systematically deceived by some demonic demiurge? We cannot. There is only one thing we can be sure of: *cogito ergo sum* ('I think, therefore I am'). From this one secure starting point Descartes sets out to prove the existence of a god who would not allow us to be systematically deceived, and then to construct an account of the natural world as consisting purely of matter in motion. The *Discourse* is a strange work because it is both autobiography and philosophy; Descartes teaches us how to think by telling us the steps he went through in learning how to think. And so he tells the reader not about the dreams, which were too private an experience, but about the day he spent in the stove-heated room, when his life as a philosopher began.

The problem with Descartes' story is that it is not true. There is no reason to doubt the stove-heated room or the dreams, but all the evidence suggests that Descartes' new philosophical life had begun exactly a year earlier, on 10 November 1618. That day, he was in Breda, in the Low Countries, serving in the army of the Protestant Maurice of Nassau. In the town someone had stuck up a poster challenging people to solve a mathematical problem. The poster was in Flemish, and so Descartes turned to the person standing next to him, also studying the poster, and asked him to translate it. This person was Isaac Beeckman, a schoolteacher and engineer, and we know about his relationship with Descartes because he kept a diary which was rediscovered in 1905 and published in four volumes between 1939 and 1953.[2]

Descartes and Beeckman chatted away in Latin and discovered that they had interests in common. 'Physico-mathematicians are very rare,' wrote Beeckman in his diary a few days later; indeed, the stranger had told him that 'he has never met anyone other than myself who pursues his studies in the way I do, combining Physics and Mathematics in an exact way. And for my part, I have never spoken with anyone apart from him who studies in this way.'[3] But Beeckman was far ahead of Descartes in his thinking. He had already decided that the universe consisted of corpuscles in motion, and that the 'laws' of motion (Beeckman's standard word for a law of nature is *pactum*: 'covenant')[4] which functioned at a microscopic level must be the same as those which functioned at a macroscopic level. He was well on the way to formulating, entirely independently, Galileo's law of fall. For two months Beeckman and Descartes worked closely together, and when Beeckman left Breda they kept up a correspondence in which Descartes told Beeckman that they were bound together 'in a bond of friendship that will never die'. Beeckman was assured by Descartes that:

> [Y]ou are the one person who has shaken me out of my nonchalance and made me remember what I had learned and almost forgotten. When my mind strayed far from serious concerns, it was you who guided it back down the correct path. If, therefore, by accident I propose something which is not contemptible, you have every right to claim it for yourself.[5]

Years later, in 1630, Beeckman did just that. In a letter to Descartes' friend Mersenne he mentioned that some of Descartes' ideas on music had come from him. Descartes was absolutely furious and denied any influence, but when Mersenne visited Beeckman and read his journal he discovered that, indeed, many of Descartes' ideas had first been formulated by Beeckman. Descartes exploded again, telling Beeckman he had learnt as much from him as he had learnt from ants and worms. This was followed on 17 October 1630 by one of the longest letters Descartes ever wrote; it runs to twelve pages in print and is full of bitter invective in which Descartes explains to Beeckman that he is mentally ill and delusional.[6]

Why could Descartes not bear the simple truth, that much of what he knew he had learnt from Beeckman? Because ever since he had awoken from his dreams on the morning of 11 November 1619 he had told himself that he was building a new philosophy single-handedly, that he was starting from scratch, that he owed nothing to anyone. The truth of his intellectual dependence on Beeckman was utterly intolerable to him. And so, in the autobiographical account which opens the *Discourse on Method* in 1637, there is no Beeckman, there is only the famous account of the stove-heated room:

> At that time I was in Germany, where I had been called by the wars that are not yet ended there. When I was returning to the army from the coronation of the Emperor, the onset of winter detained me in quarters where, finding no conversation to divert me and fortunately having no cares or passions to trouble me, I stayed all day shut up alone in a stove-heated room, where I was completely free to converse with myself about my own thoughts. Among the first that occurred to me was the thought that there is not usually so much perfection in works composed of several parts and produced by different craftsmen as in the works of one man. Thus we see that buildings undertaken and completed by a single architect are usually more attractive and better planned than those which several have tried to patch up . . .[7]

§ 2

Let us return to the table which we discussed in Chapter 3. In its case, according to an Aristotelian, the formal and final explanations are exterior to it – its form is in the mind of the carpenter and its purpose is to provide someone with a place to work. But in the case of the oak tree, the form and the purpose are in some sense *inside* the acorn. Efficient causes are external; formal and final causes are, in natural objects, internal; and material causes are, to begin with, external, although (like the water absorbed by the oak's roots, or the breakfast I have just eaten) they become internal.

A mechanical explanation, by contrast, is about external causes, not internal ones. If you are an ancient atomist – Epicurus or Lucretius, for example – you do not accept that there is any internal principle causing the formation and development of the oak tree, directing it towards the realization of its potential. Atoms are merely passive lumps of matter. An oak tree is an agglomeration of atoms which has been given a certain shape by external forces, in the same way that my house is an agglomeration of bricks that has been given a certain shape. For an ancient atomist or an early modern mechanist (such as Beeckman or Descartes) causation is always external, never internal; there are only efficient or mechanical causes.[i] There are no formal or final causes, and the material cause never varies.

For Epicurus and Lucretius, what matters about atoms is their size, shape and movement. If this is all that atoms have, then the qualities that we perceive in the world – colour, taste, smell, sound, texture, temperature – must be by-products of size, shape and movement. Size, shape and movement must be primary, and the other qualities must be secondary. If sound is produced by vibration, then it is easy to see that sound may be the result of movement. If rubbing two sticks together produces heat, then it is possible to imagine that heat may be a form of movement. One can hypothesize that smell is produced by particles entering the nose. Primary qualities are objective; secondary qualities are subjective, in that they depend on our ways of sensing. In a world without ears there would be no sounds, only vibrations; in a world without noses there would be no smells, only particles floating in the atmosphere. Galileo's example, to clarify this idea of the subjectivity of sensation, is a tickle: tickle me with a feather and I feel a distinct sensation, but there is nothing in the feather which corresponds to my sensation of being tickled. This distinction between objective reality and subjective sensation was made by Lucretius. Galileo, in *The Assayer* (1623) was the first modern author to echo it, although without mentioning Lucretius by name (for he was regarded as a dangerous atheist; but we know that Galileo owned two copies of *On the Nature of Things*).[8] After Galileo the distinction was adopted by Descartes. The terminology we now use to express this distinction, between primary

i We will return to the question of mechanism in Chapter 12.

and secondary qualities, was introduced by Boyle in 1666 and popularized by Locke in 1689.[9] (The Lockean language of primary and secondary qualities replaces an earlier Aristotelian language of primary and secondary qualities, the primary qualities being hot and cold, wet and dry.)

Although Descartes followed the ancient atomists in their distinction between primary and secondary qualities, he rejected their belief in empty space, the void. Matter, as far as Descartes was concerned, had only one fundamental characteristic, that it occupied space; it followed that there could be no vacuum, as this would be space with nothing occupying it. For Descartes the material world consists of divisible corpuscles. He avoids the term 'atoms' because the ancient atomists had insisted that atoms were not divisible and that the space between them was empty.

Matter, in Descartes' scheme of things, can interact only by direct contact; there can be no action over a distance, and when two bodies interact they can do so only by pushing against each other, so magnetism and gravity have to be explained as the result of some sort of pushing process, not a pulling one. According to Descartes, in the case of gravity this pushing process was the result of the Earth being caught up in a vast vortex of fluid swirling round the sun. This vortex both held the planets in their orbits and pressed objects down towards the surface of the Earth. The sun was just one among many stars, each surrounded by its own vortex. Similarly, magnetism worked through little streamers of corkscrew-like matter reaching out and locking on to iron: the pull of a magnet is in fact a push, just as a corkscrew pushes a cork out of a bottle. (I may pull on the corkscrew, but the corkscrew pushes on the cork.)

In Descartes' system there is only one type of matter which, by its interactions and conglomerations, produces the vast diversity of materials that we experience. The laws of its interaction are the three laws of nature. These are that 'each thing, as far as is in its power, always remains in the same state; and that consequently, when it is once moved, it always continues to move'; that 'all movement is, of itself, along straight lines'; and that 'a body, upon coming in contact with a stronger one, loses none of its motion; but that, upon coming in contact with a weaker one, it loses as much as it transfers to that weaker body.'[10]

It is easy to regard Cartesianism, with its insistence that magnets, corkscrews and even what we now call gravity always push and never pull, as something of a joke, but recent work has shown that Descartes carried out some subtle and beautiful experiments, and his vortex theory was still viable well into the eighteenth century.[11] The crucial dispute between Cartesians and Newtonians was over the shape of the Earth: Newton predicted an oblate ellipsoid, or flattened, Earth, while the Cartesians had predicted a prolate ellipsoid, or egg-shaped, Earth. French expeditions to Peru and Lapland (1735–44) discovered (to the dismay of those involved) that Newton was right and the Cartesians wrong.[ii][12]

§ 3

The modern idea of laws of nature is a by-product of Descartes' philosophy, for Descartes was the first person to treat the laws of nature as being what knowledge of nature was all about. Galileo, Harriot and Beeckman had each independently discovered what we call the law of fall; but none had used the word 'law' in this context. According to the Comte de Buffon, writing in the eighteenth century, 'Nature is the system of eternal laws established by the creator';[13] it follows that the core task of science is the identification of the laws of nature.[iii] Buffon could, if he wished, look back to the seventeenth century and identify a whole series of laws that had been discovered during the Scientific Revolution: Stevin's law of hydrostatics, Galileo's law of fall, Kepler's laws of planetary motion, Snell's law of refraction, Boyle's law of gases, Hooke's law of elasticity, Huygens' law of the pendulum, Torricelli's law of flow, Pascal's law of fluid dynamics, Newton's laws of motion and law of gravity. Most, perhaps all, of these had been given the title 'laws' by Buffon's day (only Newton used the word 'laws' when describing his own discoveries), though only a minority had already acquired eponymous labels – the rest were yet to be named after their discoverers.[iv] It is hardly surprising that there is a book on the Scientific Revolution entitled *Nature and Nature's Laws*, since the discovery of laws of nature is one of the Scientific Revolution's most remarkable achievements.[14] In 1703 Newton became President of the Royal Society and drew up a scheme to define its goals. 'Natural Philosophy,' he wrote, 'consists in

ii Are there any living Cartesians? In 1980 in Montreal I was handed a pamphlet which rejected Newton's account of gravity and defended a modified Cartesian account. So perhaps there are.

iii I take 'laws of nature' and 'scientific laws' to be synonymous, although I am aware that some philosophers want to use these as labels to distinguish two different types of law.

iv See above, p. 102.

discovering the frame and operations of Nature, and reducing them, as far as may be, to general Rules or Laws, – establishing these rules by observations and experiments, and thence deducing the causes and effects of things . . .'[15] Laws of nature were now what science was all about.

By contrast, the ancients had known, by our reckoning, only four physical laws: the law of the lever, the optical law of reflection, the law of buoyancy and the parallelogram law of velocities.[16] Or, rather, the ancients had known four principles that we call laws. The ancients did refer to the 'laws' of nature when they wanted to say that nature is regular and predictable, but they never identified any particular scientific principle as a law. The Romans talked a great deal about the law of nature (*lex naturae*), but they usually meant the moral law.

A law is an obligation ('Thou shalt not kill', for example) imposed on some creature (human beings, angels) capable of accepting or rejecting that obligation. The moral law applies to rational, language-using creatures, and the law of nature binds all human beings by virtue of their capacity to recognize that there are moral obligations that are common to them all. There are no laws in non-human nature because human beings are (as far as we know) the only rational, language-using beings in nature. To talk about those regularities that appear in the natural world as 'laws' is to speak metaphorically: this was as obvious in the first century and the seventeenth as it is now. But the metaphor is pretty straightforward. The Greeks made use of it now and again (although most of the time they liked to contrast the natural and the social), and the Romans, who were always in and out of the law courts, found it an obvious way of referring to the fact that nature is regular and predictable in its workings. It was an even more obvious metaphor for Christians, as it was easy to think of God as a legislator imposing laws on nature and to personify nature as obeying him.

So when we talk about nature's laws we can be talking about laws governing human behaviour or laws governing nature – 'natural law' and 'laws of nature', as we now say. In classical Latin there is no distinction: *lex* (or *ius*) *naturae* and *naturalis lex* (or *ius*) are synonymous, and their most frequent usage is to refer to those moral laws that all human beings have (or are supposed to have) in common. So, too, at first, in the modern languages. 'Law of nature' is the most common term in English before 1650 (Hobbes, an extreme case, uses 'natural law' twice in *Leviathan* and 'law of nature' more than a hundred times) and *loy naturelle* the most common in French (and so, too, *legge naturale* in Italian, *ley natural* in Spanish). A linguistic distinction to separate the two kinds of law, moral and scientific, emerges with Descartes, who writes of *la loy* (or *les loix*) *de la nature* and never (when discussing scientific matters) of *la loi naturelle*. Before Descartes *la loy de nature* and *la loy de la nature* were synonymous, although the first was more common. But Descartes and his authorized translator from Latin into French never write of *la loy de nature*. Thus Descartes opted for the least common phrase available to him in French as a translation for *lex naturae* in order to give his phrase a precise reference to scientific laws, not moral ones. By a similar process in German, the rarer term, *Naturgesetz*, comes to mean primarily law of nature, while the commoner term, *Naturrecht*, continues to mean natural law.

It is surely easier to give an uncommon phrase a new meaning than a common phrase. The English, however, followed Descartes in using 'law of nature', not 'natural law' to refer to scientific laws; but this had a peculiar effect, as 'law of nature' was the most common term in English for the moral law. Using the same term for both was unnecessarily confusing, and over time the moral and political philosophers and the theologians largely abandoned 'law of nature', ceding it to the scientists, and switched to 'natural law', bringing themselves into line with the French, Germans and Italians. This is a striking case of French imposing itself upon English, and of the scientists determining for the first time the language of the theologians. As a result, for us moderns, laws of nature are scientific laws and natural laws are moral laws. In this respect, we are all Cartesians.

§ 4

Long before Descartes we can find references to laws of nature in a scientific context, and scholars have struggled to disentangle the origins of the concept.[17] There is no doubt that it has multiple origins, nor that it takes on a wholly new importance with Descartes. I will distinguish three origins, of which the most important (in my view) has been largely neglected hitherto. First, nominalist philosophers from William of Ockham (1288–1348) on attacked the Aristotelian doctrine of forms. There is, they argued, no such thing as a form or essence, there are only particular objects. When we talk about forms we are using a label (or name, hence the term 'nominalism') that we have chosen to attach to certain particulars. In their view, Aristotelian forms were ghostly presences; you could never capture them, but they were always being added into the explanation. Clearly, in the case of making a table, the carpenter does have a plan: the form is an idea in his mind, and the table he makes corresponds to that form. But where is the form of the oak tree? And if you cannot locate it, how can it act in the world? If the universe is regular and predictable, this is not because there are internal forms but because God has imposed order upon it from the outside. God could have made the universe in many different ways; he has arbitrarily chosen to make it as it is, and the order it displays is an order he has chosen to impose upon it. Thus Jean Gerson (1363–1429), a nominalist, maintains that 'the law of nature as concerns created things is what regulates their movement and action and their tendency towards their goals.'[18] The term 'law' here implies external, divine causation, but the particular content of the law of nature is never specified, and there is surely scope for occasional exceptions to the law, if only monsters and miracles. Some modern commentators would like to argue that the invention of laws of nature could take place only in a monotheistic culture, where God could be conceived as an absolute legislator – thus, the Scientific Revolution owes everything to Christianity. It is certainly true that the arguments of the nominalists are theocentric, but as we shall see this is not true for other ways of thinking about laws of nature.

Secondly, in the mathematical disciplines, *lex* was often used as a synonym for *regula*, or 'rule', to refer either to natural regularities which could not be shown to be strictly necessary – in other words, where there was no full philosophical (causal) explanation – or to axioms. Thus Roger Bacon refers to the law of reflection (the angle of reflection equals the angle of incidence) and Copernicus's disciple Rheticus claimed that Copernicus had discovered 'the laws of astronomy' (Copernicus himself made no such claim). Ramus, as we have seen, writes of the 'laws' of Ptolemy and of Euclid.[19] The term 'law' implies unbroken regularity, with no exceptions, but nothing is conveyed about causation. These laws have a specifiable content.

Both traditions come together in Paris in the work of Jean Fernel (1497–1558), who started his career as an astronomer and mathematician and then turned to medicine, inventing the term 'physiology'. According to Fernel, there are eternal, immutable laws which govern the universe: these are ordained by God, and without them there would be no order in the universe. The laws of medicine fit within this wider structure of laws, and the fundamental law of medicine is the ancient Hippocratic principle that opposites cure opposites – a fever is cured by cooling the body, for example. This looks to us like a principle, maxim, or rule of thumb, not a law, because it lacks specificity.[20]

Neither the nominalist nor the mathematical usages are particularly common, and we cannot show any direct influence from either of them on seventeenth-century usages. Galileo makes only three references to the laws of nature, on each occasion when he is arguing against the theological objections to Copernicanism; there are no laws of nature in his more properly scientific works.[21] The first person to place the idea of a universal law at the heart of the attempt to understand nature, and to give some specific content to that idea, was Descartes, first in his correspondence in 1630, next in *The World* (completed by 1633 but published only posthumously: Descartes gave up all hope of publication when he heard of the condemnation of Galileo), and then in *The Principles of Philosophy* (Latin 1644, French 1647; in the earlier *Discourse of Method* Descartes uses the phrase 'principles of nature' instead of 'laws of nature').[v] Descartes, as we have seen, propounded three laws, but his laws, the first laws that matter, do not appear on modern lists of scientific discoveries: Descartes' first two laws of nature became Newton's first law of motion, and the third was refuted by Newton's laws.

More to the point, a modern list of laws would have deeply puzzled Descartes. His three laws were intended to be the only laws. From them, it should be possible to derive a complete system of knowledge embracing every aspect of the natural world, just as one can deduce the whole of Euclidean geometry from five axioms. He had no intention of seeing laws proliferate and multiply. Of course, as he worked out the implications of his laws, he drew a series of subsidiary conclusions. There were, for example, seven rules (*regulae*) which enabled one to predict what would happen in collisions between bodies travelling along the same straight line (bodies in a vacuum, although Descartes held there was no such thing; the rules for bodies travelling in a plenum were beyond even his capacity): these rules are never called 'laws'. In the course of half a century Descartes' term 'laws of nature' established itself as central to the language of science, but at the same time its meaning mutated, so that it soon ceased to resemble Descartes' original conception.

Where did Descartes' conception of a law of nature come from? Well, Lucretius had a concept of a law of nature, although he does not use the

v Earlier, in the *Novum organum* (1620), Bacon had claimed that the discovery of the law of nature was the fundamental goal of natural philosophy (Book 2, aphorism 2: Bacon, *Works* (1857), Vol. 1, 228), but for him this is a project, not an attainment.

phrase *lex naturae*; instead he uses (three times) the phrase *foedus naturae*. A *foedus* is a league, or compact, but it is often used as synonymous with *lex*, and Renaissance commentators on Lucretius interpreted him as talking about the laws of nature.[22] Bacon writes of 'the law of nature and the mutual contracts of things': he is paraphrasing Lucretius. For Lucretius the attraction of iron by a magnet takes place according to a law of nature, and species breed true, dogs producing dogs and cats cats, according to a law of nature. It seems absolutely certain that Descartes had Lucretius in mind when formulating his laws of nature because he picks up on a phrase he uses, *quantum in se est* (a phrase very difficult to translate but, roughly, 'as much as in it lies'), in his first law of motion. Lucretius uses the phrase four times in *On the Nature of Things*, twice in discussions of the way in which atoms naturally fall downwards, 'as far as in them lies', through the void, passages which prefigure Descartes' conception of inertia. The same phrase is then employed by Newton in his definition of inertia; he evidently took the phrase from Descartes, and discovered only later that it originated in Lucretius.[23]

So far, in tracing the idea of a law of nature, we have been following an established line of argument. But to understand where Descartes' preoccupation with laws of nature comes from we must consider a text that has not previously been discussed in this context. We must turn to the longest and most philosophical of Montaigne's essays, 'An Apology for Raymond Sebond', first published in 1580. The passage originally contained a single direct quotation from Lucretius, but two more were added in 1588, and we will see in a moment that it is directly inspired by Lucretius on *foedus naturae*. Here is an abbreviated version, in which, for simplicity's sake, I omit Montaigne's later additions to the 1580 text:

> Nothing of ours can be compared or associated with the Nature of God, in any way whatsoever, without smudging and staining it [i.e. the Nature of God] with a degree of imperfection . . .
>
> We wish to make God subordinate to our human understanding with its vain and feeble probabilities; yet it is he who has made both us and all we know. 'Since nothing can be made from nothing: God could not construct the world without matter.' What! Has God placed in our hands the keys to the ultimate principles of his power? Did he bind himself not to venture beyond the limits of human knowledge? . . . You only see – if you see that much – the order and government of this little cave in which you dwell;

beyond his Godhead has an infinite jurisdiction. The tiny bit that we know is nothing compared with ALL:

omnia cum coelo terraque marique
Nil sunt ad summam summaï totius omnem.

[Lucretius: 'The entire heavens, sea and land are nothing compared with the greatest ALL of all']

The laws you cite are by-laws [*une loy municipale*]: you have no conception of the Law of the Universe [*l'universelle*; i.e. *la loi universelle*]. You are subject to limits: restrict yourself to them, not God . . . [Montaigne enumerates various miracles] a material body cannot pass through a solid wall; a man cannot stay alive in a furnace . . . It is for you that he made these laws [*regles*]; it is you who are restricted by them. God, if he pleases, can be free from all of them: he has made Christians witnesses to that fact . . .

That Reason of yours never attains more likelihood or better foundations than when it succeeds in persuading you that there are many worlds . . . it seems unlikely, therefore, that God made only this one universe and no other like it . . . Now if there are several worlds, as Epicurus and almost the whole of philosophy have opined, how do we know whether the principles and laws which apply to this world apply equally to the others?[24]

Montaigne's thinking here derives from his reading of Lucretius. In his own copy, against one of the four passages where Lucretius discusses the *foedera naturae*, the laws of nature, he wrote, summarizing Lucretius, 'The order and uniformity of the conduct of nature makes evident the uniformity of her principles.'[25] It is against this position that he appears to be arguing here. Whether he is sincere, it is a little hard to say: having emphasized his belief in miracles, he proceeds to make the concept of a miracle entirely subjective only a few paragraphs later. What is important for our immediate purposes is the way in which his discussion echoes through the later literature on the law of nature, for of course Montaigne was read by all the educated.

Here is Walter Charleton more or less paraphrasing Montaigne in 1654:

[B]y the Law of Nature, every Body in the Universe is consigned to its peculiar Place, i. e. such a canton of space, as is exactly respondent to its Dimensions: so that whether a Body quiesce, or be moved, we alwayes understand the Place wherein it is Extense, to be one and the same, i. e. equal to its Dimensions.

> We say, *By the Lay* [sic] *of Nature*; because, if we convert to the Omnipotence of its Author, and consider that the Creator did not circumscribe his own Energy by those fundamental Constitutions, which his Wisedom imposed upon the Creature: we must wind up the nerves of our Mind to a higher key of Conception, and let our Reason learn of our Faith to admit the possibility of a Body existent without Extension, and the Extension of a Body consistent without the Body it self; as in the sacred mystery of our Saviours Apparition to his Apostles, after his Resurrection . . . the dores being shut [compare Montaigne: 'a material body cannot pass through a solid wall']. Not that we can comprehend the *manner* of either, i. e. the Existence of a Body without Extension, and of Extension without a Body; for our narrow intellectuals, which cannot take the altitude of the smallest effect in Nature, must be confest an incompetent measure of supernaturals: but that, whoever allowes the power of God to have formed a Body out of no praeexistent matter [compare Montaigne: 'nothing can be made from nothing'] cannot deny the same power to extend to the reduction of the same Body to nothing of matter again.[26]

And here is Boyle distinguishing, following Montaigne, between the universal laws and the municipal laws of nature (the term 'municipal laws' is, as he recognizes, an odd one to use in English; he only uses it, I feel sure, because he has Montaigne in mind):

> [W]e may sometimes usefully distinguish between the *Laws of Nature*, more properly so call'd, and the *Custom of Nature*, or, if you please, between the Fundamental and General Constitutions among Bodily Things, and the Municipal Laws, (if I may so call them,) that belong to this or that particular sort of Bodies. As, to resume and somewhat vary our Instance drawn from Water; when this falls to the Ground, it may be said to do so by virtue of the *Custom of Nature*, it being almost constantly usual for that Liquor to tend downwards, and actually to fall down, if it be not externally hinder'd. But when Water ascends by Suction in a Pump, or other Instrument, that Motion, being contrary to that which is wonted, is made in virtue of a more Catholick *Law of Nature*, by which 'tis provided, that a greater Pressure, which in our case the Water suffers from the weight of the Incumbent Air, should surmount a lesser, such as is here the Gravity of the Water, that ascends in the Pump or Pipe.[27]

Descartes also certainly read Montaigne, and he drew from him an astonishing idea: a proper law of nature would be universal in the sense that it would not just be true for this universe, it would be true for any possible universe. Nowadays we have a less stringent test: laws of nature are true for every time and every place in our universe.[28] If we take this to be a central feature of laws of nature, then it is very difficult to see how an Aristotelian could have any conception of them. In Aristotelian physics different laws apply in the sublunary and supralunary spheres.[29] In the one there is change and natural movement is vertical, while in the other there is no change and natural movement is circular. There are no physical laws common to both spheres. In the sublunary sphere it might seem easy to formulate some general laws: all living creatures die; children take after their parents. But the phoenix does not die, and monstrous births do not resemble their parents. Aristotelians therefore recognize that, in the sublunary sphere, there are no regularities which do not have exceptions; in the supralunary sphere, all is regularity without exception; there are no regularities which apply in both spheres. Consequently, there are no Aristotelian laws of nature.

Descartes, however, is not after universality in the limited sense in which we would understand the term but in the stronger sense introduced by Montaigne when he asks what laws would apply in other universes, assuming such exist. In *The Principles of Philosophy* (1644) Descartes insists that he is not describing the laws that govern our universe but a set of laws such that, if one began with pure chaos, a universe indistinguishable from ours would evolve. This is not, Descartes assures us, how our universe began: God made it, God ordered it, as we all know. But it enables us to establish the laws that would need to apply in any possible universe. Descartes gets himself into a bit of a tangle here. He wants, like the nominalists, to insist that God freely determined what the laws of nature and even of mathematics are: they seem necessary to us, but they are not necessary to him; at the same time he wants to argue that any rational God would have to opt for these laws if he wanted to create an orderly, coherent universe. As Newton's disciple Roger Cotes complained:

> He who thinks to find the true principles of physics and the laws of natural things by the force alone of his own mind, and the internal light of his reason [i.e. Descartes]; must either suppose that the world exists by necessity, and by the same necessity follows the laws proposed; or if the order of nature was established by the will of GOD, that himself, a miserable reptile, can tell what was fittest to be done.[30]

How did Descartes get into this tangle? Because he was trying to establish laws that were, in Montaigne's terms, truly universal, laws which would work both for a universe created by an omnipotent God and for an Epicurean universe created out of chaos by the random concatenation of atoms – hence his failure to use the term 'laws' for local effects.[vi]

Descartes' conception of the laws of nature was deeply influential. In the *Principia*, Newton, like Descartes, has only three laws. He took the view that Kepler's principles of planetary motion (which Kepler had never called laws) were, as propounded by Kepler, merely statistical regularities; they acquired law-like status only when they were shown, along with Galileo's law of fall, to derive of necessity from a genuinely universal principle, that of gravity.[31] (Newton evidently hesitated over whether to call gravity a law, as it does not correspond to the three Cartesian laws; he does call it a law in the *Optics*, but not in the *Principia*.) Boyle, too, evidently thought there was only a small number of 'more Catholic laws', and these were the laws of nature properly so called.

But Bacon had advocated a different approach. Beneath a single supreme law (he called it the *summa lex*, the fundamental law, but he never worked out what it was), he had sought other subordinate laws (which he sometimes thinks of as 'clauses' within the overarching law), for, after all, even Montaigne allowed for the existence of municipal laws: the law of heat is an example Bacon gives, which would define the essence of heat in all its various manifestations; while Lucretius had discussed the law of magnetism. This approach opened the way to a multiplication of laws: Boyle's hypothesis regarding gases (which he never called a law) could now be counted as one. We see this looser attitude already at work in Walter Charleton, where there are lots of other laws apart from the three '*General Laws* of Nature, whereby she produceth All Effects', such as 'the Laws of Rarity and Density' and 'the setled and unalterable Laws of Magnetical Attraction'.[32] It is this more easygoing Lucretian, Baconian, Charltonian approach which eventually became that of the Royal Society and of eighteenth-century science, in contradistinction to the far bolder one of Montaigne and Descartes.[33]

vi This tangle is reflected in the modern philosophical discussion of laws of nature. Broadly speaking, there are two schools: one holds that laws of nature are simply regularities which we identify in nature; the other that they are necessary features of the world. The regularists are the heirs of the nominalists; the necessitarians of the Epicureans. The result is that there is no agreement on how to answer the question 'What is a law of nature?'

§ 5

Descartes and his followers, who were the first to emphasize the idea of laws of nature, were faced with a series of theological difficulties; despite this, they argued that their approach was easier to reconcile with Christianity than was Aristotelianism, as Aristotle had believed the universe to be eternal and had not believed in personal immortality. There were four particular pinch points.

First, how can the soul be made to fit into a mechanistic universe? Descartes made a strict separation between mind and matter: mind was immaterial and immortal, so that mind's relationship to the sensory world of space and time was inherently problematic. Descartes solved this problem as best he could by claiming that mind acted on the body through the pineal gland. As a result, mind became 'the ghost in the machine'.[34]

Second, what is the role of God in the making of the universe? Descartes was willing to envision a universe where God set up the initial conditions and then left the machine to assemble and run itself. Others, though, argued that it was quite clear that general laws of the sort described by Descartes could never produce the perfect design that one could find in the paw of a dog. Descartes never compared the universe as a whole to a man-made machine because he had no intention of saying that the universe had been deliberately designed and constructed in the way that a man-made machine is. Robert Boyle, on the other hand, insisted that was precisely how one had to think of the universe: following Kepler, he compared the universe to a clock, and so compared God to a clockmaker. Descartes' universe is an automaton, but (at least, potentially) a self-fabricating automaton. The Cartesian universe is not made for man;[35] the Boylean universe is. We are entitled to feel at home in Boyle's universe, even if it is a mechanical apparatus; it is not clear that an immortal, immaterial soul has any business feeling at home in Descartes' universe.

Third, how do laws of nature work as causes? It is arguable that 2 + 2 will equal 4 in any universe; and, surely, levers and balances would work in the same way in any universe. But need the angle of reflection be equal to the angle of incidence in any universe? Might there be a universe in which Descartes' third law of nature actually holds? If laws of nature are something less than mathematical truths and something more than perceived regularities, then it would seem clear that they exist only because God has chosen to make them apply. This is voluntarism, and it seems to follow naturally from the idea of laws of nature. There is a puzzle here, because the standard alternative to voluntarism is rationalism, and a rationalist would hold that the laws of nature, like the laws of mathematics, exist because they are necessary. On most questions Descartes is a rationalist, but as far as the laws of nature are concerned he appears to want to have it both ways.

A related question is: What is God's role in causation? Has he simply set general rules in place, or does he step in on each and every occasion to ensure the rule is applied? On my keyboard, if I press the shift key I cannot then type a lower-case letter. There's no choice involved: the letter has to be upper case. A choice made by the manufacturer when the computer was designed determines this, and nothing can change it now. On the other hand, almost every time I type the letter 'Q' I follow it with the letter 'U'; but there is no causal connection between the 'Q' and the 'U', I simply choose to make one follow the other. The claim that, just as I choose to make 'U' follow 'Q', God acts to create what look like causal connections on each and every occasion – that there are, strictly speaking, no causal connections but only temporal coincidences – is called occasionalism. It was adopted by Malebranche and other followers of Descartes, and Newton sometimes speaks as if every act of gravitational attraction is directly willed by God. You cannot be an occasionalist without being a voluntarist, and every voluntarist has taken at least the first step on the road to occasionalism.

Some historians of science want to argue that you cannot have laws of nature without voluntarism, and you cannot have voluntarism without an omnipotent creator God.[36] Consequently, the Greeks and the Romans were incapable of formulating the idea of laws of nature, and without them they could not develop modern science. This would surely have puzzled Descartes and Newton, who found in Lucretius (in Descartes' case) an inspiration for their own ideas, or (in Newton's case) a prefiguration of them. The idea of an omnipotent creator God may help in formulating a theory of laws of nature, but it seems wrong to claim that it is a necessary precondition.

This brings us to our fourth and final problem: Does God override the laws of nature? Boyle was happy to argue that God makes miracles happen, and that in doing so he breaks his own laws. But Galileo described nature as inexorable and immutable, and it is very hard to understand how there can be exceptions of any sort to Descartes' laws.[37] French Cartesians faced an effective censorship, and so had to be careful what they said: Descartes' *Meditations* of 1641 were placed on the Catholic Church's Index of Prohibited Books in 1663 because Descartes' corpuscular philosophy (since, like Lucretian atomism, it denied that there was such a thing as substance or form) was held to be incompatible with the Catholic doctrine of transubstantiation (which declared that during the Mass there was a transformation in the substance of the bread and wine, even though they retained their original outward appearance).[38] In Protestant countries censorship was less rigorous, although there were still limits as to what could be published. Thus some disciples of Newton were prepared to follow the logic of natural law through to its conclusion and maintain that everything that happens happens in accordance with the laws of nature.[39] William Whiston (a pupil of Newton who was, like Newton, an Arian – that is, he denied that Christ had existed from all time and, consequently, denied the doctrine of the Trinity), for example, argued in 1696 that the Flood had been caused by the Earth passing through the tail of a comet.[40] Similarly, there must be natural explanations for the parting of the Red Sea, or the plagues of Egypt; what is evidence of divine providence is that God arranged for these exceptional events to coincide with the need for them.

Protestants had long argued that the modern miracles reported by Catholics were simply (where they were not frauds) misunderstandings of natural events; the same arguments were now applied to the Bible itself. It was obviously safer to apply such theories to the Old Testament than to the New but, by implication, Christ's miracles, too, even the Resurrection, were to be understood as natural events, which wonderfully coincided with the need for an appearance of divine intervention. So, too, when God answers petitionary prayer, it was argued, he does not alter the course of events in order to answer the prayer, but, as an omniscient God, he knows in advance that the prayer will be followed by an event which looks like an answer. Miracles and answers to prayer thus become entirely subjective experiences; objectively, there is nothing there, apart from a coincidence. Montaigne had already asked, 'How many things are there which we call miraculous or contrary to Nature? All men and nations do that according to the measure of their ignorance.'[41]

10

Hypotheses/Theories

> ... an accompt of a Philosophicall discovery ... being in my Judgment the oddest if not the most considerable detection[i] which hath hitherto beene made in the operations of Nature.
>
> – Isaac Newton to Henry Oldenburg, 18 January 1672

§ 1

'In the beginning of the Year 1666' Isaac Newton had just turned twenty-three (his birthday was Christmas Day). The year before, he had obtained his BA degree; a year or so later he began to develop his theory of gravity; less than four years later, in October 1669, he became Lucasian professor of mathematics (at the time the only chair in mathematics at Cambridge), and exactly four years later, in the beginning of 1670, he gave his first university lectures, on the subject of optics. In the beginning of 1666, he tells his readers, he acquired a prism. Plenty of people before Newton had used a prism to split light into the colours of the spectrum; as it happens, all of them had projected the light from the prism on to a nearby surface. Newton set up his prism in his rooms at Trinity: he made a hole in the shutter of his window to let in a thin beam of light and placed his prism near the hole so that the light from it was projected on to a wall that was 22 feet away. The sun is circular; the hole in Newton's shutter was circular, and so the patch of colours on the wall should also have been circular; but it was not, it was about five times longer than it was wide.[1]

i 'Detection' in the seventeenth century is a synonym for 'discovery' (in Latin, *detego* means 'uncover'), except that, then as now, one normally detects something that has been deliberately hidden, or someone who is deliberately hiding, so 'detection' leaves less scope for happenstance than 'discovery'.

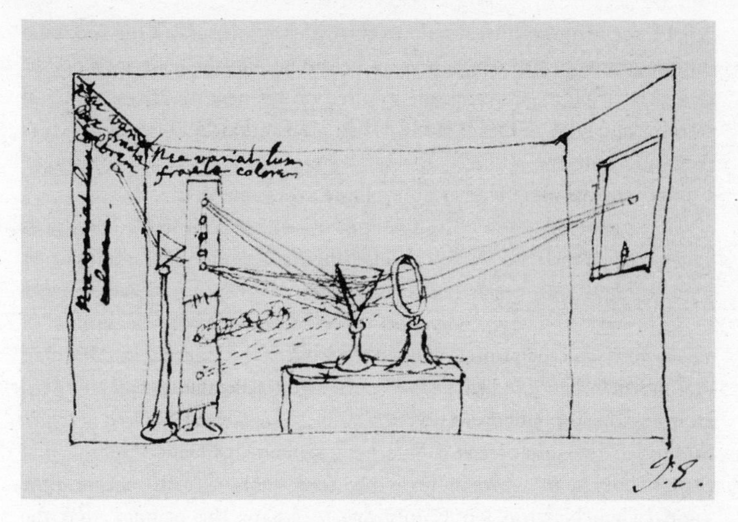

This sketch of the *experimentum crucis* was produced by Newton to provide a guide for an illustration to accompany the French translation of the *Opticks* (1720). A ray of light enters the darkened room through a hole in the shutter at the right; it passes through a lens to narrow the beam, and then through a prism, which splits it into the colours of the rainbow, which are projected in an oblong shape on to a screen. One colour passes through a hole in the screen and encounters a second prism. This colour is refracted a second time, but it remains a fine beam of light and its hue is unchanged.

Newton considered various possibilities. He established there was nothing wrong with the prism, and that the light travelled in a straight line from the prism to the wall, that it wasn't curving in some sort of odd fashion, like a tennis ball with spin on it. So he made the light pass through a smaller hole before the prism, and then passed fractions of the light that emerged through another small hole in a board, on the other side of which he placed a second prism. White light entering the first prism was split into a spectrum of colours, but each colour remained the same as it passed through the second prism, and each colour was refracted by the second prism to the same degree that it had been refracted by the first: this he calls the *experimentum crucis*. Newton had discovered that white light is not homogeneous but is made out of all the colours of the spectrum, and that each colour is refracted by a different amount as it passes through a prism. He went on to conclude that a reflecting telescope would be far superior to the standard refracting telescope because the image would not be marred by a halo of the different colours of the spectrum (although it was another two years before he had the chance to pursue this idea properly).[ii] In 1670 he lectured on his new theory of light and colours, and in 1672 this became his first publication, 'A Letter of Mr Isaac Newton, Professor of the Mathematicks in the University of Cambridge; Containing His New Theory about Light and Colors'.

The story as Newton tells it will not do. The experiment he describes cannot be performed in Cambridge at the beginning of the year: it requires the sun to reach an elevation of 40 degrees above the horizon. In any event, Newton was not in Cambridge at the beginning of 1666. In a conversation towards the end of his life he said he had bought the prism in August 1665 (corrected in the manuscript to 1663) at Sturbridge fair, but there was no fair in 1666, and he was not in Cambridge for the fair of 1665. The best we can do is say that the first experiments with a prism probably took place shortly before June 1666 (when Newton left Cambridge to escape the plague), that the prism was purchased at some other fair and that Newton then conducted further experiments, including the crucial experiment, in the summer of 1668.

The exact date hardly matters. More important is the evidence of Newton's notebooks, which suggests that he knew about the differential refraction of colours in 1664, at which date he already had a prism (perhaps purchased at Sturbridge fair in August 1663). Newton looked through his prism at a card, half of which was white, and half painted black; and at a thread, half of the length of which was red and the other half blue: in both cases the prism appeared to split the object in two, failing to line up one colour with the other. When Newton performed his experiment in 1666 it was probably deliberately designed to produce the elongated spectrum, an effect he describes as if were a complete surprise. Newton's modern biographer, Richard Westfall, concludes that we should regard Newton's claim to have been surprised by the oblong image cast by the prism 'as a rhetorical device which is not to be understood literally'.[2] Thomas Kuhn maintained that '[t]he implication of Newton's account of 1672 is wrong in that Newton did not proceed so directly or so immediately from the first prism experiment to the final

ii In a reflecting telescope the image is magnified by a curved mirror, while in a refracting telescope it is magnified by passing through a lens. Since a reflecting telescope does not rely on refraction, it does not produce a halo of different colours around the image. Producing a curved mirror, however, is far from straightforward!

version of the theory as the first paper would imply.'[3] Peter Dear goes further (too far, perhaps): Newton's account is 'spurious' for 'the event described did not actually take place'.[4]

Why would Newton redescribe what had actually happened in this way? One answer is that he wanted to pretend that he had worked from the phenomena to a theory, and not the other way round: the Royal Society admired Bacon, and this would have been the Baconian way to proceed.[5] Another is that his reference to an *experimentum crucis* is an implicit nod to Pascal's Puy-de-Dôme experiment: Pascal's experiment had been preceded by earlier experiments and earlier theorizing, but in a way that was irrelevant. Why not just cut to the chase? (Boyle would have been horrified, since he had always insisted that experimental reports must be a faithful record of what had actually happened, but in a passage cut from the published version of his paper Newton expressed impatience with lengthy historical narrations.)[iii]

We can debate when Newton conducted his prism experiments, and we can argue about the order in which he performed them and exactly when he formulated his first theory, but there is no scope for debate as to *whether* Newton conducted the experiments he describes – the events certainly took place, even if the when and why are now hard to establish. Conventional history of science tends to stop at this point. But I want to focus on something else: Newton says in his first publication that he is presenting a new 'doctrine'; the text is headed by the editor, Oldenburg, 'A Letter of Mr Isaac Newton ... containing his New Theory about Light and Colors' – this is the first article in the *Philosophical Transactions* to have the word 'theory' in its title, and it is only in subsequent correspondence that Newton himself adopts the word.[6] A critic, Ignatius Pardies, called Newton's paper 'a most ingenious hypothesis', 'an extraordinary hypothesis' which, if true, would overturn the foundations of optics.[7] Newton, replying in Latin, explained that he had decided not to consider this an insult:

> I do not take it amiss that the Rev. Father calls my theory an hypothesis, insomuch as he was not acquainted with it. But my design was quite different, for it seems to contain only certain properties of light, which, now discovered, I think easy to be proved, and which if I had not considered them as true, I would rather have them rejected as vain and empty speculation, than acknowledged even as an hypothesis.[8]

iii See the end of note v, below.

Pardies replied, insisting he had not used the word 'out of any disrespect'.[9] Newton responded that he thought of his own work as establishing the properties of light; one might, if one wanted, then hypothesize about the possible cause of those properties, but hypotheses should be subservient to the properties of things, and the only useful ones were the ones that led one to devise new experiments. He went on to complain that, in this case at least, there was no difficulty in constructing hypotheses that appeared to fit the facts: '[I]t is an easy matter to accommodate hypotheses to this doctrine. For if any one wish to defend the Cartesian hypothesis, he need only say that the globules[iv] are unequal, or that the pressures of some of the globules are stronger than others, and that hence they become differently refrangible, and proper to excite the sensation of different colours.'[10] (We know from Newton's notebooks that he had started his work on refraction with the idea that 'slowly moved rays are refracted more than swift ones', precisely the sort of hypothesis he was now rejecting as pointless.)[11] He ended his letter by returning to the subject and saying that he was sure Pardies meant no harm, 'as a practice has arisen of calling by the name hypothesis whatever is explained in philosophy', but that he felt that this practice could prove 'prejudicial to true philosophy'.[12]

He had in fact used the word 'hypothesis' himself in his original publication, but only to refer to an inaccurate mathematical rule of thumb;[13] more to the point, Oldenburg had removed a passage in which he insisted that what he was proposing was not an hypothesis, for he had proved his conclusions beyond any doubt.[v] Pardies had thus touched on a fundamental difference between Newton and the Royal Society of the 1660s and early 1670s: unlike Newton, the Royal Society favoured tentative expressions of opinion. My first goal in this chapter, then, is to establish why Newton was hostile to the word 'hypothesis' and felt that its use in the context of his own work amounted to an insult.

iv That is, the invisible corpuscles or atoms which, for a Cartesian, constitute light.

v This is the omitted passage: 'A naturalist would scearce expect to see ye science of those [*Colours*] become mathematicall, & yet I dare affirm that there is as much certainty in it as in any other part of Opticks. For what I shall tell concerning them is not an Hypothesis but most rigid consequence, not conjectured by barely inferring 'tis thus because not otherwise or because it satisfies all phænomena (the Philosophers universall Topick,) but evinced by ye mediation of experiments concluding directly & without any suspicion of doubt. To continue the historicall narration of these experiments would make a discourse too tedious & confused, & therefore . . .' Both Hooke and Newton may not have realized it had been omitted, as they later refered to it in print. Newton, *The Correspondence of Isaac Newton* (1959), Vol. 1, 96–7; see the editor's notes, 105 n. 19, 190 n. 18, 386 n. 22.

§ 2

The fashion for the word 'hypothesis' was new; it began with the publication of Descartes' *Principles* in 1644. There, in the third part, Descartes turned from the various 'hypotheses' that had been propounded to explain the movements of the planets (those of Ptolemy, Tycho and Copernicus), to discuss the task of explaining movement and change on the earth. Three crucial paragraphs (43–45) carry the following marginal glosses:

> 43: If a cause allows all the phenomena to be clearly deduced from it, then it is virtually impossible that it should not be true.
>
> 44: Nevertheless, I want the causes that I shall set out here to be regarded simply as hypotheses.
>
> 45. I shall even make some assumptions which are agreed to be false.[14]

It is not surprising that Descartes' formulations provoked confusion and controversy. First, he seemed to say that a hypothetical cause could give one true knowledge; then to backtrack and say that his arguments were *only* hypothetical; and finally to acknowledge that some of his arguments must be false. Where did this leave the new philosophy? Was it producing indisputable knowledge that could not be contested? Knowledge that might or might not be true? Or knowledge that was obviously false? From 1644 on, the use and status of hypotheses became a central issue.

In order to understand what is going on here it helps to know that 'hypothesis' had three distinct technical meanings in the Middle Ages.[15] In logic an hypothesis was something that came under the thesis ('hypo-' means 'under' in Greek, as in 'hypodermic', a needle which goes under the skin). So one might say that human beings are mortal (the thesis); Socrates is a human being; so Socrates is mortal. Here the statement that Socrates is a human being is an 'hypothesis' which follows after the thesis and generates the claim that Socrates is mortal; it can be stated in the hypothetical form '*If* Socrates is a human being, then he is mortal.' That example is straightforward, but consider this one: the apostle Peter had authority over the Church; the pope is Peter's successor; therefore the pope has authority over the Church. A Catholic would regard this as a valid syllogism, while a Protestant would maintain that the hypothesis is false; the pope may be Peter's successor as bishop of Rome, but he is not Peter's successor in the required sense.

In mathematics the word 'hypothesis' was also used to mean a supposition or postulate upon which an argument was based; in geometry, for example, one might propose to argue on the assumption that two angles were equal, even though they had not been proved to be so. But in mathematics the word 'hypothesis' also had a quite different technical meaning.[16] An hypothesis was the theoretical model which generated predictions of the future locations of the planets in the heavens. Different hypotheses might produce the same result: for example an eccentric circle will generate exactly the same movement as an epicycle upon a deferent. There might be philosophical reasons for preferring one to the other, but an astronomer could happily use either to perform calculations. Thus what mattered about an hypothesis was not that it be true, but that it produce accurate results (and what we consider to be mistaken hypotheses were quite capable of generating accurate results). Henry Savile, when invited to state a preference between Ptolemy and Copernicus, replied: 'hee cared not which were true, so the Apparences were solved, and the accompt exact: sith each way either the old of *Ptolomy*, or the new of *Copernicus*, would indifferently serve an Astronomer.'[17] In this sense, of an account that saves the phenomena, but may or may not be true, we find Hobbes using the word 'hypothesis' (in Latin) before 1640, and this is the sense in which Descartes uses the word 'hypothesis' in his discussion of cosmology.[vi] [18]

Those who held that Copernicanism was literally true, however, insisted that in this case the truth of the hypothesis mattered. Kepler distinguished between a geometrical hypothesis – the mathematical model used to generate predictions – and an astronomical hypothesis, the actual path of the planet through the heavens. As geometrical hypotheses, the Ptolemaic, Tychonic and Copernican systems were

vi 'The treatment of natural things differs greatly from that of other sciences ... In the explanation of natural causes, we must necessarily have recourse to a different kind of principle, called "hypothesis" or "supposition". For when a question is raised about the efficient cause of any event which is perceptible by the senses (what is normally called a "phenomenon"), the question consists principally in the designation or description of some motion, from which such a phenomenon necessarily follows. And since it is not impossible that dissimilar motions may produce similar phenomena, it may happen that the effect is correctly demonstrated from the supposed motion, and yet that that supposition is not true' (Hobbes, 'Tractatus opticus', quoted in Malcolm, 'Hobbes and Roberval' (2002), 183–4 – Malcolm's translation). Compare Pascal's first reply to Noel in 1647, Pascal, *Oeuvres* (1923), 98–101; but Pascal approaches Popperian falsificationism, for he stresses that though it may not be possible to prove a good hypothesis true it will often be possible to show that a bad hypothesis is false.

equivalent; but as astronomical hypotheses they were radically different. It is from this line of thinking that we get, it would seem, the first reference in English to an hypothesis as a theory that needs to be tested. In his 1576 edition of his father's *Prognostication* Thomas Digges proposed 'An Hypothesis or supposed cawse of the variation of the Cumpasse, to be mathematically wayed [i.e. weighed, meaning assessed]'.[19] The implication is that if the hypothesis passes the test it will be promoted to being a true statement. This seems to be the earliest usage of the term 'hypothesis' in its standard modern sense, at any rate in English.[vii] For the little group looking for a mathematical pattern in the variation of the compass – what Robert Norman called 'a Theorik with Hypotheses, and rules for the salving of the apparant irregularitie of the Variation'[20] – it was a simple step to adopt the astronomical language of 'hypotheses' and give it a new experimental twist.[21] But this twist effectively marks the birth of a new philosophy of science: a scientific principle is now an hypothesis that has survived the test of experience. Thus Galileo in his 'Discourse on the Flux and Reflux of the Sea' of 1616 presents his theory of the tides as an hypothesis which needs to be confirmed or disconfirmed by a systematic programme of observations.[22]

Boyle used the word 'hypothesis' in this sense over and over again, and even wrote a short paper (never published) on 'the requisites of a good hypothesis'. Boyle thought of an hypothesis as a useful step towards establishing the truth: a good hypothesis leads to novel predictions which can be tested by experiments. At its best an hypothesis is like the key which enables one to decipher a coded communication: everything now makes sense, and it is evident that this, and this only, is the right solution (exactly the view expressed by Descartes in §43).[viii] Locke wrote a section of the *Essay* on the 'true use of hypotheses'. He acknowledged that hypotheses may lead us to new discoveries, but stressed that most ('I had almost said all') hypotheses in natural philosophy were no more than very doubtful conjectures.[23]

vii The earliest date given by the *OED* for 'hypothesis' in this sense is 1646: Sir T. Browne, *Pseudoxia epidemica* ii. ii. 60: 'Irons doe manifest a verticity not only upon refrigeration . . . but (what is wonderfull and advanceth the magneticall hypothesis) they evidence the same by meer position according as . . . their extreams [are] disposed . . . unto the earth.'

viii It is a feature of an excellent hypothesis 'That it enable a skilfull Naturalist to foretell future Phenomena, by their Congruity or Incongruity to it: and especially the Events of such Expts as are aptly devised to Examine it; as things yt ought or ought not to be Consequent to it.' Westfall, 'Unpublished Boyle Papers' (1956), 69–70.

William Wotton, on the other hand, like Newton, generally used the word to refer to arguments that are false or unsatisfactory. For Wotton, to say an argument is an hypothesis is to reject it, for if it really explained all the phenomena it would no longer be an hypothesis. And we find a third use, as in Descartes' §45: the use of 'hypothesis' to refer to an argument that is acknowledged to be false but held in some way to be useful. Osiander, in his anonymous introduction to Copernicus's *On the Revolutions*, insisted that Copernicus should only be read as presenting an hypothesis, not as describing how the world really is. Bellarmine told Galileo that he could talk about Copernicanism if he did so hypothetically – Copernicanism being, as far as Bellarmine was concerned, untrue.[24] Descartes, following in this tradition, used the word in §45 to refer to principles that ought, for theological reasons, to be acknowledged as false, but that are helpful if one pretends they might be true.[ix][25]

There is a futher use of the word 'hypothesis' that we must note. In Gilbert's *On magnetism* (1600) the word is used in a purely conventional way in the main body of the text, to refer, for example, to the Copernican hypothesis. But something strange happens in the preface:

> [N]othing hath been set down in these books which hath not been explored and many times performed and repeated amongst us. Many things in our reasonings and hypotheses will, perchance, at first sight, seem rather hard, when they are foreign to the commonly received opinion; yet I doubt not but that hereafter they will yet obtain authority from the demonstrations [i.e. experiments] themselves ... [W]e but seldom quote ancient Greek authors in our support, because ... our doctrine magnetical is at variance with most of their principles and dogmas ... [O]ur age hath detected and brought to light very many things which they, were they now alive, would gladly have accepted. Wherefore we also have not hesitated to expound in

ix Descartes sometimes makes a distinction between hypotheses, which he is always prepared to disown, and suppositions, which he usually claims he can show to be true, both *a posteriori*, and, notionally, by deduction from first principles: Descartes, *Philosophical Writings* (1984), 250–1, 255–8 (hypotheses); 40–1, 150 (suppositions), but see 152–3, where the original French has *suppositions*, although the authorized Latin translation has *hypotheses* (I owe this last point, and much else, to John Schuster). When Newton objects to the feigning of hypotheses, and Locke says one should not elevate hypotheses into principles, both presumably have Descartes in mind, and particularly Descartes' 'suppositions' (which is what, presumably, Locke means by 'principles', and what Newton is objecting to when he complains that hypotheses should not be given priority over experimental evidence).

> convincing [*probabilibus*] hypotheses those things which we have discovered by long experience.[26]

Gilbert here uses the word 'hypothesis' as we would use the word 'theory'; we assume that an hypothesis awaits confirmation or disconfirmation, but Gilbert's hypotheses derive from and are confirmed by a long sequence of experiments. They are new additions to secure knowledge; they are, in our terms, theories. We find the same usage in Galileo. In his book on sunspots (1613) he refers to his claim that the moon is opaque and mountainous as a true hypothesis, confirmed by sensory experience.[27]

Thus the standard modern meaning of 'hypothesis', as an explanation that may in due course be tested and which, if confirmed, will be elevated to the status of a theory, did not become firmly established until the 1660s.[28] In 1660 Robert Boyle described an experiment proposed by Christopher Wren that 'would discover the truth or erroneousness of the *Cartesian Hypothesis* concerning the Ebbing and Flowing of the Sea';[29] Power's *Experimental Philosophy* of 1664 uses the word frequently; then in 1665 Hooke prefaced his *Micrographia* with a dedicatory letter to the Royal Society: 'The Rules YOU have prescrib'd YOUR selves in YOUR Philosophical Progress do seem the best that have ever yet been practis'd. And particularly that of avoiding *Dogmatizing*, and the *espousal* of any *Hypothesis* not sufficiently grounded and confirm'd by *Experiments*.' From this point on 'hypothesis', meaning a conjecture or query (to use Hooke's terms) which could be confirmed or disconfirmed by observation or experiment, became central to the terminology of the new science. Indeed one can say that 'hypothesis' only really acquired its modern sense after the foundation of the Royal Society.

These various meanings of 'hypothesis' serve to explain its peculiar distribution through seventeenth-century texts. Most mathematicians – Galileo, Pascal, Descartes, Newton – were familiar with the word's use in technical astronomy, and tended to avoid it in other contexts. But once it became common to refer to Copernicanism as an hypothesis, then other hypotheses – magnetical, atomic, mechanical – multiplied. These were the big theories of the new science; within them smaller hypotheses, such as Digges's account of the declination of the compass needle, or Boyle's account of the spring of the air, could be neatly packed.

But the term was not uncontroversial, particularly because Descartes had acknowledged that his hypotheses might be (indeed in some cases must be) false. Newton wrote, in the second edition of the *Principia* (1713) *hypotheses non fingo*, and we know that he himself would have translated this as 'I do not feign hypotheses' – *fingo* and 'feign' here mean 'imagine', which is the core meaning of the word 'feign' in the seventeenth century.[30x] Thus both Copernicus and Francis Bacon wrote of astronomers 'feigning' eccentrics and epicycles – they meant these are imaginary entities.[xi31] So what Newton meant was 'I do not invent imaginary entities in order to explain natural properties.'[xii] In the *Discourse on Method* (1637) Descartes had dismissed Aristotelian philosophy as 'speculative'; his own philosophy would get at the truth by proposing explanations (hypotheses, we would say) that could then be tested experimentally.[32] In the *Principles of Philosophy* (1644), however, he retreated from this position. It would often, he acknowledged, be impossible to choose between competing explanations because one could

x In a draft for the second edition he wrote: 'From the phenomena it is very certain that gravity is given and acts on all bodies according to the laws described above in proportion to the distances, and suffices for all the motions of Planets and Comets, and thus it is a law of nature, although it has not yet been possible to understand the cause of this law from the phenomena. For I avoid [*fugio*] hypotheses, whether metaphysical or physical or mechanical or of occult qualities. They are harmful and do not engender science.' (Newton, *Unpublished Scientific Papers* (1962), 353.)

xi 'aliis ante me hanc concessam libertatem, ut quos libet fingerent circulos ad demonstrandum phaenomena astrorum': Copernicus, *De revolutionibus orbium coelestium* (1543), iiii(r); 'Of Superstition' (Bacon, *The Essayes* (1625), 97), where 'feign' is used as synonymous with 'frame'; the passage was borrowed without acknowledgment in Wilkins, *A Discourse* (1640), 26. Bacon described Utopia as 'a feigned Common-wealth'; and he writes: 'Now let me put a feigned Case, (And yet Antiquity makes it doubtful, whether it were Fiction, or History,) of a *Land* of *Amazons,* where the whole *Government,* publick and private, yea the *Militia* it Self, was in the hands of *Women.*' Horrocks and Newton, writing in Latin, also use an almost identical word, *confingo* – Horrocks when discussing the imaginary epicycles of Ptolemaic astronomers (Hevelius & Horrocks, *Mercurius in Sole visus* (1662), 133), and Newton in the Latin translation of the *Opticks*, to translate 'without feigning Hypotheses' (Cohen, 'The First English Version of Newton's *Hypotheses non fingo*' (1962), 380–1). 'Feign' thus carries no necessary implication of deception; it simply means 'imagine', and Cohen is on the wrong track when he maintains that '"to feign" has also the sense of to practice dissimulation, to dissemble, to pretend, to counterfeit, to sham.' (Cohen, 'The First English Version of Newton's *Hypotheses non fingo*' (1962), 381.)

xii This line of thinking goes back to Ramus (d. 1572), whose work on logic was immensely influential, particularly among Protestants. Ramus had called for an astronomy without hypotheses, that is to say an astronomy without imaginary entities such as epicycles (a standard view was that the celestial orbs or spheres were real, physical entities, but that the epicyles that astronomers placed within them were fictitious). Kepler claimed to have produced such an astronomy by escaping from the principle of circular movement. (Granada, Mosley, and others, *Christoph Rothmann's Discourse* (2014), 55–63, 134–43.)

not see what was actually happening within the invisible world of particles out of which our visible world was constructed. Just as a clockmaker, looking at a clock from the outside, could imagine various ways in which the machinery might be configured, so the philosopher must acknowledge that there might be several equally good explanations of a natural process; it was not always possible to devise a test to choose between them.[33] It was this process of making up explanations that might or might not be true that Newton rejected when he insisted *hypotheses non fingo*. (That the word 'hypothesis' was particularly associated with his old enemy, Hooke, will not have been irrelevant.) As far as Newton was concerned the only worthwhile hypotheses were the ones that could be tested: and if they survived testing they ceased to be hypotheses. Gilbert's and Galileo's use of 'hypothesis' to refer not to a claim that *might* be true, but to one that we could be sure was true, would have made no sense to him, just as it seems peculiar to us.

§ 3

Pascal's Puy-de-Dôme experiment explained the height of the mercury in a barometer by showing that it was directly related to the weight of the air. It made visible a causal relationship: the weight of the air and the weight of the mercury balanced each other. From the point of view of a conventional seventeenth-century philosopher, this was a peculiar sort of explanation. As far as Aristotle was concerned, as we saw in Chapter 3, causal explanations had four components: the formal cause, the final cause, the material cause and the efficient cause. In Pascal's account of why the mercury does not descend in the Torricellian tube the formal and material causes are so attenuated as to be uninteresting, and the final cause has disappeared completely. You can substitute water or wine for mercury, so the exact substance is irrelevant; any liquid will do. You can substitute lead piping for glass, so again the material cause is irrelevant; any pipe sealed at one end will work. The mercury has no natural tendency to stand up in a column, so there is no final cause at work here. There is only an efficient cause: the balance of weights; and a structure or form that makes the balance possible: a sealed tube up-ended in a bath of mercury. For an Aristotelian, there is only one discipline that isolates efficient causes and structures and ignores all others, and that discipline is mechanics. Pascal's explanation is a mechanical explanation, and its oddity is that it extends the scope of mechanical explanations from the artificial world of levers and pulleys to the natural world of gases and liquids. Moreover, like any mechanical explanation, Pascal's can be expressed mathematically, either as a measurement (pounds per square inch; or, which amounts to the same thing, the height of the column of mercury) or as a ratio (since the barometer is a balance, the ratio of the two weights is 1:1, but carrying the barometer to the top of the Puy-de-Dôme shows that y metres of air equals in weight x cm of mercury). This is why Boyle's contribution to the vacuum debate was entitled *New Experiments Physico-mechanical*: mechanics is now being used to explain physics.

Pascal's Puy-de-Dôme experiment seems pretty straightforward to us, but that is because we are used to modern physics. To Aristotelians, it did not seem to offer any sort of explanation of what was happening, just as it seems to us conceptually confused to say (as Aristotelians did) that inanimate objects have goals or purposes. Pascal's explanation looks right to us; to an Aristotelian, it looked all wrong, which is why Aristotelians (and most intellectuals in Pascal's day were still Aristotelians) tried to substitute explanations in terms of nature abhorring a vacuum and so trying to prevent one coming into existence. It is hard to reconstruct in our minds a mental universe where Pascal's explanation seems obviously unsatisfactory and an explanation in terms of nature's purposes seems obviously preferable.

The problem for the Aristotelians was that they could not fashion an explanation that would successfully predict the outcome of the Puy-de-Dôme experiment. Why should nature abhor a vacuum less at the top of a mountain than at the bottom? Pascal could answer this question, and they could not. Pascal's explanation could be tested and be shown to work. But to acknowledge it as a good explanation philosophers had to change their definition of what constituted an explanation; they had to learn to satisfy themselves with the sort of explanations that mathematicians were accustomed to providing. Even people who thought Pascal's explanation was a bad one could see that he could make successful predictions (that the height of a column of water in a Torricellian tube would be fourteen times the height of a column of mercury, for example), and they could not.

Let us take another example, one familiar to Pascal: Galileo's 'law' (as we call it) of fall. Galileo showed that (in the absence of air resistance) all falling objects accelerate at the same rate, and that one can therefore predict the distance travelled in any time by a falling body, and its terminal velocity; indeed, these are related in such a way that the units of measurement are irrelevant. The distance travelled is proportional to the square of the elapsed time, whether you are measuring in feet and seconds or kilometres and Hail Marys (it is an accident that we have only one standard, universal system for measuring time and several systems for measuring distance, but early modern peoples did use informal measures of duration, such as the Hail Mary). Galileo's law of fall *describes* in mathematical terms what happens when bodies fall under ideal conditions; but it does not *explain* anything. It does not even offer (as Pascal's vacuum experiments do) a mechanical explanation. It tells you what to measure and enables you to predict, but it provides no answer to the question 'Why?'

If science explains things, this is not science. What makes it science is not that it provides an explanation but that it provides reliable predictions in the form of a mathematical model. Accepting Galileo's law of fall as good science thus involves an even more radical step away from an Aristotelian conception of science than does accepting Pascal's account of why the mercury stands tall in the barometer. You may think that this is simply because Galileo's law is incomplete: Newton's theory of gravity provides an explanation both of Galileo's law of fall and of Kepler's laws of planetary motion. That is partly true, but Newton has absolutely no explanation of what gravity is or how it works; as we have seen, he admits as much. The theory of gravity simply makes possible reliable predictions across a wider field. The problem of explanation has been moved, not solved. Consequently, Huygens' response to Newton's theory of gravity was straightforward: 'I had not thought ... of this regulated diminution of gravity, namely that it was in inverse ratio to the squares of the distances from the centre: which is a new and remarkable property of gravity, of which the reason is well worth looking for.'[34] Huygens was still looking for explanations; Newton had left the world of explanation and entered a new world, the world of theory.

Scientific explanations are not (at least not so far) complete: they come to a halt, often abruptly. A scientific law marks the point beyond which there are no explanations, although further explanations sometimes come along later. Aristotelian science was not like this: Aristotelian philosophers had no sense that their knowledge was incomplete in important respects, and they thus had a different measure of success from a Galileo or a Pascal. For them, the proof that their system of knowledge was successful was that there was nothing it could not explain, although the explanations often look circular to us now: Molière in *The Imaginary Invalid* (1673) mocked the idea that one could explain why opium puts people to sleep by saying it was because 'there is in it a dormitive power whose nature it is to lull the senses to sleep.' Such explanations look foolish after Pascal, but not before.

For a Galileo, or a Pascal, or a Newton what mattered was being able to make successful predictions where such predictions had previously been impossible. But this involved acknowledging the limits of their knowledge. Aristotelian philosophers looked backwards, assuming Aristotle had known everything that needed to be known; the new scientists looked forward, aiming to expand the limited range of topics in which they could make satisfactory predictions. One reason why the new science made progress and the old philosophy did not is that it was conscious of being imperfect and incomplete.

§ 4

What is science? James Bryant Conant, who has a good claim to be the founder of modern history of science (he was Kuhn's mentor) defined it as 'a series of concepts or conceptual schemes (theories) arising out of experiment or observation and leading to new experiments and observations'.[35] Science is thus an interactive process between *theory* on the one hand and *observation* (our old friend 'experience') on the other. In astronomy, this process really gets under way with Tycho Brahe; in physics, with Pascal. We can trace it clearly through Newton's notebooks, even if he compresses it in his first publication. It would seem evident that this extraordinary transformation in the nature of knowledge must be reflected in the language of science: and it is, although the language in which we speak about science has become so completely second nature to us that a key aspect of this linguistic adaptation has become almost completely invisible.[36] The adaptation itself is, once one has realized that it must be there, easy to identify and, once identified, its importance is evident.

A helpful way of beginning is to look up the word *théorie* in a series of French dictionaries.[37] It is not until the end of the nineteenth century that we find (in Littré's great dictionary) the obvious modern meaning, with examples given of the theories of heat and electricity. Previously, 'theory' is defined as speculative rather than practical knowledge (the etymological origin of the word is in a Greek word for looking or spectating), with one particular additional usage noted: *la théorie des planètes*, the mathematical models for the movement of the planets. If we look for the words 'theory'/*théorie*/*teoria* in Galileo, Pascal, Descartes, Hobbes, Arnauld and Locke, we find nothing,[xiii] while in Hume we find the word used in its modern sense frequently – and more and more so as time goes by.

In English, in the sixteenth century, the word 'theory' (or 'theoric'; the words are used interchangeably) is used as we would expect from our inspection of French dictionaries: on the one hand to refer to speculative or abstract knowledge, usually in opposition to practice (thus musicians learn both the theory and practice of music, and gunners learn the theory and practice of gunnery), and on the other to refer to the theory of the planets. References to the theories of Ptolemy and Copernicus are thus references to their mathematical modelling of the cosmos. The first example I can find of the word being used in the modern sense, without an implied reference to a mathematical model, is in Bacon's *Sylva sylvarum* (1627), when he is criticizing Galileo's explanation of the tides:

> *Galilaeus* noteth it well; That if an *Open Trough*, wherein *Water* is, be driven faster than the *Water* can follow, the *Water* gathereth upon an heape, towards the *Hinder End*, where the *Motion* began; Which he supposeth, (holding confidently the *Motion* of the *Earth*,) to be the *Cause* of the *Ebbing* and *Flowing* of the *Ocean*; Because the *Earth* ouer-runneth the *Water*. Which *Theory*, though it be false, yet the first *Experiment* is true.[xiv][38]

It is presumably from Bacon that this new meaning of the word spread.[xv] We find it in 1649 and 1650 in translations of and commentary on van Helmont, and in 1653 in a translation of and commentary on Descartes: in each case there is no equivalent in the original.[39] Boyle announces in 1660 that he is going to offer new experiments regarding the vacuum, but not new theories;[xvi] in 1662 he proudly announces a new 'theory' (the word is his), which we now call Boyle's law.[40] The first

xiii In Pascal and Descartes we find only the conventional distinction between practical and theoretical knowledge but never a particular theory.

xiv Galileo had, it is true, a sophisticated mathematical theory of why 'the earth over-runneth the water', but when Bacon says the theory is false I take it he is referring not to Copernicanism in general, nor to Galileo's deduction from Copernicanism, but simply to the theory that when 'the earth over-runneth the water' the tides are caused.

xv The first examples of 'theory' and 'theorize' in the relevant sense in *OED* are from 1638.

xvi 'And though I pretend not to acquaint you, on this occasion, with any store of new Discoveries, yet possibly I shall be so happy, as to assist you to *know* somethings which you did formerly but *suppose*; and shall present you, if not with new Theories, at least with new *Proofs* of such as are not yet become unquestionable.' Boyle, *New Experiments Physico-mechanical* (1660), 2 = Boyle, *The Works* (1999), Vol. 1, 157.

occurrence of the word in its new sense (that is, not as a contrast between theory and practice, nor as a mathematical model) in the *Philosophical Transactions of the Royal Society* appears to be in Oldenburg's editorial introduction to an explanation of the tides by John Wallis (Wallis writes of an hypothesis, an essay and a surmise, but not of a theory; in the index to the volume, this is 'a new theory'); the second in Robert Boyle's 'Tryals proposed to Dr Lower' regarding blood transfusions in animals.[41] In Sprat's *History* (1667) the word takes on its full, modern range of meaning: even the scholastics are now said to have had theories, and the production of new theories is now as important a part of the new science as the performance of experiments.[42] Newton's letter to the Royal Society of 1672 was, as we have seen, given by Oldenburg the title 'A Letter of Mr Isaac Newton, Professor of the Mathematicks in the University of Cambridge; Containing His New Theory about Light and Colors',[xvii] and the phrase 'new theory' runs through the titles of the ensuing correspondence: his *Opticks* (1704) declares itself to be a study in 'the Theory of Light'.[xviii][43] Optics was traditionally a branch of mathematics, and Boyle's law is a mathematical relationship, but Hooke writes not only of 'the true Theory of *Elasticity* or *Springiness*' but also of his own theory of flame, where there is no mathematics involved.[44] The word used in its new sense first appears in the title of a book in Thomas Burnet's *Telluris theoria sacra* (1681), translated as *The Theory of the Earth* in 1684, and followed in 1696 by William Whiston's *A New Theory of the Earth*. In French the new usage seems to have been adopted first by the mathematicians (Johann Bernoulli, *Nouvelle théorie du centre d'oscillation*, 1714), but it quickly spread more generally: Voltaire's *Élémens de la philosophie de Newton* (1738) discusses '*la théorie de la lumière*'. George Berkeley is translated into Italian in 1732: *Saggio d'una nuova teoria sopra la visione*.

The new sense of the word 'theory' is fundamental to an understanding of what the new science claimed to do. Traditionally, philosophy had concerned itself with *scientia*, true knowledge, but mathematicians practising astronomy had been content with mathematical models – hypotheses, theories – which might or might not correspond to reality but which fitted more or less exactly with the phenomena. Mathematical theories

xvii Newton himself calls it a doctrine, not a theory, just as Boyle entitled one of his books *A Defence of the Doctrine Touching the Spring and Weight of the Air* (1662): 'doctrine' is the old language of the schools, which 'theory' replaces.

xviii The word *theoria* occurs four times in Newton's *Principia* (1687), three times in the modern sense. 'Theory' occurs over and over again in Andrew Motte's translation of 1729.

were not explanations, they were conceptual systems for making predictions. The new theory that Boyle announced relating to the pressure of gases (1662), or Newton's new theory of light (1672) were not *explanations* – they did not answer the question *why*; they were concepts that enabled one successfully to predict the outcome of experimental procedures and to identify processes in the natural world. Moreover, the word 'theory' carried with it a useful ambiguity: it could refer either to an established truth (which is how Newton used the word) or to a viable hypothesis, thus fudging the differences between those who wanted to claim indisputable truth and those who wanted to make tentative knowledge claims.

In adopting the term 'theory' the scientists were thus freeing themselves from the philosophers' preoccupation with truth in so far as it implied knowledge of causes and of what Aristotelian philosophers called substances, or forms. Locke and Newton insisted that we could have no knowledge of substance (suppose the world is made up of atoms – we can have no idea of their size or shape); we could have knowledge only of properties (oak is hard, balsa is soft, and so forth). For knowledge of substance Newton substituted conceptual models that worked reliably and accurately. Philosophers of science right through to the present day have been preoccupied with what is called 'realism', the question of whether science is true; what they have failed to notice is that the founding of modern science was accompanied by an escape from the old notion of true knowledge (*scientia*) and its replacement by the concept of 'theory'.[xix] The adoption of the word marks the break between the classical traditions of philosophy and mathematics, which were concerned with deduction and with true knowledge of substances, and modern science, which is concerned with viable theories. Locke's *Essay* (1690) symbolizes this shift in its title. It is not a book about *knowledge* (which is now thought of as largely beyond human capacity), but it is an *Essay concerning Humane Understanding*: even the word 'essay' implies that understanding is necessarily provisional. In a crucial passage in the epistle to the reader he writes of the understanding:

> that as it is the most elevated Faculty of the Soul, so it is employed with a greater, and more constant Delight than any of the other. Its searches after Truth, are a sort of Hawking and Hunting, wherein the very pursuit makes

xix It is not surprising that this concept was taken up so wholeheartedly in the sceptical philosophy of Hume.

> a great part of the Pleasure. Every step the Mind takes in its Progress towards Knowledge, makes some[xx] Discovery, which is not only new, but the best too, for the time at least.

Thus knowledge, in so far as we have it, is not absolute but progressive, not definitive but provisional. We make progress, but, unlike those who go hawking and hunting, we may never catch our prey.

It follows that even Galileo was never more than a reluctant scientist, for he always sought the certainty of deduction; rather, modern science begins with Bacon's re-description of Galileo's demonstration of the movement of the Earth as a 'theory'. By the 1660s the standard terminology for discussing science in England included 'facts' and 'evidence' (from the law; we will discuss 'evidence' in the next chapter), and 'hypotheses' and 'theories' (from astronomy). Science had been invented. The first book to contain these four words, all used in their modern senses, along with 'experiment', also used in its modern sense, was, it would seem, Walter Charleton's paraphrase of van Helmont, the *Ternary of Paradoxes* of 1649. Charleton was a deliberate and self-conscious innovator in linguistic usage: the *Oxford English Dictionary* quotes him 151 times as the first entry under a definition (he is recorded as the first to use 'projectile', 'pathologist' and, alas mistakenly, 'erotic').[45] But none of the usages which immediately concern us were new with him, and indeed he insisted on the remarkable merits of English, on

> the Venerable Majesty of our *Mother Tongue*; out of which, I am ready to assert, may be spun as fine and fit a garment, for the most spruce *Conceptions of the Minde* to appeare in publick in, as out of any other in the World: especially, since the *Carmination* or refinement of it, by the skill and sweat of those two Heroicall Wits, the Lord St Alban [Francis Bacon], and the now flourishing Dr Browne; out of whose incomparable Writings may be selected a Volume of such full and significant *Expressions*, as if uprightly fathomed by the utmost Extent of the sublimest *Thought*, may well serve to stagger that Partiall Axiome of some Schoolemen, that the *Latin is the most symphoniacall and Concordant Language of the Rationall Soule.*[46]

Charleton's language did not meet with the approval of his contemporaries, and he opens his next publication, *Deliramenti catarrhi* (1650), with a lengthy and bitter diatribe against his thick-skulled detractors

xx 'Some' was not in the printed text of the first edition, but it is sometimes inserted in ink by the printers and appears in all later editions.

whose depraved appetites have rendered them, he declares, 'fit to digest nothing, but crude *Sallads* gathered in the *Poets Elizium*, and soft *Romances*, oyled with the effeminate *Extracts* of the *Stage*, and spiced with some new *French-English* idioms' rather than his own masculine idiolect. But Charleton was one of the most active members of the Royal Society in its early years, and his idiolect, tamed and domesticated by Boyle and Sprat, has become the language of science. Where the old philosophy had laid claim to indisputable certainties, the new one modelled itself on astronomy and the law, disciplines in which facts and evidence had long been marshalled in order to generate reliable, even incontrovertible, hypotheses and theories.

11
Evidence and Judgement

> I shook my head. 'Many men have been hanged on far slighter evidence,' I remarked.
>
> 'So they have. And many men have been wrongfully hanged.'
>
> – Arthur Conan Doyle, 'The Boscombe Valley Mystery' (1891), *The Adventures of Sherlock Holmes*

§ 1

To repeat the question: What is science? The answer: knowledge of natural processes based on evidence. In which case there can be no science without a concept of evidence. Yet if we start looking for the word 'evidence' being used by seventeenth-century scientists we discover something peculiar: they have the word, but they hardly ever use it. Bacon, for example, who is of course familiar with the use of the word 'evidence' in a legal context, never employs it when discussing natural philosophy.[1] Either they have a different concept of evidence from ours, or there is some obstacle to their use of the word.[2]

We need to start by recognizing that we use the word 'evidence' in four different senses. First, 'evidence' can refer to something that is evident. It is evident that 2 + 2 = 4. This is the original meaning of 'evidence', in that it comes directly from the Latin *evidentia*. Because this is etymologically the base meaning of the word, the *Oxford English Dictionary* lists it first, with two examples of its original use in 1665 – despite the fact that it has examples of the use of the word in other senses going back to 1300 (in its earliest sense in English, 'an evidence' is an example to imitate). One of the first examples given is from Robert Boyle: '[T]here are certain Truths, that have in them so much of native Light or Evidence that . . . it cannot be hidden.'[3] The comparison here between what is obvious to the mind and what is obvious to the eye runs through the use of 'evidence' in this sense. A passage from John Locke's *Essay Concerning Humane Understanding* (1690) illustrates this eye-language well:

> Perception of the Mind, being most aptly explained by Words relating to the Sight, we shall best understand what is meant by Clear, and Obscure in our *Ideas*, by reflecting on what we call Clear and Obscure in the Objects of Sight. Light being that which discovers to us visible Objects, we give the name of *Obscure*, to that, which is not placed in a Light, sufficient to discover minutely to us the Figure and Colours, which are observable in it, and which, in a better Light, would be discernable. Thus our *simple Ideas* are *clear*, when they are such as the Objects themselves, from whence they were taken, did, in a well-ordered Sensation or Perception, present them . . .[4]

Although Locke does use the word 'evidence' ('the degrees of its evidence', 'Certainty and Evidence') elsewhere, he much prefers to use the word 'clear', and to write of intuitive knowledge as being 'like the bright Sun-shine, [which] forces it self immediately to be perceived, as soon as ever the Mind turns its view that way'.[5] His discussion of clear and distinct ideas thus follows the example of Descartes, who maintains that only ideas which are clear can be used in argument.[i]

One reason Locke avoids using the word 'evidence' as far as he can is that the word in English has multiple meanings. Thus in 1654 Walter Charleton had offered a translation into English of two Latin phrases with which Gassendi had summarized the epistemology of Epicurus: 'That Opinion is true, to which the Evidence of Sense doth either assent, or not dissent: and that false, to which the evidence of Sense doth either not assent, or dissent.'[6] Since he is translating the Latin word *evidentia,* he ought to mean 'obviousness' or 'evidentness' by 'evidence' here, and his gloss implies that he does: by the 'Assent of the Evidence of Sense, is meant an Assurance that our Apprehension or Judgment of any Object occuring to our sense, is exactly concordant to the reality thereof; or, that the Object is truly such, as we, upon the

i This is classically expressed in the opening paragraphs of the Third of the *Meditations on First Philosophy* (1641). Very often this eye-language implied (and was intended to imply) that nothing should be taken as real unless it could be drawn (and engraved) – which meant that only the mechanical philosophy could be true. Thus Beeckman insisted in 1629: 'In philosophy I allow nothing that is not represented to the imagination as if it were observable.' Berkel, *Isaac Beeckman* (2013), 81, 173–85; Lüthy, 'Where Logical Necessity Turns into Visual Persuasion' (2006).

perception of it by our sense, did judge or opinion it to be'.[7] So the evidence of sense is not, as one might think, the testimony of the senses but our confidence that our senses have properly grasped the object. Charleton's example is a figure walking towards us from a distance: at a certain point it becomes obvious that it is Plato. The *Oxford English Dictionary* is certainly wrong to suggest that the word was not used in this sense before 1665. Here is Thomas Jackson in 1615 carefully using the word in its Latin sense:

> *Evidence*, besides cleerenes or perspicuity (directly and formally included in its prime and native signification) collaterally drawes with it a conceit of such plenary comprehension of the object knowne, as fully satiates our desire of its knowledge: (for evident wee hardly accompt that knowledge which leaves the apprehensive faculty capable of further or better information then it already hath from the particulars which we desire to know) . . .'[8]

Following Jackson, let us call this type of evidence Evidence-Perspicuity.

Second, there is 'evidence' used as a term in English (and only in English) law. Initially (from 1439), English courts considered testimony and evidence – evidence being documents relevant to the case; then (from 1503), 'evidence' became a portmanteau term, covering both testimony and documentary evidence. '*Evidence*, (*Evidentia*),' writes John Cowell in *The Interpreter* (1607), a book explaining legal terminology, 'is used in our lawe generally for any proofe, be it testimonie of men or instrument.'[9] There is no single Latin term for 'evidence' in this legal sense; documents are *instrumenta* and testimony is *testimonium*. Let us name this portmanteau legal sense Evidence-Legal. We find Charleton using 'evidence' in this sense, too, in his *Ternary of Paradoxes*: 'For we will make it our business now, for your information, to call the action of Magnetism to the bar, and by the evidence of Meridian truths, convince the ignorance and stupidity of its adversaries.'[10]

From an even earlier date 'evidence' meant anything that gives one grounds for belief or assent (Evidence-Assent). So Cowell went on to extend his definition of evidence: in a trial, he says, the accused is called on to testify. He 'telleth what he can say: after him likewise all those, who were at the apprehension of the prisoner, or who can give any Indices or tokens, which we call in our language (*Evidence*) against the malefactour'. He is quoting Sir Thomas Smith (d.1577). Smith and Cowell realize that this further sense of 'evidence' is peculiar to English. The Latin for indices and tokens is *signa* or *indicia*; the French is *preuves*. Thus we have a fourth sense of 'evidence', Evidence-Indices. In

1. A Renaissance image of Aristotle: from *The Triumph of St Thomas Aquinas* (1471) by Benozzo Gozzoli. The book Aristotle is holding is his *Metaphysics*; the text, translated, reads: 'A sign of those who know is that they can teach.' Aristotle, with Galen and Ptolemy, was taken to be the basis of all knowledge of the natural world until the Scientific Revolution, and within universities he continued to be the basis of teaching until the end of the seventeenth century.

2. Richard of Wallingford (1292–1336) constructing a mathematical instrument, probably an astrolabe. Richard, an Oxford mathematician and abbot of St Albans, made sophisticated instruments and designed an important clock. He was the closest thing the medieval world had to what we would call a scientist; however, he assumed that mathematics could be employed to interpret the heavens but not the sublunary world, and he had no notion of the experimental method. His face is spotted with what his contemporaries believed to be leprosy.

3. The earth as envisaged in Oresme's manuscript *Du ciel et du monde* (1377). Oresme took the idea of a rotating earth seriously, and thus prefigures Copernicus. The earth floats in space; one quarter of the sphere is habitable, but half of it is covered by water, and the remaining quarter represents terra or aqua incognita. Earth and water together make up what looks like a single globe, although they are in fact distinct spheres with different centres. Consequently, although Oresme can think of this globe as rotating, he cannot allow for the possibility of antipodes (areas of land separated by 180°) except along the equator.

4. The oldest surviving celestial globe, made in Valencia by Ibrâhim 'Ibn Saîd and his son Muhammad in year 478 of the Hegira (1085 of the Christian era). Arabic astronomy was highly sophisticated, and at least the equal of any Western astronomy until Copernicus – indeed, Copernicus may well have used technical solutions devised by Arab astronomers.

5. A late fifteenth-century equatorium and astrolabe. The equatorium (*top*) enables one to calculate the positions of the moon (using one of the inner circles), of Mercury and Venus (using another), and of Mars, Saturn and Jupiter (with the third). This would primarily have been used for astrology.

The astrolabe on the other side (*below*) calculates the position of the sun, tells the time from the height of the sun in the sky (if you know your latitude) or establishes your latitude from the height of the noon sun, shows which stars will be visible at any time and determines the direction of true north. This fine instrument must have belonged to a mathematician who, unless on a journey, would have used it primarily to tell the time. Such instruments, however well constructed, could not make sufficiently exact measurements or calculations to test the limits of Ptolemaic astronomy.

6. Waldseemüller's world map of 1507, the first to include the name 'America', the first to show the New World as, in effect, a new continent, and the first to show antipodes. This map was crucial in destroying the two spheres theory, and in making Copernicanism possible. At the top are the figures of Ptolemy (with a map of the Old World) and Vespucci (with a map of the New).

AMERICI VESPUCII
AQUILO
CECIAS
SUBSOLANUS
VULTURNUS EURUS
EURONOTUS
ASIA
INDIA MERIDIONALIS
CHATAY
OCCEANUS INDICUS MERIDIONALIS
TROPICUS CAPRICORNI
AMERICI VESPUCII ALIORUQUE LUSTRATIONES

PLANISPHÆRIVM
Sive
ORBIVM MVNDI
PTOLEMA
NO DI
PTOLEMAICVM,
Machina
EX HYPOTHESI
ICA IN PLA
SPOSITA.
VIA SOLIS CIRCVLVS
ORBITA IOVIS
CIRCVLVS MARTIS
SEPTEM
PLA NE TARVM
ORBES
SATVRNI
CVRRICVLVM
VENERIS

PLANISPHÆRIVM
Sive
VNIVERSI TO
EX HYPO
COPERNI
PLANO
COPERNICANVM
Systema
TIVS CREATI
THESI
CANA IN
EXHIBITVM
MARS

7. The Ptolemaic, Copernican and Tychonic systems from Andreas Cellarius, *Harmonia macrocosmica* (1660), an atlas of star maps. The Ptolemaic system (*opposite, top*) shows the earth at the centre, with, working outwards, the spheres of air, fire, the moon, Mercury, Venus, the Sun, Mars, Jupiter, Saturn, and the signs of the Zodiac. The Copernican system (*opposite, below*) shows the Sun at the centre, with, working outwards, Mercury, Venus, the Earth and the Moon, Mars, Jupiter with its moons, Saturn, and the fixed stars. The Tychonic system (*above*) shows the Moon and the Sun orbiting the earth, and the planets (with Jupiter's moons shown) orbiting the Sun; but the text suggests that the outer planets orbit not the Sun but the earth – these are in fact two alternative versions of the system. At this date the Ptolemaic system was of historical interest (which is why it has not been updated to show Jupiter's moons), but both the Copernican and Tychonic systems had their advocates.

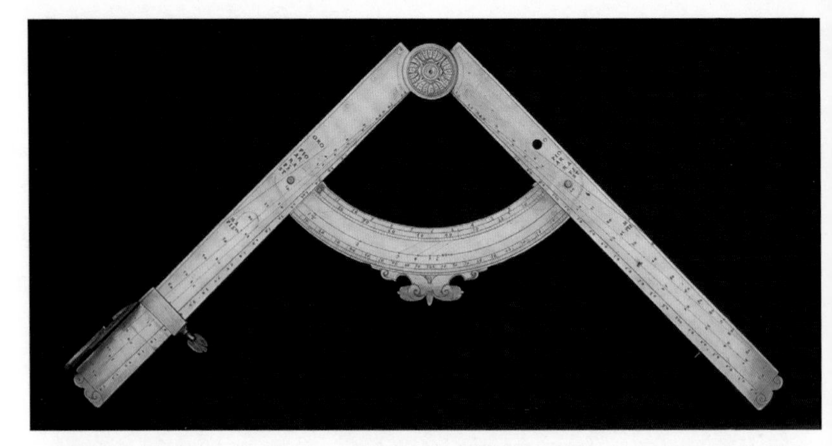

8. Galileo's *Compasso geometrico et militare*. While he was teaching at the University of Padua, Galileo made a significant income from teaching young gentlemen how to use the *compasso* (the proportional, or military, compass is commonly called a sector in English). Instruments of this general sort were common in the early seventeenth century, but Galileo's were made especially for him and were perhaps the most sophisticated. A sighting device could be attached for measuring the elevation of objects at a distance; a plumb line could be attached for measuring the angle of elevation of a cannon's barrel; and the scales on the arms could be used for mathematical calculations such as converting from one currency to another or a volume of wood into board feet. Thus Galileo's sector was a primitive theodolite, slide rule and protractor all in one. Just as the astrolabe embodies the medieval application of mathematics to the heavens, the sector embodies the new technical skills of the mathematician applying himself to the world.

9. This seventeenth-century instrument, known as Galileo's jovilabe, was used to predict the positions of Jupiter's moons. Galileo certainly invented an instrument of this sort, and such instruments were common later in the seventeenth century, when they were used by the Cassinis, Rømer and Halley in their efforts to predict accurately the positions of the moons and so calculate longitude. The jovilabe allowed the modelling of a highly complicated theoretical system without endless calculations or any loss of precision.

10. Giotto, *The Annunciation to St Anne*, 1304, from the Scrovegni Chapel in Padua. The painting depends on the successful creation of a sense of depth, but note that the roof beams do not converge towards a vanishing point, and that there is considerable ambiguity about the spaces: where would you find yourself, for example, if you entered the room by the door? Although Giotto was perfectly capable of producing a geometrically legible space, doing so is not his primary concern. He sees the world qualitatively, not quantitively.

11. Ambrogio Lorenzetti's *Annunciation*, 1344, originally in the town hall of Siena, is a very early example of the geometrical representation of space. The tiled floor establishes a spatial framework, although this is not maintained in the bodies of Mary and the angel or in the architecture. Lorenzetti portrays the moment when Mary becomes pregnant. The angel is saying, 'Nothing is impossible for the word of God.'

12. Masaccio's *Holy Trinity*, 1425, in Santa Maria Novella in Florence, is the earliest rigorous perspective painting to survive, and is evidently dependent on Brunelleschi's studies. An altar originally stood in front of the fresco, marking the transition between the upper and lower portions. The lines Masaccio drew to plot out the composition on geometrical principles are still visible in the plaster.

13. Fra Angelico, *Annunciation*, 1451. Mary is told by the angel that she is about to become pregnant. The door beyond represents both the entry to Mary's womb and the gates of Paradise. In this, 14 and 17, the vanishing point, incomprehensible in Aristotelian terms, is mysteriously obscured, showing how sensitive artists were to the clash between mathematics and philosophy.

14. Piero della Francesca, *Annunciation c.*1470, from the *Polittico di Sant'Antonio*. Piero, who was a mathematician as well as a painter, demonstrates his total command of perspectivist illusion and uses the vanishing point to convey God's incomprehensibility.

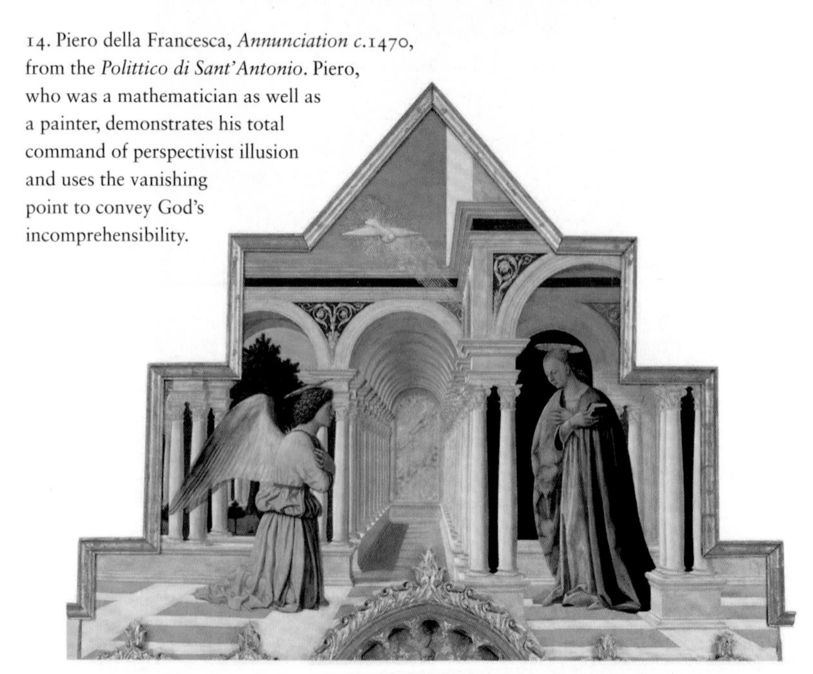

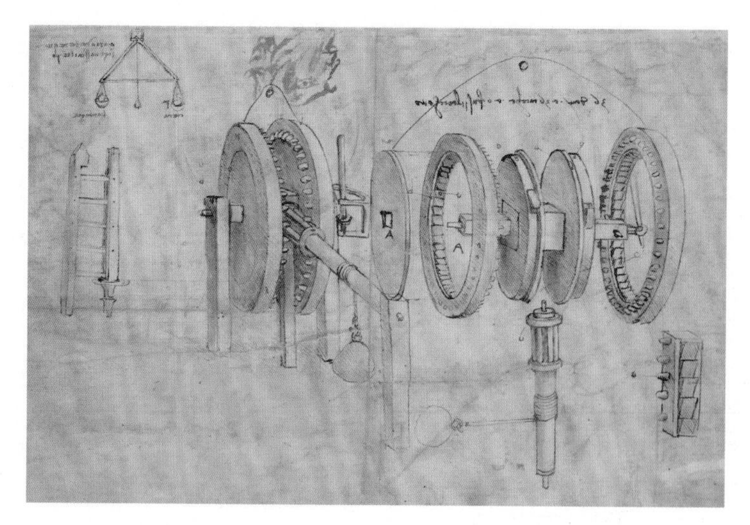

15. Leonardo da Vinci's perspective drawing of a ratchet winch with exploded construction diagram, from the *Codex Atlanticus* (1478–1519), demonstrates the capacity of perspective to transform engineering.

16. Leonardo's 'perspectograph', also from the *Codex Atlanticus*. The artist looks through a small hole at the object he wants to draw or paint. Between the hole and the object is a pane of glass on to which he can draw the outline of the object. The pane of glass thus constitutes the picture plane, and the image drawn on to it can then be copied on to another surface. This may have been Brunelleschi's technique when he produced the first accurate perspective representation.

17. A view of an ideal city, after 1470, variously attributed (perhaps Fra Carenvale or Francesco di Giorgio Martini), and apparently commissioned for the palace of Duke Federico da Montefeltro of Urbino. The whole painting focuses the viewer's attention on the vanishing point, and then playfully hides it behind a half-open door. The full mathematization of perspective is combined with a hint of the impossibility of a purely mathematical world.

18. This portrait of Luca Pacioli, often mistakenly attributed to Jacopo de' Barbari, was apparently painted in 1495 (the slip of paper on the desk gives the date). Pacioli is teaching from Euclid, and a copy of one of his own books on mathematics is in the right foreground. This is how mathematics was taught in the Renaissance and indeed for centuries afterwards.

19. One of several portraits of Kenelm Digby by Anthony van Dyck. Digby, a friend of Hobbes and a founding member of the Royal Society, played a crucial role in popularizing the word 'fact'. The sunflower is a symbol of constancy, and the portrait represents Digby in mourning for his wife, Venetia, who had died suddenly in 1633. The sunflower's ability to follow the Sun could not be explained in Aristotelian terms, and Digby presented it (along with magnetism and the weapon salve) as a paradigmatic example of the problems addressed by the new science of experiment.

English courts Evidence-Indices is considered only in so far as presented via testimony or documents, as part of Evidence-Legal.

Evidence-Indices is what we are referring to when we say that science depends upon evidence. Thus a fingerprint left at the scene of a crime is an Evidence-Index, an indication, or token, that someone in particular has been there. Cowell gives an example in explaining the word 'bankrupt':

> *Bankrupt*, (*alias brankrowte*) commeth of the French (*banque route*) and (*faire banqueroute*) with the French is as much as (*foro cedere, solum vetere*) with the Romanes: The composition of the French word I take to be this, (*banque*, i.[e.] *mensa*) & (*route*, i.[e.] *vestigium*) metaphorically taken from the sign left in the earth, of a table once fastned unto it, & now taken away. So that the originall seemeth to have sprung from those Roman (*mensarii*) which as appeareth by many wrighters, had their (*tabernas et mensas*) in certaine publique places, wherof when they were disposed to flie, & decieve men that had put them in trust with their monies, they left but the signes or carkasses behind them.[11]

You entrust some money to someone who holds a stall in the marketplace. One day you go to the market and where the stall used to be there is only a mark in the earth, a *vestigium*. This is a token, indication, or sign that the stall has gone; the fact that the stall has gone is a sign that your banker has gone out of business; and the fact that your banker has gone out of business implies that you have lost your money.

Charleton uses 'evidence' in this sense, too. He describes how if you draw a circle around a tumour with a sapphire the tumour will die, and argues that the sapphire continues to act on the tumour from a distance (magnetically): '[T]he *place* it self will afford a more certain and satisfactory evidence on the behalf of *Magnetism*; for it grows not black and torrid in the minute of, or by the affriction of the *Saphire*; but many minutes after ... the virulency does successively breathe forth, in obedience to the *Magnetical alliciency* [attractiveness] of the absent *Gem*.'[12]

Arguments of this sort were well known to the ancient Romans. These are arguments which start from things (or from what we call facts). They are discussed in Book 5 of Quintilian's *Institutes of the Orator*, a work which dates to the first century CE. Thus B is found dead, with A's knife in him. This is a sign that A murdered B; unless of course A's knife has been stolen from him, or B attacked A, so that A was acting in self-defence. Such signs therefore cannot be full proofs but only indications; they have to be interpreted in the context. Here is Quintilian. A sign

> is that from which something else is inferred, for example murder from blood. But it may be the blood of a sacrificed animal that has got on to the [suspect's] clothes, or just a nosebleed: a man whose clothes are bloody has not necessarily committed murder. But though this sign is not enough in itself, in combination with others it is taken as equivalent to a witness's statement [*ceteris adiunctum testimonii loco ducitur*] – if the man is an enemy, or has previously made threats, or was in the same place. When the sign is added to these it makes what was only suspected seem certain.[13]

We say that such Evidence-Indices is circumstantial evidence, where the legal understanding of 'circumstantial' essentially means 'contextual'. The word 'circumstantial' goes back to Quintilian, who is the only Latin author to use *circumstantia* to refer not to a spatial distribution (the sheep standing around the shepherd) but a problematic deduction which is context-dependent. Quintilian's example of an appeal to circumstance is a made-up one. Imagine there is a law which says that a high priest can pardon one person from the death penalty, and another law which says that if one person committing adultery is put to death then their partner must be put to death as well. The high priest is taken in adultery and condemned to death. 'No problem,' he says; 'I will pardon myself.' 'Not at all,' comes the reply, 'because if you pardon yourself your partner cannot be executed, and so you will have pardoned two people, and that you cannot do. So you must die.' In the specific context of adultery the high priest's right of pardon cannot be used.[14]

From Quintilian's Latin, 'circumstance' and 'circumstantial' enter English to refer to an incomplete (but not insignificant) argument for belief, one that needs to be interpreted in context. Here is the Jesuit Robert Parsons in 1590:

> For albeit the Apostle S. Paul declareth, the things which we beleeve, be not such in themselues, as may be made apparent by reason of humaine arguments: yet such is the goodnes and most sweete proceeding of our mercifull God toward us, as he wil not leave himselfe without sufficient testimonie, both inward and outward, as the same Apostle in another place doth testifie. For that inwardlie, hee testifieth the trueth of such things as we beleeve, by giving us light & understanding, with internall joy and consolation in beleeving them. And outwardly, hee giveth testimonie to the same, with so many conveniences, probabilities, and arguments of credibilitie, (as Divines

> doe call them,) that albeit the very poynt of that which is beleeved, remaine styll with some obscuritie: yet are there so many circumstaunces of likelyhoods, to induce a man to the beleefe thereof, as in all reason it may seeme against reason to denie or mistrust them.[15]

Probabilities, arguments of credibility, circumstances of likelihoods – these all derive from Evidence-Indices. They are not in themselves 'testimony', equivalent to words spoken or texts written; but when we find them in the Bible, as it were orchestrated by God, they become the equivalent of testimony. It is the Bible and the tradition of the Church which provides the textual testimony; the circumstances contextualize the testimony and accompany it. Inward light, understanding and joy, similarly, function as if they were testimony: to Parsons, it is as if they speak to the truth of our beliefs.

Between these four types of evidence (Perspicuity, Legal, Assent, Indices) there was and is plenty of scope for confusion, and for collapsing one into the others. John Wilkins in his posthumous *Principles and Duties of Natural Religion* (1675) has an extended discussion of evidence in the broadest possible sense of the grounds for belief (Evidence-Assent); consequently, he includes under the heading of 'evidence' sensation, demonstration [i.e. deduction], testimony and experience – the first two being evidence *of* and the second two evidence *for*.[16] Evidence *for* a belief (Evidence-Legal, Evidence-Indices) could be better or worse, stronger or weaker, producing different 'degrees of assent' or 'degrees of Veracity, Certainty, or Credibilitie'.[17] Similarly, the evidence *of* a belief (Evidence-Perspicuity) could be greater or less, clearer or more obscure, producing different degrees of knowledge. For Locke there are 'three degrees of Knowledge, viz. Intuitive, Demonstrative, and Sensitive: in each of which, there are different degrees and ways of Evidence and Certainty'.[18] In Locke's view Evidence-Indices – and he does use the word 'evidence' in this sense when discussing probabilities – is not a form of knowledge (which is confined to intuition, demonstration and sensation: all types of evidence *of*) but of opinion, which is measured in terms of 'degrees of assent'. Nevertheless, some Evidence-Indices can be regarded 'as if it were certain knowledge'.[19]

Obviously, different types of expert work with different types of knowledge. Mathematicians deal with demonstrative knowledge, or Evidence-Perspicuity. Aristotelian philosophers thought that all true knowledge could be expressed in syllogistic form, arguing from indisputable premises to undeniable conclusions, all based on Evidence-Perspicuity.

On the other hand, lawyers were concerned with Evidence-Legal and Evidence-Indices, and so were theologians. Since 1400 theologians had been discussing what they called 'moral certainty' – evidence good enough to rely on, even if there is a great deal at stake. Thus I can be morally certain that there is a city called Rome, even if I have never been there. There is testimony, there are documents, there are maps, there are photographs, and there is plenty of other evidence to confirm the existence of a place called 'Rome'. It is utterly implausible that all of this evidence could have been faked, so I am quite certain that Rome exists. But my certainty is not of the same sort as my knowledge that the square on the hypotenuse is equal to the sum of the squares of the other two sides, which can be rigorously proved; the evidence for Rome is an argument from experience, and so an argument from probability.[ii] A standard argument was that a Christian needed moral certainty of the truths of their faith, since the fate of their soul was at stake.

Some theologians were not content with moral certainty: in 1689 the Presbyterian propagandist Richard Baxter discussed the concept of evidence at great length, and decided that the only sort of evidence that counted was Evidence-Perspicuity. One couldn't rely, he maintained, on experience in any of its many forms. 'Even our Experimental Philosophers and Physicians find, that an experiment that hits ofttimes, quite misseth afterwards on other Subjects, and they know not why. A course of effects may oft come from unknown causes.'[20] True faith required certainty, and certainty required self-evidence, or Evidence-Perspicuity. (Baxter's efforts to prove that the Bible was self-evidently true need not detain us.)

It should be apparent at this point that one way of characterizing the Scientific Revolution is as the displacement of Evidence-Perspicuity by Evidence-Indices, as people learning to trust circumstantial or probable

ii The term 'moral certainty' was invented by Jean Gerson (1363–1429) to describe the degree of certainty required to take certain sorts of decisions without the agent being morally at fault if they prove to be wrong (Schüssler, 'Jean Gerson, Moral Certainty' (2009)). Thus a judge condemning a prisoner to death needs to be morally certain that they are indeed guilty ('beyond reasonable doubt', we would now say): but the most conscientious judge might occasionally be misled by lying witnesses without being morally at fault. Moral certainty is thus quite different from the sort of absolute certainty one might have in mathematics or in logic. Since what is often at stake in moral choices is confidence that one has got the facts right, the term was extended to other contexts in which people make judgements about questions of fact. Thus I have moral certainty that Caesar crossed the Rubicon, and moral certainty that Pascal accurately reported the Puy-de-Dôme experiment. For the later history of the concept, see Leeuwen, *The Problem of Certainty* (1963).

evidence in place of intuitive or demonstrative evidence. Thus in the Torricellian experiment you cannot see the pressure of the air, but the height of the column of mercury is an indicator of that invisible pressure. When you look at the moon through a telescope you cannot see mountains, but the ragged terminator indicates that mountains are there. When Galileo saw the moons of Jupiter he could not see that they were moons, but their patterns of movement indicated that they were in orbit around Jupiter. In each case, what you can see points beyond itself to something you can reliably infer. The mathematicians began to handle evidence in the way that lawyers and theologians had been handling it for many centuries.

§ 2

So far we have been looking at evidence mainly through English sources. In English law the jury was the judge of matters of fact: its task was to weigh the evidence and decide if the defendant was guilty or not.[21] There were no set rules on how to reach that decision. The 'beyond reasonable doubt' rule was formulated only in the eighteenth century, and the whole point of this rule is that juries have to decide for themselves what it means. Juries were thus free to reach decisions on the basis of circumstantial evidence if they chose to. As a character in a play of 1616, entitled, with obvious sarcasm, *The Honest Lawyer*, puts it:

> In case of Murder should we never judge
> By circumstanciall likelihoods and presumptions,
> No life could be secure.[22]

In Roman law jurisdictions the situation was, as we have seen, very different. There were clear rules on how to handle evidence.[23] It was the judge who collected the evidence, applied the rules and reached a verdict. A verdict of guilty in a capital case required a *complete proof* of guilt, for example the testimony of two witnesses who had seen the crime occur, or a confession.[iii] When Quintilian says that circumstantial evidence can take the place of a witness, later lawyers took him as authorizing it to be considered as half of a complete proof. In a capital case, where there was less than a complete proof, the standard procedure was to use torture in order to obtain a confession, but torture could

iii The distinction between complete proofs and half proofs goes back to the Glossators of the 1190s: Franklin, *The Science of Conjecture* (2001), 18–19.

be employed only if there were substantial grounds for suspicion amounting to half a proof. Where in English we generally use the word 'proof' to mean a demonstration (a mathematical proof), so that the idea of half a proof makes no sense, judges across the Continent and in Scotland accumulated proofs until they either had a complete proof or sufficient evidence to justify torture. Gossip, for example, represented one seventy-second of a proof.[iv]

So the French, following their Latin sources, used the world *preuve*, or 'proof', not the word 'evidence'; but they saw proofs as things that could be accumulated in exactly the same way as evidence can be accumulated until it amounts to a proof beyond doubt. But where in English we can speak of 'the evidence' as a totality – 'the evidence of his guilt was overwhelming' – in French it would be necessary to use a plural, *les preuves*. The standard French translation of 'evidence-based medicine' is *médecine fondée sur les faits*. The French were not alone in having the same word for Evidence-Indices as for proof; this was the case in all the other modern European languages apart from English and Portuguese (where *evidencia* is frequently found in the plural, as it was in eighteenth-century English).[24]

§ 3

Thus there is (or there appears to be) a fundamental continuity between the way in which Quintilian talks about signs and tokens and the way in which seventeenth-century Englishmen and women talk about Evidence-Indices and seventeenth-century continental Europeans talk about proofs in law. The claim has been made, however, that there was no concept of evidence until about 1660 (which is why I have concentrated on examples of Evidence-Indices taken from before this date).[25] The claim is based on distinguishing three types of evidence: the evidence of witnesses; the evidence of the senses; and the evidence of (for want of a better term) clues.[26] Ian Hacking distinguishes the last two from each other by quoting J. L. Austin:

> The situation in which I would properly be said to have evidence for the

iv Torture became rarer in the seventeenth century, when it became commonplace to sentence criminals in cases where the proof was incomplete to a term of service in the galleys; well-grounded suspicion could thus lead to conviction but not to execution. Circumstantial evidence on its own could now be given greater weight.

> statement that some animal is a pig is that, for example, in which the beast itself is not actually on view, but I can see plenty of pig-like marks on the ground outside its retreat. If I find a few buckets of pig food, that's a bit more evidence, and the noises and smell may provide better evidence still. But if the animal then emerges and stands there plainly in view, there is no longer any question of collecting evidence; its coming into view doesn't provide me with more evidence that it's a pig. I can now just see that it is.[27]

This concept of evidence (Evidence-Indices, it would seem) was missing in the Renaissance, so the claim goes. Instead, they had the concept of signs.[28]

This claim involves a series of errors. First, it confuses 'signs' (that is, indices, vestiges or tokens) with 'signatures'.[29] According to the Renaissance theory of signatures, some natural objects had their significance apparent in their form. Thus a kidney-shaped bean might be good for treating a disease of the kidneys. This doctrine, held by Platonists and Paracelsians, is quite different from the theory of signs (also known as indices, vestiges or tokens). Second, it is claimed that the doctrine of signs/signatures belongs to 'low' disciplines such as medicine and alchemy; no mention is made of the law or theology. Third, it is asserted that signs were 'read' as if they were texts and, consequently, no distinction was drawn between the evidence of clues and the evidence of witnesses. This is the point at which this argument becomes interesting, for, as noted earlier, Quintilian does take signs to be equivalent to testimonies, and so does Parsons. For them, testimony is the paradigm form of evidence (Evidence-Legal) to which Evidence-Indices is expected to conform.

Nevertheless, Quintilian carefully distinguishes between what he terms 'technical' and 'non-technical' proofs.[30] Non-technical proofs are things like documents, witnesses, confessions extracted under torture: they speak for themselves. Technical proofs have to be constructed by the advocate. He notes, as we have seen, that some signs virtually speak for themselves – bloodstained clothing, a cry – but others depend heavily on interpretation. *Indicia*, *vestigia* and *signa* (tokens, vestiges and signs) thus do not provide the same sort of evidence as documents and witnesses, even though they can be used to do the same work as a document or a witness. The claim that signs are read as texts is held to follow from the Renaissance belief that the universe is a book, the book of nature, but this is to place the theory of signs in the wrong context. The theory of signs originates in the law, and signs are treated as if they speak because court cases are discursive performances. It is the prosecutor's job to turn blood on the suspect's clothes into the equivalent of a witness against them – he must make the blood speak.

Pierre Gassendi, in his *Syntagma philosophicum* (1656), elaborated the classical doctrine of signs into a sophisticated theory of knowledge.[31] He identifies two types of sign. There are those that allow us to know something that we could have known by direct sensory experience if we had been present at the right time. These are vestiges: the banker's table, or the pig's trotter, or the criminal's finger leave a vestige, a track or a trace. So, too, we would say, the craters we see on the moon are vestiges of past collisions with asteroids. A sign 'leads us to the knowledge of something hidden in the way that tracks [*vestigium*] are a sort of sign indicating to a dog which way he should pursue the chase in order to catch the quarry'.[32] By 'vestige' Gassendi means Evidence-Index. On the other hand, he says, there are signs that point to something that we can never see. Thus we cannot see the pores in the skin, but the passage of sweat through the skin proves they must be there. We can never see the legs of the scabies mite, but we can tell from the fact that it moves that it must have legs or something similar. In fact, Gassendi points out, when the microscope was invented both the pores in the skin and the legs of the mite became visible, thus confirming the validity of earlier arguments to prove their existence. Such arguments depended on analogies: comparing the skin, for example, to porous earthenware. This concept of argument from analogy had been taken up by the Epicureans from medicine, but it was also familiar to lawyers. Thus Quintilian assumes the lawyer will appeal to what we might call stereotypes: 'It is easier to believe brigandage of a man, poisoning of a woman.' These are arguments by analogy, based on the circumstances of the case.[33]

Gassendi is certainly right to emphasize the importance of arguments by analogy. Robert Boyle, when, in 1660, he wanted to explain the new doctrine of the elasticity of the air, compared the air to sheep's wool, which could be compressed but which would bounce back if the pressure was removed: this analogy, he felt, made plausible the notion of elasticity. Torricelli compared the weight and pressure of the air to the weight and pressure of water. The distinction between two very different types of inference, between vestiges and analogies, is still important for Locke. Nobody disputed the reliability of vestiges (no scar without a wound, for example), but analogies were obviously much more problematic. Clearly, if much of our knowledge is analogical in origin it cannot be certain, and the real causes of events may always escape us.

Gassendi saw no need to make an explicit distinction between the evidence of vestiges and the evidence of witnesses, but he certainly did not think that the scar 'testifies' to the wound.[34] He had a perfectly good concept of evidence as distinct from testimony. However, if you mistakenly assume that the key problem lay in distinguishing clues from testimony, then you might conclude that it was only in *The Logic of Port-Royal* that Arnauld finally differentiated the two, for we are told that he drew a distinction between 'internal evidence' and 'external evidence'. The internal evidence is the evidence of clues (his knife was in the victim); the external evidence is the evidence of witnesses (his wife says he never left her side). Except, of course, Arnauld didn't use the word 'evidence', as he was writing French, not English. The word he uses is *circonstances*. For example: 'In order to judge of the truth of some event and to decide whether or not to believe in its occurrence, the event need not be considered in isolation – such as a proposition of geometry would be; rather all the circumstances of the event, both internal [the clues] and external [the testimony], should be considered.'[35]

'Circumstances', we have seen, is Quintilian's coinage. Quintilian also separates the evidence of signs or clues from the evidence of witnesses; indeed, as in *The Logic of Port-Royal*, the evidence of signs is 'internal' and the evidence of witnesses is 'external'. In the *Port-Royal* logic it is a bit hard to grasp what the evidence of signs is internal to. The knife may be in the body, but exactly what is the fingerprint *inside*? In Parsons, the distinction, as we have seen, is quite simple: internal evidence consists of my feelings, which are inside me. In Quintilian it is a little more complicated: witnesses and documents come to the lawyer from outside; technical proofs are constructed by the lawyer himself, so that they are formed within the discipline of oratory. Technical proofs are the lawyer's own contribution to the evidence. Arnauld is not copying Quintilian, but he is reworking him in order to go beyond him.[v]

Is *The Logic of Port-Royal* (1662) the first text to draw a clear distinction between testimony and Evidence-Indices? As we have seen, in earlier texts it is certainly assumed that the one can stand in for the other, and sometimes (as in Parsons) the two seem to be conflated. Nevertheless, here is Richard Hooker (d.1600): '[T]hings are made

v Quintilian is a fundamental reference point for Arnauld: 'Quintilian and all the other rhetoricians, Aristotle and all the philosophers . . .' (Arnauld & Nicole, *La Logique* (1970), 294). Quintilian may seem an obscure figure to us; but not to anyone educated when rhetoric was still a core part of the curriculum.

credible, eyther by the knowne condition and qualitie of the utterer, or by the manifest likelihood of truth which they have in themselves.'[36] This is exactly the distinction between internal ('in themselves') evidence and external (testimony) evidence made in *The Logic of Port-Royal*. Hooker implicitly recognizes that both types of evidence have to be considered in the light of the circumstances. The accused has blood on his clothes, but he is a butcher; the testimony is explicit, but the witness is unreliable. The fact that the one can substitute for the other does not mean that Hooker thinks they are one and the same thing; they are clearly different in character, one depending on the regularities of nature, the other on the veracity of people. Another example: in 1648, well before the publication of *The Logic of Port-Royal*, Wilkins was already arguing that Archimedes' discoveries (such as his famous burning mirror, with which he destroyed a fleet of ships) might seem too astonishing to be true ('such strange exploits, as ... would scarce seeme credible even to these more learned ages'; i.e. if we can't do them, how could he?), 'were they not related by so many and such judicious Authours', above all Polybius, who was either an eyewitness or at least had the opportunity to speak to eyewitnesses.[37] Here, internal evidence is set against external evidence, as it must often have been in a court of law. (The suspect had blood on his clothes, but his wife says he never left her side.)

Thus the claim that there is a new concept of evidence in the 1660s is mistaken; wherever we look what we find is nothing but the reworking of Quintilian's distinctions. What *is* new is the transfer of concepts from one discipline to another. Evidence-Indices had been the business of lawyers and theologians; in 1660 it became the business of the Royal Society. 'Moral certainty' had been the language of theologians; in 1662 we find it being used by the first statisticians, Graunt and Petty.[38] Just as facts moved out of the court room into the laboratory, so evidence made the same move at around the same time; and, as part of the same process of constructing a new type of knowledge, moral certainty moved from theology into the sciences. When it comes to evidence, the new science was not inventing new concepts but recycling existing ones.

§ 4

The search for a new concept of evidence in the mid-seventeenth century is bound to be fruitless because there was a classic context in which the question of the reliability of inferences from facts had been debated for centuries. That context was the discussion of Ptolemaic epicycles. According to the Aristotelian philosophers, epicycles ought not really to exist: all movement in the heavens ought to be circular movement around the centre of the universe. For them epicycles were useful fictions which made computation of the position of the planets possible. The mathematicians, however, interpreted the apparently irregular movement of the planets in the heavens as evidence that some invisible reality was causing the movement. They took their observations of planetary movement to be good evidence for the reality of epicycles. Here is Clavius outlining the mathematicians' view:

> [J]ust as in natural philosophy we arrive at knowledge of causes through their effects, so too in astronomy, which has to do with heavenly bodies very far away from us, we must attain to knowledge of them, how they are arranged and constituted through our senses . . . It is therefore suitable and very highly rational that, from the particular motions of the planets and the various appearances, astronomers should search out the number of particular circles that carry the planets round with such varying motions, and their arrangement and shapes . . . But our opponents try to weaken this argument, saying that they concede that all the phenomena can be saved by postulating eccentric circles and epicycles, but that it does not follow from this that the said circles are found in nature; on the contrary they are wholly fictitious; for perhaps all the appearances can be saved in a more suitable way, though it is not yet known to us . . . But by the assumption of eccentric and epicyclic circles not only are all the appearances already known preserved, but also future phenomena are predicted, the time of which is altogether unknown. Thus if I am in doubt whether, for example, the full moon will be eclipsed in January 1582, I shall be assured, by calculation from the motions of the eccentric and epicyclic circles, that the eclipse will occur, so that I shall doubt no further . . . But it is not credible that we should force the heavens (but we seem to force them if eccentrics and epicycles are fictions, as our adversaries would have it) to obey our fictions and move as we wish or as agrees with our principles.[39]

Clavius's argument here is identical with that of modern realists who claim that science must approximate to the truth, otherwise it would not be able to make successful predictions. Robert Boyle, on the other hand, sided with the philosophers and presented a line of argument that goes back to Averroes and forward to the modern pragmatists and instrumentalists:

> [A]s confidently as many Atomists, and other Naturalists, presume to know the true and genuine Causes of the Things they attempt to explicate, yet very often the utmost they can attain to in their Explications, is, That the explicated *Phaenomena* May be produc'd after such a Manner as they deliver, but not that they really Are so: For as an Artificer can set all the Wheels of a Clock a going, as well with Springs as with Weights, and may with violence discharge a Bullet out of the Barrel of a Gun, not onely by means of Gunpowder, but of compress'd Air, and even of a Spring. So the same Effects may be produc'd by divers Causes different from one another; and it will oftentimes be very difficult, if not impossible for our dim Reasons to discern surely which of those several ways, whereby it is possible for Nature to produce the same *Phaenomena* she has really made use of to exhibit them.[40]

From the outside, a clock powered by a battery and a clock powered by clockwork look the same: the fact that the hands go round does not tell you what sort of mechanism is driving them. In scholastic terms this was a debate over the reliability of *a posteriori* reasoning; in our terms it is a debate over the evidence of things, or over Evidence-Indices. The debate was not new in the second half of the seventeenth century, and neither were the arguments. Here is the humanist and philosopher Alessandro Piccolomini in 1558:

> [S]uppose we should see a stone strike a wall and with great force, and not knowing the origin of such fury we should imagine that the stone had come from a bow or a crossbow. And suppose that our theory was false and that as chance would have it the stone had come from a sling shot. Nevertheless it would have struck the wall with the same fury if it had come from the imagined bow. For the aforesaid fury of that stone could have derived from more than one cause. Similarly when we see many appearances in the planets in the sky, though the causes from which such appearances truly arise are hidden from us, nevertheless it is enough for us that, supposing these theories to be true, those appearances would derive from them just as we see them. This for us is more than enough for the calculations, predictions, and information which we need to have of the positions, places, magnitudes and motions of the planets.[41]

If this debate between realists and instrumentalists seems to echo our own debates about the nature of scientific knowledge of invisible entities such as the electron, this is because what is being put to work here is the very same concept of evidence that we use. All that is missing from these discussions is the word on which we place so much emphasis, and for which they felt no need: 'evidence'. Their vocabulary of appearances, predictions and causes is perfectly adequate for the task.

§ 5

From the 1640s, with the triumph of experiment, the sort of evidence that had been good enough for the lawyers, the doctors and the astronomers – the evidence of clues, or facts – began to be good enough for mathematicians such as Pascal when doing physics. *A posteriori* reasoning, reasoning backwards from appearances to causes, began to push to one side the *a priori* reasoning of the geometers and the scholastic philosophers, who maintained that the only reliable form of reasoning was reasoning forwards from definitions to consequences. Just as the astronomers acknowledged that, in principle, different hypotheses might do the job equally well (Clavius had no doubt that the Copernican system produced reliable predictions, but he felt sure the Earth was stationary, not moving), so many people continued to argue that there were a number of perfectly good competing explanations of the Torricellian experiment. Pascal disagreed.

It was this new sort of knowledge that Sprat had in mind when he defended the Royal Society from its critics, insisting that the starting point of the new knowledge, experimental evidence, was extremely reliable. Indeed, Sprat uses (exceptionally for a seventeenth-century author) the very word 'evidence' as we would use it:

> [T]here is not any one thing, which is now approv'd and practis'd in the World, that is confirm'd by stronger evidence, than this, which the Society requires; except onely the Holy Mysteries of our Religion. In almost all other matters of Belief, of Opinion, or of Science; the assurance, whereby men are guided, is nothing near so firm, as this. And I dare appeal to all sober men; whether, seeing in all Countreys, that are govern'd by Laws, they expect no more, than the consent of two, or three witnesses, in matters of life, and estate; they will not think, they are fairly dealt withall, in what concerns their Knowledg, if they have the concurring Testimonies of threescore or an hundred?[42]

But for the most part the new scientists avoided the word 'evidence' because it inevitably carried with it an implicit reference to the law courts – a reference that Sprat was willing to make explicit. Thus in

1660 Boyle describes one experiment 'as a plausible, though not demonstrative proof, that Water may be transmuted into Air'.[43] Here, he uses 'proof' as if he were writing *preuve* in French, where we would use 'evidence' in the sense of Evidence-Indices. Nowadays, in English, we can no more have a merely plausible proof than we can have a false fact; but Boyle is using 'proof' differently from us.

Above all, Boyle, like any seventeenth-century scientist, realizes that most of his readers will be mathematicians. And he feels he must:

> excuse my selfe to Mathematicall Readers. For some of them, I fear, will not like that I should offer for Proofs such Physical Experiments, as do not alwayes demonstrate the things, they would evince, with a Mathematical certainty and accuratenesse; and much less will they approve, that I should annex such Experiments to confirm the Explications, as if Suppositions and Schemes, well reason'd on, were not sufficient to convince any rational man about matters Hydrostaticall.[44]

In other words, he feels he must apologize for appealing to Evidence-Indices in a field where mathematical demonstration (Evidence-Perspicuity) seemed possible. This aspiration towards demonstration was not confined to what we would think of as the empirical sciences but was also commonplace in theology. Thus in 1593 the mathematician John Napier presented his interpretation of the Book of Revelation as being 'in form of proposition, as neer the analytick or demonstrative manner as the phrase and nature of holy Scripture will permit'.[45]

Mathematics in the Renaissance looked in two directions. Aristotle had distinguished between geometry and arithmetic (which dealt with purely theoretical entities) and optics, harmonics and astronomy (which dealt with physical realities). Bacon provided labels for this distinction: 'pure mathematics' and 'mixed mathematics'. (Bacon extends the list of types of mixed mathematics to include perspective, engineering, architecture, cosmography 'and divers others'.)[46] Pure mathematics deals in proofs and demonstrations; mixed in phenomena. But pure mathematics had a higher status, with the result that there was a constant hankering after the language and argumentative style of pure mathematics.

Galileo, for example, liked to keep as close to geometry as he could. He measured the length of the shadows on the moon, and used geometry to demonstrate how high the mountains that caused them were; he used geometry to prove (rather elegantly) that the spots on the sun must be on or near the surface of the sun.[47] But these proofs involved arguing from things (shadows and shapes) to the applicability of geometrical theorems. Indeed, they started as analogies. Galileo thought the patches of light and dark along the moon's terminator looked like a mountain range seen from above when the sun was rising – so like that he argued that it could only be exactly that. The spots on the sun reminded him of clouds; he knew they were not clouds, but they bore the same relationship to the sun's surface as clouds do to the Earth's. The new science often presented itself as a system of axioms and demonstrations, in Galileo's *Two New Sciences* for example, or Newton's *Principia*, but it was always grounded in facts and, less surely, in analogies.

This helps to explain the rarity of the word 'evidence' (which carried with it implications of disagreement and contention, of court-room conflict) in these seventeenth-century texts. It occurs three times (once as a verb) in Sprat's *History of the Royal-Society*. In the *Opticks* (1704), the greatest triumph of the new experimental science, Newton uses it only once. In the early volumes of the *Philosophical Transactions* it appears only once or twice a year. Even in a work as late as Desaguliers' *Course of Experimental Philosophy* (1734–44) the word appears only twice in two volumes. As we have seen, after 1660 the new scientists talked endlessly about 'facts' (though Newton avoided the term as not suitable for a mathematician), 'experience', 'experiments', 'hypotheses', 'theories' and 'laws of nature'. When they used the word 'evidence' it was usually casually and inadvertently, and often because (as in the example quoted earlier from Sprat) they wanted to imply a comparison with law and/or theology.

If there was one term that summed up the new science for those who were involved in constructing it, it was not 'evidence' but 'experience'. Pascal went further than simply insisting on the authority of experience; he argued that our knowledge of nature is capable of endless progress *because* it is founded in experience, and experience accumulates over time.[48] The emphasis on experience was thus tied to the ideas of progress and discovery. Of course, there was something deeply problematic about relying on the measurement of the height of mercury in a tube to substantiate a theoretical understanding of nature: the measurement is a particular event, conducted with particular equipment, on a particular day, under particular circumstances, while the theory has to be universally applicable. Early experimentalists such as Galileo tried to play down this problem by reporting an experiment in general terms as having been conducted over and over again, but from Pascal on the experiment becomes a local event, described as such. Such narratives do not play down the epistemological problem of moving from the specific to the universal; they accentuate it.[49] One way round this problem is to devise a series of different experiments which capture the phenomenon from different perspectives: Pascal was keen to go beyond the Torricellian experiment and devise new ones precisely in order to bridge this gap.

The correspondence between theories and facts was so crucial that Galileo and Newton were prepared to bend the facts to fit the theories, even while insisting on the priority of experience over theory. Mersenne, at any rate, was convinced that Galileo's falling-body experiments could not be precisely replicated, and we know that Newton fiddled the figures to create a perfect fit between his theoretical physical principles and the measured speed of sound.[50] We would say that Galileo's and Newton's sciences were always empirical (at least in aspiration), but that would be to use the word in a nineteenth-century sense – in the seventeenth century empirics were thought of as people who were untrained in reasoning, not people who grounded theories in evidence.[vi] Gassendi and Locke never thought of themselves as founding an empirical philosophy, although we would say that is what they were doing.

Just as the new scientists aspired to mathematical demonstration wherever possible, and claimed an overly precise fit between theory and fact when they needed to, so, in English, they avoided the word 'evidence', which was inevitably associated with the law, as far as they could. The result is that where we can identify the Typhoid Mary, or the index case, for the languages of facts (Chapter 7), of laws of nature (Chapter 9) and of hypotheses and theories (Chapter 10), there is no index case for the language of evidence – or, if there is, it lies outside science.

For the language of Evidence-Indices (as distinct from the concept) was taking off towards the end of the seventeenth century not in science but in works of natural theology, such as John Wilkins' *Of the Principles and Duties of Natural Religion* (1672; 75 occurrences) and Matthew Hale's *The Primitive Origination of Mankind* (1677; 280 occurrences, but then Hale was Chief Justice). It became commonplace in philosophy: it occurs forty-eight times in Hume's *Treatise* (1739–40). The long-term fortune of the word has a great deal to do

vi Thus Newton to Hawes, 25 May 1694: 'Experience is necessary, but yet there is the same difference between a mere practical Mechanick and a rational one, as between a mere practical Surveyer or Guager and a good Geometer, or between an Empirick in Physick [i.e. medicine] and a learned and a rational Physitian.' Newton & Cotes, *Correspondence* (1850), 284; see also Bacon, *Novum organum* Vol. 1, xcv (Bacon, *Works* (1857), Vol. 1, 201).

with William Paley's *Natural Theology; or, Evidences of the Existence and Attributes of the Deity* (1802). Slowly but surely we can see the language of Evidence-Indices moving from law, theology and philosophy into the sciences, but the language was far behind the concept, for experimentation was nothing other than an appeal to Evidence-Indices.

§ 6

It would be wrong, however, to concentrate solely on the word 'evidence' rather than the concept which it expresses, for if we do so we will miss a crucial development. For Locke and everyone before him, knowledge equals truth. Defeasible knowledge, knowledge that might be corrected in the light of new observations, was merely opinion or probability. Moral certainty had been introduced as 'hard', reliable opinion, but the whole point about morally certain knowledge was that you could commit yourself to it without reserve; it would, in practice, never be falsified, even if it could not be demonstrated as mathematical theorems could be demonstrated. The concept of evidence (meaning relevant experience) had been introduced into science from the law. In English law, once the jury had given a verdict there could be an appeal on a question of law but there could be no appeal on a question of fact. There was, until 1907, no procedure for introducing new evidence.[51] The jury thus had to be certain of their verdict. Moral certainty, grounded in probability, had, in the writings of Wilkins and Locke, joined deductive or self-evident knowledge as a form of truth. This simply conformed to the practice of the courts.

Locke had discussed a case where what looked like reliable knowledge turned out to be mistaken: that of the King of Siam, who was not willing to believe the testimony of the Dutch ambassador, who assured him that in the Netherlands water could become so hard that an elephant could walk on it. But Locke never proposed such a case as a paradigm example of how empirical knowledge (as we call it) works when it is working well. He formulated the general principle that 'as the conformity of our Knowledge, as the certainty of Observations, as the frequency and constancy of Experience, and the number and credibility of Testimonies, do more or less agree, or disagree with it, so is any Proposition in it self, more or less probable'.[52] This means that what is probable changes over time; but Locke never nerved himself up to take the further step of saying that knowledge itself changes over time. On the one hand, there is what he calls here 'our Knowledge', which is changeable. On the other, there is Knowledge, which is Truth.

The *Oxford English Dictionary* usefully distinguishes between 'knowledge' meaning 'the fact of knowing or being acquainted with a thing, person, etc.' and 'knowledge' meaning 'justified true belief' (Latin: *scientia*). 'Our Knowledge', in Locke's phrase, which must conform to observation, experience and testimony, is knowledge as acquaintance, but Locke still hankers over justified true belief. In that respect he still thinks like Hobbes, who had written:

> This taking of Signes by *experience*, is that wherein men do ordinarily think, the difference stands between man and man in *Wisdom*, by which they commonly understand a mans whole ability or *power cognitive*; but this is an *errour*: for the Signes are but *conjectural*; and according as they have often or seldom failed, so their *assurance* is more or less; but *never full* and *evident*: for though a man have always seen the day and night to follow one another hitherto; yet can he not thence conclude they shall do so, or that they have done so eternally: *Experience concludeth nothing universally.* If the Signes hit twenty times for one missing, a Man may lay a Wager of Twenty to One of the event; but may not conclude it for a Truth.[53]

A radical shift from the position of Hobbes (who clearly states Hume's problem of induction and rejects Evidence-Indices as no basis for certainty), and even from that of Locke (who may be described as clearly defining the choice between Evidence-Perspicuity and Evidence-Indices but then fudging it) is to be found in William Wotton's *Reflections upon Ancient and Modern Learning* (1694):

> The new Philosophers, as they are commonly called, avoid making general Conclusions, till they have collected a great Number of Experiments or Observations upon the Thing in hand; and, as new Light comes in the old Hypotheses, fall without any Noise or Stir. So that the Inferences that are made from any Enquiries into Natural Things, though perhaps set down in general Terms, yet are (as it were by Consent) received with this Tacit Reserve, *As far as the Experiments or Observations already made, will warrant.*[54]

Wotton's 'Tacit Reserve', which is the principle that all scientific reasoning is defeasible, is of fundamental importance.[vii] It transforms science

vii Compare Robert Boyle: 'I would have such kind of superstructures [i.e. intellectual systems] look'd upon only as temporary ones, which though they may be preferr'd before from knowledge of the truth, regarded as indisputable, into a form of progressive knowledge in which established truths may always be disputed and in which an ultimate truth is never attained. Where did Wotton get the concept of tacit reserve from? The phrase itself comes from moral philosophy, where, traditionally, every promise was accompanied by an unspoken reservation, '*If I Can; If I Ought;* or *if things Continue in the same State:* So that by the Change of Circumstances, I am discharg'd of my Obligation.'[55] But the principle that intellectual systems are merely temporary constructions which may need later to be revised and improved comes directly from the mathematical language of hypotheses and theories which Wotton invokes when he writes of 'the old Hypotheses' which 'fall without any Noise or Stir'.

Now, with Wotton's formulation of a tacit reserve, moral certainty was required to give way to a new sort of temporary knowledge, a purely provisional understanding. This tacit reserve had never been so clearly stated before. Locke's King of Siam case is one where a tacit reserve would have been appropriate, but Locke had not formulated one. What Wotton understands is that scientists can agree ('as it were by Consent'), for the moment, to treat acquaintance knowledge, or experience, as true belief, and that this is not a mistake but rather the way in which you change an ideal, unchanging knowledge-as-truth, a final unappealable verdict, into a peculiar, progressive form of knowledge, knowledge-as-incomplete-acquaintance. Induction is always imperfect, the evidence is always incomplete, but it can be good enough to be going on with. Wotton's formulation of the notion that in science all knowledge is accompanied by a tacit reserve gives him a claim to be the first to have an adequate understanding of the conceptual underpinnings of modern science; or, if you prefer, to be the first both to understand modern science and to acknowledge its limitations. It is only at this point that the modern theory of evidence becomes complete (although, of course, Wotton writes of experiences and observations and does not use the word 'evidence'). So, when we finally come to Wotton's formulation of a tacit reserve, we encounter, for the first time, a sound comprehension of the nature of scientific knowledge.[56]

---

any others, as being the least imperfect, or, if you please, the best in their kind that we yet have, yet are they not entirely to be acquiesced in, as absolutely perfect, or uncapable of improving Alterations.' Boyle, *Certain Physiological Essays* (1661), 9 = Boyle, *The Works* (1999), Vol. 2, 14.

§ 7

Nevertheless, searching for the concept of evidence pays off in unexpected ways. Indeed, it brings us to a discovery which is strange and unexpected. For when scientists began to make assessments regarding the reliability of evidence they were required to exercise what they all called 'judgement' (for example, Locke: 'Knowledge being to be had only of visible certain Truth, *Errour* is not a Fault of our Knowledge, but a Mistake of our Judgment giving Assent to that, which is not true').[57] And exercising judgement requires a specific set of virtues, the virtues you would hope to find in a jury of your peers: impartiality, assiduity, sincerity. We find these virtues everywhere in discussions of Evidence-Indices, while they are largely irrelevant in discussions of Evidence-Perspicuity. Here, for example, is a theologian in 1677 explaining the difference between certainty and faith, between Evidence-Perspicuity and Evidence-Indices:

> A Mathematical Demonstration brings so strong a Light that the Mind cannot suspend its assent, but is presently overcome by the naked propounding of the Object: And hence it is that in Mathematical matters, there are neither Infidels nor Hereticks. But the motives of Faith are such, that although the Object be most certain, yet the Evidence is not so clear and irresistible, as that which flows from Sense, or a Demonstration. And 'tis the excellent observation of Grotius, God has wisely appointed this way of perswading Men the truth of the Gospel, that Faith might be accepted as an act of Obedience from the reasonable Creature. For the Arguments to induce belief, though of sufficient certainty, yet do not so constrain the mind to give its assent, but there is prudence and choice in it.[58]

Mathematicians don't need prudence; Christians, lawyers and scientists do. So Sprat describes his ideal philosopher: 'The True Philosophy must be first of all begun, on a scrupulous, and severe examination of particulars: from them, there may be some general Rules, with great caution drawn'; 'Let us then imagin our Philosopher, to have all slowness of belief, and rigor of Trial, which by some is miscall'd a blindness of mind, and hardness of heart.'[59] The scientist must be slow, scrupulous, severe, rigorous. Here is Wilkins: 'The word Moderation is a Quality, a Habit, an Affection of intellectual virtue, whereby we are concerned for any truth according to a due measure, not more or less then the evidence and importance of it doth require, to which the notion of fierceness or fanaticalness is opposed as the deficient extreme.'[60] The scientist must be moderate: this is a new type of intellectual virtue. Of its novelty Wilkins is quite clear: we have no choice, he says, but to adopt the view which we think is best supported by the evidence:

> yet must it withal be granted to be a particular virtue and felicity to keep the mind in such an equal frame of judging. There are some men, who have sufficient abilities to discern betwixt the true difference of things; but what through their vicious affections and voluntary prejudices, making them unwilling that some things should be true; what through their inadvertency or neglect to consider and compare things together, they are not to be convinced by plain Arguments; not through any insufficiency in the evidence, but by reason of some defect or corruption in the *faculty* that should judg of it. Now the neglect of keeping our minds in such an equal frame, the not applying of our thoughts to consider of such matters of moment, as do highly concern a man to be rightly informed in, must needs be a vice. And though none of the Philosophers (that I know of) do reckon this kind of Faith (as it may be styled), this teachableness and equality of mind in considering and judging of matters of importance, among other intellectual virtues; yet to me it seems, that it may justly challenge a place among them.[61]

Impartiality is also now an intellectual virtue. Here is Locke:

> Nevertheless, I do not yet Question, but that Humane Knowledge, under the present Circumstances of our Beings and Constitutions, may be carried much farther than it hitherto has been, if Men would sincerely, and with freedom of Mind, employ all that Industry and labour of Thought, in improving the means of discovering Truth, which they do for the colouring or support of Falshood.[62]

Sincerity and industry are intellectual virtues, too.

So as knowledge ceases to be a question of Evidence-Perspicuity and becomes a question of Evidence-Indices a whole new set of intellectual virtues are demanded of the knower. Eventually, with the concept of a tacit reserve, a principle from moral philosophy is introduced directly into epistemology, establishing a limit to knowledge claims. We might want to say that these virtues and this limit can be summed up in the word 'objectivity'; but objectivity is a nineteenth-century concept and implies new ways of observing nature and recording information.[63] It would be wrong to read it back into the Scientific Revolution; before the precision instrumentation of the Industrial Revolution impartiality and judgement were virtues, not ways of re-describing professional competence.

Discovery implies individualism and competition. Scientists have to be buccaneering and entrepreneurial. But, as Robert K. Merton kept pointing out, science isn't only about individual success. Scientists are required by the culture of their profession to declare their allegiance to a quite different set of values, which he summarized as communism (knowledge is shared: often renamed 'communalism'; we have seen the first community of experimental scientists emerging in 1640s France), universalism (knowledge must be impersonal and unbiased), disinterestedness (scientists must help each other) and organized scepticism (ideas must be tested and tested again).[64] This set of values is sometimes referred to by the acronym CUDOS. Thus every scientist is subject to two competing and conflicting imperatives: they are obliged to be both competitive and cooperative. Scientists are required to be Janus-faced, both self-effacing and self-assertive, and Merton saw his task as a sociologist of science as being to work out how scientists negotiated this conflict, which he regarded as being constitutive of science as a social enterprise.[viii]

How did this conflict come about? The answer is very simple. It is the result of combining discovery with the moral virtues associated with Evidence-Indices. The result is a structural conflict in the nature of science, a conflict whose origins are historical. You won't find Copernicus, or Kepler, or Galileo praising moderation, impartiality, industry; but then Copernicus, Kepler and Galileo are primarily mathematicians. The generation after Galileo had to acknowledge its dependence on Evidence-Indices and so, willy-nilly, it had to adopt the virtues of the judiciary.

A great deal of play has been made with the idea that the Royal Society was committed to the pursuit of facts, to impartiality, to moderation, because it was founded in the immediate context of the Restoration.[65] For twenty years people had been killing each other in the name of the truth; now they must learn to manage their disagreements in a different way. The new science has to be placed in this local context. I don't deny that there is truth in this, but it hardly explains why Merton found the same virtues being admired by scientists in the 1940s as were admired

viii Some people have tried to rewrite CUDOS by making the O stand for originality, but that misses the point – CUDOS represents one set of interlocking values, originality or discovery a quite different and conflicting set.

by scientists in the 1660s. The new virtues go deeper than the immediate context of the Restoration.

Where else do we find a similar conflict between competition and cooperation? In the legal profession. The adversarial system means that lawyers want to win, and the better they are at winning, the more they get paid. At the same time, every lawyer is an officer of the court. They are bound by a code of professional standards. They must never lie on behalf of a client. They must never withhold evidence from the other side. They must be simultaneously competitive and cooperative. What happened when Evidence-Indices moved from the court room to the laboratory was that the contradictory characteristics of any evidence-based legal system (there are other legal systems – trial by ordeal, for example – which don't have these characteristics) were imported into science, and scientists became divided against themselves just as lawyers had always been within adversarial legal systems, right back to Quintilian, who is looking all the time both for good arguments and for winning arguments, knowing full well that the two aren't always the same thing.

With the Renaissance revival of Stoicism the word 'philosophical' acquired a new meaning: philosophers, it was held, were able to moderate their passions and to be unmoved by the blows of fortune.[ix] They could stand back from their immediate experience and contemplate the bigger picture. Wilkins, Sprat and Locke are looking for a quite different sort of person, one who exemplifies CUDOS. This chapter began with the search for a new type of evidence; it ends with the Janus face of a new type of intellectual, one produced by philosophers being required to engage with an old type of evidence: circumstantial evidence.

§ 8

There is a further step in this argument. In 1976 Thomas Kuhn published an article entitled 'Mathematical versus Experimental Traditions in the Development of Physical Science'.[66] In England, Kuhn claimed, experimental science flourished from the late seventeenth century onwards, science conducted in the tradition of Bacon. On the continent, a much more deductive style of science was preferred, science conducted

ix The earliest date given by the *OED* for 'philosophical' in this sense is 1638, but it is already used in this way by Montaigne (e.g. *Essais* (1580), Book 2, Ch. 7), and this usage is taken over in Florio's translation of 1603.

in the style of Descartes. The English were preoccupied with facts, the French (for it is primarily the French he has in mind) with theories. As Kuhn states it, this argument, which sets the experimentalists against the mathematicians, seems wrong. The English had Halley and Newton, who were simultaneously experimentalists and mathematicians. The French had Pascal, the Cassinis and Huygens (the Cassinis and Huygens not being French by origin, but by choice), ditto.[x] It was a Frenchman, Claude Bernard, not an Englishman, who wrote *An Introduction to the Study of Experimental Medicine* (1865).

But we might restate Kuhn's argument in a different way. The English had the common law, which relied on the jury system. The jury system allowed a significant role to circumstantial evidence, as long as it was introduced in testimony. It gave prosecutors plenty of scope to argue from analogy. When the scientists reorganized science around Evidence-Indices they imported into it the virtues of the jury system, at least in an idealized form: a willingness to listen to both sides, a desire to mix proof with persuasion, an appeal to common sense. (Newton, of course, is a great exception here.) The French, by contrast, had a Roman law system. Their new science was organized around the virtues of a *juge d'instruction*: intellectual rigour, a set of formalized procedures, a quest for a complete proof, a confidence that one need only answer to other professionals. If there are two different scientific traditions, as Kuhn claimed, perhaps they should be seen as reflecting the importation into science of two different legal traditions, of two different ways of handling Evidence-Indices, rather than as embodying a conflict between mathematicians and experimentalists. Perhaps it is therefore wrong to say that one can safely translate the English 'evidence' with the French *preuve*, for the two terms reflect two different and incommensurable forensic cultures, trial by jury and trial by inquisition. French Evidence-Legal was different from English Evidence-Legal, and so French Evidence-Indices has always been different from English Evidence-Indices.[xi]

A friend of mine was once in hospital in Paris. The doctors told him that they had an hypothesis regarding the nature of his illness which

x In 1756 d'Alembert celebrated what he took to be the triumph of the new experimental physics in France in the '*Expérimentale*' article of the *Encyclopédie*.

xi Strangely, Cassin, Rendall & Apter (eds.), *Dictionary of Untranslatables* (2014) (itself a translation from French), contains no discussion of evidence/proof: on this showing at least, it appears the French are unaware that they lack the term 'evidence'. The discussion of experience/experiment claims that the English vocabulary fosters empiricism, while the French vocabulary is at odds with it; this would also seem to be true in this case.

they intended to prove, where in England they would have told him that he had certain symptoms which suggested a diagnosis which they would run tests to confirm. Two cultures: one emphasizes the difference between Evidence-Indices and Evidence-Perspicuity; the other minimizes it. But, nevertheless, they have one common enterprise, that of transmuting signs and symptoms into knowledge.

What I have offered here is what Boyle would call a plausible proof – a good argument (I hope) that falls short of being conclusive. What I hope to have shown is that science modelled its new concern with Evidence-Indices on legal processes, and for as long as it could it sought to downplay the extent to which Evidence-Indices differed in character from Evidence-Perspicuity. Even David Hume, in his essay 'Of Miracles' (1748), still slipped and slid between the two senses of the word.[xii]

I also want to stress two things that it is easy for us to overlook, so used are we to handling Evidence-Indices and finding them convincing. First, there have been plenty of systems of knowledge that have rejected Evidence-Indices and relied on something else – on geometrical proof, for example, or on signatures, or on the identification of concealed meaning (astrology, for instance). Evidence-Indices may always have been used in an unthinking way by people going about their daily business; but to elevate them into being a reliable basis for theoretical knowledge, as happened in England in the mid-seventeenth century, was to make a claim that was culturally peculiar and far from obviously true.

Second, it is not obvious even now that relying on Evidence-Indices was bound to be a winning strategy. The whole point of Hume's formulation of the problem of induction is that we cannot explain *why* induction tends to work quite well, why nature seems extremely regular in its proceedings (or at least seems regular to us, who have been trained to look for regularities). Even if one intended to rely on Evidence-Indices,

xii 'All the Objects of human Reason or Enquiry may naturally be divided into two kinds, *viz. Relations of Ideas*, and *Matters of Fact*. Of the first Kind are the Propositions in Geometry, Algebra, and Arithmetic; and in short, every Proposition which is either intuitively or demonstratively certain ... Tho' there never were a true Circle or Triangle in Nature, the Propositions, demonstrated by *Euclid*, would for ever retain all their Truth and Evidence. 'Matters of Fact, which are the second Objects of human Reason, are not ascertain'd to us in the same Manner; nor is our Evidence of their Truth, however great, of a like Nature with the foregoing. The contrary of every Matter of Fact is still possible; because it can never imply a Contradiction, and is conceiv'd by the Mind with equal Distinctness and Facility, as if ever so conformable to Truth and Reality.' (Hume, *Philosophical Essays* (1748), 47–8.)

how can one tell what constitutes a good argument? Doctors treat a medicine as being of demonstrated effectiveness if the results of a trial are better than anything one would obtain by chance nineteen times out of twenty; nuclear physicists say there is evidence for something if there is no more than a one in 741 chance of a false positive result – they regard something as proved if the odds are 3.5 million to one. The early scientists did not even know how to perform a test for statistical significance.

The point about evidence is not that it was a natural type of argument to rely on, nor that it was a sort of argument that was obviously bound to be successful; the point is that relying on evidence just happened to work rather well. As Evidence-Indices replaced Evidence-Perspicuity the new scientists were able to claim more and more successes (the Puy-de-Dôme experiment, Boyle's Law, Newton's new thory of light), and these successes further boosted the attractions of Evidence-Indices. The intellectual apparatus of the new science – facts, experiments, theories, laws of nature, evidence – did not establish its worth by philosophical arguments; its success depended upon the fact that, in practice, it produced good results. There may be inhabited worlds in which no culture ever becomes evidence-based; and there could, for all we know, be universes in which looking for evidence simply doesn't pay off – in which the sceptics not only win the arguments but have the facts on their side. Reality and the new science happened, in certain areas of physics, to dovetail quite neatly together. This was, in the end, good luck. Locke doubted whether our sensory capacities were sufficient to allow us to develop an adequate knowledge of corporeal substances.[67] It turns out he was wrong, but he could easily have been right.

PART FOUR

Birth of the Modern

Natural Philosophy is therefore but young.
– Thomas Hobbes, *Elements of Philosophy*, 1656

Part Four addresses two very different consequences of the Scientific Revolution. The first and last chapters explore the scientific origins of the Industrial Revolution, which turn out to be earlier and closer than previously suspected. The middle chapter looks at belief in supernatural agency – witches, demons, poltergeists. Initially, key figures engaged in the new science hoped that it would help prove the reality of supernatural activity; after the publication of Newton's *Principia* (1687) the result was quite the opposite: the new science seemed to legitimize a new scepticism.

12
Machines

> The Renaissance view of nature as a machine . . . is based on the human experience of designing and constructing machines. The Greeks and Romans were not machine-users, except to a very small extent: their catapults and water-clocks were not a prominent enough feature of their life to affect the way in which they conceived the relation between themselves and the world. But by the sixteenth century the Industrial Revolution was well on the way. The printing-press and the windmill, the lever, the pump, and the pulley, the clock and the wheel-barrow, and a host of machines in use among miners and engineers were established features of daily life. Everyone understood the nature of a machine, and the experience of making and using such things had become part of the general consciousness of European man. It was an easy step to the proposition: as a clockmaker or mill-wright is to a clock or mill, so is God to Nature.
>
> – R. G. Collingwood, *The Idea of Nature* (1945)[1]

§ 1

The great philosopher and archaeologist R. G. Collingwood proposed a fairly straightforward technological determinism: new machines encourage new ways of thinking. There are two problems with his argument. The first is that the only machine that he lists that was new in the Renaissance was the printing press. The Middle Ages, it has long been maintained, saw a technological revolution, with the invention of the clock, the widespread diffusion of the water mill and the wheelbarrow, and the development of the various hoists and windlasses required to build the cathedrals; the view of nature as a machine should have come not in the sixteenth century but the fourteenth.[2] The second problem is even more fundamental: although Collingwood devotes a long chapter to the Renaissance idea of nature, and returns to the claim that the Renaissance viewed nature as a machine, he never gives a single example of someone describing nature as being like a machine. Collingwood was so sure that the Renaissance thought of nature as a machine that he failed to notice that he had produced no evidence to support his claim.

With this cautionary tale in mind, let us begin by asking a question which seems too obvious to need asking but is in fact an essential preliminary: What is a machine? First, in conceptual terms at least, there are the 'simple machines'. Archimedes studied three elementary tools that could be used to move weights: the lever, the pulley and the screw. Hero of Alexandria (10–70 CE) added to these the windlass and the wedge, and in the late sixteenth century Simon Stevin included the inclined plane. All these simple machines provide a mechanical advantage in moving a load. The modern science of mechanics was crystallized in Galileo's *On Mechanics* (1600 in manuscript; first published by Mersenne in 1634).[3] Galileo was the first to demonstrate that the work performed by a machine could never be greater than the work put into it, so that machines can never trick nature into doing something that breaks its normal rules. (Thus a lever enables a light weight to lift a heavier weight, but the light weight travels further than the heavier weight, so the work done on each side of the fulcrum is the same.) Galileo thus established a new equivalence between natural processes and artificial ones. Because Galileo thought about machines in this narrow, technical way, he never says that the universe is a machine or that all natural processes can be understood in mechanical terms; neither does he ever compare the universe to a clock, which he certainly could have done had he wanted to.

What Galileo did discuss was atomism. The atomism of Democritus, Epicurus and Lucretius implied that the universe is made up of building blocks that function through their size, shape and solidity. As Democritus put it, 'By convention sweet, by convention bitter, by convention hot, by convention cold, by convention colour: but in reality atoms and void.'[4] In a world of atoms and the void all natural processes result from the ways in which atoms jostle together. In 1618, as a result of a conversation with Isaac Beeckman, the young Descartes came up with an alternative to ancient atomism: where the ancients had thought of atoms bumping into each other in empty space, Descartes rejected the possibility of empty space and thought in terms of corpuscles filling all the available space, as water fills the ocean. The next year Descartes formulated his famous doctrine *cogito ergo sum*, 'I think therefore I am'; consequently, there is something, one thing, I know for certain. On this secure foundation he set out to build a new philosophy to replace that of Aristotle, and he began to publish elements of his new system in 1637. Since the publication of a book-length article by Marie Boas in 1952 it has become customary to refer to these two alternatives to the Aristotelian theory of forms and qualities – the atomic philosophy of the ancients (revived by Galileo, Gassendi and others) and the corpuscular philosophy of Descartes – together, as 'the mechanical philosophy'.[5] The term was certainly widely used in the late seventeenth century, but it is more misleading than helpful.

Descartes, who was taken to be the founder of the mechanical philosophy, never described himself in print as a mechanical philosopher; he says that all mechanical laws are physical or natural laws (which Galileo had shown), but not that all laws of nature are mechanical laws: he does not describe nature as a mechanical system. He does use the term 'mechanical philosophy' once in a letter (in 1637), where he refers to 'the rather greasy and mechanical philosophy' – in other words, the sort of philosophy a cart-maker would have. He was replying to a critic who had described his philosophy as 'coarse and rather greasy' and 'excessively gross and mechanical' – that is, too physical (as we might put it) to count as a philosophy at all. He wrote, 'If my philosophy seems to him excessively gross because it considers shapes, sizes, and motions, as happens in mechanics, he is condemning what I think deserves praise above all else, and in which I take particular pride.'[6] (Leonardo, too, had decided that being a mechanical philosopher, in this sense, should be a matter of pride.)[7]

The term 'mechanical philosophy' was coined by Henry More (a Cambridge don and life-long admirer of Plato) in 1659, after Descartes' death, in the course of an attack on Cartesianism, which he had once enthusiastically supported.[8] More wanted to defend the idea that spirit and purpose are active in nature and reject the Cartesian claim that natural processes are soulless, that matter is passive and that everything that happens (leaving aside the free choices of God, angels and men) happens of necessity. Outside England the term was slow to catch on: the first reference to it in Latin is in Samuel Parker's *Disputationes* of 1678, and in French in Pierre Bayle's *Nouvelles de la république des lettres* (1687).[9] In English there was an alternative: Robert Boyle invented the term 'the corpuscularian philosophy' in 1662 to cover both ancient atomism and Descartes' new corpuscular theory.[10] 'The corpuscular philosophy' and 'the mechanical philosophy' are thus two competing terms for exactly the same thing: indeed, the first occurrence of either term in French is in a reference to *la philosophie mécanique ou corpusculaire* (1687; when Boyle was translated two years later the phrase used was '*la philosophie des corpuscules*').[11]

This is how Walter Charleton, writing in 1654, summarized what would soon be called the mechanical philosophy. Everything he says could have been said by Descartes:

> Consider we, that the *General Laws* of Nature, whereby she produceth All Effects, by the Action of one and Passion of another thing, as may be collected from sundry of our praecedent Discertations, are these: (1.) That every Effect must have its Cause; (2) That no Cause can act but by Motion; (3) That Nothing can act upon a Distant subject, or upon such whereunto it is not actually Praesent, either by it self, or by some instrument, and that either Conjunct, or Transmitted; and consequently, that no body can move another, but by contact Mediate, or Immediate, *i.e.* by the mediation of some continued Organ, and that a Corporeal one too, or by it self alone.

Having set up his definition, Charleton goes on to attack the traditional concepts of sympathy and antipathy, and to argue that they must be reconceptualized in mechanical terms:

> Which considered, it will be very hard not to allowe it necessary, that when two things are said either to *Attract* and *Embrace* one the other by mutual *Sympathy*, or to *Repell* and *Avoid* one the other, by mutual *Antipathy*; this is performed by the same wayes and means, whereby we observe one Body to Attract and hold fast another, or one Body to Repell and Avoid conjunction with another, in all Sensible and Mechanique Operations. This small Difference only allowed, that in Gross and *Mechanique* operations, the Attraction, or Repulsion is performed by *Sensible* Instruments: but, in those finer performances of Nature, called *Sympathies* and *Antipathies*, the Attraction or Repulsion is made by Subtle and *Insensible*.

This means that he now knows in principle how sympathy and antipathy work:

> The means used in every common and Sensible Attraction and Complection of one Bodie by another, every man observes to be Hooks, Lines, or some such intermediate Instrument continued from the Attrahent to the Attracted; and in every Repulsion or Disjunction of one Bodie from another, there is used some Pole, Lever, or other Organ intercedent, or

> somewhat exploded or discharged from the Impellent to the Impulsed. Why therefore should we not conceive, that in every Curious and Insensible Attraction of one bodie by another, Nature makes use of certain slender Hooks, Lines, Chains, or the like intercedent Instruments, continued from the Attrahent to the Attracted, and likewise that in every Secret Repulsion or Sejunction [pushing apart], she useth certain small Goads, Poles, Levers, or the like protruding Instruments, continued from the Repellent to the Repulsed bodie? Because, albeit those Her Instruments be invisible and imperceptible; yet are we not therefore to conclude, that there are none such at all.[12]

This mechanical philosophy, as described by Charleton, would have made perfect sense to Lucretius, but he would have found the label itself deeply puzzling, because the Romans did not think of poles and hooks as machines, any more than we do – these are the simple machines of the mathematicians. But the Roman idea of a machine was neither Charleton's nor ours. The key source for knowledge of Roman machinery is Vitruvius's *On Architecture*, which describes machines used in construction and for warfare. When Vitruvius wrote about a machine (Latin: *machina*) he used the word to means something quite different from what we mean. Scaffolding is a machine. A scaling ladder is a machine. A tower on wheels built to enable you to approach the enemy's walls and scale them is a machine. A platform on which spectators stand is a machine. Roman machines do not necessarily do work in the sense of moving things, nor do they necessarily have moving parts. Their common characteristic is that they are substantial structures designed to be stable. Thus a hoist with a pulley is a machine, but what makes it a machine seems to be the fact that it is solidly supported. A trebuchet is a machine, but what makes it a machine is not that it throws large rocks but that it is made out of great solid baulks of timbers lashed together. The nearest synonym to *machina* was *fabrica*, a word that can often be translated as 'structure'. When Lucretius talks about the machine of the world (*machina mundi*) he does so in the context of discussing the dissolution of our universe. When our world ends its structure will come apart. The machine of the world is thus the stable structure of our universe: the heavens, the earth and the four elements. All these will disappear when the universe comes to an end and a new one is born.[13]

The phrase *machina mundi* is echoed by Tertullian (160–225) and Augustine (354–430), and thus appears throughout medieval philosophy (in Sacrobosco, for example), even though the text of Lucretius had been lost and was not rediscovered until 1417,[i] but it does not imply an interlocking set of moving parts, a geared system or a power train. To translate it as 'machine of the world' is to mistranslate. The best translation into English is perhaps a phrase of John Wilkins's of 1675 which is surely meant to be the English equivalent: 'this visible frame which we call the World'.[14]

§ 2

With time, of course, the original meaning of Lucretius's phrase was lost; as machines changed, the meaning of Lucretius's phrase changed with them. Crucial here is the clock. One of the primary purposes of the first clocks was to model the movement of the heavens, not just to tell the time. Thus in 1364 (about sixty years after the invention of the escapement mechanism which made the mechanical clock possible) Giovanni de' Dondi constructed in Padua an *astrarium* (a star-arium, or planetarium) which showed the time, the movements of the sun, moon and other planets, and the religious feast days. Part of his purpose was to prove that the Ptolemaic system was an accurate representation of how the heavens really work and not just a mathematical model.[15] It was natural therefore to claim that, since a clock models the heavens, the heavens are like a clock. So far as we know, this claim was first made by Oresme in 1377, seven years after the erection of a clock on the Palais Royal in Paris: the movement of the spheres, he said, was perhaps like 'a man making a clock and letting it go and be moved by itself'.[16] Implicitly, he was saying that God might be rather like a clockmaker. Collingwood is right: '[I]t was an easy step to the proposition: as a clockmaker . . . is to a clock . . . so is God to Nature.' But no medieval author compared the universe to anything so coarse as a mill; and Oresme's comparison was very carefully limited: he was comparing the circular movement of the heavens to the turning wheels of a clock, not the whole universe to a clock; he did not think of clocks as machines and he does not use the clock metaphor to prove the existence of God. Oresme had no intention of expounding a mechanical philosophy, for

i According to Tertullian, Dionysius, when he witnessed the miraculous eclipse of the sun which coincided with the crucifixion, exclaimed, '*Aut deus naturae patitur, aut dissolvitur machina mundi*' ('God suffers, or the world breaks into pieces'), and the phrase then appears in the Latin breviary; Augustine: *moles et machina mundi* ('the vast bulk of the world').

he lived in a world of Platonic and Aristotelian forms; indeed, he ended up accepting the conventional view that the heavenly spheres were governed by spiritual intelligences.

Around 1550, however, commentators on Vitruvius (writing in Latin) began to express dissatisfaction with his account of what a machine is.[17] They wanted to include water-wheels and clocks among the machines (for the first time) and give prominence to powered machinery. (The Greeks and Romans had very few water-wheels and no clocks, hence their lack of interest in powered machinery.) Thus the modern idea of the machine was born by giving a new meaning to the Latin term *machina*. This new understanding of what a machine is meant that automata – that is to say, devices that move of themselves (including clocks) – were now for the first time classified as machines.

Clocks had long had little statues that came out of the clock or moved to mark the time: the bell indicating the hour was often struck by the figure of a man with a hammer: a *jacquemart*, or, in English a 'jack'. Sometimes statues of the Virgin Mary and child appeared and the three kings paraded by them; or mechanical heralds emerged and blew on trumpets. The clock inside Strasbourg Cathedral (first built in 1352–4) was the most famous example of such an elaborated clock. On its top, for example, stood a gilded cock that flapped its wings, opened its mouth, stuck its tongue out and crowed to mark midday.[18] The modern cuckoo clock is a simplified version of these 'automatic' mechanisms. One of the most sophisticated was the repeater mechanism invented in 1676: if you pulled on a string the clock would chime the last set of hours and quarters: in other words, the clock would answer you if you asked it the time (a capacity that was primarily intended for use in the dark).

A particularly important example of the new concept of the machine is provided by *The Relations of Moving Forces* by the French Protestant Salomon de Caus (1615).[19] De Caus was interested only in moving mechanisms, whether they were driven by air pressure (he devised a primitive steam engine), by flowing water or by descending weights. He invented, for example, a player organ: an organ which, like a player piano, automatically played music according to the information conveyed by pins on a turning drum. He constructed elaborate fountains, with grottos containing mechanical singing birds. But he also described machines for pumping water, sawing wood and performing other industrial tasks. In making ornate fountains and singing birds he was following classical precedents; but the Greeks and Romans had nothing to say about powered hammers and saws, about automata that performed mechanical tasks beyond the strength of a human being.

De Caus is important because when Descartes wrote about machines it is particularly his machines that he had in mind, and it is de Caus who transmits the new terminology from Latin into French, and through Descartes into English.[20] Before they read Descartes in 1637 the English had called complex machines 'engines', not 'machines'.[21] (The word 'engine' comes from the Latin *ingenium*, meaning intelligence, from which we derive 'ingenious'; in English the word meant 'cunning' before it came to be used for a cunning device.) So when More invented the term 'mechanical philosophy' he was playing his part in importing the new terminology into English.[22] 'Engine' and 'machine' still have overlapping meanings as a consequence of Descartes' influence on the English language.

Descartes himself designed, and perhaps built, automata: he designed an improvement to the mechanism of clocks but also a tightrope walker, powered by magnets, and an installation in which a dog springs at a partridge, which flies away. There was even a story that he had made himself a woman; convinced that this living machine must be inhabited by a devil, she was thrown overboard by a ship's captain when his ship, on which Descartes was a passenger, was caught in a storm.[23]

Descartes' remarkable and novel claim, first stated in the *Discourse on Method*, was that animals are automata, that is, complex, self-moving machines. They appear to have some extra quality that we call 'life', or 'intelligence' but are in fact simply performing predetermined routines, like the cock on the Strasbourg Cathedral clock. The soul, Descartes claimed, is unique to rational human beings; animals have no soul and no capacity to reason. (The Aristotelians had distinguished three different types of soul – vegetative, animal and rational – and so had no problem in acknowledging that animals had a sort of soul.) When Descartes describes something in nature as a machine it is biological entities which he always has in mind. He denied that animals were designed; but they do move (like de Caus's machines), and they do reproduce themselves, so, once they have come into existence, their complex structures do not have to be put together again from scratch.

If animals are machines and nothing but machines, then for Cartesians it must follow that the human body, which is evidently similar to the body of an ape, works like a machine, and Cartesian doctors were eager to study human anatomy as an example not of a geared mechanical system but of a hydraulic system of the sort that powered de Caus's fountains and player organs. If the human body is a machine it must have a power source, which is perhaps why Descartes wanted to think of the heart as a heat engine rather than as a pump (de Caus had powered fountains by heating water by shining sunlight through lenses); to describe it as a pump simply begs the question of what powers the pump. But, of course, once animals have been claimed to be machines, it is a small step to arguing that human beings are also machines, and so to the adoption of a systematic materialism of a sort which would have been anathema to Gassendi, Descartes, Boyle and Newton. Julien Offray de La Mettrie's *Man the Machine* (1748) is a logical development of this sort of uncompromising mechanistic thinking.[24] The challenge set by Descartes was, of course, that of building an automaton that could behave like an animal. A hundred years later Jacques de Vaucanson (1709–82) made a mechanical duck which could walk, quack, eat and defecate.[25]

Descartes does not think of the universe as being like a clock because in his view outer space is filled not with the crystal spheres of Ptolemaic astronomy, nor with the gears and levers of de Caus's machines, but with liquid vortices which carry the planets in their orbits around the stars.[26] However, he does say that understanding the universe is comparable to the problem of understanding a clock. If you look at a clock or a watch from the outside you can tell that there is a mechanism which turns the hands. You may conclude that the hands are driven around by a descending weight. But they might equally be driven by a spring (or be regulated by a pendulum – but Descartes died before the invention of the pendulum clock). You can only tell exactly what is going on if you can take the clock apart.[27] Descartes thinks that much of our understanding of nature is like this: we can come up with a convincing account of how things might work, but we cannot know for sure that that is how they really work because the mechanism is invisible to us; hidden not because it is hidden inside a box but because it is too small to see. The initial hope was that the microscope might make the invisible visible, and it did, for example, when it showed how a fly can walk up a pane of glass. But it could not show the mechanism which causes light to reflect or refract, or the particles that cause smells.[28] Descartes thought that sometimes experiments could be constructed to enable a secure choice between possibilities (thus you can experiment with a spherical bottle filled with water and show how a rainbow is produced), but this might not always be possible: in Descartes' view, Pascal's vacuum experiments did not serve to eliminate the possibility of a

AVEC PERMISSION
DU MAGISTRAT DE LA VILLE,

On expoſera a la vüe du Publique les 3. chefs d'Oeuvres Mechaniques du Celébre Monſieur VAUCANSON, *Mémbre de l'Academie Royale des Sciences de Paris*, *qui conſiſtent en trois Figures Automates.*

SCAVOIR:

LA premiére, Un homme de Grandeur naturelle habillè en SAUVAGE qui jouë Onze airs ſur la Flûte traverſiére par les mémes mouvements des Levres des doits & le ſouffle de ſa bouche comme l'homme vivant.

LA ſeconde, un homme auſſi de Grandeur naturelle, habillé en BERGER PROVENCAL qui jouë 20. airs differens ſur le Flûtet de Provence d'une main & du Tambourin de l'autre avec toute la preçiſion & perfection de même qu'un habile joueur.

LA troiſiéme un CANARD artifiçiel en Cuivre d'oré qui Bois, Mange, Croüaſſe Barbote dans l'eau & fait la digeſtion commo un Canard vivant.

CEs 3. Pieces qui ont fait meriter une Récompenſe a l'Autheur d'une Penſion de 8. mille & 5. cent Livres par le Roy, & qui ont engagè un grand nombre des Perſonnes de diſtinction a des longs & penibles Voyages pour les voir, marque mieux leur mérite qu'un plus long detail. On Eſpere que dans cette Ville un chacun ſera charmè de profiter de l'occaſion de les voir & qu'ils en feront la difference du nombre des bagatels, que l'on fait voir tous les jours au publique. Comme le Proprietaire doit ſe trouver le 12. a Francfort il donnera pendant 8. jours a commencer ce jourd'huy 2. Répreſentations par jour a 3. & 5. heures apres midy au Poil du Miroir, l'on payera 24. Sols au premiére, 16. au ſecond & 8. au troiſieme place, & comme il n'y a aucune tricherie dans ces beaux ouvrages l'on enfera voir l'interieur a decouvert en payant 24. Sols par perſonne, l'on vend auſſi dans la méme Sale le mémoire preſentè par l'Auteur a Meſſieurs de l'Academie Royale qui contient un ample detail des pieces contenües dans ces ouvrages & auſſi l'Approbation des Meſſieurs de l'Academie.

Les Compagnies particulieres pourront les voir a tout heure, en avertiſſant d'avance & payeront 3. Livres par Perſonne etant au moin au nombre des huits.

An undated eighteenth-century handbill announcing an exhibition of three of Vaucanson's automata: the flute player, the drummer and the digesting duck. The inner workings of the duck are unknown, despite various attempts to construct replicas.

plenum. The clock metaphor is thus used by Descartes to make an epistemological argument about the limits of our understanding, rather than as an analogy as to how the universe actually works.[ii]

Once a clock has started, it performs its tasks automatically, but it cannot regulate itself. If it is running fast, it cannot slow itself down, nor catch up if it is running slow. One of de Caus's machines, however, has a sophisticated feedback mechanism.[29] The machine is designed to use the weight of water to lift half a tank of water while the other half balances it by going downwards. It consists of three tanks on different levels. In order to work the machine two valves have to be closed when the lower tank is full; this is done by having an overflow which fills a hopper; when the hopper empties, a weight closes the valves. Self-regulating machines were very rare in the seventeenth century (a few years later Cornelis Drebbel designed an incubator for chicken's eggs with a thermostat to regulate the temperature) and de Caus's may have been the first to be invented since Hero of Alexandria designed the float regulator, which we still use in water tanks and toilet cisterns.[30] They become common only in the late eighteenth century: methods for automatically turning valves on and off at the right moment would later prove crucial to the workings of the steam engine (the fan-wheel which turns a windmill into the wind is another simple example). The self-regulating machine is not just a crucial step towards a whole series of more advanced technologies. It is the founding concept of modern social science: David Hume's theory of the balance of trade and Adam Smith's conception of the market depend on the concept of a feedback mechanism.[31] There is an old debate about why the Greeks and the Romans failed to develop a general theory of economic behaviour; one good answer is that they had no machines with feedback mechanisms, so they lacked an essential tool for thinking about social processes.[32]

§ 3

So classical atomism, as revised and reconstructed in the seventeenth century by Gassendi, Descartes and others, explained nature in terms of interacting particles. After 1659 the English usually called this 'the mechanical philosophy', although Boyle was soon to introduce the much less

ii The argument itself was not new: it had already been formulated in the debate between Averroists and their opponents over the reality of epicycles. See p. 413.

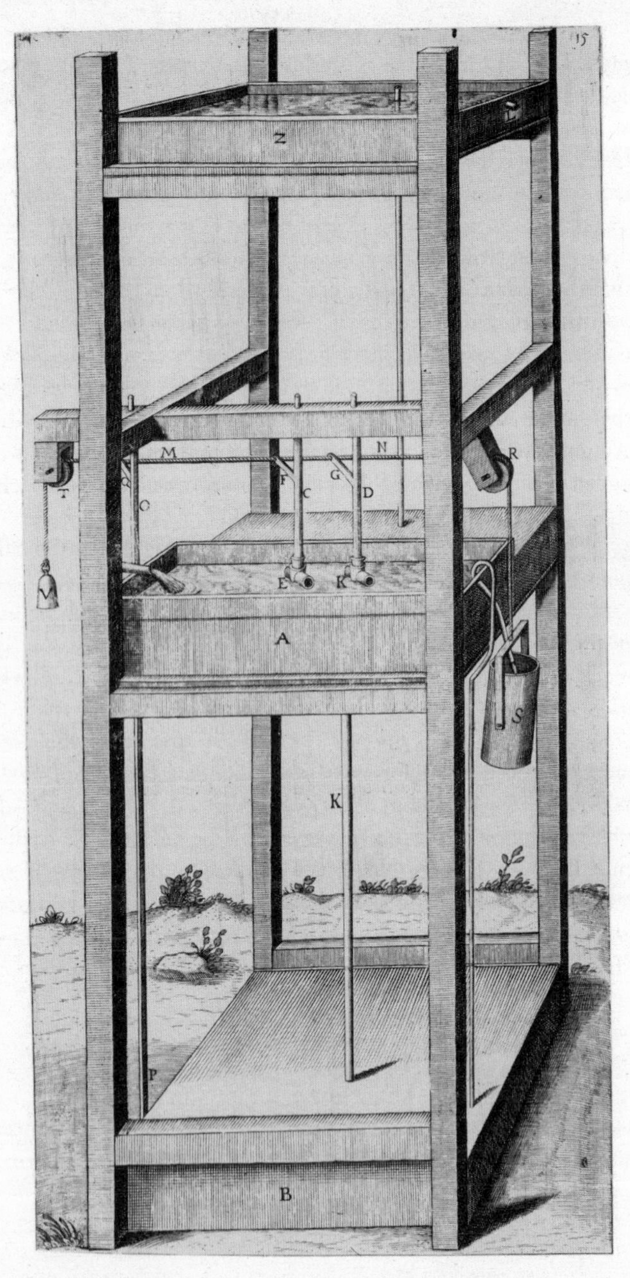

De Caus's self-regulating machine for raising water, from *La Raison des forces mouvantes* (1615).

confusing terms 'the corpuscularian philosophy' and 'the corpuscular philosophy'. (Eventually, the term 'mechanical philosophy' was exported into French, spreading confusion even among Cartesians.) Atoms and corpuscles are not machines, but their interaction is determined by size, shape and hardness, just as the interaction of the parts of a clock are. The disciples of Gassendi (such as Charleton in England) and of Descartes (such as Henry Power) disagreed about lots of things, but they agreed that one should privilege corpuscular explanations for natural processes. Boyle and Newton followed them in this, although they did not insist that everything was susceptible to corpuscular explanation (indeed, Newton's theory of gravity proved to be the great exception, destroying in the end the corpuscular philosophy out of which it had been born).

A quite separate argument was the claim that the universe has been designed to serve a purpose, like a clock or any other complex machine, and that it therefore demonstrates the existence of God. Descartes does not use this argument. His universe can barely be described as 'designed', being the result of allowing very basic laws to work themselves out in practice, and it is not so much mechanical as fluid: it involves vortices and other liquid flows, not cogs and gears. It has been claimed that the modern argument from design first appears in John Wilkins, one of the founders of the Royal Society, in *Of the Principles and Duties of Natural Religion*, published posthumously in 1675.[33] According to that argument, we can tell that the universe has a creator because only an external agent could have designed and constructed it so that its different parts serve their specific functions. An Aristotelian could never argue like this: Aristotle himself did not believe the universe had a creator, and his medieval successors thought that purposiveness was part of the very fabric of nature.[iii] Wilkins, however, is not the first to use this argument, for it can be found in Henry More in 1668, and before More in an early-seventeenth-century Dutch Jesuit with an interest in probability, Leonardus Lessius, in 1631.[34]

Tracing the argument back through these texts, it becomes apparent that it is a variation on a much older argument, which is sometimes called the infinite monkey theorem – the claim that a monkey randomly tapping keys on a typewriter for all eternity would eventually produce

iii Aquinas's fifth proof of the existence of God is that we can see that the parts of the universe act as if they were being guided by an intelligence, and so an intelligence must be guiding them: Aquinas is proving the existence not of a clockmaker God but of a God who inhabits the universe and directs it 'as the arrow is directed by the archer'.

the works of Shakespeare. A refutation of this argument, in essence, can be found in Cicero, although of course without either typewriters or Shakespeare. Cicero rejected the claim of the atomists that the universe is the product of chance, saying:

> I cannot understand why he who considers it possible for this to have occurred should not also think that, if a countless number of copies of the one-and-twenty letters of the alphabet, made of gold or what you will, were thrown together into some receptacle and then shaken out on to the ground, it would be possible that they should produce the *Annals* of Ennius, all ready for the reader. I doubt whether chance could possibly succeed in producing even a single verse![35]

Similarly, Lessius and his successors say that if you throw a quantity of bricks down on the ground you will never get a palace. A book requires an author; a palace requires an architect; a clock requires a clockmaker; and the universe requires a creator. We have already seen Kepler suggesting that even a salad requires a cook.[36] (This argument will, of course, later be disputed by Hume and by Darwin, but for a long time it seemed almost irresistible.)

Boyle has his own version of this argument, which he uses not against the atomists but against the scholastics. They, he maintains, do not have an adequate conception of an omnipotent god. Indeed, he may well have had Aquinas in mind:

> [T]he difference betwixt their Opinion of God's Agency in the World, and that which I would propose, may be somewhat adumbrated by saying, That *they* seem to imagine the World to be after the nature of a Puppet, whose Contrivance indeed may be very Artificial, but yet is such, that almost every particular motion the Artificer is fain (by drawing sometimes one Wire or String, sometimes another) to guide, and oftentimes over-rule, the Actions of the Engine; whereas, according to us, 'tis like a rare Clock, such as may be that at *Strasbourg*, where all things are so skillfully contriv'd, that the Engine being once set a Moving, all things proceed, according to the Artificers first design, and the Motions of the little Statues, that at such hours perform these or those things, do not require, like those of Puppets, the peculiar interposing of the Artificer, or any Intelligent Agent employed by him, but perform their functions upon particular occasions, by vertue of the General and Primitive Contrivance of the whole Engine.[37]

You cannot substitute a text or a palace for the clock in Boyle's version of the argument (as you can in Wilkins's) because texts and palaces are static while clocks move. His argument is that it is astonishing that in our universe everything carries on according to the same general laws without intervention being necessary to correct faults in the machinery, without an archer directing the arrow. To mount this argument he needs not just a complex contrivance but one which is in continuous movement. Only a clock will do: indeed, only a pendulum clock, since earlier clocks were constantly in need of adjustment. In that respect, Boyle's argument is a new and distinctively mechanical one.

But the legacy of the mechanical philosophy was not simply modern versions of the argument from design, which are still widely defended in the form of Intelligent Design. The future lay not only with the new philosophies, whether mechanical or Newtonian; it also lay with the new machines. De Caus in 1615 played with a very simple steam engine, and working steam engines require simple feedback mechanisms. Vaucanson not only made a mechanical duck, he also devised a machine that would automatically weave brocades. Friedrich von Knauss (1724–89) invented a mechanical hand that wrote on a piece of paper just like a living hand; he also constructed the first typewriter.[38] The Industrial Revolution would depend on the skills of such men, skills that would have been familiar to the craftsmen who built the first Strasbourg Cathedral clock. The Scientific Revolution started as a revolution of the mathematicians; it would eventually turn into a revolution of the mechanics. There is a direct line of descent from the Strasbourg clock to the spinning jenny.

This brings us back to the problem with which we began. The Strasbourg clock was built in the middle of the fourteenth century – but the mechanical philosophy was invented three centuries later. Machines did not change much in the meantime, but philosophers did. Once Lucretius was available (he was rediscovered in 1417), his concept of the *machina mundi* could be turned into a quite new idea, the idea of a clockwork universe. In order for this to happen, however, the text of Lucretius was not enough. What was needed was not just new machines but also a new language for discussing machinery. Before this new language, clocks could be used to understand the heavens but not terrestrial physics or biology. It was engineers such as de Caus who, by generalizing the concept of a moving mechanism, made the clockwork universe and the mechanical man possible.

Geography had been remade at the beginning of the sixteenth century by mariners; the philosophy of nature was remade in the seventeenth by the 'mathematicians and engineers'.[39] Natural philosophy was no longer an enterprise to be conducted simply with pen and paper. Boyle's air pump and Huygens' pendulum clock were philosophical machines – machines made by philosophers (with the assistance, of course, of technicians) – the first to tackle a scientific problem, and the second to embody a scientific theory. They helped transform the way in which philosophers thought about machinery, just as Descartes' obsession with automata resulted in a new mechanical philosophy. Already in the seventeenth century the mathematicians' revolution was becoming indistinguishable from a mechanical revolution. Collingwood's claim that the Industrial Revolution was 'well on the way' by the sixteenth century seems to me misconceived, for no new power sources had been brought to bear, but in Chapter 14 I will argue that it was indeed well under way by the end of the seventeenth century, thanks to the appearance of a new type of expert, the engineer-scientist.

§ 4

It will now be apparent that Descartes and Boyle both have what we may term mechanical philosophies, but that they are very different. Of the core three arguments we have distinguished – the corpuscular philosophy, animals as automatons and the clockwork universe – they agree on the first, but each picks one and only one of the other two. Animal automata lead to atheism if humans are held to be little different from animals, but not if one can prove (as Descartes thought he could) the existence of an immaterial mind. The corpuscular philosophy leads to atheism if it is combined with the claim that the universe arises from chance, but not if this further step is blocked, as Boyle sought to block it, by the argument from design. Descartes and Boyle are confident they can protect themselves from atheism, the first by distinguishing mind from matter, and the second by regarding the natural world as providing proof of God's design.[40] Boyle's argument was to prove rather robust, and he was to inspire a long tradition of Christian theologians, such as William Paley (1743–1805); until Darwin, there was no good answer to it, though Hume did his best to blunt it in his posthumous *Dialogues Concerning Natural Religion* (1779). Descartes' argument did not fare so well; even Locke thought that there might be such a thing as thinking matter.[41] Newton proved to be right to ask:

> If we say with Descartes that extension is body, do we not manifestly offer a path to Atheism, both because extension is not created but has existed eternally, and because we have an absolute idea of it without any relationship to God, and so in some circumstances it would be possible for us to conceive of extension while imagining the non-existence of God? Nor is the distinction between mind and body in this [Cartesian] philosophy intelligible, unless at the same time we say that mind has no extension at all, and so is not substantially present in any extension, that is, exists nowhere; which seems the same as denying the existence of mind . . . And hence it is not surprising that Atheists arise ascribing that to corporeal substances which solely belongs to the divine.[42]

Many eighteenth-century atheists, such as d'Holbach and Diderot, were to take their inspiration from Descartes' mechanism and turn it into a systematic materialism with no room for God.

The corpuscular philosophy was absolutely crucial to the Scientific Revolution in that it provided an alternative to the Aristotelian doctrine of forms or substances, of immaterial essences; consequently, it excluded teleology from the nature of things.[43] It was helpful for the generation of the theoretical models which explained air pressure and the vacuum, even if those models were unacceptable to Descartes. But the Newtonian revolution (which we will come to in Chapter 13) ensured that it quickly ceased to be a core part of science as that concept was inherited in the eighteenth and nineteenth centuries. Modern physics, chemistry and biology do not emerge out of the corpuscular philosophy, but out of its collapse. The corpuscular philosophy is, in the end, a parenthesis between scholasticism and Newtonianism.

If Newtonianism destroyed the corpuscular philosophy, it greatly strengthened the argument from design. Only an omnipotent God, creating laws of nature and ensuring that they constantly took effect in the world, could explain the working of gravity, since gravity could not be explained in either Aristotelian or corpuscular terms. Newton's God did not direct individual arrows to their targets; he established the laws that determined the flight of each and every arrow. Newtonianism was thus only conceivable within a culture which had elaborated and become dependent upon the argument from design. That culture was peculiarly English for, as we have seen, Descartes studiously avoided appeals to design. In this respect, Newton is Boyle's heir, and exporting Newtonianism to the Continent depended not just on persuading scientists abroad to accept the possibility of action over a distance but also on persuading them to adopt the design argument.

Voltaire said in 1733 that the English and the French lived in two different worlds, the world of Newton and the world of Descartes:

> A Frenchman, who arrives in *London*, will find Philosophy, like every Thing else, very much chang'd there. He had left the World a *plenum*, and he now finds it a *vacuum*. At *Paris* the Universe is seen, compos'd of Vortices of subtile Matter; but nothing like it is seen in *London*. In *France* 'tis the Pressure of the Moon that causes the Tides; but in *England* 'tis the sea that Gravitates towards the Moon . . . The very Essence of Things is totally chang'd. You neither are agreed upon the Definition of the Soul, nor on that of Matter . . . How furiously contradictory are these Opinions![44]

Voltaire serves to remind us that there was more than one world in which scientists could live.

But in both the world of Descartes and the world of Newton the laws of nature were inexorable, and human beings inhabited a universe which, far from reflecting back to them an image of themselves, macrocosm to microcosm, appeared utterly indifferent to their existence. For both, the sun was just one star among an uncountable multitude, and the universe was, if not infinite, at least without any known limit. 'When I consider the short duration of my life, swallowed up in the eternity before and after, the little space which I fill, and even can see, engulfed in the infinite immensity of spaces of which I am ignorant, and which know me not, I am frightened . . . The eternal silence of these infinite spaces frightens me,' wrote Pascal, who had helped bring this new world into being.[45] Meanings inscribed by God in the forms of things, the great chain of being, sympathy and antipathy, natural magic, had been replaced by blind mechanisms and inexorable laws. Even animals, in Descartes' view, were just automata. Blake painted Newton playing at God, measuring out the universe; preoccupied with the simplified, mathematical language of the laws of nature, he can no longer see the complexity and variety that surround him; quality has been reduced to mere quantity. This was the beginning of what Weber called 'the disenchantment of the world'.[46]

13
The Disenchantment of the World

> Thus the growing process of intellectualization and rationalization does *not* imply a growing understanding of the conditions under which we live. It means something quite different. It is the knowledge or the conviction that if *only we wished* to understand them we *could* do so at any time. It means that in principle, then, we are not ruled by mysterious, unpredictable forces, but that, on the contrary, we can in principle *control everything by means of calculation*. That in turns means the disenchantment of the world.
>
> – Max Weber, 'Science as a Vocation' (1918)[1]

§ 1

In March of 1661 a gentleman called John Mompesson, of Tedworth in Wiltshire, had a busker who was drumming arrested (beggars had to have a licence, and this beggar's licence was forged) and his drum taken from him. For the next two years or so Mompesson's house was haunted by a poltergeist.[2] There were drumming noises but also strange levitations of objects and alarming noises. Here is a typical report:

> On the Fifth of *Novemb.* 1662. it kept a mighty noise, and a servant observing two Boards in the Childrens Room seeming to move, he bid it give him one of them. Upon which the Board came (nothing moving it that he saw) within a yard of him. The Man added, Nay let me have it in my Hand; upon which it was shov'd quite home to him. He thrust it back, and it was driven to him again, and so up and down, to and fro, at least twenty times together, till Mr *Mompesson* forbad his Servant such familiarities. This was in the day-time, and seen by a whole Room full of People. That morning it left a sulphurous smell behind it, which was very offensive. At

> night the Minister one Mr *Cragg*, and divers of the Neighbours came to the House on a visit. The Minister went to Prayers with them, Kneeling at the Childrens Bed-side, where it was then very troublesome and loud. During Prayertime it withdrew into the Cock-loft, but returned as soon as Prayers were done, and then in sight of the Company, the Chairs walkt about the Room of themselves, the Childrens shooes were hurled over their Heads, and every loose thing moved about the Chamber. At the same time a Bedstaff was thrown at the Minister, which hit him on the Leg, but so favourably that a Lock of Wool could not have fallen more softly, and it was observed, that it stopt just where it lighted, without rolling or moving from the place.[3]

There have always been plenty of stories of the weird and the wonderful. This story comes from *Saduscismus triumphatus*, written by a clergyman, Joseph Glanvill, one of the chief propagandists for the new science, and a fellow of the Royal Society from 1664. Glanvill began publishing in defence of the reality of witchcraft in 1666, and his first version of the Mompesson story appeared the next year, in *A Blow at Modern Sadducism*, Sadducism being understood to be the denial of the reality of spirits. (The version just quoted comes from the posthumous work *Saducismus triumphatus*, or *The Saducee Triumphed Over* (1681), seen through the press by Glanvill's friend the Platonist philosopher Henry More, which went through a further five editions.) The purpose of *Saducismus triumphatus* was simple: Glanvill sought to produce unimpeachable testimony (including his own), what he called 'a choice Collection of modern Relations', which would establish witches, poltergeists and demons as matters of fact (the use of the language of contemporary science was deliberate); thus he would prove the reality of a spirit world, and thereby refute atheistical materialism.[4] A physician, John Webster, wrote *The Displaying of Supposed Witchcraft* against him in 1677: Webster had trouble getting a licence to publish, but finally obtained one from the vice-president of the Royal Society. In the conflict between Glanvill and Webster the Society hedged its bets; Webster, however, was never elected a fellow. For Glanvill, and others like him, the new science was intended to serve as a bulwark against materialism and atheism; being modern and believing in witchcraft went hand in hand.[5]

§ 2

The word 'modern' (*modernus*) dates to the sixth century.[i] It postdates the sack of Rome by the Visigoths (410) and the establishment of a new, Christian order under Theodoric (493–526). Then, the modern age was an age of restoration, after a lengthy period of catastrophe, crisis and collapse. What 'modern' means has shifted century by century, and discipline by discipline. For a millenium or so there were ancients and moderns, which corresponded roughly to pagans and Christians. As early as 1382 the Florentine chronicler Filippo Villani refers to 'ancient, middle, and modern times'; in 1604 the term *medium aevum* (the forerunner of 'medieval') was introduced, setting up the distinction between ancient, medieval and modern history which is still standard.[6] Other terms come and go: 'the Renaissance' and 'the Enlightenment' (with the definite article) are nineteenth-century terms, which for the last fifty years have been giving ground to 'early modern'; all three reflect a reluctance to think of all history since 1453 (the fall of Constantinople) as 'modern'.[7] A nineteenth-century traveller, Baedeker in hand, embarking on a train at one of the great railway stations of Europe, no longer felt much in common with Erasmus, who had crossed Europe on horseback in the early sixteenth century; in the Enlightenment the only advance since Erasmus's day was the introduction of the coach. 'Modernity' (an eighteenth-century word), for the late-nineteenth-century historian, began not with the fall of Rome or the fall of Constantinople but with the railway timetable.[ii] And there it seems it has (at least for now) stuck, for we have invented the term 'postmodern' to mark out the differences between our world (the world of the last fifty years or so) and that of our parents, grandparents and great-grandparents.

Shakespeare used the word 'modern' to mean both 'ordinary' and 'contemporary'. He did not have a strong enough sense of historical change to want to emphasize the peculiar features of the modern world, and the peculiar feature he was most acutely aware of – the Reformation – he was obliged to refer to only obliquely, for fear of being accused of Catholicism. So when Lafeu in *All's Well that Ends Well* says, 'They

i The first recorded usage is in the *Institutiones* of Cassiodorus, *c.*530s: '*antiquorum diligentissimus imitator, modernorum nobilissimus institutor*'.

ii According to a Google n-gram, the word 'modern' first becomes more common than 'ancient' in 1894.

say miracles are past; and we have our philosophical persons, to make modern and familiar, things supernatural and causeless. Hence is it that we make trifles of terrors, ensconcing ourselves into seeming knowledge, when we should submit ourselves to an unknown fear' (II.iii.891), he is attacking the new, Protestant doctrine that miracles are past; but his use of 'modern' as a synonym for 'ordinary' obscures, rather than clarifies, his purpose. In the fifth century the sack of Rome marked the end of one world and the beginning of another; so did the railway in the nineteenth century. Shakespeare is not aware of living in a distinctly 'modern' world, despite the compass, the printing press, gunpowder and the discovery of America. He was keen to elide the differences between ancient Rome and his own London, as well as those between Verona and Canterbury.

In some distinct disciplines a sense of the modern exists in the Renaissance: in painting, in music, in warfare, in literature (where modern literature, Dante being the supreme example, is written in the vernacular, not in Latin).[iii] But the idea that there was something that might be called the 'modern age', or the 'modern world', or 'modern times' established itself only after Shakespeare's death (1616).[iv] Take, for example, the lengthy comparison between the achievements of the ancients and moderns published by Alessandro Tassoni in 1620. Tassoni is well aware of all sorts of things that the ancients did not have: falconry, for example, or silk, or perspective painting. He thinks certain modern technologies – the clock, the compass, gunpowder, the telescope – represent a real advance over anything achieved by the ancients. But his view of history is fundamentally cyclical: the gains of one age can all too easily be lost by the next. Above all, he has no conception of a decisive transition in the natural sciences. In his discussion of natural philosophy he praises the moderns for not accepting anything on Aristotle's mere authority and for making numerous discoveries (mainly as a by-product of the discovery of the New World), but he regards the superiority of the Greeks over the moderns as indisputable. In his discussion of astronomy he shows himself well aware of Brahe's demonstration that comets are in the supralunary world and of Galileo's discoveries with the telescope, but he classifies Sacrobosco, along

iii See above, pp. 36–7.

iv These three phrases occur in EEBO (see p. 26, note xxiv) three times before 1600 (first in 1593); 49 times between 1600 and 1624; 88 times between 1625 and 1649; 179 times between 1650 and 1674; and 162 times between 1675 and 1699. The peak decade is the 1650s.

with Copernicus, as a modern (just as he classifies the clock, along with the telescope, as modern). And he praises Cremonini, who had refused to look through Galileo's telescope. It has been claimed that Tassoni expresses a sense of liberation from the schoolroom of antiquity, but his claim is merely that the moderns must find a place in the schoolroom alongside the ancients. It does not occur to him that they might supplant them.[8]

This chapter is about the birth of the modern in two senses: first, there is the emergence of a new sense of the word 'modern' in the 1660s to refer to post-Galilean science. Thus in Glanvill's *Plus ultra* (1668) the first section is entitled 'Modern Improvements of Useful Knowledge', and he uses the word 'modern' frequently ('the modern world', 'modern times', 'the modern way of Philosophy', 'modern Experimenters', 'the modern Discoveries') to refer to the post-Columbus age.[v] This is the same sense of 'modern' which underlies Butterfield's title *The Origins of Modern Science*, and reflects a usage established by the contemporaries of Newton. Our understanding of the word 'modern', when talking about science, thus still corresponds to theirs, and their use of the term 'the modern' is their way of acknowledging what we call the Scientific Revolution.

Second, there is the decline of belief in magic and witchcraft, already hinted at in Lafeu's speech, if we take 'modern' to mean something more than just 'ordinary'. At the time this was seen as something new and unparallelled; this, too, was modern. Early-eighteenth-century England represents a key moment in Weber's 'disenchantment of the world'. It is Weber's conception of modernity which particularly draws our attention to this aspect of the Scientific Revolution.[9]

v See also, for works with 'modern' in the title, Bartholin, Walaeus & others, *Bartholinus Anatomy: Made from the Precepts of His Father, and From the Observations of All Modern Anatomists* (1662); Dary, *The General Doctrine of Equation Reduced into Brief Precepts: In III Chapters. Derived from the Works of the Best Modern Analysts* (1664); Salusbury (ed.), *Mathematical Collections and Translations in Two Parts from the Original Copies of Galileus and Other Famous Modern Authors* (1667); Croll, Hartmann & others, *Bazilica Chymica & Praxis Chymiatricae; or, Royal and Practical Chymistry in Three Treatises. Wherein All Those Excellent Medicines and Chymical Preparations are Fully Discovered, from whence All Our Modern Chymists Have Drawn Their Choicest Remedies* (1670); Barbette, *The Chirurgical and Anatomical Works . . . Composed according to the Doctrine of the Circulation of the Blood and Other New Inventions of the Moderns* (1672).

§ 3

In 1704 Jonathan Swift, later the author of *Gulliver's Travels*, published a little satire entitled *The Battle of the Books*. It described a war between books in a library, a war between the Ancients and the Moderns. Swift had written this squib by 1697, and by the time it appeared the conflict it satirized appeared to be over. Swift's text was teasingly incomplete, so that it was impossible to tell who had emerged victorious. The conflict, in its English version, had broken out in 1690 when the distinguished politician and diplomat Sir William Temple (who employed Swift as his secretary on and off from 1688 until his death in 1699) published an essay defending the ancients against the moderns.[10] Temple was responding to a dispute which had begun a few years earlier in France, where it had been claimed that the writings of French authors of the seventeenth century (which the French now call *l'âge classique*; the usage is apparently new in the twentieth century) were superior to anything the Greeks or Romans had to offer. In England this debate turned on the relative merits of authors such as Milton and Dryden on the one hand and Vergil and Homer on the other (Shakespeare had not yet established a claim to be the greatest of all poets). The 'moderns' had acquired a new self-confidence.

In this dispute the question of the relative merits of ancient and modern science was at first entirely secondary. Temple touched on it only tangentially in his essay; it recedes into the background again in Swift's *Battle of the Books*.[11] But it was a central topic when Fontenelle took up the defence of the moderns against the ancients in France (1686),[12] and it again played a central role in the major reply to Temple, *Reflections upon Ancient and Modern Learning* (1694; with an expanded second edition in 1698), by a young clergyman, William Wotton, who had managed to get himself elected a Fellow of the Royal Society, though science was far from being his primary interest. (Wotton was commissioned to write the first life of Robert Boyle; he began work but never finished as he fell, for a time, into a life of drunkenness and debauchery.)[13] Temple knew very little about science, much less than Wotton, and lacked any inclination to make good this deficiency. He left an unfinished reply to Wotton when he died in 1699 – missing was the discussion of science which needed to be the crux of his argument. He had evidently hoped that someone else, perhaps Swift, would draft this section for him.

Swift published this incomplete text posthumously in 1701.[14] According to Swift, the following passage was in Temple's handwriting, but the information in it, which implies a good knowledge of the work of Godwin, Wilkins and others, was surely supplied by Swift, who was well informed on questions of science, and indeed on almost every subject under the sun:[15]

> What has been produced for the Use, Benefit, or Pleasure of Mankind, by all the airy Speculations of those, who have passed for the great Advancers of Knowledge and Learning these last fifty Years, (which is the date of our modern Pretenders) I confess I am yet to seek, and should be very glad to find. I have indeed heard of wondrous Pretensions and Visions of Men, possess'd with Notions of the strange Advancement of Learning and Sciences, on foot in this Age, and the Progress they are like to make in the next: As, The Universal Medicine, which will certainly cure all that have it: The Philosopher's Stone, which will be found out by Men that care not for Riches: The Transfusion of young Blood into old Men's Veins, which will make them as gamesom as the Lambs, from which, 'tis to be derived: An Universal Language, which may serve all Mens Turn, when they have forgot their own: The Knowledge of one anothers Thoughts, without the grievous Trouble of Speaking: the Art of Flying, till a Man happens to fall down and break his Neck: Double-bottom'd Ships, whereof none can ever be cast away, besides the first that was made: The admirable Virtues of that noble and necessary Juice call'd Spittle, which will come to be sold, and very cheap in the Apothecarys Shops: Discoveries of new Worlds in the Planets, and Voyages between this and that in the Moon, to be made as frequently as between *York* and *London*: Which, such poor Mortals as I am, think as wild as those of *Ariosto*, but without half so much Wit, or so much Instruction; for there, these modern Sages may know, where they may hope in Time to find their lost Senses, preserved in Vials, with those of *Orlando*.[16]

All this (the transfusions, the double-bottomed ships, the communication without speaking, even the cure by spittle) is accurate enough. The knowledge is Swift's, not Temple's.

The publication of Temple's *Defence* in 1701 had not provoked Wotton into a reply, for his adversary was dead, and in any case his essay confessedly lacked the lengthy discussion of the new science which could alone have enabled it to sustain its argument. But he did not take the publication of Swift's *Battle of the Books*, which made mocking references to his own work, so stoically, particularly, perhaps, as he may now have suspected that Swift had had more of a role in the *Defence* than had at first been apparent. So in 1705 he published his own *Defense*, along with a savage attack on Swift's *Tale of a Tub* (which had been published alongside *The Battle of the Books*), an allegory which he interpreted as an attack upon the core beliefs of Christianity.

Temple was a man of noble family, who treated the upstart Wotton with barely concealed disdain.[17] Even Swift, whose background was modest indeed, dismissed Wotton as someone of unknown parentage.[18] It would have astonished Temple to think that one day his name would be remembered principally because he had been for a while Swift's employer, and that he and Wotton would be remembered only for having given occasion to Swift's *Battle of the Books*.[vi] For Wotton has sunk into an obscurity even deeper than Temple's. No one now reads his scholarly disquisitions on the Tower of Babel, or on the Scribes and the Pharisees. But, unlike Temple, he deserves to be better remembered. He had an excellent grasp of the Scientific Revolution; indeed, he was the first to survey the field. He saw that he had to give an account of the differences between ancient and modern sciences; that he had to analyse the contribution of the printing press, and of the telescope and the microscope, to the new sciences; and that he had to describe the way in which a new critical attitude combined with better dissemination of information had led to greater reliability in both facts and theories.[vii] His enquiry into the prehistory of the idea of the circulation of the blood was the beginning of the history of science as a learned enterprise.[19] It used to be thought that his judgement was clearly defective, in that he makes no mention of Copernicus; now that Tycho Brahe has come to be seen as the true founder of the new astronomy, this looks less culpable.[viii][20] And it is Wotton who was the first to articulate the view that the founding of the Royal Society marks the true beginning of modern science, for he held that the achievements of the sixteenth century had been primarily destructive ('It was the Work of one Age to remove the Rubbish'), while it was only in the last forty or fifty years that 'the new Philosophy had gotten Ground in the World'.[21]

Nowadays, Wotton says, in his final summary:

> (1.) No Arguments are received as cogent, no Principles are allowed as current, among the celebrated Philosophers of the present Age, but what are in

vi Temple has also long been remembered for the letters written to him by his future wife, Dorothy Osborne, during their prolonged courtship; there were numerous editions between 1888 and 2002, and the story of their relationship is still being told: Dunn, *Read My Heart* (2008).

vii See above, pp. 281–2, 305 and 420–1.

viii Neither the blame nor the credit properly belongs to Wotton, as he had Halley write the chapter on astronomy for him: it is, though, of considerable interest that Halley, a great astronomer, did not think Copernicus worth mentioning.

themselves intelligible ... Matter and Motion, with their several Qualities, are only considered in Modern Solutions of Physical Problems. *Substantial Forms, Occult Qualities, Intentional Species, Idiosyncrasies, Sympathies and Antipathies of Things*, are exploded ... because they are only empty Sounds, Words whereof no Man can form a certain and determinate Idea.

(2.) Forming of Sects and Parties in Philosophy ... is, in a manner, wholly laid aside. *Des Cartes* is not more believed upon his own Word, than *Aristotle*: Matter of Fact is the only Thing appealed to ...

(3.) Mathematicks are joyned along with Physiology [i.e. natural science], not only as Helps to Men's Understandings, and Quickners of their Parts; but as absolutely necessary to the comprehending of the Oeconomy of Nature, in all her Works.

(4.) The new Philosophers, as they are commonly called, avoid making general Conclusions, till they have collected a great Number of Experiments or Observations upon the Thing in hand; and, as new Light comes in the old Hypotheses, fall without any Noise or Stir.[22]

Thus Wotton had a sophisticated analysis of the Scientific Revolution, although he did not use the term. It had been made possible by the printing press and the invention of the telescope; it depended on mathematics and the mechanical philosophy; and it relied on a new experimental method and the establishment of matters of fact. The new science was different in kind from anything that had gone before because it was based on experiment and observation, not on empty theorizing, and because it recognized that scientific understanding would continue to change over time. By 1694 Newton's *Principia* had been published, and Wotton had some grasp of its significance; by 1705 he was able to present Newton's *Opticks* as the exemplary text of the new science. It was already possible to look back over the Scientific Revolution, to identify its leading protagonists and sketch out its main characteristics. This present book stands squarely in a tradition established by my namesake, William Wotton.

That last sentence is not a whimsical one, as it might appear, for by 1700 a conception of science had developed which has remained largely unchanged ever since, and with it came a fundamentally reliable account of what had changed over the previous two hundred years. In 1650 nobody quite knew how to study the physical world. By 1700 the idea that the study of the physical world is all about facts, experiments, evidence, theories and laws of nature had become well established. Later scientific revolutions have transformed our knowledge, but they have not melted down and recast our idea of science.

§ 4

The idea of modern science also raises, however, a set of further questions which are tied together in Weber's phrase 'the disenchantment of the world'. As Wotton interpreted Swift, it was Swift, the critic of the moderns, who was the sceptical unbeliever, while Wotton presented himself as an orthodox Protestant.[23] He shows no anxiety that science might be associated with unbelief. Reading Wotton, we may gather that there is no conflict between the new philosophy and Christian faith, but we obtain no insight into the nature of their relationship to each other, or of the relationship between science and a range of beliefs which we have now rejected as incompatible with science, particularly magic and witchcraft. The one topic Wotton does touch on is alchemy, where he makes his own scepticism clear, although he expresses himself very cautiously, perhaps because he was aware that Boyle and Newton were adepts; Temple is thus given an opportunity to attack Wotton for being far too sympathetic to the alchemists.[24] One might be forgiven for thinking that it is Temple and Swift, the critics of modern science, who live in a disenchanted world, not Wotton, its advocate.

On the important subject of science and the decline of magic the work done by the last generation of historians is curiously unhelpful. In many respects the key text remains Keith Thomas's *Religion and the Decline of Magic* (1971).[25] Thomas believed that there was a double foundation to magic. On the one hand, magic represented an attempt to gain control over nature, an attempt which was inevitable in societies which were incapable of protecting their members from bad harvests, fires, disease, pain and sudden death. Thus, in principle, belief in magic ought to decline with improvements in technology, particularly medicine, and the beginnings of insurance policies and other methods of reducing the impact of unforeseen disasters. On this account belief in magic ought not to have declined until the nineteenth century, or even later. (It is true that Nicholas Barbon's Fire Office insurance company was founded in 1680, but few benefitted from its services; similarly, friendly societies, such as the Freemasons, go back to the beginning of the eighteenth century, but they had relatively few members until the nineteenth.)

Secondly, Thomas held that individuals were identified as practising witchcraft and demonic magic as a result of social tensions, particularly with regard to the distribution of charity. By this argument, belief in demonic magic ought not to have declined until there was a general improvement in living standards, and perhaps not until the development of the welfare state. Certainly, both the beliefs and the tensions remained prevalent in early-eighteenth-century society. Joseph Addison, in the *Spectator* (1711), insists that in every village there are people believed to be witches, and this belief straightforwardly reflects social tensions:

> When an old Woman begins to doat, and grow chargeable to a Parish, she is generally turned into a Witch, and fills the whole Country with extravagant Fancies, imaginary Distempers, and terrifying Dreams. In the mean time the poor Wretch that is the innocent Occasion of so many Evils begins to be frighted her self, and sometimes confesses secret Commerces and Familiarities that her Imagination forms in a delirious old Age. This frequently cuts off Charity from the greatest Objects of Compassion, and inspires People with a Malevolence towards those poor decrepid Parts of our Species, in whom Human Nature is defaced by Infirmity and Dotage.[26]

Yet it is clear that belief in witchcraft among the educated elite diminished rapidly in the early eighteenth century. Addison himself claimed to be neutral on the subject: he believed in witchcraft in principle, but not in the validity of any particular accusation of witchcraft. Neatly reflecting this ambivalence, Jane Wenham was convicted of witchcraft in 1712, but pardoned and set free. She was not, however, the last to face a capital charge: Mary Hickes and her nine-year-old daughter were executed in 1716 for raising a storm. The anti-witchcraft legislation was, however, abolished in 1736.[27] Somewhat surprisingly, contemporaries (including Addison) insisted that the clergy were at the forefront of the new scepticism with regard to witchcraft accusations.[28]

There would appear to be a straightforward solution to this puzzle. There may be no technological or sociological explanation for the decline in belief in witchcraft in the early eighteenth century, but there is an alternative explanation ready to hand. The new science must be responsible. Thomas, who generally avoids intellectualist explanations, falls back on this. He has been criticized for doing so, on the grounds that this is not really an explanation at all, since the new science and scepticism towards witchcraft represent merely two sides of the same coin.[29] I cannot see that this criticism holds. Think of the boats on a tidal river; at low tide they are stuck in the mud. One can certainly explain the lifting of the boats by the tide, even if the lifting of the boats is itself the best evidence that the tide is rising.

However, if the new science is responsible, the mechanism is far from straightforward.[30] In the years after the founding of the Royal Society in 1660 it was very important to a significant group of leading members, and to Boyle above all, to establish that the new philosophy was favourable to Christianity, not hostile to it. This was one of the central tasks assigned to Sprat's so-called *History* (which appeared only seven years after the Society was founded; Sprat was a clergyman and future bishop, and he wrote under the supervision of Wilkins, who was promoted to a bishopric the year after the *History* appeared), just as it was the key objective of Glanvill's *Philosophia pia* (1671, which, despite the Latin title, was written in English) and of Boyle's *Christian virtuoso* (1690).[31] It is not difficult to understand the reasons for this. As we have seen, the 'corpuscularian philosophy', as Boyle called it, was based on the teachings of Epicurus and Lucretius, opponents of all religion. Thomas Hobbes, although he was not an atomist, had developed a materialist, Epicurean philosophy which was universally understood to be hostile to religion. Irreligious thinking was apparently widespread in the coffee houses of London (which rapidly increased in number after the Restoration in 1660), although it had little expression in print. There were evidently unbelievers even among the members of the Royal Society: it was said of Halley 'that he would not so much as pretend to believe the Christian religion', and it was for this reason that he was denied the chair of astronomy at Oxford in 1691; he was somewhat irritated, consequently, to see that Nicholas Saunderson's reputation for unbelief (he later became the model for the blind atheist in Diderot's *Letter on the Blind*) was no bar to his obtaining the Lucasian chair of mathematics at Cambridge in 1710.[32]

Moreover, since the founding of the universities in the twelfth century, Christian theology had been taught within a framework established by the philosophy of Aristotle; it was natural for defenders of Aristotelian natural philosophy to accuse the new philosophy of being contrary to good theology as well as good philosophy. Thus members of the Royal Society were bound to think it was of strategic importance that the new philosophy should demonstrate that it was favourable to Christian faith; but, of course, for many of them this was not merely a matter of calculation. Boyle was a deeply religious man who gave money to pious causes and by the terms of his will founded the Boyle Lectures for the conversion of unbelievers. His insistence on the compatibility of the new science and Christian faith was an expression of his deepest convictions.[33]

In the years immediately before the founding of the Royal Society two new sets of arguments for Christian faith were pioneered. First, there were the Cartesian arguments that the mind must be immaterial, that rational beings must therefore have an immortal soul, and that our knowledge of God as a being superior to ourselves must come from outside ourselves. In other respects, Cartesianism sat uncomfortably with traditional belief, since Descartes was prepared to imagine a universe which was completely unplanned, once the fundamental laws of nature had been established, and various forms of irreligious argument were developed out of Cartesianism, above all by Spinoza. But at the core of Descartes' philosophy there was this set of arguments in favour of belief. Second, there was the argument from design: the mechanical philosophers, by describing the universe as like a clock (see Chapter 12), were able to argue that it was incomprehensible except as the production of an omnipotent clockmaker. Boyle laid great stress on this argument, which ran counter to the Epicurean and Lucretian insistence that the universe was the result of a random swerve which had brought two atoms into contact and set off a chain reaction.

Both these arguments were fundamentally new. In traditional medicine 'spirits' did the work in the body that we would attribute to electrical impulses travelling along the nerves; these spirits were, we might say, barely material. And in theology, angels and demons occupied space, even if they did not have bodies in the conventional sense. The spirit world was thus a blurred zone between the material and the immaterial.[34] This is how John Webster describes it in 1677:

> [A]s we know not the intrinsick nature of body, so also we are ignorant of the highest degree of the purity and spiritualness of bodies, nor do we know where they end, and therefore cannot tell where to fix the beginning of a meer spiritual and immaterial being. For there are of Created bodies in the Universe, so great a diversity, and of so many sorts and degrees of purity and fineness, one exceeding another, that we cannot assign which of them cometh nearest to incorporeity, or the nature of spirit . . . So the vital part in the bodies of men are by Physicians called Spirits in relation to the bones, ligaments, musculous flesh and the like . . . and yet still are contained within the limits of body, and are as really Corporeal as any of the rest, and so are the air and aether. And those visible species of other bodies that are carried in the air and represented unto our Eyes, by which we distinguish the shape, colour, site and similitude of one body from another, though by the Schools passed over with that sleight title of qualities, as though they were either simply nothing, or incorporeal things, are notwithstanding really Corporeal . . . So that if we have bodies of so great

purity, and near approach unto the nature of spirit, we cannot tell where spirit must begin, because we know not where the purest bodies end.[35]

Cartesianism, by making an unambiguous divide between the material and the immaterial, left it unclear how angels and demons might be present in the world. Long before Descartes, Reginald Scot had drawn the conclusion that there could be no place (except within the mind) for an immaterial being in a material world. But Webster, who saw himself as following Descartes, was happy to argue that angels and demons were material entities, capable of appearing to the sight and of communicating, but, like air, too spiritual to be touched or held.[36] Even human beings had not only an immaterial mind but a material sensitive soul, capable of a physical presence after death. Moreover, a severe blow was dealt to Descartes' argument for an immaterial mind when John Locke, in his *Essay Concerning Humane Understanding* (1690), acknowledged that there was no logical impossibility in the idea of thinking matter.[37] Thus the Cartesian distinction between mind and matter turned out to be less sharp and decisive than at first it had seemed.

As for the argument from design, it was fundamentally different from the traditional Thomistic argument, which held that the universe was imbued with purpose, and that the ultimate purpose was to be found in God. The new philosophers regarded matter as the passive recipient of a divine shaping and denied the existence of Aristotelian forms. For them, as we saw in Chapter 9, the argument from design depended on envisaging the universe as manufactured, rather than on showing nature itself to be purposive. This argument was much more robust than the mind/matter distinction, and came under systematic attack only in Hume's posthumous *Dialogues*. Both these arguments thus depended on prior acceptance of the mechanistic or corpuscular philosophy, according to which matter is passive and always acted on from outside.

Alongside them, in the years between 1653 and 1691, a third argument developed, an argument which was based on the new language of 'matter of fact'.[38] The idea was simple: Christianity depended on belief in a spiritual world beyond the material world. To deny the existence of spirits in the form of angels and demons was a key step towards denying the existence of the immortal soul; to demonstrate the existence of spirits would prove the reality of the spiritual world. Although battle was to be engaged on the question of the existence of spirits, the assumption was that it was really the existence of God that was at stake. As Glanvill put it, '[*T*]*hose that dare not bluntly say,* There is NO GOD, *content themselves (for a fair step, and* Introduction) *to deny there are* SPIRITS *or* WITCHES.'[39] This emphasis on grounding faith in indisputable matter of fact was not peculiar to the Protestant world, and has its origins before the new language of facts had established itself. In Rome the *advocatus diaboli*, or devil's advocate, had been established as early as 1587 to test the evidence adduced in support of the miracles claimed for those proposed for canonization.

The new strategy for the refutation of unbelief begins with Henry More's *Antidote against Atheism* (1653), which includes an extensive study of witchcraft cases. More's disciple Glanvill became its chief exponent. Glanvill made certain cases famous, above all 'The Drummer of Tedworth'.[40] It was as a contribution to this literature that Boyle arranged for the translation from French of *The Devil of Mascon* (1658) and Méric Casaubon published the record of John Dee's conversations with angels, or, as Casaubon would have it, with devils (1659). Boyle went on to carry out extensive research into the phenomenon of second sight, which one might describe as the beginnings of parapsychology.[41] The last significant work to be published in this tradition is Richard Baxter's *The Certainty of the World of Spirits* (1691).

There was a simple problem with this strategy of establishing the existence of the spirit world by accumulating testimonies from reliable witnesses. It assumed that the sort of evidence which would be persuasive in the report of a laboratory experiment or in a court of law when dealing with a murder or a theft could be convincing when dealing with cases of demonic possession or levitation. This view was shared by Glanvill's chief opponent, John Webster. Indeed, it is striking (given that we rarely find the word 'evidence' in Boyle's texts) how frequently they use the word 'evidence': thirty-two times in Webster, sixty-six in the 1681 edition of Glanvill. Webster wanted to deny that there was reliable evidence for pacts with the devil, copulation with the devil, familiars such as black cats who sucked on protuberances on the witch's skin, witches flying through the air or being turned into wolves or hares. He defined reliable evidence as one would in a court of law: more than one witness, and the witnesses must be of sound mind, not partial or prejudiced. Relying on these criteria, he accepted the reliability of evidence for apparitions, for the body of the murdered bleeding in the presence of the murderer, for alchemy, and so forth. Thus the debate between Glanvill and Webster was not over the reality of demons, but only over the limits of their actions in the world; and Webster believed many

The frontispiece to the second part of Joseph Glanvill's *Saducismus triumphatus* (1681). Clockwise from top left, the following are represented: the drummer of Tedworth (see pp. 449–50); the Somerset witch Julian Cox; a rendezvous of witches at Trister Gate; a celestial apparition in Amsterdam; the Scottish witch Margaret Jackson; and the levitation of Richard Jones at Shepton Mallet.

things that seem ridiculous to us on the basis of evidence that, in the eyes of Boyle and Glanvill, was no stronger than the evidence for witchcraft.[42]

However, the *Logic of Port-Royal* (1662) had acknowledged that the more unlikely an event the stronger the evidence in favour of it would have to be in order to ensure that it was more unlikely that the evidence should be false than that the event should not have occurred. This obviously presented a problem for evidence for miracles, one which was rapidly glossed over by the *Logic* and by Locke; indeed, it is presumably for this reason that the argument was slow to be taken up. But in the early years of the eighteenth century this argument from probability was applied with devastating force. It was at the heart of Francis Hutchinson's *Historical Essay Concerning Witchcraft* (1718) (though Hutchinson, a future bishop, carefully affirmed his belief in angels), from where it was adopted, in 1722, by Trenchard and Gordon in *Cato's Letters*.[43] And it was put to work in one of the pamphlets generated by the famous case of Mary Toft, who claimed in 1726 to have given birth to seventeen rabbits:

> Suppose one were to see a Letter from *Battersea*, importing that a Woman there had been delivered of five Cucumbers, or indeed a hundred Letters, would that lead a man of Sense to believe any Thing, but, either that the People who wrote those Letters had been grossly impos'd upon themselves, or intended to impose upon him. Either of these two Things may, and do happen every Day; but it was never known, that ever any Creature brought forth any one Creature of a Species in all Respects different from it self, much less five or seventeen such Creatures; for which therefore, a Man of common Sense, much more a penetrating and quicksighted Anatomist, should look upon all such Letters with the utmost Contempt.[44]

In a more sophisticated form it became the argument of David Hume's essay 'Of Miracles' in *An Enquiry Concerning Human Understanding* (1748).[45]

The Christian churches could not abandon their belief in miracles and angels, but Christians could certainly retreat from their insistence on the reality of witchcraft, demonic possession, poltergeists, levitation and second sight. Indeed, as we have seen, the clergy, who had been in the vanguard of the army of those advocating belief in spirits, were among the leaders in this retreat, which was well under way by the trial of Jane Wenham in 1712. What made this retreat possible was the development of a new and powerful argument for religious faith.

§ 5

In 1687 Newton published his *Principia*, which established his theory of gravity. Gravity involved action over a distance, something impossible according to the mechanical philosophy. On the Continent resistance to Newtonianism continued into the 1740s and involved key intellectual figures such as Huygens, Leibniz and Fontenelle.[46] In England the importance of the *Principia* was slow to establish itself simply because the book was so technical: it is said that only ten people could properly understand it in the period between its first publication and Newton's death in 1727.[47]

The key moment for the popular transmission of Newton's discovery came in 1692, when Richard Bentley delivered the first set of Boyle Lectures. Bentley was the greatest classical scholar of his age, but he would also become a Fellow of the Royal Society.[48] He had joined the battle of the ancients versus the moderns on his friend Wotton's side by providing for the second edition of Wotton's *Reflections* a lengthy demonstration that the letters of Phalaris, which Temple had singled out as one of the jewels of classical literature, were a later forgery. He had not started out on a clerical career, but he had been ordained deacon in 1690 and was later ordained as a priest. In preparation for his lectures, he wrote to Newton, who replied, 'When I wrote my treatise about our systeme, I had an eye upon such principles as might work with considering men for the beliefe of a Deity; and nothing can rejoyce me more than to find it useful for that purpose.'[49]

Bentley's eight lectures were published under the title *The Folly and Unreasonableness of Atheism Demonstrated from the Advantage and Pleasure of a Religious Life, the Faculties of Human Souls, the Structure of Animate Bodies, & the Origin and Frame of the World* (1693). The first lecture dealt with the social and psychological benefits of religion; the second with the Cartesian argument against thinking matter; the third, fourth and fifth with the design of the human body; and the last three with the Newtonian account of the universe. Bentley, following Newton, made a simple case: gravity required that God constantly 'inform and actuate' the universe; gravity was 'the immediate *Fiat* and Finger of God, and the Execution of the Divine Law . . . which at once, if it be proved, will undermine and ruine all the Towers and Batteries that the Atheists have raised against Heaven'. This was 'a New and Invincible Argument for the Being of God'.[50] Moreover, our solar system could not have come into existence by chance but required the deliberate organization of its component parts in order to create a stable system. A benevolent deity could thus be shown to have created both the universe and humankind.

It is these new arguments which made possible a profound shift in the culture of both scientists and theologians in the years after 1692. The old arguments from the spirit world were discarded (in all the published Boyle Lectures from the eighteenth century I can find only one passing reference to witchcraft) and a new rationalized (as we would see it) theology put in their place. Bentley's Newtonian Christianity was presented as an alternative not only (implicitly, if not explicitly) to belief in demonic activity but also to the excessive rationalism of the Cartesians. Bentley's target here was Thomas Burnet, who, in his *Sacred Theory of the Earth* (Latin, 1681–9; English, 1684–90) had sought to give a scientific account of Noah's Flood.

As we have seen, Aristotelian philosophers believed the sphere of water was ten times larger than the sphere of Earth, so that for them the puzzle was not why did the waters cover the land during the Flood, but why did they not always do so. Once the sphere of waters and the sphere of land had been merged into one, it seemed evident that there was not enough water to cover the whole surface of the globe. Moreover, Cartesians held that the universe was a plenum: it was full. So if God chose temporarily to create more water he would have simultaneously to destroy the matter that currently occupied the space into which he was going to place the water. Burnet found this implausible. Instead, he hypothesized that the Earth was once a perfectly smooth shell completely surrounding the water; a crisis provoked the shell to crack and large parts of it then fell beneath the waves, thus creating the Earth as we know it today. Burnet's argument was met with widespread horror; as Herbert Croft, the bishop of Hereford, put it, 'This way of philosophizing all from natural causes, I fear, will make the whole world turn scoffers.'[51] Bentley, drawing on works such as John Ray's *The Wisdom of God Manifested in His Works of Creation* (1691), insisted that the Earth had been made from the start with oceans and harbours for the benefit of mankind. Consequently, the Flood had to be seen as a true miracle, not a mere natural event which happened to coincide with an excess of human depravity.[52]

So, from one point of view, Bentley's argument was a restatement of traditional Christianity, much more conservative, for example, than William Whiston's *A New Theory of the Earth* (1696), which developed

Newton's account of the cosmos to argue that the Flood was the result of a close pass by a comet. But in arguments about devils and witches Bentley was on the side of the radicals, as is apparent from his later attack on one of the great irreligious texts of the early eighteenth century, Anthony Collins's *Discourse of Freethinking* (1713). In his *Remarks upon a Late Discourse of Freethinking* (1713), Bentley, hiding behind the pseudonym of Phileleutherus (lover of freedom), makes clear that he no more believes in witchcraft than does Collins, referring with approval to Balthasar Bekker's *The Enchanted World* (1691), an influential attack on belief in witchcraft and demonic possession originally published in Dutch, and to a work by Samuel Harsnett, probably the *Declaration of Egregious Popish Impostures* of 1605. Harsnett had been much influenced by Reginald Scot, and had set out to prove that cases of supposed demonic possession were in fact deliberate frauds. Bentley is thus, as his choice of name indicates, prepared to meet the freethinkers on common ground at least when it comes to witchcraft, while defending the established Church.

The chief importance of Bentley's *Remarks* for present purposes lies in the explanation he offers for the decline of belief in witchcraft:

> In the dark times before the Reformation, not because they were Popish, but because Unlearn'd, any extraordinary Disease attended with odd Symptoms, strange Ravings or Convulsions, absurd Eating or Egestion, was out of ignorance of *natural* Powers ascrib'd to *Diabolical*. This Superstition was universal, from the Cottages to the very Courts; nor was it ingrafted by Priestcraft, but is implanted in Human Nature: no Nation is exempted from it; not our Author's *Paradise* of *New Jersey*, where no *Priests* have yet footing: and if the next Ages become unlearn'd, That Superstition will, I will not say return, but spring up anew. What then has lessen'd in *England* your stories of Sorceries? Not the *Growing Sect* [of Freethinkers], but the Growth of Philosophy and Medicine. No thanks to Atheists, but to the Royal Society and College of Physicians; to the *Boyle's* and *Newton's*, the *Sydenham's* and *Ratcliff's*. When the People saw the Diseases they had imputed to Witchcraft, quite cur'd by a course of Physic, they too were cur'd of their former Error: they learn'd Truth, by the *Event*; not by a false Position *a priori*, That there was neither Witch, Devil, nor God.[53]

Note the care with which Bentley formulates his view: systematic denial of belief in witch, devil and God is the false position of the atheist; rejection of the superstitious belief that witchcraft causes diseases is the correct position of the philosopher. The atheist argues '*a priori*'; the philosopher argues 'by the *Event*', in other words from experience. Bentley surely did not think that Boyle, Newton, Sydenham and Ratcliffe had directly attacked belief in witchcraft; and he probably knew that Boyle had been all in favour of such belief. Rather he meant that, whatever their intentions, the new sciences had undermined credulous belief. The new attitude to evidence, which Sprat had praised as 'this Inquiring, this scrupulous, this incredulous Temper', had encouraged a general scepticism of miracles, providences and witchcraft. As God was increasingly assumed to work by his 'known, and standing Laws', not by miracles, so, too, the Devil was held to operate through the ordinary temptations of vice, not the extraordinary means of possession and incantation.[54] Bentley is thus an advocate of the Thomas thesis: an improved technology for dealing with disease, combined with better scientific knowledge, undermined belief in magic and witchcraft. On the one hand he places learning, and on the other superstition. (Of course, there was no improved technology for treating disease, but Bentley evidently thought that medicine was making great strides, even though this belief now seems unjustified.)

He was not alone in assuming that the progress of science destroyed superstitious beliefs. Indeed, the erosion of traditional beliefs had been going on for quite some time. First under attack had been belief in fairies and hobgoblins (which, according to Reginald Scot, had largely disappeared among the educated by the 1580s); then had come the doctrine of sympathies;[ix] and this had been followed by the belief in prodigies – strange shapes in the clouds, double and triple suns, comets, monstrous births – which were held, as in ancient Rome, to herald catastrophe. In *A Discourse Concerning Prodigies* of 1663 John Spencer had argued that natural philosophy was the proper cure for superstition:

> Its the nature of all knowledge to give a kinde of strength and presence of minde to a man, but especially of *Philosophy:* this will secure us, as from the rocks of *Atheism* because leading us into a notice of some First cause, into which all the second doe gradually ascend and finally resolve; so also from the shelves of *superstition*, because acquainting us with the second causes: for fancy is apt to suggest very monstrous and superstitious notions of those things of whose causes and natures we are unresolv'd; all which

ix Above, pp. 243, 269, 275–7, 281, 434 and 457.

> flie (like the shadows of the twilight) before the approaching beams of knowledge. Philosophy leads us (as men doe horses) close up to the things we start at, and gives us a distinct and through view of what frighted us before, and so shames the follies and weakness of our former fears.[55]

The attack on prodigies was part of a larger post-Restoration programme to undermine 'Enthusiasm' (particularly the belief in immediate inspiration by the Holy Spirit), in the conviction that it could only lead to civil conflict.[56] Thus Sprat had wanted to emphasize the tendency of the new science to moderate the 'extravagances' of those who believed in providences and wonders:

> Let us then imagin our *Philosopher*, to have all slowness of belief, and rigor of Trial, which by some is miscall'd a blindness of mind, and hardness of heart. Let us suppose that he is most unwilling to grant that any thing exceeds the force of *Nature*, but where a full evidence convinces him. Let it be allow'd, that he is alwayes alarm'd, and ready on his guard, at the noise of any *Miraculous Event*; lest his judgment should be surpriz'd by the disguises of *Faith*.[57]

Enthusiasm, in Sprat's view, by making false claims of divine intervention in the world, simply offered hostages to unbelief. It was necessary to pare down faith to a set of core beliefs in order to make it defensible.

The key value that took the place of credulous piety was politeness, which was the great preoccupation of writers in the late seventeenth and early eighteenth centuries. Spencer already reflects this new concern when he redescribes Christianity as follows:

> But they which talk of and look for any such vehement expressions of Divinity now [as occurred in the Old Testament], mistake the temper & condition of that Oeconomy which the appearance of our Saviour hath now put us under; wherein all things are to be managed in a more sedate, cool, and silent manner, in a way suited to, and expressive of the temper our Saviour discover'd in the world, *Who caused not his voice to be heard in the streets*; and to the condition of a Reasonable Being made to be manag'd by steady and calm arguments, and *the words of Wisdom heard in quiet*; the mysteries of the Gospel come forth cloth'd in sedate and intelligible forms of speech; the minds of men are not now drawn into ecstasie by any such vehement and great examples of Divine Power and Justice as attended the lower and more servile state of the World. The miracles our Saviour wrought were of a calm and gentle nature [curing the blinde,

restoring the sicke and lame, not causing of thunder and storms, as *Samuel*, but appeasing them].[58]

It would be easy to think of witchcraft, too, as belonging to a more primitive dispensation, inappropriate in this new sedate, cool, quiet and reasonable age. Spencer's argument is of course ambiguous – has to be ambiguous – as to just when the 'lower and more servile state of the World' came to an end. Was it really with the birth of our Saviour? Or was it perhaps with the Reformation, or even with the Restoration?

Between 1653 and 1692 many of the new philosophers were concerned to assert their orthodoxy by demonstrating their belief in angels and demons, even though these arguments fitted uncomfortably with the sedate, polite world which, in other respects, they aspired to occupy. After 1692 Newtonianism offered a viable alternative argument for faith, the argument from the balance of probabilities, which was set loose, first against witchcraft and then eventually against miracles (with Middleton's *Free Enquiry* of 1747 and Hume's essay of 1748). Arguments for belief in magic and witchcraft were largely abandoned. But, over time, the middle ground that people like Sprat and Bentley had sought to occupy between superstition and rationalism became increasingly embattled, and the pendulum began to swing the other way. As the gospel miracles came (at least implicitly) under attack, what had recently been regarded as superstition became respectable again. Hogarth represents this new world in *Credulity, Superstition and Fanaticism* (1762), which, as well as satirizing contemporary events, looks back to the beginning of the century, when such views had last been voiced: Mary Toft is giving birth to rabbits; a copy of Glanvill is piled with Wesley's sermons under the thermometer of fanaticism, and on the top of the thermometer stands the drummer of Tedworth. An era of scepticism was giving way to a new species of Enthusiasm.

§ 6

Thus, in simple terms, Bentley was right: the new science did undermine belief in magic and witchcraft, just as it undermined belief in astrology and alchemy. But this process was not straightforward. Between 1653 and 1692 belief in witchcraft and the practice of alchemy often went hand in hand with the new science and, if the new science eventually proved incompatible with both, this was for many an unintended,

Hogarth's *Credulity, Superstition and Fanaticism. A Medley* (1762). Around a dozen cases of witchcraft and superstition are represented: among them, the drummer of Tedworth stands on top of the thermometer on the right (the thermometer itself stands on a copy of Glanvill), while Mary Toft gives birth to rabbits in the foreground.

not an intended, consequence of the new philosophy. Only after 1692 did a new rationalism begin to take a secure hold; and when that rationalism later came under sustained attack from John Wesley, the result was not its defeat but the emergence, for the first time, of two cultures, of science on the one hand and faith on the other.

For the remarkable thing about Wotton and Bentley is that they were simultaneously theologians and advocates of the new science, while Swift, who mocked them both in *The Battle of the Books*, was a clergyman who knew as much science as they did. In England scientists and clergymen still inhabited a common culture, and there was no division between them when it came to belief or disbelief in magic and witchcraft. In being clergymen with an interest in science, Wotton and Bentley were typical among the early supporters of Newtonianism. Many of the leaders of the Newtonian party were clergymen: John Harris, author of the *Lexicon technicum* (1704); Samuel Clarke, Boyle Lecturer and Newton's champion against Leibniz; James Bradley, Savilian professor of astronomy at Oxford and astronomer royal; William Derham, author of *Physico-Theology* (1713), which went through numerous editions and translations; William Whiston, Newton's successor as Lucasian professor of mathematics at Cambridge; and so on. Unbelievers – and there were plenty of them – were usually more interested in classical learning than in contemporary science; they published works with titles such as *The Two First Books of Philostratus, Concerning the Life of Apollonius Tyaneus: Written Originally in Greek, and now Published in English: Together with Philological Notes upon each Chapter* (Charles Blount, 1680); the bulk of Bentley's attack on Collins is given over to a dispute over the interpretation of a passage in Cicero. The battle lines of the nineteenth century were yet to be drawn.

Making and remaking the common culture that bound together clergymen, mathematicians, instrument makers and aristocrats such as James Brydges, Duke of Chandos, and George Parker, second Earl of Macclesfield, required constant effort.[59] We can distinguish four components to the project. First, there was the need to provide a Newtonian education at the universities: the first Newtonian textbook in physics was John Keill's *Introductio ad veram physicam* (1701), which competed with Samuel Clarke's adaptation of Rohault's *Physica* (1697), in which Rohault's Cartesianism was steadily swamped by Clarke's Newtonian commentary as one edition succeeded another. Newton himself was simplified by Willem 's Gravesande, first in English (1720) and then in Latin (1723); and made even simpler by John Pemberton in his

A View of Sir Isaac Newton's Philosophy (1728). Second, Newtonian Christianity had to be defended against its critics, with works such as James Jurin's *Geometry No Friend to Infidelity* (1734). Then, Newtonianism had to be made popularly accessible. Whiston's *New Theory* (1696) was the first detailed popular account of the arguments of the *Principia*, but it was rapidly followed by works such as Nehemiah Grew's *Cosmologia sacra* (1701), Edward Wells's *The Young Gentleman's Astronomy* (1718), John Harris's *Astronomical Dialogues between a Gentleman and a Lady* (1729), Voltaire's *The Elements of Sir Isaac Newton's Philosophy* (1738) and Francesco Algarotti's *Sir Isaac Newton's Philosophy Explain'd for the Use of the Ladies* (1739; the book went through thirty editions in six languages).

The seeming obsession with the education of ladies derives partly from the imitation of Fontenelle's *Dialogues* (in France women had a central place in the culture of the *salon*) and partly through the example of Émilie du Châtelet, Voltaire's companion, who was a competent mathematician and translated the *Principia* into French (1756).[60] Even Voltaire, who avoided the dialogue between a philosopher and a lady, a form popularized by Fontenelle, liked to imagine his book being read by a sophisticated woman at her dressing table. And surely he had some readers of this sort; Voltaire corresponded with Laura Bassi, the first woman to obtain a degree from the University of Bologna (1732), and the first to teach there. Bassi held a chair in physics, and naturally taught the physics of Newton.[61] In England, it was the playwright Aphra Behn who translated Fontenelle and the poet Elizabeth Carter who translated Algarotti, so the female audience was more than fictional.[62]

These three components of the campaign on behalf of Newtonianism developed momentum over time. A simple measure of this is to count books in which Newton's name appears in the title: the peak period clearly runs from 1715 to 1745. When Samuel Johnson, in his essay on 'The Vanity of Authors' (1751), wrote, 'Every new system of nature gives birth to a swarm of expositors, whose business is to explain and illustrate it, and who can hope to exist no longer than the founder of their sect preserves his reputation,' he was taking for granted a state of affairs that was entirely new.[63] No one had popularized systems of nature before Fontenelle's *Dialogues on the Plurality of Worlds* (1686); the Newtonians adopted and adapted the techniques of the Cartesians in order to make a much more abstruse and complex intellectual system available to a mass audience. In the process they sought not only to preserve the idea of a common culture shared by all the educated but also to adapt that idea to a new era of cheap books, mass communication and nearly universal literacy.

This was not all. The best way of communicating the experimental philosophy was through making it possible for people to see experiments performed. John Harris gave public lectures, accompanied by experiments, in London from 1698 to 1707, teaching 'the principles of true Mechanick Philosophy'. He was soon competing with James Hodgson, Francis Hauksbee the elder and Humphrey Ditton. In 1713 William Whiston (who had been expelled from Cambridge in 1710 for his heretical views) began lecturing and demonstrating in London. In January he was lecturing from his home, and also with the elder Francis Hauksbee; in the spring he was lecturing and demonstrating with the younger Francis Hauksbee (nephew to Francis Hauksbee the elder) in Crane Court, on mathematics at Douglas's Coffee House, and at the Marine Coffee House. The greatest of the popular lecturers was John Theophilus Desaguliers (another Newtonian clergyman, although he paid little attention to his clerical duties and showed few signs of religious conviction), who began lecturing and demonstrating in London in the spring of 1713, publishing his *Physico-Mechanical Lectures* in 1717. By 1734 he had delivered 121 courses, not only in London but also on provincial tours and in the Low Countries, and could boast that, of the dozen or so professional lecturers on the circuit, eight had been educated by him. Indeed, courses of lectures were available far and wide: in Newcastle, in Spalding, in Scarborough, in Bath.[64]

It would be easy to imagine that Newton, because of the intellectual level at which he operated, brought about a new professionalisation of science, so that it became an esoteric activity in which only an elite could participate.[65] But the reverse is the case. From the end of the seventeenth century, through sermons and lectures, through popular textbooks and dramatic dialogues, the new science was disseminated to a wider audience than ever before. If it played a role in disenchanting the world, it did so precisely because it was effectively inculcated among the educated, clerical and lay, male and female. The real historical puzzle, we might think, is not this eighteenth-century loss of belief in witches and demons, but the progressive re-enchantment of the world in the nineteenth century.

14

Knowledge is Power

> I should not have neer so high a value as I now cherish for Physiology, if I thought it could onely teach a Man to discourse of Nature, but not at all to master Her; and served onely, with pleasing Speculations, to entertain his Understanding without at all increasing his Power.
>
> – Robert Boyle, *Some Considerations* (1663)[1]

§ 1

What is the relationship between the Scientific Revolution and the Industrial Revolution, between the mathematicians' revolution and the mechanical revolution? The claim with which this book opened, that the Scientific Revolution is the most important event since the Neolithic Revolution, depends on our answer to this question; for if the Scientific Revolution was merely an event in the world of ideas, its importance is relatively limited, while if it opened the way to a new control over nature, then the Industrial Revolution can be seen as merely an extension of the Scientific Revolution, the extension of the procedures, language and culture of the new science to a wider social stratum of technicians and engineers. There is no doubt that Bacon and his followers aspired to transform the world through their new science. In the mid-eighteenth century Birch's *History of the Royal Society* carried as its epigraph a quotation from Bacon: 'Natural philosophy as I understand it does not slip away into sublime and subtle speculations, but is applied effectively to relieve the inconveniences of the human condition.'[2] The motto of the French Academy of Sciences, founded in 1666, was *naturae investigandae et perficiendis artibus* ('the investigation of nature and the improvement of technology') changed in 1699 to the snappier *invenit et perficit* ('progress through discovery').

It is easy now to find some of the first scientists' expressions of enthusiasm naïve: thus Ambroise Sarrotti, who had come to England to accompany his father, Paolo, the Venetian ambassador (1675–81), returned home to organize a scientific society which conducted experiments with vacuums.[i] At the end of the first year he proudly announced to his colleagues: 'If, since the beginning of the world, all mankind united together could have done every year, as much as you alone have done this year last past, they would live now as happy in this world, as in a terrestrial paradise.'[3] This, despite the fact that they had discovered nothing of any use whatsoever. It is not surprising then that not everyone was convinced of the practical utility of the new science.[4] Jonathan Swift wrote Book 3 of *Gulliver's Travels* (1726) with the sole aim of denying it. Yet his attack suggests that he hesitated as he tried to define the nature of the enemy. Laputa is an airborne island, ruled by scientists who are so obsessed with matters mathematical that they are unable to pay any attention to the world around them: they rely on the services of flappers, who strike their ears and mouths with inflated bladders in order to remind them when to listen and when to speak. But down below in Balnibarbi, the colony over which they rule, an academy has been formed, in imitation of Laputa, where scientists pursue practical objectives in the most impractical ways, making sunbeams from cucumbers, and thread from spider's webs. The governor general, who alone disapproves of the new inventions, tells Gulliver:

> That he had a very convenient Mill within Half a Mile of his House, turned by a Current from a large River, and sufficient for his own Family, as well as a great Number of his Tenants. That about seven Years ago, a Club of those Projectors came to him with Proposals to destroy this Mill, and build another on the Side of that Mountain, on the long Ridge whereof a long Canal must be cut, for a Repository of Water, to be conveyed up by Pipes and Engines to supply the Mill: Because the Wind and Air upon a Height agitated the Water, and thereby made it fitter for Motion: And because the Water, descending down a Declivity, would turn the Mill with half the Current of a River whose Course is more upon a Level. He said, that being then not very well with the Court, and pressed by many of his Friends, he complyed with the Proposal; and after employing an Hundred Men for two Years, the work miscarryed, the Projectors went off, laying

i Giovanni Ambrosio Sarotti (or Sarrotti) was elected a Fellow of the Royal Society on 1 Dec. 1679: Hunter, *The Royal Society and Its Fellows* (1982), no. 356.

> the Blame intirely upon him, railing at him ever since, and putting others upon the same Experiment, with equal Assurance of Success, as well as equal Disappointment.[5]

He does not say what sort of 'engines' were employed, but Swift surely had in mind the first steam engines, which were generally used for raising water. So on Swift's account the new science is both totally impractical and at the same time obsessed with practicality. This is not an impossible combination – indeed, it seems to describe Sarrotti rather well – but it is certainly a puzzling one.

Historians of science have not advanced our understanding of the relationship between the new science and technological progress much beyond Swift. Naturally, Marxist historians have wanted to argue that the new science was the result of new social relations. As the Russian Boris Hessen (who was executed in 1936, an early victim of Stalin's Great Purge) put it in 1931, 'Step by step, science flourished along with the bourgeoisie. In order to develop its industry, the bourgeoisie required a science that would investigate the properties of material bodies and the manifestations of the forces of nature.' But Marxists were not alone in assuming that the new science was motivated by its possible practical applications: Robert K. Merton in his classic study of 1938, *Science, Technology and Society in Seventeenth-century England*, in which he emphasized the role of Puritanism in encouraging useful knowledge, followed Hessen in arguing that seventeenth-century science was indeed intended, through and through, to have practical applications, despite his own rejection of Hessen's Marxist assumptions.[6]

A series of studies, however (those of Alfred Rupert Hall being particularly influential), have claimed to show that, whatever the intentions of scientists may have been, in practice, the new science had virtually no influence on technological progress. A key case-study was provided by Watt's steam engine (1765). Watt developed his new engine in Glasgow, where Joseph Black had proposed the concept of latent heat (*c*.1750). Black later collaborated with Watt and invested in his new engine. Was Watt familiar with the concept of latent heat when he devised his new engine, and did the new theory inform his new technology? He insisted that he was not, and historians came (almost reluctantly) to take him at his word.[7] Lawrence Joseph Henderson is frequently quoted as saying (apparently in 1917), 'Science owes more to the steam engine than the steam engine owes to science.'[8] After all, Sadi Carnot finally produced a satisfactory theory of the steam engine only in 1824, more than a hundred years after Newcomen's first engine, and sixty years after Watt's. Hall thought it was 'not quite' but very nearly true to say that 'engineering owed nothing to science' until very late in the eighteenth century. Thomas Kuhn thought that science and technology were antithetical to each other, at least until the 1870s.[9]

One might think that the historians of technology would have wanted to question this disjuncture between theory and practice – but at first they were the same people as the historians of science.[10] The major attack on the established orthodoxy has come only very recently, and from an unexpected quarter: the new economic historians of the Industrial Revolution, who emphasize the importance of skills and technical innovation, of what they call 'the knowledge economy'.[11]

On this question the new economic historians are (as will become apparent) in the right. But those who argue that science played a key role in the Industrial Revolution need to have an answer to a simple and by now classic question: What role did science play in the invention of the steam engine? Before tackling this problem, however, we need to unpack the apparently straightforward notion of practical knowledge. The key issue here is one of timescale: How long should one wait before dismissing a theoretical achievement or a technological advance as having little or no practical relevance? Does, as Hall assumed, the new science have to be *contemporary* with the technology that derives from it?[12]

Take ballistics: Galileo initially hoped that his discovery of the law (as we say) of fall, and with it of the parabolic path of projectiles, would revolutionize gunnery. When his disciple Torricelli entered into practical tests to see if Galileo's theory described how cannon balls actually fly, he discovered that it didn't: he insisted that the theory remained sound, even though it could not be applied to fast-moving projectiles because the effects of air resistance were not adequately understood (it could, it turned out, be applied to mortar shells fired short distances at low speeds).[13] Ballistics was finally revolutionized between 1742 and 1753 by Robins and Euler, with the discovery of the sound barrier and an understanding of the effects of rotation in flight (deliberately induced, of course, by rifling; but Torricelli's cannon balls were tumbling as they flew), and, as a result, the production of equations for reliably calculating trajectories. Galileo's physics was intended to be practical but turned out to be of little practical use in its most obvious field of application. Nevertheless, his idealized parabolic trajectory in a vacuum was an essential precondition for the much more sophisticated analysis by Robins and Euler of actual trajectories. Galileo's theory *was* practical; it just took a full century to pay off. For the young Napoleon, who was exceptionally good at mathematics, the problems which had defeated the great Torricelli were by the 1780s mere school exercises – the school, of course, being the École Militaire.[14]

Or take the challenge that preoccupied Galileo through a large part of his working life: that of establishing longitude at sea. Degrees north and south (latitude) are easy to calculate, providing one knows the date, from the height of the sun at midday; degrees east and west (longitude) are much harder to establish, since there is no obvious reference point that one can use. Galileo theorized that one could use eclipses of the moons of Jupiter (which he had discovered in 1610) as a sort of universal clock. With reliable tables predicting future eclipses, one ought to be able to tell the exact time wherever one was in the world; and, if one knew the local time (time elapsed since midday, for example) one could then compare local time with the time at the place for which the tables were calculated, and then easily calculate degrees west or east of the reference point. The theory was fine. Calculating the movement of the moons was less straightforward, but Galileo and his associates made strenuous efforts, and Galileo even constructed a little mechanical model, the Jovilabe, which enabled him to work out the location of the moons without doing complex calculations; he would have done better, of course, if he had known that it was also necessary to make an allowance for the speed of light, since the time at which an eclipse seems to take place varies depending on how far away Jupiter is from the Earth.

The central problem, however, was simple: How could one look through a powerful telescope at a tiny, distant object and make reliable observations while on a ship bouncing about on the waves? Galileo devised a pair of powerful binoculars which were clamped to the head, since it was hard to hold a telescope still enough on a moving boat, in what amounted to a gimballed chair in which one could sit to observe (compasses were already mounted in gimbals). Solving the problem of longitude was a universally recognized challenge: indeed, governments had promised enormous rewards for anyone who could succeed. Galileo hoped to establish his immortal fame by this discovery more than any other: he tried to claim the reward offered by the Spanish government, but failed (his pupil Castelli went to sea, but became hopelessly seasick); and in his last years he entered into clandestine negotiations with the Dutch government in the hope that they would take up his ideas and make them work in practice, but he failed.[15]

Or did he? By 1679 the Cassini family (who had emigrated from Italy to France, where they became famous as astronomers and cartographers) were using the moons of Jupiter to calculate longitude, if not at sea then at least on dry land. Such measurements guided their recalculation of the size of France (France turned out to be a full 20 per cent smaller than previously believed), and their calculation of the shape of the globe (which turned out to be good news for Newtonians, and a devastating blow for Cartesians). Galileo was right: the moons of Jupiter were a promising way of measuring longitude. It just took sixty years to make his proposal work in practice, and then it worked only if one had one's feet on solid ground.[16]

There were alternative projects for calculating longitude. For a long time it was hoped that measuring compass deviation and dip would enable sailors to establish their coordinates. Despite generations of effort, this proved illusory, because dip and deviation change unpredictably over time.[17] The simplest scheme in the end turned out to be the best: all one needed to do was to take a reliable timepiece on the journey and use it to measure the difference between local time (local noon, for example) and the time at the reference point (the Greenwich meridian, for example).

Galileo believed that he had shown that pendulums tell perfect time, and devised a pendulum clock (though he did not build it; by the time he turned his attention to the question he was blind, and his son, who tried to help him, lacked the necessary manual skills). Huygens, without knowing of Galileo's work, went on to build the first pendulum clock (1656) and to refine the law of the pendulum (1673). Meanwhile, Robert Hooke, Huygens and Jean de Hautefeuille devised between 1658 and 1674 ways of controlling a balance wheel (which had been invented in the fourteenth century, and was more stable than a pendulum for a travelling timepiece) with a spring so that small clocks and watches would tell time reliably. Still, the task of making a seagoing clock or watch was far from solved: such a timepiece had to remain accurate despite changes in temperature and humidity, and despite the movement of the waves. The problem was not solved until John Harrison produced the first reliable marine chronometer in 1735.[18] Were the discoveries of Galileo, Hooke and Huygens irrelevant? Certainly not, but they were insufficient. It took more than a century to solve the problem, but in the course of that century steady progress was made towards a solution.

Clockwork, of course, was not a seventeenth-century innovation. As we have seen, the first mechanical clocks date to the late thirteenth century, and their geared machinery derived from water-wheels and

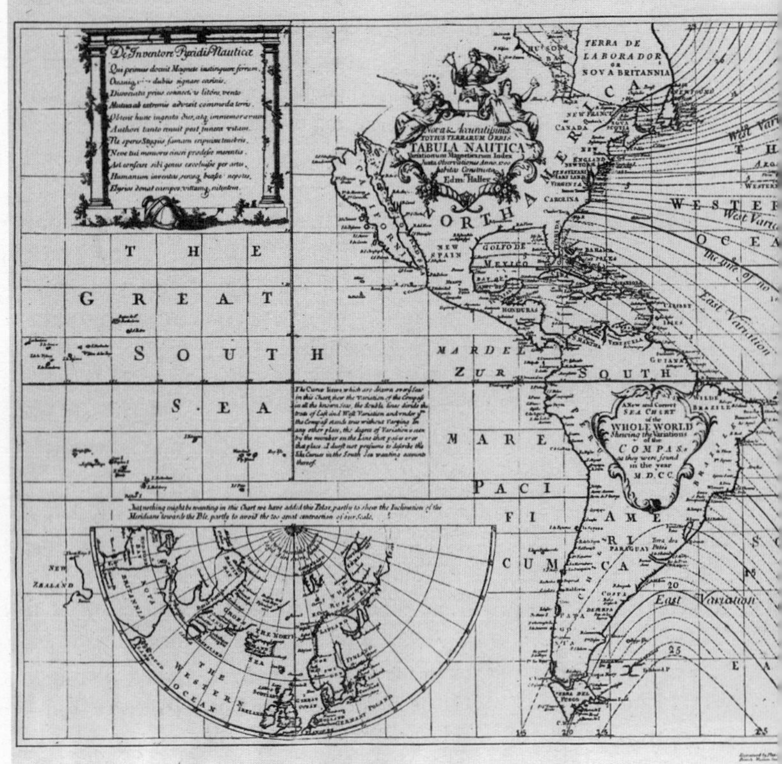
HALLEY'S M
De Inventore Pyxidis Nauticæ
TERRA DE LABORADOR OR NOVA BRITANNIA
NEW SPAIN
CANADA
VIRGINIA
CAROLINA
GOLFO DE MEXICO
HONDURAS
GUIANA
NORTH AMERICA
Nova & Accuratissima Totius Terrarum Orbis TABULA NAUTICA
Edm. Halley
West Variation
East Variation
WESTERN OCEAN
THE GREAT SOUTH SEA
MAR DEL ZUR
SOUTH
MARE PACIFICUM
AMERICA
A New and Correct SEA CHART of the WHOLE WORLD Shewing the Variations of the COMPASS as they were found in the year M.D.CC.
BRAZILE
PARAGUAY
PATAGONIA
TERRA DEL FUEGO
NEW ZEALAND
THE WESTERN OCEAN
THE NORTH SEA
IRELAND
POLAND
GERMANY
NOVA BRITANNIA
20
25
15

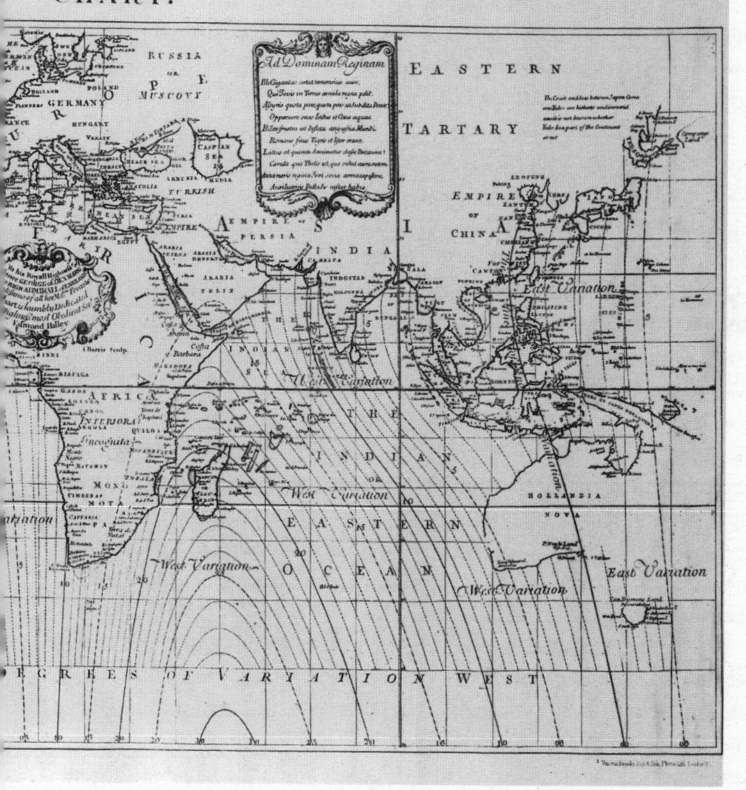

Halley's isogonic map of magnetic variation, published in 1701. Each line on the map is like a contour line but, instead of marking a uniform measurement of height, it marks a uniform measurement of magnetic variation. Halley had conducted two expeditions to make the measurements on which the map was based, and the hope was that this would open the way to using magnetic variation to measure longitude.

windmills. Water-wheels were known to the ancient Greeks and Romans but were far from common; thanks to an early medieval proto-Industrial Revolution, they quickly became widespread around the end of the first millennium CE. The Domesday Book records more than six thousand mills driven by water-wheels in England in 1086. Vertical windmills followed quickly: the first securely dated one was in Weedley, Yorkshire, in 1185. Given that the greatest concentration of medieval watermills was in England, it is surely not a coincidence that it was in England that we find both the first recorded vertical windmill and the first recorded clock. Steam did not overtake water and wind as a source of power until after 1830;[19] in Swift's Laputa, as in eighteenth-century England, steam power did not replace water power but supplemented it.

Nevertheless, it has been claimed that the innovations of Galileo, Hooke and Huygens made possible the geared machinery of the Industrial Revolution.[20] Before the mid-seventeenth century, gears were laid out and cut by hand; Hooke designed the first machine to produce identical gears, thus making possible the mass production of machinery. Inevitably, eighteenth- and nineteenth-century engineers turned to clockmakers to build their machines (Richard Arkwright, for example, worked with the clockmaker John Kay to produce the spinning frame in 1769), and the quality of what they could accomplish had been greatly enhanced as a result of the revolution in clockmaking that had taken place in the years after 1656.[21]

Mechanical clocks give us a valuable chance to think comparatively, because we can see how other cultures responded when they were introduced to them by European travellers in the sixteenth century. The Japanese were soon manufacturing their own clocks (just as they were quick to manufacture their own guns); while the Chinese showed little interest in using clocks for telling time, nor in making their own, despite the fact that Su Song had produced a sophisticated water-driven clock for astronomical purposes in the eleventh century. As far as they were concerned, clocks were merely delightful but useless luxury objects – rather like musical boxes. (The Chinese were similarly slow to adopt the technology of the military revolution, despite the fact that gunpowder originated in China.) There was thus nothing automatic about the widespread adoption of the clock which took place in medieval Europe.[22]

Yet clocks spread rapidly in the fourteenth and fifteenth centuries: first, because Europeans were already mechanically minded (all those water-wheels and windmills); second, because their circular geared movements reflected in miniature the movements of the Ptolemaic heavens (early clocks often measured astronomical time – the phases of the moon, the signs of the zodiac – as well as diurnal time); and third, because clocks provided an impersonal mechanism for the coordination of community activities (the saying of the offices in monasteries and cathedrals, the opening and closing of markets in towns and cities). Egalitarian communities (cities, monasteries and cathedral chapters all chose their leaders through elections) are governed by the clock, while despotisms are not; clocks were given prominent, public places in monasteries, cathedrals and town halls, but they were slower to establish themselves in royal palaces. (Even now, my university campus, built in the 1960s, is dominated by a clock tower which is there not to tell the time but to convey the impression that ours is a disciplined, egalitarian community.) These factors – cultural, technological, conceptual, political – were absent in China, and hence the Chinese admired clockwork but had no use for it.

Clockwork obviously fostered the notion that the universe could be understood as a complex mechanism, and Copernicans were committed to the view that the same physical principles were at work in the heavens and on Earth. Thus Kepler could write in 1605, inspired by his reading of Gilbert on magnetism:

> My aim is this, to show that the celestial machine is not like a divine creature, but like a clock (he who believes the clock to be animate assigns the glory of the artificer to the work), insofar as nearly all the diversity of motions are caused by a simple, magnetic and corporeal force, just as all the motions of a clock are caused by a most simple weight.[ii] I will also show how this physical account is to be brought under mathematics and geometry.[23]

But medieval and Renaissance clocks were so imperfect that not only did the weight which drove them have to be raised every day, the time also had to be corrected, and it was only the improved clocks of Huygens which made it possible to think of the universe as a perfect, clock-like mechanism which required no tending by the divine clockmaker. We find the new post-Huygens imagery grafted on to Descartes' imagery of 'automata' as early as 1662, six years after the first

ii Most large early clocks are driven by a weight which slowly descends: grandfather clocks had to be tall in order to give the weight space in which to descend. The descent of the weight is regulated by an escapement mechanism and, before the pendulum, it was the inaccuracy of this regulation that was the source of error.

pendulum clock, in this passage by Simon Patrick, an advocate of the new science:

> Then certainly it must be the Office of Philosophy to find out the process of this Divine Art in the great automaton of the world, by observing how one part moves another, and how these motions are varied by the severall magnitudes, figures, positions of each part, from the first springs . . .[iii][24]

Clockwork, by providing a fruitful metaphor, encouraged the Scientific Revolution, and, by fostering the development of sophisticated, geared machinery, it facilitated the Industrial Revolution, but it was not itself the product of either of those revolutions, nor was it a necessary precondition for either one, for there were other types of geared machinery.

There is another important example of the delayed pay-off in technological progress. Hydraulic engineering was a major concern for the first engineers, such as Leonardo, and consequently an immediate one for Galileo and his disciples. Galileo advised on drainage projects; his pupil Castelli advised the papacy on the management of rivers and published a major treatise on the subject (*Della misura delle acque correnti*, 1628); his pupil Torricelli made a major theoretical breakthrough when he formulated what is now known as Torricelli's law (1643), which enables one to establish the velocity of flow given a particular head of water (or the head given a known velocity of flow), and he also did practical work on the flow of the River Chiana, a tributary of the Arno; his pupil Famiano Michelini, who was also his successor as philosopher to the Grand Duke, also published on hydraulics (*Trattato della direzioni de' fiumi*, 1664).[25]

Yet a hundred years went by before John Smeaton in England, relying on Torricelli's work, set out to perform a systematic programme of experiments with model water-wheels in order to establish which designs were the most efficient (comparing undershot and overshot) and how each design could be made to work best: How big should the wheel be, and how fast should it revolve for maximum efficiency? How deep into the water should the paddles of an undershot wheel go? Smeaton discovered, to his surprise, that overshot wheels (where the water enters

iii Patrick has in mind a small clock driven by a spring, not a weight, but the plural suggests that he has particularly in mind a clock (or watch) with both a mainspring and a balance spring: the balance spring adapts the principle of the pendulum to a smaller space and to movement on the wrist and was invented by Hooke in 1658.

John Smeaton's model water-wheel in 'undershot' configuration: the wheel is two feet in diameter (from *An Experimental Enquiry*, 1760).

the wheel at the top) were twice as efficient as undershot wheels (where the water flows along the bottom of the wheel): theory had led him to expect them to perform equally (though Desaguliers had rightly suspected that, in practice, overshot wheels were superior),[26] and he had difficulty explaining why they performed so differently. Smeaton thus developed a series of practical rules of thumb to guide water-wheel construction, and was widely influential in instigating a shift from undershot wheels to overshot or, where this was not practical, to breast wheels (where the water strikes the wheel halfway up). It is at this point – and at only this point – that we can say that the work of Galileo and his pupils on water flow had finally paid off by facilitating a markedly improved practical technology.[27]

The case of water-wheels is a particularly interesting one because the technology had developed extremely slowly for almost a thousand years. Millwrights had learnt by trial and error what worked and what did not, but rapid advance required systematic experimentation, and this only occurred after the experimental method had been elevated to a new intellectual status. Smeaton himself had trained in the law and then apprenticed as a machine-maker before becoming an engineer (he was the first to call himself a 'civil engineer' – as opposed to a military engineer – and he went on to found a society of civil engineers)[28] and a Fellow of the Royal Society. He combined practical and theoretical knowledge, as Hooke had in watchmaking. And, of course, he was responding to an economic situation where the demand for power was growing rapidly. He went on to build steam engines, harbours, bridges and canals (including the Calder Navigation, a set of cuts and locks which made and still make the River Calder navigable).

What was the obstacle to performing Smeaton's experiments in the 1680s or even the 1580s?[iv] Smeaton's work depended on two intellectual preconditions being satisfied. First, it was well known that working with scale models could often be misleading, because full-size machines often performed quite differently. The conceptual apparatus for thinking about this problem had been provided by Galileo in his *Two New Sciences*, and Smeaton addressed one aspect of it, the fact that friction tends to be greater in scale models than in full-size machines, by cleverly measuring the amount of friction generated in his models and then compensating for it. Second, Smeaton's work depended on the systematic application of Torricelli's law. We might want to add a third precondition: in calculating the efficiency of a water-wheel by comparing the output of the wheel with the input of the stream, Smeaton was assuming a Newtonian law of the conservation of energy. In that sense his work was post-Newtonian. But he could have compared the output from different types of water-wheel without having an absolute measure of efficiency. Moreover, in defining force Smeaton steered clear of the conflict between the followers of Newton and the followers of

iv In 1627 Isaac Beeckman and his associates planned a programme of research using models to ascertain the relative merits of vertical and horizontal windmills. But there is no evidence that they actually implemented their plans: Berkel, *Isaac Beeckman* (2013), 38–9.

Leibniz over the definition of 'force' (a conflict now resolved by distinguishing momentum from kinetic energy): this conflict did not have to be resolved for his work to succeed.

It would thus seem clear that it would have been impossible to carry out Smeaton's experiments in the 1580s but perfectly possible in the 1650s, and straightforward once the arguments of Newton's *Principia* (1687) began to be widely understood. Nor was there anything new about working with models: Desaguliers was building model steam engines by the 1720s, and he was surely not the first. Yet it was not until the middle of the eighteenth century that Smeaton and Watt both used models to work out how to transform the efficiency of powered machinery. Those who think that modern science derives from the empirical experimental enquiries of craftsmen and artisans need to account for the extraordinarily slow evolution of water-wheel technology prior to the introduction of Smeaton's scientific method. To employ the experimental method systematically and self-consciously, as Smeaton and Watt did, one needed both a certain amount of sound theory and a confidence that experimentation, though it might be laborious, presented an excellent prospect of making major progress. The theory was not new in the 1750s, but the confidence was. The source of that confidence was a sustained programme of advertising the new science through public lectures and books conducted by the disciples of Newton, above all by Desaguliers.[29]

In the end, early modern science defeated two of the most difficult practical problems it had set itself: the calculation of the path of projectiles under real-life conditions, and the measurement of longitude. If seventeenth-century scientists did not see these problems solved, nevertheless they prepared the ground for their eighteenth-century successors, who did. In addition, in the mid-century, Smeaton and Watt transformed the efficiency with which water and steam power were harnessed to drive machinery; in the short term, Smeaton's achievement was the more important; in the long term, Watt's. In 1726, when these practical problems remained as yet unsolved, Swift's case against the utility of science seemed sound; it would have been much harder to make the case against in 1780, or even in 1750. Strangely, historians have remained stuck in Swift's world, and when they read texts such as Smeaton's they read them naïvely, as if they simply reflect a programme of tinkering around with models, and as if all the terminology used is commonsensical; they remain oblivious to the fact that it was the new science that discovered the relationship between the head of water and the speed of the stream.

§ 2

The first great, practical achievement of the new science was Newcomen's steam engine of 1712 – the very engine Swift was presumably mocking when he complained about mills being built where there were no rivers. It is important to put Newcomen's achievement into perspective. By 1800 only 2,200 steam engines had been built in Britain, some two thirds of which were Newcomen engines, and a quarter Boulton and Watt engines.[30] Between 1760 and 1800 almost twice as much new water power (much of it the result of Smeaton's work) as steam power became available.[31] The great age of steam still lay ahead: when Mary Shelley published *Frankenstein* in 1818 her vision of the horrifying power of the new science scarcely included steam (a single reference to 'the wonderful effects of steam' was added, probably by Percy Shelley, as the book went to press), although Blake was already writing about 'dark satanic mills' in 1804 (he probably had in mind the Albion Flour Mills, the first large factory in London, powered by a Boulton and Watt steam engine and built in 1786).[32] In 1807 Fulton's steamboat began a regular passenger service between New York City and Albany, the state capital; in 1819 the SS *Savannah*, a ship that combined sail and steam, crossed the Atlantic; Stephenson's *Rocket* rattled along the rails in 1829. By 1836 it was possible to describe steam as marking 'a new era in the history of the world'. It had multiplied the powers of mankind 'beyond calculation'.[33]

In 1712 the Industrial Revolution and the age of steam lay far in the future; by 1836 they were a reality. They had been summoned into existence by a new culture of technical expertise, by men like Watt and Smeaton and by England's high wages (for many of the new inventions were profitable only in a high-wage economy).[34] The steam engine did not make the Industrial Revolution inevitable, but it did make it possible. There had been high-wage economies before (after the Black Death, for example), but no Industrial Revolution. It is true that many of the new inventions that were central to the Industrial Revolution – Arkwright's spinning frame, for example – owed nothing to science; but without Smeaton's improved water-wheels and Boulton's and Watt's improved steam engines, the factories in which they were made could never have been powered.

In order to understand steam engines it may be helpful to think about methods of making coffee. Some people make it by dripping water through a filter containing coffee grounds: they are relying on gravity. Others use an espresso pot, which uses steam to drive water up through the grounds: the pot is a pressurized steam system, which is why it requires a safety valve. And some use the vacuum method, where water is driven up into a higher container by steam (at low pressure, for it has only to overcome the weight of the water), but then, when the heat is removed and the steam condenses, it creates a vacuum, sucking the water back down through the grounds. The vacuum method relies on atmospheric pressure.

The steam engine was the product of seventeenth-century science, which had experimented with vacuums, and with air and steam pressure.[35] A simple example of air pressure is the air gun, called in the seventeenth century 'the wind-gun'. Mersenne described one in 1644, which is also the first year in which one is referred to in English; Boyle published a design for one in 1682.[36] It worked by compressing air into a container with a bellows and using the compressed air to power a dart or pellet. Steam in a confined space could also be used to create pressure. This principle was employed by della Porta in 1606, and in 1625 Salomon de Caus devised a steam fountain. It worked just like a stove-top espresso machine: the pressure of steam in a chamber from which there was only one exit drove the water in the container upwards and out. Boyle's law provided a theoretical account of how pressure could be used to produce powerful force, if one could only find a way of harnessing that force for a useful purpose.

But there was an alternative to constructing some sort of high-pressure mechanism. That alternative – a low-pressure mechanism – derives from von Guericke's work with his air pump. Von Guericke had shown that, if one pumped the air out of a cylinder, the pressure of the atmosphere would drive a piston down in the cylinder and the force would be such that even a team of strong men would be unable to resist it.[37] In 1680 Huygens came up with an alternative way of harnessing atmospheric pressure. He used an explosion to drive the air out of a cylinder through a valve; then, when the hot gases cooled, a piston was sucked downwards, raising a weight.

This idea was taken up by Denis Papin, a medical doctor who had started his scientific career as an assistant to Huygens, performing air-pump experiments. He had then moved to England: Papin was a Protestant and life was becoming increasingly uncomfortable for Protestants in France. Here, he worked as an assistant to Boyle; on Boyle's own testimony, Papin devised many of the experiments published in

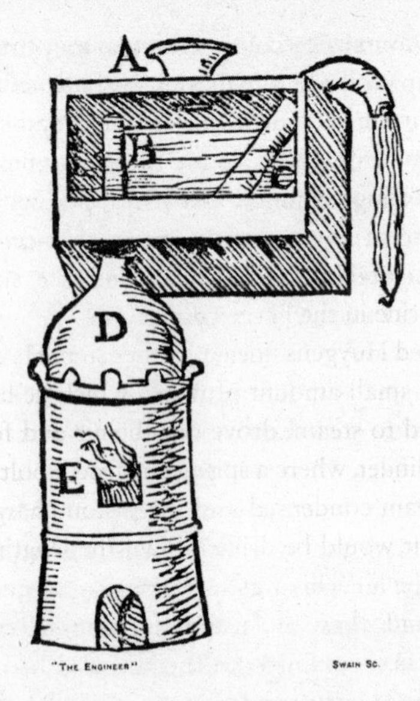

Giovanni Battista della Porta's steam pressure pump, from *Tre libri de' spiritali* (1606).

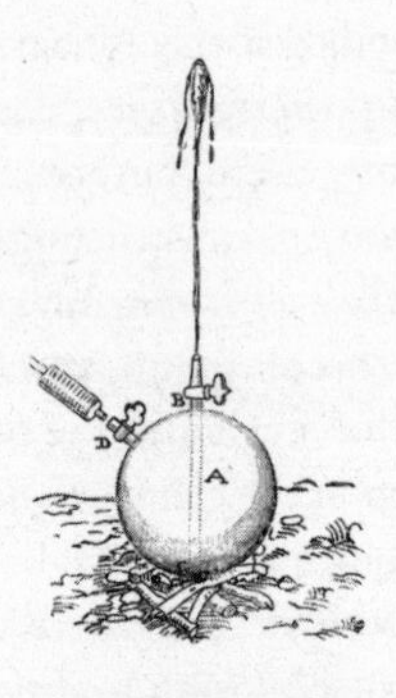

De Caus's steam-powered fountain, from *La Raison des forces mouvantes* (1615).

Boyle's *A Continuation of New Experiments* (Latin, 1680; English, 1682) and performed all of them. Indeed, the book was not really by Boyle at all, since it was written by Papin.[38] Papin was elected a Fellow of the Royal Society in 1680 (his social status was quite different from that of a mere technical assistant), but his financial position was precarious (he was exempted from paying fees); from 1681 to 1684 he was employed in Venice and, although he returned to England, he left again in 1687, first becoming a professor of mathematics in Marburg (where he fell out with his fellow academics, who saw no need for a mathematics professor, and his co-religionists, who excommunicated him), and then from 1695 working as an engineer advising the land-grave (or count) of Hesse in Kassel. There he successfully tested a primitive submarine in the River Fulda.[39]

Papin advanced Huygens' idea a further step. He constructed a cylinder containing a small amount of water, which he heated over a flame. The water turned to steam, drove out the air and forced the piston to the top of the cylinder, where a spring engaged a bolt. The heat was then removed, the steam condensed and the piston charged; as soon as the bolt was pulled it would be driven down the length of the cylinder by the pressure of the air. This was, in effect a wind gun driven by atmospheric pressure and where the piston replaced the bullet. Papin went on to imagine a series of such pistons turning gears to drive a boat and so saving on the cost of oarsmen (galleys were still in widespread use, particularly in the Mediterranean and on rivers), and he thought an engine of this sort could be used to pump water out of a mine if there were no river to drive a water-wheel nearby.[40] Unfortunately, he had no mechanism for rapidly charging and discharging the cylinders, or (at this point) for getting them to discharge in an orderly fashion.

During these years he engaged in a number of different steam-engine experiments, the high point of which was the construction of a steam-powered carriage which ran around the floor of his parlour.[41] He even looked forward to the time when steam-driven armoured cars would travel faster than the cavalry. His enemies, mocking him, put it about that he was working on a flying machine, and indeed he admitted that the thought had crossed his mind.[42] He designed, as his personal contribution to the war against Louis XIV (who had driven Protestants, including Papin, out of France), a mortar to throw grenades 90 yards at the rate of two hundred an hour (or even five hundred, he later claimed). The design was straightforward: by pulling on a lever, a piston was drawn down in a cylinder, creating a vacuum; when the piston was released it was driven up the cylinder, creating the propulsive force to lob the mortar towards the enemy. This was, in other words, an adaptation of his atmospheric steam engine, or rather a reversion to his earlier plan for a wind-gun powered by atmospheric pressure.[43]

Papin was still working on his atmospheric steam engine in March of

1704. How much progress did he make? The answer to this question is to be found in a notebook belonging to an English lawyer, musician and literary figure, Roger North.[44] There North described and sketched a two-cylinder atmospheric steam engine which he said he had seen 'onely in modell'. In this period the word 'model' is ambiguous: it can carry the modern meaning but more often it refers to a graphic representation, a plan or drawing.[45] The phrase 'in model' is extremely rare, but a single sheet published in 1651 is described in its title as a summary of Christian doctrine 'in model': it is a wall-chart.[46] So North probably saw not a working model, or even a maquette, but a drawing – hence his insistence that he had only seen it in model. *When* he saw this drawing we cannot be sure: an earlier entry was written in 1701, which gives us an approximate date. This, presumably, is the engine which powered the little steam carriage which ran around Papin's parlour.

The engine sketched by North is a development of Papin's atmospheric engine; now the cylinders have automatic valve gear and operate reciprocally (in 1676 Papin had designed an air pump with precisely

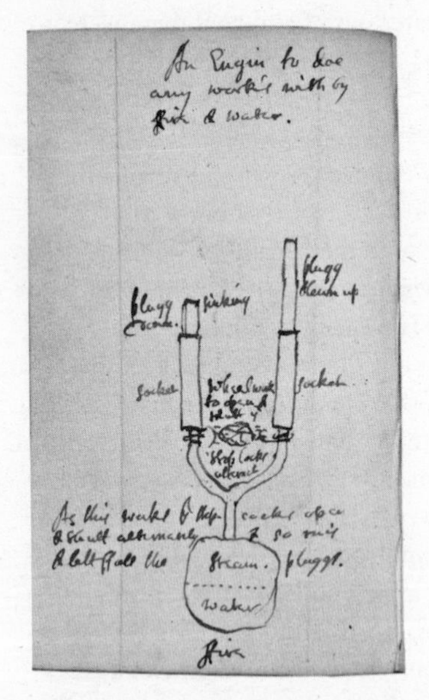

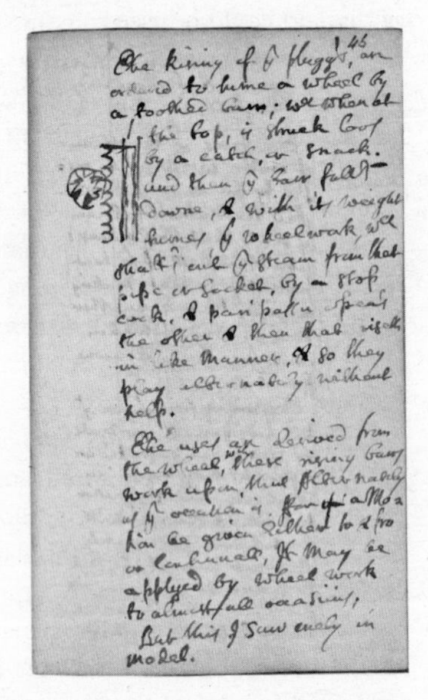

Roger North's notebook entry showing his drawing of a two-cylinder steam engine and the rack-and-pinion mechanism by which the pistons turn an axle. (From British Library Add. MS 32504.)

these features). The drive mechanism is evidently very similar to the one Papin had illustrated when he published an account of his steam-engine experiments in French in 1695, and which he claimed was modelled on mechanisms to be found in watches, although now the pinion moves away from the rack when the drive stroke is completed, rather than the other way round: the rack-and-pinion with a ratchet mechanism is distinctive because it is far from the best solution to the problem of how to drive a wheel by a piston (a crank being far better). Boyle's first air pump had used a rack-and-pinion mechanism to drive the piston in the pump (the opposite of its use here, where the piston is driving the rack-and-pinion mechanism), but there was no ratchet to allow the rack to draw back without turning the pinion. It is possible it is the work of someone following in Papin's footsteps, but it seems much more likely to be the work of Papin himself; evidently, he had sent a drawing of his latest engine to one of his friends in England, and this drawing had been shown to North. But there is no sign of Papin working on an atmospheric steam engine after 1704, or of news of the version of his engine recorded by North being disseminated. The advances Papin had made between 1695 and 1704 had no influence, and but for North's sketch we would have no knowledge of them. Papin's real contribution, as we shall see, lay elsewhere.

§ 3

In 1698 Thomas Savery, a military engineer and Fellow of the Royal Society, obtained a patent for a steam-driven pump that used both atmospheric and steam pressure to raise water (some suspected he had simply copied an earlier design by Edward Somerset, the Marquess of Worcester (d.1667), who had devised a steam-powered pump).[47] Steam was introduced into a cylinder, which was then cooled by having water sprayed on to it. The steam then condensed, drawing water up a pipe into the cylinder. A valve was closed, the water was heated, and the steam generated drove the water upwards out of the cylinder. Thus Savery's engine sucked and blew, just like a bellows, but the sucking was caused by condensing steam and the blowing was caused by expanding steam. Apart from the valves, this engine had no moving parts. Because the suck was driven by atmospheric pressure it could raise water no more than 30 feet or so, while the blow could push water upwards any distance, providing the pressure in the cylinder was high enough. Savery

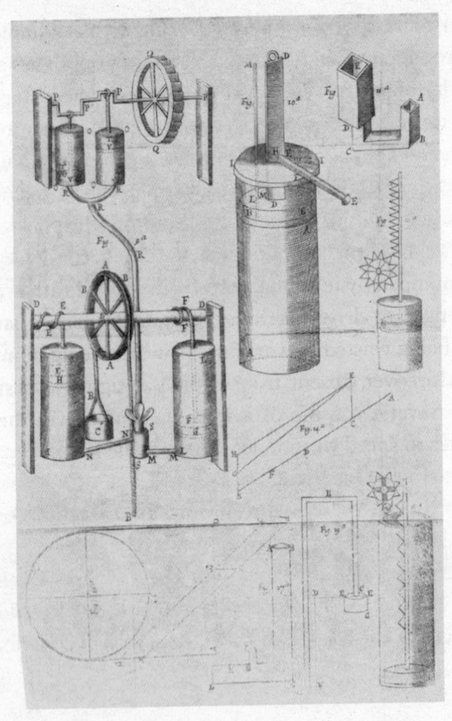

Papin's 1695 illustration of various pneumatic engines. The system on the left uses a water-wheel to drive pistons which pump air, which, by driving a second set of pistons, raises and lowers a bucket. At the centre top there is a representation of Papin's piston powered by the pressure of the atmosphere: once the steam has condensed, removal of the pin labelled E causes the piston to descend. On the right there are two images of his rack-and-pinion ratchet mechanism.

thus proposed mounting his device near the bottom of a mine and using it to pump water to the surface. In practice, the engine was used to power ornamental fountains, but not to pump water out of mines, because Savery could not build boilers and cylinders which would sustain a sufficiently high pressure.[48]

News of Savery's engine reached the landgrave of Hesse, and Papin was put to work devising a high-pressure steam pump. His initial efforts were apparently not very successful, and Savery was consulted on how to improve his design. In time, Papin successfully used a steam engine to pump water for an ornamental fountain (Louis XIV's waterworks at Versailles had made ornamental fountains a highly competitive field for rulers and aristocrats). One of his engines blew up (despite the fact that Papin had invented the first safety valve), almost killing the landgrave, while the boiler of another had burst when it froze in the winter. Papin's pump is often, and with good reason, described as a modification of Savery's, though Papin claimed he had devised it independently.[49] It was significantly different from Savery's pump in that it raised water only on the blowing cycle, and it separated the water used to power the system (which was being turned into steam and then condensed) from the water being pumped by using a float (which looked rather like a piston, but was not used to drive machinery), the idea being that this would prevent heat being wasted in warming the water being pumped through the engine. Moreover, absent from Papin's design was a simple device employed by Savery: the use of a spray of water over the cylinder to cool it in order to speed the condensation of the steam.[50]

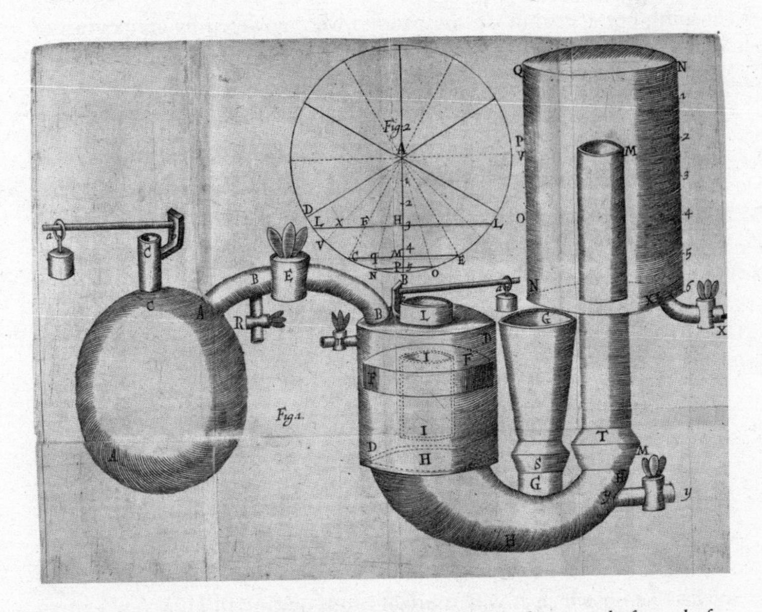

Papin's steam pump, from *Nouvelle manière pour élever l'eau par la force du feu* (1707). The boiler is on the left and the tank that is being filled on the right; there needs to be a steady supply of water into the hopper labelled G. Figure 2 is concerned with the design of the water-wheel that the pump is intended to drive. The pump is a modification of Savery's engine, with a float introduced to separate the steam from the water being pumped. It is fitted with two of Papin's safety valves.

Papin became increasingly discontented in Hesse, where the landgrave failed to give his researches the support he felt they warranted, and so he determined to return to England. It is commonly said that he built a steam-powered boat on to which he loaded all his belongings. He set out from Kassel on the River Fulda bound, eventually, for England. Unfortunately, he had gone 15 miles when he came to the junction of the Weser, and so to a stretch of river over which a guild of boatmen held a monopoly. He had made efforts without success to obtain an official exemption. The boatmen, determined to enforce their rights, seized his craft and destroyed it. And this was the end of steam-powered transportation for almost a century.

But the story of the steam-powered boat is based on a misunderstanding: Papin had built a boat powered neither by sail nor oars, and the boat was indeed destroyed. But it was not (as is clear from his correspondence) powered by a functioning steam engine. Papin had built a paddleboat (and not the first paddleboat – here, too, Savery was ahead of him), not a steamboat: the paddles were powered by cranks turned by hand.[51] It is puzzling that this story is so often repeated without any sign of scepticism: after all, if it was possible to build a working steamboat in 1707, why would it take a further century to establish reliable steam propulsion on water? One author has not hesitated to reach the obvious conclusion that a dastardly conspiracy must have been at work. Yet the evidence to refute this myth was printed as long ago as 1880.[52]

Papin, having lost most of his belongings with the wreck of his boat, and having been separated from his wife, finally arrived in England in 1707 and proposed to the Royal Society that they fund him to build his steam-powered craft. The Society submitted his proposal to Savery, who was not only the leading expert in the field but also held a patent which was so broadly worded that it covered any steam-powered engine. Savery insisted that the float/piston would produce too much friction to be workable. Newton, as President of the Society, dismissed the whole project as far too expensive.[53] Of course, Newton may have been biased against Papin because Papin was a friend of Leibniz, with whom Newton was increasingly coming into conflict. The Royal Society, after years of decline, in which it was short of funds and performing few experiments, was showing signs of a new enthusiasm for experimental science, but Papin did not benefit.[54]

Certainly, Newton was right: Papin's scheme was hopelessly expensive. The reason is apparent from the illustrations of Papin's engine. It requires a supply of water into the pumping mechanism, and the source of that supply must be higher than the top of the cylinder.[55] If the engine were to be installed in a boat, and the water taken from the river or sea, then the whole of the engine would have to be below the waterline, which requires a very large boat with a very deep draught.[56] Papin was well aware of this: he proposed to the Royal Society a ship of eighty tons, perhaps 100 feet long, costing 'but four hundred pounds' to build.[57] How much was £400 worth? Fifty thousand pounds in modern money, using a retail price index, but £725,000 in modern money using an average wage multiplier. A more helpful measurement perhaps is that it was four times the salary of the Lucasian professor of mathematics at Cambridge: so let's say £400,000.[v]

Nineteenth-century illustrations of Papin sailing his steam-powered boat are thus completely misleading in that they show an engine mounted on top of the deck of a small craft, not a large sea-going vessel with an engine below decks. It is impossible to get round the problem that Papin's steam-pressure engine could not be made to work to power a boat unless it was implemented on a large scale.[vi] The scheme is simply impractical. Papin, who was constantly coming up with new schemes (he, like so many others, thought he could make a clock accurate enough to measure longitude) could find no one to back him. His final years were ones of failure and poverty. We last hear of him on 23 January 1712. 'I am,' he writes, 'in a sad case.'[58] We do not know where, when, or how this great engineer-scientist died.[59]

§ 4

Only five years after Papin's failure, Newcomen produced the first commercially viable steam engine. The great merit of Newcomen's engine was its simplicity of conception and modesty of ambition. It consisted

v The Lucasian chair had been established in 1663, but monetary values had changed little in the intervening period. Modern salaries, of course, are supplemented by the employer's pension contributions and other charges, so they cannot be directly compared with seventeenth-century salaries.

vi Impossible if the water-wheel driven by the engine is also going to function as the paddle which powers the boat through the water. In principle, one could separate the two functions and connect the water-wheel and the paddle wheel by a drive belt, and one could recycle the water from the water-wheel back into the engine, but if the water-wheel was up on deck it would catch the wind and act as a sail, with the result that the boat would be impossible to steer and likely to capsize – unless, again, it was a very large boat with a very deep draught.

of a single piston, powered by the pressure of the atmosphere. When the piston is driven downwards it pulls on a large beam which works a force pump. The weight of the pump mechanism ensures that the piston rests in the up position. Air is driven out of the cylinder by filling it with steam; the steam is then condensed by injecting water into the cylinder (Newcomen discovered how to do this by accident), and the atmosphere pushes the piston down; steam is then reintroduced into the cylinder at atmospheric pressure; this releases the piston, which is raised by the weight of the pump. The engine ran slowly, at about fifteen cycles per minute. The engine is simple because, like Papin's first steam engine of 1690, it consists of a single cylinder powered by atmospheric pressure only. Savery's engine, and Papin's second engine, needed to build up a high pressure in order to be effective but, in practice, boilers and cylinders could not be made that withstood such pressure. On the other hand, Newcomen, unlike Savery, had to construct a moving piston, with all the difficulties of potential friction and leaking that entailed.

Newcomen had little formal education. Born in 1664, he was an ironmonger in Dartmouth, Devon, and an elder in the local Baptist Church. Yet, almost singlehandedly (we know of one assistant, Mr Cawley, a glazier), he brought a new technology into existence. How was this possible? It puzzled contemporaries just as it puzzles us. The first possibility is that he worked in complete isolation, without knowledge of anything that had gone before. One only has to formulate this possibility to see that it must be wrong. For a start, Newcomen could not have devised his engine without knowing about the pressure of the atmosphere, since this provides its driving force. It is true that knowledge of air pressure was widespread by 1712, and any explanation of the workings of a barometer would have conveyed to Newcomen the discoveries of Torricelli and Pascal. But he would have needed this as an absolute minimum.

We have almost no direct information about Newcomen prior to 1712 but, from what he told his associates in later life, two things seem evident. First, he began work on his steam engine around the same time as Savery began work on his, so no later than 1698. Second, he worked in complete independence of Savery.[60] Nevertheless, some scholars have felt this makes no sense. Newcomen must have benefitted from either Savery's or Papin's expertise. One scholar has boldly, and against all the evidence, claimed that Newcomen was simply an employee of Savery.[61] Another, going as he himself puts it, 'in the face of all the evidence', has suggested that Newcomen and Savery could have met in January

The Newcomen Engine, as illustrated in John Theophilus Desaguliers, *A Course of Experimental Philosophy* (1734–44; taken from the 1763 reprinting). The boiler is on the left, with the piston rising vertically out of it and connected to the rocking arm.

1707 or soon after, when we know that Savery went to Dartmouth – but this is far too late (and so is soon altered by a sleight of hand into 'by 1705', which is still too late).[62] A late-eighteenth-century scholar 'solved' the problem by claiming that Hooke (who died in 1703) had written to Newcomen describing Papin's first steam engine. This story still gets repeated, despite the fact that the documents which are supposed to support it do not exist, and have been known not to exist since 1936.[63] Another scholar says that 'Thomas Newcomen surely must have seen Papin's sketches of his models of proto-engines and pumps, published in various issues of *Philosophical Transactions* between 1685 and 1700' – glossing over the fact that none of Papin's publications in the

Transactions dealt with steam power; they all relied on water-power or man-power.[64]

Papin's engines were the closest in conception to Newcomen's. The great historian Joseph Needham said, very sensibly, 'I find it almost impossible to believe that Newcomen did not know of Papin's steam cylinder.'[65] But Papin had made and operated his first engine in Germany. No Englishman, as far as we know, had ever seen it. He described it several times in print, in Latin and in French, but never in English. A single paragraph appeared describing it in English in a review of one of Papin's publications which appeared in the *Philosophical Transactions* for 1697:

> The fourth letter shows a method of draining mines, where you have not the conveniency of a near river to play the aforesaid [pumping] engine [by means of a water-wheel]; where having touched on the inconveniency of making a vacuum in the cylinder for this purpose with gunpowder [as Huygens had done], he proposes the alternately turning a small surface of water into vapour, by fire applied to the bottom of the cylinder that contains it, which vapour forces up the plug [i.e. piston] in the cylinder to a considerable height, and which (as the vapour condenses as the water cools when taken from the fire) descends again by the air's pressure, and is applied to raise the water out of the mine.[66]

It is highly unlikely that Newcomen ever had access to the *Philosophical Transactions*, but even if he did this single paragraph, without any supporting illustration, would have left him with a tremendous amount of work to do. As for the more advanced Papin engine sketched by North, this would obviously have been of great interest to Newcomen had he known of it, but it probably postdates the beginning of Newcomen's programme of experimentation, and its design is much more complicated than Newcomen's – indeed, we may doubt that Papin ever got it to work properly.

It may be helpful to list some of the things that Newcomen would have had to invent to make a successful steam engine, or which he would have needed, even before that, to conduct a programme of experiments. Much of what he needed was straightforward. The pump bucket and pump handle, for example, were straightforward applications of existing technologies, and the boiler was basically a large brewer's copper. But other things were far from straightforward. First, though the idea of using a cylinder and piston went back to Guericke, there was no recent experience of combining this with steam in England. Second,

Newcomen needed a means of making the piston air-tight. He sealed his piston with a leather washer and a layer of water injected into the cylinder. (John Morland had designed pumps using pistons in the 1680s – his seal was quite different.)[67] Third, it would have been nice to have a pressure gauge: the barometer is the first pressure gauge, but Boyle and Papin had described a sophisticated pressure gauge in the *Continuation of New Experiments* of 1682. Crucially, it was essential to have a safety valve, which is in itself a form of pressure gauge: Papin had invented one, and had incorporated one in his 1707 design (though perhaps not on the version which had exploded). Newcomen used a version of Papin's safety valve (called the Puppet Clack) in his engine.[68] In addition, he needed a technique for having the valves in the piston open and close by the action of the machine itself.[69]

Lastly, there is a further prerequisite. It is a feature of Savery's engine that it runs better on a small scale than when scaled up: as the cylinders get bigger, the volume of the cylinders increases more rapidly than their surface area, and so cooling becomes less efficient. So when Savery built a model he would have been misled into thinking he had made a breakthrough. Newcomen's engine is the opposite: the ratio of power to friction is most unfavourable on a small scale, and becomes more favourable as the engine is scaled up because the volume of the cylinder (which determines the power of the engine) increases more rapidly than the piston's circumference (which determines the amount of friction).[70] Desaguliers and a friend later built models of both the Savery and the Newcomen engines: despite his extraordinary expertise, Desaguliers was plainly taken aback to see the Savery engine outperform the Newcomen engine.[71] Newcomen must have understood this scaling issue from the beginning, otherwise he would never have persisted when his first models performed (as they must have) very poorly. He must have obtained this knowledge from some source.

Of course, Newcomen could have invented all this and more; after all, he worked on his new engine for some fourteen years before he was ready to put it into public operation. But it is worth knowing that only a few years ago an attempt to build a one-third-scale replica Newcomen engine ran into a great deal of difficulty. Even with good plans, even with plenty of technical expertise, even with a knowledge of what the end product was supposed to be and an absolute certainty that it could be made to work, it turned out that it took many months of tinkering and fiddling to get the engine to run properly.[72] Ideally, Newcomen needed a source of information which would have supplied him with all sorts of bits and pieces of key information, so that he could then concentrate on working out how to assemble a working machine. That would have been quite enough to keep him busy in his spare time for a decade or more.

§ 5

There was indeed such a source, one that Newcomen is much more likely to have come across than the *Philosophical Transactions*. It is a source that historians of the steam engine have overlooked because it does not discuss steam engines. Indeed, it has been generally overlooked: there is not a single citation of it in Google Scholar, or in Thomson Reuters Web of Science. One could easily form the impression that no one has read it in the past century, despite the fact that its author is well known and, judging by the number of surviving copies, the book sold well when first published. I refer to *A Continuation of the New Digester of Bones*, published by Denis Papin in 1687.[73]

Papin published the first account of his New Digester in 1681. It was, quite simply, the first pressure cooker, a sealed bain-marie. Because the pressure cooker turns water into steam under pressure it cooks at a higher heat than normal boiling water, and so much more quickly, or (in the case of digesting bones) it cooks until hard materials have been reduced to soft pulp. (Papin's Digester has a special place in history because in 1761 or 1762 Watt carried out his first experiments with steam by attaching a syringe to the safety valve of a Papin Digester, thus making a primitive steam engine.) The *New Digester* and the *Continuation* were often bound together, and it is easy to imagine Newcomen acquiring either the *Continuation* alone or both volumes together in 1687 or at any time in the decade before he began work on his steam engine. His motive would have been straightforward: there was a possible profit to be made making and selling Papin's device and, since Papin had insisted that anyone was free to copy it, and had not protected it with any patent, there was no obstacle to Newcomen trying to make money out of it.

The short title is, however, a poor guide to the contents of Papin's book. The full title is more informative: 'A continuation of the new digester of bones: its improvements, and new uses it hath been applyed to both for sea and land: together with some improvements and new uses of the air-pump, tryed both in England and Italy'. This is, in part, a

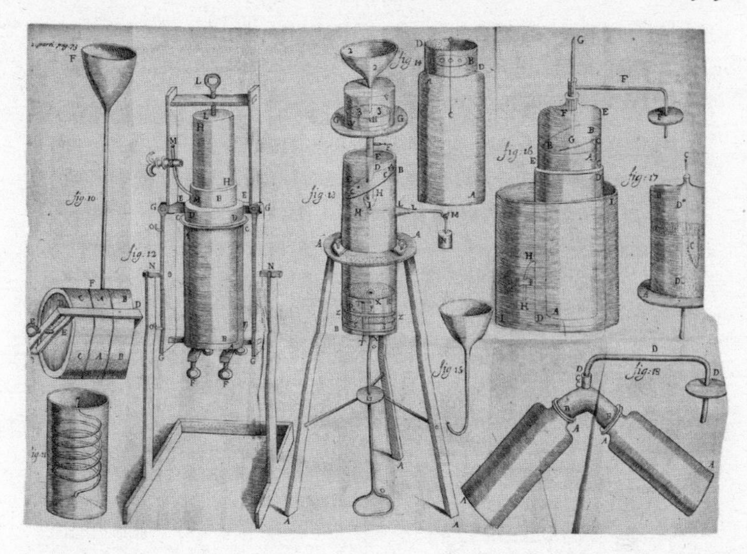

Papin's 1687 air pump, from *A Continuation of the New Digester*.

book about the air pump (although historians of the air pump and of vacuum experiments have failed to read it),[74] and it provides an illustration and description of Papin's most recent (and last) model.[75] Papin's pump consists of a cylinder with a piston; the piston is sealed by a layer of water, and Papin describes with care how to achieve this.[76] The method used corresponds to the method initially used by Newcomen – though he later found a better one.[77] The cylinder, like the piston of Newcomen's steam engine, has a number of valves and inlets which open and close with the action of the piston. (Papin was the first to make an air-pump in which the valve action was automatic.) There is one valve closed by a weight, although it is not in this case a safety valve; however, Papin describes the operation of such a valve. Thus the basic technology of the steam engine's piston is laid out because that technology overlaps with the technology of the air pump – it is precisely because of these overlaps that Papin could go on three years later to build the first steam engine.[vii]

vii Hall writes, 'Boyle's airpump (1658) . . . was the antecedent of all later machines depending on a piston accurately fitting in a true cylinder . . .' (Hall, 'Engineering and the Scientific Revolution' (1961), 337).

But the *Continuation* does more than that. It provides the reader with the line of thinking that led Papin to the invention of the steam engine. Here is what he says:

> I might also reckon among the uses of this Engine [the air-pump] the strength it can afford to produce great effects without the encumbrance of great weights: For a tube very even and well workt may be made very light and yet being emptyed of Air it will endure the pressure of the Atmosphere: Nevertheless a plug very exact at one end of that tube would be pressed towards the other with a very great strength, at least if the tube was of a pretty great Diameter: for example if it was a foot Diameter the plug would be press't with the strength of about 1800 pounds. The Renouned Mr *Guerike* hath the first tryed to apply that strength to shoot leaden bullet's [*sic*] through a gun, as may be seen in the description he hath given of it in his book of the Pneumatick engine: I have also endeavoured since to add some thing to his invention, as may be seen in the *Philosophical Transactions* for the mounth of January 1686: since that I have calculated that a bullet of lead of an Inche Diameter being thus shot through a barrel 4 foot Long should acquire the swiftness to fly about 128 feet in a second; but if the same swiftness was to be given to a bullet of a foot diameter it should be made of Iron hollow within, so that it should weigh but about 37 pound and half: For if it was made solid of Lead, it would weigh about 450 pounds, and so in passing the length of about 4 foot in the barrel, it would acquire but the swiftness of 32 feet in a second . . . The end of the barrel through which the bullet passeth must be stop't with something strong enough to uphold the pressure of the Atmosphere, and that being met in the way of the bullet takes away also some of its strength.[78]

What Papin is describing here is a wind-gun powered by atmospheric pressure; but it is clear that the device, which requires the bullet to punch a hole to escape from the barrel, is not fit for any practical purpose.

What he is also describing is a piston powered by atmospheric pressure; he was on the point of inventing the atmospheric steam engine, but here the vacuum is created by his pump, not by the condensing of steam. He does, however, repeatedly describe how to use water applied to the outside of a container full of steam in order to bring about rapid condensation (although he never used this technique in his own steam engines) and so a vacuum.[79] If he read him, Newcomen had only to put two and two together in exactly the way Papin was to go on to do to have the basic design of a steam engine. If Papin could do it, why could

Newcomen not do it, too? Moreover, Papin here introduces the reader to the problem of scale: as you scale up the gun, it becomes less efficient, because the weight of the bullet increases faster than the surface area of its end. Someone who thought hard about this might grasp that as the diameter of the tube increased, the increase in the weight of the bullet would be partially compensated for by a decrease in the proportion of energy lost to friction.

Newcomen's steam engine is a bit like a locked-room plot in a detective story. Here is a dead body in a locked room: How did the murderer get in and out, and what did he use as a weapon? Our puzzle is that we have Newcomen in Dartmouth in or around 1698, and we can see no way in which knowledge of the steam engine can reach him. As with the locked-room mystery, if we can find *one* solution, then we have found *the* solution. Of course, we cannot exclude the possibility that Newcomen went to London and met Papin in 1687; Papin indeed advertized that he would be available at a certain time each week to demonstrate his digester, although in fact he soon left the country. But we do not need to imagine such a meeting. With a copy of the *Continuation* in his hands, Newcomen would have known almost everything that Papin knew about how to harness atmospheric pressure to build an engine. All the bits and pieces were there; all he had to do was recognize how they could be assembled to serve a new purpose, not to create a gun but to power a pump. And the *Continuation*, with its instructions on how to build a revised model of the Papin Digester, is precisely the sort of book that a provincial ironmonger and small-time manufacturer would have been looking out for. The last thing Newcomen would have expected to find in it, the last thing he would have been looking for, is a description of a new type of power capable of producing great effects without the encumbrance of great weights. From this unintended encounter, I believe, the steam engine was born.

Desaguliers, in the first major study of the steam engine, insisted that all the great advances in steam-engine design had been made by chance:

> If the Reader is not acquainted with the History of the several Improvements of the Fire-Engine since Mr N*ewcomen* and Mr C*awley* first made it go with a Piston, he will imagine that it must be owing to great Sagacity and a thorough Knowledge of Philosophy, that such proper Remedies for the Inconveniencies and difficult Cases mention'd were thought of: But here has been no such thing; almost every Improvement has been owing to Chance . . .[80]

Desaguliers chose his words with care. He said that the improvements were owing to chance, but he left his readers to make up their own minds as to whether the action of first making the steam engine go with a piston required a thorough knowledge of philosophy or not. Certainly it required some philosophy, and some inherited technology. Both, I suggest, were provided by Papin's *Continuation*.

Indeed, when Desaguliers explains the workings of Newcomen's engine, he makes a remarkable move. He asks you to imagine an engine in which 'a philosopher' uses an air pump to create a vacuum in a piston, before going on to describe Newcomen's actual design, in which steam is condensed in a piston to make a vacuum. This philosopher is certainly not Newcomen; but Desaguliers had, I think, correctly intuited the only plausible route to the invention of the Newcomen engine. Where Newcomen had put two and two together to make the steam engine, Desaguliers, in order to explain its working, has taken them apart again, reinventing Papin's atmospheric wind-gun.[81]

Historians have long debated the extent to which science contributed to the Industrial Revolution. The answer is: far more than they have been prepared to acknowledge. Papin had worked with two of the greatest scientists of the day, Huygens and Boyle. He was a Fellow of the Royal Society and a professor of mathematics. In the twenty years between 1687 and 1707 he worked towards the construction of a viable steam engine, but in the end he failed. Newcomen picked up, I suggest, not where Papin ended, with his modified Savery engine, but where Papin began. In doing so he inherited some of the most advanced theories and some of the most sophisticated technology produced in the seventeenth century. It was this that made the Industrial Revolution possible. First came the science, then came the technology.[viii]

viii Although the steam engine itself also provoked new science, as its workings, particularly in the advanced form of the Boulton and Watt engine, were not fully explained until Sadi Carnot founded the science of thermodynamics with his *Reflections on the Motive Power of Fire*, 1824.

CONCLUSION

The Invention of Science

> How is it possible for a historical activity, such as scientific activity, to produce trans-historical truths, independent of history, detached from all bonds with both place and time and therefore eternally and universally valid?
>
> Bourdieu, *Science of Science* (2004), 1

The Conclusion stands back and asks what the consequences are of recognizing the reality of the Scientific Revolution. Chapter 15 looks at the key arguments on which the relativists depend and shows that they do not do what is claimed for them. Chapter 16 addresses the claim that any history of the Scientific Revolution must be Whig or teleological history, by arguing that the opponents of Whig history have defined history in such a way that change cannot be discussed. Chapter 17 ends the book by looking at Montaigne's scepticism and asking whether we are entitled to claim that we know more than he did.

15
In Defiance of Nature

> Salviati: If this question we are arguing about was some point of law or of one of the other disciplines in the humanities, where there is neither truth nor falsehood, then one could justifiably rely on intellectual subtlety, on verbal fluency, and on breadth and depth of reading, and hope that whoever had the advantage in these respects would succeed in making his argument seem, and be accepted as, the stronger; but in the natural sciences [*scienze naturali*], the conclusions of which are true and necessary, and in which the opinions of human beings are irrelevant, you have to be careful not to give your support to error, because a thousand Demosthenes and a thousand Aristotles will find themselves defeated by a mediocre intellect who has the good fortune to attach himself to the truth. Therefore, Signore Simplicio, give up on that idea and that hope that you have, that there can be men so much better educated, so much more sophisticated, and with so much more book-learning than the rest of us that they can, in defiance of nature, turn falsehood into truth.
>
> Galileo, *Dialogue Concerning the Two Chief World Systems* (1632)[1]

§1

Shakespeare had – to return to Borges's comment with which this book began – no sense of history. He read classical authors as if they were his contemporaries. He had plenty of experience of change, sometimes for better and sometimes for worse, but he had no notion of irreversible change, and he had no notion of progress. And this is hardly surprising, for there was, in his world, little evidence of progress; when Shakespeare retired from the stage in 1613 Bacon had published only one book on the new science, *The Advancement of Learning* (1605), and it was only three years since Galileo had published his telescopic discoveries. But since then progress has been uninterrupted. I see no reason to revise John Stuart Mill's opinion that one of the main drivers of economic development has been 'the perpetual, and so far as human foresight can extend, the unlimited, growth of man's power over nature', and that (as we saw in Chapter 14) this power results from increasing scientific knowledge.[2]

All sorts of taboos have developed around the use of the word 'progress'; indeed, it has become a word that can no longer be used in the humanities without its use being penalized by what Pierre Bayle called 'the law of opinion', a heavy sanction indeed in the academic world since it means the denial of tenure and of promotion.[i] So let me stress that my views on this question coincide with those of many of the sharpest critics of the idea of progress. Here is the philosopher John Gray in a book subtitled *Against Progress and Other Illusions*: 'In science progress is a fact, in ethics and politics it is a superstition. The accelerating advance of scientific knowledge fuels technical innovation, producing an incessant stream of new inventions; it lies behind the enormous increase in human numbers over the past few hundred years. Post-modern thinkers may question scientific progress, but it is undoubtedly real.'[3]

This view used to be entirely conventional. As George Sarton, the founder of the [American] History of Science Society and of its journal, *Isis*, put it in 1936: '[T]he history of science is the only history which can illustrate the progress of mankind. In fact, progress has no definite and unquestionable meaning in other fields than the field of science.'[4] Such statements have led to Sarton becoming someone who is quoted only to show how naive we once were. Alexandre Koyré's reputation has survived better than Sarton's, but he said exactly the same thing the year before: the history of science, he claimed, is the 'only history (along with that, linked to it, of technology) which gives any sense to the idea, so often glorified and so often decried, of progress.'[5]

Sarton and Koyré were right. A history of modern science without progress fails to capture science's unique feature. Moreover, the best of the supposed 'relativists' know this. Kuhn denied that science made

i As Glanvill put it, 'He that offers to dissent, shall be out-law'd in his reputation: and the fear of guilty Cain, shall be fulfilled on him, who ever meets him shall slay him.' (Glanvill, *The Vanity of Dogmatizing* (1661), 130.)

progress *towards the truth*,[ii] or could ever claim to have grasped the truth, but he always insisted that a place had to be found for the idea of progress in science, even if he had great difficulty explaining how this might be the case.[6] The last chapter of *Structure* was entitled 'Progress through Revolutions'. In it he writes: '[A] sort of progress will inevitably characterize the scientific enterprise so long as such an enterprise survives'; and he goes on to argue that progress must be understood in evolutionary terms.[7] Richard Rorty, the boldest defender of pragmatism, insisted that there are no epistemological foundations on which we can build an incontrovertible knowledge, but he was also an admirer of Kuhn, and like Kuhn acknowledged that science makes progress within its own terms: 'To say that we think we are heading in the right direction is just to say, with Kuhn, that we can, by hindsight, tell the story of the past as a story of progress.'[8] A form of knowledge which aims at prediction and control gets better at prediction and control. Progress is part of the story. This book is not aimed at the mitigated relativism of a Kuhn or a Rorty but at the strong relativism which presents progress in science as an illusion, the consequence of a misunderstanding of what is in fact taking place when scientists disagree with each other. The public – and the scientists themselves – imagine that the quality of the evidence determines the outcome; actually, we are told, it is the status, power and rhetorical skill of the combatants.

§2

This emphasis on the contingent and local character of scientific knowledge is supported by what many take to be a philosophical argument of profound importance, the so-called Duhem-Quine thesis, named after Pierre Duhem (1861–1916), a physicist and historian of science, and W. V. O. Quine (1908–2000), an American philosopher.[9] The thesis is misnamed, because, as it is usually formulated, Duhem did not hold it and Quine abandoned it, but it is the fundamental

ii In order to understand the idea that one might make progress without it being progress towards the truth, think of someone climbing a mountain in a fog. She can tell that she is going upwards. But is she approaching the summit, or is she actually on a slope that leads to a lower peak, from which she will eventually have to retreat and retrace her steps? She can only answer this question if the fog clears. In the case of scientific knowledge the fog never clears: the summit is invisible and truth always remains out of sight. But there can be progress without truth.

conceptual underpinning of much modern history and philosophy of science.[iii]

The thesis takes two forms. First, it is argued that a scientific theory cannot be refuted by experiment: not just that it cannot be refuted by a single experiment, no matter how often repeated, but that it cannot be refuted by a whole series of different experiments. Scientific theories are complex things, for they are constituted by interlocking bundles of theories, facts, and equipment. If an experiment produces a result which is at odds with the theory then something is wrong; but you cannot simply say that the theory is wrong. Some other theory that this theory depends on may be at fault, some fact that has been taken for granted may be mistaken, or some piece of equipment may not operate as envisaged. Consequently the results of experiments cannot refute a theory. This is called 'holism'.

But let us take the example of sailing to America. This was in effect an experiment, and it was a crucial experiment: it straightforwardly refuted the two spheres theory. The only way of rescuing the theory in the light of the new evidence would have been to say that all the navigators were wrong: America was not where they thought it was. Nobody thought it worth pursuing this line of argument. This would not have puzzled Duhem, who specifically formulated his thesis to address modern physics; he acknowledged that it did not apply, for example, to nineteenth-century biology.

Second, it is argued that theories have a very loose relationship to the facts. Given any set of facts, there are countless theories that can account for them, just as there are countless lines that can be drawn through any set of dots to join them together. This means that scientists, although they may not be aware of it, are never obliged to adopt any particular theory: there are always alternatives that would work as well, indeed, for all we know, better.[iv] And of course facts and theories have an intim-

iii To be exact, Quine acknowledged that the thesis is 'untenable if taken nontrivially': Quine, 'A Comment on Grünbaum 's Claim' (1976). For an example of a trivial sense, suppose the defenders of Ptolemaic astronomy had responded to the threat posed by Copernicanism by simply renaming the sun 'the earth', and the earth 'the sun'. They could then have continued to claim that the earth stands still at the centre of the universe. In principle any theory could be saved by redefining terms in this way; but the enterprise would be pointless.

iv This follows, some believe, from Thomas Kuhn's *Structure of Scientific Revolutions* (1962), which itself cites Quine in the Preface. The argument is much more boldly stated in a book Kuhn had read, Ludwik Fleck, *Genesis and Development of a Scientific Fact* (1979), first published in German in 1935. Consequently, particularly since Kuhn delivered his ate relationship: what counts as a fact depends on the theories you hold, and whether a theory appears to be valid depends on the facts you acknowledge. This loose, slippery, and incestuous relationship between facts and theories is called the principle of underdetermination.

Again, the example of sailing to America presents problems for the principle of underdetermination: we have seen that although Bodin proposed an alternative to the terraqueous globe theory, it was never viable; not a single person came to its support. The terraqueous globe theory was not underdetermined; in this case the relationship between the theory and the facts was a tight one, not a loose one. The same is true of the phases of Venus: once their existence was acknowledged, the conclusion that Venus orbited the Sun was inescapable.

These two principles – holism and underdetermination – are what is being invoked when appeal is made to the Duhem-Quine thesis. The standard argument is that this thesis proves that evidence does not determine what scientists hold to be true; consequently, it is claimed, scientific beliefs are primarily shaped by cultural and social factors. If science is largely culturally determined, then the now familiar conclusion follows: its procedures and conclusions will reflect a purely local consensus.[v] This conviction provides yet another reason for insisting that there was no such thing as 'the Scientific Revolution of the seventeenth century'. Science, we are to understand, was one thing in Florence, something quite different in Paris or London. Historians of the book seek to block the obvious objection – that scientists in London were reading books written by scientists in Florence and Paris, and so belonged to a single intellectual community – by maintaining that books mean different things to different readers, so that reading Galileo in Florence in the 1640s was a quite different enterprise from reading him in London in the 1660s.[10]

This contrast between local meanings and cosmopolitan messages is a perfectly sound one. One just has to get the balance right. Galileo may never have left Italy, but he had English and Scottish students, and his *Two New Sciences* was first published in Leiden; William Harvey, the

---

1991 lecture against the sociology of science, much contemporary history of science takes Fleck rather than Kuhn as its model: e.g. the statement of Labinger and Collings that Fleck not Kuhn 'is the true precursor of the sociology of scientific knowledge' (Labinger and Collins (eds.), *The One Culture?* (2001), 3n).

v A phrase often used in this context is 'local knowledge' – Foucault uses it, but it is particularly associated with Geertz, *Local Knowledge* (1983). For a critique, Jacob, 'Science Studies After Social Construction' (1999).

discoverer of the circulation of the blood, studied medicine in Padua; René Descartes moved from France to Holland, Christiaan Huygens from Holland to France, Thomas Hobbes from England to France; Robert Boyle may have spent his working life in Oxford and London, but he went to Italy and learnt Italian; his associate, Denis Papin, worked in France, England, Italy and Germany. And of course almost all the early scientists shared a common language. Galileo published his *Dialogue Concerning the Two Chief World Systems* in Italian in 1632, but it appeared in Latin in 1635; Boyle published his *New Experiments Physico-Mechanical Touching the Spring of the Air* in English in 1660, but it appeared in Latin in 1661; Newton published his *Opticks* in English in 1704, but it appeared in Latin in 1706. Of the first 550 fellows of the Royal Society, elected between 1660 and 1700, seventy-two were foreigners (and the proportion of foreigners rose in the eighteenth century, until it reached one third).[11] The new science knew no boundaries of language or nationality, at least not within western Europe, within the world of the printing press, gunpowder weaponry, the telescope and the pendulum clock.

A moderate interpretation of the Duhem-Quine thesis leads to a mixed constructivism, in which evidence and culture each have a part to play in the construction of scientific beliefs.[12] Kuhn's *Copernican Revolution* (1959) provides an example of this. According to Kuhn, Copernicanism won out over the alternative systems (the Ptolemaic and Tychonic) before the invention of the telescope, but this cannot be explained simply in terms of the mathematical elegance of the Copernican system; other cultural factors, such as neo-Platonism, which might have encouraged people to revere the Sun, were also important.[vi] Another example would be Newton's willingness to accept the idea of action over a distance. To Cartesians, Newton's theory of gravity made no sense; but in England, where Cartesianism had never been adopted without reservations, and where arguments from design were widely accepted, resistance to the theory was much weaker. But once a science establishes itself it becomes, to a remarkable degree, autonomous, immune to influence from other fields.[13] This is not to say that it is not shaped by culture as well as evidence; but the culture which shapes it is, first and foremost, the culture of science itself. Thus Kepler, because he

vi See the summary in Kuhn, *Structure* (1970), 69. The claim that Copernicus was influenced by neo-Platonism has been contested: Rosen, 'Was Copernicus a Neoplatonist?' (1983). Bruno certainly was.

was familiar with Gilbert's work on magnets, could use magnetic force as a model, and on the basis of it identify his laws of planetary motion: Gilbert made it possible for Kepler to conceive of an astronomy grounded in physics rather than mere geometry. Thus Newton, in England, could propose his theory of gravity, but only because he already had (unlike the Cartesians) the notion of *a theory*: something more than an hypothesis but distinct from a proof.

An uncompromising interpretation of the Duhem-Quine thesis leads to the conclusion that science is *entirely* a social construction, or at least should be studied as if it is, and that reality (the source of the Danube, the existence of America, the phases of Venus) is of no concern to historians and sociologists. If this is correct, then then there is no way of distinguishing good science from bad science because all theories are, or are to be regarded as being, equally adequate (this has been called 'cognitive egalitarianism'), and therefore it is meaningless to discuss progress in science.[vii] I call this relativism.[viii] For some considerable period this uncompromising interpretation has been the dominant position in the history of science. It is precisely because the Duhem-Quine thesis, so interpreted, cannot handle either the demise of the two spheres theory or of Ptolemaic geocentrism that these crucial historical events have been invisible to those who are convinced of the truth of the thesis. Its proponents behave just like the philosopher Cesare Cremonini when he refused to look through Galileo's telescope: they hold to their

vii This may seem astonishing, but is documented in Laudan, 'Demystifying Underdetermination' (1990). It is important to stress that what Laudan calls cognitive egalitarianism continues to be widely accepted among historians of science for the simple reason that they believe that if they question this claim they will no longer be thinking historically (cf. Shapin, 'History of Science and Its Sociological Reconstructions' (1982), 196–7). In fact, acceptance of cognitive egalitarianism means they *cannot* think historically, for a cognitive egalitarian cannot explain why, to take but one example, we no longer believe that swallows overwinter in ponds.

viii The term 'relativism' is contentious, but I use it, rather than 'constructivism', because I want to refer to those who hold that reality does not constrain the beliefs we can hold about the real world. I take all those who think that science is purely and simply a *social* construction to be relativists; there are others (including myself) who think that science is constrained by both nature and society, and they may reasonably be called constructivists in a broad sense, but they are not relativists. For a fuller discussion of relativism see the 'Notes on Relativism and Relativists' (pp.580–92). In summary, I find members of the Edinburgh and Bath schools to be relativists (with the exception of Andrew Pickering), and so are the proponents of Actor Network Theory. Ian Hacking, Larry Laudan and Andrew Pickering represent significant examples of people trying to avoid the pitfalls of both realism and relativism; I share that enterprise.

convictions even when the evidence shows they are wrong by the simple device of ignoring anything that does not fit their theory.

§3

Exactly the same relativist approach is employed with regard to facts as to theories; indeed, the two are often indistinguishable. According to Ian Hacking there is no necessary agreement on basic measurements, such as the speed of light.[14] He claims to show this by pointing out that the first person to measure the speed of light came up with a very different figure from the one that we now rely on, and so feels entitled to dismiss the argument that agreement on the speed of light was inevitable as 'dreadful'. In fact the 'dreadful argument' is (quite uncharacteristically) Hacking's, and in order to see this we need only look at the evidence.

Hacking, following established convention, presents the astronomer Ole Rømer (1644–1710) as the first to measure the speed of light. In fact, Rømer never calculated a figure for the speed of light.[15] His goal was to establish accurate figures for the periods of the satellites of Jupiter (the eclipses of the satellites were to be used to set a standard time against which to measure the longitude of different places on the earth's surface). Rømer concluded, on the basis of a very small set of observations, that when the earth is at its maximum distance from Jupiter the moment of an eclipse appears to be delayed by twenty-two minutes compared to when it is at its minimum distance. Thus it takes light twenty-two minutes to cross the diameter of the earth's orbit, or eleven minutes to cross the radius (i.e. to travel from the sun to the Earth). Claims that Rømer measured the speed of light depend on then introducing a figure for this distance, which Rømer never did (and there is no reason to think he would have regarded any figure he could have used as reliable).[ix] Direct measurements of the speed of light come much later, in the nineteenth century. The two tables opposite summarize the history of these two sorts of measurement.[16]

ix It should be obvious that if I say my flight from London to New York will take seven hours and thirty minutes I have not estimated the speed of my aeroplane; to do that I need to add in the information that the distance from London to New York is approximately 3,500 miles. Rømer was measuring elapsed time, not speed.

Table of time taken by light from the sun to Earth

Date	Author	Method	Time
1676	Rømer	Moons of Jupiter	11 mins
1687	Newton[1]	Moons of Jupiter	10 mins
1693	Cassini[2]	Moons of Jupiter	7 mins 5 secs
1704	Newton	Moons of Jupiter	7 or 8 mins
1726	Bradley	Stellar aberration	8 mins 12.5 secs
1809	Delambre	Moons of Jupiter	8 mins 13.2 secs
Modern figure			8 mins 19 secs[3]

1 Newton's figures are not based on independent measurements; they are the best available, in his judgement, at the time.
2 Cassini never accepted that light was not projected instantaneously but, like Rømer, he came up with a figure for the correction to be employed in calculating the timing of eclipses of the moons of Jupiter, and this figure was adopted by others as a figure for the speed of light.
3 This figure is an average, as the Earth's orbit is an ellipse.

Table of speed of light

Date	Investigator	Method	Result (km/s)
1849	Fizeau	Rotating toothed wheel	313,000
1850	Foucault	Rotating mirror	298,000
1875	Cornu	Rotating mirror	299,990
1880	Michelson	Rotating mirror	299,910
1883	Newcomb	Rotating mirror	299,860
1907	Rosa and Doresey	Electromagnetic constants	299,790
1926	Michelson	Rotating mirror	299,796
1928	Mittelstaedt	Kerr cell shutter	299,778
1932	Pease and Pearson	Rotating mirror	299,774
1940	Huttel	Kerr cell shutter	299,768
1941	Anderson	Kerr cell shutter	299,776
1950	Essen and Gordon-Smith	Cavity resonator	299,792
1951	Bergstrand	Kerr cell shutter	299,793
1958	Froome	Radio inferometry	299,792
Modern figure			299,792[1]

1 This is now true by definition as, since 1983, the length of a meter has been set in relation to the speed of light rather than vice versa.

There are two conclusions to be reached from these tables. The first is that a pretty reliable figure for the time light takes to travel from the sun to the Earth was available seventeen years after Rømer's first estimate of eleven minutes. And the second is that measurements of the speed of light improved steadily until 1928; then for about twenty years they settled around a figure which now seems slightly inaccurate; and then progress resumed in 1950, since when measurements have remained more or less identical.[x]

What makes Hacking's argument 'dreadful' is that he takes one figure on its own, and that figure is the first in a long succession of attempts to measure the speed of light. Of course Rømer's figure for the time light takes to travel from the sun to the earth was a very crude approximation! But science is not an enterprise conducted by isolated individuals; it is, as argued in Chapter 8, a collective undertaking in which progress is driven by competition (and cooperation).[17] Indeed John Flamsteed, the Astronomer Royal, noted that there was 'emulation', and even bad blood, between Cassini, the established authority, and Rømer, the upstart.[18] Over time, competition ensures progress. And of course competition is imperfect, and for a time scientists may set out on the wrong path; but over time good results will drive out bad ones.[xi] The claim that an alien society, if it had a sufficiently advanced technology, would come up with almost exactly the same figure as us for the speed of light is perfectly sound. The question of the speed of light is not just an arbitrary one invented to entertain a theoretical physicist: it would impose itself on anyone doing astronomy designed to predict the future apparent location of planetary bodies to a high degree of accuracy, no matter whether their purpose was astrology, chronometry (as in Rømer's case), or space navigation.[xii]

x This tendency to shared error has been labelled 'the bandwagon effect': Mirowski, 'A Visible Hand' (1994), 574. Pinch appears to think that this is the sole cause of agreement on measurements (Labinger and Collins (eds.), *The One Culture?* (2001), 223) but that can't be right, or agreement once established would never break down.

xi Under particular circumstances there may be an economic or institutional investment in a bad solution that allows it to persist. The English-language QWERTY keyboard is an example (David, 'Clio and the Economics of QWERTY' (1985)); geocentrism, for the Catholic Church after 1616, is also an example.

xii The issue arose, entirely predictably, shortly after the invention of the pendulum clock (1656), which made possible new standards of accuracy, exposing previously invisible anomalies (Cohen, 'Roemer and the First Determination of the Velocity of Light (1676)' (1940), 338). Even so, Rømer's timepiece was out by an average of about 30 seconds (Shea, 'Ole Rømer, the Speed of Light' (1998)). How do we know? Because we can compare his

§4

A relativist would respond to this argument by saying that there is no reason to think that scientists were getting better at measuring the speed of light; they were simply getting better at *agreeing* on how to measure the speed of light. This is obviously false, because one test of measurements of the speed of light is whether, when used in conjunction with Kepler's laws of planetary motion, they enable you to predict the apparent position of planets in the sky. Rømer's figure fails that test, and modern figures pass it. Still, a classic example of this line of argument is Simon Schaffer's essay 'Glass Works: Newton's Prisms and the Uses of Experiment' (1989). Schaffer's claim is that Newton had not, as is traditionally claimed, demonstrated by experiment that white light is made of rays of differently coloured light which are differently refrangible, for his experiments could not be successfully replicated except under arbitrary and unreasonable conditions (the use of prisms made in England, for example). Newton's supposed discovery only imposed itself on the scientific community because Newton acquired 'control over the social institutions of experimental authority'. His authority became 'overwhelming'. We believe Newton's theory of colour not because of the experimental evidence, but despite it; we believe it because Newton successfully imposed himself on the scientific community, and experiments were then 'staged' to produce the required results.[19] Nowadays, Newton's experiments come pre-packaged, so that they can be reliably reproduced to educate children in school; but that is because the equipment has been devised to produce the required result.

One might think that Schaffer or his readers would have rejected these arguments as inherently implausible. One might think they would have wondered about the various and varied technologies that depend on Newton's theories of refraction and of colours: the reflector telescopes that Newton himself designed, which avoid the problem that different colours of light are differently refracted, causing a penumbra of colours around objects seen through a lens; or the colour televisions which had been in widespread use for twenty years when Schaffer published his article, which make a full range of colours out of red, green and blue. On the contrary, Schaffer's claims were accepted as confirming a by-then-well-established theory of how science works – that it works not through evidence, but

observations with our retrodiction of the location of the moons of Jupiter when he was observing them.

through power and persuasion. His essay was admired because it seemed to demonstrate that the strong theory could be put into practice: one could write the history of what we now think of as good science (Newton's new theory of light) using exactly the same intellectual moves as the ones one would use for writing about what we now think of as bad science (say, alchemy). Unfortunately, it was Schaffer's evidence, not Newton's, that was 'staged' to produce the required results. Step by step, detail by detail Schaffer's argument was refuted by Alan Shapiro in 1996: it turned out that plenty of people had successfully replicated Newton's experiment without any particular difficulty, and without any need to fiddle the results. But since the year 2000 Schaffer's paper has been cited seven times for every two times that Shapiro's has, and the gap is growing, not closing: over the last four years Schaffer has had ten citations for every two to Shapiro. Bad knowledge, at least temporarily, has driven out good.[20]

Schaffer's essay is not an isolated case. There is a significant group of intellectuals, working within the same tradition as Schaffer, who claim that experiments can never be replicated in a straightforward way. Whenever experiments are repeated genuinely independently, they would argue, divergent results are obtained; to learn to get the 'right' results you have to be trained to do the experiment under very peculiar and particular conditions, and at first this involves learning it directly from people who have conducted it successfully in the past. Eventually an experiment can be reliably mass-produced by manufacturing special equipment designed to obtain precisely that result – the equipment and the results are interdependent. This is what is called 'black boxing'. Once an experiment has been black boxed the experiment is no longer a test of the result; rather, getting the right result has become a test of the reliability of the equipment. So the whole notion of replication, in the view of those who argue in this way, is misleading, and there is nothing transparent about experimental knowledge.[21] Building consensus around what an experiment demonstrates is thus primarily a social process of persuading people to act and think in the way you want them to, not an impartial process of discovering an objective aspect of the real world. These claims, of course, run into trouble if one tries to apply them to the experiment that had more influence on the behaviour of scientists than any other, the Torricellian experiment – or, indeed, to Newton's prism experiments, or to measurements of the speed of light.

Robert Boyle summed up the alternative view of science from that propounded by the devotees of holism and underdetermination and by deniers of the independent replication of experimental results:

> Experience has shown us, that divers very plausible and radicated Opinions, such as that of the Unhabitableness of the Torrid Zone, of the Solidity of the Celestial part of the World, of the Blood's being convey'd from the Heart by the Veins (not the Arteries) to the outward parts of the body, are generally grown out of request, upon the appearing of those new Discoveries with which they are inconsistent, and would have been abandon'd by the Generality of Judicious Persons, though no man had made it his business purposely to write Confutations of them: so true is that Vulgar saying, that *Rectum est Index sui & Obliqui*. ['The line which shows itself to be straight shows also what line is crooked.'][22]

In other words, as with the terraqueous globe and the phases of Venus, very often a new theory triumphs rapidly and without resistance, because new evidence simply eliminates the viability of all known alternatives.

§5

If the relativist account of science were correct, every major paradigm shift ought to be accompanied by bitter disputes between competing intellectual communities: indeed it was Kuhn's view that this is exactly what happens. Some are, but others take place silently, as Boyle says, without anyone bothering to write confutations of the old theory. One army abandons the field almost immediately after the first blow has been struck; its opponents declare victory and are rapidly joined by deserters from the other side. What causes this sudden transformation? When Vadianus maintained in 1507 that Aristotle did not know everything, that he was a fallible human being (the question immediately at issue was the source of the Danube, but of course the two-spheres theory was also at stake), the claim seems so obvious to us as to be trivial; it would not have been at all obvious to Vadianus's contemporaries. Why did Aristotle make mistakes? Because of *experientiae penuria*, insufficient experience.[23] The victory of the terraqueous globe theory following the discovery of America is the first great triumph of experience over philosophical deduction, and thus the beginning of a revolution.[xiii]

xiii This revolution went through virtually without opposition. If Clavius was still insisting on it fifty-five years later, it was not because it was contentious; it was simply that he was providing a commentary for Sacrobosco's *Sphere*, which was still the textbook in every university. Clavius was really arguing with the dead, not the living. Why then did

But it would be dangerous to rely on examples such as this to support an oversimplified view of the role of experience. Experience, we might say, comes in three kinds. Sometimes, as we have just seen, it falsifies beliefs and in doing so immediately imposes an alternative; sometimes it confirms beliefs that are already held (the measurement of the shape of the Earth by French expeditions to Peru and Lapland (1735–44) confirmed Newtonianism); and sometimes it represents only one step along a path that leads to an outcome which cannot be foreseen. Of this third kind, there are both answers to scientific questions that might have been right but turn out to be wrong, but nevertheless are a crucial step towards a successful answer; and right answers whose full significance only slowly becomes apparent in the light of further experience. Kuhn held that the fact that the outcome of a revolutionary crisis is unpredictable while it is going on means that it cannot be explained with the benefit of hindsight. On the contrary, there is often only one path through the debate which is capable of producing a stable outcome. Getting it right can be like finding the way out of a maze.

In the late Middle Ages, for example, the Venetians became wealthy importing spices from Asia; the spices were carried overland from the Red Sea to Alexandria, which meant that the Venetian merchants, who bought them in order to ship them across the Mediterranean, had to pay a high price for them. The Portuguese, hoping to undercut the Venetians, sought a route to the spice islands by sailing around Africa, and in the end they succeeded. They were followed by the Dutch, who made the spice trade the foundation of a great commercial empire. Columbus sought a route to the West, but his successors discovered that circumnavigating South America was arduous and time-consuming; as a commercial route to Asia his discovery proved to be a failure, but for this there was more than compensation in the discovery of gold and silver in South America. In the early 1600s the French explorer Samuel Champlain thought he might find a route by water across Canada – up

Sacrobosco continue to be the standard text after the discoveries of Vespucci? Because the university curriculum was geared to classic texts (Sacrobosco was for astronomy the equivalent of Euclid for geometry); and because as long as the Ptolemaic cosmology was viable the problems with Sacrobosco's text were manageable – and were neatly managed in the texts which followed the example set by the Wittenberg edition of 1538. Still, those problems were accumulating over time: Barozzi, *Cosmographia* (1585) contains a list of eighty-four errors in Sacrobosco; Barozzi aspired to replace Sacrobosco, but his hopes were misplaced. Sacrobosco retained his position until, after 1611, Ptolemaic geocentrism was no longer intellectually respectable. And after 1611 no astronomical textbook would ever again have the status that Sacrobosco had had.

the St Lawrence, through the Great Lakes, and on.[24] He carried Chinese court dress with him in his canoe in case he should meet Chinese agents coming from the East. His transcontinental route too was a failure. Until 1794 ships sought a Northwest Passage, but none was to be found.[25]

Here we have a series of attempts to answer the same question, against a background of changing geographical knowledge: indeed it was the attempt to find a better route to Asia which was the main driver in improving that knowledge. As it turned out, between the end of the fifteenth century and the end of the nineteenth century efforts to find a new route failed over and over again. It was impossible to know in advance that this would be the outcome; but we can be sure, as they could not, that Columbus would not reach China, that Champlain was never going to meet an emissary from the Chinese court, and (until the advent of global warming) that the search for a commercially viable Northwest Passage was doomed to failure. By 1800 all the possible alternatives had been eliminated, and the question of the best route to Asia had finally been settled (at least until the opening of the Suez Canal in 1869).

Such examples of path dependency are the rule and not the exception. Once Copernicus had suggested that the earth was not the centre of the universe but a planet orbiting the sun, people were bound to puzzle over what sort of planet it could be. In the Aristotelian universe the earth had been the recipient of light but had given no light. It was easy to imagine looking down on the earth, but what one would see would be a miniature earth. There was, surprisingly, an extensive discussion among Aristotelian philosophers as to what the earth would look like seen from the heavens, but no one envisaged it as one of the brightest stars in the night sky. Nicholas of Cusa had turned the earth into a true star, but only at the price of turning the sun into an earth – hardly anyone was prepared to follow him.[26] For Digges and Benedetti, even though they were Copernicans, from a vast distance the Earth, which received light but did not transmit it, would become a dark star. Leonardo, Bruno and Galileo realized that the Earth would look like a vast moon if seen from the moon, and recognized that when the moon was new you could see that it was being illuminated by the Earth – Harriot, after reading Galileo, named this 'earthshine', which is the term we use today.[27] Galileo devised some elementary experiments to show that the Earth would reflect light, and that the land would reflect more light than the sea (which is why the moon shines so brightly by reflected light). Turning his new telescope towards Venus in 1610 he found that it had phases – proof that it too shone by reflected light. Moreover, it had a full set of phases, proof that it orbited the sun, as in the systems of Copernicus and Tycho Brahe.[28] At this point it became apparent that seen from Venus the Earth would be one of the brightest stars in the sky.

Thus Copernicanism posed a straightforward question: What sort of planet is the earth? A full range of answers to that question was canvassed. Only one answer, that all the planets shine by reflected light, proved robust and stable. It took seventy years for that answer to establish itself, but once the telescope had been invented and turned into a scientific instrument it was the only answer with any chance of survival. The answer was entirely unpredictable in 1543, but entirely inevitable after 1611.

Once a scientific question is on the agenda of a community of scientists – once it becomes 'alive' – then over a period of time we may expect a range of possible answers to be explored; indeed, sometimes all possible answers may be proposed for consideration.[29] In this early period it may be impossible to get agreement about which is the right answer. But over time a stable consensus will emerge that one answer is right and all the others are wrong. This consensus does not just depend on a rhetorical or political process of winning agreement; it also depends on the capacity over time for the supporters of one particular theory to rebut criticism and provoke fruitful lines of enquiry.[30] A 'robust' or 'stable' answer comes to be regarded as quite simply the right answer. This does not mean that its rightness was always apparent, although careless historians and scientists often imply as much; it does mean that its rightness becomes incontestable, at least for a period of time.

The discovery of antipodes led straight to the concept of the terraqueous globe; but Copernicanism did not lead directly to the view that all planets shine by reflected light – the telescope had to intervene. Between the recognition that the Torricellian tube is a pressure gauge and the invention of the atmospheric steam engine there was no intervention of an extraneous factor. Boyle's law was a natural development of Pascal's Puy de Dôme experiment, and the atmospheric steam engine was a natural development of Boyle's law (even if the technical problem of building an engine that worked was considerable). Torricelli did not for a moment imagine the steam engine, any more than Columbus imagined America; the path that led from the barometer to the steam engine was not as straight or as short as the path that led from Palos de la Frontera to the Bahamas; but the path was there, waiting to be found.

The search for the Northwest Passage, or the attempt to explain planetary motion on the model of magnetism, were mistaken but useful. There are plenty of other examples of scientific enterprises which have been doomed from their inception, but whose practitioners have simply refused to learn from experience: the attempts to turn base metal into gold or to cure infectious diseases by letting blood both lasted for more than two millennia, but neither could be made to work, and indeed neither produced valuable new knowledge, as both the quest for the Northwest Passage and Kepler's new astronomy did. And it is this which is the fundamental problem with a relativist approach – either *anything* can be made to work, in which case the philosopher's stone may yet be found, or some things can never work, in which case there is an external reality that constrains which beliefs are viable and which are not. Of course 'made to work' is a slippery concept: plenty of alchemists thought they had seen base metal turn into gold, and plenty of doctors thought they had cured their patients through bloodletting. People delude themselves in all sorts of ways. The idea that America was really Asia died within a generation, while the alchemists' project was much more resilient.

§6

Naive realists, those who think that science always establishes incontrovertible truths about the world (a view difficult to sustain, given the evidence that scientific theories change radically as the evidence they are based upon is revised),[31] assume that scientific enquiry is always going to ask similar questions and produce identical answers; relativists assume that both the questions and the answers are infinitely variable. In truth the questions may be variable, but sometimes the answers are not. You do not have to sail west, but if you do you will end up in America. And once you have found America, if you were trying to get to Asia, then the search will begin for ways round it. One question leads to another; scientific enquiry is path-dependent.[32]

There is a common-sense view that takes this idea to an extreme. It holds that once you set out to answer a question the answer you will arrive at is entirely predetermined – just like Columbus's discovery of America. Just as any two carpenters will agree on the length of a table, though one may measure in inches and the other in centimetres, so a Martian and an Earthling will agree on the speed of light, though they will surely have different units of measurement.

Thus according to the common-sense view the science of extraterrestrial beings, if they exist and if they are intelligent, must, where it overlaps with our science, agree with it. Steven Weinberg, a Nobel prize-winning physicist, expressed this view when he wrote 'when we make contact with beings from another planet we will find that they have discovered the same laws of physical science as we have.'[33] Science is thus a cross-cultural language which any culture can in principle learn to speak, and which any technologically sophisticated culture will already have learnt to speak. This was the assumption underlying a message that was broadcast into space by the Arecibo radio telescope in 1974. The message consisted of the numbers one through to ten, the atomic numbers of hydrogen, carbon, nitrogen, oxygen and phosphorus, the formulas for the sugars and bases in the nucleotides of DNA, the number of nucleotides in DNA, the double helix structure of DNA, a figure of a human being and its height, the population of Earth, a diagram of our solar system, and an image of the Arecibo telescope with its diameter. The assumption was that any extraterrestrial intelligence capable of receiving the message would recognize the maths and the science and quickly make sense of the Earth-specific information. The great mathematician Christiaan Huygens discovered the law of the pendulum in 1673; he believed that there were inhabited planets scattered across the universe; and by the time he died in 1695 he had persuaded himself that this law was known throughout the universe.[34]

The opposing view is that science is shaped by a whole series of cultural and social factors which ensure that no two societies would produce the same knowledge, just as no two societies produce the same religious beliefs. Scientific knowledge is not, in reality, unalterable truth, even if it appears to be. Consequently two different scientific communities will always produce two radically different bodies of scientific fact and theory, and science is not a cross-cultural form of knowledge but a local consensus, specific to a particular community. You might think that Boyle's law is a bit like the New World – it was there waiting to be discovered. But the cultural determinists do not accept this. They think it is more like a signature dish – Escoffier's peach melba, for example – the product of a very specific, local technology and culture (the Savoy Hotel in 1892).[35] Just as certain dishes – peach melba, prawn cocktail – have managed to spread across continents and survive over time, so some scientific doctrines are successfully disseminated, while others remain firmly tied to their time and place of origin.

This book seeks to recognize that proponents of both positions have some good arguments and some bad arguments. Its criticisms are not only directed at relativists (who are given more space simply because their views are more prevalent in history of science); they are also directed at realists who fail to take to heart the evidence on which relativism is founded, the evidence of cultural difference across time and space – the evidence of history and anthropology. It is a standard ploy of opponents of relativism to argue that we all share certain commonsensical capacities for reasoning, and thus we can all recognize better knowledge when we see it.[36] They take the view that science is, in essence, common sense applied systematically, or, as Karl Popper put it, '*common-sense knowledge writ large*, as it were'.[xiv] In my view, explaining science in terms of common sense is simply to go round in a circle. Evidently there *are* certain fundamental experiences and modes of reasoning that are universal and can be regarded as valid in any and every human culture. If there weren't, cross-cultural communication would often be impossible.[37] For example, there is some experience of hunting wild animals in every human culture. We have seen that Roman lawyers regarded *vestigia*, which originally meant the tracks left by an animal, as a form of evidence. The word 'investigate' has as its root meaning to follow tracks when hunting. Our word 'clue', in its meaning, new in the nineteenth century, of something that detectives are on the lookout for, is a metaphor derived from Ariadne's thread which allowed Theseus to escape from the Minotaur's labyrinth, an indication that there is nothing new about following the evidence and seeing where it will lead – the detective too is following a track, just as Theseus retraced his own steps. All human beings are capable of the sort of intellectual activity that tracking requires, and when we investigate a problem we are engaged, as the realists claim, in a sophisticated version of the same activity.[38]

But, and it is a big 'but', although there are certain cross-cultural uniformities common to all human societies, this doesn't help us very much, most of the time. Firstly, although there may be a very few truly

xiv Popper, *The Logic of Scientific Discovery* (1959), 22 (italics in original). Popper might have hesitated if he had remembered that Copernicus explicitly described his new theory as being (almost) at odds with common sense (Copernicus, *On the Revolutions* (1978), 4). Moreover, Popper also held that the peculiar feature of scientific knowledge is that it grows (18), while common sense ought to be the same at different times and different places. For an example of the common-sense view see Christie, 'Nobody Invented the Scientific Method' (2012). However, what we think of as common sense is itself a cultural attainment and has preconditions, such as literacy: Luria, *Cognitive Development, Its Cultural and Social Foundations* (1976) is the classic (and eye-opening) demonstration that abstract thinking (such as syllogistic reasoning) does not come naturally.

universal experiences and modes of thought, each culture naturalizes the experiences and ways of thinking that its members have in common as their own, local version of common sense – common to us, if not to those other people over there. Thus G. E. Moore thought it was common sense to believe in an external reality, but not in a creator God or a life after death, while these beliefs have seemed commonsensical to many people.[39] In practice, the range of beliefs that *can* come to be generally shared and effectively unquestioned within a society is enormous. Discussions of common sense generally fail to distinguish adequately universal and local definitions of the concept.

Take just one example: through the Middle Ages and the Renaissance, scholastic philosophers followed Aristotle in thinking that earth is naturally heavy, and tends downwards, while fire is naturally light (it has negative weight, we might say) and tends upwards. Air and water were heavy or light depending on the circumstances in which they found themselves – depending on whether they were above or below their proper location. Aristotelians also held that solids are denser and heavier than liquids. Thus ice is heavier (because it is 'denser') than water. So why does it float? It floats because ice on a pond is flat, and the water resists its sinking. Wood, they claimed, was heavier than water, since the element out of which it was primarily made was earth. You can make a boat out of wood, they would say, but only if it has a flat (or at least flattish) bottom.[40]

The works of Archimedes were widely known in the Middle Ages, but the philosophers simply didn't accept that his account of buoyancy was correct: Archimedes argued as if all substances have weight and tend downwards, and as far as the philosophers were concerned this represented a fundamental misconception. When they dismissed Archimedes as irrelevant they thought common sense was on their side; and their own experience of the world corresponded exactly to their claims. They were aware of no significant anomalies where their theories failed to correspond to reality. They sailed in ships, walked on pontoon bridges, and stepped across frozen puddles without ever experiencing anything that seemed to oblige them to rethink their theories. Nevertheless, as Galileo pointed out, it would have been easy to test their claim that ice floats because it is flat by breaking it up into small pieces and seeing if the pieces still floated.[xv] And in any case, why does a flat sheet of ice bob back to the surface if you push it under?

xv See above, pp. 71–2.

A reading of the philosophers who replied to Galileo will quickly establish that they thought they had reason, common sense, and the authority of Aristotle all on their side, although they were far from sharing a consistent position. One argued that Galileo's ice bobbed to the surface because, although pure ice would be heavier than water (which would not prevent it floating, but would prevent it surfacing once submerged), his ice had air trapped within it, which explained why it would not stay submerged.[41] Others pointed out that Galileo's theory presented its own problems. Galileo appealed to sensory experience, but sensory experience established, or so they claimed, that boats float higher in the water when far from shore, and lower as they approach port. Neither Archimedes nor Galileo could explain this basic fact, but an Aristotelian, they were happy to report, could. (A larger body of water pushes more firmly against the bottom of the boat, and so the puzzle is solved.)[42] Meanwhile they disagreed on the most fundamental questions: one thought wood heavier than water, another lighter; one held that water had no weight in its proper place, another denied it; one acknowledged that water expands when it freezes, another disputed it. One thing, though, they agreed on: Aristotle was always right.

There is a key difference between Galileo and his opponents: both appealed to experience, but Galileo had engaged in a programme of experimentation. His was an applied knowledge and theirs was not. Ultimately, the difference between him and them lies in the fact that they were prepared to make claims that they were sure were true but which they had never tested (wood is heavier than water; boats float higher in the water as they move away from the shore), while Galileo had tested each and every one of his claims. Their failure to conduct tests was reflected in their inability to agree on the most basic points. For the most part they did not dispute the truth of what Galileo stated as a matter of fact; but they expected him to accept their own factual claims, which of course he was not prepared to do. Thus they reported, on the authority of Seneca, that there was a lake in Syria where the water was so thick that bricks floated on top of it; invited to explain this, Galileo dismissed it as nothing but a tall tale, so that no explanation was necessary; at which they grew indignant, insisting that one should believe 'authors worthy of faith, such as Seneca, Aristotle, Pliny, Solinus, and so forth'.[43] In other words, the basic difference between Galileo and his opponents was that they were philosophers, while he was a mathematician in the process of becoming a scientist (rather than being, as one opponent claimed, a mathematician falsely claiming to be a competent philosopher).[44]

It seems pointless to maintain that Aristotle and his followers were deficient in common sense, or deficient in experience of how the world goes. They had plenty of both, according to the standards of their own time. What they lacked was the right intellectual tool-kit: in this case, a procedure for devising tests to confirm (or rather falsify) theoretical claims. They lacked this procedure for the fundamental reason that they thought it unnecessary: what could go wrong if one argued from uncontested premises to conclusions that necessarily followed? What could go wrong if one relied on the common experiences that we all share? So we come to a dilemma. Either common sense is amorphous and malleable, to the extent that all the beliefs that are shared within a community are compatible with common sense. Or, if you claim that many societies are lacking in common sense because they hold beliefs that could easily be refuted, then it seems there are communities in which, much of the time, nobody exhibits any common sense at all. The idea of 'common sense' either proves too much or too little. Either all societies have enough of it to get by, in which case the concept hardly helps one understand what makes for reliable knowledge; or only the arguments that accord with our own exhibit common sense, in which case across history and across cultures common sense has been in very short supply.

When Susan Haack writes, 'Our standards of what constitutes good, honest, thorough inquiry and what constitutes good, strong, supportive evidence are not internal to science. In judging where science has succeeded and where it has failed, in what areas and at what times it has done better and in what worse, we are appealing to the standards by which we judge the solidity of empirical beliefs, or the rigor and thoroughness of empirical inquiry, generally,' she is, I submit, confusing two distinct issues.[45] Yes, in our society certain standards of enquiry (including an emphasis on the acquisition of empirical information) are not peculiar to science, but are widely shared – that is only because the Scientific Revolution, and the wider cultural shifts that made it possible, have shaped the whole of our culture. It would however be simply wrong to think that we and Aristotle share the same views about what constitutes a justified true belief, and about how to acquire such beliefs. So the problem is: Who is 'we' here? Is it, to use her examples, we (contemporary) historians and we (contemporary) detectives? Or is it we as human beings who share a certain capacity for common sense with all other human beings? The first interpretation is largely true, but not very significant; the second would be significant, but is untrue.

Moreover, notions of how to judge empirical evidence vary from one enterprise to another. Boyle and Newton believed in the transmutation of base metals into gold; but they were excellent judges when dealing with other empirical questions. In the sixteenth century Jean Bodin wrote a book entitled *Method for the Easy Comprehension of History* (1566), which is often regarded as a founding text of modern historical reasoning; he also wrote a book, *On the Demon-mania of Witches* (1580), arguing that there are witches everywhere and that human beings regularly metamorphose into wolves. Both sets of views seemed to him equally commonsensical. A century later, Thomas Browne campaigned against the epidemic of false beliefs (that elephants have no knees, for example); but he continued to believe that Lot's wife had been turned into a pillar of salt, and (in his professional capacity as a doctor) he appeared in a witchcraft trial to certify that supernatural forces were at work, thus ensuring the conviction of the accused.[46] Browne was a pretty sensible chap, at least by the standards of the day, and we will not get far by trying to argue that his views on witchcraft show that he was incompetent to make judgements about empirical beliefs. If he had views about what constitutes evidence that were different from ours, then that simply shows that 'common sense' varies from one culture to another.

§7

To watch our modern conception of common sense (as applied to science and other types of empircal enquiry) being born we must turn again to the example of Galileo's revision of Archimedes' principle. Galileo was pretty much the same person when he got the right results as he had been only a few days previously when he was still getting the wrong results. What had changed? The answer is straightforward: he had begun an experimental programme of testing and refining his theories. He had started with one anomaly, the floating chip of ebony, and discovered another, the floating needle; as a consequence he had done his best to make sense of surface tension. He had also set out to illustrate Archimedes' principle in action, and as a consequence had discovered a further anomaly, that a weight of water could lift a weight greater than itself. Boldly, he had gone back, reanalysed Archimedes' principle, and revised it. He had then tested his new theory with a quite different experiment. This back and forth movement between theory and evidence, hypothesis and experiment has come to seem so familiar that it is hard for us to grasp that Galileo was doing something fundamentally new. Where his predecessors had been doing mathematics or philosophy, Galileo was doing what we call science. The difference between Galileo at the start of this process and Galileo at the end is that at the end evidence was being used to impose much tighter constraints on argument than at the beginning; evidence and argument were interacting in a new way. In arguing for such an interaction Galileo could indeed appeal to what both he and we would think of as universal principles of common sense, even though both his experimental practice and his conclusions were novel; but such appeals are always problematic when they run against convictions so entrenched that it has come to be assumed that they are indisputable truths. All parties to a dispute always claim that common sense is on their side.[xvi] Galileo seems to have had absolutely no success in his efforts to convince the philosophers that he understood why bodies float better than they did.

The whole of the Scientific Revolution is encapsulated in Galileo's little treatise on floating bodies. The subject had been discussed by brilliant philosophers and mathematicians for 2,000 years. The views of the philosophers and mathematicians, although they were sharply at odds with each other, corresponded satisfactorily to daily experience, or at least so they believed. Neither theory was in crisis. Before Galileo, no competent mathematician had doubted that Archimedes provided a complete explanation of what happens when bodies float, and no competent Aristotelian philosopher had disputed the view that shape is crucial in determining which bodies float. And yet Galileo realized, not only that ice was lighter than water, but also – it must have been an extraordinary shock – that bodies often float without displacing their own weight in water. Everybody had been wrong.

We cannot explain the timing of this intellectual revolution simply by appealing to some local, contingent event, an argument between Galileo and some Aristotelian philosophers (though such an argument did indeed take place). Nor was Galileo, when he invented the hydraulic press, working on some new, practical problem that only a competent engineer could tackle. He was simply addressing an age-old question: Why do some bodies float and others sink? If he came up with new

xvi Descartes' *Discourse on Method* (1637) begins: 'Right understanding [*le bon sens*; often translated as *common sense*] is the most equally divided thing in the World; for every one beleevs himself so well stor'd with it, that even those who in all other things are the hardest to be pleas'd, seldom desire more of it then they have.' (Descartes, *A Discourse* (1649), 3.)

answers it was because he employed new methods and was evolving a new intellectual tool-kit. Pascal went on to clarify and systematize Galileo's new understanding. His situation was a little different: he was studying pressure in fluids in order to understand how the weight of air supports a column of mercury in a barometer. (Actually, he was inventing the concept of pressure – Galileo had thought in terms of weights, not pressures.)[47] Pascal's new hydraulics is an extension of his vacuum experiments, but Galileo's work on floating bodies is not path-specific in this way. It springs, not from a new problem, but from a new type of practice and a new way of thinking.

Aristotle might have had difficulty in following Galileo's argument that the size of the container matters when trying to understand floating bodies, but Archimedes would not. So what separates Galileo from Archimedes? In what sense is this new thinking if Archimedes would have had no difficulty in understanding it? First, Galileo lived in a culture where even the most authoritative beliefs could be questioned – this was the legacy of Columbus. He lived in an age of discovery. Second, Galileo was constructing a new science in which everything was, at least in principle, subject to measurement, even the rise in the level of a pond when a duck enters the water, or of an ocean when a ship is launched. This principle of ever-more-exact measurement derives from Tycho Brahe; in tightening up the relationship between evidence and theory in physics, Galileo was extrapolating from the practices of the astronomers.[xvii] Third, Galileo had the example of Gilbert, who had used the careful manipulation of experimental apparatus to establish new and unsuspected truths. Columbus, Brahe and Gilbert did not supply arguments that Galileo used; they supplied role-models that he was bold enough to follow. They did not contribute directly to his new hydraulics, but the new hydraulics was made possible by the intellectual culture they had helped to shape – specifically, Galileo's intellectual culture, for most of his contemporaries were still happy to defer to Aristotle and were uninterested in challenging orthodox beliefs. It is only, as we have seen, with Pascal that this peculiar, idiosyncratic culture began to become widely adopted and generally respected.

Archimedes had been convinced that the real world is mathematically legible, even though oceans and ships are not shaped like circles,

xvii His insistence on being more precise than Archimedes goes back into his youth, when he had devised an instrument, the *balincetta*, which would make more exact measurements of specific gravities than any made before.

triangles and squares; he had been sure that Euclid's geometry is more powerful than Aristotle's syllogisms. There is no way of proving that in every possible world the mathematicians will be better equipped to understand the world than the philosophers: it is something that can only be established through the success of mathematical practice, and that requires a culture in which mathematicians are permitted to challenge the claims of philosophers and are rewarded for their successes.[48] But Galileo was not simply a second Archimedes; he was also an experimental scientist. Nothing guarantees that mathematicians will be interested in taking this further step. In fourteenth-century Oxford, mathematically inclined philosophers had hypothesized that falling bodies accelerate continuously; but they had made no effort to confirm this theoretical possibility through experiment. That was left to Galileo. Indeed, the Oxford philosophers had hypothesized that colour and temperature might in principle be quantified and might change with increasing or decreasing rapidity; they had no way of measuring colour or temperature, just as for all practical purposes they had no way of measuring the speed of falling bodies. Their speculations were purely abstract and theoretical; they applied to any possible world and not to our actual world. They studied mechanics in theory but they had no practical interest in machines.[xviii]

Archimedes had been available in Latin since the twelfth century, and in print since 1544: before Galileo every mathematician was happy to think about ships floating in a boundless ocean, but no one had floated model boats in model oceans and studied exactly what happened. The finest mathematicians were entirely satisfied with Archimedes' principle, which seemed to them both coherent and complete. Galileo was the first to turn Archimedes' account of how bodies float into a theory to be tested with an experimental apparatus; at which point the theory proved incomplete.

Galileo's determination to embody mathematical theory in a corresponding apparatus is fundamental. He even hoped to build a mechanical model that would illustrate his theory of the tides.[49] From it follows the

xviii With one exception: Richard of Wallingford (1292–1336) designed a wonderfully elaborate and sophisticated clock. But clocks were philosophical instruments – they modelled the movements of the sun and the other planets; and the sophisticated measuring instruments that Richard made with his own hands were to be used to further the liberal arts, not the mechanical arts. (North, *God's Clockmaker* (2005).) Richard was the son of a blacksmith, but he had no interest in designing load-bearing machines such as watermills or windmills.

great advantage of Galileo's analysis over Aristotle's and even over Archimedes', which is that Galileo's analysis gives you better prediction and more control: the proof of the pudding is in the eating. And then, of course, success once achieved and disseminated is redescribed as common sense, and we are assured that any sensible person can see that there is only one way – the right way, in this case Galileo's way – of tackling the problem, by moving back and forth between induction and deduction, by devising thought experiments and real experiments, by never trusting a theory until it has been tested and attempts have been made to falsify it.[xix] 'Common sense', if we mean the practices which *we* think embody it, did not exist before the publication of Galileo's *Discourse on Floating Bodies* in 1612.[xx] Galileo is the first person who is one of us in the sense of 'us' that is required when Susan Haack writes of 'our standards of what constitutes good, honest, thorough inquiry and what constitutes good, strong, supportive evidence'.[50] In seeking to understand a case like Galileo's study of floating bodies we can, in short, have too much realism, in which case we will never understand Galileo's originality or the opposition he faced; but we can equally have too much relativism, in which case we will never acknowledge that he was right, and his opponents were wrong.

Thus we are obliged to recognize that both realists and relativists have a good point. We must run with the hare and hunt with the hounds. Like the relativists, we must acknowledge the dangers of arguing from universal standards of human rationality (which is not to say that such arguments are always invalid – Galileo's knowledge of buoyancy was much more reliable than those of his opponents, as they would have discovered if they had embarked on a programme of experimentation). Like the realists, we must insist that it is not so difficult to tell good

xix There have been claims, dating back at least to Randall, 'The School of Padua' (1940), that Aristotelian philosophers understood how to do this; but there are no convincing examples of their actually doing it before opponents of Aristotle, such as Galileo, had shown them how. In the first half of the seventeenth century there was a lively scientific tradition among the Jesuits, one which involved adapting Aristotelianism to meet the challenges of the new science, but this was slowly smothered by the obligation members of the order were under to conform to the traditional teachings of the Church: Feingold, *Jesuit Science* (2002).

xx I am arguing here from dates of publication: Galileo had already developed a sophisticated theory of falling bodies, backed up by exquisite experiments; but he had not published, and would not until 1638. The treatise on floating bodies is the first published experimental science to tackle the core principles of Aristotelian physics head on: Gilbert's work on magnetism was a brilliant demonstration of the experimental method, but it did not clash directly with the arguments of the philosophers.

science from bad, as long as you agree that knowledge should be carefully and systematically tested against experience.

I started the previous section of this chapter by presenting two alternative views of science, and I have argued that both have merit. From one point of view, the knowledge we end up with seems to be culturally relative, contingent, peculiar; from another it seems to be commonsensical, predictable, inevitable. Kuhn tried to maintain both these views, despite the tension between them, by distinguishing revolutionary science from normal science. The outcome of a revolution is, he argues, culturally relative, contingent, peculiar; but it leads to a period of stability during which progress is the norm. Kuhn's distinction between two types of science was too neat, but his basic approach was sound: sometimes, as we have seen, one discovery leads to another in a way which is, with hindsight, entirely inevitable. Between Copernicus and Newton a number of revolutions took place, and there is no simple path that leads from one to the other; but from Torricelli to Newcomen the route is fairly straightforward, and once the barometer had been invented and experimentation had become accepted as the best route to new knowledge, the steam engine could straightforwardly follow. In the history of science there is no simple solution to the apparently binary choice between radical contingency and predictable evolution. For the choice is a false one. The answer is always somewhere between the two extremes, and the balance between the two has to be struck afresh with each new topic.

§8

We have been studying the origins of science. We have seen that science is an enterprise we have invented and agreed to play by certain rules. There are plenty of enterprises where we invent the rules and we change them when we want to. The age of majority used to be twenty-one; now it is eighteen. Women used to be denied the vote; now they have it. There are other enterprises where our ability to change the rules is constrained by factors over which we have no control.

Gardeners, for example, create a peculiar micro-environment for their plants. If the gardener stops working, nature takes over. A garden is thus both natural and artificial, and it is both of these things entirely and simultaneously. It is easy to assume that the law of the excluded middle requires something to be *either* natural *or* artificial: thus a shirt is made of a material which is either natural (cotton, linen, wool) or synthetic (nylon, polyester). But rayon is both natural and artificial, being made from wood. So, too, a boat capable of sailing into the wind uses natural forces to achieve a result that would not occur in nature. In gardening, in cooking, in naval architecture there are lots of culturally specific choices that we can make, but there are plenty of other things that simply cannot be done. Plants die, mayonnaises curdle, ships sink. Wishing and willing will not make it otherwise.[xxi] Such activities depend on a complex collaboration between the natural and the social. Thus it must be wrong to say, as Andrew Cunningham does, that science is 'a human activity, wholly a human activity, and nothing but a human activity':[51] it is wholly a human activity, but it is not 'nothing but a human activity'. Poetry and Scrabble are nothing but human activities. Science belongs to the very extensive class of activities which combine the natural and the artificial, which are constrained by both reality and culture.

The peculiar feature of science is that it claims not simply to co-operate with nature (as gardeners, cooks, and naval architects do), but to discover a truth that existed before that cooperation began. It is not surprising that the history of science is a problematic activity, because science itself constantly claims to escape from its own temporal specificity, its own artificiality. In claiming to escape from its own process of production, science presents itself as being natural, not artificial; it is only to be expected that, in opposition to this obvious misrepresentation, some want to claim that science is entirely artificial and not at all natural. But the simple truth is that it is both, and that the scientists are right to claim that this artificial enterprise can discover what happens in nature.[xxii]

Some would simply deny that it is possible to escape culture and discover nature in the way that scientists claim. Bruno Latour maintained

xxi In a notorious essay, Stanley Fish, a famous literary critic (assuming that is not, like 'military intelligence', an oxymoron), argued that science and baseball are basically alike, as both are social constructions and both are 'real' (Fish, 'Professor Sokal's Bad Joke' (1996)). Fish acknowledges that the rules of baseball can be changed by a vote; he then goes on to argue that the facts of nature also change, and that funding agencies decide what they are. He's wrong. Ice was never heavier than water. 'Even if two activities are alike social constructions,' he writes, 'if you want to take the measure of either, it is the differences you must keep in mind.' Quite so: science is not like baseball. My own argument agrees at certain points with Fish's, but it is the differences that the reader should keep in mind.
xxii Denying the validity of the nature/artifice distinction was a key move which made the Scientific Revolution possible: see above, pp. 332–4.

that the fact that the bacterium that causes tuberculosis has been discovered in the lungs of an ancient Egyptian pharaoh, Ramses II, did not mean that Ramses died of tuberculosis. Tuberculosis was only discovered in the nineteenth century. Before its discovery there was no such thing as tuberculosis, and consequently nobody could die of it. This is just wrong: of course Ramses II did not know he was dying of tuberculosis, but nevertheless we now know that it was tuberculosis that killed him. Latour's historicism misses a key point about science, which is that it is about matters which are the case whether we believe them to be so or not. The bacterium which causes tuberculosis was *discovered* not *invented* by Robert Koch in 1882. Latour says that Ramses II could no more die of tuberculosis than he could be killed by a Gatling gun (Richard Gatling invented the Gatling gun in 1861), making clear that he thinks discovery and invention are the same thing. They are not. The Gatling gun involves a new sort of cooperation between nature and society; but the bacterium that causes tuberculosis requires no deliberate cooperation on our part, even if identifying it and killing it do require techniques which originate in the laboratory and involve a complex cooperation between nature and society.[xxiii][52]

Science, as a method and practice, is a social construct. But science as a system of knowledge is more than a social construct because it is successful, because it fits with reality.[53] This fit cannot be shown to be necessary or inevitable, which is what the realists do not understand. Aristotle thought his method was necessarily trustworthy; he was wrong. If our method works better than his it is because it fits better with the world as it is, not because the world was bound to be like that.[xxiv] Nevertheless, wherever this fit is established (and it has to be established anew in each new scientific discipline) it establishes a positive feedback loop. Outside the narrowly mathematical disciplines (including astronomy and optics), that loop was first closed in 1600. Consequently, we need to think of science as being the result of an evolutionary process where good science has had, over the last five centuries, a better prospect of

xxiii See 'Notes on Relativism and Relativists', No. 7' (p. 587). Compare, for a similar confusion, the idea that facts are made (above, p. 256n); contrasted with my argument that the concept of a fact is an invention, but particular facts are not invented but established. So, too, science itself is an invention, but Newton did not invent gravity, he discovered its law.

xxiv Contrast, for example, Boghossian, *Fear of Knowledge* (2006), in which the idea of an objective fact is treated as if it were transparent and unproblematic; the notion that this is an idea which itself had to be constructed, and which has evolved over time, seems to escape Boghossian.

survival than bad science. As Kuhn rightly put it, 'Scientific development is like Darwinian evolution, a process driven from behind rather than pulled toward some fixed goal towards which it grows ever closer.'[54]

§9

The problem with the relativists is that they explain bad science and good science, phrenology and nuclear physics, in exactly the same way – advocates of 'the strong programme' explicitly insist on this equivalence.[xxv] The problem with the realists is that they assume there is nothing peculiar about the method and structure of science. According to them the scientific method is somehow natural, like walking, not artificial, like a watch. This book will look, I trust, realist to relativists and relativist to realists: that is how it is meant to look. It stands in the tradition of Kuhn's 1991 lecture 'The Trouble with the Historical Philosophy of Science'. There Kuhn criticized the relativists (who had taken much of their inspiration from his own work), saying that their mistake was

> taking the traditional view of scientific knowledge too much for granted. They seem, that is, to feel that traditional philosophy of science was correct in its understanding of what *knowledge* must be. Facts must come first, and inescapable conclusions, at least about probabilities, must be based upon them. If science does not produce knowledge in that sense, they conclude, it cannot be producing knowledge at all. It is possible, however, that the tradition was wrong not simply about the methods by which knowledge was obtained but about the nature of knowledge itself. Perhaps knowledge, properly understood, is the product of the very process these new studies describe.[55]

The task, in other words, is to understand how reliable knowledge and scientific progress can and do result from a flawed, profoundly contingent, culturally relative, all-too-human process.

One of the obstacles to understanding *knowledge* (to echo Kuhn) lies in the vocabulary we use to discuss our difficulties. There is a satisfactory label for people who think there is not really such a thing as *knowledge* at all (but merely belief systems that pass as *knowledge*): they are relativists. But there is no collective term for all the different positions which have in common a recognition that some forms of knowledge of nature

xxv See 'Notes on Relativism and Relativists', No. 8 (pp. 587–8).

are more successful than others, and that consequently knowledge can progress. One could of course adopt the term 'progressivist', but that would elide all the difficulties that are associated with the idea of progress. For progress often comes to a halt, and in many fields of life one step forward results in two steps back. Progress is not linear or incremental, and it is often difficult to reach agreement on the standard by which it should be measured. Nevertheless, it happens.

What all these groups – whether they call themselves realists, pragmatists, instrumentalists, fallibilists, or whatever – have in common, apart from a willingness to recognize progress when they see it, is a recognition that nature (or reality, or experience) establishes practical constraints on what can pass as successful prediction or control – that nature 'pushes back'.[56] These people recognize that scientific knowledge is not fully determined, nor is it *un*determined; it is *semi*-determined. It is not possible to be a full-blooded relativist and to acknowledge that nature pushes back; but it is possible to be a constructivist (to say that we make knowledge out of the cultural resources available to us) and to acknowledge the resistance of nature. Indeed scientific knowledge, properly understood, must be seen to be both constructed and constrained. Hasok Chang has proposed the label 'active realism' for this double recognition.[xxvi][57]

Anyone who tries to occupy this 'best of both worlds' position needs to face up to a further challenge, that of fleshing out the notion that nature pushes back. Kuhn saw this challenge, but he misdescribed it. He complained about those who 'freely acknowledge that observations of nature do play a role in scientific development. But they remain almost totally uninformative about that role – about the way, that is, in which nature enters the negotiations that produce beliefs about it.'[58] If you try to grasp Kuhn's meaning you will find it slips through the fingers. For there is a sense in which science itself is the account of how nature enters into negotiations that produce beliefs about it; in which case the question that Kuhn is asking is not really an historical or philosophical question, but a request to have some bit of science explained to him.

So we have to turn Kuhn's formulation around. In order to understand the way in which the physical world enters the negotiations that produce beliefs, we must look at the ways in which we communicate with and about it. On one level, this is a question about equipment: the telescope transformed the way in which astronomers negotiated with

xxvi See 'Notes on Relativism and Relativists', No. 8 (pp. 587–8).

nature. Secondly, it is a question about intellectual tools: the concept of laws of nature, for example, shapes the sorts of questions scientists ask and the sorts of answers nature provides. In the dialogue between the scientist and the physical world, the physical world (by and large) stays the same, while what scientists bring to the dialogue changes, and this transforms the role that the physical world plays. The ways in which nature pushes back alter as we alter. Hence the need for an historical epistemology which allows us to make sense of the ways in which we interact with the physical world (and each other) in the pursuit of knowledge. The central task of such an epistemology is not to explain *why* we have been successful in our pursuit of scientific knowledge; there is no good answer to that question. Rather it is to track the evolutionary process by which success has been built upon success; that way we can come to understand *that* science works, and *how* it works.

Earlier in this chapter I quoted Popper's claim in 1958 that science 'is *common-sense knowledge writ large*, as it were'. In that, as we have seen, he was wrong. A few months later he added a new epigraph to the second printing of *The Logic of Scientific Discovery*, a quotation taken from the papers of the historian Lord Acton (1834–1902): 'There is nothing more necessary to the man of science than its history . . .'[59] And just what should the scientist and the citizen learn from the history of science? That nothing endures. That just as the theories of Ptolemy and Newton seemed perfectly satisfactory for centuries, so too our most cherished theories will one day be supplanted. As Kuhn pointed out over and over again, one of the central purposes of an education in the sciences is to hide this basic truth from the next generation of scientists.[60] Science replicates itself by indoctrination, since scientific communities work most efficiently when they are agreed about what they are trying to do.

But, as Kuhn also grasped, the fact that even the best-established scientific theories may not endure does not mean that they are unreliable, and it does not mean that science does not progress. Ptolemy gave the astrologers the information they needed, and Newton explained Kepler's laws of planetary motion. We demonstrate the reliability of modern science at every moment of every day. Recognizing both science's limitations and its strengths requires a peculiar mixture of scepticism and confidence; relativists overdo the scepticism and realists overdo the confidence.

16

These Postmodern Days

'History is not the study of origins . . .'

Herbert Butterfield, *The Whig Interpretation of History* (1931)[1]

This knowledge [of what happened afterwards] makes it impossible for the historian to do merely what the history-minded say he should do – consider the past in its own terms, and envisage events as the men who lived through them did. Surely he should try to do that; just as certainly he must do more than that simply because he knows about those events what none of the men contemporary with them knew; he knows what their consequences were.

Jack Hexter, 'The Historian and His Day' (1954)[2]

§1

This book is called *The Invention of Science*. It is about a process whose significance can only be fully grasped with hindsight. Something close to our sense of what had happened had already been attained by William Wotton in 1694 and by Diderot in 1748, but it is striking that the modern historiography really begins immediately after the Second World war. James B. Conant, then president of Harvard, who had played a central role in the Manhattan Project to build the atomic bomb, began teaching an undergraduate course 'On Understanding Science' in 1948 (Kuhn's work sprang directly out of this initiative). As we have seen, Herbert Butterfield's lectures on 'The Origins of Modern Science' were delivered the same year: science had won the war in the Pacific, and its history now became properly the concern of every educated person.

Given that this book relies in part on hindsight, many historians will feel justified in condemning it as an example of Whig history. Way back in 1931 when Butterfield attacked Whig history his target was the view that history had a purpose and a goal, which was to produce our values, our institutions, our culture. His attack was on those who wrote history to ratify and glorify the present, particularly present political arrangements.[3] Butterfield's solution was to advise historians to make the past their present.[4]

In the last half century, though, the term 'Whig history' has subtly changed its meaning, and the very historians who have made the most strenuous efforts to understand the past in its own terms have found themselves accused of being guilty of this intellectual crime.[5] The key confusion that has bedevilled discussion of this question is expressed in the statement – made by people who present themselves as opponents of Whig history – that 'the historian's point of vantage on the past must necessarily be in the present' and that consequently the choice of what to study is 'in the end . . . not simply historiographic, but political'.[6] If this were true then it would become extremely difficult to see how historians could avoid writing Whig history. But it is no more than a half truth.

Historians necessarily rely on the evidence which has survived from the past: in that sense their vantage point is a collection of material remains which exist in the present. Historians also necessarily write in their own language, and rely on intellectual tools and procedures which (unless they are writing contemporary history) they do not share with the people they are studying. Their choice of what to write about will necessarily be shaped by their own interests and concerns. In all these respects history is written by an historian who, inescapably, lives in the present.

But the perspective adopted by the historian need not be that of the present at all. The late Tom Mayer wanted to write a book entitled *Galileo was Guilty*. By that he did not mean that *he* thought Galileo was guilty, or that *we* should think he was guilty; he meant that according to the established procedures of the Inquisition which tried and condemned him in 1633 he was guilty. Mayer's goal was to understand the trial of Galileo as just one of a vast number of trials conducted by the Inquisition, and to put himself in a position where he could assess whether the normal procedures had been followed or whether the treatment of Galileo was in any way exceptional. His project was to make the inquisitors' past his present. His conclusion was that Galileo was guilty according to the law as then established.[7]

The notion that we must understand the past in its own terms, though, can place historians under a perverse pressure to find someone in the past who can act as spokesperson for their own views. Sometimes this manoeuvre can be justified; but often it gives rise to a new sort of Whig history. A striking example of this is to be found in Shapin and Schaffer's *Leviathan and the Air-pump*.

Hobbes distinguishes between the knowledge we have of things we have ourselves made (geometry, the state) from the knowledge we have of things which are not of our making (natural philosophy). In the case of geometry and the state, on Hobbes's view, we can know for sure why something is the case because neither geometry or the state would exist if we had not constructed them. But in the case of natural philosophy he argues that there is a limit to what we can know as we study nature, because various different mechanisms can produce the same effect, so, as we try to reason backwards from effects to causes, all we can do is produce a reasonable conjecture as to the cause which is likely to have produced a particular effect.[8]

Shapin and Schaffer, however, want to turn Hobbes into a seventeenth-century Wittgensteinian, someone who believes that all knowledge is conventional and constructed.[9] Their only evidence to support this claim is a quotation from the English translation of *De cive* (1642), Chapter 16, §16:

> But what some man may object against Kings, that for want of learning, they are seldome able enough to interpret those books of antiquity in the which Gods word is contained, and that for this cause it is not reasonable that this office should depend on their authority, he may object as much against the Priests, and all mortall men, for they may erre; and although Priests were better instructed in nature, and arts then other men, yet Kings are able enough to appoint such interpreters under them; and so, though Kings did not themselves interpret the word of God; yet the office of interpreting them might depend on their authority; and they who therefore refuse to yeeld up this authority to Kings, because they cannot practise the office it selfe, doe as much as if they should say that the authority of teaching *Geometry* must not depend upon Kings, except they themselves were *Geometricians*.

From this they conclude that according to Hobbes 'the force of logic ... is the delegated force of society, working on the natural reasoning capacities of all men' and that 'the force which lies behind geometrical inferences' is the force of Leviathan.[10] This is part of a broader argument that 'Solutions to the problem of knowledge are solutions to the problem of social order' and that 'the history of science occupies the same terrain as the history of politics.'[11] Both Hobbes and Wittgenstein, they want to argue, grasped that a form of knowledge implies a form of social order and vice versa.

Unfortunately, their argument depends on a profound misunderstanding of Hobbes's text. Its argument is straightforward. We all constantly employ experts to do jobs for us, and we don't have to be experts ourselves to make a reasonable choice of an architect, or a builder, or a car mechanic, or a surgeon. In the case of the king, he makes a particular kind of choice of expert, in that he licenses people to practice – above all, in Hobbes's world, he licenses the clergy to preach. To do this he does not need to be an expert himself, he just needs to take sensible advice. Hobbes's point is not that expertise is whatever the sovereign says it is; it is that non-experts can be competent to select experts.

In the England of Hobbes's day you needed to be licensed to teach in a school, and you could only teach Latin from the authorized textbook, *Lily's Grammar*. This does not mean that everyone thought that Latin grammar was whatever the sovereign decreed it to be; it means that the government had consulted experts and chosen a single grammar book so that students who changed school would not be confused by having to start learning from a different book. When Hobbes says that geometricians have their authority to teach from the king, he means the king licenses people to teach geometry, and only people with a licence can teach; he does not mean that royal decree establishes what counts as a good argument in geometry. Of course the king might make a bad decision; in modern terms, he might give practitioners of homeopathy the same legal status as practitioners of the germ theory of disease, or, in seventeenth-century terms, Catholic clergy the same legal status as Protestant clergy. But his bad decision would not make bad logic or bad geometry or bad medicine, or indeed bad theology, into good logic, geometry, medicine or theology; it would simply have the effect of giving the wrong people the right to practise.

What Hobbes is writing about here is not truth, but authorization. The website of the General Medical Council of the United Kingdom states: 'To practise medicine in the UK all doctors are required by law to be registered and hold a licence to practise. The licence to practise gives a doctor the legal authority to undertake certain activities in the UK, for example prescribing, signing death or cremation certificates and holding certain medical posts (such as working as a doctor in the

NHS).' It does not follow that the capacity to heal the sick is delegated to doctors by the GMC, or that good medicine is whatever the GMC says it is, or that doctors cannot seek to improve medicine by introducing new treatments. It simply means that you cannot practise without a licence. Hobbes was no more committed to the supposed Wittgensteinian view of truth than the GMC are.

Shapin's and Schaffer's misunderstanding serves a purpose: by making Hobbes's view of truth correspond to that of Wittgenstein (as they understand him), they are able to turn the dispute between Hobbes and Boyle into a dispute which aligns with our own contemporary disputes about the nature of science and thus to make it serve their own polemical purposes. Hobbes stands for the relativists and Boyle for the realists. Had they not been able to find a spokesperson for their own views in the past they would have been open to the charge that they were making sense of the past by relying on our categories rather than understanding it in its own terms; luckily Hobbes, on their account, employs the very same categories as they want to, so past and present merge seamlessly together. Thus they legitimize a profoundly anachronistic reading of the dispute between Hobbes and Boyle by placing their own view of that dispute into the mouth of Hobbes.

Shapin and Schaffer end *Leviathan and the Air-pump* with the statement: 'As we come to recognize the conventional and artifactual status of our forms of knowing, we put ourselves in a position to realize that it is ourselves and not reality that is responsible for what we know. Knowledge as much as the state is the product of human actions. Hobbes was right.'[12] My argument has been that responsibility for what we know lies both with ourselves *and* reality. Science is not like the state, which is entirely of our own making, though unmaking it would, like unmaking money, be far from easy. Columbus was not 'responsible' for the existence of America, nor Galileo for the moons of Jupiter, nor Halley for the return of Halley's comet, although certainly the credit for these discoveries belongs to them. Reality had its own part to play. And this was precisely Hobbes's point when he maintained that natural philosophy depended on both 'appearances, or apparent effects' (we would say 'reality') and 'true ratiocination'.[13]

But for the moment let us concentrate on the last phrase: 'Hobbes was right.' Shapin and Schaffer would never say, 'Boyle was right.' They would never say, 'Galileo was right' or, 'Newton was right.' That would be Whig history. What, then, makes it permissible for them to say 'Hobbes was right'? The answer is simple. Hobbes was not right about science: in science, as far as they are concerned, there is no right or wrong. He was right about the conventional nature of knowledge. Here their relativism breaks down and they offer us their own version of Whig history. How do they know Hobbes was right? Because they think they can make him into a supposed seventeenth-century Wittgensteinian.

§2

Making the past our present, as Butterfield advocated, is one thing. It is quite another to condemn all use of hindsight and to insist that the past should only be presented 'in its own terms'. Thus we are told 'the one and only error of general principle which the historian can commit . . . is to read history not forwards, as it happened, but backwards.'[i][14] As we saw in Chapter 2, this is nonsense, in that we have to read history both forwards and backwards. Robespierre had no intention of bringing Napoleon into existence, but if we want to understand Napoleon we have to look backwards and see how the combination of French absolutism and the French Revolution made Napoleon possible. What has happened here is that Butterfield's strictures against a particular sort of hindsight (hindsight used to glorify the present) have been turned into a condemnation of hindsight *in general*.

Failure to recognize that history can, indeed must, be read backwards leads to a whole series of historical questions being outlawed. Thus Quentin Skinner, the most influential historian of ideas of the second half of the twentieth century, has apologized for writing his landmark book *The Foundations of Modern Political Thought* (1978). He was, he said in an interview in 2008,

> wrong . . . in using a metaphor that virtually commits one to writing teleologically. My own book is far too much concerned with the origins of our present world when I ought to have been trying to represent the world I was examining in its own terms as far as possible. But the trouble with writing early-modern European history is that, although their world and our world are vastly different from each other, our world nevertheless

i But see the classic essay by Hexter, 'The Historian and His Day' (1954), reprinted in his *Reappraisals in History* (1961). It is perhaps to Hexter that we owe the popularity of the phrase, now ubiquitous, that the task of the historian is 'to understand the past in its own terms' – a view that Hexter sensibly takes to be only one part of what the historian does (221, 231).

> somehow emerged out of theirs, so that there's a very natural temptation to write about origins, foundations, evolutions, developments. But it's not a temptation to which I would ever think of yielding in these postmodern days.[15]

Note that what is being condemned here is not just any history which relates the past to the present (a temptation apparently always to be resisted), but any history which deals with origins, evolutions, developments, any history which is written with the outcome in mind.

It might be thought that this is just a slip (Skinner was, after all, speaking not writing), but it follows inevitably from the project of representing the past 'in its own terms'. The fundamental feature of all human history is that people cannot see into the future, and this is because the future, although it is the result of innumerable deliberate actions, is, for everyone, an unintended outcome.[16] Nobody gets exactly what they planned, expected, intended. To write a history which presents the past in its own terms must inevitably be to write a history in which the process of change is utterly incomprehensible, for the simple reason that it cannot be foreseen.[ii] It is important here to distinguish between teleological history – the notion that history has a purpose or goal – and retrospective history, which seeks to study history as a process of development. Human history has no purpose or goal; but there are plenty of origins, foundations, evolutions and developments, and if you leave them out then you leave out any possibility of understanding change.[iii]

This enterprise of reading the past backwards need not involve any assumption that the participants knew where they were going or that the end result was foreordained. A. J. P. Taylor (*The Origins of the Second World War*, 1961) did not imagine that it was possible for German politicians to see the Second World War coming – indeed he was trying to escape from the assumption that the war was the result of deliberate planning and was looking for a better explanation. Skinner,

ii Walter Benjamin wrote in 1940: 'To historians who wish to relive an era, Fustel de Coulanges [1830–1889] recommends that they blot out everything they know about the later course of history. There is no better way of characterizing the method with which historical materialism has broken.' (Benjamin & Arendt, *Illuminations* (1986), 247–8.) You do not, I would add, need to be an historical materialist to see that historians need to think about change, development, origins.

iii The claim that any search for origins must be dismissed as teleological derives from Foucault's '*Nietzsche, l'histoire, la généalogie*' (1971), in Foucault, *Dits et écrits* (2001), Vol. 1, 1004–24.

when he wrote *Foundations* (if I may take the liberty of defending early Skinner from late Skinner) did not imagine that Machiavelli, Bodin and Hobbes were deliberately trying to lay the foundations of modern liberalism or of the modern theory of the state. Donald Kelley (*The Beginning of Ideology*, 1981) did not for a moment imagine that sixteenth-century French intellectuals foresaw the modern -isms.[iv] The writing of retrospective history is (I know some historians will find this shocking) a perfectly sensible intellectual enterprise.[17] Historians who refuse to engage in it narrow the intellectual scope of history unnecessarily and arbitrarily; indeed history written without the benefit of hindsight (if such a thing were possible) would not be history at all, but rather, to adopt Foucault's term, 'genealogy'.

The source of much of this confusion derives from two apparently problematic features of the historian's enterprise. The first is that people in the past do, to a considerable degree, understand what is going on and react intelligently in response to it. It is thus easy to think that one could write history from Galileo's point of view. But it is an elementary truth that the full significance of his actions was hidden from Galileo. Galileo, for example, despite performing exquisite experiments, never fully grasped the power of the experimental method, dismissing Gilbert's work as not sufficiently philosophical. Correcting for his limitation in this respect need not involve adopting a present-centered view; we need only look on Galileo from the vantage point of Mersenne, who quickly set about replicating Galileo's experiments with a view to producing more accurate measurements. (Of course there is always the danger here that we will simply turn Mersenne into the spokesperson for our own views, as Shapin and Schaffer do with Hobbes, so we have to proceed carefully.)

The second apparently problematic feature of the historian's enterprise is that some important developments are quite simply invisible to the participants.[v] Sometimes people really do not have any sense of the significance of what they are doing, or if they do they fail to put their thoughts on paper. Newton kicked up a fuss about the use of the word 'hypothesis'; but there was no comparable debate about theories, facts, or laws of nature. The new terminology was adopted quietly, casually,

iv See 'Notes on Relativism and Relativists', No. 10 (pp. 589–90).

v See, for example, Christopher Clark, *The Sleepwalkers: How Europe Went to War in 1914* (London: Allen Lane, 2012) and Chris Wickham, *Sleepwalking into a New World: The Emergence of Italian City Communes in the Twelfth Century* (Princeton: Princeton University Press, 2015).

carelessly. Yet it marks the birth of a new way of thinking, a way of thinking that continues to be our own. Identifying that way of thinking and identifying it as still ours are both proper tasks for the historian – this can be called intellectual archaeology, if one wishes, in so far as it deals with developments which were not the result of conscious deliberation.[18] I don't think that either of these two types of historical perspective ought to be problematic, although they must seem so to anyone who seeks to present the past in its own terms, nor do I think they deserve the title of Whig history.

Let us now turn to an example of Whig history as that term is generally understood. When Sir George Cayley published 'On Aerial Navigation' (1809–10), an analysis of the physics of heavier-than-air flight, he could not imagine a modern aeroplane; but he had invented the cambered aerofoil and what we would now call the propeller,[vi] and he was confident that heavier-than-air flight was possible; indeed he went on to successfully construct a glider that could carry a person. What he lacked of course was a suitable power source: he tried to imagine an aeroplane powered by a steam engine, but the difficulties were obvious.

It is perfectly sensible to say that Cayley laid the foundations of modern aeronautics. Cayley was convinced that he had made an important breakthrough; but it would be more than a century before the significance of what he had achieved would become apparent. Thus we find in the journal *Science* for 1912 an article on 'The Problem of Mechanical Flight' which begins 'The scientific period in aviation began in 1809 when Sir George Cayley published . . . the first complete mechanical theory of the aeroplane,' but then quickly adds '[t]his memoir passed unnoticed until unearthed some sixty years later.'[19] What are historians to do with Cayley? I see no reason why they should pretend to ignore his existence, even though any discussion of him is liable to be dismissed as Whig history. There is nothing peculiarly present-centered about regarding Cayley as important, unless the present is extended to include the period of the first flying machines. Nor does writing about Cayley involve glorifying modern air travel or (for example) denying its contribution to global warming. Cayley is a minor but not insignificant figure in the

vi The earliest usage of this word, in the sense of a device for propelling a ship (in reality a paddle wheel), given by the *OED* is 1809, and the first practical screw propeller on a boat apparently dates to 1829.

history of science and technology; but there is no doubt that our sense of his significance derives entirely from hindsight.

§3

Such is the fear of being accused of writing Whig or teleological history that it is difficult to find an historian making simple and elementary points of this sort. Fortunately the philosopher Richard Rorty can come to our assistance. Rorty pounced on a remark by Steven Weinberg, who had written:

> What Herbert Butterfield called the Whig interpretation of history is legitimate in the history of science in a way that it is not in the history of politics or culture, because science is cumulative, and permits definite judgements of success or failure.[20]

Here Weinberg had inadvertently confused three separate issues: cumulation (all history, indeed all human activity, is cumulative); success or failure (there are many human activities that permit definite judgements of success or failure); and progress (a unique feature of modern science and technology). And so Rorty attacked:

> Does Weinberg really want to abstain from definite judgements of the success or failure of, say, the constitutional changes brought about by the Reconstruction Amendments and by the New Deal's use of the interstate commerce clause? Does he really want to disagree with those who think that poets and artists stand on the shoulders of their predecessors, and accumulate knowledge about how to write poems and paint pictures? Does he really think that when you write the history of parliamentary democracy or of the novel that you should *not*, Whiggishly, tell a story of cumulation? Can he suggest what a non-Whiggish, legitimate history of these areas of culture would look like?[21]

History is a cumulative record of success and failure, and the pretense that it can be something other than that is a peculiar shibboleth of historical writing over the last fifty years. (Rorty and Weinberg would have had, one may imagine, little difficulty in agreeing with each other if only they could have agreed on the language to use.)

The important thing about the science of Galileo and Newton, Pascal and Boyle is that it was, in part, successful, and that it laid the foundations for future successes. They did not know what the future would hold; but they did have a clear sense of what they were trying to achieve. They were confident that they were making progress, and we cannot leave that progress out of our history, any more than we can leave out the influence they had on those who came after them. Nor can we leave success out of the history of parliamentary democracy or the novel; but democracies sometimes fail, and sometimes the novels get worse not better. The remarkable thing about science is that the process is not only *cumulative* but, it would seem (to make a distinction the dictionary does not recognize), *accumulative*. The past not only shapes the present; in science, gains made in the past are only ever given up (except where there is censorship or religious or political interference) in order to be exchanged for greater gains made in the present.[vii] It is this peculiar feature of modern science which makes the history of science since 1572 uniquely a history of progress, and makes it inappropriate to write history of science in the same sceptical way that one might write the history of democracy or of the novel.

§4

This book has therefore been deliberately written in opposition to certain conventions which have become established in 'these postmodern days'. Soon, I trust, those conventions will be every bit as mysterious as the ones that governed the writing of Whig political history. What is the driving force behind relativism and postmodernism? Some think it is fundamentally a political commitment to multiculturalism, a commitment that needs re-examination when cultures clash. Broadly speaking, this seems to me true. An alternative view is that postmodernists do not want to recognize the existence of 'reality'. The problem with this view is that it takes the concept of 'reality' for granted, whereas what we need is a history of the changing nature of reality. But still, this second view has also, I think, a kernel of truth. To insist, as postmodern

vii The process whereby the triumph of a new paradigm in science involves the loss of some perfectly satisfactory theories which have to be abandoned because they are inseparable from the old paradigm has been called 'Kuhn loss' (the term originates with Post, 'Correspondence, Invariance and Heuristics' (1971)). Graney, *Setting Aside All Authority* (2015), for example, studies the claims of those who maintained that the losses that followed from Copernicanism outweighed the gains, and shows that their arguments were far from negligible. Nevertheless, I would argue that in science one step back always results in two steps forward and there are no clear cases where the loss outweighs the gain.

historians of science do, on radical contingency, on the idea that there is no such thing as path dependency, that we might still be practising alchemy or riding penny-farthings, is to claim that there was never a reality-based logic which led certain theories and technologies to succeed while others failed.[22] Alongside the political commitment to multiculturalism, honourable in intention though profoundly problematic in practice, we must also acknowledge a powerful fantasy, the fantasy that we can remake the world in any way we choose, and, equally powerful, the fantasy that no one can tell us that what we are trying to do can never be done. Multiculturalist politics had a real referent in post-colonialism and immigration. But postmodernist epistemology also has a fantasy referent in what we may call the politics of wish-fulfillment, according to which there are no obstacles to our remaking the world as we choose, apart from the ideas in our minds. The world can be anything we want it to be, because thinking makes it so. When Shapin and Schaffer say 'it is ourselves . . . that is responsible for what we know' they seem to imply that knowledge can be whatever we choose to make it; and if we do not like science as we find it, then all we need do is wish for it to be otherwise.

Concealed within relativism there thus lies a dream of omnipotence, a fantasy recompense, perhaps, for the impotence and irrelevance of academic life. During the years 1919–20 Antonio Gramsci, the Italian Marxist, adopted as his own the maxim 'pessimism of the intellect, optimism of the will'.[23] Foucauldian politics is the opposite of this: optimism of the intellect, pessimism of the will. It declares that we are trapped in a world not of our making, while insisting that the obstacles to our remaking the world are indeed all of our own making. This is a vision of politics first invented by Montaigne's beloved friend Étienne de la Boëtie. Montaigne, although his love for his friend knew no limits, was never taken in by it.

17
'What Do I Know?'

'What must the world be like in order that man may know it?'
Thomas Kuhn, *The Structure of Scientific Revolutions* (1962)[1]

No theory of knowledge should attempt to explain why we are successful in our attempts to explain things . . . there are many worlds, possible and actual worlds, in which a search for knowledge and for regularities would fail.

Karl Popper, *Objective Knowledge* (1972)[2]

§1

In 1571 Montaigne retired from his professional life as a judge. He was 37, still young by our standards, but on the threshold of old age by those of the sixteenth century. He was mourning – still mourning – the death of la Boëtie in 1563, and he was preoccupied with thoughts of dying. He intended to spend time with his books – he owned a thousand volumes, a vast collection. On the beams of his library he had painted sixty or so quotations from the classics, all emphasizing the vanity of human life and of human aspirations to knowledge. They were, in effect, an epitome of his reading. He had a medal struck which bore the words '*Que sçay-je?* – What do I know?' – over the image of a pair of scales. The scales did not represent justice, for they teetered. They represented uncertainty.

Montaigne found no happiness in his new life, and so he turned to writing as a form of therapy, a way of keeping himself company. The result was to be the *Essays*, of which the first volume, containing Books One and Two, was published in 1580 (a third book was added in 1588, and Montaigne went on revising his essays until his death in 1592). The word *essays* has come to seem normal and natural to us – students write essays all the time. But when Montaigne used the word it meant an *assay* or a test. Montaigne was testing himself, exploring himself, studying himself, trying to make sense of himself. In the *Essays* Montaigne was making a fundamental claim about our knowledge of the world, that knowledge is always subjective, personal. He was also inventing a new literary genre.

In the first edition of his *Essays* two were of particular importance. At the centre of Book One was an essay on friendship, a prelude to what was originally intended to be the first publication of a remarkable work by La Boëtie, *The Discourse of Voluntary Servitude*, a work now often regarded as the first anarchist text.[3] In the end, Montaigne was unable to publish *The Discourse* because it had already been published by Protestant rebels and condemned as seditious. La Boëtie wanted to know why we obey authority, and his answer was that we shouldn't.

At the heart of Book Two (though not this time at the centre – the central essay is entitled 'On Freedom of Conscience') was the longest of all the essays, 'An Apology for Raimond Sebond', a passage from which, as we saw in Chapter 9, was crucial to later thinking about the laws of nature. Sebond (1385–1436), a Catalan theologian, had written, in Latin, a book providing a rational demonstration of the truths of Christianity, and Montaigne had been requested by his dying father to translate it into French (Montaigne dated the dedicatory epistle to the translation, addressed to his father, to the day of his father's death, 18 June, 1568). Thus the origin of the 'Apology' was every bit as private and personal as that of the essay on friendship, and here again we have a pairing of texts – Sebond's defence of Christianity with Montaigne's 'Apology'. But this time it is Montaigne who is the author of the revolutionary text, for the 'Apology' was only in outward appearance a defence of Sebond; on closer examination it is a devastating attack on everything he stands for, a sustained critique of religion. Evidently Montaigne's argument had to be expressed with exquisite care. Even Sebond's work had fallen foul of the censors, not for its basic thrust, but for the extravagant claims made on behalf of it by Sebond in his preface. Since Sebond had harnessed faith and reason together, Montaigne's critique set out to undermine faith by showing that all claims to knowledge are overstated. What was at stake in the 'Apology' was not just the reasonableness of Christian faith, but the reliability of all the claims made by philosophers. The subjects we would now call 'science' formed, in the sixteenth century, part of philosophy,[i] so

i Or of subjects subordinated to philosophy. See above, pp. 23–4.

Montaigne's 'Apology' is, among other things, an attack on the science of his day.

The sources of Montaigne's scepticism are not difficult to identify. The bitter conflict between Protestants and Catholics, which had led to prolonged civil war in France, to the most terrible massacres and brutalities, had made all claims to truth seem partisan. Humanist learning (Montaigne had been brought up to speak Latin as his first language, so that he would have the learning his father lacked) had brought back to life the beliefs of the pagan Greeks and Romans, offering a real alternative to Christianity. The philosophical disputes of the medieval universities (between the Aristotelianism of Avicenna and the Aristotelianism of Averroes, and between realists and nominalists) had been made to seem parochial by the publication of two texts unknown to the Middle Ages: *On the Nature of Things* by Lucretius, a work of materialist atheism which Montaigne had studied with great care (his copy, heavily annotated, has recently been identified); and the *Outline of Pyrrhonism* of Sextus Empiricus (rediscovered in the 1420s but only published in 1562).[4] The discovery of the New World had fatally undermined any claim that there are some things on which all human beings can agree – here were societies practising nudity and cannibalism.

Montaigne's scepticism had its limits. He did not doubt that you can make wine out of grapes, or find your way from Bordeaux to Paris. Someone had once tried to persuade him that the ancients did not understand the winds in the Mediterranean. Montaigne was impatient with such an argument: did they try to sail East and end up going West? Did they set out for Marseille and find themselves in Genoa? Of course not. He gave no indication of doubting that two plus two make four, or that the angles of a triangle add up to two right angles (although he found a geometrical proof that two lines could approach each other for ever but never meet paradoxical).[5] What he doubted is that you can prove the truth of the Christian religion or of any religion. He doubted that the universe was created to provide a home for human beings any more than a palace is built for rats to live in.[6] He doubted that there is any principle of morality which can command universal assent, and he doubted that any of our sophisticated intellectual systems make sense of how the world is. Doctors, he felt sure, were more likely to kill their patients than cure them. For nearly a millennium and a half Ptolemy had seemed an entirely reliable expert on all questions to do with geography and astronomy; and then the discovery of the New World had shown that his geographical knowledge was hopeless, and

Copernicus had shown that there was at least a viable alternative to his cosmology.[7] Our claims to knowledge, Montaigne said, are generally misconceived because we will not acknowledge our limits as human beings. We need to remember that the wisdom of Socrates consisted in acknowledging his own ignorance.[8]

Montaigne ended (or almost ended) 'The Apology' with a quotation from Seneca: 'Oh, what a vile and abject thing is Man if he does not rise above humanity.' 'A pithy saying;' he commented, 'a most useful aspiration, but absurd withal. For to make a fistful bigger than the fist, an armful larger than the arm, or to try and make your stride wider than your legs can stretch, are things monstrous and impossible. Nor may a man mount above himself or above humanity: for he can see only with his own eyes, grip only with his own grasp.' Of course he could not quite stop there, for the heretical implications were too clear. And so he went on: 'He will rise if God proffers him – extraordinarily – His hand; he will rise by abandoning and disavowing his own means, letting himself be raised and pulled up by purely heavenly ones.'[9] Was this a reluctant addition? Readers of Montaigne are – and always have been – sharply divided between those who think his protestations of Catholic orthodoxy were genuine, and those who think they were merely concessions to the censor. My own sympathies will already be clear.[10] After all, Montaigne never gave an example of heavenly inspiration, of divine intervention, without hedging it about with doubts and difficulties. He pointed out that, far from our being made in the image of God, we make our gods in our own image: 'we forge for ourselves the attributes of God, taking ourselves as the correlative.'[11] One moment he insisted he believed in miracles, the next he doubted his own belief. In the end he grounded the obligation to be a Christian in the obligation to obey the laws of the country in which one finds oneself – and, from the point of view of a reasonable person, the content of those laws is entirely arbitrary.[12]

There is no need to resolve this issue here. What matters for present purposes is Montaigne's rejection not of the practical knowledge, of wine-making and bread-baking, of his day, but of the learned knowledge, of medicine, geography, astronomy. Montaigne called these various branches of knowledge 'sciences'. Montaigne's scepticism, when applied to the sciences of the times, was entirely justified: for there is not a single natural philosophical principle taught in the universities in 1580 that a student in the sciences would still learn today. Montaigne's arguments against religious belief and against conventional moral certainties are still as sharp as ever they were; but his arguments against the sciences of his day have no purchase against the sciences of our day. Science is now something utterly different from what it was then.

§2

Human beings, Montaigne argued, are imperfect, and so human knowledge is necessarily unreliable. Galen had claimed that the hand of a healthy doctor was the perfect instrument for judging hot and cold, wet and dry – the four qualities that made up the world. If the patient was hotter than the doctor's hand they were, in absolute terms, hot, and that was that. The world had been divinely ordered so that our sensations of hot and cold corresponded to real qualitative differences. Montaigne would have none of this. We have five senses, but who knows how many we ought to have if we want to know what is really going on? Who knows what we are missing? And of course he is right: bats experience the world fundamentally differently from the way in which we experience it, and it is wrong to assume that echo location merely enables them to know what we know by a different means, for it may give them insights which we will never have.[13] Diderot, in the *Letter on the Blind* (1749), a work every bit as subversive as Montaigne's 'Apology', was to formulate the view that a blind philosopher would of necessity be an atheist, for they would be quite unable to perceive order and harmony in the universe.[14] What we know about the world and what we think we know depend entirely on how we perceive it.

Part of the great transformation that we know as the Scientific Revolution, a transformation which began in earnest the year after Montaigne retired to his library, consisted in improving our senses. The compass enabled sailors to perceive the earth's magnetic field. The telescope and the microscope enabled scientists to see previously invisible worlds. The thermometer replaced Galen's hand as a measure of temperature. The barometer displayed the pressure of the air on the skin. The pendulum clock provided an objective measure of a subjective experience – the passage of time. New instruments meant new perceptions, and with them came new knowledge.

All of these instruments relied, at least in part, on glass manufacturing skills and provided visual information. Alongside these we can put the mechanical reproduction of text and images by means of the printing press, which transformed the communication of knowledge and established new types of intellectual community. Montaigne's *Essays*, which he wrote in his library surrounded by serried ranks of printed books, are themselves testimony to the emergence of a new bookish culture; disseminated in print they showed each and every reader how to engage in their own project of self-exploration.

There is a tendency to think of the telescope as a scientific instrument and the printing press as something external to science: but the first telescopes were not made by or for scientists, and the printing press transformed the intellectual aspirations of scientists because it was now possible to work with detailed images alongside text. Both began as practical technologies and became scientific instruments. The new science was thus dependent on a few key technologies which functioned, to use Elizabeth Eisenstein's phrase, as 'agents of change'.[15]

The printing press had a further crucial consequence which we can also see reflected in Montaigne's *Essays*: it fostered a new critical attitude to authority which led to the insistence that knowledge must be tested and retested. In Montaigne's case this resulted in a peculiar emphasis on the subjectivity of what we know, its dependence on our personal experience. Inherited knowledge could no longer be accepted without question. But as new knowledge accumulated, the printing press instead of fostering scepticism began to make possible a new type of confidence. Facts could be checked, experiments replicated, authorities could be set side by side and compared. Intellectual scrutiny could be much more intensive and extensive than ever before. The printing press was the precondition for this new insistence that knowledge, no longer authoritative, might at last become reliable.

The new instruments and the oceans of printed books opened up new experiences and destroyed old authorities. The old history of science, the history of science of Burtt, Butterfield and Koyré, rejected the idea that the new science of the seventeenth century was primarily the consequence of this new evidence; what mattered were new ways of thinking. The new history of science, beginning with Kuhn, tried to ground these new ways of thinking in intellectual communities: the success of new ideas depended upon conflict and competition within and between communities of thinkers. By problematizing the idea that experiments could be successfully replicated, the generation after Kuhn, the generation of Shapin and Schaffer, sought to demonstrate that experience itself is unpredictable, malleable, socially constructed. On their account (and here they parted company with Kuhn) the social history of knowledge is not just one aspect of the history of science; rather the social history of knowledge is the only history that can be written.

Recognizing the inadequacies of postmodern history of science does not mean that we should simply go back to Kuhn or to Koyré. The problem with concentrating on the paradigm shifts that interested them is that you lose sight of the wider environment within which those shifts took place: thus Kuhn gave an account of Copernicanism in which discovery was taken for granted, the telescope barely appeared, and the language in which science was conducted was never mentioned. Kuhn's approach took the scientific enterprise as given, and so inevitably it missed the process of its formation, which was critical to the belated triumph of Copernicanism. Kuhn failed to see what he was missing because he assumed that science had been invented long before 1543 and because he seriously underestimated the obstacles to the adoption of Copernicanism, obstacles which came from the subordination of astronomy to philosophy. Such an approach might explain local revisions: how Pascal developed a theory of pressure, or Boyle came up with Boyle's law; it cannot explain the long series of vacuum experiments from Berti to Papin (Newcomen's atmospheric steam engine was not so much a new beginning as the final conclusion of that extended enterprise), for during that sequence a new culture was constructed, one which sought to resolve intellectual disputes through experimentation. That culture was itself founded in imitation of an earlier enterprise, which sought to resolve disputes regarding the structure of the universe through ever more exact observation, the enterprise of the new astronomy founded by Tycho Brahe. As the mathematicians turned their attention from observation to experiment, from astronomy to physics, they found they required a new set of intellectual tools, a new language. Some of that language – hypotheses and theories – came from astronomy; some of it – facts, and later evidence – came from the law. This new vocabulary was crucial to explaining the status of the new knowledge, and yet it is a language that we have come to take so completely for granted that its invention has become invisible. The presumption has been, either that thinking comes naturally, or that the required intellectual tools for thinking about natural science had all been developed by the ancient Greeks. As we have seen, this is not the case.

Enquiries into science have tended to assume that there are basically three variables to take into account: experience (facts, experiments), scientific thought (hypotheses, theories), and society (social status, professional organizations, journals, networks, textbooks). Kuhn's concept of a paradigm, which he presented as an amalgam of a practice, a theory, and an educational programme, represented a particular way of interlocking these three variables. This fundamental schema might have come into question with the publication of Ian Hacking's *The Emergence of Probability* (1975), which argued that probability thinking provided a powerful intellectual tool which had not existed until the 1660s.[ii] But it should now be apparent that probability was just one of a series of key intellectual tools that appeared in the course of the seventeenth century: the materials out of which one could construct such a new history of science were not to hand in 1975.

Hacking's identification of probability theory as a particular mode of thought could, however, have served to clarify the intellectual alternatives that were available before the emergence of probability. No lines written by Galileo are more frequently quoted than these:

> Philosophy is written in this very great book which always lies open before our eyes (I mean the universe), but one cannot understand it unless one first learns to understand the language and recognize the characters in which it is written. It is written in mathematical language and the characters are triangles, circles and other geometrical figures; without these means it is humanly impossible to understand a word of it; without these there is only clueless scrabbling around in a dark labyrinth.[16]

In Galileo's view the intellectual tools provided by geometry were the only tools required by a scientist. This was a reasonable view, since they were the only tools needed for Copernican astronomy and for Galileo's two new sciences, the science of projectiles and the science of load-bearing structures.[iii] In insisting these were the only tools required Galileo was dismissing Aristotelian logic as an irrelevance. Since Galileo, of course, all sorts of new languages have been invented with which to do science, including algebra, calculus and probability theory.

ii On *Emergence*, see Daston, 'The History of Emergences' (2007). I have not stressed the importance of probability thinking in this book because its main impact comes later: Gigerenzer, Swijtink and others, *The Empire of Chance* (1989). Hacking saw his own work as a development of Michel Foucault's studies in the archaeology of knowledge, but I take *Emergence* to be a work that is much less Foucauldian than Hacking believed; Hacking also acknowledged a considerable debt to Alistair Crombie. *Emergence* fits, at least for my purposes, in the tradition of historical studies of intellectual tools which for historians begins with Febvre, *Le Problème de l'incroyance* (1942). Febvre's own concern with intellectual tools was apparent in the historical encyclopaedia he planned, Rey, Febvre, and others (eds.), *L'Outillage mental* (1937); its source lies in the anthropology of Lucien Lévy-Bruhl, for example Lévy-Bruhl, *How Natives Think* (1925).

iii Modern books (e.g. Heilbron, *Galileo* (2010)) often present Galileo's discoveries as algebraic formulae; such presentations are profoundly anachronistic.

It is easy to think that new knowledge comes from new types of apparatus – Galileo's telescope, Boyle's air pump, Newton's prism – not from new intellectual tools.[iv] Often this is a mistaken view: in a hundred years time the randomized clinical trial (streptomycin, 1948) may look much more significant than the X-ray (1895) or even the MRI scanner (1973). New instruments are plain as pikestaffs; new intellectual tools are not. As a result we tend to overestimate the importance of new technology and underestimate the rate of production and the impact of new intellectual tools. A good example is Descartes' innovation of using letters from near the end of the alphabet (x, y, z) to represent unknown quantities in equations, or William Jones's introduction of the symbol π in 1706. Leibniz believed that the reform of mathematical symbols would improve reasoning just as effectively as the telescope had improved sight.[17] Another example is the graph: graphs are now ubiquitous, so it comes as something of a shock to discover that they only began to be put to use in the natural sciences in the 1830s, and in the social sciences in the 1880s. The graph represents a powerful new tool for thinking.[18] An absolutely fundamental concept, that of statistical significance, was first propounded by Ronald Fisher in 1925. Without it, Richard Doll would not have been able to prove, in 1950, that smoking causes lung cancer.

Physical tools work very differently from intellectual tools. Physical tools enable you to act in the world: a saw cuts through wood, and a hammer drives home nails. These tools are technology-dependent. The screwdriver only came into existence in the nineteenth century, when it became possible to mass produce identical screws; before that the few handmade screws that were used were turned with the tip of a knife blade.[19] Telescopes and microscopes depended on pre-existing techniques for making lenses, and thermometers and barometers depended on pre-existing techniques for blowing glass. Telescopes and thermometers do not change the world around them as saws and hammers do, but they change our awareness of the world. They transform our senses. Montaigne said that people can see only with their own eyes; when they look through a telescope (which of course Montaigne never did) they still see only with their own eyes, but they see things they could never see with their unaided eyesight.

iv I presume a distinction between new thinking and new tools for thinking; Galileo's new sciences were certainly new, but the intellectual tools he used to construct them were familiar to any mathematician; calculus was a new tool for thinking, and without it Newton would not have been able to formulate his account of gravitational attraction.

Intellectual tools, by contrast, manipulate ideas, not the world. They have conceptual preconditions, not technological preconditions. Some instruments are both physical and intellectual tools. An abacus is a physical tool for carrying out complicated calculations; it enables you to add and subtract, multiply and divide. It is perfectly material, but what it produces is a number, and a number is neither material nor immaterial. An abacus is a physical tool for performing mental work. So too are the Arabic numerals we take for granted. I write 10, 28, 54, not, as the Romans did, x, xxviii, liv. Arabic numerals are tools which enable me to add and subtract, multiply and divide on a piece of paper far more fluently than I could with Roman numerals. They are tools that exist as notations on the page and in my mind; like the abacus, they transform the way I operate on numbers. The number zero (unknown to the Greeks and the Romans), the decimal point (invented by Christoph Clavius in 1593), algebra, calculus: these are intellectual tools which transform what mathematicians can do.[20]

Modern science, it should now be apparent, depends on a set of intellectual tools which are every bit as important as the abacus or algebra, but which, unlike the abacus, do not exist as material objects, and which, unlike arabic numerals, algebra, or the decimal point, do not require a particular type of inscription. They are, at first sight, merely words ('facts', 'experiments', 'hypotheses', 'theories', 'laws of nature', and indeed 'probability'); but the words encapsulate new ways of thinking. The peculiar thing about these intellectual tools is that (unlike the intellectual tools employed by mathematicians) they are contingent, fallible, imperfect; yet they make possible reliable and robust knowledge. They imply philosophical claims which are difficult, perhaps impossible, to defend, yet in practice they work well. They served as a passage between Montaigne's world, a world of belief and misplaced conviction, and our world, the world of reliable and effective knowledge. They explain the puzzle that we still cannot make a fistful bigger than a fist, or a stride longer than our legs can stretch, but that we can now know more than Montaigne could know. Just as the telescope improved the capacities of the eye, these tools improved the capacities of the mind.

During the seventeenth century the meaning of key words shifted and changed, and a modern scientific – or rather metascientific – vocabulary slowly took shape. This both reflected and gave rise to new styles of thinking.[21] These changes were rarely the subject of explicit debate within the intellectual community, and have generally been overlooked by historians and philosophers (partly because the terms themselves were not new – 'probability' is typical in this respect – even if they were now being used in a new way), but they transformed the character of knowledge claims.[22]

Alongside these intellectual tools we can see the emergence of a community accustomed to using them: the new language of science and the new community of scientists are two aspects of a single process, since languages are never private. What held this community together was not just the new language, but a set of competitive and cooperative values which were expressed in the language used to describe the scientific enterprise (rather than in scientific arguments themselves), expressed in terms of discovery and progress and eventually institutionalized in eponymy. What is striking about these intellectual tools and cultural values is that they have proved to have a history quite unlike that of paradigms. Paradigms flourish; some then die, and others get relegated to introductory textbooks. The new language and the new values of science have now survived for 300 years (500 if we go back to their common origin in 'discovery'), and there is nothing to suggest they are likely to go out of fashion soon. Just like algebra and calculus, these tools and these values represent acquisitions which are too powerful to be discarded, and which remain not as museum pieces but are in constant use. Why? Because the new language and culture of science still constitute (and I believe will always constitute) the basic framework within which the scientific enterprise is conducted. Their invention is part and parcel of the invention of science.

§3

The Scientific Revolution was a single transformative process, the cumulative consequence, not of one sort of change repeated many times, but of several distinct types of change overlapping and interlocking with each other. First, there was the cultural framework within which science was invented. This framework consisted of concepts such as discovery, originality, progress, authorship and the practices (such as eponymy) associated with them. An older school of historians and philosophers took this framework for granted, while a newer school has wanted to debunk or deconstruct the concepts rather than explain their significance and trace their origin. This culture emerged at a particular moment in time: before it came into existence there could be no science as we understand the term. Of course the critics are right in that concepts such as discovery are problematic: discoveries are rarely made by a single individual at a precise moment in time. But just like plenty of other problematic concepts (democracy, justice, transubstantiation), they provided and still provide a framework within which people made sense and make sense of their activities and decided and still decide how to live their lives. We cannot understand science without studying the history of these foundational concepts.

Alongside this new framework, the printing press was transforming the nature of intellectual communities, the knowledge they could exchange, and the attitude to authority and to evidence that came naturally to them. Next there came new instruments (telescopes, microscopes, barometers, prisms), and new theories (Galileo's law of fall, Kepler's laws of planetary motion, Newton's theory of light and colour). Finally the new science was given a distinctive identity by a new language of facts, theories, hypotheses and laws. Five fundamental changes thus interacted and interlocked in the course of the seventeenth century to produce modern science. Changes in the wider culture, in the availability of and attitude to evidence, in instrumentation, in scientific theories narrowly defined, and in the language of science and the community of language users all operated across different time-scales, and were driven by different, independent factors. But the cumulative effect was a fundamental transformation in the nature of our knowledge of the physical world, the invention of science.

Since each of these changes was necessary for the construction of the new science we should be wary of trying to rank them. But, if one looks closely, it is apparent that the new science was about one thing more than anything else, and that is the triumph of experience over philosophy. Each and every one of these changes weakened the position of the philosophers and strengthened the position of the mathematicians, who, unlike the philosophers, welcomed new information. The new language of science was above all a language which gave the new scientists tools for handling evidence, or, as it was called at the time, experience. Leonardo, Pascal and Diderot (and Vadianus, Contarini, Cartier, and all the others) were right: it was experience that marked the difference between the new sciences and the old.

§4

Montaigne, too, was right – right to think that the men and women of his day were hopelessly fallible when it came to understanding the world. Since then, the claims of the postmodernists notwithstanding, we have learnt to develop reliable knowledge, even though we as human beings have continued to be as fallible as ever. Of course, our present-day knowledge will prove to be incomplete and limited in the eyes of future generations; we cannot even begin to guess what will one day be known. But there is no prospect of it proving simply unreliable. We can calculate reliably the path a rocket will take as it flies from the earth to Mars. We can sequence human DNA, and identify genetic mutations that cause, for example, diabetes. We can build a particle accelerator. We could not do these things if our knowledge was entirely misconceived – anyone who suggests we could should be met with the same impatience as that with which Montaigne greeted the claim that the Romans did not understand the Mediterranean wind system.

Hilary Putnam claimed in 1975 that realism, the belief that science gets at the truth, 'is the only philosophy that doesn't make the success of science a miracle'.[23] The thinking is simple: science is very good at explaining what happens and predicting what is going to happen. If scientific knowledge is true this state of affairs needs no further explanation; but if scientific knowledge is not true then only a miracle could bring about such a perfect coincidence between the predictions of scientists and what actually happens. Putnam's argument was demolished by Larry Laudan, who objected to the claim that successful scientific theories were likely to be true, and he was quite right to do so.[24] Plenty of theories which we now regard as plain wrong have been successful in the past. By this I do not mean the theories that always were defective, were recognized as being defective by some people at the time, but nevertheless acquired a widespread following: Hippocratic (humoral) medicine, or alchemy, or phrenology. I mean rather theories which became well-established within the science of their time, were based upon significant evidence, appeared to provide robust explanations, and were successfully used to make novel predictions: theories such as the Ptolemaic system, phlogiston (a substance held, from 1667 until late in the eighteenth century, to be released by combustible substances when set alight), caloric (an elastic fluid which was supposed, in the first half of the nineteenth century, to be the material basis of heat) and the electromagnetic aether (which was held, in the second half of the nineteenth century, to be the medium for the propagation of light).

These cases differ from that, for example, of Newtonian physics. Using Einstein's theory of relativity you can construct a world – the world of our daily experience – in which Newtonian laws very closely correspond to what actually happens. Astrophysicists still use Newton not Einstein to plot the orbits of space craft, because although the Newtonian calculations are based on what we would now regard as misconceptions, the differences between them and calculations that recognize the relativity of space and time are too small to be worth worrying about. Einsteinian physics can thus be regarded as inheriting the results of Newtonian physics while going far beyond them. But in the cases of caloric or the electromagnetic aether there is no inheritor theory, and we would not now say that these theories, which at one time seemed perfectly well established, were useful approximations of the truth. Nevertheless, it does not follow from the fact that we no longer regard these theories as true, or even useful, that they were never associated with reliable experimental practices; like Ptolemaic astronomy, they were well-founded within certain limits. Laudan's arguments tell against Putnam's claim that science gets at the truth, not against the claim that what marks science out is that it is reliable.[25] As Margaret Cavendish put it in 1664, comparing the search for the truth to the futile search for the philosopher's stone which would turn base metal into gold:

> although Natural Philosophers cannot find out the absolute truth of Nature, or Natures ground-works, or the hidden causes of natural effects; nevertheless they have found out many necessary and profitable Arts and Sciences, to benefit the life of man ... Probability is next to truth, and the search of a hidden cause finds out visible effects.[26]

Of course reliability is a slippery concept. We only need to turn to the doctors of Montaigne's day for a cautionary example. They thought they were using their knowledge to cure patients. In fact, their preferred remedies (bleedings and purges) did no good at all.[27] They mistook the patients' spontaneous recovery (thanks to the workings of their immune systems), combined with the placebo effect, for cures brought about by medical therapy (and intelligent bystanders such as Montaigne suspected as much).[v] In medicine there were no reliable methods of measuring success until the nineteenth century.

But the Ptolemaic astronomers of Montaigne's day were very different from the Hippocratic doctors. Clavius claimed that eccentrics and

v Modern trials of the efficacy of therapies depend on the idea that an effective therapy must perform better than a placebo. As it happens, Denis Papin, who had qualified as a doctor, was the first to describe the placebo and the placebo effect, in a letter to Leibniz, 11 or 12 August 1704. As we have seen (above, p. 292), the need to measure success against outcomes in a control group had been grasped by Walter Charleton.

epicycles must exist, otherwise the success of the predictions made by astronomers were inexplicable:

> But by the assumption of eccentric and epicyclic circles not only are all the appearances already known preserved, but also future phenomena are predicted, the time of which is altogether unknown. . . . it is not credible that we should force the heavens (but we seem to force them, if eccentrics and epicycles are fictions, as our adversaries would have it) to obey our fictions and to move as we wish or as agrees with our principles .[28]

Clavius was wrong – there are no eccentrics and epicycles – but he was right to claim that he could predict the future movements of the heavenly bodies with a high degree of reliability. Like Clavius, we test our knowledge by doing things with it, which is the fundamental difference between our knowledge and most of the sciences of Montaigne's day. In comparison to sixteenth-century philosophy, all our sciences are applied sciences, and all our scientific knowledge is robust enough to withstand real-world application, if only in the form of experiment. We can summarize this in two words: Science Works.

If you learn to navigate a boat you will be taught to work with a Ptolemaic system, with a stationary earth and a moving sun, not because this is true, but because it makes for an easy set of calculations. So a false theory can be perfectly reliable when used in the appropriate context. If we no longer use epicycles, phlogiston, caloric, or aether it is not because no reliable results can be obtained with those theories; it is that we have alternative theories (theories we take to be true) which are just as easy to use and have a wider range of applications. There are no good grounds for thinking that one day our physical sciences will prove, like Hippocratic medicine, to be learned nonsense; but it is perfectly possible that where they are right they are, like Ptolemy's epicycles, right for entirely the wrong reasons. Science offers reliable knowledge (that is, reliable prediction and control), not truth.[29]

One day we may discover that some of our most cherished forms of knowledge are as obsolete as epicycles, phlogiston, caloric, the electromagnetic aether and, indeed, Newtonian physics. But it seems virtually certain that future scientists will still be talking about facts and theories, experiments and hypotheses. This conceptual framework has proved remarkably stable, even while the scientific knowledge it is used to describe and justify has changed beyond all recognition. Just as any progressive knowledge of natural processes would need a concept akin to 'discovery', so as further advances occurred it would need a way of representing knowledge as both reliable and defeasible: terms that do the work done by 'facts', 'theories' and 'hypotheses' would have to play a role in any mature scientific enterprise.

We should end by acknowledging that we have the scientific knowledge we have against all the odds. There is no evidence that the universe was made with us in mind, but by good fortune we seem to have the sensory apparatus and the mental capacities required to make a start on understanding it; and over the last 600 years we have fashioned the intellectual and material tools needed to make progress in our understanding. Robert Boyle asked:

> And how will it be prov'd, that the Omniscient God, or that admirable Contriver, Nature, can exhibit *Phaenomena* by no wayes, but such as are explicable, by the dim Reason of Man? I say, Explicable rather than Intelligible; because there may be things, which though we might understand well enough, if God, or some more intelligent Being then our own, did make it his Work to inform us of them, yet we should never of our selves finde out those Truths.[30]

God, angels, and extraterrestrials have yet to come to our assistance; yet more and more phenomena have proved explicable by the dim reason of human beings.

Science – the research programme, the experimental method, the interlocking of pure science and new technology, the language of defeasible knowledge – was invented between 1572 and 1704. We still live with the consequences, and it seems likely that human beings always will. But we do not just live with the technological benefits of science: the modern scientific way of thinking has become so much part of our culture that it has now become difficult to think our way back into a world where people did not speak of facts, hypotheses and theories, where knowledge was not grounded in evidence, where nature did not have laws. The Scientific Revolution has become almost invisible simply because it has been so astonishingly successful.

Some Longer Notes

A NOTE ON GREEK AND MEDIEVAL 'SCIENCE'

The whole of this book is an argument against the continuity thesis (exemplified by Lindberg, *The Beginnings of Western Science* (1992)), but I want in this note to present some general arguments and to offer some crucial concessions.

The argument that there were no sciences before Tycho saw his nova in 1572 is open to some obvious (but mostly mistaken) objections. Kuhn thought Ptolemaic astronomy was a mature science (Kuhn, *Structure* (1970), 68–9): it certainly had functioning paradigms and a capacity for progress. Although some of its central arguments – that all movements in the heavens are circular, that there is no change in the heavens, that the earth is at the centre of the universe, that there can be no vacuum – derived from philosophy (Kuhn, *The Copernican Revolution* (1957) calls these 'blinders' (86) and 'entanglements' (90)), they corresponded rather well with experience. And it made possible, not only Copernicanism, but also Tycho's research programme. Astronomy, though, was a peculiar discipline because it accepted unquestioningly the Aristotelian distinction between the sublunary and supralunary worlds. That distinction only began to break down in 1572, and with it went the notion that there might be different principles governing different parts of the universe, that there could be different sciences for different places. 1572 thus really is a crucial moment of change.

There are strong arguments for thinking that Aristotelian biology was a science (Leroi, *The Lagoon* (2014)). But Aristotle established no tradition of biological enquiry. In the seventeenth century William Harvey saw himself as an Aristotelian biologist, but he recognized only one person between himself and Aristotle who had understood how to conduct biological research, and that was his own teacher (and Galileo's friend), Girolamo Fabrizi d'Acquapendente (Lennox, 'The Disappearance of

Aristotle's Biology' (2001); Lennox, 'William Harvey' (forthcoming)). Similarly, there are strong arguments for thinking that Archimedes was a scientist (Russo, *The Forgotten Revolution* (2004)), but his science had little influence in the Middle Ages except in so far as it could be integrated into Aristotelianism; it is only late in the sixteenth century that the mathematicians begin to imagine an Archimedean science which might supplant Aristotle (Clagett, 'The Impact of Archimedes on Medieval Science' (1959); Laird, 'Archimedes among the Humanists' (1991)). Thus the Scientific Revolution recuperated the lost sciences of Aristotelian biology and Archimedean mathematics; but very quickly it moved away from its sources: Harvey had no followers who claimed, as he did, to be true Aristotelians, and Galileo had no followers who claimed, as he did, to be disciples of Archimedes.

As far as Kuhn was concerned, Aristotelian dynamics was itself a mature science (Kuhn, *Structure* (1970), 10; see also Kuhn, *The Copernican Revolution* (1957), 77–98; Kuhn, *The Essential Tension* (1977), 24–35, 253–65; Kuhn, *The Road since Structure* (2000), 15–20). Although he refused to recognize that optics was a science before Newton, because there were always competing schools (and so no 'normal' science), he presented Aristotelian dynamics as a successful paradigm which was supplanted in the late Middle Ages by impetus theory, which in its turn led to Galileo's new physics (Kuhn, *Structure* (1970), 118–25). The test here is that 'the successive transition from one paradigm to another via revolution is the usual developmental pattern of mature sciences' (Kuhn, *Structure* (1970), 12). But the medieval theory of impetus produced no such transition. Aristotle continued to be the textbook, and although the theory of impetus was used to patch and mend problems within Aristotle's theory, there were no separate treatises devoted to impetus theory (Sarnowsky, 'Concepts of Impetus' (2008)). Impetus theory was used to handle some anomalies, not to bring about a revolution; indeed, medieval natural philosophers were incapable of imagining a revolution that would supplant Aristotle. Because they were not conducting normal science, they never finally resolved the problems that puzzled them. There are two characteristic forms that natural philosophy takes in the Middle Ages: one is the commentary on Aristotle; the other is the collection of *quaestiones*, of problems to which there is no agreed solution. Over time new problems were added; old ones were never eliminated.

Of course one reason why Aristotelian natural philosophy survived virtually unchallenged through the Middle Ages was that outside three very restricted areas (the magnet, the rainbow, alchemy) experiments were not conducted, and where appeals to experience were made these never involved measurement. Thus in the vast bulk of Clagett, *The Science of Mechanics in the Middle Ages* (1959), the first proper experiments are those conducted by Galileo. Turn to the even vaster bulk of Grant (ed.), *A Source Book in Medieval Science* (1974), and we find, for example, a section entitled by its editor 'Experiments Demonstrating that Nature Abhors a Vacuum' (327–8), translated from Marsilius of Inghen (1340–1396). But these are *experientiae* or experiences: Marsilius has collected examples of phenomena which seem best explained by the claim that nature abhors a vacuum (one can suck water up through a straw, for instance). He has not conducted any experiments. When we turn to William Gilbert (*On the Magnet*, 1600), on the other hand, we find not only specially designed experiments, but also (something we do not find in his predecessors, such as Garzoni) experiments that require measurements.

A very powerful intellectual tradition has been dedicated to showing that medieval philosophy was a precondition for modern science (e.g. Grant, *The Foundations of Modern Science* (1996); Hannam, *God's Philosophers* (2009)). This work builds on the pathbreaking studies of Pierre Duhem (1861–1916), Annalise Maier (1905–1971), and Marshall Clagett (1916–2005). It is no part of my argument to dispute the claim that we only have the sciences we have because Aristotle and the medieval philosophers opened up certain lines of enquiry; the first scientists inherited a set of problems from their predecessors, but their procedures for resolving those problems were new, and the intellectual tools they constructed to facilitate those procedures were drawn not from philosophy but from astronomy and the law. No medieval natural philosopher had a view of natural science as making progress, and no medieval natural philosopher was engaged in research, if we understand that to mean the gathering of relevant new information. Tycho, on the other hand, had a research programme which he conducted systematically over many years, and which he believed would resolve fundamental problems in contemporary astronomy; and with the idea of a research programme came, necessarily, the idea of progress.

A NOTE ON RELIGION

Rethinking an important subject like the Scientific Revolution involves a complex process of recalibration and revaluation; topics which once seemed central become marginal, and topics which once seemed of merely antiquarian interest take on a new importance. There is a very extensive literature devoted to the relationship between Christianity and science in the early modern period.[i] Some argue that belief in a creator God was a fundamental prerequisite for modern science, as it made possible the idea of laws of nature, an idea unknown in ancient Greece and Rome, or in China. Others claim that there is a particular affinity between one or other particular sort of Christianity (Puritanism, for example) and the new science.[ii] I do not find these arguments convincing, although they are certainly intriguing. If monotheism was what counted, there would have been a scientific revolution in the Islamic and Orthodox worlds. If Protestantism was what counted, Galileo would not have been a great scientist. The idea of laws of nature represents a crucial test case, and theological questions do not prove to be fundamental: indeed, the key source for the concept appears to be Lucretius; and, as for the religious convictions of the first scientists, the only safe conclusion is that generalization is impossible. There are Jesuits and Jansenists, Calvinists and Lutherans, and some who have little or no belief. The first scientists appear, as far as their religious beliefs are concerned, to be a more or less random sample of the intellectuals of seventeenth-century Europe. Many of the scientists I have discussed were profoundly pious, but their religious faith was not what they had in common. To grasp this point one only has to think of Pascal and Newton, the first a Jansenist and the second an Arian.[iii] What they had in common was not religion but mathematics and, of course, a need for freedom of expression. '*Me tenant comme je suis, un pied dans un pays et l'autre en un autre, je trouve ma condition très heureuse, en ce qu'elle est libre*,' wrote Descartes to Elizabeth of Bohemia in the summer of 1648 ('Carrying on as I do, with one foot in one country [France] and

i Merton, 'Science, Technology and Society' (1938); Hooykaas, *Religion and the Rise of Modern Science* (1972); Lindberg & Numbers (eds.), *God and Nature* (1986); Webster (ed.), *The Intellectual Revolution of the Seventeenth Century* (1974); Funkenstein, *Theology and the Scientific Imagination* (1986); more recently, Harrison, *The Bible, Protestantism and the Rise of Natural Science* (1998), and Harrison, *The Fall of Man and the Foundations of Science* (2007). Doubting that Christianity or Protestantism was a precondition for the Scientific Revolution leaves plenty of scope for studying the interaction between faith and science: e.g. Picciotto, *Labors of Innocence* (2010).

ii The Merton thesis, that Puritanism fostered science, had very little influence among historians until it was taken up in Hill, *Intellectual Origins* (1965); the defects of that book were immediately apparent: e.g. Rabb, 'Religion and the Rise of Modern Science' (1965).

iii On Newton's theology, see, for example, Snobelen, 'Isaac Newton, Heretic' (1999); Snobelen, '"God of Gods, and Lord of Lords"' (2001).

the other in another [the Netherlands], I find my situation very happy, in that I am free.'

WITTGENSTEIN: NO RELATIVIST

The conviction that Wittgenstein was a relativist is entrenched in the literature on sociology and the history of science, although philosophers are far from united on the question (Kusch, 'Annalisa Coliva on Wittgenstein and Epistemic Relativism' (2013); see also Pritchard, 'Epistemic Relativism, Epistemic Incommensurability and Wittgensteinian Epistemology' (2010)). It seems to me to be at odds with a number of passages in which Wittgenstein expresses a quite different view of science. In a note from 1931 he wrote: 'As simple as it sounds: the distinction between magic and science can be expressed by saying that in science there is progress, but in magic there isn't. Magic has no tendency within itself to develop' (Wittgenstein, 'Remarks on Frazer's Golden Bough' (1993), 141). The fact that an enterprise makes progress does not necessarily mean I should adopt it: athletes run faster every year, but that is no reason why I should take up athletics. But science is a special case: if science gets better at understanding nature, gets better at prediction and control, then it is very difficult to see how I can remain indifferent in the face of such progress.

A remark from 1931 can easily be dismissed as unrepresentative, but we find essentially the same views in Wittgenstein's last set of notes, *On Certainty* (1969). Consider the following passage:

> 131. No, experience is not the ground for our game of judging. Nor is its outstanding success.
>
> 132. Men have judged that a king can make rain; we say this contradicts all experience ...

I take Wittgenstein to be saying that we cannot ground induction in experience, just as Hume showed that we cannot ground our notion of causation in experience; but even though we cannot ground a particular procedure in a philosophical justification, we should certainly go on using it if it is outstandingly successful. The magical claim that a king can make rain is not an 'outstanding success'; and when we say that it 'contradicts all experience' what we have is a clash between their magic and our science in which our science is superior to their magic.

Compare:

> 170. I believe what people transmit to me in a certain manner. In this

way I believe geographical, chemical, historical facts etc. That is how I *learn* the sciences. Of course learning is based on believing.

If you have learnt that Mont Blanc is 4000 metres high, if you have looked it up on the map, you say you *know* it.

And can it now be said: we accord credence in this way because it has proved to pay?

Again, the argument would seem to be that I cannot *prove* that Mont Blanc is 4000 metres high, but believing it, on the authority of a map, has 'proved to pay'. In other words, the social procedures we have for establishing certain types of fact cannot be *justified*, but they are *successful*, they pay, and this is why we employ them.

And (to take one of a series of notes dealing with the idea of going to the moon: 106, 108, 111, 117, 171, 226, 238, 264, 269, 286, 327, 332, 337, 338, 661, 662, 667):

286. What we believe depends on what we learn. We all believe that it isn't possible to get to the moon; but there might be people who believe that that is possible and that it sometimes happens. We say: these people do not know a lot that we know. And, let them be never so sure of their belief - they are wrong and we know it.

If we compare our system of knowledge with theirs then theirs is evidently the poorer one by far.

It is easy to assume that Wittgenstein's point here is a relativist one: we say their knowledge is inferior to ours; but *they* say the same about us. But suppose a society which believes that one can travel to the moon by leaving one's body, as shamans do (cf. §§106, 667), and compare it with Wittgenstein's own world in 1950: isn't it fair to say that the scientific knowledge of 1950, which made possible the jet engine and the atom bomb, was superior to (more successful than) the magical knowledge of a shamanistic culture? (see Child, *Wittgenstein* (2011), 207–12)

The same sort of point is made again:

474. This game [assuming the stability of things as the norm] proves its worth. That may be the cause of its being played, but it is not the ground.

Thus I assume that this table will continue to exist if I get up from it and leave the room. I cannot *justify* this belief, but believing works out well (it pays, it is successful), and this is why I continue to act as if this belief were true (this is the cause of this game's being played).

Lastly:

> 617. Certain events would put me into a position in which I could not go on with the old language-game any further. In which I was torn away from the *sureness* of the game.
>
> Indeed, doesn't it seem obvious that the possibility of a language-game is conditioned by certain facts?

Take the language-game represented by Ptolemaic astronomy; that game ceased to be possible when the telescope showed that Venus has a full set of phases. Thus language-games do not simply succeed, progress, pay, or prove their worth; they can also become unsustainable if the facts change.

Taken together these passages suggest that there are some types of knowledge which are superior to others because *they work*, *they pay*, *they are superior*, *they make progress*, and they are not at odds with the known facts. We cannot provide a satisfactory philosophical justification for these types of knowledge (broadly, 'the sciences'), but we can tell that they work, and other cultures interested in understanding, predicting, or controlling natural phenomena (and all cultures must be interested in these activities) should be able to recognize the utility of our knowledge (of our maps, or of our weather forecasts), just as indigenous Americans could recognize the advantages of horses and guns for hunting buffalo. This amounts to an anti-foundationalist but far from relativist view of science. It would follow that when scientific views are abandoned and replaced by new ones it is because the new ones are thought to be better at succeeding, paying, etc. In other words, science evolves, and it does so because theories that fail to develop, or that are unable to adapt in the face of new discoveries, are eliminated.

This is (as it happens) the view of science put forward in this present book, which, it would therefore seem, is authentically in the tradition established by Wittgenstein. But Wittgenstein's texts are puzzling, problematic, and unfinished. They are open to more than one reading. I have no great quarrel with those who wish to read Wittgenstein as a relativist, providing they do not use this reading to justify a relativist history of science. If pointing out that Wittgenstein himself was not a relativist in his understanding of science helps persuade historians to abandon their hostility to what they (misleadingly) call 'Whig history' then it is worth debating what Wittgenstein really meant. For note that to say that a practice pays, succeeds, proves its worth is, of necessity, to make a retrospective judgement: we can only distinguish good science from bad science, on Wittgenstein's account, with the benefit of hindsight.

And we cannot opt simply to ignore the distinction between good and bad science, because if we do we will miss one of science's peculiar characteristics, that it makes progress.

The question of what Wittgenstein really thought must, in any event, be kept separate from the question of his influence: *On Certainty* was not published until 1969, by which point a view of Wittgenstein as an uncompromising relativist was firmly established. And so his texts played a decisive role in legitimizing the new post-Kuhnian history of science because they were wrongly read as endorsing a thoroughgoing relativism. (See also below, pp. 586–7.)

NOTES ON RELATIVISM AND RELATIVISTS

This book is directed against three types of relativism. First, there is the claim that history must be written without benefit of hindsight. This claim, which dates back to Butterfield's book *The Whig Interpretation of History* (1931), had no discernible influence in the history of science until the 1960s. It can't be right: it is only hindsight, for example, which enables one to identify Columbus's discovery of America as a key moment in the development of modern science (see, in general, MacIntyre, 'Epistemological Crises' (1977)). Second, there is the claim that the concept of rationality is always culturally relative. This claim derives from Wittgenstein but began to have a major impact on history and philosophy of science after the publication of Peter Winch's *The Idea of a Social Science* (1958). It is, I maintain, incompatible with any grasp of the achievements of modern science. And, third, there is the claim that, in science, successful claims and failed claims should be understood and explained in exactly the same way, an argument which originates with David Bloor's *Knowledge and Social Imagery* (1976) and which Bloor named 'the strong programme'. This argument involves denying that scientific claims are ever adopted because they fit the evidence better than the alternatives. Its consequences for the history of science have been, it seems to me, pernicious. Each of these arguments, of course, has become part of a larger intellectual movement which may loosely be labelled 'postmodernism'. Postmodernism has, I believe, a great deal to teach naïve realists, but as naïve realism hardly gets a look in among historians of science these days I have concentrated here on its defects, not its merits.

1. See Shapin and Schaffer on truth as an actor's judgement (i.e.

truth is what you think it is): Shapin & Schaffer, *Leviathan and the Air-pump* (1985), 14 (compare Bloor, *Knowledge and Social Imagery* (1991), 37–45, and Shapin, *A Social History of Truth* (1994), 4: 'For historians, cultural anthropologists, and sociologists of knowledge, the treatment of truth as accepted belief counts as a maxim of method, and rightly so'). Truth is only an actor's judgement for statements which are necessarily subjective: e.g. 'That is the funniest joke I have ever heard' is true if and only if I think it so. It hardly helps to make rationality an actor's judgement either (Garber, 'On the Frontlines of the Scientific Revolution' (2004), 158), since the whole point of the concept is that it can be (and was) used to show that actors can be and often are mistaken. There is a difference between checkmate and death: changing the rules of chess may alter who wins and who loses, but we cannot bring ourselves back to life by changing our concepts (and to believe we can is a form of madness). If anything and everything is to be treated as an actor's judgement then the ideas of truth, rationality and objective reality become meaningless, and we can all be immortal if we choose. But at least those who make this move avoid truly puzzling formulations, such as Newman's and Principe's claim that Starkey's belief in the philosopher's stone was 'not unwarrantable' (Newman & Principe, *Alchemy Tried in the Fire* (2005), 176) – thus they avoid claiming it was sensible, and also avoid acknowledging it was foolish.

2. Barnes & Bloor, 'Relativism, Rationalism' (1982), 23, formulate the core doctrine of the strong programme as the 'equivalence postulate': 'Our equivalence postulate is that all beliefs are on a par with one another with respect to the causes of their credibility. It is not that all beliefs are equally true or equally false, but that regardless of truth and falsity the fact of their credibility is to be seen as equally problematic.' And so Simon Schaffer insists that it would be mistaken to 'account for the establishment of one version of natural philosophy [rather than an opposing one] through the superiority of its grasp over nature' (Schaffer, 'Godly Men and Mechanical Philosophers' (1987), 57). But it should be obvious that not all beliefs are on a par with one another, and that the causes of their credibility vary greatly. The Galilean belief that ice is lighter than water is not on a par with the Aristotelian belief that ice is heavier than water; the modern

belief that magnets are indifferent to garlic is not on a par with the classical belief that garlic disempowers magnets. In these cases the first belief has the facts on its side and the second doesn't; one version of natural philosophy established itself over its opponent through the superiority of its grasp over nature. To insist that the issue of validity must be separated from the issue of credibility is to insist that well-founded beliefs be treated as if they are unfounded beliefs. Enquiries based on this premise are bound to conclude that the claims made on behalf of well-founded beliefs are excessive because that conclusion is built into the methodology.

Of course, the question of how best to interpret the strong-programme approach is much disputed: see the marvellous exchange of fire between Bloor, 'Anti-Latour' (1999) and Latour, 'For David Bloor' (1999): I find Latour's reading of Bloor entirely convincing. For an effective critique, see Laudan, 'The Pseudo-science of Science?' (1981).

3. Secord, 'Knowledge in Transit' (2004), 657. The intellectual context within which *Leviathan and the Air-pump* was written is conveniently established by Shapin, 'History of Science and Its Sociological Reconstructions' (1982). For the strong programme, see Bloor, *Knowledge and Social Imagery* (1991); for other works by Barnes and Bloor: Bloor, *Wittgenstein* (1983); Barnes, *T. S. Kuhn and Social Science* (1982). The strong programme explicitly advocates 'methodological relativism', a term of art which means 'All beliefs are to be explained in the same general way regardless of how they are evaluated' (Bloor, *Knowledge and Social Imagery* (1991), 158: i.e. it is identical with the symmetry principle, on which see above, pp. 43–4 and below, No. 7, and a restatement of the equivalence postulate, on which see above, No. 2).

Harry Collins, founder of the Bath School, whose work is closely related to that of the Edinburgh School, is happy, at least on occasion, to employ the word 'relativism' unequivocally and identify those he regards as fellow relativists and aids to relativism: Collins, 'Introduction' (1981). But 'relativism' is rather like the word 'atheism' in the seventeenth century: lots of people attack it but few own up to it; and when they do they insist in defining the word in their own peculiar way (Bloor, 'Anti-Latour' (1999), 101–3). The result is a certain amount of confusion over who can fairly be called a relativist and who cannot. I have been repeatedly told, for instance, by people who should know better that Shapin is not a relativist, and certainly he rarely uses the word: nevertheless, he has recently explicitly identified himself as a 'methodological relativist', in other words as a supporter of the strong programme (which is not, from a sociological point of view, surprising, as he was a member of the Edinburgh Science Studies Unit from 1973 to 1989). Shapin practises what he preaches by giving an explanation for his own belief in the reliability of scientific knowledge which could equally be applied (in a different culture) to a belief in witchcraft: 'My confidence in science is very great: that is just to say that I am a typical member of the overall overeducated culture, a culture in which confidence in science is a mark of normalcy and which produces that confidence as we become and continue to be members of it.' (Shapin, 'How to be Antiscientific' (2010), 42 = Labinger & Collins (eds.), *The One Culture?* (2001), 111; compare Collins's claim that those who believe in astrology are making a *social* mistake (in Labinger & Collins (eds.), *The One Culture?* (2001), 258–9); see also Shapin's exposition of the 'equivalence postulate': Shapin, 'Cordelia's Love' (1995), and the description of the 'relativist genre' in Ophir & Shapin, 'The Place of Knowledge' (1991), 5, which is evidently, on Shapin's part, a self-description). I address Shapinesque relativism in Chapter 15.

I agree with Bricmont and Sokal in their contributions to Labinger & Collins (eds.), *The One Culture?* (2001), that 'methodological relativism cannot be *justified* unless one also adopts philosophical relativism or radical skepticism' (244). It is important to distinguish between methodological relativism (which is the adoption of relativism as a method) and a very different position with which it can easily be confused, methodological agnosticism, the claim that one cannot know *a priori* which method will work and which will not – a position I would defend – which is perfectly compatible with the claim that *ex post facto* one can see that one method is more successful than another (a claim that methodological relativists are committed to denying): see Kuhn, *The Structure of Scientific Revolutions* (1996), 173.

4. Shapin, *A Social History of Truth* (1994). Shapin advocates a 'liberal' rather than a 'restrictive' view of truth (4). Such an

approach involves holding that garlic does disempower magnets, or did (for Pliny, Abertus Magnus, van Helmont, etc.). The claim that garlic does not disempower magnets becomes simply an alternative truth, not a discovery; the experimental method becomes one way of making truths, not a reliable way; and Boyle's policy of disciplined suspicion becomes a new way of trusting others.

Shapin also advocates 'a methodological disposition towards charity' (4). In *Leviathan and the Air-pump* he and Schaffer wrote, 'Following Gellner, we shall be offering a "charitable interpretation"' of Hobbes, and cited an article by Gellner which had first appeared in 1962, and the use made of it by Harry Collins (on whom, see below, No. 9). In fact, Collins was quite clear that he was not following, but rather going against, Gellner (Collins, 'Son of Seven Sexes' (1981), n. 15), for Gellner's article was, in his own words, a 'plea against charity' (Gellner, 'Concepts and Society' (1970), 48); he held that 'Excessive indulgence in contextual charity blinds us to what is best and what is worst in the life of societies. It blinds us to the possibility that social change may occur through the replacement of an inconsistent doctrine or ethic by a better one . . . It equally blinds us to . . . the employment of absurd, ambiguous, inconsistent or unintelligible doctrines' (42–3). It is my book, not *Leviathan and the Air-pump*, which advocates following Gellner. Indeed, Gellner stated my overall argument precisely: 'In recent centuries, there has been an important shift from the use of merely social to genuinely cognitive concepts: this is normally known as the Scientific Revolution. Wittgensteinianism makes it impossible to ask any questions about this event, for on its terms nothing of the kind could ever occur, could make any sense' (Gellner, *Relativism and the Social Sciences* (1985), 185). No wonder Shapin insists that '[t]here was no such thing as the Scientific Revolution'! (Shapin, *The Scientific Revolution* (1996), 1).

It should be stressed that those who were trying to build the new science were acutely aware of how one might claim to provide a sociology of knowledge but wanted to escape from a world in which knowledge was entirely socially determined: cf. Bacon on the Idols (Bacon, *Instauratio magna* (1620), 53–80 (Book 1, §§23–68) = Bacon, *Works* (1857), Vol. 4, 51–69), and Glanvill, *The Vanity of Dogmatizing* (1661), esp. 125–35,

194–5. For reviews of Shapin's *A Social History of Truth*, see Feingold, 'When Facts Matter' (1996) and Schuster & Taylor, 'Blind Trust' (1997).

5. Thomas Kuhn is often credited with introducing the word 'paradigm' into English-language philosophy of science in Kuhn, *Structure* (1962) (e.g. Lehoux, *What Did the Romans Know?* (2012), 227, and Hacking, 'Introductory Essay' (2012), xvii–xxi), but in fact the word is used repeatedly in Hanson, *Patterns of Discovery* (1958): 16, 30, 91, 150, 161; some, though not all, of these seem distinctly proto-Kuhnian. Kuhn's first use of the term 'paradigm' was in a conference paper delivered in 1959, after the appearance of Hanson's book ('The Essential Tension', reprinted in Kuhn, *The Essential Tension* (1977), 225–39). Hanson also preceded Kuhn in stressing the importance of gestalt psychology and in laying emphasis on the philosophy of Wittgenstein. He is cited four times in *Structure*, and Kuhn later stressed the extent to which he was influenced by him (Kuhn, *The Road since Structure* (2000), 311; Nye, *Michael Polanyi and His Generation* (2011), 242).

This raises a larger issue in the interpretation of Kuhn. Joel Isaac has claimed that the apparent similarities between Kuhn's work and a number of nearly contemporary works are a retrospective construction (Isaac, *Working Knowledge* (2012), 232), but he does not consider the influence some of these works had on Kuhn. Thus he says that Kuhn 'hit upon' the concept of a paradigm in 1958–9 (234), ignoring the possibility that Hanson had an influence on Kuhn. (Feyerabend, when he read *Structure* in draft, found it altogether too reminiscent of Hanson: Hoyningen-Huene, 'Two Letters' (1995).) Isaac also thinks that the apparent similarities between Kuhn's *Structure* and Polanyi's *Personal Knowledge* (1958) are misleading, despite the fact that in *Structure* Kuhn refers to Polanyi's book as 'brilliant' (44; it is sometimes said that Kuhn plagiarized many of his ideas from Polanyi. Thus MacIntyre wrote that Kuhn's view of natural science 'seems largely indebted to the writings of Michael Polanyi (Kuhn nowhere acknowledged any such debt)' (MacIntyre, 'Epistemological Crises' (1977), 465). The bracketed statement is simply false: the acknowledgement is there from the first edition, though it is overlooked in the index to the third and later editions.) So, too, with the similarities between Kuhn and

Feyerabend, despite the fact that they were in close communication during 1960 and 1961 (Hoyningen-Huene, 'Three Biographies' (2005)). Isaac claims that reading Kuhn, alongside these other authors, as an opponent of positivism 'conflates the reception of Kuhn's book with the historical context of its composition' (4; the classic reception text is Shapere, 'The Structure of Scientific Revolutions' (1964)). But Kuhn himself approved the interpretation Isaac seeks to overturn (Kuhn, *The Road since Structure* (2000), 90–1).

Thus Isaac understates the significance of *Structure*'s direct attack on positivism, which, according to Kuhn, underpinned 'the most prevalent contemporary interpretation of the nature and function of scientific theory' (Kuhn, *Structure* (1996), 98–103; Isaac, *Working Knowledge* (2012), 231–2; for a summary account of this contemporary interpretation, see Hesse, 'Comment' (1982), 704), and misrepresents the context of *Structure*'s composition. Isaac's 'local' reading of Kuhn in a Harvard context is valuable, but Kuhn left Harvard in 1956; the key text for a Harvard reading should therefore be not *Structure* but *The Copernican Revolution*, published in 1957, a text which Isaac largely ignores – and *Structure* is (as Isaac occasionally seems to acknowledge) correctly read as an engagement in a much wider, international, anti-positivist debate.

6. Not everyone would agree that the truths of mathematics are necessary. Wittgenstein held that we 'make' or 'invent' mathematical truths, we do not 'discover' them (http://plato.stanford.edu/entries/wittgenstein-mathematics/ revised 21/02/2011), and the strong programme seeks to extend this principle from mathematics to science (Bloor, 'Wittgenstein and Mannheim' (1973)). The question I ask is not 'Were Regiomontanus and Hobbes right about mathematics?' but 'How did their understanding of mathematics help lay the groundwork for reliable scientific knowledge?' Even Wittgenstein held that there is a reality which corresponds to mathematical truths, but 'the reality which corresponds to them is that we have a use for them' (Conant, 'On Wittgenstein's Philosophy of Mathematics' (1997), 220). Science is one of the uses we have for our mathematics, and our mathematics and our science mutually support each other. Bloor, when discussing the utility of mathematics, tacitly assumes that it is useful in making possible certain types of social

relationship – thus he thinks it might be right to call mathematics an ideology, like monarchism (189); but mathematics also involves what Wittgenstein calls 'our practical requirements' (188), and if 2 + 2 = 4 is a norm, it is not like the divine right of kings but rather like 'when making a mayonnaise you should add the oil drop by drop.'

7. One way of escaping from the standard relativist argument that good science and bad science cannot be distinguished while avoiding an appeal to an independent reality is to argue that reality itself changes, so that one can then treat nature and society 'symmetrically' as part of the same history. This is the approach of Actor Network Theory (ANT); for an impressive example, see Law, 'Technology and Heterogeneous Engineering' (1987), and, for the thinking behind this approach, Latour, 'The Force and the Reason of Experiment' (1990), and Latour, 'One More Turn after the Social Turn' (1992). This approach is admirable in that it rejects the methodological relativism of the Edinburgh and Bath schools, but it leads to a radical historicism ('My solution . . . is to historicize more not less': Latour, *Pandora's Hope* (1999), 169), according to which Tasmania did not exist before Tasman 'discovered' it in 1642 and tuberculosis did not exist before Koch 'discovered' it in 1882. All facts are thus artefacts (see above, pp. 256n and 540), which is not true. Nature and reality are also held to be artefacts, which brings us back to relativism by a different route: according to Latour, the laws of nature hold only where there are scientists and scientific instruments, just as frozen fishfingers can be found only where there are freezers and freezer trucks (Latour, *We Have Never Been Modern* (1993), 91–129).

8. Bloor, *Knowledge and Social Imagery* (1991): for a critique, see Slezak, 'A Second Look' (1994). A striking example of Bloor's inability to acknowledge that nature constrains science is to be found on p. 39 (though the concession in the last sentence – 'Doubtless we are fully justified in preferring our theory [to Priestley's] because its internal coherence can be maintained over a wider range of theoretically interpreted experiments and experiences' – would seem devastating because incompatible with the equivalence postulate). It is important to distinguish between the principle of symmetry (that good and bad science should be explained in the same way) and the principle of impartiality (that failed science

should be studied as carefully as successful science – a principle stated by Alexandre Koyré as early as 1933: Zambelli, '*Introduzione*' (1967), 14). Thus Bertoloni Meli, *Equivalence and Priority* (1993), 14, appeals to a symmetry principle, but his argument requires only an impartiality principle. Indeed, his account of Leibniz's conflict with Newton is not symmetrical, since Leibniz was a plagiarist and Newton was not.

9. My view is similar to that of Pickering, *The Mangle of Practice* (1995), although Pickering avoids the word 'constraint' because he thinks it implies *social* constraint (65–7), preferring 'resistance'. Contrast Harry Collins's defence of his assumption that 'the natural world in no way constrains what is believed to be' (Collins, 'Son of Seven Sexes' (1981), 54; Collins says his position became less extreme in 1980 (Labinger & Collins (eds.), *The One Culture?* (2001), 184n), so it is worth remarking that I am quoting here from statements of his mature or moderated position). If this were true, Columbus would have reached China, garlic would disempower magnets, and pigs could fly. It is important to grasp that Collins's relativism (like that of the strong programme) is not the outcome of an empirical programme of enquiry (even though he calls it 'the Empirical Programme of Relativism' (Collins, 'Introduction' (1981)) but its premise: his whole enterprise 'rests on the prescription, "treat descriptive language as though it were about imaginary objects"' (Collins, *Changing Order* (1985), 16). Clearly, if this is your premise your only conclusion must be that science involves some sort of 'artful trick' (6), the trick of persuading people that imaginary objects actually exist. Even Collins, of course, succumbs to this trick (see Collins, 'Son of Seven Sexes' (1981), 34, 54), while insisting that it is wrong to do so: thus the empirical enterprise exists only to illustrate and not to test Collins's relativist premises, and is profoundly implausible in that it requires one to think that things that can't be true (are 'literally incredible') might be true. Some readers may think that Collins can't be real and that I have made him up (after the Sokal hoax – Sokal, *Beyond the Hoax* (2008) – such a thought would not be unreasonable). I assure them he does exist and is not a crank: cranks do not get elected as Fellows of the British Academy.

For a more cautiously formulated rejection of 'constraint' talk, see Shapin, 'History of Science and Its Sociological

Reconstructions' (1982), 196–7. What Shapin offers is essentially a circular argument: talk of constraint is incompatible with relativism, but historians are committed to relativism, consequently they must not talk of constraint. Second, he relies on the Duhem–Quine thesis to claim that what constrains scientists is not reality but a particular description of reality; but it is wrong to assume, as this claim does, that the outcome of scientific disputes is always open-ended. When Galileo saw the phases of Venus there was no alternative way of describing what he had seen; nor could there be, unless one was prepared to question assumptions that everyone, with good reason, held in common (that light travels in straight lines, for example).

10. It is easy to add to these examples: Mornet, *Les Origines intellectuelles de la Révolution française* (1933); Lefebvre, *The Coming of the French Revolution* (1947); Bailyn, *The Ideological Origins of the American Revolution* (1967); Trevor-Roper, 'The Religious Origins of the Enlightenment' (1967); Stone, *The Causes of the English Revolution* (1972); Weber, *Peasants into Frenchmen* (1976); Baker, *Inventing the French Revolution* (1990); Chartier, *The Cultural Origins of the French Revolution* (1991); Skinner, 'Classical Liberty and the Coming of the English Civil War' (2002); Bayly, *The Birth of the Modern World* (2003). Equally retrospective in character are books on decline, such as Thomas, *Religion and the Decline of Magic* (1997), or failure, such as MacIntyre, *After Virtue* (1981).

One of the reasons, of course, for abandoning the old retrospective stories is that they were profoundly unsatisfactory, as Elton and a whole series of scholars after him showed for the English Civil War (Elton, 'A High Road to Civil War?' (1974)), and as Cobban and a whole series of scholars after him showed for the French Revolution (Cobban, *The Social Interpretation of the French Revolution* (1964)). But the fact that a job has been done badly does not mean it cannot be done better, and it is hard to imagine how a situation in which we have no explanation for the English Civil War, other than that it was an unfortunate accident (which simply begs the question of why it was impossible to put Humpty together again) can be regarded as satisfactory. Nor can I see why historians should cede many of the most interesting questions to other disciplines – politics, philosophy, sociology – simply because they require a consideration of beginnings and endings.

The simple truth is that the definition of Whig history has got tighter and tighter year by year. But in history of science the issue of so-called Whig history is particularly vexed because it is used to censor any acknowledgement that there is progress in science, and so to entrench the equivalence postulate as a principle of historical method. Here, too, attitudes have become more restrictive with each passing year. In 1996 Roy Porter, an historian as opposed to Whig history as anyone, published a work (evidently written earlier, perhaps in 1989) in which he referred to the Scientific Revolution as resulting in 'substantial and permanent achievements, full of future promise' and wrote of 'the advancement of science' (Porter, 'The Scientific Revolution and Universities' (1996), 538, 560; compare Porter, 'The Scientific Revolution' (1986), 302). It is time to release the ratchet.

11. A fine analysis of the state of play when I began work on this book is provided by Daston, 'Science Studies and the History of Science' (2009); the difference between us is one of emphasis, for in my judgement Daston underestimates the extent to which fear of anachronism has debilitated history of science and overestimates the extent to which history of science has distanced itself from the symmetry principle. (For an earlier acknowledgement that history of science had lost its sense of direction, Secord, 'Knowledge in Transit' (2004), 671.) Golinski, 'New Preface' (2005), xi, summed up the situation immediately after the Science Wars: 'Constructivism may have lost some of the bloom of its early promise ... but it still informs much historical scholarship at the level of tacit assumptions.'

Golinksi is perhaps also typical in his muddle-headed view that the relativism of the strong programme can and should be used 'as a tool rather than as the expression of a totalizing skepticism' and in suggesting that constructivism can be regarded as 'complementary to a range of other approaches' (x–xi). It is true that sophisticated supporters of the strong programme insist that they are not relativists when going about their daily lives, but they do not suggest that you can be a part-time relativist when studying science as an historian or sociologist; their relativism cannot be picked up and put down like a tool because it is a methodological postulate which rules non-relativist questions out of order – in that respect, they are relativists through and through.

Golinski is also wrong to suggest that it is only with the outbreak of the Science Wars that the constructivist enterprise lost its sense of direction. In fact, by the time of the Sokal hoax (1996), the enterprise was already in deep trouble. From outside, it had been subjected to a devastating critique: Laudan, 'Demystifying Underdetermination' (1990). The gathering sense of crisis was marked by Bruno Latour's statement, as an insider: 'After years of swift progress, social studies of science are at a standstill' (Latour, 'One More turn After the Social Turn' (1992), 272). They were, and (despite Pickering, *The Mangle of Practice* (1995), which represents an important attempt to get back on the right road) they still are.

It is now fifteen years since Victoria E. Bonnell and Lynn Hunt published a collection of essays entitled *Beyond the Cultural Turn* in which they sought – but did not find – a way out of what they called 'our current predicament' (Bonnell & Hunt, 'Introduction' (1999), 6). Alas, there are still plenty of people who think, with Nick Wilding, that 'social constructivism does not go nearly far enough.' Wilding suspects that in the seventeenth century 'scientific practice was so localized and nontransferable that the idea of a norm belongs to an Enlightenment, rather than early modern, epistemological landscape' (Wilding, *Galileo's Idol* (2014), 136–7). Such an approach inevitably renders the Scientific Revolution totally invisible. It implies that Galileo's claim that it is impossible, in defiance of nature, to turn falsehood into truth, was entirely misconceived; that Hobbes was wrong to admire Galileo as the founder of a new type of knowledge; and that Diderot's dream represents the beginning, not the end, of the story of the birth of the scientific enterprise. And, of course, it is wrong: the transferability of Galileo's new science is straightforwardly demonstrated by a list of the cities in which his work was published in the fifty years after his condemnation in 1633: Strasbourg (1634, 1635, 1636), Leiden (1638), Paris (1639, 1681), Padua (1640, 1649), Lyon (1641), Ravenna (1649), London (1653, 1661, 1663, 1665, 1667, 1682, 1683), Bologna (1655–6, 1664), Amsterdam (1682), to which may be added the popularizing works of Mersenne, Danese, Wilkins, and others. If this is localism, what would its opposite look like?

A NOTE ON DATES AND QUOTATIONS

I give dates of publication as on the title page: Locke's *Essay* came out in 1689, but the title page says 1690; Popper's *Logik der Forschung* came out in the autumn of 1934, but the title page says 1935; Koyré's *Études* are dated 1939 but came out in 1940. The exception is Walter Charleton's *Ternary of Paradoxes*; there are two distinct editions dated 1650, one of which actually appeared in 1649, so I give 1649 as the publication date in order to show which edition I have used.

I date years as beginning on 1 January. Newton's first publication is dated 6 February 1671/72: I call this 1672.

I have preserved original spelling and punctuation in quotations, except that I have regularized 'u' and 'v', and 'i' and 'j'.

A NOTE ON THE INTERNET

Over the last decade the scholarly enterprise has been transformed by the internet. All the early modern texts cited here are to be found on the internet, some at subscription services (Early English Books Online (EEBO); Eighteenth Century Collections Online (ECCO)) but many at open-access sites (Google Books, Gallica).

In particular, my research on the history of words has been internet based. The key sources are as follows: 1. For English, the *Oxford English Dictionary*, supplemented by the search facilities in EEBO and ECCO. EEBO searches all titles and, apparently, some 25 per cent of texts (but, in reality, rather more than that, as many texts are duplicated in several editions), while ECCO searches (very erratically) all texts on the database (which is nearly complete). One can also search early-modern-English dictionaries at http://leme.library.utoronto.ca. 2. For French, the public-access dictionary collection at http://artfl-project.uchicago.edu/contentdictionnaires-dautrefois. 3. For Italian, the *Vocabolario degli accademici della Crusca* (1612) at http://vocabolario.signum.sns.it/. 4. For all languages, and particularly for Latin, the resources at Google Books and other collections of ebooks (such as archive.org and gallica.bnf.fr). I have not noted the date of these searches, but the bulk of the book was written in 2012–14: outcomes will change, of course, as further materials come on line and as the *OED* is revised.

But this is only part of my debt to the internet: day after day the postman has brought packages of books to my door, acquired from far-flung corners of the world. Seventeenth-century scholars sometimes felt they were drowning in an ocean of books. As the piles on and beside my desk have grown I, too, have had this feeling, but mostly I have felt as though I were far out at sea, unsure where or when I would make landfall but delighted to be on my own voyage of discovery.

Acknowledgements

This book was born out of a sense that for the most part, and with some honourable exceptions, historians of science were not doing their subject justice.[i] I do not expect them to agree with this assessment of their profession; it will, inevitably, seem to them to be based, at best, on a misunderstanding. Nevertheless, my greatest debt is to those with whom I disagree. In the words of Alexandre Koyré, 'Human thought is polemic; it thrives on negation. New truths are foes of the ancient ones which they must turn into falsehoods.'[1] Without disagreement, sometimes sharp disagreement, there would be no thriving.

But I have not sought either disagreement or novelty for their own sake; rather, I have come to a difference of opinion slowly and reluctantly, and only because crucial features of science and of the Scientific Revolution are (as it seems to me) overlooked or dismissed in the accounts that now pass for sound scholarship. As Pascal said when he announced that nature was indifferent to the existence of a vacuum, '[I]t is not without regret that I abandon opinions so generally received. I only yield to the compulsion of truth. I resisted these new ideas as long as I had any reason for clinging to the old.'[2]

It will be apparent that my own intellectual development owes a great deal to Lucien Febvre. His *The Problem of Unbelief in the Sixteenth Century* (1942) is still the most important book on the transition from medieval to modern ways of thinking; I spent the first decade of my academic career attacking that book, and so it is a nice example of the complicated way in which we struggle with our predecessors that I now find myself, years later, defending it.[3] Another book from the same era that has provided me with a model of how to think is Bruno Snell's *The Discovery of the Mind* (1946).

But old books are not my only source of inspiration. I have learnt

i 'Notes on Relativism and Relativists', No. 11 (pp. 590–2).

from Ian Hacking and Lorraine Daston how to practise historical epistemology; from Jim Bennett that the Scientific Revolution is not many revolutions but one, for the simple reason that the inspiration for all the different revolutions that make it up came from the mathematicians; and I have taken particular encouragement from Larry Laudan, 'Demystifying Underdetermination' (1990), Andrew Pickering, *The Mangle of Practice* (1995) and John Zammito, *A Nice Derangement of Epistemes* (2004).

Chapter 7 first appeared in public as the 2011 Emden Lecture at St Edmund Hall, Oxford, and then as an interdisciplinary lecture at the University of Sheffield and as a lecture to the York Philosophical Society. The core arguments of the book have been presented in the 2014 Aylmer Lecture at the University of York and in a lecture at the Illinois Institute of Technology. Some arguments, particularly from Chapters 3 and 7, were first trialled in review essays for the *Times Literary Supplement*: I am grateful to my editors there for the opportunities they have given me. I also owe a great debt to my department and students at the University of York: my department for allowing me to concentrate on history of science for the past decade, and my students for being both clever and industrious.

A number of friends and colleagues – Jim Bennett, Sabine Clark, Michael Kubovy, Rachel Laudan, Paolo Palmieri, Klaus Vogel, Tom Welch – have read parts of the book and have made helpful criticisms. Alan Chalmers, Stephen Collins, Christopher Graney, John Kekes, Alan Sokal and Sophie Weeks read a draft right through and out-argued me on crucial issues. John Schuster has, with extraordinary generosity, read more than one draft, and provided a perfect mixture of encouragement and criticism. Julia Reis has provided invaluable help, particularly with German texts. A large number of individuals have provided me with guidance and saved me from error: Fabio Acerbi, Adrian Aylmer, Mike Beaney, Marco Bertamini, Pete Biller, Ann Blair, Stuart Carroll, H. Floris Cohen, Stephen Clucas, Simon Ditchfield, Toby Dyke, John Elliott, Mordechai Feingold, Felipe Fernández-Armesto, Pierre Fiala, Arthur Fine, Mary Garrison, Alfred Hiatt, Mark Jenner, Stephen Johnston, Harry Kitsikopoulos, Larry Laudan, Steven Livesey, Michael Löwy, Noel Malcolm, Saira Malik, Adam Mosley, Jamie Newell, Eileen Reeves, Chris Renwick, Stuart Reynolds, Richard Serjeantson, Alan Shapiro, Barbara Shapiro, William Shea, Mark Smith, Shelagh Sneddon, Rick Watson, Nick Wilding, Albert van Helden, David Womersley. I want particularly to thank Owen Gingerich and Michael Hunter, who read the book for the publishers: an author could not wish for better readers, and I have come back to them again and again with queries.

The original project for this book was constructed in close collaboration with my marvellous agent, Peter Robinson. Stuart Proffitt at Allen Lane has given the book the exquisite care and attention for which he is justly famous – it is a much better book than it would have been without him. It is also much longer: from the beginning he wanted a big book and somehow or other he has managed to get one. At the same time, pulling in the opposite direction, my American agent, Michael Carlisle, and publisher, Bill Strachan, have been keen to see me actually finish; and at last I have. Susannah Stone has done a wonderful job sourcing illustrations. Sarah Day has been a lynx-eyed copyeditor. The index is, as an index on this scale should be, signed by its author. I have used Mellel as my wordprocessor and Sente as my bibliography programme: I cannot praise them enough.

None of the above bears any responsibility for my errors and omissions.

As before, my thinking developed during conversations with Matthew Patrick. Above all, I am indebted to Alison Mark, without whom nothing, with whom everything.

Theddingworth, Leicestershire
Spring 2015

Endnotes

INTRODUCTION

1. Harrison, 'Reassessing the Butterfield Thesis' (2006), 7, argues that the concept of the Scientific Revolution is incoherent because there is no way of knowing when it began and when it ended. I disagree: the concept would be coherent even if the dates were uncertain (compare the 'Industrial Revolution'), but actually the dates are fairly easy to specify.

CHAPTER 1

1. Borges, *The Total Library* (2001), 465. 2. Barker, *The Agricultural Revolution in Prehistory* (2006). 3. Stein, *Everybody's Autobiography* (1937), 289. 4. Turgot's *A Philosophical Review of the Successive Advances of the Human Mind* was written in 1750 but not published until the nineteenth century (Turgot, *Turgot on Progress* (1973)); Condorcet, *Outlines of an Historical View of the Progress of the Human Mind* (1795) – original French edition the same year; Bury, *The Idea of Progress* (1920). 5. II.1.813–15. 6. II.3.1440. 7. MacGregor, *Shakespeare's Restless World* (2012), Ch. 18: 'London becomes Rome.' 8. Borges, *The Total Library* (2001), 472 ('The Enigma of Shakespeare', 1964). 9. Kassell, *Medicine and Magic in Elizabethan England* (2005). 10. Donne, *The Epithalamions, Anniversaries and Epicedes* (1978). 11. Wootton, *Galileo* (2010), 5–6. 12. Jacquot, 'Thomas Harriot's Reputation for Impiety' (1952). 13. Hill, *Philosophia epicuraea* (2007). 14. Brown, '*Hac ex consilio meo via progredieris*' (2008), 836–8. I don't think it can originally have come from the college library, for library books are not interleaved with blank pages. 15. Trevor-Roper, 'Nicholas Hill, the English Atomist' (1987), 11 (quoting Robert Hues), 13 (quoting Thomas Henshaw). 16. Trevor-Roper, 'Nicholas Hill, the English Atomist' (1987), 3–4. 17. Trevor-Roper, 'Nicholas Hill, the English Atomist' (1987), 28–34. 18. Kepler, *Kepler's Conversation with Galileo's Sidereal Messenger* (1965), 34–6, 38–9. 19. Trevor-Roper, 'Nicholas Hill, the English Atomist' (1987), 11. 20. Lynall, *Swift and Science* (2012). 21. Letter to Thomas Poole, 23 March 1801. 22. Gingerich, 'Tycho Brahe and the Nova of 1572' (2005); McGrew, Alspector-Kelly & others, *The Philosophy of Science* (2009), 120–2. For the view that Brahe, not Copernicus, marks the beginning of the revolution in astronomy,

Donahue, *The Dissolution of the Celestial Spheres* (1981); Lerner, *Le Monde des sphères* (1997); Grant, *Planets, Stars and Orbs* (1994); Randles, *The Unmaking of the Medieval Christian Cosmos* (1999). **23.** Wesley, 'The Accuracy of Tycho Brahe's Instruments' (1978). **24.** Thoren, *Lord of Uraniborg* (2007); Christianson, *On Tycho's Island* (2000); Mosley, *Bearing the Heavens* (2007).

CHAPTER 2

1. Weinberg, *To Explain the World: The Discovery of Modern Science* (2015), xi. **2.** Mayer, 'Setting Up a Discipline (2000); the first appointment of an historian to study and teach history of science came later, in 1948. Butterfield, *The Origins of Modern Science* (1950); and Bentley, *The Life and Thought of Herbert Butterfield* (2011), 177–203. **3.** Snow, *The Two Cultures* (1959). See also Leavis, *Two Cultures?* (2013). **4.** Cohen, *The Scientific Revolution: A Historiographical Inquiry* (1994), 21, 97–121; and, for example, Porter, 'The Scientific Revolution and Universities' (1996), 535. **5.** In addition to Snow, *The Two Cultures* (1959); and Ashby, *Technology and the Academics* (1958). **6.** Butterfield, *The Origins of Modern Science* (1950), viii. **7.** Laski, *The Rise of European Liberalism* (1936). Ornstein in 1913, Preserved Smith in 1930 and Bernal in 1939 had also used the term 'the Scientific Revolution' (Cohen, *The Scientific Revolution: A Historiographical Inquiry* (1994), 389–96), but that would not justify describing the concept as popular. **8.** Greeley, 'The Age We Live In' (1848), 51: 'Lowell, Manchester, Lawrence, are but types of the Industrial Revolution which is rapidly transforming the whole civilized world.' **9.** Koyré, *The Astronomical Revolution* (1973) (French original, 1961). **10.** Cunningham & Williams, 'De-Centring the "Big Picture"' (1993). **11.** Shapin, *The Scientific Revolution* (1996), 3. **12.** Koyré's source when he first introduced the notion of the Scientific Revolution in 1935 was Bachelard, *Le Nouvel Esprit scientifique* (1934), later translated as Bachelard, *The New Scientific Spirit* (1985). In later editions he introduced a reference to the classic work, Bachelard, *La Formation de l'esprit scientifique* (1938), translated as Bachelard, *The Formation of the Scientific Mind* (2002). **13.** Butterfield, *The Whig Interpretation of History* (1931). For its continuing significance, see, for example, Wilson & Ashplant, 'Whig History' (1988). **14.** Elton, 'Herbert Butterfield and the Study of History' (1984), 736. *Origins* is, says B. J. T. Dobbs, 'the most Whiggish history of science imaginable': Dobbs, 'Newton as Final Cause' (2000), 30. See Westfall, 'The Scientific Revolution Reasserted' (2000), 41–3, for a defence. **15.** Shapin, *The Scientific Revolution* (1996). Key authorities on the Scientific Revolution are Dijksterhuis, *The Mechanization of the World Picture* (1961); Cohen, *The Birth of a New Physics* (1987); Lindberg & Westman (eds.), *Reappraisals of the Scientific Revolution* (1990); Cohen, *The Scientific Revolution: A Historiographical Inquiry* (1994); Applebaum, *Encyclopedia of the Scientific Revolution* (2000); Osler (ed.), *Rethinking the Scientific Revolution* (2000); Dear, *Revolutionizing the Sciences* (2001); Rossi, *The Birth of Modern Science* (2001); Henry, *The Scientific Revolution* (2002); Wussing, *Die* *grosse Erneuerung* (2002); Hellyer (ed.), *The Scientific Revolution* (2003); Cohen, *How Modern Science Came into the World* (2010); Principe, *The Scientific Revolution* (2011). For a survey of recent trends in scholarship, Smith, 'Science on the Move' (2009). **16.** Wilson & Ashplant, 'Whig History' (1988), 14. **17.** Wagner, *The Seven Liberal Arts* (1983). **18.** Thus Milliet de Chales, *Cursus seu mundus mathematicus* (1674), I, †3r: *Plebeiae sunt caeterae disciplinae, mathesis Regia*; ††1r: *Primum inter naturales scientias locum, sibi iure vendicare Mathematicas disciplinas*; and in the expanded posthumous edition, Milliet de Challes, *Cursus seu mundus mathematicus* (1690), Vol. 1, 1–2: *Quòd si hoc praesertim saeculo, assurgere non nihil videtur Physica, fructúsque edidisse non poenitendos, si multa scita digna, jucunda, Antiquis etiam incognita decreta sunt; ideò sane quia Mathematici philosophantur, rebúsque physicis Mathematices placita admiscent.* Bennett, 'The Mechanics' Philosophy and the Mechanical Philosophy' (1986), is significant in this regard; as is the table in Gascoigne, 'A Reappraisal of the Role of the Universities' (1990), 227; and there is an important article on the collaborations between mathematicians and anatomists: Bertoloni Meli, 'The Collaboration between Anatomists and Mathematicians in the Mid-seventeenth Century' (2008). Robert Boyle is an interesting (and partial) exception to the rule that the new scientists are nearly always either mathematicians or doctors: Shapin, 'Boyle and Mathematics' (1988). Recognizing the clash between mathematicians and philosophers helps clarify the role of universities in the Scientific Revolution: for a positive view of their role, see Gascoigne, 'A Reappraisal of the Role of the Universities' (1990) (but note table 5.2, which shows that only one third of scientists born between 1551 and 1650 held university positions); and Porter, 'The Scientific Revolution and Universities' (1996). **19.** Leonardo da Vinci, *Treatise on Painting* (1956), no. 1. For later puzzlement, see Leonardo da Vinci, *Trattato della pittura* (1817), 2. For an extended argument that mathematics is the foundation of all true knowledge, Aggiunti, *Oratio de mathematicae laudibus* (1627), esp. 8, 26, 33. There is no reason to think this text is by Galileo (*pace* Peterson, *Galileo's Muse* (2011)), but he certainly approved of it. **20.** Biagioli, 'The Social Status of Italian Mathematicians, 1450–1600' (1989). **21.** '*Galilaeus, non modo nostri, sed omnium saeculorum philosophus maximus*', Hobbes, *De mundo* (1973), 178. **22.** Hooke, *The Posthumous Works* (1705), 3–4. **23.** Baxter, *A Paraphrase on the New Testament* (1685), annotations on 1 Corinthians, Ch. 2 (misquoted in *OED* s.v. physic); and Harris, *Lexicon technicum* (1704), quoted in *OED* s.v. physiology (I quote from the second edition, 1708). See also Hooke, *The Posthumous Works* (1705), 172: 'the Science of Physicks, or of Natural and Experimental Philosophy'. Wotton thinks that in English 'physick' and 'physical' are properly restricted to medicine (Wotton, *Reflections upon Ancient and Modern Learning* (1694), 289), but in practice he uses 'physical' to refer to physics in general. **24.** For 'physiology' used as synonymous with 'physical science', see the full title of Gilbert, *De magnete* (1600) (*physiologia nova*); and Charleton, *Physiologia Epicuro-Gassendo-Charletoniana* (1654); also Parker, above, p. 40, and Wotton, *Reflections upon Ancient and Modern Learning*

(1694), 457. **25.** Andrew Cunningham has been particularly insistent that the correct category for the early modern period is 'natural philosophy', and that natural philosophy differs from science in that it is God-centred. See his debate with Edward Grant: Cunningham, 'How the *Principia* Got Its Name' (1991); Grant, 'God, Science and Natural Philosophy' (1999); Cunningham, 'The Identity of Natural Philosophy' (2000); and Grant, 'God and Natural Philosophy' (2000). Grant seems to me to be in the right. See also Dear, 'Religion, Science and Natural Philosophy' (2001). A much more challenging argument is being developed by John Schuster (see, at the moment, Schuster, *Descartes-Agonistes* (2013), 31–98), but I don't agree with his view that natural philosophy is *the* category for thinking about the Scientific Revolution and that the Scientific Revolution is a civil war *within* natural philosophy. See also on an alternative category, physico-mathematics, which seems to me much more helpful: Dear, *Discipline and Experience* (1995), 168–79, Schuster, 'Cartesian Physics' (2013), 57–61; and Schuster, *Descartes-Agonistes* (2013), 10–13, 56–9. The question of whether the new science should be seen as, willy-nilly, a break from natural philosophy, or a struggle within it, depends in part on whether one thinks the mature Descartes is typical or atypical. **26.** Kuhn, *The Road since Structure* (2000), 42–3. *Discipline and Experience* (1995), 151–2 on Mary Hesse's 'network model'. **27.** For the French, see Schaffer, 'Scientific Discoveries' (1986), 408. **28.** Boyle, *The Christian Virtuoso* (1690), title page = Boyle, *The Works* (1999), Vol. 11, 281. **29.** Quoted from Secord, *Visions of Science* (2014), 105. **30.** 1831 is from Google Books; *OED* gives 1835–6. **31.** Hannam, *God's Philosophers* (2009), 338. **32.** Hill, 'The Word "Revolution" in Seventeenth-century England' (1986), 149, on how 'things precede words'. **33.** Benveniste, *Problèmes de linguistique générale II* (1974), 247–53; English dates from *OED*, checked against EEBO and Google Books; I owe the 1895 date to Pierre Fiala, who kindly searched the Frantext database. **34.** An exception is Bruno: Bruno, *The Ash Wednesday Supper* (1995), 139. **35.** An example taken at random: Denton, *The ABC of Armageddon* (2001), 84–5. **36.** Shapiro, *John Wilkins* (1969), 192. **37.** Laslett, 'Commentary' (1963). **38.** On Digges, Johnson & Larkey, 'Thomas Digges, the Copernican System' (1934); Ash, 'A Perfect and an Absolute Work' (2000); and Collinson, 'The Monarchical Republic' (1987). On Harriot, Fox (ed.), *Thomas Harriot* (2000); Schemmel, *The English Galileo* (2008); and Greenblatt, 'Invisible Bullets' (1988). **39.** Laslett's failure to recognize that mathematics was a science-related profession (he mentions only medicine) suggests he had never encountered Taylor, *The Mathematical Practitioners* (1954). Such a mistake would hardly be made now, one would hope, thanks to work such as Dear, *Discipline and Experience* (1995). **40.** A key figure who published on almost all of these is the Dutch mathematician Simon Stevin: Dijksterhuis, *Simon Stevin: Science in the Netherlands around 1600* (1970). **41.** Snow, 'The Concept of Revolution' (1962). Hill, 'The Word "Revolution" in Seventeenth-century England' (1986), argues for an earlier date, but most of his examples are, at best, ambiguous. **42.** Hull, 'In Defence of Presentism' (1979). **43.** Hooke, *Micrographia, or Some Physiological Descriptions of Minute Bodies* (1665), a4. **44.** Sprat, *The*

History of the Royal-Society (1667), 327, 363. **45.** Sprat, *The History of the Royal-Society* (1667), 328–9. **46.** Cohen, 'The Eighteenth-century Origins of the Concept of Scientific Revolution' (1976); and Baker, *Inventing the French Revolution* (1990). **47.** Hunter & Wood, 'Towards Solomon's House' (1986), 81. Compare Sprat, *The History of the Royal-Society* (1667), 29: 'one great Fabrick is to be pull'd down, and another to be erected in its stead.' **48.** On the word 'modern' in English, Withington, *Society in Early Modern England* (2010), 73–101. **49.** Galilei, *Dialogue on Ancient and Modern Music* (2003). **50.** Kuhn, *Structure* (1970), 161; Feyerabend, *Farewell to Reason* (1987), 143–61. There was soon (1587–95) an attempt by Bernardino Baldi to write a history of modern mathematics modelled on Vasari's Lives: Swerdlow, 'Montucla's Legacy' (1993), 301; and Rose, 'Copernicus and Urbino' (1974). **51.** For some two hundred examples, see Thorndike, 'Newness and Craving for Novelty' (1951). **52.** For example, Thorndike, *A History of Magic and Experimental Science* (1923), II, 451–527; and Crombie, *Styles of Scientific Thinking* (1994), 345 **53.** Gilbert, *De magnete* (1600), Ch. 1. Gilbert also identifies 'more modern' authors, i.e. Renaissance authors. **54.** Filarete, *Trattato di architettura* (1972), Bk 13; Panofsky, *Renaissance and Renascences* (1970), 28, quoted from Greenblatt & Koerner, 'The Glories of Classicism' (2013). See http://fonti-sa.sns.it/TOCFilarete TrattatoDiArchitettura.php, 380. **55.** Swift, *A Tale of a Tub* (2010), 153. **56.** Rapin, *Reflexions upon Ancient and Modern Philosophy* (1678), 189 (first French edition, 1676). **57.** Boyle, *Hydrostatical Paradoxes* (1666), A7r = Boyle, *The Works* (1999), Vol. 5, 195; and Glanvill, *Plus ultra* (1668), 1. For Glanvill, Bacon, Galileo, Descartes and Boyle are *the* moderns. **58.** Harvey, *The Vanities of Philosophy and Physick* (1699), 10. **59.** Glanvill, *Plus ultra* (1668); Le Clerc, *The History of Physick* (1699) (first published in French, 1696). **60.** '*Audendum est, et veritas investiganda; quam etiamsi non assequamur, omnino tamen propius, quam nunc sumus, ad eam pervenivemus.*' (The root meaning of *investigare* is 'to follow in the tracks of'.) Boyle, *The Origine of Formes and Qualities* (1666) = Boyle, *The Works* (1999), Vol. 5, 281; Boyle, *A Free Enquiry* (1686) = Boyle, *The Works* (1999), Vol. 10, 437. See Eamon, *Science and the Secrets of Nature* (1994), 269–300. **61.** Wotton, *Reflections upon Ancient and Modern Learning* (1694), unpaginated preface, 91, 105, 146, 169, 341. For early examples of the phrase 'progress of science' (and variations), see: Jarrige, *A Further Discovery of the Mystery of Jesuitisme* (1658); Borel, *A New Treatise* (1658), 2 – Borel is under the misapprehension that Bacon had written a book *de progressu Scientiarum* (92); Naudé, *Instructions Concerning Erecting of a Library* (1661); Bacon, *The Novum organum . . . Epitomiz'd* (1676), 11; and Le Clerc, *The History of Physick* (1699), 'To the Reader' (unpaginated). **62.** Quoted in Gingerich & Westman, 'The Wittich Connection' (1988), 19. **63.** Wootton, *Galileo* (2010), 96, 123, 286 n. 53. **64.** Hunter & Wood, 'Towards Solomon's House' (1986), 87. **65.** Glanvill, *The Vanity of Dogmatizing* (1661), 178, 181–3. **66.** Hobbes, *Elements of Philosophy* (1656), B1r (first Latin edition, 1655). **67.** Power, *Experimental Philosophy* (1664), 192. **68.** Wallis, 'An Essay of Dr John Wallis' (1666), 264. **69.** Parker, *A Free and Impartial Censure* (1666), 45. **70.** Dryden, *Of*

Dramatic Poesie (1668), 9. 71. Kuhn, *Structure* (1970), 162–3. 72. Winch, *The Idea of a Social Science* (1958); Hanson, *Patterns of Discovery* (1958); Kuhn, *Structure* (1962). Wittgenstein was a key influence on David Bloor and the 'Edinburgh school': Bloor, *Knowledge and Social Imagery* (1991); and Bloor, *Wittgenstein* (1983). For a devastating assessment of the project of using Wittgenstein to ground a relativist sociology, Williams, 'Wittgenstein and Idealism' (1973). 73. Wittgenstein, *Philosophical Investigations* (1953), §43. 74. E.g. Phillips, *Wittgenstein and Scientific Knowledge* (1977), 200–1. I have started with Wittgenstein, but William James thought the concept of truth as something other than a human device died in 1850: James, 'Humanism and Truth (1904)' (1978), 40–1. 75. Biagioli (ed.), *The Science Studies Reader* (1999), provides an introduction to what used to be called Science Studies and is now called Science and Technology Studies. 76. Russell, 'Obituary: Ludwig Wittgenstein' (1951). 77. Wittgenstein, *On Certainty* (1969), §612. 78. Feyerabend, *Against Method* (1975); Feyerabend, *Farewell to Reason* (1987). 79. Wilson (ed.), *Rationality* (1970); and Hollis & Lukes (eds.), *Rationality and Relativism* (1982). 80. Galilei, *Le opere* (1890), Vol. 5, 309–10. 81. Shapin & Schaffer, *Leviathan and the Air-pump* (1985), 67. 82. Hamblyn, *The Invention of Clouds* (2001); and Gombrich, *Art and Illusion* (1960), 150–2. 83. Hooke, *The Posthumous Works* (1705), 3. 84. Gilbert, *On the Magnet* (1900), iii. 85. Tuck, *Natural Rights Theories* (1979), 1–2. 86. Burtt, *The Metaphysical Foundations of Modern Physical Science* (1924). 87. Butterfield, *The Origins of Modern Science* (1950), 5; Burtt, *The Metaphysical Foundations of Modern Physical Science* (1924) (on which, see Daston, 'History of Science in an Elegiac Mode' (1991)); and Koyré, 'Galileo and the Scientific Revolution of the Seventeenth Century' (1943), 346. 88. Diderot, *The Indiscreet Jewels* (1993), 136.

PART ONE

1. Copernicus, *On the Revolutions* (1978), 7.

CHAPTER 3

1. Hanson, 'An Anatomy of Discovery' (1967), 352. 2. Columbus, *The Journal* (2010), 35–6. 3. Lester, *The Fourth Part of the World* (2009). 4. Grafton, Shelford & others, *New Worlds, Ancient Texts* (1992), 80; Galilei, *Le opere* (1890), Vol. 3, 57; above, p. 56. 5. Galilei, *The Essential Galileo* (2008), 47; Giordano da Pisa: '*Non é ancora venti anni che si trovó l'arte di fare gli occhiali, che fanno vedere bene, ch'é una de le migliori arti e de le piú necessaire che 'l mondo abbia, e é così poco che ssi trovò: arte novella che mmai non fu. E disse il lettore: io vidi colui che prima la trovó e fece, e favvellaigli*' (quoted from Ilardi, *Renaissance Vision* (2007), 5); and Filarete: '*Pippo di ser Brunelleschi inventò la prospettiva, la quale precedentemente non si era mai usata . . . Benché gli antichi fossero acuti e sottili, essi non conobbero la prospettiva*' (quoted from Camerota, *La prospettiva*

del Rinascimento (2006), 61). **6.** On the meaning of *descobrir*, Morison, *Portuguese Voyages to America* (1940), 5–10, 43 (the 1484 usage, translated by Morison as 'explore'), 45–6 (1486, translated by Morison as 'discover'). See also Randles, '*Le Nouveau Monde*' (2000), 10, for the word *descubre* occurring in Spanish to mean 'discover' in 1499. **7.** Caraci Luzzana, *Amerigo Vespucci* (1999), 321–83; searchable text at http://eprints.unifi.it/archive/00000533/02/Lettera_al_Soderini.pdf. The earlier *Mundus novus* does not contain *discooperio*, but the original Italian text is lost. Waldseemüller is far too good a Latinist to copy Vespucci's usage in the *Cosmographiae introductio*. O'Gorman argues that Waldseemüller's *invenio* should be translated as 'conceive', not 'discover', which is to ignore the fact that Waldseemüller is working from a Latin text of Vespucci in which *invenio* is already a translation of *discooperio* (O'Gorman, *The Invention of America* (1961), 123 and n. 117). **8.** See, for example, Wolper, 'The Rhetoric of Gunpowder' (1970). **9.** Watson, *The Double Helix* (1968), 197. **10.** On the discovery of discovery: Fleming (ed.), *The Invention of Discovery* (2011); and Margolis, *It Started with Copernicus* (2002), Ch. 3 – neither explores the new terminology. On curiosity: Huff, *Intellectual Curiosity and the Scientific Revolution* (2011); Harrison, 'Curiosity, Forbidden Knowledge' (2001); Ball, *Curiosity* (2012); Daston, 'Curiosity in Early Modern Science' (1995); and Daston & Park, *Wonders and the Order of Nature* (1998), 303–28. An interesting account of the cultural foundations of modern science is to be found in Muraro, *Giambattista della Porta, mago e scienziato* (1978), 171–9. **11.** Bury, *The Idea of Progress* (1920), 44–9. **12.** Leroy, *Variety of Things* (1594), fol. 127rv. On Le Roy's secular historical philosophy, see Huppert, 'The Life and Works of Louis Le Roy, by Werner L. Gundersheimer' (1968). **13.** For a firm statement of the principle of progress in knowledge based on the new geographical discoveries, see Piccolomini, *De la sfera del mondo* (1540), 39v. **14.** I owe this point to Stuart Carroll. **15.** 1829 comes from a search on Google Books; *OED* gives 1853. **16.** On 'nostalgia', see *OED*. For the first English use: Harle, *An Historical Essay on the State of Physick in the Old and New Testament* (1729) (*OED* gives 1756); for an early French usage of *maladie du pays*, Constantini, *La Vie de Scaramouche* (1695). **17.** I take the term from Dunn, *Modern Revolutions* (1972), 226; for the underlying issues, Skinner, *Visions of Politics*, Vol. 1 (2002), 128–44; and Shapin & Schaffer, *Leviathan and the Air-pump* (1985), 14. **18.** Leroy, *Variety of Things* (1594), sig. A4v. **19.** Leroy, *De la vicissitude* (1575), '*Sommaire de l'oeuvre*'. **20.** Vergil, *On Discovery* (2002); Copenhaver, 'The Historiography of Discovery in the Renaissance' (1978); and Atkinson, *Inventing Inventors in Renaissance Europe: Polydore Vergil's 'De inventoribus rerum'* (2007). **21.** Hay, *Polydore Vergil* (1952), 74. **22.** Zhmud, *The Origin of the History of Science* (2006), 299–301. **23.** Vergil, *A Pleasant and Compendious History* (1686), 149. See also Vergil, *An Abridgement* (1546); and Vergil, *The Works* (1663). **24.** For an introduction, Bodnár, 'Aristotle's Natural Philosophy' (2012); also Kuhn, *The Road since Structure* (2000), 15–20. **25.** See below, 315–17, 530–1. **26.** Thorndike, *A History of Magic and Experimental Science* (1923), Vol. 5, 37–49. **27.** Westman, *The Copernican Question* (2011), 99. **28.** Thorndike,

Science and Thought in the Fifteenth Century (1929), 209 (translation modified). Thorndike had not seen the first edition: Achillini, *De elementis* (1505), 84v–85r. **29.** Thorndike, *Science and Thought in the Fifteenth Century* (1929), 209. **30.** See the extended argument that experience must take precedence over authority, especially in questions of geography, in Piccolomini, *Della grandezza della terra et dell'acqua* (1558), 7v–10r. **31.** Thorndike, *Science and Thought in the Fifteenth Century* (1929), 210. **32.** See, for example, the preface to Book 1 of Machiavelli's *Discourses* (Machiavelli, *Selected Political Writings* (1994), 82–4); Montaigne, *The Complete Essays* (1991), 605–6; and Schmitt, 'Experience and Experiment' (1969) on Zabarella. **33.** Quoted in Eamon, *Science and the Secrets of Nature* (1994), 272. **34.** Thorndike, *A History of Magic and Experimental Science* (1923), Vol. 5, 581–2; and Taisnier, *Opusculum* (1562), 16–17. It is sometimes said that Ramus defended the thesis that everything that Aristotle said is false, but this is a mistranslation: Ong, *Ramus* (1958), 36–46. **35.** Galilei, *Dialogue Concerning the Two Chief World Systems* (1967), 107–8; for a discussion of this and similar examples, and of the adage that it was better to err with Plato/Aristotle/Galen than be right, Maclean, *Logic, Signs and Nature* (2002), 191–3. **36.** Muir, *The Culture Wars of the Late Renaissance* (2007), 15–18; compare Pascal, *Oeuvres* (1923), 9 – Pierre Guiffart on Pascal's experiments. **37.** Glanvill, *Plus ultra* (1668), 65–6. **38.** Harvey, *Anatomical Exercitations* (1653), preface, fol. 4r; and Charleton, *Physiologia Epicuro-Gassendo-Charletoniana* (1654), 183. **39.** Guicciardini, *Maxims and Reflections (Ricordi)* (1972), 76. **40.** Montaigne, *The Complete Essays* (1991), 648. **41.** Montaigne, *The Complete Essays* (1991), 644 (the 1588 text); Borges, *The Total Library* (2001), 'The Doctrine of Cycles' (115–22), 'Circular Time' (225–8), with Vanini 'quoted' on 225 (this is, in fact, a Borgesian improvement on what Vanini actually says: see Vanini, *De admirandis* (1616), 388); and Trompf, *The Idea of Historical Recurrence* (1979). **42.** Zhmud, *The Origin of the History of Science* (2006), 299. **43.** Righter, *Shakespeare and the Idea of the Play* (1962), 15, 23. **44.** Bacon, *Instauratio magna* (1620), Vol. 1 §84, 99 = Bacon, *Works* (1857), Vol. 1, 191; and Browne, *Pseudodoxia epidemica* (1646), 20; see also Pascal, '*Préface sur le traité du vide*' (Pascal, *Oeuvres complètes* (1964), 772–85); and Glanvill, *The Vanity of Dogmatizing* (1661), 140–1. **45.** Johnson, 'Renaissance German Cosmographers' (2006), 34–5. **46.** Alberti, *On Painting and On Sculpture* (1972), 33 (dedication to Brunelleschi; translation modified). **47.** Alberti, *On Painting and On Sculpture* (1972), 57–8; above, p. 58 and n; and Serlio, *Libro primo [-quinto] d'architettura* (1559), Book 2, 1r (1537 is the date of the first edition). **48.** *Discourses on Livy*, introduction to Book 1 (a paragraph missing from the Neville translation of 1675); Book 2, Ch. 17; and Machiavelli, *Art of War*, preface. **49.** Copernicus, *On the Revolutions* (1978), 5; and Gingerich, 'Did Copernicus Owe a Debt to Aristarchus?' (1985). **50.** Rheticus, *Narratio prima* (1540); and Rosen (ed.), *Three Copernican Treatises* (1959), 135. **51.** Digges & Digges, *A Prognostication Everlasting* (1576), fol. 43. **52.** Galilei, *Dialogue Concerning the Two Chief World Systems* (1967), 274, 276, 318, 328. **53.** Eamon, *Science and the Secrets of Nature* (1994); and Long, *Openness,*

Secrecy, Authorship (2001). **54.** Quoted from Minnis, *Medieval Theory of Authorship* (1988), 9. **55.** See Guillaume de Testu's defence of the concept of *chozes nouvelles* in 1556: Lestringant, *L'Atelier du cosmographe* (1991), 187. **56.** Rosenthal, '*Plus ultra, non plus ultra*' (1971); and Rosenthal, 'The Invention of the Columnar Device' (1973). **57.** Randles, 'The Atlantic in European Cartography' (2000), 15. **58.** Galilei, *Le opere* (1890), Vol. 3, 253. **59.** Galilei, *Le opere* (1890), Vol. 15, 155. **60.** Norman, *The New Attractive* (1581), Aiirv. **61.** Lodovico delle Colombe, quoted Wootton, *Galileo* (2010), 7; compare the same author in 1612: Galilei, *Le opere* (1890), Vol. 4, 317. **62.** Galilei, *Le opere* (1890), Vol. 13, 345. **63.** Quoted from Eamon, *Science and the Secrets of Nature* (1994), 272. **64.** Thorndike, *A History of Magic and Experimental Science* (1923), Vol. 7, 430; or, for example, Thevet, *Cosmographie universelle,* 1575: '*en ces matieres cy, les plus sçavans n'y voient pas si clairement, que font les Matelots et ceux qui ont par cy devant long temps voiagé en ces terres, d'autant que l'experience est maistresse de toutes choses*': quoted in Lestringant, *L'Atelier du cosmographe* (1991), 25; see also 27–35, 45–6, 50. **65.** Glanvill, *The Vanity of Dogmatizing* (1661), 140. **66.** On *experientia magistra rerum*, n. 64 above; Gilbert, *Machiavelli and Guicciardini* (1965), 39; Tedeschi, 'The Roman Inquisition and Witchcraft' (1983); Gerson, *Opera omnia* (1706), Vol. 1, 76; and Himmelstein, *Synodicon herbipolense* (1855), 207. Erasmus, however, thought that only fools need to learn from experience: Vaughan, 'An Unnoted Translation of Erasmus in Ascham's "Schoolmaster"' (1977). 67. Cooper, *Inventing the Indigenous* (2007). **68.** Ashworth Jr, 'Natural History and the Emblematic World View' (1990). **69.** 'Discovery': *OED*; 'discover': Münster, *A Treatyse of the Newe India* (1553), sig. H7r; 'voyage of discovery': Bourne, *A Regiment for the Sea* (1574), 35v. **70.** Phillips, 'The English Patent' (1982), 71. **71.** Bacon, *The Advancement of Learning* (1605), 48v = Bacon, *Works* (1857), Vol. 3, 384. For a general discussion of this theme, see Gascoigne, 'Crossing the Pillars of Hercules' (2012). Later, Hooke, in his 'The Present State of Natural Philosophy', sought to give rules for how to make discoveries, and thus ensure that hard work, rather than genius, would be all that was required to advance knowledge: Hooke, *The Posthumous Works* (1705), 1–70. **72.** Serjeantson, 'Francis Bacon and the "Interpretation of Nature" in the Late Renaissance' (2014). **73.** Weeks, 'Francis Bacon and the Art–Nature Distinction' (2007), 105, quoting *Novum organum* CIX (translation modified). **74.** Weeks, 'The Role of Mechanics in Francis Bacon's "Great Instauration"' (2008). Compare della Porta, *Natural Magick* (1658) [1589], 2. **75.** Compare Pascal, *Oeuvres* (1923), 136–41; and above, p. 36. **76.** Galilei, *Le opere* (1890), Vol. 3, 59. **77.** De Bruyn, 'The Classical Silva' (2001). **78.** Fattori, '*La diffusione di Francis Bacon nel libertinismo francese*' (2002). For a discussion of his ideas by Mersenne, see Thorndike, *A History of Magic and Experimental Science* (1923), Vol. 7, 430. **79.** Bartholin, *Anatomicae institutiones* (1611), 449. **80.** Wotton & Bentley, *Reflections upon Ancient and Modern Learning. The Second Part* (1698), 45–6; Thorndike, *A History of Magic and Experimental Science* (1923), Vol. 5, 44–5; and Park, 'The Rediscovery of the Clitoris' (1997). **81.** Laqueur, *Making Sex* (1990). **82.**

Bartholin, *Anatomicae institutiones* (1611), 174. **83.** Gingerich & van Helden, 'From Occhiale to Printed Page' (2003), 251–4. **84.** Galilei & Scheiner, *On Sunspots* (2008). **85.** Kuhn, 'Historical Structure of Scientific Discovery' (1962). **86.** Schaffer, 'Scientific Discoveries' (1986); and Schaffer, 'Making Up Discovery' (1994). **87.** This is the argument of O'Gorman, *The Invention of America* (1961). O'Gorman complicates matters by distinguishing 'discovery' from 'invention' in a peculiar way (9), but his basic claim is that Waldseemüller discovered America (123). **88.** Schaffer, 'Making Up Discovery' (1994), 13. **89.** Broughton, 'The First Predicted Return of Comet Halley' (1985); and Yeomans, Rahe & Freitag, 'The History of Comet Halley' (1986). **90.** There is a difference of opinion as to when Bessel may be said to have made this prediction. Compare Bamford, 'Popper and His Commentators on the Discovery of Neptune' (1996), 216, who says 1823, and Smith, 'The Cambridge Network' (1989), 398–9, who says 1840. Morando, 'The Golden Age of Celestial Mechanics' (1995), 216, has 'after 1835'. **91.** Wittgenstein, *Philosophical Investigations* (1953), §§66–8. **92.** Merton, 'Priorities in Scientific Discovery' (1957); Merton, 'Singletons and Multiples' (1961); Merton, 'Resistance' (1963); and Merton, *The Sociology of Science* (1973) (which collects the earlier articles); see also Lamb & Easton, *Multiple Discovery* (1984); and Stigler, 'Stigler's Law of Eponymy' (1980). **93.** Merton, *On the Shoulders of Giants* (1965); Merton & Barber, *The Travels and Adventures of Serendipity* (2006); and Sills & Merton, *International Encyclopedia of the Social Sciences: Social Science Quotations* (1991). **94.** Koyré, *Études Galiléennes* (1966), 80–158; and Schemmel, *The English Galileo* (2008). **95.** Schaffer, 'Scientific Discoveries' (1986), 400–6. **96.** Hanson, *Patterns of Discovery* (1958), 4–30 (which goes a long way towards stating Kuhn's 'different worlds' thesis); Putnam, *Meaning and the Moral Sciences* (1978), 22–5; and Lehoux, *What Did the Romans Know?* (2012), 226–9. **97.** Burkert, *Lore and Science* (1972), 307. **98.** Galilei, *Le opere* (1890), Vol. 10, 296; see also, for example, 372. **99.** Wilding, *Galileo's Idol* (2014), 108–11. **100.** For *concurrence*, see Leroy, *De la vicissitude* (1575); the *Vocabolario delli Accademici della Crusca* does not give the modern sense of *concorrente* when defining the word but uses it in the modern sense when defining *rivale*. For English usages, see *OED* (also 'emulation', 1552), but for the first use of 'competition', see Stubbes, *The Discoverie of a Gaping Gulf* (1579), E5r. **101.** Quoted from Hobbes, *Examinatio et emendatio* (1660), in Malcolm, 'Hobbes and Roberval' (2002), 164–5 (Malcolm's translation). **102.** Hall, *Philosophers at War* (1980); Bertoloni Meli, *Equivalence and Priority* (1993). On priority disputes: Iliffe, 'In the Warehouse' (1992). **103.** Westfall, *Never at Rest* (1980), 446–53, 471–2, 511–12. **104.** For some reason, Merton, although he started out as an historian of science, never developed the line of argument I am about to present. But see Merton, *Science, Technology and Society* (1970) [1938], 169, n. 30. **105.** Jardine, *The Birth of History and Philosophy of Science* (1984). **106.** Clark & Montelle, 'Priority, Parallel Discovery and Pre-Eminence' (2012). **107.** For a possible earlier example, Van Brummelen, *The Mathematics of the Heavens* (2009), 182. **108.** Hellman, *Great Feuds in Mathematics* (2006); Toscano, *La formula segreta*

(2009). **109.** Biagioli, 'From Ciphers to Confidentiality' (2012). **110.** Mattern, *Galen and the Rhetoric of Healing* (2008); Lehoux, *What Did the Romans Know?* (2012), 6–8, 10–11, 132. **111.** Merton, *The Sociology of Science* (1973), 273–5. **112.** Park, 'The Rediscovery of the Clitoris' (1997). **113.** Ambrose, 'Immunology's First Priority Dispute' (2006). **114.** Serrano, 'Trying Ursus' (2013). **115.** Ruestow, *The Microscope in the Dutch Republic* (1996), 47–8; Cobb, *Generation* (2006), 155–87. **116.** Röslin, *De opere Dei creationis* (1597). (I owe this reference to Adam Mosley.) Compare the three medical systems in Severinus, *Idea medicinae philosophicae* (1571). As far as I can tell, Brotton is wrong to claim that Brahe immodestly named his system after himself: Brotton, *A History of the World in Twelve Maps* (2012), 266. **117.** http://www-history.mcs.st-andrews.ac.uk/Curves/Limacon.html. **118.** On Waldseemüller's map 'America' does not appear to refer to the continent as a whole, but his collaborator Matthias Ringmann, in a book accompanying the map, clearly intended the new name to refer to the continent as a whole, and by 1520 it appears on maps with this meaning: Meurer, 'Cartography in the German Lands, 1450–1650' (2007), 1205. On other maps 'Asia' continued to be used until at least 1537: Rosen, 'The First Map to Show the Earth in Rotation' (1976), 174, reprinted in Rosen, *Copernicus and His Successors* (1995). On the naming of America, Johnson, 'Renaissance German Cosmographers' (2006) – although, unfortunately, she takes eponymy for granted. **119.** The earliest usage of the adjective 'Alphonsine' (in Latin) that I have found is 1483, but there may well be earlier occurrences. **120.** Randles, 'Bartolomeu Dias' (2000), 26. **121.** McIntosh, *The Johannes Ruysch and Martin Waldseemüller World Maps* (2012), 17. **122.** Galilei, *The Essential Galileo* (2008), 46. **123.** On Galileo's success in elevating the Medici among the pagan gods, Aggiunti, *Oratio de mathematicae laudibus* (1627), 20. **124.** Ramazzini & St Clair, *The Abyssinian Philosophy Confuted* (1697). **125.** Bailey, *An Universal Etymological English Dictionary* (1721). **126.** Ippocratista: Siraisi, *Taddeo Alderotti* (1981), 40; Scotista: Gerson, *Opera* (1489), index, s.v. *Distinctionis*; the rest from *OED*. **127.** Little has been written about eponymy, but there is an interesting dictionary of medical eponyms at http://www.whonamedit.com, and Stigler's Law states that the wrong person always get the credit: Stigler, 'Stigler's Law of Eponymy' (1980). When Pascal writes (Pascal, *Oeuvres complètes* (1964), 523), '*[Q]uand nous citons les auteurs, nous citons leurs démonstrations, et non pas leurs noms; nous n'y avons nul égard que dans les matières historiques*,' he is distinguishing between two ways of using an author's name. If one refers to Copernicus (using the name to refer to a book), this is a shorthand way of referring to heliocentrism; but if one refers to the nova of 1604 (a matter of historical fact), then the credibility of its existence depends on the authority of Kepler and other observers of astronomical phenomena. **128.** See *OED*, 'algorism'. **129.** Proclus & Euclid, *In primum Euclidis* (1560), 207; and van Brummelen, *The Mathematics of the Heavens* (2009), 56. **130.** Proclus & Euclid, *In primum Euclidis* (1560), 198, 200. **131.** Proclus & Euclid, *In primum Euclidis* (1560), index (under *admirabile*) – compare 134, 270; Drayton, *Poly-Olbion* (1612), A3rv, offers various accounts of the origin of the name. It is true that there was some uncertainty as to whether Pythagoras was indeed the originator of the theorem: Proclus had been cautious in reporting the attribution (and had gone on to present the theorem as being of limited significance). Vitruvius, *Zehen Bücher* (1548). **132.** Ruby, 'The Origins of Scientific "Law"' (1986), 357. **133.** Devlin, *The Man of Numbers* (2011), 145. **134.** Pascal, *Oeuvres* (1923), 478–95; Dear, *Discipline and Experience* (1995), 186–9. (Koyré found these protestations disingenuous: Koyré, *Études d'histoire de la pensée scientifique* (1973), 378.) Similarly, Pascal claimed that since he had invented the void-in-the-void experiment, he deserved the credit for the discoveries made by others with modified versions of it; for a similar claim made by Leibniz, see Bertoloni Meli, *Equivalence and Priority* (1993), 6. **135.** 'Plagiary': there is a 1585 occurrence in EEBO, unrecorded in *OED*, but this is in the etymologically correct sense of 'kidnapper'. **136.** Browne, *Pseudodoxia epidemica* (1646), 22. **137.** *OED*, s.v. 'Ptolemean'. **138.** Starkey, *Nature's Explication and Helmont's Vindication* (1657). **139.** Bartholin, Walaeus and others, *Bartholinus Anatomy* (1662). **140.** Stubbe, *An Epistolary Discourse Concerning Phlebotomy* (1671). **141.** Dates from *OED* unless otherwise noted; in Latin, *boyliano* originates with Line, *Tractatus de corporum inseparabilitate* (1661). **142.** Harris, *Lexicon technicum* (1704). **143.** Reynolds, *Death's Vision* (1713). **144.** Voltaire, *Letters Concerning the English Nation* (1733). **145.** Zhmud, *The Origin of the History of Science* (2006). **146.** Galilei, *The Essential Galileo* (2008), 45. **147.** Bacon, *Sylva sylvarum* (1627), 45–6 = Bacon, *Works* (1857), Vol. 3, 165–6. **148.** Charleton, *Physiologia Epicuro-Gassendo-Charletoniana* (1654), 3. **149.** Huff, *Intellectual Curiosity and the Scientific Revolution* (2011). **150.** Harris, *Lexicon technicum* (1704). **151.** Phillips, 'The English Patent' (1982); Long, 'Invention, Authorship, "Intellectual Property"' (1991). **152.** May, 'The Venetian Moment' (2002). **153.** Wootton, 'Galileo: Reflections on Failure' (2011); on Baliani, Wallis, 'An Essay of Dr John Wallis' (1666), 270. Wallis thinks that Galileo's theory produces two high tides a day, but see Palmieri, 'Re-examining Galileo's Theory of Tides' (1998), 242. **154.** McGuire & Rattansi, 'Newton and the "Pipes of Pan"' (1966), 109. For a discussion which would have delighted Newton, see Russo, *The Forgotten Revolution* (2004), 365–79.

CHAPTER 4

1. Adams, *The Hitchhiker's Guide* (1986), 15, 274, 463. **2.** Kuhn, 'Dubbing and Redubbing: The Vulnerability of Rigid Designation' (1990), 299. **3.** Kuhn, *Structure* (1970), 171. **4.** Russell, *Inventing the Flat Earth* (1991). **5.** Columbus, *The Four Voyages* (1969), 217–19. **6.** O'Gorman, *The Invention of America* (1961), 98–101. **7.** Biro, *On Earth as in Heaven* (2009); Schuster & Brody, 'Descartes and Sunspots' (2013); also Johnson, *The German Discovery of the World* (2008), 51–7; the first book with 'cosmology' in the title is, I think, Mizauld, *Cosmologia: Historiam coeli et mundi* (1570). (Worldcat shows a couple of earlier entries, but both are likely to be ghosts.) 8. Aristotle, *On the Heavens* (1939). 9. To be exact, the problem seems to have originated with the last of the pagan philosophers, Olympiodorus of Alexandria: Duhem, *Le Système du monde*, Vol. 9 (1958), 97–8. 10. Pliny the Elder, *Natural History* (1938), Book 2, cap. 65; the translator has been unable to make sense of the passage: compare Pliny the Elder, *L'Histoire du monde* (1562). Pliny's claim would seem to be that the distance from the centre of the Earth to the ocean shore is less than the distance from the deepest point of the ocean to the high seas. 11. The key text, on which the following paragraphs are largely based, is Duhem, *Le Système du monde*, Vol. 9 (1958), 79–235 (available online at www.gallica.fr). Since discussions of this subject are usually based on an incomplete acquaintance with the literature, I offer here an attempt at a more complete bibliography, in chronological order, although Rosen, 'Copernicus and the Discovery of America' (1943), is a necessary preliminary; Boffito, *Intorno alla 'Quaestio'* (1902) (which amounts to an anthology of sources) – available online at www.archive.org; Thorndike, *A History of Magic and Experimental Science* (1923), Vol. 4, 161, 166, 176, 233; Vol. 5, 9, 24–5, 156, 321, 389, 427–8, 552–3, 569, 591, 614; Vol. 6, 10, 12, 27, 34, 50, 60, 83, 380; Vol. 7, 50, 54–5, 339, 385, 395–6, 404, 481, 601, 644, 692; Wright, *The Geographical Lore of the Time of the Crusades* (1925), 186–7, 258; Thorndike, *Science and Thought in the Fifteenth Century* (1929), 200–16; Duhem, *Le Système du monde*, Vol. 9 (1958), 79–235 (available online at www.gallica.fr); O'Gorman, *The Invention of America* (1961), especially 56–8; Goldstein, 'The Renaissance Concept of the Earth' (1972); Randles, *De la terre plate au globe terrestre* (1980); Grant, 'In Defense of the Earth's Centrality and Immobility' (1984), 20–32 (the best starting point); Hooykaas, *G. J. Rheticus's Treatise on Holy Scripture and the Motion of the Earth* (1984), 127–32; Margolis, *Patterns, Thinking and Cognition* (1987), 235–43; Russell, *Inventing the Flat Earth* (1991) (although he rather misses the point of Randles' work); Wallis, 'What Columbus Knew' (1992); Vogel, '*Das Problem der relativen Lage von Erd- und Wassersphäre im Mittelalter*' (1993); Randles, 'Classical Models of World Geography' (1994) (reprinted in Randles, *Geography, Cartography and Nautical Science in the Renaissance* (2000)); Grant, *Planets, Stars and Orbs* (1994), 622–37; Vogel, *Sphaera terrae* (1995); Headley, 'The Sixteenth-century Venetian Celebration of the Earth's Total Habitability' (1997); Margolis, *It Started with Copernicus* (2002), 96–102; Besse, *Les Grandeurs de la terre* (2003), 65–110; Vogel, 'Cosmography' (2006); Lester, *The Fourth Part of the World* (2009); Biro, *On Earth as in Heaven* (2009) (the thesis on which Biro's book is based is available online at unsworks.unsw.edu.au/fapi/datastream/unsworks:993/SOURCE02); and Schuster & Brody, 'Descartes and Sunspots' (2013). A convenient point of entry to the medieval debates is provided by Alighieri, *La Quaestio de aqua et terra* (1905) (facsimile and translations), available at www.archive.org (there is also a translation by Philip Wicksteed at http://alighieri.scarian.net/translate_english/alighieri_dante_a_question_of_the_water_and_of_the_land.html). It is remarkable that the relevant issues are not even mentioned in Westman, *The Copernican Question* (2011), although

Westman is familiar with the two works by Grant, and with Goldstein (Margolis, *Patterns, Thinking and Cognition* (1987), 314), and historians of cartography are generally unacquainted with the issues: e.g. Brotton, *A History of the World in Twelve Maps* (2012); Simek, *Heaven and Earth in the Middle Ages* (1996); and Woodward, 'The Image of the Spherical Earth' (1989). The vast bulk of Woodward (ed.), *Cartography in the European Renaissance* (2007) contains three sentences on the subject (59, and 327 – where Randles' argument is misrepresented), but a key contribution, on globes, shows no knowledge of it (136–7). **12.** Oresme, *Le Livre du ciel et du monde* (1968), 397, 562–73. **13.** Duhem, *Le Système du monde*, Vol. 9 (1958), 91–6. In 1505 Alessandro Achillini expressed doubt as to the validity of the traditional ratios, but (if I understand him correctly) he did not go so far as to question the two-spheres theory: Achillini, *De elementis* (1505), 84v–85r. **14.** Hiatt, *Terra incognita* (2008), 100–4. **15.** Thorndike, *The Sphere of Sacrobosco and Its Commentators* (1949), provides text and translation. **16.** There are two listings, which differ significantly: Roberto de Andrade Martins at http://www.ghtc.usp.br/server/Sacrobosco/Sacrobosco-ed.htm; and Hamel, *Studien zur 'Sphaera'* (2014), 68–133. **17.** For example, Taylor, *The Haven-finding Art: A History of Navigation from Odysseus to Captain Cook* (1971), 154; Russell, *Inventing the Flat Earth* (1991), 19; Lester, *The Fourth Part of the World* (2009), 28–9. **18.** Hiatt, *Terra incognita* (2008), 142 (quoting British Library MS Cotton Julius D.VII, 46r (for a photograph of the passage see 123). **19.** Hiatt, *Terra incognita* (2008), 133, quoting Petrarch, *Le familiari (Familiarum rerum libri)*, ed. V. Rossi (4 vols., Florence: Sansoni, 1933–42) Vol. 2, 248. **20.** Wright, *The Geographical Lore of the Time of the Crusades* (1925), 86–7, 259–61; Arim: Oresme, *Le Livre du ciel et du monde* (1968), 24, 330–5; and Sen, 'Al-Biruni on the Determination of Latitudes and Longitudes in India' (1975). **21.** Duhem, '*Un précurseur français de Copernic*' (1909); Duhem, *Le Système du monde*, Vol. 9 (1958), 202–4, 329–44; Sarnowsky, 'The Defence of the Ptolemaic System' (2007), 35–41; Grant, *Planets, Stars and Orbs* (1994), 642–7; and Oresme, *Le Livre du ciel et du monde* (1968). **22.** For the conventional view, see Thorndike, *The Sphere of Sacrobosco and Its Commentators* (1949), 274–5, 296 (the commentary is attributed to Michael Scot). **23.** Johnson, *The German Discovery of the World* (2008), 57–71 (though the thrust of her argument is at odds with mine). **24.** On the first representations of the terraqueous globe, see Helas, '"*Mundus in rotundo et pulcherrime depictus*"' (1998); and Helas, '*Die Erfindung des Globus durch die Malerei – Zum Wandel des Weltbildes im 15. Jahrhundert*' (2010). Medieval representations of a globe, such as the *globus cruciger*, should be understood as representations of either the sphere of earth or the sphere of the heavens – in other words, the universe as a whole (Vogel, *Sphaera terrae* (1995), 360). **25.** Colón, *The Life of the Admiral Christopher Columbus* (1992), 15–40 (the globe is on 19); Dalché, 'The Reception of Ptolemy's Geography' (2007), 329; and Randles, 'The Evaluation of Columbus' "India" Project' (1990). **26.** Besse, *Les Grandeurs de la terre* (2003), 62–3. **27.** Vogel, 'America' (1995), 14. **28.** For example, Guillaume Fillastre, writing 1414–18: 'I say that supposing the shape of the earth [*terra*: the land mass] to be spherical they who live in the furthest parts of the east are antipodeans to those who live in the furthest parts of the west.' Hiatt, *Terra incognita* (2008), 158. **29.** Ezekiel 7:2; Isaiah 11:12. On the habitable earth's four corners, Oresme, *Traitié de l'espère* (1943), Ch. 31. **30.** Donne, Holy Sonnets VII. **31.** Leurechon, *Selectae propositiones* (1629), 19. An advance on the earliest date (1646) known to Randles: Randles, *Geography, Cartography and Nautical Science in the Renaissance* (2000), article 1, 74. The first known English usage (predating the *OED* date of 1658) is Charleton, *The Darkness of Atheism Dispelled* (1652), 8. **32.** Trutfetter, *Summa in tota[m] physicen* (1514), Book 2, Ch. 2 (sig. liii–miiv). In Trutfetter, *Summa philosophiae naturalis* (1517), an abbreviated version of the previous edition, the issue is not addressed. **33.** *Habes lector* was reprinted in 1518, 1522 and 1557, but it also appears in the apparatus to Pomponius Mela, *De orbis situ* (1518, 1522, 1530, 1540, 1557), sometimes bound separately and catalogued as a separate publication. See Randles, 'Classical Models of World Geography' (1994) (reprinted in Randles, *Geography, Cartography and Nautical Science in the Renaissance* (2000)), 66–7 (note that the key passage from Vadianus quoted on 67 varies in different editions – compare Agricola & Vadianus, *Habes lector* (1515), sig. B iii(r) with Mela, *De orbis situ libri tres. Adiecta sunt praeterea loca aliquot ex Vadiani commentariis* (1530), sig. X5v). **34.** Mela, *De orbis situ libri tres. Adiecta sunt praeterea loca aliquot ex Vadiani commentariis* (1530), V2v, V3r, V4r, X2r, X6r, Y3v. On Tannstetter's role as editor of the 1518 Sacrobosco, see Hayton, 'Instruments and Demonstrations' (2010), 129. For an illustration from 1524, see Margolis, *Patterns, Thinking and Cognition* (1987), 236. Oronce Fine's 1528 *La Theorique des cielz*, which is not a commentary on Sacrobosco, also adopts the new image of the globe: Cosgrove, 'Images of Renaissance Cosmography' (2007), 62–3. **35.** Similar views had already been expressed by Fernández de Enciso (1519), Margalho (1520) and Fernel (1528): Randles, 'Classical Models of World Geography' (1994), 65–9. **36.** Gingerich, 'Sacrobosco as a Textbook' (1988). **37.** Hamel, *Studien zur 'Sphaera'* (2014), 42–50. **38.** Gingerich, 'Sacrobosco Illustrated' (1999), 213–14. **39.** I have seen the later reprint: Sacrobosco, *Sphaera ... in usum scholarum* (1647). **40.** See, for example, Beyer, *Quaestiones novae* (1551) and Sacrobosco, *Sphaera* (1552) (I have used the edition of 1601). Piccolomini, *La Prima parte delle theoriche* (1558) claims to be new and shocking, but he is evidently writing for an inexpert audience, and he treats the one contemporary he has found upholding the old views with contempt. **41.** Schott, *Anatomia physico-hydrostatica* (1663). Similar issues are discussed in Carpenter, *Geographie Delineated* (1635). **42.** Berga & Piccolomini, *Discorso* (1579); and Benedetti, *Consideratione* (1579); the two texts were then published together in Latin (Berga's text being translated by another because Berga was already dead or dying): Berga & Benedetti, *Disputatio* (1580). (The texts have separate title pages but continuous pagination.) **43.** Madeleine Alcover claims (Cyrano de Bergerac, *Les États et empires de la lune et du soleil* (2004), 27), on the basis of an article by Maurice Laugaa, that Vincent Leblanc denied the existence of a second hemisphere in 1634. I have not seen Laugaa's article; but Leblanc's text, at least in translation, does not support the claim: Leblanc, *The World Surveyed* (1660), 171–3. **44.** Bataillon, '*L'Idée de la découverte de l'Amérique*' (1953), 31. The phrase originates with Peter Martyr in 1493: O'Gorman, *The Invention of America* (1961), 84–5. Columbus himself held that his third voyage had discovered a new world, even though his first two voyages had discovered part of Asia: O'Gorman, *The Invention of America* (1961), 94–104. Equally, the new lands were sometimes said to be *extra orbem*, outside the sphere of land: Randles, '*Le Nouveau Monde*' (2000), 31. **45.** Bodin, *Universæ naturæ theatrum* (1596), 183–93; Bodin, *Le Théatre de la nature universelle* (1597), 252–65; Blair, *Annotations in a Copy of Jean Bodin, Universae naturae theatrum* (1990). **46.** Schott, *Anatomia physico-hydrostatica* (1663), 245–8. **47.** Cesari, *Il trattato della sfera* (1982), 144–7. **48.** Copernicus, *On the Revolutions* (1978). **49.** Rosen (ed.), *Three Copernican Treatises* (1959). **50.** Goldstein, 'The Renaissance Concept of the Earth' (1972), is the key text, echoed in Grant, 'In Defense of the Earth's Centrality and Immobility' (1984), 27 n. 90 and Grant, *Planets, Stars and Orbs* (1994), 636 n. 66. **51.** Rosen, 'Copernicus and the Discovery of America' (1943). **52.** Compare Swerdlow's translation (Swerdlow, 'The Derivation and First Draft of Copernicus's Planetary Theory' (1973), 444): 'And thus the earth rotates together with the water that flows around it and the nearby air'; Swerdlow's translation depends on emending *circumfluis* as *circumflua*, against the manuscript evidence. **53.** For the apple, Mela, *De orbis situ libri tres. Adiecta sunt praeterea loca aliquot ex Vadiani commentariis* (1530), X5(v); Gaspar Peucer, *Elementa doctrinae* (1551), quoted in Besse, *Les Grandeurs de la Terre* (2003), 110; and Hooykaas, *G. J. Rheticus's Treatise on Holy Scripture and the Motion of the Earth* (1984), 86, 128–31. Note, too, that *circumfluere* is the verb used in this context by Vadianus. Rheticus had evidently read Vadianus, and perhaps he learned of this text from Copernicus: Hooykaas, *G. J. Rheticus's Treatise on Holy Scripture and the Motion of the Earth* (1984), 87. **54.** If this is correct, Swerdlow's commentary misleads in certain respects. Thus Copernicus's second postulate cannot be taken to be simply a consequence of postulates 3 and 6, and Copernicus's phrase *centrum gravitatis* means more than just 'center towards which heavy things move' (Swerdlow, 'The Derivation and First Draft of Copernicus's Planetary Theory' (1973), 437–8). Similarly, the claim that 'to understand Copernicus's work properly, as he understood it, one must completely remove it from natural philosophy . . . and terrestrial physics' (440) cannot be right. **55.** Swerdlow takes the earliest possible date to be 1500. For a series of strong arguments for a date *c.*1508 (but omitting any reference to the crucial evidence discussed here), see Goddu, 'Reflections on the Origin of Copernicus's Cosmology' (2006). Goddu (37–8, following Rosen) discusses Corvinus's poem. **56.** Digges & Digges, *A Prognostication Everlasting* (1576), M2r. **57.** Swerdlow, 'The Derivation and First Draft of Copernicus's Planetary Theory' (1973), 425–9. **58.** Besse, *Les Grandeurs de la Terre* (2003), 91–6. **59.** Copernicus, *On the Revolutions* (1978), 4. **60.** Shank, 'Setting up Copernicus?' (2009). **61.** A set of arguments against a moving earth are already to be found in Albert of Saxony, who is responding to Oresme: Sarnowsky, 'The Defence of the Ptolemaic System' (2007),

35–8. **62.** Swerdlow, 'The Derivation and First Draft of Copernicus's Planetary Theory' (1973), 425, 442, 474, 477. For a recent discussion of these issues (one which at least acknowledges the work of Margolis), see Clutton-Brock, 'Copernicus's Path to His Cosmology' (2005), 209 and n. 27 (echoed in Goddu, 'Reflections on the Origin of Copernicus's Cosmology' (2006), n. 55) **63.** Rheticus, *Narratio prima* (1540), D3v, D4v; Rosen (ed.), *Three Copernican Treatises* (1959), 14 ('like a ball on a lathe'), 149; Calcagnini, *Opera aliquot* (1544), 389 (where the surrounding elements serve to turn the Earth into a perfect sphere, *pilae absolutae rotunditatis*); and Hooykaas, *G. J. Rheticus's Treatise on Holy Scripture and the Motion of the Earth* (1984), 49 (*totum globum ex terrâ et aquâ, cum adiacentibus elementis*), 54–5. **64.** On Bruno in England, Massa, 'Giordano Bruno's Ideas in Seventeenth-century England' (1977); McMullin, 'Giordano Bruno at Oxford' (1986); Ciliberto & Mann (eds.), *Giordano Bruno, 1583–85* (1997); Feingold, 'Giordano Bruno in England, Revisited' (2004); and Rowland, *Giordano Bruno* (2008), 139–87. **65.** Rowland, *Giordano Bruno* (2008), 145–6. **66.** McNulty, 'Bruno at Oxford' (1960), 302–3. **67.** Goldstein, 'Theory and Observation' (1972), 43. The question of whether Copernicus's initial motivation in adopting heliocentrism was to avoid equants or to fix the order of the planets is a vexed one: see Westman, 'The Copernican Question Revisited' (2013). In favour of the view that the key issue was equants, it seems to me, is the evidence discussed in the next paragraph. **68.** Gingerich, *An Annotated Census* (2002); see also Gingerich & Westman, 'The Wittich Connection' (1988); and Gingerich, *The Book Nobody Read* (2005). **69.** Bruno, *The Ash Wednesday Supper* (1995). Bruno's English world is wonderfully described in Bossy, *Giordano Bruno and the Embassy Affair* (1991), but its central claim, that Bruno was a spy, needs to be corrected in the light of Bossy, *Under the Molehill* (2001). **70.** Rowland, *Giordano Bruno* (2008), 149–59. **71.** Copernicus, *On the Revolutions* (1978), 16. **72.** Grant, *Planets, Stars and Orbs* (1994), 395–403. **73.** Singer & Bruno, *Giordano Bruno* (1950); Gatti, 'Bruno and the Gilbert Circle' (1999). **74.** Koyré, *From the Closed World to the Infinite Universe* (1957), 6–23; Montaigne, *The Complete Essays* (1991), 505; and Montaigne, *Oeuvres complètes* (1962), 429. **75.** Redondi, '*La nave di Bruno e la pallottola di Galileo*' (2001); Granada, 'Aristotle, Copernicus, Bruno' (2004). **76.** McMullin, 'Bruno and Copernicus' (1987), who criticizes Yates, *Giordano Bruno and the Hermetic Tradition* (1991); on which, see also Westman & McGuire, *Hermeticism and the Scientific Revolution* (1977); Gatti, *Essays on Giordano Bruno* (2011), Ch. 2. **77.** It was first named in a radio programme, and in print the next year. **78.** Digges & Digges, *A Prognostication Everlasting* (1576). **79.** Johnson & Larkey, 'Thomas Digges, the Copernican System' (1934). **80.** This was already apparent to Dreyer, although he had seen only the 1592 edition: Dreyer, *History of the Planetary Systems* (1906), 347. **81.** Duhem, *Le Système du monde*, Vol. 10 (1959), 247–347; and Koyré, *From the Closed World to the Infinite Universe* (1957), 6–24. **82.** Digges, *Alae* (1573); Pumfrey, 'Your Astronomers and Ours Differ Exceedingly' (2011). **83.** Westman, *The Copernican Question* (2011). **84.** Swerdlow, 'Copernicus and Astrology' (2012), 373. Rather oddly, Swerdlow appears to hold both that there are equants in Copernicus and that getting rid of equants was the main motivation for Copernicus's adoption of heliocentrism (Westman, 'The Copernican Question Revisited' (2013), 104–15). **85.** Ragep, 'Copernicus and His Islamic Predecessors' (2007); Saliba, *Islamic Science and the Making of the European Renaissance* (2007), 193–232. **86.** Although Oresme had already developed such a theory with considerable care: Oresme, *The 'Questiones de Spera'* (1966), Q. 8, and Oresme, *Le Livre du ciel et du monde* (1968), 518–39: he may be a common source for Digges and Bruno. **87.** This argument continued to be important until the mid-seventeenth century; Riccioli thought it was the key argument against Copernicanism: Graney, 'The Work of the Best and Greatest Artist' (2012); and Graney, 'Science Rather than God' (2012). It was strengthened by the fact that telescope lenses turn stars from points into discs, so that, if Copernicanism required them to be at a great distance, the telescope required them to be of an even more enormous size. Flamsteed, for example, thought that some stars were as much bigger than the sun (itself now assumed to be a star) as the sun is than the Earth: Hunter, 'Science and Astrology' (1995), 280. **88.** Johnson & Larkey, 'Thomas Digges, the Copernican System' (1934), 102, and (for a reference in the earlier *Alae*) 111; and Digges & Digges, *A Prognostication Everlasting* (1576), M2r, N4r. **89.** Palingenius, *The Zodiake of Life* (1565); Koyré, *From the Closed World to the Infinite Universe* (1957), stresses the importance of Digges (35–9), but is aware of his close dependence on Palingenius (24–7, 38–9). The dark star is already in Oresme: Oresme, *Le Livre du ciel et du monde* (1968), 515. **90.** Harvey, *Gabriel Harvey's Marginalia* (1913), 161. **91.** Bacchelli, 'Palingenio' (1999); see also, Palingenius, *The Zodiake of Life* (1947); and Granada, 'Bruno, Digges, Palingenio' (1992). **92.** Ariew, 'The Phases of Venus before 1610' (1987), assumes that the heavenly bodies must a) shine by their own light, b) be translucent or c) reflect light. The fourth option, that they could be 'dark', does not appear in his discussion. **93.** Benedetti, *Diversarum speculationum* (1585), 195. Dreyer, *History of the Planetary Systems* (1906), 350, does not fully understand this passage. I have not seen it discussed elsewhere (it is not discussed, for example, in Di Bono, '*L'astronomia Copernicana nell'opera di Giovan Battista Benedetti*' (1987), where it is mistakenly claimed that Benedetti thinks the moon and the Earth are similiar bodies; the passage quoted on 293–4 is not at odds with the interpretation presented here: '*if* the earth were to shine like the sun . . .' – but it doesn't). **94.** Gatti, 'Bruno and the Gilbert Circle' (1999). Gilbert, *De mundo nostro sublunari philosophia nova* (1651), 173. **95.** Pumfrey, 'The Selenographia of William Gilbert' (2011); and Bacon, *Works* (1857), Vol. 2, 80. **96.** Pumfrey, 'Your Astronomers and Ours Differ Exceedingly' (2011).

PART TWO

1. Bartholin, *The Anatomical History* (1653), 127.

CHAPTER 5

1. Galilei, *Le opere* (1890), Vol. 6, 232; translation from Sharratt, *Galileo: Decisive Innovator* (1994), 140. 2. Gleeson-White, *Double Entry* (2011). 3. Galilei, *Dialogue Concerning the Two Chief World Systems* (1967), 207–8. 4. For an historiographical survey, see Baldasso, 'The Role of Visual Representation' (2006). 5. On the dating, see Kemp, *The Science of Art* (1990), 9; Camerota, *La prospettiva del Rinascimento* (2006), 60; Tanturli, '*Rapporti del Brunelleschi con gli ambienti letterari fiorentini*' (1980), 125. 6. White, *The Birth and Rebirth of Pictorial Space* (1987), 119; (and, for a cautionary note) Raynaud, *L'Hypothèse d'Oxford* (1998), 7. Manetti, *Vita di Filippo Brunelleschi* (1992). (The description is largely reproduced in White, *The Birth and Rebirth of Pictorial Space* (1987), 113–17). 8. Key texts are Edgerton, *The Renaissance Rediscovery of Linear Perspective* (1975); Arnheim, 'Brunelleschi's Peepshow' (1978); Kemp, 'Science, Non-Science and Nonsense' (1978); and Kubovy, *The Psychology of Perspective and Renaissance Art* (1986). 9. cf. Leonardo, '[I]t is impossible for a painting to look as rounded as a mirror image . . . except if you look at both with one eye only.' Quoted from Gombrich, *Art and Illusion* (1960), 83. 10. 'Realism' and 'naturalism' take a number of different forms in art (see, for example, Smith, 'Art, Science and Visual Culture in Early Modern Europe' (2006); Smith, *The Body of the Artisan* (2006); and Ackerman, 'Early Renaissance "Naturalism" and Scientific Illustration' (1991)). What seems to me particularly important is what Ivins called 'a rigorous two-way, or reciprocal, metrical relationship between the shapes of objects as definitely located in space and their pictorial representations' (Ivins, *On the Rationalization of Sight* (1975), 9). A correspondence theory of truth is already to be found in Aquinas (*De veritate*, Q.1, A.1–3; cf. *Summa theologiae*, Q.16); although some would claim there are classical antecedents, I do not find them convincing. We will return to the question of 'external' reality. 11. Yiu, 'The Mirror and Painting' (2005), is helpful here. Schechner, 'Between Knowing and Doing' (2005), who works from surviving mirrors, would seem unduly pessimistic regarding the quality of mirrors in view of the material discussed by Yiu. 12. Vasari, *Lives of the Artists* (1965); Alberti, *On Painting and On Sculpture* (1972) (Latin and English); and Alberti, *On Painting* (1991) (English only). 13. Tanturli, '*Rapporti del Brunelleschi con gli ambienti letterari fiorentini*' (1980). 14. Belting, *Florence and Baghdad* (2011). 15. Raynaud, *L'Hypothèse d'Oxford* (1998). 16. Boccaccio quoted from Gombrich, *Art and Illusion* (1960), 53. 17. Hahn, 'Medieval Mensuration' (1982). 18. See the appendix in Kemp, *The Science of Art* (1990), 344–5; and Camerota, *La prospettiva del Rinascimento* (2006), 63–7. 19. Additional elements are the representation of the three dimensions in two on astrolabes (Aiken, 'The Perspective

Construction of Masaccio's *Trinity* Fresco' (1995)), sundials (Lynes, 'Brunelleschi's Perspectives Reconsidered' (1980)) and Ptolemy's third method for mapping the Earth on a flat surface (Edgerton, *The Heritage of Giotto's Geometry* (1991), 152–3). 20. Filarete, *Trattato di architettura* (1972) (also on the Web at http://fonti-sa.sns.it/TOCFilareteTrattatoDiArchitettura.php). 21. Melchior-Bonnet, *The Mirror* (2002), 18–19. 22. Gombrich, *Art and Illusion* (1960), 5. Gombrich is often misinterpreted: for a careful analysis, see Bertamini & Parks, 'On What People Know about Images on Mirrors' (2005). Although several authors have discussed this issue in this context (for example, Lynes, 'Brunelleschi's Perspectives Reconsidered' (1980), 89), only Rotman illustrates the effect of this in his representation: Rotman, *Signifying Nothing* (1993), 15. Strangely, Camerota's simulation, which appears to use a mirror, does not show this effect, and he seems to assume in the text that the mirror image and the original scene would appear to be the same size: Camerota, *La prospettiva del Rinascimento* (2006), 62. I can only assume that his images are not mirror images but printed reproductions, and that they are misleading. 23. The established view is that the Latin text preceded the Italian one, in which case Alberti apparently withdrew the claim in a text intended to be read by Brunelleschi (which might support my case). On the other hand, recently, scholars have argued that the Italian text preceded the Latin one, in which case Alberti apparently added the claim, perhaps after discussions with Brunelleschi (which might tell against my case): see Alberti, *On Painting* (2011). 24. Alberti, *De pictura* §§31, 32: Latin in Alberti, *On Painting and On Sculpture* (1972); Italian in Alberti, *De pictura* (1980) (available on the Web). This revision is not noted in Alberti, *On Painting* (2011), whose source text, strangely, is not the Italian (supposed) original but the Basle Latin *editio princeps*. 25. Panofsky, *Perspective as Symbolic Form* (1991), 75–6 n. 3. 26. Camerota, *La prospettiva del Rinascimento* (2006), 66–7. 27. Field, *The Invention of Infinity* (1997), 43–61. 28. Vasari, *Lives of the Artists* (1965), 136. 29. Niceron, *La Perspective curieuse* (1652). See Massey, *Picturing Space* (2007). 30. Mackinnon, 'The Portrait of Fra Luca Pacioli' (1993). 31. Vergil, *On Discovery* (2002), 245. 32. Baxandall, *Painting and Experience in Fifteenth-century Italy* (1972). 33. Gleeson-White, *Double Entry* (2011). I hope to return elsewhere to the influence of double-entry bookkeeping on ideas of rationality in the early modern period. 34. Panofsky, *Perspective as Symbolic Form* (1991), 143: translating Palladio, Panofsky remarks that '*orizzonte* ... in the older terminology always means "vanishing point".' 35. Alberti, *On Painting* (1991), 54 (§19). 36. Hintikka, 'Aristotelian Infinity' (1966); it is instructive to read Charleton, *Physiologia Epicuro-Gassendo-Charletoniana* (1654), 62–71, which atttempts to formulate a concept of space. 37. Rotman, *Signifying Nothing* (1993). 38. Vitruvius Pollio, *De architectura* (1521). Koyré regarded the concept of infinity as the key distinction between Aristotelian and modern physics: Koyré, *Études d'histoire de la pensée scientifique* (1973), 165. 39. Moffitt, *Painterly Perspective and Piety* (2008); Parronchi, '*Un tabernacolo brunelleschiano*' (1980). 40. Song of Songs 4:12. 41. Edgerton, *The Heritage of Giotto's Geometry* (1991), 108–47; Long, 'Power, Patronage and the Authorship of Ars' (1997); Galluzzi,

The Art of Invention (1999); Ackerman, 'Art and Science in the Drawings of Leonardo da Vinci' (2002); Lefèvre, 'The Limits of Pictures' (2003); and Long, 'Picturing the Machine' (2004). **42.** Chapman, 'Tycho Brahe in China' (1984). **43.** Thorndike, *A History of Magic and Experimental Science* (1923), Vol. 5, 498–514. **44.** Carpo, *Architecture in the Age of Printing* (2001), 16–22. Cunningham, *The Anatomical Renaissance* (1997), which argues that Renaissance anatomy is a continuation of classical anatomy, misses the fundamental transformation which resulted from the mechanical reproduction of illustrations. For an excellent case study of the difficulty of transmitting visual information within a manuscript culture, see Eagleton, 'Medieval Sundials and Manuscript Sources' (2006). **45.** Ogilvie, *The Science of Describing* (2008); and Kusukawa, 'The Sources of Gessner's Pictures for the *Historia animalium*' (2010). **46.** Quoted from Ackerman, 'Early Renaissance "Naturalism" and Scientific Illustration' (1991), 202. **47.** Ivins, *Prints and Visual Communication* (1953), is the classic text. Not everyone agreed with Fuchs and Vesalius on the value of images: Kusukawa, *Picturing the Book of Nature* (2011), 124–31, on opposition to Fuchs, and 233–7 on opposition to Vesalius. **48.** Swerdlow, 'Montucla's Legacy' (1993), 299; Byrne, 'A Humanist History of Mathematics?' (2006). **49.** Swerdlow, 'Montucla's Legacy' (1993), 299. **50.** Swerdlow, 'Montucla's Legacy' (1993), 188 (translation modified). **51.** Wootton, *Galileo* (2010), 22, 138, 165–6, 210. Compare Boyle, below, p. 416. Thus in its first phase the Scientific Revolution amounts to a rediscovery of Greek mathematical science: Russo, *The Forgotten Revolution* (2004). **52.** The original text is reproduced in Jervis, *Cometary Theory in Fifteenth-century Europe* (1985), 170–93, along with her translation, 96–112. **53.** Jervis, *Cometary Theory in Fifteenth-century Europe* (1985), 108–10. **54.** Bennett, *The Divided Circle* (1987). **55.** Regiomontanus is misleadingly described as generalizing from Ptolemy's method of calculating the distance of the moon by Barker & Goldstein, 'The Role of Comets in the Copernican Revolution' (1988), 311. Ptolemy's method involved only one measurement, not two: van Helden, *Measuring the Universe* (1985), 16; and Newton, 'The Authenticity of Ptolemy's Parallax Data – Part 1' (1973). They may well be right that Regiomontanus's method and the idea of applying it to comets had already occurred to Levi ben Gerson, but this part of his work was not known in the Renaissance. **56.** Jervis, *Cometary Theory in Fifteenth-century Europe* (1985), 114–20. **57.** Jervis, *Cometary Theory in Fifteenth-century Europe* (1985), 125. **58.** Gingerich, 'Tycho Brahe and the Nova of 1572' (2005). **59.** Barker & Goldstein, 'The Role of Comets in the Copernican Revolution' (1988), argues that this is an oversimplification, in that an alternative theory of comets, which saw them as lenses focusing the sun's rays, had been propounded, and this theory was agnostic as to the location of comets. But, in the first place, this theory did not provide an adequate account of change in the heavens, and, in the second, had it provided an account of the path of comets through the heavens, that account would have been incompatible with the crystalline-spheres theory. They are right to argue that cometary theory is not the cause of Copernicanism (as I have argued, the one-sphere theory of the Earth is a crucial precondition), and right to argue that Copernicanism itself preserves much of the old astronomy; wrong to argue that it would have been possible to continue to make ad hoc adjustments to the Aristotelian–Ptolemaic system to account for cometary parallax, and wrong to claim that the idea of a consistent cosmological system is itself new with Kepler and Galileo. **60.** Gingerich & Voelkel, 'Tycho Brahe's Copernican Campaign' (1998). **61.** There is a French translation: Brahe, *Sur des phénomènes plus récents du monde éthéré, livre second* (1984). On Brahe, Thoren, *Lord of Uraniborg* (2007); Mosley, *Bearing the Heavens* (2007); and Christianson, *On Tycho's Island* (2000); on the comet, Hellman, *The Comet of 1577* (1971). **62.** Donahue, *The Dissolution of the Celestial Spheres* (1981); Randles, *The Unmaking of the Medieval Christian Cosmos* (1999); and Lerner, *Le Monde des sphères* (1997). **63.** I am grateful to Christopher M. Graney for confirming this to me. On the decisive role of the telescope in resolving philosophical and astronomical debates, Aggiunti, *Oratio de mathematicae laudibus* (1627), 20; naturally, in view of the condemnation of Copernicanism in 1616, he avoids going into detail. **64.** Bogen & Woodward, 'Saving the Phenomena' (1988). **65.** Klein, *Statistical Visions in Time* (1997), 149–51. **66.** Hellman, 'A Bibliography of Tracts and Treatises on the Comet of 1577' (1934); and Hellman, 'Additional Tracts on the Comet of 1577' (1948). **67.** Eisenstein, *The Printing Press as an Agent of Change* (1979); Estienne, *The Frankfurt Book Fair* (1911); and Kepler in Jardine, *The Birth of History and Philosophy of Science* (1984), 277–80. **68.** Barker, 'Copernicus, the Orbs and the Equant' (1990) (and see his note 4 for earlier literature). **69.** See Cohen, *The Scientific Revolution: A Historiographical Inquiry* (1994), 59–97; and Cohen, *How Modern Science Came into the World* (2010), xvii–xviii, 201. The phrase originates with Koyré, 'Galileo and the Scientific Revolution of the Seventeenth Century' (1943), 347, although it is only a restatement of ideas present in the *Études* of 1939. As it happens, it had already been used in Needham, *The Sceptical Biologist* (1929), 91. **70.** Wootton, *Galileo* (2010), 58. A similar argument is already found in Calcagnini, *Opera aliquot* (1544), 389. **71.** See above, p. 164. **72.** Vergil, *On Discovery* (2002), Bk 1, Ch. 18, para. 3. **73.** Panofsky, *Perspective as Symbolic Form* (1991), 57–8. **74.** Hale, 'The Early Development of the Bastion' (1965); Henninger-Voss, 'Measures of Success' (2004); and Gerbino & Johnston, *Compass and Rule* (2009), 31–44. **75.** *Othello*, I, i, 19. **76.** Alberti, *On Painting* (2011). **77.** Cuomo, 'Shooting by the Book' (1997). **78.** Brook, *Vermeer's Hat* (2008), 102. **79.** For example, Edgerton, *The Renaissance Rediscovery of Linear Perspective* (1975), 91–123. **80.** Wootton, *Bad Medicine* (2006), 73–93. **81.** For example, Harley, 'Maps, Knowledge and Power' (2001). **82.** Donne, 'First Anniversary', ll. 278–82. **83.** Parker, *The Army of Flanders* (1972), 42–90; Hale, 'Warfare and Cartography' (2007). **84.** Cipolla, *European Culture and Overseas Expansion* (1970). **85.** For example, Long, 'Power, Patronage and the Authorship of Ars' (1997). **86.** Jesseph, 'Galileo, Hobbes and the Book of Nature' (2004), 193. For further praise of mathematics, see, for example, the inaugural lecture of Galileo's pupil Niccolò Aggiunti (unconvincingly attributed to Galileo by Peterson, *Galileo's Muse* (2011)): Aggiunti, *Oratio de mathematicae laudibus* (1627). **87.** Tuck, 'Optics and Sceptics' (1988).

CHAPTER 6

1. 1610 from Georg Kepler's letter to Kepler: see Kepler, *The Six-cornered Snowflake* (1966), 65 n. 1. **2.** Kepler, *The Six-cornered Snowflake* (2010), 99. **3.** Kepler, *The Six-cornered Snowflake* (2010), 31. Kepler apparently has in mind a creature smaller than the scabies mite: Kepler, *L'Étrenne* (1975), 88 n. 21. **4.** Kepler, *Kepler's Conversation with Galileo's Sidereal Messenger* (1965), 9–11. **5.** Kepler, *Dissertatio cum Nuncio sidereo* (1993) (for an English text, Kepler, *Kepler's Conversation with Galileo's Sidereal Messenger* (1965)). **6.** Mario Biagioli has claimed that Galileo *had* seen a telescope. His evidence is a letter by Sarpi to Francesco Castrino (21 July 1609) in which Sarpi says a telescope has arrived 'in Italy'. 'In Italy' need not mean 'in Venice', as is apparent from the context (Biagioli, 'Did Galileo Copy the Telescope? A "New" Letter by Paolo Sarpi' (2010)). Bucciantini, Camerota & others, *Galileo's Telescope* (2015), 35–6, accept Biagioli's line of argument at least in so far as they claim, as a fact, that Sarpi had handled a telescope before writing this letter, but this goes beyond what Sarpi says. Sarpi's 'new' letter was first published in 1833. Biagioli puzzles over why Favaro did not include it in Galileo's *Opere*. The explanation is simple: Favaro read it as a news report, not a description of Sarpi's personal experience; on this reading (which seems to me perfectly legitimate), it is irrelevant to the question of Galileo's knowledge of the telescope. **7.** Wootton, *Galileo* (2010), 87–92; van Helden, 'The Invention of the Telescope' (1977). **8.** Alexander, 'Lunar Maps and Coastal Outlines' (1998); and Pumfrey, 'Harriot's Maps of the Moon' (2009). **9.** Wootton, *Galileo* (2010), 130. **10.** Freedberg, 'Art, Science and the Case of the Urban Bee' (1998), at 298; Power, *Experimental Philosophy* (1664), had few and poor illustrations. **11.** Wootton, *Bad Medicine* (2006), 110–38; Wilson, *The Invisible World* (1995); Ruestow, *The Microscope in the Dutch Republic* (1996). **12.** Plutarch, 'The Face of the Moon' (1957), §§21–2, 133–49. **13.** Kepler, *Kepler's Somnium* (1967); Kepler, *Kepler's Dream* (1965). See Aït-Touati, *Fictions of the Cosmos* (2011), 17–44; and Campbell, *Wonder and Science* (1999), 133–43. **14.** Quoted from Wootton, *Galileo* (2010), 65. **15.** Kepler, *Kepler's Conversation with Galileo's Sidereal Messenger* (1965), 11, 34–9, 44–5; Campanella to Galileo, 13 Jan. 1611, Galilei, *Le opere* (1890), Vol. 11, 21–2. **16.** Donne, *Devotions Upon Emergent Occasions* (1624), 98–9. **17.** Kepler, *Kepler's Conversation with Galileo's Sidereal Messenger* (1965), *passim*, but especially 38: 'Therefore, Galileo, you will not envy our predecessors their due praise. What you report as having been quite recently observed by your own eyes, they predicted, long before you, as necessarily so.' **18.** Alexander, 'Lunar Maps and Coastal Outlines' (1998), 346–7. **19.** Gingerich & van Helden, 'From Occhiale to Printed Page' (2003), 260–1. **20.** My discussion here and in the following paragraphs draws extensively on Palmieri, 'Galileo and the

Discovery of the Phases of Venus' (2001). **21.** Shank, 'Mechanical Thinking' (2007), 22–6, on Regiomontanus; Ragep, 'Copernicus and His Islamic Predecessors' (2007), discusses the Islamic tradition, which he characterizes as a failure (71). **22.** Galilei, *Le opere* (1890), Vol. 10, 483. **23.** Galilei, *Le opere* (1890), Vol. 10, 409–505, Vol. 11, 11–12; Kepler, *Dioptrice* (1611), 11–12; and Kepler, *Dioptrice* (1611), 21–3. **24.** Lattis, *Between Copernicus and Galileo* (1994), 199–202. **25.** Lattis, *Between Copernicus and Galileo* (1994), 186–93. **26.** Lattis, *Between Copernicus and Galileo* (1994), 193–5; Lattis does not seem to have grasped that no one (except perhaps Galileo) had yet seen Venus 'as a circle like the full moon' – superior conjunction had last occurred in May 1610 and would not reoccur until December 2011. **27.** Lattis, *Between Copernicus and Galileo* (1994), 205–16. **28.** Galilei & Scheiner, *On Sunspots* (2008), 173. **29.** Galilei & Scheiner, *On Sunspots* (2008), 93. **30.** Galilei & Scheiner, *On Sunspots* (2008), 196. **31.** Galilei, *Le opere* (1890), Vol. 11, 177; Galilei & Scheiner, *On Sunspots* (2008), 265. **32.** Hooke, *Micrographia, or, Some Physiological Descriptions of Minute Bodies* (1665), 234. **33.** Milton, *Paradise Lost*, Book 2, 1052; Pascal, *Pensées*, no. 199; and Locke, *An Essay* (1690), 277, 296. **34.** Ball, *Curiosity* (2012), 215–55; and Cressy, 'Early Modern Space Travel' (2006). **35.** Act II, scene 4. The text was not published until 1623. **36.** Godwin, *The Man in the Moone* (2009). **37.** Empson, *Essays on Renaissance Literature* (1993), 220–54; Aït-Touati, *Fictions of the Cosmos* (2011), 45–55; Campbell, *Wonder and Science* (1999), 155–71; and Campbell, 'Speedy Messengers' (2011). **38.** Aït-Touati, *Fictions of the Cosmos* (2011), 56–63; and Chapman, 'A World in the Moon – Wilkins and His Lunar Voyage of 1640' (1991). **39.** Cyrano de Bergerac, *Les États et empires de la lune et du soleil* (2004); Darmon, *Le Songe libertin* (2004); Aït-Touati, *Fictions of the Cosmos* (2011), 63–71; and Campbell, *Wonder and Science* (1999), 171–80. **40.** Borel, *A New Treatise* (1658), 93–4. **41.** Hunter, 'Science and Astrology' (1995), 280–1. Hunter, in his notes, mistakes Borel (Borellus) for Giovanni Alphonso Borelli (Borellius). **42.** Fontenelle, *Entretiens sur la pluralité des mondes* (1955); Aït-Touati, *Fictions of the Cosmos* (2011), 79–94; and Campbell, *Wonder and Science* (1999), 143–9. **43.** Aït-Touati, *Fictions of the Cosmos* (2011), 95–125. **44.** Bentley, *The Folly and Unreasonableness of Atheism* (1692), 241–2. **45.** Chang, *Inventing Temperature* (2004), 10. **46.** Crease, *World in the Balance* (2011). **47.** Aït-Touati, *Fictions of the Cosmos* (2011), 139. **48.** Griffith, *Mercurius Cambro-Britannicus, or, News from Wales* (1652), preface (*2r). **49.** Voltaire, '*Micromégas*': *A Study* (1950); but on the date of the composition, Barber, 'The Genesis of Voltaire's "Micromégas"' (1957). **50.** Ball, *Curiosity* (2012), 222. **51.** Power, *Experimental Philosophy* (1664), preface (c2v–c3r). **52.** Ball, *Curiosity* (2012), 318; but 'may' is an important word here, as Hooke does not actually mention the microscope in this context (Hooke, *The Posthumous Works* (1705), 140). **53.** Bertoloni Meli, *Mechanism, Experiment, Disease* (2011). **54.** Pinto-Correia, *The Ovary of Eve* (1997). **55.** Pascal, *Pensées* (1958), no. 72. **56.** Cyrano de Bergerac, *Les États et empires de la lune et du soleil* (2004), 116–17; Pascal, *Pensées* (1958); and Borges, *Other Inquisitions* (1964), 'Pascal', 100. Borges thinks this idea was already present in Anaxagoras, but this seems wrong: see Vlastos, '*Wege und Formen frühgriechischen Denkens, Hermann Fränkel*' (1959). **57.** Swift, *Gulliver's Travels* (2012), 158–9. **58.** Cyrano de Bergerac, *The Comical History* (1687), 13–14. **59.** Pascal, *Pensées* (1958), no. 206. **60.** Malcolm, 'Hobbes and Roberval' (2002), 170. (Malcolm argues that the view may be as much Hobbes's as Roberval's.) **61.** Cyrano de Bergerac, *The Comical History* (1687), 41. **62.** Locke, *An Essay* (1690), 46. **63.** Aggiunti, *Oratio de mathematicae laudibus* (1627), 19–20. **64.** See, among many examples, Shapin & Schaffer, *Leviathan and the Air-pump* (1985), 6–7.

PART THREE

1. Popper, *Objective Knowledge* (1972), 23.

CHAPTER 7

1. Barthes, '*Le Discours de l'histoire*' (1967). **2.** Kuhn, *The Trouble with the Historical Philosophy of Science* (1992), 6; reprinted in Kuhn, *The Road since Structure* (2000). **3.** Kepler, *Epitome of Copernican Astronomy, Books 4 & 5* (1995), 5. **4.** The major works are: Poovey, *A History of the Modern Fact* (1998); Shapiro, *A Culture of Fact* (2000); and Daston & Park, *Wonders and the Order of Nature* (1998), 215–53. There are a number of significant essays by Daston: 'The Factual Sensibility' (1988); 'Marvellous Facts and Miraculous Evidence' (1991); 'Baconian Facts' (1994); 'Strange Facts, Plain Facts' (1996); 'The Cold Light of Facts' (1997); 'The Language of Strange Facts' (1997); and '*Perché i fatti sono brevi?*' (2001). Highly influential are Shapin & Schaffer, *Leviathan and the Air-pump* (1985), 22–79; and Shapin, *A Social History of Truth* (1994), 193–242. **5.** This is presumably what misled Lorraine Daston into stating that the modern usage of the word 'fact' is contemporary with Bacon. *OED* gives two early usages of 'fact' under impersonal definitions, but they are, in fact, agency usages. The earliest impersonal usage it records is from Evelyn, discussed above. Daston and I also offer different accounts of how fact-establishing originates: she turns to the establishing of strange and wonderful facts (monstrous births, for example), while I suggest Kepler was establishing facts. **6.** Paula Findlen thinks that Pliny the Elder used the word *factum* to refer to a singular piece of information; his *Natural History* would then be the source of the modern concept of the fact. She cites Pliny, *Natural History*, preface, 17–18 in the Loeb edition: Findlen, 'Natural History' (2008), 437–8; Pliny, *Natural History* (1938), Vol. 1, 12–13. This is an uncharacteristic lapse. The word *factum* does not appear in the passage cited, and the word which Rackham translates as 'fact' is, as one would expect, *res*. **7.** An early example is Bossuet, *Quakerism À-la-Mode* (1698), 91: 'To what purpose are his Arguments, when Matters of Fact speak?' **8.** Hume, *Philosophical Essays* (1748), 47 (First Enquiry, Part 4, section 1). **9.** Hume, *Political Discourses* (1752), 211 ('Of the Populousness of Ancient Nations'). **10.** Browne,

Pseudodoxia epidemica (1646), a5r ('To the Reader'). **11.** Barnhart, *The American College Dictionary* (1959). **12.** Latour, 'The Force and the Reason of Experiment' (1990), 63–5. **13.** Searle, *The Construction of Social Reality* (1995), 1–30, 121. **14.** Galilei, *Le opere* (1890), Vol. 10, 226–7. **15.** See below, p. 332. **16.** Wootton, 'Accuracy and Galileo' (2010). **17.** Hume, *Political Discourses* (1752), 155–261. For two examples of exceptionally early attempts at statistical accuracy, see Giovanni Villani's *Cronica* for 1338 (discussed in Biller, *Measure of Multitude* (2000), 406–14), and Giovanni Botero's contribution to Tolomei, Guicciardini & others, *Tre discorsi appartenenti alla grandezza delle citta* (1588); the most recent modern edition is in Botero, *On the Causes of the Greatness and Magnificence of Cities, 1588* (2012). **18.** McCormick, *William Petty* (2009); Holmes, 'Gregory King' (1977); Slack, 'Measuring the National Wealth in Seventeenth-century England' (2004); and Slack, 'Government and Information in Seventeenth-century England' (2004). **19.** Kepler, *New Astronomy* (1992), 210–11. **20.** Goldstein & Hon, 'Kepler's Move from Orbs to Orbits' (2005). **21.** Kepler, *New Astronomy* (1992), 405, 410–16. **22.** Gingerich, 'Johannes Kepler' (1989), 63. **23.** Gingerich, 'Circles of the Gods' (1994), 23. **24.** Shapin & Schaffer, *Leviathan and the Air-pump* (1985), 22–79. **25.** Westman, *The Copernican Question* (2011), 401. (I have modified Westman's translation; there is a French translation, Kepler, *L' Étoile nouvelle dans le serpentaire* (1998)). **26.** Barthes, 'The Reality Effect' (1986). **27.** Kepler, *New Astronomy* (1992), 27. **28.** Owen, '*Tithenai ta phainomena*' (1975). For seventeenth-century English attacks on the notion that consensus could provide a reliable basis for knowledge, see Skinner, *Reason and Rhetoric* (1996), 257–67. **29.** For an example of someone wrestling with this, see Piccolomini, *La prima parte delle theoriche* (1558), 29r–30v, which argues that public opinion is to be trusted in questions of morality but not natural philosophy. **30.** Gingerich, 'Johannes Kepler' (1989), 63. **31.** Duhem, *To Save the Phenomena* (1969). **32.** Lehoux, *What Did the Romans Know?* (2012), 136–54, 209–17. **33.** Lehoux, *What Did the Romans Know?* (2012), 140; the whole discussion of garlic and magnets covers 136–54. **34.** Browne, *Pseudodoxia epidemica* (1646), 67. **35.** Della Porta, *Natural Magick* (1658), 212. **36.** Lehoux, *What Did the Romans Know?* (2012), 143. This is Lehoux's translation of Plutarch, *Quaestiones convivales*, Bk 2, Ch. 7. **37.** Lehoux, *What Did the Romans Know?* (2012), 145–6. **38.** Della Porta, *Natural Magick* (1658), 10. **39.** Shea, *Designing Experiments* (2003), 116–17; Augst, 'Descartes' Compendium on Music' (1965). Descartes also explained why a corpse bleeds in the presence of its murderer: Daston & Park, *Wonders and the Order of Nature* (1998), 241. **40.** Charleton, *Physiologia Epicuro-Gassendo-Charletoniana* (1654), 358. **41.** Balbiani, *La magia naturalis* (2001), 20. **42.** Eamon, *Science and the Secrets of Nature* (1994), 194–229; Tarrant, 'Giambattista della Porta and the Roman Inquisition' (2013), which simplifies Porta's response to censorship; and Valente, '*Della Porta e l'Inquisizione*' (1999). **43.** Clubb, *Giambattista della Porta, Dramatist* (1965), 23–4, 26, 51. **44.** Della Porta, *Natural Magick* (1658), Book 6; della Porta, *La Magie naturelle* (1678), Book 3, Ch. 4. **45.** Della Porta, *La Magie naturelle*

(1678), *préface aux lecteurs* (A4v). For some reason the prefaces are missing from the Italian translation, della Porta, *De i miracoli* (1560); della Porta, *Natural Magick* (1658), preface to the reader. **46.** Garzoni, *Trattati della calamità* (2005); Ugaglia, 'The Science of Magnetism before Gilbert' (2006). **47.** Muraro, *Giambattista della Porta, mago e scienziato* (1978), 143–71. **48.** Hobson, 'A Sale by Candle in 1608' (1971); Grendler, 'Book Collecting in Counter-Reformation Italy' (1981). **49.** Garzoni, *Trattati della calamità* (2005), 81–2. **50.** Gilbert as the first modern: Zilsel, 'The Origin of William Gilbert's Scientific Method' (1941); Monica Ugaglia argues that Gilbert had both direct access to Garzoni and to a lost manuscript by Paolo Sarpi. The evidence seems to me less than conclusive (and Sarpi's manuscript may well have post-dated Gilbert's book). Garzoni, *Trattati della calamità* (2005), 60–79. **51.** Della Porta, *Natural Magick* (1658), 190. **52.** Della Porta, *Natural Magick* (1658), 212. **53.** Della Porta, *Natural Magick* (1658), 213. **54.** Della Porta, *Natural Magick* (1658), 214. **55.** Della Porta, *Natural Magick* (1658), preface to the reader. **56.** Della Porta, *Natural Magick* (1658), 8–10. **57.** Cesi, *Mineralogia* (1636), 40 (Lehoux's translation), 534 (the major discussion, which is not cited by Lehoux). **58.** Lehoux, *What Did the Romans Know?* (2012), 144. **59.** An almost identical point is made by Lehoux himself in an earlier version of his argument: Lehoux, 'Tropes, Facts and Empiricism' (2003), 13 n. 12, where he says that Alessandro Vicentini 'is really making an old-fashioned generalized appeal to experiences of the world rather than to experiment as we understand it'. cf. Dear, *Discipline and Experience* (1995), 149. **60.** Quoted from Garzoni, *Trattati della calamità* (2005), 91 n. 4. **61.** Browne, *Pseudodoxia epidemica* (1672), 70. (I quote the 1672 edition as the 1646 text appears corrupt.) **62.** Rohault, *Traité de physique* (1671), 234. **63.** De Boodt, *Gemmarum et lapidum historia* (1609), 225. The authority of the sailors presumably derived from the traditional principle that Boyle summarized as 'the *Logicians* Rule, the Skilfull *Artists* should be Credited in their own Art' (Serjeantson, 'Testimony and Proof' (1999), 218; Browne, *Pseudodoxia epidemica* (1646), 26). **64.** De Boodt, *Gemmarum et lapidum historia* (1609), 222, 234–5. **65.** De Boodt, *Gemmarum et lapidum historia* (1609), 60. **66.** Jonkers, *Earth's Magnetism in the Age of Sail* (2003), 166. **67.** Helmont & Charleton, *A Ternary of Paradoxes* (1649), 40–1; Fletcher & Fletcher, *Athanasius Kircher* (2011), 150; and Thorndike, *A History of Magic and Experimental Science* (1923), Vol. 2, 310–11. **68.** Midgeley, *A New Treatise of Natural Philosophy* (1687), 31. **69.** Ross, *Arcana microcosmi* (1652), 110. **70.** Starkey, *Nature's Explication and Helmont's Vindication* (1657), b7v ('Epistle to the Reader'). For Galileo's version of this argument, see Wootton, *Galileo* (2010), 164. **71.** Nor do I wish to overstate de Boodt's modernity: he had an interest in alchemy, on which see Purs, 'Anselmus Boëtius de Boodt' (2004). **72.** Boyle, *Certain Physiological Essays* (1669), 33 = Boyle, *The Works* (1999), Vol. 2, 29–30; and Boyle, *Certain Physiological Essays* (1661), 31, 27, 7 = Boyle, *The Works* (1999), Vol. 2, 28 (where 'barely' is mistranscribed as 'basely'), 27, 13. Note that Boyle is not particularly concerned about the social status of his witnesses, as he would be if authority were what was at issue (contra Shapin, *A Social History of*

Truth (1994)); what matters is that they have direct knowledge and enough expertise not to misinterpret their experience. Note also that we are not dealing here with a peculiarly English empiricism. 73. Boyle, *Experimenta et observationes physicæ* (1691), 30 = Boyle, *The Works* (1999), Vol. 11, 386. 74. The best introductions to the world that existed before the scientific method became common sense remain Febvre, *The Problem of Unbelief* (1982), first published 1942; Koyré, '*Du monde de "l'à-peu-près" à l'univers de la précision*' (1971), first published 1948; and Febvre, '*De l'à peu près à la précision*' (1950). 75. Wotton, *Reflections upon Ancient and Modern Learning* (1694), 233–4. 76. Wotton, *Reflections upon Ancient and Modern Learning* (1694), 24; see also Glanvill, *Plus ultra* (1668), 77–9. 77. On the new culture of eyewitness testimony, see Frisch, *The Invention of the Eyewitness* (2004). 78. Della Porta, *De telescopio* (1962). 79. Garzoni, *Trattati della calamità* (2005), 94. 80. Adorno, 'The Discursive Encounter of Spain and America' (1992). 81. Lessing, *Gotthold Ephraim Lessings Leben* (1793), Vol. 3, 177–8 ('*Ueber das Wörtlein "Thatsache"*'); *Grimms Wörterbuch* gives 1756 as the date for the first use of '*Thatsache*'. 82. Browne, *Pseudodoxia epidemica* (1646), 3. 83. III.vii.44–7. 84. Daston and Park are thus wrong to see innovation in Bacon's distinction between matters of fact and matters of judgement: Daston & Park, *Wonders and the Order of Nature* (1998), 230. 85. Bartlett, *Trial by Fire and Water* (1986). 86. Bacon, *Sylva sylvarum* (1627), 243 = Bacon, *Works* (1857), Vol. 2, 642. 87. Johnson, in his *Dictionary*, relies on an earlier passage: 'Those *Effects*, which are wrought by the *Percussion* of the *Sense*, and by *Things* in *Fact*, are produced likewise, in some degree, by the *Imagination*.' Bacon, *Sylva sylvarum* (1627), 206 = Bacon, *Works* (1857), Vol. 2, 598. 88. Biggs, *Mataeotechnia medicinae praxeos* (1651), 37. 89. Ross, *Arcana microcosmi* (1652), 132. 90. Evelyn, *The Diary* (1955), Vol. 2, 38. 91. Daston, 'Baconian Facts' (1994). But Daston does not attribute the word 'fact' to Bacon, unlike Shapiro, 'The Concept "Fact"' (1994), 15–16, which fails to distinguish Latin and English usages. 92. Wootton, *Galileo* (2010), 99–100. See also Galilei, *Le opere* (1890), Vol. 5, 389. 93. Shapiro, *A Culture of Fact* (2000), 133–5. 94. See also Roberval's use of '*fait*' to mean 'fact' twice in an undated fragment: Pascal, *Oeuvres* (1923), 49–51 (Roberval died in 1675, which establishes a terminus *ante quem*). The text may date to 1648. 95. http://artfl-project.uchicago.edu/content/dictionnaires-dautrefois; one has to search under '*faire*'. 96. Serjeantson, 'Testimony and Proof' (1999), n. 84, gives two examples: Bacon, *Works* (1857), Vol. 1, 402, Vol. 3, 736. To these may be added: Vol. 3, 775, and perhaps Vol. 1, 210 (where the usage is evidently metaphorical). 97. Digby, *Two Treatises* (1644), 330. 98. 'Betwixt our *Divine* and *Physician*, there is at all no dispute *de facto*, about the verity of the *fact*; for both unanimously concede the cure to be wrought upon the wounded person: The contention lies onely in this, that the *Physician* asserts this Magnetical Cure to be purely *Natural*, but the *Divine* will needs have it *Satanical*' (Helmont & Charleton, *A Ternary of Paradoxes* (1649), 4); '*Inter theologum & medicum non est quaestio facti*' (Helmont, *Ortus medicinae* (1652), 595b); 'I know an *Herb*, commonly obvious, which if it be rubbed, and cherished in thy hand, until it wax warm, you may hold fast the hand of another person, until that also grow warm, and he shall continually burn with an ardent love, and fixt dilection of thy person, for many days together. I held in my hand, first bathed in the steam of this love procuring plant the foot of a *dog*, for some few minutes: The dog, wholly renouncing his old Mistress, instantly followed me, and courted me so hotly, that in the night he lamentably howled at my Chamber door, that I should open and admit him. There are some now living in *Bruxels*, who are witnesses to me, and can attest the truth of this fact' (14); '*Adsunt Bruxellae mihi hujus facti testes*' (599a) ; 'Since in earnest I have held forth examples of the *Fact*, in *Sublunaries*, and brought upon the stage very many and very apposite *instances*, as that of the insititious or engrafted *Nose*, of the *Saphire*, of *Arsmarte*, *Asarum*, and most other *Herbs*' (35); '*Siquidem in sublunaribus exempla facti*' (604b); 'For it is an action of insolent petulancy for any, therefore to deny the *contingence* of that *fact*, which is every where so trivial and frequent, that it can hardly escape the observation of any' (35); '*Idcirco inseolentis est petulantiae, negare facti esse*' (604b) **99.** Hobbes, *Leviathan* (1651), 21, 30–1, 40, 200; Hobbes, *Of Libertie and Necessitie* (1654), 75. **100.** EEBO assumes he is George Wither, but this seems unlikely to me. **101.** Hobbes, *Humane Nature* (1650), 31–41; cf. Hacking, *The Emergence of Probability* (2006), 31, 47–8; and Glanvill, *The Vanity of Dogmatizing* (1661), 189–93. **102.** Pascal, *Les Provinciales, or, The Mysterie of Jesuitisme* (1657); and Pascal, *Les Provinciales, or, The Mystery of Jesuitisme* (1658). **103.** Keynes, *John Evelyn, a Study in Bibliophily* (1937), 119–24; Jansen, *De Blaise Pascal à Henry Hammond* (1954). **104.** Digby, *A Late Discourse* (1658), 4. **105.** Della Porta, *Natural Magick* (1658), 229. See Hedrick, 'Romancing the Salve' (2008), 162 n. 5; and McCord, 'Healing by Proxy' (2009). **106.** Helmont & Charleton, *A Ternary of Paradoxes* (1649), d1r. Charleton's change of mind is often attributed to a conversion to mechanism: this is disputed by Lewis, 'Walter Charleton and Early Modern Eclecticism' (2001). **107.** Charleton, *Physiologia Epicuro-Gassendo-Charletoniana* (1654), 380–2. **108.** Weld, *A History of the Royal Society, with Memories of the Presidents* (1848), Vol. 2, 527. **109.** Sprat, *The History of the Royal-Society* (1667), 47–8, 70; also 73, 99, 359. **110.** Pomata, 'Observation Rising: Birth of an Epistemic Genre, 1500–1650' (2011). **111.** De Bils, *The Coppy of a Certain Large Act* (1659), A2v = Boyle, *The Works* (1999), Vol. 1, 43; Boyle, *The Correspondence of Robert Boyle, 1636–1691* (2001), Vol. 1, 396. (I owe this reference to Michael Hunter.) **112.** For example, Poovey, *A History of the Modern Fact* (1998), 112–15. **113.** Glanvill, *The Vanity of Dogmatizing* (1661), 159–68. **114.** Malcolm, *Aspects of Hobbes* (2002), 317–35. **115.** Hedrick, 'Romancing the Salve' (2008), 184. **116.** Sprat, *The History of the Royal-Society* (1667), 36. **117.** Glanvill, *The Vanity of Dogmatizing* (1661), 207. **118.** Salusbury (ed.), *Mathematical Collections* (1661), Vol. 1: 240, 413, 428, 445 (twice), 455; Vol. 2: 57; and *de facto*, which, like 'fact' has acquired a new sense: Vol. 1, 21, 161, 367, 376, 401, 455. **119.** Boyle, *New Experiments Physico-mechanical* (1660), 229–32 = Boyle, *The Works* (1999), Vol. 1, 238–9; Galilei, *Discorsi e dimostrazioni matematiche* (1638), 12. **120.** Hunter, *Establishing the New Science* (1989), article 14, 42 (but the upper part shows not 'the earth and planets' but Jupiter and its moons, discovered by Galileo). **121.** This revolution is best described in Serjeantson, 'Testimony and Proof' (1999). **122.** Pierre Du Moulin in 1598, quoted from Serjeantson, 'Testimony and Proof' (1999), 203, with a revision to Serjeantson's translation. **123.** Pascal, *Oeuvres complètes* (1964), 772–85; Browne, *Pseudodoxia epidemica* (1646), 25–6. **124.** Browne, *Pseudodoxia epidemica* (1646), 26. **125.** Glanvill, *The Vanity of Dogmatizing* (1661), 143; Sprat, *The History of the Royal-Society* (1667), 25, 29. **126.** In this respect, the argument of Shapin, *A Social History of Truth* (1994), is incontrovertible. **127.** Daston, 'Strange Facts, Plain Facts' (1996); and Daston, 'The Language of Strange Facts' (1997). **128.** Berkel, *Isaac Beeckman* (2013), 144–5. **129.** Clark, *Thinking with Demons* (1997). **130.** Arnauld & Nicole, *La Logique* (1970), Part 4, Ch. 14. **131.** Accademia del Cimento, *Saggi di naturali esperienze* (1667), 146; Accademia del Cimento, *Essayes of Natural Experiments* (1684), 77. **132.** Westrum, 'Science and Social Intelligence about Anomalies' (1978). **133.** Nield, *Incoming!* (2011), 67–72. **134.** Pantin, 'New Philosophy and Old Prejudices' (1999), 260. Gilbert uses the phrase '*libere philosophare*' in the preface to the reader of *De magnete*. **135.** Goulding, 'Henry Savile and the Tychonic World-system' (1995), 175. **136.** Jacquot, 'Thomas Harriot's Reputation for Impiety' (1952), 167. **137.** Pascal, *Oeuvres complètes* (1964), 779. **138.** Sutton, 'The Phrase "*Libertas Philosophandi*"' (1953); Broman, 'The Habermasian Public Sphere' (1998); Daston, 'The Ideal and Reality of the Republic of Letters' (1991). **139.** Latour, 'Visualization and Cognition' (1990). **140.** Stigler, 'John Craig and the Probability of History' (1986). **141.** Montaigne, *Essayes* (1613), A3v (prefatory poems). **142.** Gilbert, *On the Magnet* (1900), ii. **143.** Galilei, *Le opere* (1890), Vol. 10, 441; Schneider, *Disputatio physica de terrae motu* (1608), seems the best candidate. It is worth remarking that some Renaissance scientists held a theory according to which the material out of which the Earth is made is in imperceptible movement, as it transmutes into other elements and encounters heat and water, as a result constantly shifting the Earth's centre of gravity (for example, Prosdocimo de' Beldomandi). This theory of the movement of the Earth is presumably the subject of the lost treatise on the subject by Galileo's friend and mentor Guidobaldo del Monte, as there is no reason to think that he was a Copernican (Grant, *Planets, Stars and Orbs* (1994), 624–6; Thorndike, *A History of Magic and Experimental Science* (1923), Vol. 4, 239, Vol. 7, 230, 601). **144.** Reiss & Hinderliter, 'Money and Value in the Sixteenth Century' (1979). **145.** Bayle, *Projet* (1692); Joubert and Primerose were known to Browne: Browne, *Pseudodoxia epidemica* (1646), a5r. **146.** Grafton, *The Footnote* (1997). **147.** Charleton, *Physiologia Epicuro-Gassendo-Charletoniana* (1654), 3. **148.** For example, Grafton, 'Review: The Importance of Being Printed' (1980); for a manifesto laying out an alternative approach to those of Eisenstein and Latour, Johns, 'Science and the Book in Modern Cultural Historiography' (1998). **149.** Leonardo da Vinci, *Trattato della pittura* (1651) = *Traitté de la Peinture* (2012): 382–3. **150.** Mosley, *Bearing the Heavens* (2007); and Gingerich & Westman, 'The Wittich Connection' (1988). **151.** Buringh & van

Zanden, 'Charting the "Rise of the West"' (2009). 152. G. W., *The Modern States-man* (1653), 21–3. 153. Culverwel, *An Elegant and Learned Discourse* (1652), 171, 138. 154. Below, p. 457. 155. Desaguliers, *A Course of Experimental Philosophy* (1734), Vol. 1, preface (b3r). 156. Quoted in Carpenter, *John Theophilus Desaguliers* (2011), 70.

CHAPTER 8

1. Antinori, '*Notizie istoriche*' (1841), 27. 2. On barometer/vacuum experiments, Waard, *L'Expérience barométrique* (1936); Middleton, *The History of the Barometer* (1964); and Shea, *Designing Experiments* (2003). The primary sources for Pascal's experiments are in Pascal, *Oeuvres complètes* (1964), Vol. 2; key texts are translated in Pascal, *The Physical Treatises of Pascal* (1937). 3. Boyle, *A Defence* (1662), 48 = Boyle, *The Works* (1999), Vol. 3, 50. As Hunter and Davis note, Boyle acknowledges the phrase as coming from Bacon, but Bacon's phrase was *instantia crucis* (he was thinking of an observation rather than an experiment), so that later uses of the phrase *experimentum crucis* by Hooke and Newton evidently derive from Boyle. (Boyle's role in establishing the phrase and, consequently, his reference to Pascal are missed in the earlier literature: for example, Dear, *Discipline and Experience* (1995), 22.) 4. Newton, 'A Letter of Mr Isaac Newton' (1672), 3078. 5. For a slightly earlier example that lacks only rapid dissemination, Graney, 'Anatomy of a Fall' (2012); and Koyré, *Études d'histoire de la pensée scientifique* (1973), 289–319; for another alternative candidate, see Gassendi's 1640 public experiments to prove the relativity of motion: Koyré, *Études d'histoire de la pensée scientifique* (1973), 329. For an excellent survey of seventeenth-century experimentation, see Bertoloni Meli, 'Experimentation in the Physical Sciences of the Seventeenth Century' (2013). 6. On Roberval's experiments and arguments, see Auger, *Un savant méconnu* (1962), 117–33; and Malcolm, 'Hobbes and Roberval' (2002), 193–6. The difficulty in reaching agreement about the meaning of the Torricellian experiments is well illustrated by Hale, *Difficiles nugae* (1674). 7. Dear, 'The Meanings of Experience' (2006), 106; Crombie, *Styles of Scientific Thinking* (1994), 331–2, 349. The key article on experiment is Schmitt, 'Experience and Experiment' (1969), whose scope is much broader than its title would suggest. 8. Bacon, *Works* (1857), Vol. 8, 100–1. 9. Hobbes, *Humane Nature* (1650), 35–6; see Maclean, *Logic, Signs and Nature* (2002), for an earlier example of this distinction. 10. Daston & Lunbeck (eds.), *Histories of Scientific Observation* (2011). 11. The earliest occurrence I can find of *expérimentation* (apart from in Italian–French dictionaries and a charter dating from 1639) is in Scarpa, *Réflexions et observations anatomico-chirurgicales sur l'anéurisme* (1809), 3, a translation from Italian. 12. Valente, '*Della Porta e l'Inquisizione*' (1999), 422. 13. For lively debates on this subject, see Pesic, 'Proteus Rebound' (2008); Merchant, 'The Violence of Impediments' (2008); Vickers, 'Francis Bacon, Feminist Historiography and the Dominion of Nature' (2008); and Park, 'Response to Brian Vickers' (2008). The fact that nature is

female does not seem to me particularly relevant in this case – men and women were tortured indiscriminately. Compare the two legal mottos *magister rerum usus* and *magistra rerum experientia*: the fact that experience is a mistress, and practice (*usus*) a master, is irrelevant. Or our own tradition of referring to boats as female. On the other hand, it does matter to Machiavelli that fortune is a woman who can consequently be overpowered, and so it *might* be relevant that nature is female. **14.** Gilbert, *De magnete* (1600), '*Verborum quorundam interpretatio*' (*vi[r]). There are two translations of Gilbert, of which the first is to be preferred: Gilbert, *On the Magnet* (1900); Gilbert, *De magnete* (1951). On pointing, Gilbert, *On the Magnet* (1900), ii. Shapin introduced the concept of 'virtual witnessing' in the context of Boyle's experimental reports: Shapin, 'Pump and Circumstance' (1984), but there is virtual witnessing before Boyle; in particular, Pecquet, *New Anatomical Experiments* (1653), a work read by Boyle, uses all the literary techniques identified by Shapin in its description of vivisections. (I owe this point to Jamie Newell.) **15.** The discussion that follows is much influenced by Palmieri, 'The Cognitive Development of Galileo's Theory of Buoyancy' (2005), although my interpretation differs from his. The key source in translation is in Drake, *Cause, Experiment and Science* (1981), and a crucial discussion is Shea, *Galileo's Intellectual Revolution* (1972), 16–22. **16.** Galilei, *Le opere* (1890), Vol. 4, 52–4. **17.** Galilei, *Le opere* (1890), Vol. 4, 54–5. **18.** Lehoux, *What Did the Romans Know?* (2012), 143 n. 22; Lindberg, 'Alhazen's Theory of Vision' (1967). The Latin translation has been newly edited: Ibn al-Haytham, *Alhacen's Theory of Visual Perception* (2001) and subsequent volumes. **19.** King, 'Medieval Thought-experiments' (1991). **20.** Boyle, *Hydrostatical Paradoxes* (1666), 5–6 = Boyle, *The Works* (1999), Vol. 5, 206; on Galileo, Wootton, *Galileo* (2010), 78 (the mistaken view originates with Koyré, '*Galilée et l'expérience de Pise*' (1973), first published in 1937). **21.** Westfall, *Never at Rest* (1980), 60–4. **22.** Sacrobosco, Peuerbach & others, *Textus sphaerae* (1508), 87v. **23.** Eastwood, 'Robert Grosseteste's Theory of the Rainbow' (1966). **24.** Crombie, *Styles of Scientific Thinking* (1994), 348. For an example of someone falsely claiming to have conducted an experiment, see ibid., 380–1. Crombie's evolving interpretation of Grosseteste can be traced from his earliest texts, in which Grosseteste is a founder of experimental science (Crombie, 'Grossesteste's Position in the History of Science' (1955)), through his later texts, to the very cautious formulations with which he ended his career: see Eastwood, 'On the Continuity of Western Science' (1992). For further criticism, Eastwood, 'Grosseteste's "Quantitative" Law of Refraction' (1967); Eastwood, 'Medieval Empiricism' (1968); Serene, 'Robert Grosseteste on Induction' (1979); and Southern in Crombie (ed.), *Scientific Change* (1963), 305. **25.** Cranz, *Reorientations of Western Thought* (2006). **26.** Eastwood, 'Medieval Empiricism' (1968), 306–11; and Serene, 'Robert Grosseteste on Induction' (1979), 103. **27.** The issues here are complex, and I do not pretend to have a full grasp of them. As I see it, there are three distinct phases: for the Greeks, the knower is one with the known; for the medieval philosophers, the knower can have true sensory knowledge of the known; and for the early modern scientists, sensation comes to be an unreliable instrument. See: Tachau, *Vision and Certitude in the Age of Ockham* (1988); Smith, 'Knowing Things Inside Out' (1990); the opening section of Buchwald & Feingold, *Newton and the Origin of Civilization* (2013); Tuck, 'Optics and Sceptics' (1988); and Cranz, *Reorientations of Western Thought* (2006). **28.** Newton, *The Mathematical Principles of Natural Philosophy* (1729) (Cotes's preface, unpaginated). An alternative view, of course, is that much that happens in nature is not regular and predictable: Céard, *La Nature et les prodiges* (1996); and Daston & Park, *Wonders and the Order of Nature* (1998). **29.** Quoted from Crombie, *Styles of Scientific Thinking* (1994), 1102 (for the original text, see Sarpi, *Pensieri naturali* (1996), 3). **30.** Webster, 'William Harvey's Conception of the Heart as a Pump' (1965). **31.** Pérez-Ramos, *Francis Bacon's Idea of Science* (1988). **32.** Weeks, 'Francis Bacon and the Art–Nature Distinction' (2007). **33.** Dear, *Discipline and Experience* (1995), 153–61. **34.** See further discussion in Chapter 12. **35.** Grant (ed.), *A Source Book in Medieval Science* (1974), 435–41; and Crombie, *Robert Grosseteste* (1953), 233–59. **36.** Boyer, *The Rainbow from Myth to Mathematics* (1959), 125; and Topdemir, 'Kamal al-Din al-Farisi's Explanation of the Rainbow' (2007). **37.** Crombie, *Robert Grosseteste* (1953), 233. **38.** Boyer, *The Rainbow from Myth to Mathematics* (1959), 141. **39.** Trutfetter, *Summa in tota[m] physicen* (1514); Trutfetter, *Summa philosophiae naturalis* (1517). **40.** Buchwald, 'Descartes' Experimental Journey' (2008). **41.** Sabra, 'The Commentary that Saved the Text' (2007). **42.** Kant, *Critique of Pure Reason* (1949), preface to the second edition (1787). **43.** To be exact, it first appeared in print before 1520, but the edition is extraordinarily rare, and there is no evidence of anyone ever having read it. The standard edition is Peregrinus, *Opera* (1995). **44.** Pumfrey, *Latitude: The Magnetic Earth* (2001). **45.** Digges & Digges, *A Prognostication Everlasting* (1576), O3v–O4r. **46.** See the references to him in Melchior Inchofer's justification for the condemnation of Galileo in Blackwell, *Behind the Scenes at Galileo's Trial* (2006). **47.** Pumfrey, '*O tempora, O magnes!*' (1989). **48.** Zilsel, 'The Sociological Roots of Science' (1942), and Zilsel, 'The Origin of William Gilbert's Scientific Method' (1941) provide the classic Marxist account. For de Maricourt, Radelet de Grave & Speiser, '*Le "De magnete" de Pierre de Maricourt*' (1975), 203 (French translation). **49.** Gilbert, *On the Magnet* (1900), 7–8. **50.** Wootton, *Galileo* (2010), 91–2, 102–3. **51.** On Harriot, Schemmel, *The English Galileo* (2008); and Shirley, *Thomas Harriot* (1983). **52.** Wootton, *Galileo* (2010), 36–42; Wootton, 'Accuracy and Galileo' (2010), 49; Dear, *Discipline and Experience* (1995), 67–85, 124–44; Bertoloni Meli, 'The Role of Numerical Tables in Galileo and Mersenne' (2004); Sarasohn, 'Nicolas-Claude Fabri de Peiresc' (1993); and Palmerino, 'Experiments, Mathematics, Physical Causes' (2010). **53.** Wootton, *Galileo* (2010), 168–9; Wootton, 'Galileo: Reflections on Failure' (2011), 16–18; Shea, *Designing Experiments* (2003), 17–24. **54.** Shea, *Designing Experiments* (2003), 24–39; Shank, 'Torricelli's Barometer' (2012). **55.** Pascal, *Oeuvres* (1923), 486; and Shea, *Designing Experiments* (2003), 41–7. The public performance was not a peculiarly French phenomenon: Valeriano Magni performed his own version of the Torricellian experiment in Warsaw in July

1647 in front of the king, queen and courtiers (Dear, *Discipline and Experience* (1995), 187–8). **56.** Koyré, '*Galilée et l'expérience de Pise*' (1973) (first published 1937); Wootton, *Galileo* (2010), 273–4 n. 10; Devreese & Vanden Berghe, '*Magic is No Magic*' (2008), 152–4. **57.** Charleton, *Physiologia Epicuro-Gassendo-Charletoniana* (1654), table of contents for Ch. 5; and Shank, 'Torricelli's Barometer' (2012), 162; see also Glanvill, *Plus ultra* (1668), 94; and Boyle, *Certain Physiological Essays* (1661), 189 = Boyle, *The Works* (1999), Vol. 2, 155. **58.** Shea, *Designing Experiments* (2003), 47–127. **59.** Webster, 'The Discovery of Boyle's Law' (1965); Pecquet, *New Anatomical Experiments* (1653). **60.** Bertoloni Meli, 'The Collaboration between Anatomists and Mathematicians in the Mid-seventeenth Century' (2008), 672 (where the concept is attributed to Pecquet rather than Roberval). **61.** Boyle, *A Defence* (1662), 63–4. The question of the nature of the various contributions made by Boyle's collaborators is a vexed one. Agassi, 'Who Discovered Boyle's Law?' (1977), makes some valuable arguments but also contains some basic errors (regarding the date, for example, of Hooke's first experiments); Pugliese, 'The Scientific Achievement of Robert Hooke' (1982) is reliable. **62.** Hunter, *Boyle* (2009), 190; Boyle, *A Continuation of New Experiments* (1682), a3v, a4r = Boyle, *The Works* (1999), Vol. 9, 128–9. Boyle had, it is true, laid out the principle of acknowledging others in 1661 (Boyle, *Certain Physiological Essays* (1661), 32 = Boyle, *The Works* (1999), Vol. 2, 29), but he certainly did not live by it then. **63.** Hale, *Difficiles nugae* (1674), 8. **64.** Ugaglia, 'The Science of Magnetism before Gilbert' (2006), 72–3; Duhem, '*Le Principe de Pascal*' (1905); and Pascal, *Oeuvres complètes* (1964), Vol. 2, 1037 – the work was published only posthumously, but it seems to have been ready for the press in 1654; there is nothing to suggest that Pascal intended to add any acknowledgement of his sources. **65.** Abercromby, *Academia scientiarum* (1687). Sprat distinguished authors from 'Finishers': Iliffe, 'In the Warehouse' (1992), 32. **66.** Mersenne describes his virtual academy as being 'entirely mathematical': Garber, 'On the Frontlines of the Scientific Revolution' (2004), 156. **67.** Pascal, *Oeuvres* (1923), 161–2. **68.** Schott, *Mechanica hydraulico-pneumatica* (1657). **69.** Webster, 'New Light on the Invisible College' (1974). **70.** Pascal, *Oeuvres complètes* (1964), 777–85; and Shea, *Designing Experiments* (2003), 187–207. **71.** Merton, *On the Shoulders of Giants* (1965). **72.** Lehoux, *What Did the Romans Know?* (2012), 10–11. **73.** The first occurrence of the phrase is apparently 1635: Anstey, 'Experimental versus Speculative Natural Philosophy' (2005), 217. **74.** Ibn al-Haytham, *The Optics: Books I–III, on Direct Vision* (1989), Book 2, 15–19. **75.** Schmitt, 'Experience and Experiment' (1969), 115–22. **76.** Siraisi, *Communities of Learned Experience* (2013). **77.** Shea, *Designing Experiments* (2003), 43; and Power, *Experimental Philosophy* (1664), 88. **78.** Collins, *Changing Order* (1985); and Pinch, *Confronting Nature* (1986). **79.** Shapin & Schaffer, *Leviathan and the Air-pump* (1985). **80.** Secord, 'Knowledge in Transit' (2004), 657. **81.** Webster, 'Henry Power's Experimental Philosophy' (1967), 169. **82.** Shapin & Schaffer, *Leviathan and the Air-pump* (1985), 225–82. **83.** Shapin & Schaffer, *Leviathan and the Air-pump* (1985), 254. **84.** There is an interesting discussion of anomalous suspension, not mentioned by Shapin and Schaffer, in Papin, *A Continuation of the New Digester* (1687). Boyle was already aware of the problem of replication in 1661: see the two essays on unsuccessful experiments (mainly chemical) in Boyle, *Certain Physiological Essays* (1661); the problem was later discussed in Sprat, *The History of the Royal-Society* (1667), 243–5. **85.** Charleton, *Physiologia Epicuro-Gassendo-Charletoniana* (1654), 35. **86.** The recent literature on alchemy is vast. Particularly noteworthy are: Dobbs, *The Foundations of Newton's Alchemy* (1975); Smith, *The Business of Alchemy* (1994); Principe, *The Aspiring Adept* (1998); Newman, *Gehennical Fire* (2003); Newman, *Promethean Ambitions* (2004); and Newman & Principe, *Alchemy Tried in the Fire* (2005). **87.** Hunter & Principe, 'The Lost Papers' (2003). **88.** Principe, *The Aspiring Adept* (1998); Hunter, 'Alchemy, Magic and Moralism' (1990), esp. 404–5 (repeal of the act); and Hunter, *Boyle* (2009). **89.** Principe, *The Aspiring Adept* (1998), 98–113, 190–201. **90.** Principe, *The Aspiring Adept* (1998), 115–34; Malcolm, 'Robert Boyle, Georges Pierre des Clozets and the Asterism' (2004); and Principe, 'Georges Pierre des Clozets' (2004). On the question of fraud and alchemy, Nummedal, 'On the Utility of Alchemical Fraud' (2007); and Nummedal, *Alchemy and Authority in the Holy Roman Empire* (2007), 147–75. **91.** Newman & Principe, *Alchemy Tried in the Fire* (2005), 189. **92.** Principe, *The Aspiring Adept* (1998), 110, 159; and Starkey, *Alchemical Laboratory Notebooks* (2004), xxii–xxiii, 2–41 (assuming that the gold which was turned into a black powder was the same gold he had previously made). **93.** Du Chesne, *The Practise of Chymicall, and Hermeticall Physicke* (1605), K1v, K2r, K3r. **94.** Newman & Principe, *Alchemy Tried in the Fire* (2005), 175–6. **95.** Vickers, 'The "New Historiography"' (2008), 127. For Newman's reply to this essay, Newman, 'Vickers on Alchemy' (2009). **96.** Vickers, 'The "New Historiography"' (2008), 132; and Principe & DeWitt, *Transmutations* (2002). **97.** Hunter, 'Alchemy, Magic and Moralism' (1990), 403–4. **98.** For example, Newman & Principe, 'Alchemy versus Chemistry' (1998); Newman, *Atoms and Alchemy* (2006); and Newman, 'Recent Historiography' (2011). **99.** For emphasis on this chronology, see, for example, Principe, 'Alchemy Restored' (2011), 306. **100.** Powers, '*Ars sine arte*' (1998), 176 (translation altered). **101.** Powers, '*Ars sine arte*' (1998), 177. **102.** Klein, 'Origin of the Concept of Chemical Compound' (1994); Chalmers, *The Scientist's Atom and the Philosopher's Stone* (2009); Newman, 'How Not to Integrate the History and Philosophy of Science' (2010); Chalmers, 'Understanding Science through Its History' (2011); and Chalmers, 'Klein on the Origin of the Concept of Chemical Compound' (2012). **103.** Hunter, 'The Royal Society and the Decline of Magic' (2011), 105. **104.** Hunter, 'Alchemy, Magic and Moralism' (1990), 407. **105.** Wootton, 'Galileo: Reflections on Failure' (2011). **106.** Newman & Principe, 'Alchemy versus Chemistry' (1998), 60–1. **107.** Glanvill, *Plus ultra* (1668), 12. **108.** Quoted from Newman & Principe, 'Alchemy versus Chemistry' (1998), 62. **109.** Popper, *The Open Society and Its Enemies* (1945).

CHAPTER 9

1. Baillet, *La Vie de Monsieur Des-Cartes* (1691), Vol. 1, 77–86, is the key source; for modern discussions, see Gaukroger, *Descartes: An Intellectual Biography* (1995), 104–11; and Clarke, *Descartes: A Biography* (2006), 58–63. 2. Beeckman, *Journal* (1939); Berkel, *Isaac Beeckman* (2013); Gaukroger, *Descartes: An Intellectual Biography* (1995), 68–103, 222–4; and Clarke, *Descartes: A Biography* (2006), 46–52, 142. 3. Beeckman, *Journal* (1939), Vol. 1, 244; translation from Gaukroger, *Descartes: An Intellectual Biography* (1995), 69. 4. Beeckman, *Journal* (1939), Vol. 1, 101, 253, 260–1, 265, Vol, 3, 104, but he also uses *theorema* (Vol. 1, 256) and *ratio naturalis* (Vol. 3, 104); see also Berkel, *Isaac Beeckman* (2013), 238 n. 52, on his use of *modus*. 5. Gaukroger, *Descartes: An Intellectual Biography* (1995), 90. 6. French translations of the Latin letters in Descartes, *Oeuvres philosophiques* (1963), Vol. 1, 270–84. 7. Descartes, *Philosophical Writings* (1984), Vol. 1, 116. 8. Camerota, '*Galileo, Lucrezio e l'atomismo*' (2008); Favaro, '*Libreria di Galileo Galilei*' (1886), nos. 353, 354. 9. Boyle, *The Origine of Formes and Qualities* (1666), 10, 43 = Boyle, *The Works* (1999), Vol. 5, 308, 317. 10. Descartes, *Principles of Philosophy*, Vol. 2, 37–40. 11. For example, Schuster, 'Waterworld' (2005); and Buchwald, 'Descartes' Experimental Journey' (2008). 12. Hoare, *The Quest for the True Figure of the Earth* (2004). 13. Quoted in Wilson, 'From Limits to Laws' (2008), 13. 14. Boas Hall, *Nature and Nature's Laws* (1970). What we call Boyle's law, reported by Boyle in 1662, had become a law by 1676 (Dear, *Discipline and Experience* (1995), 207). 15. Westfall, *Never at Rest* (1980), 632. 16. Boyer, 'Aristotelian References to the Law of Reflection' (1946), 92. 17. Key discussions are: Zilsel, 'The Genesis of the Concept of Scientific Progress' (1945); Needham, 'Human Laws and Laws of Nature in China and the West (I)' (1951); Needham, 'Human Laws and Laws of Nature in China and the West (II)' (1951); Oakley, 'Christian Theology and the Newtonian Science' (1961); Milton, 'The Origin and Development of the Concept of the "Laws of Nature"' (1981); Ruby, 'The Origins of Scientific "Law"' (1986); Steinle & Weinert, 'The Amalgamation of a Concept' (1995); Milton, 'Laws of Nature' (1998); Henry, 'Metaphysics and the Origins of Modern Science' (2004); Oakley, *Natural Law, Laws of Nature, Natural Rights* (2005); Joy, 'Scientific Explanation' (2006); and Harrison, 'The Development of the Concept of Laws of Nature' (2008). 18. '*Lex naturae est in rebus creatis regulatio motuum et operationum et tendentiarum in suos fines.*' Quoted from Oberman, 'Reformation and Revolution' (1975), 425 n. 47. 19. Ruby, 'The Origins of Scientific "Law"' (1986), 342–3, 353–5, 357. 20. Fernel, *Therapeutice, seu medendi ratio* (1555), 1r–6r. See also Fernel, *On the Hidden Causes of Things* (2005), 30 n. 90. I owe these references to Sophie Weeks. 21. Steinle & Weinert, 'The Amalgamation of a Concept' (1995), 320–1. 22. Hine, 'Inertia and Scientific Law in Sixteenth-century Commentaries on Lucretius' (1995). 23. Lehoux, *What Did the Romans Know?* (2012), 49–54; Cohen, '"*Quantum in se est*"' (1964); '*Nos in jure naturae enucleando et rerum foederibus interpretandis paulo*

diligientiores erimus ...': 'The History of the Sympathy and Antipathy of Things', Bacon, *Works* (1857), Vol. 2, 81. **24.** Montaigne, *The Complete Essays* (1991), 585–7; and Montaigne, *Oeuvres complètes* (1962), 504–5. **25.** Screech (ed.), *Montaigne's Annotated Copy of Lucretius* (1998), 229. **26.** Charleton, *Physiologia Epicuro-Gassendo-Charletoniana* (1654), 263. **27.** Boyle, *A Free Enquiry* (1686), 256–7 = Boyle, *The Works* (1999), Vol. 10, 524. **28.** http://www.iep.utm.edu/lawofnat/. **29.** I am grateful to Sophie Weeks for making this point to me. **30.** Newton, *The Mathematical Principles of Natural Philosophy* (1729), Mr Cotes's Preface, A8r (a translation of the Latin edition of 1713). **31.** Russell, 'Kepler's Laws' (1964) and Wilson, 'From Kepler's Laws, So-called, to Universal Gravitation' (1970), on the reception of Kepler's 'laws'. **32.** Charleton, *Physiologia Epicuro-Gassendo-Charltoniana* (1654), 343, 258, 395. **33.** Steinle, 'Negotiating Experiment, Reason and Theology' (2002). **34.** Ryle, *The Concept of Mind* (1949). **35.** Descartes is explicit on this: Descartes, *Principia philosophiæ* (1644), Part 3, § iii; and Descartes, *Les Principes de la philosophie* (1668),114–15. **36.** Harrison, 'The Development of the Concept of Laws of Nature' (2008); see also the debate between Harrison and Henry on voluntarism: Harrison, 'Voluntarism and Early Modern Science' (2002); Henry, 'Voluntarist Theology at the Origins of Modern Science' (2009); and Harrison, 'Voluntarism and the Origins of Modern Science' (2009). **37.** Galilei, *Le opere* (1890), Vol. 5, 283. **38.** Henry, *The Scientific Revolution* (2002), 92. **39.** Harrison, 'Newtonian Science, Miracles, and the Laws of Nature' (1995). **40.** Snobelen, 'William Whiston, Isaac Newton' (2004). **41.** Montaigne, *The Complete Essays* (1991), 588; and Montaigne, *Oeuvres complètes* (1962), 506.

CHAPTER 10

1. Newton, *Papers & Letters* (1958), prints the key published texts and letters; Kuhn's analysis is on 27–45. Westfall's accounts are: Westfall, 'The Development of Newton's Theory of Color' (1962); and Westfall, *Never at Rest* (1980), 156–74. My own chronology is based on Westfall, Shapiro, 'Introduction' (1984); Guerlac, 'Can We Date Newton's Early Optical Experiments?' (1983); and Hall, *All was Light: An Introduction to Newton's Opticks* (1993), 33–59. **2.** Westfall, 'The Development of Newton's Theory of Color' (1962), 352. **3.** Newton, *Papers & Letters* (1958), 34 n. 11. **4.** Dear, *'Totius in verba'* (1985), 155. **5.** Lohne, 'Isaac Newton: The Rise of a Scientist, 1661–1671' (1965), 138. **6.** Newton, *Papers & Letters* (1958), 47, 53, 93. **7.** Newton, *Papers & Letters* (1958), 79. **8.** Newton, *Papers & Letters* (1958), 92. **9.** Newton, *Papers & Letters* (1958), 105. **10.** Newton, *Papers & Letters* (1958), 108. **11.** Westfall, 'The Development of Newton's Theory of Color' (1962), 350. **12.** Newton, *Papers & Letters* (1958), 109. **13.** Newton, *Papers & Letters* (1958), 49. **14.** Descartes, *Philosophical Writings* (1984), Vol. 1, 255–6. **15.** The best general discussion is Koyré, 'Concept and Experience in Newton's Scientific Thought' (1965). 16. Blake, Ducasse, and others, *Theories of Scientific Method*

(1960), 22–49; Gingerich, 'From Copernicus to Kepler' (1973); Martens, *Kepler's Philosophy and the New Astronomy* (2000), 60–8; Granada, Mosley, and others, *Christoph Rothmann's Discourse* (2014), 55–64. **17.** Carpenter, *Geographie Delineated* (1635), 143. **18.** Malcolm, 'Hobbes and Roberval' (2002), 167. **19.** For the details of Digges's hypothesis see Johnston, 'Theory, Theoric, Practice' (2004). Borough, *A Discours of the Variation of the Cumpas* (1581), Giiir/v, uses the word 'hypothesis' in exactly the same way. Both Borough and Digges had been taught by John Dee, whose somewhat eccentric use of 'hypothesis', to mean, apparently, a true claim unsupported by argument, is worth noting: Dee, *General and Rare Memorials* (1577), 41. **20.** Norman, *The New Attractive* (1581), Aiiirv. **21.** The same move is to be found in Scaliger, *Opuscula varia ante hac non edita* (1610), 424–5 = Hues, *Tractatus de globis* (1617), 111 = Hues, *A Learned Treatise of Globes* (1659), 142. **22.** Galilei, *Le opere* (1890), Vol. 5, 395. See also Galilei, *Le opere* (1890), Vol. 7, 485. **23.** Locke, *Essay*, iv, 12, §13. On Locke on hypotheses, Laudan, 'Nature and Sources' (1967); Osler, 'John Locke' (1970) (though, strangely, Osler had not read Laudan); Farr, 'The Way of Hypotheses' (1987). For a wide-ranging discussion of hypothetical arguments in this period, Roux, '*Le scepticisme et les hypothèses de la physique*' (1998). **24.** Finocchiaro, *The Galileo Affair: A Documentary History* (1989), 67, and references in index, s.v. 'hypothesis'; also, for example, Dini to Galileo, 2 May 1615: Galilei, *Le opere* (1890), Vol. 12: no. 1115. **25.** Descartes, *Principia philosophiae* (1644), especially Vol. 3, 44, 45, 47; Descartes, *Philosophical Writings* (1984), Vol. 1, 255–8, 267; Martinet, '*Science et hypothèses chez Descartes*' (1974); Clarke, *Occult Powers and Hypotheses* (1989), 131–63. For the impact of the trial of Galileo on Descartes, Finocchiaro, *Retrying Galileo, 1633–1992* (2007), 43–51. **26.** Gilbert, *De magnete* (1600), *iii[r]; translation modified from Gilbert, *On the Magnet* (1900), iii–vi. **27.** Galilei, *Le opere* (1890), Vol. 5, 225. **28.** Wilkins, *A Discourse* (1640), 19, describes Copernicanism as an hypothesis confirmed by later 'inventions' (meaning discoveries). **29.** Boyle, *New Experiments Physico-Mechanicall* (1660), 133, 382. **30.** Cohen, 'The First English Version of Newton's *Hypotheses non fingo*' (1962); Koyré, 'Concept and Experience in Newton's Scientific Thought' (1965); Cohen, 'Hypotheses in Newton's Philosophy' (1966); Sabra, *Theories of Light* (1967), 231–50; Hanson, '*Hypotheses fingo*' (1970); McMullin, 'The Impact of Newton's *Principia* on the Philosophy of Science' (2001); Anstey, 'The Methodological Origins of Newton's Queries' (2004); Anstey, 'Experimental versus Speculative Natural Philosophy' (2005). **31.** On Bacon's astronomy, Jalobeanu, 'A Natural History of the Heavens' (2015). **32.** Descartes, *A Discourse* (1649), 100, 103–5. **33.** Laudan, 'The Clock Metaphor and Probabilism' (1966). **34.** Sabra, *Theories of Light* (1967), 168–9, translating from Huygens, *Oeuvres completes*, 21:472. **35.** Conant, *Robert Boyle's Experiments in Pneumatics* (1950), 4. **36.** On 'theory' before 1600, see Westman, *The Copernican Question* (2011), 38–43; on hypothesis, Brading, 'Development of the Concept of Hypothesis' (1999); and Ducheyne, 'The Status of Theory and Hypotheses' (2013). Ducheyne (188) thinks Boyle and Hooke use 'hypothesis' and 'theory' interchangeably. This is not so: Hooke, for example, repeatedly uses the phrase 'true theory'; he never writes of a 'true hypothesis'. See also Anstey, 'Experimental versus Speculative Natural Philosophy' (2005). **37.** Conveniently available at: https://artfl-project.uchicago.edu/content/dictionnaires-dautrefois. **38.** Bacon, *Sylva sylvarum* (1627), 204–5 = Bacon, *Works* (1857), 2:596. **39.** Helmont & Charleton, *A Ternary of Paradoxes* (1649); Helmont, *Deliramenta catarrhi* (1650); and Descartes, *Excellent Compendium of Musick* (1653). **40.** Boyle, *A Defence* (1662), 63–8 = Boyle, *The Works* (1999), Vol. 3, 61–5. **41.** Wallis, 'An Essay of Dr John Wallis' (1666); Boyle, 'Tryals Proposed by Mr Boyle to Dr Lower' (1667) = Boyle, *The Works* (1999), Vol. 5, 554–6. **42.** Sprat, *The History of the Royal-Society* (1667), 18, 155. **43.** Newton, 'A Letter of Mr Isaac Newton' (1670); and Newton, *Opticks* (1704), Advertisement, Book 1, 12, Book 2, 1, 78, 102, 111. **44.** Hooke, *Lectiones Cutlerianæ* (1679), 31; and 'Lampas', 9. **45.** 'Erotick' is to be found in Powell, *The Passionate Poet* (1601), and in Ferrand, *Erotomania, or, A Treatise Discoursing of the Essence, Causes, Symptomes, Prognosticks and Cure of Love or Erotic Melancholy* (1645). **46.** Helmont & Charleton, *A Ternary of Paradoxes* (1649), c1rv.

CHAPTER 11

1. Serjeantson, 'Testimony and Proof' (1999), 211. On the history of the word 'evidence' in English, Wierzbicka, *Experience, Evidence and Sense* (2010), 94–148. **2.** Thus Buchwald & Feingold, *Newton and the Origin of Civilization* (2013), provides an admirable study of Newton's use and understanding of evidence, but no account of why he very rarely uses the word. **3.** Boyle, *Occasional Reflections* (1665), 156 (3rd pagination) = Boyle, *The Works* (1999), Vol. 5, 154. **4.** Locke, *An Essay* (1690), 163. **5.** Locke, *An Essay* (1690), 264. **6.** Charleton, *Physiologia Epicuro-Gassendo-Charletoniana* (1654), 19; see Gassendi, *Animadversiones* (1649), 158: '*Opinio illa vera est, cui vel suffragatur, vel non refragatur Sensus evidentia . . . Opinio illa falsa est, cui vel refragatur, vel non suffragatur Sensus evidentia.*' **7.** Compare Glanvill, *The Vanity of Dogmatizing* (1661), 24, 77, 90, 109, where it seems clear that 'evidence of sense' is used in the modern sense. **8.** Jackson, *Justifying Faith* (1615), 13. **9.** Cowell, *The Interpreter* (1607), s.v. 'evidence'. **10.** Helmont & Charleton, *A Ternary of Paradoxes* (1649), 37. **11.** Cowell, *The Interpreter* (1607), s.v. 'bankrupt'. **12.** Helmont & Charleton, *A Ternary of Paradoxes* (1649), 18. **13.** Quintilian, *The Orator's Education* (2001), 362–3 (translation modified). **14.** Quintilian, *The Orator's Education* (2001), 419. **15.** Parsons, *The Seconde Parte of the Booke of Christian Exercise* (1590), 157–8. **16.** Wilkins, *Natural Religion* (1675), 1–11. Ian Hacking claims that there is a new concept of evidence (Evidence-Indices) present in this text, but this is a misinterpretation of what Wilkins says. Wilkins is discussing experience, and none of his examples is of Evidence-Indices. Experience does not mix demonstration and testimony, but sensation and demonstration (Hacking, *The Emergence of Probability* (2006), 83; the pagination is the same in the first, 1984, edition). The mixed character of experience had already been discussed by

Gassendi: Gassendi, *Opera omnia* (1727), 72b. **17.** Jackson, *Justifying Faith* (1615), 14. **18.** Locke, *An Essay* (1690), 268. **19.** Locke, *An Essay* (1690), 336 **20.** Baxter, *A Treatise of Knowledge and Love Compared* (1689), 59. **21.** Weiner, 'The Civil Jury Trial and the Law–Fact Distinction' (1966). **22.** Sheppard, *The Honest Lawyer* (1616), J4v. **23.** Franklin, *The Science of Conjecture* (2001), 12–63; and Langbein, *Torture and the Law of Proof* (1977). **24.** I owe a number of these points about linguistic usage to Alan Sokal. **25.** Hacking, *The Emergence of Probability* (2006), 32–5, 79, 83–4 **26.** Hacking calls this the evidence of things, but this seems misleading, as others (Wilkins, for example) refer to sensation as the evidence of things. 'Clues' is also problematic, as it is a nineteenth-century term: Ginzburg, *Myths, Emblems, Clues* (1990). **27.** Quoted, Hacking, *The Emergence of Probability* (2006), 32, from Austin, *How to Do Things with Words* (1962), 115. **28.** Hacking, *The Emergence of Probability* (2006), 39–48. **29.** Maclean, 'Foucault's Renaissance Episteme' (1998). **30.** Quintilian, *The Orator's Education* (2001), 325–7, 355–9. **31.** Hacking, *The Emergence of Probability* (2006), 46–7; LoLordo, *Pierre Gassendi* (2007), 94–9; on the classical doctrine, Allen, *Inference from Signs* (2001). **32.** Quoted from Eamon, *Science and the Secrets of Nature* (1994), 283; Latin text at Gassendi, *Opera omnia* (1727), Vol. 1, 108. **33.** Quintilian, *The Orator's Education* (2001), 375–415 (at 379), 461, 483, 501. **34.** For Gassendi on witnesses, Gassendi, *Opera omnia* (1727), 86. **35.** Quoted in Hacking, *The Emergence of Probability* (2006), 79 (my interpolations). **36.** Hooker, *Ecclesiasticall Politie* (1604), 100. **37.** Wilkins, *Mathematicall Magick* (1648), 120–1. **38.** Graunt, *Natural and Political Observations ... Made upon the Bills of Mortality* (1662), 20; and Petty, *A Treatise of Taxes* (1662), 27. Graunt also uses the phrase 'in all probability', *pace* Daston, *Classical Probability in the Enlightenment* (1988), 12. **39.** Clavius, *In sphaeram* (1585), 450–2, translation from Crombie, *Styles of Scientific Thinking* (1994), Vol. 1, 535–6. **40.** Boyle, *Some Considerations* (1663), 81 = Boyle, *The Works* (1999), Vol. 3, 255–6 (see Crombie, *Styles of Scientific Thinking* (1994), Vol. 2, 1175–6). **41.** Piccolomini, *La prima parte delle theoriche* (1558), 22v; translation from Crombie, *Styles of Scientific Thinking* (1994), Vol. 1, 532. **42.** Sprat, *The History of the Royal-Society* (1667), 100. **43.** Boyle, *New Experiments Physico-mechanical* (1660), 176 (Boyle, *The Works* (1999), Vol. 1, 218, where 'transmuted' is misread as 'transmitted'). But see Boyle's references to judges and witnesses (above, p. 280) and the discussion in Sargent, *The Diffident Naturalist* (1995), 42–61, esp. 54. I cannot find an early modern scientist who uses the word 'evidence' as freely as, for example, Glanvill in his *Vanity of Dogmatizing*. **44.** Boyle, *Hydrostatical Paradoxes* (1666), a1r = Boyle, *The Works* (1999), Vol. 5, 196. **45.** Quoted in Buchwald & Feingold, *Newton and the Origin of Civilization* (2013), 140–1. **46.** Bacon, *The Advancement of Learning* (1605), 31; and Brown, 'The Evolution of the Term "Mixed Mathematics"' (1991). **47.** Galileo's second letter on sunspots (1612), in Galilei & Scheiner, *On Sunspots* (2008), 107–70. **48.** '*Préface sur le traité du vuide*', in Pascal, *Oeuvres complètes* (1964), Vol. 2, 772–85. **49.** Dear, *Discipline and Experience* (1995), 15, 180; for an example published by Riccioli in 1651,

78. 50. Palmerino, 'Experiments, Mathematics, Physical Causes' (2010); and Westfall, 'Newton and the Fudge Factor' (1973). 51. The legal issues and their history were recently summarized in the House of Lords judgement on Regina *v.* Pendleton, 13 Dec. 2001. 52. Locke, *An Essay* (1690), 333. 53. Hobbes, *Humane Nature* (1650), 38–9; quoted in Hacking, *The Emergence of Probability* (2006), 48 54. Wotton, *Reflections upon Ancient and Modern Learning* (1694), 301. 55. Seneca, *Seneca's Morals Abstracted* (1679), Part 3, 99–100. 56. I thus have some sympathy with Feyerabend's claim that there is a radical incoherence in earlier accounts of the validity of empirical knowledge: Feyerabend, 'Classical Empiricism' (1970). 57. Locke, *An Essay* (1690), 353; on Locke on judgement, Laudan, 'Nature and Sources' (1967), 214–16. For an example of how judgement was exercised in day-to-day scientific procedure, Buchwald & Feingold, *Newton and the Origin of Civilization* (2013), 66–71: Hevelius would make repeated measurements to determine the position of a star; he would not then average the results but make a judgement as to which result was the right one. 58. Bates, *The Divinity of the Christian Religion* (1677), 41–2. 59. Sprat, *The History of the Royal-Society* (1667), 31, 360. 60. Wilkins, *An Essay towards a Real Character* (1668), 289–90. 61. Wilkins, *Natural Religion* (1675), 35–6. 62. Locke, *An Essay* (1690), 269. 63. Daston & Galison (eds.), *Objectivity* (2007). However, the discussion of 'objectivity' in Gaukroger, *The Emergence of a Scientific Culture* (2006), 239–45, is in part at least a discussion of 'judgement'. 64. Merton, 'Science and Technology in a Democratic Order' (1942), reprinted as Merton, 'The Normative Structure of Science' (1973). 65. Shapin & Schaffer, *Leviathan and the Air-pump* (1985), 72–6. 66. Kuhn, 'Mathematical versus Experimental Traditions' (1976), reprinted in Kuhn, *The Essential Tension* (1977). This line of argument is already sketched out in Conant, *Robert Boyle's Experiments in Pneumatics* (1950), 67–8 under the heading 'The Two Traditions'. 67. Locke, *An Essay*, Book 3, Ch. 6, §2.

CHAPTER 12

1. Collingwood, *The Idea of Nature* (1945), 8–9. 2. Cipolla, *Clocks and Culture, 1300–1700* (1967); White, 'The Medieval Roots of Modern Technology' (1978); Gimpel, *The Medieval Machine* (1976); Reynolds, *Stronger than a Hundred Men* (1983); North, *God's Clockmaker* (2005); and Walton, *Wind and Water* (2006). Some would argue that it did: Kaye, *Economy and Nature* (1998). On Renaissance and early modern machinery, Sawday, *Engines of the Imagination* (2007); and Rossi, *Philosophy, Technology and the Arts* (1970). 3. Drabkin & Drake (eds.), *Mechanics in Sixteenth-century Italy* (1969). 4. Kirk, Raven & others, *The Presocratic Philosophers* (1983), 410. 5. On the mechanical philosophy, Boas, 'The Establishment of the Mechanical Philosophy' (1952); and Dijksterhuis, *The Mechanization of the World Picture* (1961), translated from a Dutch original first published in 1950. 6. Berkel, *Isaac Beeckman* (2013), 83 and n. 42. 7. See above, p. 24. 8. More, *The Immortality of the Soul* (1659), preface

(b7r). More had earlier welcomed Cartesianism as a 'Fortification about Theology' to defend it from the assaults of the atheists: McGuire & Rattansi, 'Newton and the "Pipes of Pan"' (1966), 131. The classic study of More and Descartes is Webster, 'Henry More and Descartes, Some New Sources' (1969). **9.** Parker, *Disputationes de Deo et providentia divina* (1678), 64; and Bayle (ed.), *Nouvelles* (1684), Vol. 2, 753 **10.** Boyle, *A Defence* (1662), preface (*1v) = Boyle, *The Works* (1999), Vol. 3, 9; Boyle, *Experiments and Considerations Touching Colours* (1664), preface (A4r) = Boyle, *The Works* (1999), Vol. 4, 7: 'corpuscular philosophy'. See also Power, *Experimental Philosophy* (1664), preface (b2r): 'the Atomical and Corpuscularian Philosophers'. **11.** Boyle, *Nouveau traité* (1689), title page. **12.** Charleton, *Physiologia Epicuro-Gassendo-Charletoniana* (1654), 343–4. **13.** Popplow, 'Setting the World Machine in Motion' (2007): an article on which I draw extensively. **14.** Wilkins, *Natural Religion* (1675), 62. See John Dee's 1563 formula 'the huge frame of this world', quoted in Bennett (1986), 10. **15.** Shank, 'Mechanical Thinking' (2007), 19–22. On planetaria, see King & Millburn, *Geared to the Stars* (1978). **16.** Crombie, *Styles of Scientific Thinking* (1994), Vol. 1, 404. **17.** Popplow, 'Setting the World Machine in Motion' (2007), 57. **18.** Bedini, 'The Role of Automata' (1964), 29–30 **19.** De Caus, *Les Raisons des forces mouvantes* (1615). **20.** Baltrusaitis, *Anamorphoses* (1955), 37. **21.** Power still prefers 'engine' to 'machine'; and Boyle uses 'engine' much more often than he uses 'machine', as in 'Pneumatic Engine' (the air-pump) and 'living Engine' (biological creatures). The awkwardness of the linguistic shift is reflected in Boyle's frequent use of the phrase 'Mechanical Engine', as if there were any other sort. On 'engine', see Carroll, *Science, Culture and Modern State Formation* (2006), 30–2. **22.** *OED* engine 6a is 1538, and mechanical 5a (2nd example) is 1579–80; but machine 6b is 1659 (a translation from the French of Cyrano de Bergerac); mechanism 2a is 1665; and automaton 2b is 1664; both these last two examples are evidently influenced by Descartes. **23.** Mayr, *Authority, Liberty & Automatic Machinery* (1986), 63; and Gaukroger, *Descartes: An Intellectual Biography* (1995), 1. **24.** La Mettrie, *La Mettrie's 'L'homme machine': A Study* (1960). **25.** Riskin, 'The Defecating Duck' (2003); Schaffer, 'Enlightened Automata' (1999); and Standage, *The Turk: The Life and Times of the Famous Eighteenth-century Chess-playing Machine* (2002). **26.** Mayr, *Authority, Liberty & Automatic Machinery* (1986), 64; Schuster, 'Waterworld' (2005). **27.** Laudan, 'The Clock Metaphor and Probabilism' (1966). **28.** *pace* Power, *Experimental Philosophy* (1664), preface (b2r), who thought he might see atoms. **29.** De Caus, *Les Raisons des forces mouvantes* (1615), Book 1, problem 6. **30.** Mayr, *Authority, Liberty & Automatic Machinery* (1986), 190–3, which misses de Caus. **31.** Mayr, *Authority, Liberty & Automatic Machinery* (1986), 155–80; Wootton, 'Liberty, Metaphor and Mechanism' (2006). **32.** Finley, 'Aristotle and Economic Analysis' (1970). **33.** Hacking, *The Emergence of Probability* (2006), 82–3. **34.** More, *Divine Dialogues* (1668), 20–8; and Lessius, *Rawleigh, His Ghost* (1631), 27–41; on Lessius, Franklin, *The Science of Conjecture* (2001), 244–5. **35.** Cicero, *De natura deorum* (1933), 213. **36.** Above, p. 265. **37.** Boyle, *A Free Enquiry* (1686), 11–12 = Boyle, *The Works*

(1999), Vol. 10., 448. 38. Bedini, 'The Role of Automata' (1964), 37–9; Riskin, 'The Defecating Duck' (2003), 625–9. 39. See above, p. 27n. 40. Mayr, *Authority, Liberty & Automatic Machinery* (1986), 57, 64–5, 92. 41. Yolton, *Thinking Matter* (1983). 42. Newton, *Unpublished Scientific Papers* (1962), 142–4 (Latin on 109–10). 43. Cook, 'Divine Artifice and Natural Mechanism' (2001). The revolt against teleology, while it made possible the Scientific Foundation, simultaneously destroyed the traditional framework of ethical discourse: MacIntyre, *After Virtue* (1981). For the claim that even this overstates the importance of the mechanical philosophy: Chalmers, 'The Lack of Excellency of Boyle's Mechanical Philosophy' (1993); Chalmers, *The Scientist's Atom and the Philosopher's Stone* (2009); and Chalmers, 'Intermediate Causes and Explanations' (2012). 44. Voltaire, *Letters Concerning the English Nation* (1733), 109–11. Compare 'After Copernicus astronomers lived in a different world' (Kuhn, *Structure* (1962), 117). 45. Pascal, *Pensées* (1958), 61. 46. 'Science as a Vocation' (1918): Weber, *The Vocation Lectures* (2004), 13.

CHAPTER 13

1. Weber, *The Vocation Lectures* (2004), 12–13. 2. Hunter, 'New Light on the "Drummer of Tedworth"' (2005). 3. Glanvill, *Saducismus triumphatus* (1681), 93–4. 4. Glanvill, *Saducismus triumphatus* (1681), 81. 5. Jobe, 'The Devil in Restoration Science' (1981). 6. McLaughlin, 'Humanist Concepts of Renaissance and Middle Ages in the Tre- and Quattrocento' (1988), 135; Considine, *Dictionaries in Early Modern Europe* (2008), 259–61. 7. Thomas, *The Ends of Life* (2009), 4, on 'early modern'. 8. Tassoni, *Dieci libri di pensieri diversi* (1627); Rossi, *Philosophy, Technology and the Arts* (1970), 91–3; and Hale, *The Civilization of Europe in the Renaissance* (1993), 589–90. 9. Walsham, 'The Reformation and "The Disenchantment of the World" Reassessed' (2008). 10. Jones, *Ancients and Moderns* (1936); and, primarily on the literary aspects, Levine, *The Battle of the Books* (1991). On Wotton, Hall, 'William Wotton and the History of Science' (1949); and on Swift, Elias, *Swift at Moor Park* (1982). 11. Jones, *Ancients and Moderns* (1936), treated the debate over science as the central issue in the battle between the ancients and the moderns; the emphasis was corrected by Levine, *The Battle of the Books* (1991) and Levine, *Between the Ancients and the Moderns* (1999), but in the process the debate over science, and with it Wotton's reply to Temple, were lost to view. 12. Fontenelle, *Entretiens sur la pluralité des mondes* (1955). 13. The surviving fragment is printed in Hunter (ed.), *Robert Boyle: By Himself and His Friends* (1994), 111–48. 14. Swift, *A Tale of a Tub* (2010), 199–200; Wotton, *A Defense of the Reflections* (1705), 14–15, 45–7; and Elias, *Swift at Moor Park* (1982), 76–7. Note especially Wotton's suggestion that Swift's use of the hiatus was in imitation of Temple's, which slyly implies that Swift may have had a hand in Temple's text. For an echo of Temple's (or Swift's) passage in the *Battle*, see Elias, *Swift at Moor Park* (1982), 298 (but note that spittle was an important part of the healing performances of

Valentine Greatrakes, who was endorsed by Boyle). **15.** Lynall, *Swift and Science* (2012). **16.** Temple, *Miscellanea. The Third Part* (1701), 281–3. **17.** Elias, *Swift at Moor Park* (1982), 116–20, 191. **18.** Swift, *A Tale of a Tub* (2010), 155. **19.** Wotton, *Reflections upon Ancient and Modern Learning* (1694), 206–18; and Wotton & Bentley, *Reflections upon Ancient and Modern Learning. The Second Part* (1698), 46–53. **20.** Hall, 'William Wotton and the History of Science' (1949), 1061–2. **21.** Wotton, *Reflections upon Ancient and Modern Learning* (1694), 3. **22.** Wotton, *Reflections upon Ancient and Modern Learning* (1694), 300–1. The numbering is Wotton's, but the paragraph divisions are mine. **23.** Wotton may well have been right: Womersley, 'Dean Swift Hears a Sermon' (2009). **24.** Wotton, *Reflections upon Ancient and Modern Learning* (1694), 128–30; and Temple, *Miscellanea. The Third Part* (1701), 292–5. **25.** Thomas, *Religion and the Decline of Magic* (1997); see Macfarlane, 'Civility and the Decline of Magic' (2000). **26.** *Spectator*, no. 117, 14 July 1711: Addison and Steele (eds.), *The Spectator* (1712), Book 2, 189. **27.** Bostridge, *Witchcraft and Its Transformations* (1997). **28.** For example, Bentley, *Remarks upon a Late Discourse of Free-thinking* (1713), 33. **29.** Macfarlane, 'Civility and the Decline of Magic' (2000). **30.** Hunter, 'Science and Heterodoxy' (1990); Hunter, 'The Royal Society and the Decline of Magic' (2011); and Hunter, 'The Decline of Magic' (2012) are crucial. **31.** On Sprat, Wood, 'Methodology and Apologetics' (1980). **32.** Schaffer, 'Halley's Atheism and the End of the World' (1977); *ODNB*, 'Saunderson, Nicholas'; and Tunstall & Diderot, *Blindness and Enlightenment* (2011), 41–6. **33.** Hunter, *Boyle* (2009). **34.** Schaffer, 'Godly Men and Mechanical Philosophers' (1987); and Webster, 'Henry Power's Experimental Philosophy' (1967), 173–6. **35.** Webster, *The Displaying of Supposed Witchcraft* (1677), 203–4. **36.** Webster, *The Displaying of Supposed Witchcraft* (1677), 147–8, 197–215. **37.** Yolton, *Thinking Matter* (1983). **38.** Hunter, 'The Decline of Magic' (2012), 405–8. Hunter explores the extent to which this tradition survived after 1691, but the conclusion would seem to be: barely. **39.** Glanvill, *Saducismus triumphatus* (1681), preface (F3r). **40.** Hunter, 'New Light on the "Drummer of Tedworth"' (2005). **41.** Hunter, *The Occult Laboratory* (2001). **42.** Jobe, 'The Devil in Restoration Science' (1981). **43.** Wootton, 'Hutchinson, Francis' (2006); and Trenchard & Gordon, *Cato's Letters, or, Essays on Liberty, Civil and Religious, and Other Important Subjects* (1995), Vol. 3, no. 79, 2 June 1722. **44.** Gulliver, *The Anatomist Dissected* (1727), 5–6. **45.** Wootton, 'Hume's "Of Miracles"' (1990). **46.** Shank, *The Newton Wars* (2008). **47.** Valenza, *Literature, Language* (2009), 58. **48.** Haugen, *Richard Bentley* (2011). **49.** Newton to Bentley, 10 December 1692: Bentley, *The Correspondence* (1842), 47; see also Stewart, *The Rise of Public Science* (1992), 31–59. **50.** Bentley, *The Folly and Unreasonableness of Atheism* (1692), 225, 102, 277. **51.** Croft, *Some Animadversions* (1685), 40–1. **52.** Stewart, *The Rise of Public Science* (1992), 33–7, 41, 67–73. **53.** Bentley, *Remarks upon a Late Discourse of Free-thinking* (1713), 33–4. **54.** Sprat, *The History of the Royal-Society* (1667), 362, 360. **55.** Spencer, *A Discourse Concerning Prodigies* (1663), 76. **56.** Burns, '"Our Lot is Fallen into an Age of Wonders"' (1995); Burns, *An*

Age of Wonders (2002). **57.** Sprat, *The History of the Royal-Society* (1667), 360. **58.** Spencer, *A Discourse Concerning Prodigies* (1663), 11–12 (square brackets Spencer's). **59.** Obviously, this process can be read in Habermasian terms: Broman, 'The Habermasian Public Sphere' (1998). **60.** Du Châtelet, *Selected Philosophical and Scientific Writings* (2009). **61.** Findlen, *A Forgotten Newtonian* (1999); and Cieslak-Golonka & Morten, 'The Women Scientists of Bologna' (2000). **62.** Valenza, *Literature, Language* (2009), 78–86; and Feingold, *The Newtonian Moment* (2004), 119–41. **63.** Johnson, 'The Vanity of Authors' (1752), 53. **64.** Wigelsworth, *Selling Science in the Age of Newton* (2011), 147–74; Jacob & Stewart, *Practical Matter* (2004), 61–92; Stewart, *The Rise of Public Science* (1992), 94–182; and Carpenter, *John Theophilus Desaguliers* (2011). **65.** Valenza, *Literature, Language* (2009), argues that 'difficulty itself was translated by his followers into a commercial product' (55).

CHAPTER 14

1. Boyle, *Some Considerations* (1663), Vol. 2, 3 = Boyle, *The Works* (1999), Vol. 2, 64. **2.** Bacon, *Works* (1857), Vol. 1, 500 (my translation). **3.** Quoted in Papin, *A Continuation of the New Digester* (1687), 105. **4.** See, for example, Thomas Shadwell's *The Virtuoso* (1676); and Carroll, *Science, Culture and Modern State Formation* (2006), 40–3. **5.** Swift, *Gulliver's Travels* (2012), 257–8. **6.** Hessen (1931) reprinted in Hessen & Grossman, *The Social and Economic Roots of the Scientific Revolution* (2009) (quotation on 56); Merton, 'Science, Technology and Society' (1938); for a Mertonian approach, see Webster, *The Great Instauration* (1975); and for a more recent study, Westfall, 'Science and Technology' (1997) (although Westfall should not be taken to be a follower of Hessen: see Ravetz & Westfall, 'Marxism and the History of Science' (1981)). If the old, Mertonian sociology of science assumed science served practical purposes, the new, post-Foucauldian sociology/history of science has had surprisingly little to say about technology in the Scientific Revolution. Thus Shapin, 'Understanding the Merton Thesis' (1988), offers a valuable re-reading of Merton's 1938 text but offers no judgement on either the Hessen thesis or the Merton thesis. **7.** The classic article is Fleming, 'Latent Heat and the Invention of the Watt Engine' (1952); for recent scholarship, see Miller, *James Watt, Chemist* (2009). **8.** This is generally quoted as 'attributed'. On Google Books the quotation appears unattributed from 1957; the attribution to Henderson and the date first appear in 1963. **9.** Hall, 'Engineering and the Scientific Revolution' (1961), 337; Hall, 'What Did the Industrial Revolution in Britain Owe to Science?' (1974); and Kuhn, 'The Principle of Acceleration: A Non-dialectical Theory of Progress: Comment' (1969). Hall's argument is already sketched out in Conant, *Robert Boyle's Experiments in Pneumatics* (1950), 69–70. **10.** Hall, for example, was one of the editors of the standard Oxford history of technology. His position was first articulated in Hall, *Ballistics in the Seventeenth Century* (1952); see especially Hall, 'What Did the Industrial Revolution in Britain Owe to Science?'

(1974); Hall, 'Engineering and the Scientific Revolution' (1961) (contrast Kerker, 'Science and the Steam Engine' (1961)); and Singer, Hall & others, *A History of Technology* (1954). A classic essay is Layton Jr, 'Technology as Knowledge' (1974). I have found Wengenroth, 'Science, Technology and Industry' (2003), helpful. **11.** Mokyr, *The Enlightened Economy* (2009); Mokyr, *The Lever of Riches* (1990); Mokyr, *The Gifts of Athena* (2004); Mokyr, 'The Intellectual Origins of Modern Economic Growth' (2005); van Zanden, *The Long Road to the Industrial Revolution* (2009); and Allen, *The British Industrial Revolution* (2009); for a comparison of Mokyr and Allen, see Crafts, 'Explaining the First Industrial Revolution' (2011). For an earlier study, Musson & Robinson, *Science and Technology* (1969). For a Marxisant approach, Jacob, *Scientific Culture and the Making of the Industrial West* (1997). H. Floris Cohen is almost alone among contemporary historians of science in emphasizing the contribution of science to industry: Cohen, 'Inside Newcomen's Fire Engine' (2004). And yet his major work, Cohen, *How Modern Science Came into the World* (2010), does not discuss the topic. **12.** Hall, 'What Did the Industrial Revolution in Britain Owe to Science?' (1974), 136. **13.** Segre, 'Torricelli's Correspondence on Ballistics' (1983). **14.** Steele, 'Muskets and Pendulums' (1994). Hall, with characteristic pessimism, pushes this revolution back into the nineteenth century: Hall, 'Engineering and the Scientific Revolution' (1961), 334; and Hall, *Ballistics in the Seventeenth Century* (1952), 159. **15.** Bedini, *The Pulse of Time* (1991); and Wootton, *Galileo* (2010), 130–1, 167, 169. **16.** Brown, *Jean Domenique Cassini and His World Map of 1696* (1941), 39, 47, 58–60; Brotton, *A History of the World in Twelve Maps* (2012), 306. **17.** Pumfrey, 'O *tempora, O magnes!*' (1989); see also Waters, 'Nautical Astronomy and the Problem of Longitude' (1983). **18.** Sobel, *Longitude* (1995). **19.** Allen, *The British Industrial Revolution* (2009), 173. **20.** Allen, *The British Industrial Revolution* (2009), 204–6. On the introduction of interchangeable parts, see Alder, 'Making Things the Same' (1998). **21.** This line of argument is foreshadowed in Koyré, '*Du monde de l'à-peu-près à l'univers de la précision*' (1971), first published in 1948. **22.** Landes, 'Why Europe and the West?' (2006). **23.** Latin text of letter to Herwart von Hohenburg quoted in Koyré, *The Astronomical Revolution* (1973), 378; translation from Snobelen, 'The Myth of the Clockwork Universe' (2012), 177 n. 18. **24.** Patrick, *A Brief Account of the New Sect of Latitude-Men* (1662), 19. **25.** Maffioli, *Out of Galileo* (1994); and Maffioli, *La via delle acque* (2010). **26.** Desaguliers, *A Course of Experimental Philosophy* (1734), 532. **27.** Smeaton, *An Experimental Enquiry* (1760); Schaffer, 'Machine Philosophy' (1994); and Reynolds, *Stronger than a Hundred Men* (1983). **28.** *OED*, s.v. 'civil'. **29.** Stewart, 'A Meaning for Machines' (1998), 272–6. **30.** Rolt & Allen, *The Steam Engine of Thomas Newcomen* (1977), 145. **31.** Allen, *The British Industrial Revolution* (2009), 173. **32.** Ketterer, 'The Wonderful Effects of Steam' (1998). **33.** Galloway & Hebert, *History and Progress of the Steam Engine with a Practical Investigation of Its Structure and Application* (1836), preface (i). **34.** Allen, *The British Industrial Revolution* (2009), 25–56. **35.** Allen, *The British Industrial Revolution* (2009), 157–8. **36.** *OED* s.v.

'wind-gun'; Wilkins, *Mathematicall Magick* (1648), 153; Bertoloni Meli, *Thinking with Objects* (2006), 130; and Boyle, *A Continuation of New Experiments* (1682), 16–18 = Boyle, *The Works* (1999), Vol. 9, 147–9. **37.** A survey of the early history of the steam engine is provided by Dickinson, *A Short History of the Steam Engine* (1963), 1–17. Von Guericke designed a vacuum-powered gun of which Papin made a version: Papin, 'Shooting by the Rarefaction of the Air' (1686); Papin then calculated the speed at which air would enter the exhausted cylinder, Papin, 'A Demonstration' (1686). **38.** Boyle, *A Continuation of New Experiments* (1682), preface to the Latin edition (A3v–a1r) = Boyle, *The Works* (1999), Vol. 9, 124–5. I take it that Papin was responsible not just for the experimental notes (in French) but also for the Latin text: it is hard to see what else Boyle's statement that the 'style' and 'choice of words' were his can mean (cf. Shapin, 'Boyle and Mathematics' (1988), 35). **39.** Tönsmann, '*Wasserbauten und Schifffahrt in Hessen*' (2009). Papin makes a major appearance in Shapin, 'The Invisible Technician' (1989) and Shapin, *A Social History of Truth* (1994) as an anomalous case of a 'technician' whom Boyle identifies by name. Shapin nowhere mentions that Papin was elected FRS (presumably in acknowledgement of his contribution to the *Continuation*), which was a clear recognition that he was more than a technician: Hunter, *The Royal Society and Its Fellows* (1982), 87, 133 n. 3, F369. **40.** Dickinson, *A Short History of the Steam Engine* (1963), 9–11. Further details on Papin in Galloway, *The Steam Engine and Its Inventors* (1881); Ernouf, *Denis Papin* (1883); Wintzer, *Dénis Papins Erlebnisse in Marburg, 1688–1695* (1898); Schaffer, 'The Show that Never Ends' (1995), 13–14 (but Schaffer confuses the atmospheric piston with the Hessian pump, a centrifugal pump or bellows described in Papin, *Recueil de diverses pièces* (1695); and he repeats the old myth that Papin built a steam-powered boat, on which, see below); Stewart, *The Rise of Public Science* (1992), 24–7 (but 'Hellish Bellows' on p. 25 is a misreading for 'Hessian Bellows'), 131–2, 175–8; Shapin, previous note; Boschiero, 'Translation, Experimentation and the Spring of the Air: Richard Waller's "Essayes of Natural Experiments"' (2009); and Ranea, 'Theories, Rules and Calculations' (2015), which I saw too late to take account of. The Latin text with German translation of Papin's *Nova methodus* (1690) is in Tönsmann & Schneider (eds.), *Denis Papin* (2009), 136–41, and there is a French translation in Ducoux, *Notice sur Denis Papin* (1854), 56–63 and in Figuier, *Exposition et histoire* (1851), Vol. 3, 419–23; there is an English translation of Papin's *Nouvelle manière pour lever l'eau par la force du feu* in Smith, 'A New Way of Raising Water by Fire' (1998). Papin's works were collected in Papin, *La Vie et les ouvrages de Denis Papin* (1894); the first volume reprints Péan & La Saussaye, *La Vie et les ouvrages* (1869) (the only volume printed). Of an intended eight volumes, six and and a half were, it seems, printed and bound, but all (I suspect) are missing the planned plates. This extremely rare work is apparently available on the internet in the USA (at http://www.hathitrust.org), but not currently in Europe. Worldcat lists one 'complete' set, the one copied for the Web, at Wisconsin, and there is another in the municipal library at Blois: Smith, 'A New Way of Raising Water by Fire' (1998), 178–9. (The copy at Oklahoma is missing Vol. 1, and all the other copies listed are missing several volumes. There is not a single complete copy listed in COPAC (Consortium of Online Public Access Catalogues). The copy I cite is my own, seemingly one of four extant complete sets.) The *Works* contains the complete Papin–Leibniz correspondence in Vols. 7 & 8 – a fact which has escaped authors on the *vis viva* controversy, most recently Rey, 'The Controversy between Leibniz and Papin' (2010) (whose dating of the letters sometimes varies from that of the *Works*). An extensive selection from the Papin–Leibniz correspondence, along with other letters, was published in Leibniz, Huygens & others, *Leibnizens und Huygens' Briefwechsel mit Papin* (1881), of which there is a modern reprint. This edition was unknown to the editors of the *Works*. To Papin's works must be added 'Dr Pappins Letter containing a Description of a Wind-fountain', an account of a fountain powered by compressed air, in Hooke, *Lectures de potentia restitutiva* (1678), 25–8. **41.** Letter to Leibniz, 25 July 1698 (Papin, *La Vie et les ouvrages de Denis Papin* (1894), Vol. 8, 17–19 = Leibniz, Huygens & others, *Leibnizens und Huygens' Briefwechsel mit Papin* (1881), 233–4). There is, unfortunately, no way of knowing how this worked; but Papin was considering pistons driven by atmospheric pressure in 1704: letter to Leibniz, 13 March (Papin, *La Vie et les ouvrages de Denis Papin* (1894), Vol. 8, 151–4 = Leibniz, Huygens & others, *Leibnizens und Huygens' Briefwechsel mit Papin* (1881), 284–7: a reply, it seems to me, to Leibniz's undated letter at Papin, *La Vie et les ouvrages de Denis Papin* (1894), Vol. 8, 215–19 and at Leibniz, Huygens & others, *Leibnizens und Huygens' Briefwechsel mit Papin* (1881), 276–80). **42.** Letter to Leibniz, 13 March 1704, and Leibniz's undated letter, as above. **43.** Letter to Leibniz, 7 September 1702 (Papin, *La Vie et les ouvrages de Denis Papin* (1894), Vol. 8, 126–9 = Leibniz, Huygens & others, *Leibnizens und Huygens' Briefwechsel mit Papin* (1881), 264–7; and subsequent correspondence until 23 March 1705 (Papin, *La Vie et les ouvrages de Denis Papin* (1894), Vol. 8, 223–6 = Leibniz, Huygens & others, *Leibnizens und Huygens' Briefwechsel mit Papin* (1881), 342–4). **44.** The interpretation of this sketch is controversial. Dickinson thought, correctly in my view, that it was an atmospheric-pressure engine, but see the note of his editors, who argue it was a high-pressure engine: Dickinson, *Sir Samuel Morland* (1970), 79. North evidently thought that the engine worked on the up-stroke, but such a misunderstanding is hardly surprising if he only saw a drawing or maquette. In any case, as Papin worked with both atmospheric-pressure and high-pressure systems, the issue may not be important for present purposes. Rhys Jenkins, Dickinson & Rolt attribute the engine to Morland (Rolt & Allen, *The Steam Engine of Thomas Newcomen* (1977), 18–19), which implies that North was sketching from memory an engine he had seen many years earlier. They do not note the evident similarity to Papin's drive mechanism, or the ambiguity of the word 'model'. Wallace attributes the engine to Savery (Wallace, *The Social Context of Innovation* (1982), 58–60), despite Savery's conviction that pistons were impractical because there would be too much friction to overcome. The attribution to Papin, or, at the limit, someone working from Papin's publications, seems to me secure. **45.** It would appear, however, to carry the modern meaning when Savery contrasts a model with a draft (or drawing): Smith, 'A New Way of Raising Water by Fire' (1998), 172. **46.** Gibbon, *A Summe or Body of Divinitie Real* (1651). **47.** Desaguliers, *A Course of Experimental Philosophy* (1734), 465–6. **48.** Dickinson, *A Short History of the Steam Engine* (1963), 18–27. **49.** Gaulke, '*Die Papin–Savery-Kontroverse*' (2009). **50.** Papin, *Nouvelle manière pour élever l'eau* (1707). Newcomen discovered by chance that it was better to inject the water into the cylinder: Desaguliers, *A Course of Experimental Philosophy* (1734), 533. **51.** Savery, *Navigation Improv'd* (1698); Papin, *La Vie et les ouvrages de Denis Papin* (1894), Vol. 1, 206–7; letters to Leibniz 13 March 1704, 7 July 1707 (Papin, *La Vie et les ouvrages de Denis Papin* (1894), Vol. 8, 280–2 = Leibniz, Huygens & others, *Leibnizens und Huygens' Briefwechsel mit Papin* (1881), 378–80); and letter of Drost von Zeuner to Leibniz, 29 September 1707 (Papin, *La Vie et les ouvrages de Denis Papin* (1894), Vol. 8, 294 = Leibniz, Huygens & others, *Leibnizens und Huygens' Briefwechsel mit Papin* (1881), 385): '*sa petite machine d'un vaisseau à roues*'; Leibniz to Sloane, quoted in Tönsmann, '*Wasserbauten und Schifffahrt in Hessen*' (2009), 99. **52.** The myth originates with Figuier, *Exposition et histoire* (1851), Vol. 3, 70–106, 419–32 (which becomes Vol. 1 in later editions). Gerland, '*Das sogenannte Dampfschiff Papin's*' (1880) gets the facts right. The false story continues to be told (http://en.wikipedia.org/wiki/Denis_Papin, accessed 3 June 2014); see also Gerth, '*Der Dampfkochtopf = Digestor – Eine Erzählung*' (1987); and above, note 40. http://www.schillerinstitute.org/educ/pedagogy/steam_engine.html for the conspiracy. **53.** Smith, 'A New Way of Raising Water by Fire' (1998), 169–77. **54.** cf. Leibniz to Papin, 24 or 27 June 1699 (Papin, *La Vie et les ouvrages de Denis Papin* (1894), Vol. 8, 101–2, 303–4 = Leibniz, Huygens & others, *Leibnizens und Huygens' Briefwechsel mit Papin* (1881), 248–9); Stewart, *The Rise of Public Science* (1992), 14–15; Boas Hall, *Promoting Experimental Learning* (1991), 122; and Heilbron, *Physics at the Royal Society* (1983), esp. 14, 21, 31, 43. **55.** Savery identified this as a problem (Smith, 'A New Way of Raising Water by Fire' (1998), 174). **56.** Hence Papin's view that his steamboat could not run on the Fulda but needed to be tried at a seaport: letter to Leibniz, 7 July 1707. See also his letter to Leibniz, 15 September 1707: '*Je suis persuadé que si Dieu me fait la grâce d'arriver heureusement à Londres et d'y faire des vaisseaux de cette construction qui aient assez de profondeur pour appliquer la machine à feu à donner le mouvement aux rames* [paddles], *je suis persuadé, dis-je, que nous pourrions produire des effects qui paroïtront incroyables* . . .' (Papin, *La Vie et les ouvrages de Denis Papin* (1894), Vol. 8, 291–3 = Leibniz, Huygens & others, *Leibnizens und Huygens' Briefwechsel mit Papin* (1881), 383–5). Tönsmann, '*Wasserbauten und Schifffahrt in Hessen*' (2009), correctly identifies the claim that Papin had built a working steamboat as a legend, but mistakenly assumes that Papin's plans for a steamboat relied on his atmospheric-pressure-driven piston of 1690 rather than his Savery-type engine of 1707. **57.** Smith, 'A New Way of Raising Water by Fire' (1998), 169. **58.** Royal Society archives; mistranscribed at Papin, *La Vie et les ouvrages de Denis Papin* (1894), Vol. 7, 74. **59.** De la Saussaye, following Bannister, *Denis Papin: Notice sur sa vie et ses écrits* (Blois: F Jahyer, 1847), 23 (worldcat lists only one copy of this work; another is in this author's collection), quotes undated passages from Leibniz's correspondence which he claims establish that Papin was in Hesse in 1714: Papin, *La Vie et les ouvrages de Denis Papin* (1894), Vol. 1, 251–2. But see Leibniz, Huygens & others, *Leibnizens und Huygens' Briefwechsel mit Papin* (1881), 114, 256–60. **60.** Rolt & Allen, *The Steam Engine of Thomas Newcomen* (1977), 39. **61.** Wallace, *The Social Context of Innovation* (1982), 60–1. **62.** Rolt & Allen, *The Steam Engine of Thomas Newcomen* (1977), 38–9. **63.** Rosen, *The Most Powerful Idea in the World* (2010), 31 and note; compare Rolt & Allen, *The Steam Engine of Thomas Newcomen* (1977), 36. **64.** Compare Mokyr, 'The Intellectual Origins of Modern Economic Growth' (2005), 298 note, with Wallace, *The Social Context of Innovation* (1982), 55–6. **65.** Quoted Wallace, *The Social Context of Innovation* (1982), 56. **66.** Anon, 'Account of Books' (1697). **67.** Dickinson, *Sir Samuel Morland* (1970), 57–8. **68.** Desaguliers, *A Course of Experimental Philosophy* (1734), 472; Hills, *Power from Steam* (1989), 33; Savery's engine lacked a safety valve (steam could blow straight through it if the water was not being raised too high); Desaguliers introduced one as an improvement in 1717. **69.** This is sometimes said to be a later development, but see Rolt & Allen, *The Steam Engine of Thomas Newcomen* (1977), 79–80. The valves could be opened and closed by hand in a test run, but the machine would have to operate too slowly to be practical without automation. **70.** Leibniz had grasped this principle: letter to Papin, 28 August 1698 (Papin, *La Vie et les ouvrages de Denis Papin* (1894), Vol. 8, 28–32 = Leibniz, Huygens & others, *Leibnizens und Huygens' Briefwechsel mit Papin* (1881), 239). **71.** Desaguliers, *A Course of Experimental Philosophy* (1734), Vol. 2, 489–90. **72.** Hills, *Power from Steam* (1989), 21–2. **73.** The English Short Title Catalogue lists twenty-eight copies, compared, for example, to thirty-nine copies of Robert Boyle's *Continuation of New Experiments* of 1682; and we can be confident a much higher proportion of Boyle's book survived than of this one, which has some obvious characteristics of an ephemeral publication. A French translation of both the original *New Digester* and the *Continuation* appeared in 1688: Papin, *La Manière d'amollir les os* (1688). **74.** No mention, for example, in the literature on the air-pump: Andrade, 'The Early History of the Vacuum Pump' (1957); van Helden, 'The Age of the Air-pump' (1991); and Schimkat, '*Denis Papin und die Luftpumpe*' (2009); or the classic study, Wilson, 'On the Early History of the Air-pump in England' (1849). **75.** This is presumably the model used for the experiments in Boyle, *A Continuation of New Experiments* (1682), which Boyle tells us in the preface differed from his own air-pump (also designed by Papin), which is described and illustrated at the beginning of Boyle, *Experimentorum novorum* (1680) and Boyle, *A Continuation of New Experiments* (1682) (= Boyle, *The Works* (1999), Vol. 9, 134; the copy of Boyle, *A Continuation of New Experiments* (1682) on EEBO is defective in that it lacks the illustrations). However, it incorporates an improvement dating to 1684: Papin, *La Vie et les ouvrages de Denis Papin* (1894), Vol. 5, 7–11. **76.** Papin, *A Continuation of the New Digester* (1687), 45. **77.** Desaguliers, *A Course of Experimental Philosophy* (1734), Vol. 2, 470, 482–3, 533. **78.** Papin, *A Continuation* *of the New Digester* (1687), 54–5. **79.** Papin, *A Continuation of the New Digester* (1687), 41, 48, 116. **80.** Desaguliers, *A Course of Experimental Philosophy* (1734), Vol. 2, 474 (see also 532–3). **81.** Desaguliers, *A Course of Experimental Philosophy* (1734), Vol. 2, 468.

CHAPTER 15

1. Galilei, *Le opere* (1890), Vol. 7, 78 (my translation). **2.** Mill, *Principles of Political Economy* (1909), Bk 4, Ch. 1, §2. And yet Chunglin Kwa writes: 'The idea that scientific progress has led to the constant expansion of our power over nature is a romantic myth.' Kwa, *Styles of Knowing* (2011), 11. **3.** Gray, *Heresies* (2004), 3. **4.** Sarton, *The Study of the History of Science* (1936), 5. **5.** Koyré, *Études Galiléennes* (1966), 11. **6.** Kuhn, *The Essential Tension* (1977): see index under 'progress of science'. **7.** Kuhn, *Structure* (1970), 170. **8.** Rorty, 'Science as Solidarity' (1991), 39. See also Rorty, 'Thomas Kuhn, Rocks and the Laws of Physics' (1999), at 179–80. For a swift and efficient critique of Rorty on science, see Williams, *Essays and Reviews, 1959–2002* (2014), 204–15. **9.** Quine, 'Two Dogmas of Empiricism' (1951). That the Duhem-Quine thesis is not universally true, and that its applicability must be demonstrated for each particular case, is proven by Grünbaum, 'The Duhemian Argument' (1960). Moreover, Duhem never held the Duhem-Quine thesis: Ariew, 'The Duhem Thesis' (1984). Laudan, 'Demystifying Underdetermination' (1990), provides a devastating critique of misconceived applications of the thesis. Popper had already addressed the Duhem thesis in 1935: Popper, *The Logic of Scientific Discovery* (1959), 42, 78–84. **10.** Biagioli, 'Scientific Revolution, Social Bricolage and Etiquette' (1992); Johns, *The Nature of the Book* (1998) (and see Eisenstein, 'An Unacknowledged Revolution Revisited' (2002); Johns, 'How to Acknowledge a Revolution' (2002)); Livingstone & Withers (eds.), *Geography and Revolution* (2005); Ogborn & Withers, 'Book Geography, Book History' (2010). Localism and contingency are fundamental to the strong programme: *cf.* Bloor, 'Anti-Latour' (1999). For an expression of concern that localism has not been balanced by a study of how science travels, Secord, 'Knowledge in Transit' (2004), 660; and, for a bold but misconceived attempt to escape from the localism/globalism binary, see Latour, *We Have Never Been Modern* (1993). **11.** Hunter, *The Royal Society and Its Fellows* (1982), 107. **12.** In this respect the approach of Horton, *Patterns of thought* (1997) seems to me exemplary. **13.** cf. Bourdieu, *Science of Science* (2004). **14.** Hacking, 'How Inevitable are the Results of Successful Science?' (2000), 64–6, and Hacking, *The Social Construction of What?* (1999), 163–5. **15.** Cohen, 'Roemer and the First Determination of the Velocity of Light (1676)' (1940); Van Helden, 'Roemer's Speed of Light' (1983); Kristensen & Pedersen, 'Roemer, Jupiter's Satellites and the Velocity of Light' (2012). **16.** The figures in the first table are from Boyer, 'Early Estimates of the Velocity of Light' (1941), and in the second from Fowles, *Introduction to Modern Optics* (1989), 6 and https://en.wikipedia.org/wiki/Speed_of_light#First_measurement_attempts

(accessed 8 December 2014). See also MacKay & Oldford, 'Scientific Method, Statistical Method and the Speed of Light' (2000). **17.** Hasok Chang avoids the word 'competition', preferring 'epistemic iteration': Chang, *Inventing Temperature* (2004), 44–8, 212–17, 226–31. **18.** Willmoth, 'Römer, Flamsteed, Cassini and the Speed of Light' (2012), 49. **19.** Schaffer, 'Glass Works' (1989), 100, **20.** Shapiro, 'The Gradual Acceptance of Newton's Theory of Light' (1996). A valuable alternative to Schaffer's account of the opposition to Newton's first publication is provided by Bechler, 'Newton's 1672 Optical Controversies' (1974); for Bechler the central issue was not replication, but Newton's claim to certainty. **21.** Whitley, 'Black Boxism' (1970); Callon, '*Boîtes noires*' (1981); Pinch, 'Opening Black Boxes' (1992); the paradigmatic usage is established by Pinch, *Confronting Nature* (1986). **22.** Boyle, *Certain Physiological Essays* (1661), 27–8 = Boyle, *The Works* (1999), 2:26 (which I follow in adopting '(not the Arteries)' from the second edition). **23.** Mela, *De orbis situ libri tres. Adiecta sunt praeterea loca aliquot ex Vadiani commentariis* (1530), S2(rv). **24.** Brook, *Vermeer's Hat* (2008), 26–53. **25.** Williams, *Voyages of Delusion* (2002); Fleming, *Barrow's Boys* (1998). **26.** Alessandro Achillini claims that *if* the earth were luminous, then from a distance the earth would shine like the moon, or even one of the planets: Achillini, *De Elementis* (1505), 85r. But immediately before he maintains that the earth is *not* luminous. **27.** Wootton, *Galileo* (2010), 64. **28.** Palmieri, 'Galileo and the Discovery of the Phases of Venus' (2001). **29.** Jardine, *The Scenes of Inquiry* (2000). **30.** Lakatos, *The Methodology of Scientific Research Programmes* (1978). **31.** For reasonably sophisticated defences of realism see Leplin (ed.), *Scientific Realism* (1984). **32.** For an influential critique of path-dependency, Pinch & Bijker, 'The Social Construction of Facts and Artefacts' (1987); for an insistence on its importance, Pickering, *The Mangle of Practice* (1995), 185, 209. It is necessary here to distinguish between the actual path followed by someone in exiting from a maze – which is likely to be highly erratic and contingent – and the path which allows one to exit, which is predetermined. On questions and answers, Collingwood, *An Autobiography* (1939). My argument here is very different from that of Hacking, 'The Self-Vindication of the Laboratory Sciences' (1992); clearly in the case of some laboratory sciences there is an interplay between the equipment used and the results obtained, so that the two become mutually supporting. But this does not seem to me the case with the examples I discuss below. **33.** Weinberg, 'Sokal's Hoax' (1996); for hostile commentary see Weinberg himself, Labinger and Collins (eds.), *The One Culture?* (2001), 238, Brown, *Who Rules in Science?* (2001), 19, Rorty, 'Thomas Kuhn, Rocks and the Laws of Physics' (1999), 182–7. **34.** Aït-Touati, *Fictions of the Cosmos* (2011), 105. **35.** Hacking, 'How Inevitable are the Results of Successful Science?' (2000); see also Jardine, *The Scenes of Inquiry* (2000); Stanford, *Exceeding Our Grasp* (2010). Mathematics provides an interesting case study. Against path dependency, Heeffer, 'On the Curious Historical Coincidence of Algebra and Double-entry Bookkeeping' (2011); but Pascal seems to have reinvented Euclidean geometry from scratch, and Srinivasa Ramanujan reinvented much of modern mathematics, although he also produced numerous results that were distinctive and unparalleled. 36. Sokal, *Beyond the Hoax* (2008), 234–5, but note that Sokal's and Bricmont's formulation is cautious and ambiguous. 37. O'Grady, 'Wittgenstein and Relativism' (2004), 328–9. 38. Ginzburg, *Myths, Emblems, Clues* (1990), 96–125, is fundamental for thinking about these issues. Wittgenstein might well have been able to recognize the force of this argument: O'Grady, 'Wittgenstein and Relativism' (2004). 39. Moore, *A Defence of Common Sense* (1925). For an historian's approach, Rosenfeld, *Common Sense: A Political History* (2011). 40. Galilei, *Le opere* (1890), Vol. 4, 154, 217. 41. Galilei, *Le opere* (1890), Vol 4, 218–19. 42. Galilei, *Le opere* (1890), Vol. 4, 364–5, 391. 43. Galilei, *Le opere* (1890), Vol. 4, 393. 44. Galilei, *Le opere* (1890), Vol. 4, 385 45. Haack, *Manifesto of a Passionate Moderate* (1998), 94 – but see 105, where this circle is acknowledged. 46. Geis & Bunn, *A Trial of Witches* (1997). 47. Chalmers, 'Qualitative Novelty in Seventeenth-century Science' (2015) 48. Biagioli, 'The Social Status of Italian Mathematicians, 1450–1600' (1989). 49. Galilei, *Le opere* (1890), Vol. 5, 386. 50. Above, note 45. 51. Cunningham, 'Getting the Game Right' (1988), 370 52. Latour, 'On the Partial Existence of Existing and Non-existing Objects' (2000); even Hacking finds this 'irresponsibly playful': Hacking, *Historical Ontology* (2002), 11. 53. Compare Lehoux, *What Did the Romans Know?* (2012), 232–3, 237; Kuhn, *The Trouble with the Historical Philosophy of Science* (1992), 9. 54. Kuhn, *The Trouble with the Historical Philosophy of Science* (1992), 14 55. Kuhn, *The Trouble with the Historical Philosophy of Science* (1992), 9. Pinch, 'Kuhn – The Conservative and Radical Interpretations' (1997) (originally published in 1982) provides a valuable guide to the ambiguities of Kuhn's legacy. 56. Lehoux, *What Did the Romans Know?* (2012), 232–3. An interesting example is the speed of sound: for over a century there was a major discrepancy between theoretical and experimental values for the speed of sound. Nature kept pushing back (Finn, 'Laplace and the Speed of Sound' (1964)). 57. Chang, *Is Water H_2O?* (2012), 203–51, especially 215–24. Brown and Sokal uses the term 'objectivism' to serve the same purpose: Brown, *Who Rules in Science?* (2001), 92; Sokal, *Beyond the Hoax* (2008), 229. But, as Brown himself shows (101–4), the concept of objectivity is liable to cause much confusion. 58. Kuhn, *The Trouble with the Historical Philosophy of Science* (1992), 9. 59. Popper, *The Logic of Scientific Discovery* (1959), 22 (from the 1958 preface). The new epigraph (14) was added when the book was first reprinted (see the acknowledgement on 23). The pagination differs in later editions. 60. Kuhn, *Structure* (1970), 135–42.

CHAPTER 16

1. Butterfield, *The Whig Interpretation of History* (1931), 47. 2. Hexter, 'The Historian and His Day' (1954), 231. 3. Butterfield, *The Whig Interpretation of History* (1931), v. 4. Butterfield, *The Whig Interpretation of History* (1931), 16. 5. Wilson & Ashplant, 'Whig History' (1988). 6. Ashplant & Wilson,

'Present-centred History and the Problem of Historical Knowledge' (1988), 274. 7. Mayer, *The Roman Inquisition: A Papal Bureaucracy* (2013); Mayer, *The Roman Inquisition: Trying Galileo* (2015). 8. Malcolm, 'Hobbes's Science of Politics and His Theory of Science' (2002); Malcolm, 'Hobbes and Roberval' (2002), 187–9; Hull, 'Hobbes and the Premodern Geometry of Modern Political Thought' (2004), especially 121–2 9. Shapin & Schaffer, *Leviathan and the Air-pump* (1985), 150. 10. Hobbes, *Philosophicall Rudiments Concerning Government and Society* (1651), 284; Shapin & Schaffer, *Leviathan and the Air-pump* (1985), 153–4. 11. Shapin & Schaffer, *Leviathan and the Air-pump* (1985), 332. 12. Compare Bloor, *Wittgenstein* (1983), 3. 13. Shapin & Schaffer, *Leviathan and the Air-pump* (1985), 148, quoting the translation of *De Corpore*, 65–6. 14. Laslett, 'Commentary' (1963), 863 (the original text has an obvious typographical error: it reads 'the historians'); see the discussion in Jardine, 'Whigs and Stories' (2003). 15. Goldie, 'The Context of the Foundations' (2006), 32. 16. Merton, 'Unanticipated Consequences' (1936). 17. Mayr, 'When is Historiography Whiggish?' (1990); Alvargonzález, 'Is the History of Science Essentially Whiggish?' (2013). For the claim that historians who express views such as these should be driven out of the historical profession, see Shapin, 'Possessed by the Idols' (2006); and my response in the letters page of the next issue. 18. Foucault, *L'archéologie du savoir* (1969). 19. James, 'The Problem of Mechanical Flight' (1912). 20. Weinberg, 'Sokal's Hoax' (1996). 21. Rorty, 'Thomas Kuhn, Rocks and the Laws of Physics' (1999), 186. 22. For a striking, but unconvincing, relativist approach to the history of the bicycle, see Pinch & Bijker, 'The Social Construction of Facts and Artefacts' (1984), 411–19. 23. The phrase originates with Romain Rolland, and was used on the masthead of Gramsci's newspaper, *L'ordine nuovo*.

CHAPTER 17

1. Kuhn, *Structure* (1970), 163. 2. Popper, *Objective Knowledge* (1972), 23. 3. La Boëtie, *De la servitude volontaire* (1987). 4. Greenblatt, *The Swerve* (2011); Screech (ed.), *Montaigne's Annotated Copy of Lucretius* (1998); Popkin, *The History of Scepticism* (1979). 5. Montaigne, *The Complete Essays* (1991), 643–4. 6. Montaigne, *The Complete Essays* (1991), 502–4, 594–5 (an argument repeated by Voltaire in *Candide*). 7. Montaigne, *The Complete Essays* (1991), 642–4. 8. Montaigne, *The Complete Essays* (1991), 555, 567. 9. Montaigne, *The Complete Essays* (1991), 683. 10. Scholarly views on early modern unbelief have advanced greatly over the course of the last thirty years, but Montaigne continues to be read as a Christian. For my approach to these questions, see, for example, Wootton, 'Lucien Febvre and the Problem of Unbelief' (1988); for a more recent discussion of Renaissance humanism see Brown, *The Return of Lucretius* (2010), preface and Chapter 1. Montaigne's annotations of his Lucretius demonstrate his close engagement with Lucretius's critique of all religious beliefs. 11. Montaigne, *The Complete Essays* (1991), 595. 12. Montaigne,

The Complete Essays (1991), 652–3. **13.** Nagel, 'What is It Like to be a Bat?' (1974). **14.** Tunstall & Diderot, *Blindness and Enlightenment* (2011) – though my reading of this text differs from Tunstall's, for obvious reasons. **15.** Eisenstein, *The Printing Press as an Agent of Change* (1979); Eisenstein, *The Printing Revolution* (1983); Baron, Lindquist & Shevlin (eds.), *Agent of Change* (2007). In general I prefer Eisenstein to her critics (e.g. McNally (ed.), *The Advent of Printing* (1987)): see also above, pp. 60, 197–8 and 302–6. **16.** Sharratt, *Galileo* (1994), 140; Galilei, *Le opere* (1890), Vol. 6, 232 (Sharratt's translation). **17.** Mazur, *Enlightening Symbols* (2014); Padoa, *La Logique déductive* (1912), 21. **18.** Tilling, 'Early Experimental Graphs' (1975); Maas & Morgan, 'Timing History' (2002). **19.** Rybczynski, *One Good Turn* (2000). **20.** Ginsburg, 'On the Early History of the Decimal Point' (1928). **21.** Hacking, *The Emergence of Probability* (2006), xvi, quoting Butterfield, *The Origins of Modern Science* (1950); and xx, referring to Crombie, *Styles of Scientific Thinking* (1994); Hacking, 'Language, Truth and Reason' (1982) (reprinted in Hacking, *Historical Ontology* (2002)); Hacking, '"Style" for Historians and Philosophers' (1992) reprinted in Hacking, *Historical Ontology* (2002); Hacking, 'Inaugural Lecture' (2002). Hacking is following in the footsteps of Crombie, *Styles of Scientific Thinking* (1994), a work that Crombie had described as 'forthcoming' as early as 1980 (Crombie, 'Philosophical Presuppositions' (1980)). For a critique see Kusch, 'Hacking's Historical Epistemology' (2010). For an approach based on Crombie, see Kwa, *Styles of Knowing* (2011). **22.** The study of the history of intellectual tools or foundational concepts has been called 'historical epistemology'. The term originates with Gaston Bachelard. See Daston, 'Historical Epistemology' (1994). She describes the term as 'Hackinqesque', but Ian Hacking prefers Foucault's term 'archaeology of knowledge': Hacking, *The Emergence of Probability* (2006) and 'Historical Ontology' in Hacking, *Historical Ontology* (2002). Daston has recently offered an alternative: 'history of emergences': Daston, 'The History of Emergences' (2007). **23.** Putnam, *Mind, Language, and Reality* (1975), 73. **24.** Laudan, 'A Confutation of Convergent Realism' (1981). Laudan, in my view, overstates his case; but attempts to demolish it piecemeal (such as Psillos, *Scientific Realism* (1999), 101–45) seem to me to involve too much special pleading to be entirely convincing. **25.** Chang, *Is Water H2O?* (2012), 224–7. **26.** Newcastle, *Philosophical Letters* (1664), 508. **27.** Wootton, *Bad Medicine* (2006). See the comments of Papin in his letter to Leibniz, 10 July 1704 (Papin, *Le Vie et les ouvrages de Denis Papin* (1894), Vol. 8, 190–94 = Leibniz, Huygens and others, *Leibnizens und Huygens' Briefwechsel mit Papin* (1881), 317–21. The placebo: Papin, *Le Vie et les ouvrages de Denis Papin* (1894), Vol. 8, 206–8 = Leibniz, Huygens and others, *Leibnizens und Huygens' Briefwechsel mit Papin* (1881), 328–30. **28.** See above, p. 413. **29.** My position corresponds to that of Hacking, 'Five Parables' (1984), and in Hacking, *Historical Ontology* (2002), 43–5. **30.** Boyle, *Some Considerations* (1663), 84 = Boyle, *The Works* (1999), Vol 3, 257; see, for similar sentiments expressed in secular terms, Kuhn, *Structure* (1970), 173.

ACKNOWLEDGEMENTS

1. Koyré, *Newtonian Studies* (1965), 65. 2. Shea, *Designing Experiments* (2003), 116. 3. Febvre, *Le Problème de l'incroyance* (1942); Febvre, *The Problem of Unbelief* (1982); and Wootton, Lucien Febvre and the Problem of Unbelief (1988).

Bibliography

Abercromby, David. *Academia scientiarum: Or the Academy of Sciences.* London: HC for J Taylor, 1687.

Accademia del Cimento. *Essayes of Natural Experiments.* Trans. R Waller. London: B Alsop, 1684.

———. *Saggi di naturali esperienze.* Florence: G Cocchini, 1667.

Achillini, Alessandro. *De elementis.* Bologna: J Antonius, 1505.

Ackerman, James S. 'Art and Science in the Drawings of Leonardo da Vinci'. In *Origins, Imitation, Conventions: Representation in the Visual Arts.* Cambridge, Mass: MIT Press, 2002: 143–73.

———. 'Early Renaissance "Naturalism" and Scientific Illustration'. In *Distance Points: Essays in Theory and Renaissance Art and Architecture.* Cambridge, Mass: MIT Press, 1991: 185–210.

Adams, Douglas. *The Hitchhiker's Guide to the Galaxy: A Trilogy in Four Parts.* London: Heinemann, 1986.

Addison, Joseph and Richard Steele (eds.). *Spectator.* 8 vols. London: S Buckley and J Tonson, 1712–15.

Adelman, Janet. 'Making Defect Perfection: Shakespeare and the One-sex Model'. In *Enacting Gender on the English Renaissance Stage.* Ed. V Comensoli. Urbana: University of Illinois Press, 1999: 23–52.

Adorno, Rolena. 'The Discursive Encounter of Spain and America: The Authority of Eyewitness Testimony in the Writing of History'. *The William and Mary Quarterly* 49 (1992): 210–28.

Agassi, Joseph. 'Who Discovered Boyle's Law?' *Studies in History and Philosophy of Science Part A* 8 (1977): 189–250.

Aggiunti, Niccolò. *Oratio de mathematicae laudibus.* Rome: Mascardus, 1627.

Agricola, Rudolf and Joachim Vadianus. *Habes lector: hoc libello. Rudolphi Agricolae ivnioris Rheti, ad Joachimum Vadianum Heluctiu(m) Poeta(m) Laureatu(m), Epistolam, qua de locor(um) non nullorum obscuritate quaestio sit et percontatio.* Vienna: J Singrenues, 1515.

Aiken, Jane Andrews. 'The Perspective Construction of Masaccio's "Trinity" Fresco and Medieval Astronomical Graphics'. *Artibus et historiae* 16 (1995): 171–87.

Aït-Touati, Frédérique. *Fictions of the Cosmos: Science and Literature in the Seventeenth Century.* Trans. S Emanuel. Chicago: University of Chicago Press, 2011.

Alberti, Leon Battista. *De pictura*. Ed. C Grayson. Rome: Laterza, 1980.
———. *On Painting*. Ed. M Kemp. Trans. C Grayson. London: Penguin, 1991.
———. *On Painting: A New Translation and Critical Edition*. Ed. R Sinisgalli. Cambridge: Cambridge University Press, 2011.
———. *On Painting and on Sculpture: The Latin Texts of De pictura and De statua*. Ed. C Grayson. London: Phaidon, 1972.
Alder, Ken. 'Making Things the Same: Representation, Tolerance and the End of the Ancien Régime in France'. *Social Studies of Science* 28 (1998): 499–545.
Alexander, Amir. 'Lunar Maps and Coastal Outlines: Thomas Harriot's Mapping of the Moon'. *Studies in History and Philosophy of Science Part A* 29 (1998): 345–68.
Alighieri, Dante. *La Quaestio de aqua et terra*. Ed. A Müller and SP Thompson. Florence: LS Olschki, 1905.
Allen, James V. *Inference from Signs: Ancient Debates about the Nature of Evidence*. Oxford: Clarendon Press, 2001.
Allen, Robert C. *The British Industrial Revolution in Global Perspective*. Cambridge: Cambridge University Press, 2009.
Alvargonzález, David. 'Is the History of Science Essentially Whiggish?' *History of Science* 51 (2013): 85–100.
Ambrose, Charles T. 'Immunology's First Priority Dispute – An Account of the 17th-century Rudbeck–Bartholin Feud'. *Cellular Immunology* 242 (2006): 1–8.
Andrade, EN da C. 'The Early History of the Vacuum Pump'. *Endeavour* 16 (1957): 29–35.
Anon. 'An Accompt of Some Books'. *Philosophical Transactions* 10 (1675): 505–14.
———. 'Account of Books'. *Philosophical Transactions* 19 (1697): 475–84.
———. 'An Advertisement Concerning the Invention of the Transfusion of Bloud'. *Philosophical Transactions* 2 (1666): 489–90.
Anstey, Peter R. 'Experimental versus Speculative Natural Philosophy'. In *The Science of Nature in the Seventeenth Century*. Ed. P Anstey and J Schuster. Berlin: Springer, 2005: 215–42.
———. 'The Methodological Origins of Newton's Queries'. *Studies in History and Philosophy of Science Part A* 35 (2004): 247–69.
Antinori, Vincenzo. '*Notizie istoriche*'. In *Saggi di naturali esperienze fatte nell'Accademia del cimento*. Florence: Tip. Galileiana, 1841: 1–133.
Applebaum, W. *Encyclopedia of the Scientific Revolution: From Copernicus to Newton*. New York: Garland, 2000.
Ariew, Roger. 'The Duhem Thesis'. *British Journal for the Philosophy of Science* 35 (1984): 313–25.
———. 'The Initial Response to Galileo's Lunar Observations'. *Studies in History and Philosophy of Science Part A* 32 (2001): 571–81.
———. 'The Phases of Venus before 1610'. *Studies in History and Philosophy of Science Part A* 18 (1987): 81–92.
Aristotle. *On the Heavens*. Ed. WKC Guthrie. Cambridge, Mass.: Harvard University Press, 1939.

Arnauld, Antoine. *Première Lettre apologétique de Monsieur Arnauld Docteur de Sorbonne.* [S.l.]: [s.n.], 1656.

Arnauld, Antoine and Pierre Nicole. *La Logique, ou l'art de penser.* Paris: Flammarion, 1970.

———. *Response au P. Annat, provincial des Jésuites, touchant les cinq propositions attribuées à M. l'Evesque d'Ipre, divisée en deux parties.* [s.l.]: [s.n.], 1654.

Arnheim, R. 'Brunelleschi's Peepshow'. *Zeitschrift für Kunstgeschichte* 41 (1978): 57–60.

Ash, Eric H. '"A Perfect and an Absolute Work" – Expertise, Authority and the Rebuilding of Dover Harbor, 1579–1583'. *Technology and Culture* 41 (2000): 239–68.

Ashby, Eric. *Technology and the Academics: An Essay on Universities and the Scientific Revolution.* London: Macmillan, 1958.

Ashworth Jr, William B. 'Natural History and the Emblematic World View'. In *Reappraisals of the Scientific Revolution.* Ed. DC Lindberg and RS Westman. Cambridge: Cambridge University Press, 1990: 303–32.

Atkinson, Catherine. *Inventing Inventors in Renaissance Europe: Polydore Vergil's De inventoribus rerum.* Tübingen: Mohr Siebeck, 2007.

Auger, Léon. *Un savant méconnu, Gilles Personne de Roberval, 1602–1675; son activité intellectuelle dans les domaines mathématique, physique, mécanique et philosophique.* Paris: A Blanchard, 1962.

Augst, Bertrand. 'Descartes's Compendium on Music'. *Journal of the History of Ideas* 26 (1965): 119–32.

Aurelius, Marcus. *The Meditations of the Emperor Marcus Aurelius.* Ed. ASL Farquharson. Oxford: Clarendon Press, 1968.

Austin, John Langshaw. *How to Do Things with Words.* Oxford: Clarendon Press, 1962.

Bacchelli, Franco. '*Palingenio e la crisi dell'aristotelismo*'. In *Sciences et religions: De Copernic à Galilée.* Rome: École Française de Rome, 1999.

Bachelard, Gaston. *The Formation of the Scientific Mind: A Contribution to a Psychoanalysis of Objective Knowledge.* Trans. M McAllester-Jones. Manchester: Clinamen Press, 2002.

———. *La Formation de l'esprit scientifique: contribution à une psychanalyse de la connaissance objective.* Paris: J. Vrin, 1938.

———. *The New Scientific Spirit.* Boston: Beacon Press, 1985.

———. *Le Nouvel Esprit scientifique.* Paris: Librairie Félix Alcan, 1934.

Bacon, Francis. *The Essayes or Counsels, Civill and Morall.* London: J Haviland, 1625.

———. *Instauratio magna.* London: J Bill, 1620.

———. *The Novum organum . . . Epitomiz'd.* Trans. MD. London: T Lee, 1676.

———. *Of the Proficience and Aduancement of Learning, Divine and Humane.* London: H Tomes, 1605.

———. *Sylva sylvarum, or A Naturall Historie.* London: W Lee, 1627.

———. *Works.* Ed. J Spedding, RL Ellis and DD Heath. 14 vols. London: Longman 1857–74.

Bailey, Nathan. *An Universal Etymological English Dictionary.* London: E Bell, 1721.

Baillet, Adrien. *La Vie de Monsieur Des-Cartes.* 2 vols. Paris: D Horthemels, 1691.

Bailyn, Bernard. *The Ideological Origins of the American Revolution.* Cambridge, Mass.: Harvard University Press, 1967.

Baker, Keith Michael. *Inventing the French Revolution.* Cambridge: Cambridge University Press, 1990.

Balbiani, Laura. *La magia naturalis di Giovan Battista della Porta.* Bern: Lang, 2001.

Baldasso, Renzo. 'The Role of Visual Representation in the Scientific Revolution: A Historiographic Inquiry'. *Centaurus* 48 (2006): 69–88.

Ball, Philip. *Curiosity: How Science became Interested in Everything.* London: Bodley Head, 2012.

Baltrusaitis, Jurgis. *Anamorphoses, ou Perspectives curieuses.* Paris: O Perrin, 1955.

Bamford, Greg. 'Popper and His Commentators on the Discovery of Neptune: A Close Shave for the Law of Gravitation?' *Studies in History and Philosophy of Science Part A* 27 (1996): 207–32.

Bannister, Saxe. *Denis Papin: Notice sur sa vie et ses écrits.* Blois: F Jahyer, 1847.

Barber, William H. 'The Genesis of Voltaire's "Micromégas"'. *French Studies* 11 (1957): 1–15.

Barbette, Paul. *The Chirurgical and Anatomical Works . . . Composed according to the Doctrine of the Circulation of the Blood, and Other New Inventions of the Moderns.* London: J Darby, 1672.

Barker, Graeme. *The Agricultural Revolution in Prehistory: Why Did Foragers become Farmers?* Oxford: Oxford University Press, 2006.

Barker, Peter. 'Copernicus and the Critics of Ptolemy'. *Journal for the History of Astronomy* 30 (1999): 343–58.

———. 'Copernicus, the Orbs and the Equant'. *Synthèse* 83 (1990): 317–23.

Barker, Peter and Bernard R Goldstein. 'The Role of Comets in the Copernican Revolution'. *Studies in History and Philosophy of Science Part A* 19: 299–319 (1988).

Barnes, Barry. *T. S. Kuhn and Social Science.* London: Macmillan, 1982.

Barnes, Barry and David Bloor. 'Relativism, Rationalism and the Sociology of Knowledge'. In *Rationality and Relativism.* Ed. M Hollis and S Lukes. Oxford: Blackwell, 1982: 21–47.

Barnhart, Clarence Lewis. *The American College Dictionary.* New York: Random House, 1959.

Baron, Sabrina, Eric Lindqvist and Eleanor Shevlin (eds.). *Agent of Change: Print Culture Studies after Elizabeth L. Einstein.* Amherst, Mass.: University of Massachusetts Press, 2007.

Barozzi, Francesco. *Cosmographia in quatuor libros distributa summo ordine.* Venice: G Perchacinus, 1585.

Barthes, Roland. '*Le Discours de l'histoire*'. *Social Science Information* 6 (1967): 63–75.

———. 'The Reality Effect'. In *The Rustle of Language.* Trans. R Howard. Oxford: Blackwell, 1986: 141–8.

Bartholin, Caspar. *Anatomicae institutiones corporis humani utriusque sexus historiam*. Wittenberg: Raab, 1611.

Bartholin, Caspar, Thomas Bartholin and Johannes Walaeus. *Institutiones anatomicae, novis recentiorum opinionibus & observationibus, quarum innumerae hactenus editae non sunt*. Leiden: Hackius, 1641.

Bartholin, Thomas. *The Anatomical History of Thomas Bartholinus, Doctor and Kings Professor, Concerning the Lacteal Veins of the Thorax, Observ'd by Him Lately in Man and Beast*. London: O Pulleyn, 1653.

Bartholin, Thomas, Johannes Walaeus and others. *Bartholinus Anatomy: Made from the Precepts of His Father, and from the Observations of All Modern Anatomists*. London: P Cole, 1662.

Bartlett, Robert. *Trial by Fire and Water: The Medieval Judicial Ordeal*. Oxford: Oxford University Press, 1986.

Barton, Ruth. '"Men of Science": Language, Identity and Professionalization in the Mid-Victorian Scientific Community'. *History of Science* 41 (2003): 73–119.

Bataillon, Marcel. '*L'idée de la découverte de l'Amérique chez les Espagnols du XVIe siècle (d'après un livre récent)*'. *Bulletin hispanique* 55 (1953): 23–55.

Bates, William. *The Divinity of the Christian Religion*. London: JD, 1677.

Baxandall, Michael. *Painting and Experience in Fifteenth-century Italy*. Oxford: Oxford University Press, 1972.

Baxter, Richard. *A Paraphrase on the New Testament*. London: B Simmons, 1685.

———. *A Treatise of Knowledge and Love Compared*. London: T Parkhurst, 1689.

Bayle, Pierre (ed.). *Nouvelles de la république des lettres*. Amsterdam: Desbordes, 1684–1709.

———. *Projet et fragmens d'un dictionnaire critique*. Rotterdam: R Leers, 1692.

Bayly, Christopher. *The Birth of the Modern World: Global Connections and Comparisons*. Oxford: Blackwell, 2007.

Bechler, Zev. 'Newton's 1672 Optical Controversies: A Study in the Grammar of Scientific Dissent'. In *The Interaction between Science and Philosophy*. Ed. Y Elkana. Atlantic Highlands, NJ: Humanities Press, 1974: 115–42.

Bedini, Silvio A. *The Pulse of Time: Galileo Galilei, the Determination of Longitude, and the Pendulum Clock*. Florence: LS Olschki, 1991.

———. 'The Role of Automata in the History of Technology'. *Technology and Culture* 5 (1964): 24–42.

Beeckman, Isaac. *Journal tenu par Isaac Beeckman de 1604 à 1634*. Ed. C de Waard. 4 vols. The Hague: M Nijhoff, 1939–53.

Belting, Hans. *Florence and Baghdad: Renaissance Art and Arab Science*. Cambridge, Mass.: Harvard University Press, 2011.

Benedetti, Giovanni Battista. *Consideratione di Gio. Battista Benedetti, filosofo del Sereniss. S. Duca di Sauoia, intorno al Discorso della grandezza della terra, & dell'acqua, del Excellent. Sig. Antonio Berga, filosofo nella Vniuersità di Torino*. Turin: Bevilacqua, 1579.

———. *Diversarum speculationum mathematicarum et physicarum liber*. Turin: N Bevilacqua, 1585.

Benjamin, Walter. *Illuminations*. Ed. Hannah Arendt. New York: Schocken Books, 1986.

Bennett, James A. *The Divided Circle: A History of Instruments for Astronomy, Navigation and Surveying*. Oxford: Phaidon, 1987.

———. 'The Mechanics' Philosophy and the Mechanical Philosophy'. *History of Science* 24 (1986): 1–28.

Bentley, Michael. *The Life and Thought of Herbert Butterfield*. Cambridge: Cambridge University Press, 2011.

Bentley, Richard. *The Correspondence*. Ed. JH Monk, C Wordsworth and J Wordsworth. London: J Murray, 1842.

———. *The Folly and Unreasonableness of Atheism*. London: H Mortlock, 1692.

———. *Remarks upon a Late Discourse of Free-Thinking: In a Letter to F. H.D.D. By Phileleutherus Lipsiensis*. London: J Morphew, 1713.

Benveniste, Émile. *Problèmes de Linguistique Générale II*. Paris: Gallimard, 1974.

Berga, Antonio. *Discorso di Antonio Berga della grandezza dell'acqua & della terra contra l'opinione dil S. Alessandro Piccolomini*. Turino: Bevilacqua, 1579.

Berga, Antonio and Giovanni Battista Benedetti. *Disputatio de magnitudine terræ et aquæ (contra Alex. Piccolomineum conscripta)*. Trans. FM Vialardi. Turin: IB Raterius, 1580.

Berkel, Klaas van. *Isaac Beeckman on Matter and Motion: Mechanical Philosophy in the Making*. Baltimore: Johns Hopkins University Press, 2013.

Bertamini, Marco and Theodore E Parks. 'On What People Know about Images on Mirrors'. *Cognition* 98 (2005): 85–104.

Bertoloni Meli, Domenico. 'The Collaboration between Anatomists and Mathematicians in the Mid-seventeenth Century'. *Early Science and Medicine* 13 (2008): 665–709.

———. *Equivalence and Priority: Newton versus Leibniz*. Oxford: Oxford University Press, 1993.

———. 'Experimentation in the Physical Sciences of the Seventeenth Century'. In *The Oxford Handbook of the History of Physics*. Ed. JZ Buchwald and R Fox. Oxford: Oxford University Press, 2013: 199–225.

———. *Mechanism, Experiment, Disease: Marcello Malpighi and Seventeenth-century Anatomy*. Baltimore: Johns Hopkins University Press, 2011.

———. 'The Role of Numerical Tables in Galileo and Mersenne'. *Perspectives on Science* 12 (2004): 164–89.

———. *Thinking with Objects: The Transformation of Mechanics in the Seventeenth Century*. Baltimore: Johns Hopkins University Press, 2006.

Besse, Jean-Marc. *Les Grandeurs de la terre: Aspects du savoir géographique à la Renaissance*. Lyon: ENS Éditions, 2003.

Beyer, Hartmann. *Qvaestiones novae in libellum de sphaera Joannis de Sacro Bosco*. Paris: G Cauellat, 1551.

Biagioli, Mario. 'Did Galileo Copy the Telescope? A "New" Letter by Paolo Sarpi'. In *The Origins of the Telescope*. Ed. A van Helden, S Dupré, R van Gent and H Zuidervaart. Amsterdam: KNAW Press, 2010: 203–30.

———. 'From Ciphers to Confidentiality: Secrecy, Openness and Priority in Science'. *British Journal for the History of Science* 45 (2012): 213–33.

——— (ed.). *The Science Studies Reader.* New York: Routledge, 1999.

———. 'Scientific Revolution, Social Bricolage and Etiquette'. In *The Scientific Revolution in National Context.* Ed. R Porter and M Teich. Cambridge: Cambridge University Press, 1992: 11–54.

———. 'The Social Status of Italian Mathematicians, 1450–1600'. *History of Science* 27 (1989): 41–95.

Biggs, Noah. *Mataeotechnia medicinae praxeos: The Vanity of the Craft of Physick.* London: E Blackmore, 1651.

Biller, Peter. *The Measure of Multitude: Population in Medieval Thought.* Oxford: Oxford University Press, 2000.

de Bils, Lodewijk. *The Coppy of a Certain Large Act . . . Touching the Skill of a Better Way of Anatomy of Mans Body.* London: [s.n.], 1659.

Biro, Jacqueline. *On Earth as in Heaven: Cosmography and the Shape of the Earth from Copernicus to Descartes.* Saarbrücken: VDM Verlag Dr Müller, 2009.

Blackwell, Richard J. *Behind the Scenes at Galileo's Trial.* Indiana: University of Notre Dame Press, 2006.

Blair, Ann. *Annotations in a copy of Jean Bodin, 'Universae naturae theatrum'. Frankfurt: Wechel, 1597.* 1990. http://history.fas.harvard.edu/files/history/files/blair–theaterofnature.pdf.

———. 'Annotating and Indexing Natural Philosophy'. In *Books and the Sciences in History.* Ed. M Frasca-Spada and N Jardine. Cambridge: Cambridge University Press, 2000: 69–89.

Blake, Ralph M, Curt J Ducasse and Edward H Madden. *Theories of Scientific Method: The Renaissance through the Nineteenth Century.* Seattle: University of Washington Press, 1960.

Bloor, David. 'Anti-Latour'. *Studies in History and Philosophy of Science Part A* 30 (1999): 81–112.

———. *Knowledge and Social Imagery.* 2nd edn. London: Routledge & Kegan Paul, 1991.

———. *Wittgenstein: A Social Theory of Knowledge.* London: Macmillan, 1983.

———. 'Wittgenstein and Mannheim on the Sociology of Mathematics'. *Studies in History and Philosophy of Science Part A* 4 (1973): 173–91.

Blundeville, Thomas. *A Briefe Description of Universal Mappes and Cardes, and of Their Use: And Also the Use of Ptholemey His Tables.* London: T Cadman, 1589.

Boas, Marie. 'The Establishment of the Mechanical Philosophy'. *Osiris* 10 (1952): 412–541.

Boas Hall, Marie. *Nature and Nature's Laws: Documents of the Scientific Revolution.* London: Macmillan, 1970.

———. *Promoting Experimental Learning: Experiment and the Royal Society 1660–1727.* Cambridge: Cambridge University Press, 1991.

Bodin, Jean. *Le Théatre de la nature universelle.* Trans. F de Fougerolles. Lyons: J Pillehotte, 1597.

———. *Universæ naturæ theatrum in quo rerum omnium effectrices causæ & fines quinque libris discutiuntur.* Lyons: I Roussin, 1596.

Bodnár, István. 'Aristotle's Natural Philosophy'. *The Stanford Encyclopedia of Philosophy.* 2012. http://plato.stanford.edu/archives/spr2012/entries/aristotle-natphil/ (accessed 14 December 2014).

Boffito, Giuseppe. *Intorno alla 'Quaestio de aqua et terra' attribuita a Dante.* Turin: C Clausen, 1902.

Bogen, James and James Woodward. 'Saving the Phenomena'. *Philosophical Review* 97 (1988): 303–52.

Boghossian, Paul Artin. *Fear of Knowledge: Against Relativism and Constructivism.* Oxford: Clarendon Press, 2006.

Bonnell, Victoria E and Lynn Hunt. 'Introduction'. In *Beyond the Cultural Turn: New Directions in the Study of Society and Culture.* Ed. VE Bonnell and L Hunt. Berkeley: University of California Press, 1999: 1–32.

Boodt, Anselm Boèce de. *Gemmarum et lapidum historia.* Hanover: C Marnius, 1609.

Borel, Pierre. *A New Treatise Proving a Multiplicity of Worlds.* Trans. D Sashott. London: J Streater, 1658.

Borges, Jorge Luis. *Other Inquisitions, 1937–1952.* Austin: University of Texas Press, 1964.

———. *The Total Library: Non-Fiction 1922–1986.* Ed. E Weinberger. Trans. E Allen and SJ Levine. London: Penguin, 2001.

Borough, William. *A Discours of the Variation of the Cumpas, or Magneticall Needle.* R Ballard: London, 1581.

Boschiero, Luciano. 'Translation, Experimentation and the Spring of the Air: Richard Waller's "Essayes of Natural Experiments"'. *Notes and Records of the Royal Society* (2009)

Bossuet, Jacques. *Quakerism A-la-Mode, Or A History of Quietism, Particularly That of the Lord Arch-Bishop of Cambray and Madam Guyone.* London: J Harris, 1698.

Bossy, John. *Giordano Bruno and the Embassy Affair.* New Haven: Yale University Press, 1991.

———. *Under the Molehill: An Elizabethan Spy Story.* New Haven: Yale University Press, 2001.

Bostridge, Ian. *Witchcraft and Its Transformations, c.1650–c.1750.* Oxford: Clarendon Press, 1997.

Botero, Giovanni. *On the Causes of the Greatness and Magnificence of Cities, 1588.* Trans. G Symcox. Toronto: University of Toronto Press, 2012.

Bourdieu, Pierre. *Science of Science and Reflexivity.* Trans. R Nice. Chicago: University of Chicago Press, 2004.

Bourne, William. *A Regiment for the Sea.* London: T Hacket, 1574.

Boyer, Carl B. 'Aristotelian References to the Law of Reflection'. *Isis* 36 (1946): 92–5.

———. 'Early Estimates of the Velocity of Light'. *Isis* 33 (1941): 24–40.

———. *The Rainbow from Myth to Mathematics.* New York: T Yoseloff, 1959.

Boyle, Robert. *Certain Physiological Essays and Other Tracts*. London: H Herringman, 1669.

———. *Certain Physiological Essays Written at Distant Times, and on Several Occasions*. London: H Herringman, 1661.

———. *The Christian Virtuoso Shewing, that by being Addicted to Experimental Philosophy, a Man is Rather Assisted, than Indisposed, to be a Good Christian*. London: J Taylor, 1690.

———. *A Continuation of New Experiments Physico-mechanical*. Oxford: R Davis, 1682.

———. *The Correspondence of Robert Boyle, 1636–1691*. Ed. MCW Hunter, A Clericuzio and L Principe. 6 vols. London: Pickering & Chatto, 2001.

———. *A Defence of the Doctrine Touching the Spring and Weight of the Air*. London: FG, 1662.

———. *Experimenta et observationes physicæ: Wherein are Briefly Treated of Several Subjects Relating to Natural Philosophy in an Experimental Way*. London: J Taylor, 1691.

———. *Experimentorum novorum physico-mechanicorum continuatio secunda*. Geneva: S de Tournes, 1680.

———. *Experiments and Considerations Touching Colours*. London: H Herringman, 1664.

———. *A Free Enquiry into the Vulgarly Receiv'd Notion of Nature*. London: J Taylor, 1686.

———. *Hydrostatical Paradoxes*. Oxford: R Davis, 1666.

———. *New Experiments Physico-Mechanical, Touching the Spring of the Air*. Oxford: H. Hall, 1660.

———. *Nouveau traité*. Lyons: J Certe, 1689.

———. *Occasional Reflections upon Several Subjects*. London: H Herringman, 1665.

———. *The Origine of Formes and Qualities*. Oxford: R Davis, 1666.

———. *Some Considerations Touching the Usefulnesse of Experimental Naturall Philosophy*. Oxford: R Davis, 1663.

———. 'Tryals Proposed by Mr Boyle to Dr Lower, to be Made by Him, for the Improvement of Transfusing Blood out of One Live Animal into Another'. *Philosophical Transactions* 1 (1667): 385–8.

———. *The Works of Robert Boyle*. Ed. M Hunter and EB Davis. 14 vols. London: Pickering & Chatto, 1999–2000.

Brading, Katherine. 'The Development of the Concept of Hypothesis from Copernicus to Boyle and Newton'. *Revista de Filozofie KRISIS* 8 (1999): 5–16.

Brahe, Tycho. *Sur des phénomènes plus récents du monde éthéré, livre second*. Trans. J Peyroux. Paris: A Blanchard, 1984.

Brannigan, Augustine. *The Social Basis of Scientific Discoveries*. Cambridge: Cambridge University Press, 1981.

Broman, Thomas. 'The Habermasian Public Sphere and "Science *in* the Enlightenment"'. *History of Science* 36 (1998): 123–50.

Brook, Timothy. *Vermeer's Hat: The Seventeenth Century and the Dawn of the Global World*. London: Profile, 2008.

Brotton, Jerry. *A History of the World in Twelve Maps*. London: Allen Lane, 2012.

Broughton, Peter. 'The First Predicted Return of Comet Halley'. *Journal for the History of Astronomy* 16 (1985): 123–32.

Brown, Alison. *The Return of Lucretius to Renaissance Florence*. Cambridge, Mass.: Harvard University Press, 2010.

Brown, Gary I. 'The Evolution of the Term "Mixed Mathematics"'. *Journal of the History of Ideas* 52 (1991): 81–102.

Brown, James Robert. *Who Rules in Science? An Opinionated Guide to the Wars*. Cambridge, Mass.: Harvard University Press, 2001.

Brown, Lloyd A. *Jean Domenique Cassini and His World Map of 1696*. Ann Arbor: University of Michigan Press, 1941.

Brown, Piers. '*Hac ex consilio meo via progredieris*: Courtly Reading and Secretarial Mediation in Donne's "The Courtier's Library"'. *Renaissance Quarterly* 61 (2008): 833–66.

Browne, Thomas. *Pseudodoxia epidemica, or Enquiries into Very Many Received Tenents, and Commonly Presumed Truths*. London: E Dod, 1646.

———. *Pseudodoxia epidemica: Or, Enquiries into Very Many Received Tenents and Commonly Presumed Truths*. London: N Ekins, 1672.

Brummelen, Glen van. *The Mathematics of the Heavens and the Earth: The Early History of Trigonometry*. Princeton: Princeton University Press, 2009.

Bruno, Giordano. *The Ash Wednesday Supper = La Cena de le Ceneri*. Ed. EA Gosselin and LS Lerner. Toronto: University of Toronto Press, 1995.

De Bruyn, Frans. 'The Classical Silva and the Generic Development of Scientific Writing in Seventeenth-century England'. *New Literary History* 32 (2001): 347–73.

Bucciantini, Massimo, Michele Camerota and Franco Giudice. *Galileo's Telescope: A European Story*. Cambridge Mass.: Harvard University Press, 2015.

Buchwald, Jed Z. 'Descartes' Experimental Journey Past the Prism and through the Invisible World to the Rainbow'. *Annals of Science* 65 (2008): 1–46.

Buchwald, Jed Z and Mordechai Feingold. *Newton and the Origin of Civilization*. Princeton: Princeton University Press, 2013.

Buringh, Eltjo and Jan Luiten van Zanden. 'Charting the "Rise of the West": Manuscripts and Printed Books in Europe, a Long-term Perspective from the Sixth through Eighteenth Centuries'. *Journal of Economic History* 69 (2009): 409–45.

Burkert, Walter. *Lore and Science in Ancient Pythagoreanism*. Cambridge, Mass.: Harvard University Press, 1972.

Burns, William E. *An Age of Wonders: Prodigies, Politics and Providence in England, 1657–1727*. Manchester: Manchester University Press, 2002.

———. '"Our Lot is Fallen into an Age of Wonders': John Spencer and the Controversy Over Prodigies in the Early Restoration'. *Albion* 27 (1995): 237–52.

Burtt, Edwin A. *The Metaphysical Foundations of Modern Physical Science: A Historical and Critical Essay*. London: Routledge, 1924.

Bury, John Bagnell. *The Idea of Progress: An Inquiry into Its Origin and Growth*. London: Macmillan, 1920.

Butterfield, Herbert. *The Origins of Modern Science, 1300–1800*. London: Bell, 1950.

———. *The Whig Interpretation of History*. London: Bell, 1931.

Byrne, James Steven. 'A Humanist History of Mathematics? Regiomontanus's Padua Oration in Context'. *Journal of the History of Ideas* 67 (2006): 41–61.

Calcagnini, Celio. *Opera aliquot*. Basle: H Frobenius, 1544.

Callon, Michel. *'Boïtes noires et opérations de traduction'*. *Économie et humanisme* 262 (1981): 53–9.

Camerota, Filippo. *La prospettiva del Rinascimento: arte, architettura, scienza*. Milano: Electa, 2006.

Camerota, Michele. '*Galileo, Lucrezio e l'atomismo*'. In *Lucrezio, la natura, la scienza*. Ed. F Beretta and F Citti. Florence: LS Olschki, 2008: 141–75.

Campbell, Mary Baine. 'Speedy Messengers: Fiction, Cryptography, Space Travel and Francis Godwin's "The Man in the Moone"'. *Yearbook of English Studies* 41 (2011): 190–204.

———. *Wonder and Science: Imagining Worlds in Early Modern Europe*. Ithaca: Cornell University Press, 1999.

Caraci Luzzana, Ilaria. *Amerigo Vespucci*. Nuova Raccolta Colombiana. Rome: Istituto poligrafico e Zecca dello Stato, 1999.

Cardano, Gerolamo. *De subtilitate libri XXI*. Basle: L Lucius, 1554.

Carpenter, Audrey T. *John Theophilus Desaguliers*. London: Continuum, 2011.

Carpenter, Nathanael. *Geographie Delineated Forth in Two Bookes, Containing the Spherical and Topicall Parts Thereof*. Oxford: J Lichfield, 1635.

———. *Philosophia libera, triplici exercitationum decade proposita: In qua, adversus huius temporis philosophos, dogmata quædam nova discutiuntur*. Oxford: J Lichfield, 1622.

Carpo, Mario. *Architecture in the Age of Printing*. Cambridge, Mass.: MIT Press, 2001.

Carroll, Patrick. *Science, Culture and Modern State Formation*. Berkeley: University of California Press, 2006.

Cassin, Barbara, Steven Rendall and Emily S Apter (eds.). *Dictionary of Untranslatables: A Philosophical Lexicon*. Princeton: Princeton University Press, 2014.

De Caus, Salomon. *Les Raisons des forces mouvantes*. Frankfurt: J Norton, 1615.

Cavendish, Margaret. *The Description of a New World, Called the Blazing-World*. London: A Maxwell, 1666.

Céard, Jean. *La Nature et les prodiges: L'Insolite au XVIe siècle*. Geneva: Droz, 1996.

Cesari, Anna Maria. *Il trattato della sfera di Andalò di Negro nelle Zibaldone del Boccaccio*. Milan: AM Cesari, 1982.

Cesi, Bernardo. *Mineralogia, sive, Naturalis philosophiæ thesauri*. Louvain: J & P Prost, 1636.

Chalmers, Alan. 'Intermediate Causes and Explanations: The Key to Understanding the Scientific Revolution'. *Studies in History and Philosophy of Science Part A* 43 (2012): 551–62.

———. 'Klein on the Origin of the Concept of Chemical Compound'. *Foundations of Chemistry* 14 (2012): 37–53.

———. 'The Lack of Excellency of Boyle's Mechanical Philosophy'. *Studies in History and Philosophy of Science Part A* 24 (1993): 541–64.

———. 'Qualitative Novelty in Seventeenth-century Science: Hydrostatics from Stevin to Pascal'. *Studies in History and Philosophy of Science Part A* 51 (2015): 1–10.

———. *The Scientist's Atom and the Philosopher's Stone How Science Succeeded and Philosophy Failed to Gain Knowledge of Atoms*. Dordrecht: Springer, 2009.

———. 'Understanding Science through Its History: A Response to Newman'. *Studies in History and Philosophy of Science Part A* 42 (2011): 150–3.

Chang, Hasok. *Inventing Temperature: Measurement and Scientific Progress*. Oxford: Oxford University Press, 2004.

———. *Is Water H2O?: Evidence, Pluralism and Realism*. Dordrecht: Springer, 2012.

Chapman, Allan. 'Tycho Brahe in China: The Jesuit Mission to Peking and the Iconography of European Instrument-making Processes'. *Annals of Science* 41 (1984): 417–43.

———. 'A World in the Moon – Wilkins and His Lunar Voyage of 1640'. *Quarterly Journal of the Royal Astronomical Society* 32 (1991): 121.

Charleton, Walter. *The Darknes of Atheism Dispelled by the Light of Nature. A Physico-Theologicall Treatise*. London: W Lee, 1652.

———. *Physiologia Epicuro-Gassendo-Charletoniana, or A Fabrick of Science Natural upon the Hypothesis of Atoms*. London: T Heath, 1654.

Chartier, Roger. *The Cultural Origins of the French Revolution*. Durham, NC: Duke University Press, 1991.

Châtelet, Émilie du. *Selected Philosophical and Scientific Writings*. Ed. JP Zinsser. Chicago: University of Chicago Press, 2009.

Chesne, Joseph du. *The Practise of Chymicall, and Hermeticall Physicke*. Trans. T Timme. London: T Creede, 1605.

Child, William. *Wittgenstein*. London: Routledge, 2011.

Christianson, John Robert. *On Tycho's Island: Tycho Brahe, Science and Culture in the Sixteenth Century*. Cambridge: Cambridge University Press, 2000.

Christie, Thony. 'Nobody Invented the Scientific Method'. 29 August 2012. http://thonyc.wordpress.com/2012/08/29/nobody-invented-the-scientific-method/ (accessed 10 December 2014).

Cicero, Marcus Tullius. *De natura deorum: Academica*. Ed. H Rackham. Cambridge, Mass.: Harvard University Press, 1933.

Cieslak-Golonka, Maria and Bruno Morten. 'The Women Scientists of Bologna'. *American Scientist* 88 (2000): 68–73.

Ciliberto, Michele and Nicholas Mann (eds.). *Giordano Bruno, 1583–1585: The English Experience*. Florence: LS Olschki, 1997.

Cipolla, Carlo M. *Clocks and Culture, 1300–1700*. London: Collins, 1967.

———. *European Culture and Overseas Expansion*. Harmondsworth: Penguin, 1970.

Clagett, Marshall. 'The Impact of Archimedes on Medieval Science'. *Isis* 50 (1959): 419–29.

———. *The Science of Mechanics in the Middle Ages*. Madison: University of Wisconsin Press, 1959.

Clark, Kathleen M and Clemency Montelle. 'Priority, Parallel Discovery, and Pre-eminence: Napier, Bürgi and the Early History of the Logarithm Relation'. *Revue d'histoire des mathématiques* 18 (2012): 223–70.

Clark, Stuart. *Thinking with Demons: The Idea of Witchcraft in Early Modern Europe*. Oxford: Clarendon Press, 1997.

Clarke, Desmond M. *Descartes: A Biography*. Cambridge: Cambridge University Press, 2006.

———. *Descartes' Philosophy of Science*. Manchester: Manchester University Press, 1982.

———. *Occult Powers and Hypotheses: Cartesian Natural Philosophy under Louis XIV*. Oxford: Clarendon Press, 1989.

Clavius, Christoph. *In sphaeram Ioannis de Sacro Bosco commentarius, nunc tertio ab ipso auctore recognitus*. Rome: D Basa, 1585.

———. *Opera mathematica*. 5 vols. Mainz: Hierat, 1611–12.

Clubb, Louise George. *Giambattista della Porta, Dramatist*. Princeton: Princeton University Press, 1965.

Clutton-Brock, Martin. 'Copernicus's Path to His Cosmology: An Attempted Reconstruction'. *Journal for the History of Astronomy* 36 (2005): 197–216.

Cobb, Matthew. *Generation: The Seventeenth-century Scientists who Unravelled the Secrets of Sex, Life and Growth*. New York: Bloomsbury, 2006.

Cobban, Alfred. *The Social Interpretation of the French Revolution*. Cambridge: Cambridge University Press, 1964.

Cohen, H Floris. *How Modern Science Came into the World: Four Civilizations, One 17th-century Breakthrough*. Amsterdam: Amsterdam University Press, 2010.

———. 'Inside Newcomen's Fire Engine: The Scientific Revolution and the Rise of the Modern World'. *History of Technology* 25 (2004): 111–32.

———. *The Scientific Revolution: A Historiographical Inquiry*. Chicago: University of Chicago Press, 1994.

Cohen, I Bernard. *The Birth of a New Physics*. New York: Norton, 1987.

———. 'The Eighteenth-century Origins of the Concept of Scientific Revolution'. *Journal of the History of Ideas* 37 (1976): 257–88.

———. 'The First English Version of Newton's *Hypotheses non fingo*'. *Isis* 53 (1962): 379–88.

———. 'Hypotheses in Newton's Philosophy'. *Physis* 8 (1966): 163–83.

———. '*Quantum in se est*: Newton's Concept of Inertia in Relation to Descartes and Lucretius'. *Notes and Records of the Royal Society of London* 19 (1964): 131–55.

———. 'Roemer and the First Determination of the Velocity of Light (1676)'. *Isis* 31 (1940): 327–79.

Collingwood, Robin George. *An Autobiography*. London: Oxford University Press, 1939.

———. *The Idea of Nature*. Oxford: Clarendon Press, 1945.

Collins, Harry M. *Changing Order: Replication and Induction in Scientific Practice*. London: Sage, 1985.

———. 'Introduction: Stages in the Empirical Programme of Relativism'. *Social Studies of Science* 11 (1981): 3–10.

———. 'Son of Seven Sexes: The Social Destruction of a Physical Phenomenon'. *Social Studies of Science* 11 (1981): 33–62.

———. 'Tacit Knowledge, Trust and the Q of Sapphire'. *Social Studies of Science* 31 (2001): 71–85.

———. 'The TEA Set: Tacit Knowledge and Scientific Networks'. *Social Studies of Science* 4 (1974): 165–85.

Collinson, Patrick. 'The Monarchical Republic of Queen Elizabeth I'. *Bulletin of the John Rylands University Library of Manchester* 69 (1987): 394–424.

Colón, Fernando. *The Life of the Admiral Christopher Columbus*. Ed. B Keen. New Brunswick: Rutgers University Press, 1992.

Columbus, Christopher. *The Four Voyages*. Trans. JM Cohen. Harmondsworth: Penguin, 1969.

———. *The Journal of Christopher Columbus (During His First Voyage, 1492–93)*. Ed. CR Markham. Cambridge: Cambridge University Press, 2010.

Conant, James. 'On Wittgenstein's Philosophy of Mathematics'. *Proceedings of the Aristotelian Society* 97 (1997): 195–222.

Conant, James Bryant. *Robert Boyle's Experiments in Pneumatics*. Cambridge, Mass.: Harvard University Press, 1950.

Condorcet, Marquis de. *Outlines of an Historical View of the Progress of the Human Mind . . . Translated from the French*. London: J Johnson, 1795.

Considine, John. *Dictionaries in Early Modern Europe: Lexicography and the Making of Heritage*. Cambridge: Cambridge University Press, 2008.

Constantini, Angelo. *La Vie de Scaramouche*. Paris: C Barbin, 1695.

Cook, MG. 'Divine Artifice and Natural Mechanism: Robert Boyle's Mechanical Philosophy of Nature'. *Osiris* 16 (2001): 133–50.

Cooper, Alix. *Inventing the Indigenous: Local Knowledge and Natural History in Early Modern Europe*. Cambridge: Cambridge University Press, 2007.

Copenhaver, Brian P. 'The Historiography of Discovery in the Renaissance: The Sources and Composition of Polydore Vergil's *De inventoribus rerum*, I–III'. *Journal of the Warburg and Courtauld Institutes* 41 (1978): 192–214.

Copernicus, Nicolaus. *De revolutionibus orbium coelestium*. Nuremberg: J Petreius, 1543.

———. *On the Revolutions*. Ed. J Dobrzycki. Trans. E Rosen. Baltimore: Johns Hopkins University Press, 1978.

Cosgrove, Denis E. 'Images of Renaissance Cosmography'. In *The History of Cartography*. 6 vols. Vol. 3: *Cartography in the European Renaissance*. Ed. D Woodward. Chicago: University of Chicago Press, 2007: 55–98.

Costabel, Pierre. '*Sur l'origine de la science classique*'. *Revue philosophique de la France et de l'étranger* 137 (1947): 208–21.

Cowell, John. *The Interpreter, or Booke Containing the Signification of Words*. Cambridge: J Legate, 1607.

Crafts, N. 'Explaining the First Industrial Revolution: Two Views'. *European Review of Economic History* 15 (2011): 153–68.

Cranz, F Edward. *Reorientations of Western Thought from Antiquity to the Renaissance*. Ed. NS Struever. Aldershot: Ashgate, 2006.

Crease, Robert P. *World in the Balance: The Historic Quest for an Absolute System of Measurement*. New York: WW Norton, 2011.

Cressy, David. 'Early Modern Space Travel and the English Man in the Moon'. *The American Historical Review* 111 (2006): 961–82.

Croft, Herbert. *Some Animadversions upon a Book Intituled, the Theory of the Earth*. London: C Harper, 1685.

Croll, Oswald, Georg Eberhard Hartmann and Johann Hartmann. *Bazilica Chymica, & Praxis Chymiatricae, or Royal and Practical Chymistry in Three Treatises*. London: J Starkey, 1670.

Crombie, Alistair Cameron. 'Grosseteste's Position in the History of Science'. In *Robert Grosseteste, Scholar and Bishop*. Ed. DA Callus. Oxford: Clarendon Press, 1955: 98–120.

———. 'Philosophical Presuppositions and Shifting Interpretations of Galileo'. In *Theory Change, Ancient Axiomatics and Galileo's Methodology*. Ed. J Hintikka, D Gruender and E Agazzi. Dordrecht: Reidel, 1980: 271–86.

———. *Robert Grosseteste and the Origins of Experimental Science, 1100–1700*. Oxford: Oxford University Press, 1953.

———. *Scientific Change*. New York: Basic Books, 1963.

———. *Styles of Scientific Thinking in the European Tradition*. 3 vols. London: Duckworth, 1994.

Culverwell, Nathaniel. *An Elegant and Learned Discourse of the Light of Nature: With Other Treatises*. London: J Rothwell, 1652.

Cunningham, Andrew. *The Anatomical Renaissance: The Resurrection of the Anatomical Projects of the Ancients*. Aldershot: Ashgate, 1997.

———. 'Getting the Game Right: Some Plain Words on the Identity and Invention of Science'. *Studies in History and Philosophy of Science Part A* 19 (1988): 365–89.

———. 'How the *Principia* Got Its Name, or Taking Natural Philosophy Seriously'. *History of Science* 29 (1991): 377–92.

———. 'The Identity of Natural Philosophy: A Response to Edward Grant'. *Early Science and Medicine* 5 (2000): 259–78.

Cunningham, Andrew and Perry Williams. 'De-centring the "Big Picture": "The Origins of Modern Science" and the Modern Origins of Science'. *British Journal for the History of Science* 26 (1993): 407–32.

Cuomo, Serafina. 'Shooting by the Book: Notes on Niccolò Tartaglia's *Nova scientia*'. *History of Science* 35 (1997): 155–88.

Cyrano de Bergerac, Hercule-Savinien de. *The Comical History of the States and Empires of the Worlds of the Moon and Sun*. London: H Rhodes, 1687.

———. *Les États et empires de la lune et du soleil, avec le fragment de physique*. Ed. M Alcover. Paris: H Champion, 2004.

Dalché, Patrick Gautier. 'The Reception of Ptolemy's Geography'. In *The History of Cartography*. 6 vols. Vol. 3: *Cartography in the European Renaissance*. Ed. D Woodward. Chicago: University of Chicago Press, 2007: 285–364.

Daneau, Lambert. *Physique françoise, comprenant ... le discours des choses naturelles, tant célestes que terrestres, selon que les philosophes les ont descrites*. Geneva: E Vignon, 1581.

Darmon, Jean-Charles. *Le Songe libertin: Cyrano de Bergerac d'un monde à l'autre*. Paris: Klincksieck, 2004.

Dary, Michael. *The General Doctrine of Equation Reduced into Brief Precepts: In III Chapters. Derived from the Works of the Best Modern Analysts*. London: N Brook, 1664.

Daston, Lorraine J. 'Baconian Facts, Academic Civility and the Prehistory of Objectivity'. In *Rethinking Objectivity*. Ed. A Megill. Durham, NC: Duke University Press, 1994: 37–63.

———. *Classical Probability in the Enlightenment*. Princeton: Princeton University Press, 1988.

———. 'The Cold Light of Facts and the Facts of Cold Light: Luminescence and the Transformation of the Scientific Fact, 1600–1750'. In *Signs of the Early Modern II*. Ed. DL Rubin. Charlottesville, VA: Rookwood Press, 1997: 17–45.

———. 'Curiosity in Early Modern Science'. *Word and Image* 11 (1995): 391–404.

———. 'The Factual Sensibility'. *Isis* 79 (1988): 452–67.

———. 'Historical Epistemology'. In *Questions of Evidence: Proof, Practice and Persuasion across the Disciplines*. Ed. J Chandler, AI Davidson and H Harootunian. Chicago: University of Chicago Press, 1994: 282–9.

———. 'The History of Emergences: The Emergence of Probability'. *Isis* 98: 801–8 (2007).

———. 'History of Science in an Elegiac Mode: E. A. Burtt's *Metaphysical Foundations of Modern Physical Science Revisited*'. *Isis* 82 (1991): 522–31.

———. 'The Ideal and Reality of the Republic of Letters in the Enlightenment'. *Science in Context* 4 (1991): 367–86.

———. 'The Language of Strange Facts in Early Modern Science'. In *Inscribing Science: Scientific Texts and the Materiality of Communication*. Ed. T Lenoir. Stanford: Stanford University Press, 1997: 20–38.

———. 'Marvelous Facts and Miraculous Evidence in Early-Modern Europe'. *Critical Inquiry* 18 (1991): 93–124.

———. '*Perché i fatti sono brevi?*' *Quaderni storici* 36 (2001): 745–70.

———. 'Science Studies and the History of Science'. *Critical Inquiry* 35 (2009): 798–813.

———. 'Strange Facts, Plain Facts and the Texture of Scientific Experience in the Enlightenment'. In *Proof and Persuasion: Essays on Authority, Objectivity and Evidence*. Ed. S Marchand and E Lunbeck. Turnhout: Brepols, 1996: 42–59.

Daston, Lorraine J and Peter Galison (eds.). *Objectivity*. New York: Zone Books, 2007.

Daston, Lorraine J and Elizabeth Lunbeck (eds.). *Histories of Scientific Observation*. Chicago: University of Chicago Press, 2011.

Daston, Lorraine J and Katharine Park. *Wonders and the Order of Nature, 1150–1750*. New York: Zone Books, 1998.

David, Paul A. 'Clio and the Economics of QWERTY'. *American Economic Review* 75 (1985): 332–7.

Davies, Richard. *Memoirs of the Life and Character of Dr Nicholas Saunderson: Late Lucasian Professor of the Mathematics in the University of Cambridge.* Cambridge: Cambridge University Press, 1741.

Dear, Peter. *Discipline and Experience: The Mathematical Way in the Scientific Revolution.* Chicago: University of Chicago Press, 1995.

———. 'The Meanings of Experience'. In *The Cambridge History of Science.* Vol. 3: *Early Modern Science.* Ed. K Park and LJ Daston. Cambridge: Cambridge University Press, 2006: 106–31.

———. 'Religion, Science and Natural Philosophy: Thoughts on Cunningham's Thesis'. *Studies in History and Philosophy of Science Part A* 32 (2001): 377–86.

———. *Revolutionizing the Sciences: European Knowledge and Its Ambitions, 1500–1700.* Princeton: Princeton University Press, 2001.

———. '*Totius in verba*: Rhetoric and Authority in the Early Royal Society'. *Isis* 76 (1985): 144–61.

Dee, John. *General and Rare Memorials Pertayning to the Perfect Arte of Navigation.* London: J Daye, 1577.

Della Porta, Giambattista. *De i miracoli et maravigliosi effetti dalla natura prodotti libri IV.* Venice: L Avanzi, 1560.

———. *De telescopio.* Florence: LS Olschki, 1962.

———. *La Magie naturelle en quatre livres.* Lyons: A Olier, 1678.

———. *Natural Magick in Twenty Books . . .: Wherein are Set Forth All the Riches and Delights of the Natural Sciences.* London: T Young, 1658.

Denton, Peter H. *The ABC of Armageddon: Bertrand Russell on Science, Religion and the Next War, 1919–1938.* Albany, NY: State University of New York Press, 2001.

Desaguliers, John Theophilus. *A Course of Experimental Philosophy.* 2 vols. London: Senex, 1734–44.

Descartes, René. *A Discourse of a Method for the Well Guiding of Reason, and the Discovery of Truth in the Sciences.* London: T Newcombe, 1649.

———. *Excellent Compendium of Musick with Necessary and Judicious Animadversions Thereupon.* London: T. Harper, 1653.

———. *Oeuvres philosophiques.* Ed. F Alquié. 3 vols. Paris: Garnier, 1963–73.

———. *The Philosophical Writings of Descartes.* Ed. J Cottingham, D Murdoch and R Stoothoff. 2 vols. Cambridge: Cambridge University Press, 1984.

———. *Les Principes de la philosophie.* Paris: T Girard, 1668.

———. *Principia philosophiæ.* Amsterdam: Elzevir, 1644.

Deutscher, Guy. *Through the Language Glass: Why the World Looks Different in Other Languages.* London: William Heinemann, 2010.

Devlin, Keith J. *The Man of Numbers: Fibonacci's Arithmetic Revolution.* New York: Walker, 2011.

Devreese, J T and Guido Vanden Berghe. *'Magic is No Magic' : The Wonderful World of Simon Stevin.* Southampton: WIT, 2008.

Dewey, John. *German Philosophy and Politics*. New York: H Holt, 1915.

Di Bono, Mario. '*L'astronomia Copernicana nell'opera di Giovan Battista Benedetti*'. In *Cultura, scienze e tecniche nella Venezia del Cinquecento: Atti del convegno internazionale di studio Giovan Battista Benedetti e il suo tempo*. Venice: Istituto veneto di scienze, lettere e d'arti, 1987: 288–300.

Dickinson, Henry Winram. *A Short History of the Steam Engine*. London: F Cass, 1963.

———. *Sir Samuel Morland: Diplomat and Inventor, 1625–1695*. Cambridge: Heffer, 1970.

Diderot, Denis. *Les Bijoux indiscrets*. 2 vols. [n.l.]: Au Monomotapa, 1748.

———. *The Indiscreet Jewels*. New York: Marsilio, 1993.

Digby, Kenelm. *A Late Discourse Made in a Solemne Assembly of Nobles and Learned Men at Montpellier in France*. London: R Lownes, 1658.

———. *Two Treatises . . . in Way of Discovery of the Immortality of Reasonable Soules*. Paris: G Blaizot, 1644.

Digges, Leonard and Thomas Digges. *A Prognostication Everlasting*. London: T Marshe, 1576.

Digges, Thomas. *Alae seu scalae mathematicae*. London: T Marsh, 1573.

Dijksterhuis, Eduard Jan. *The Mechanization of the World Picture*. Oxford: Clarendon Press, 1961.

———. *Simon Stevin: Science in the Netherlands around 1600*. The Hague: M Nijhoff, 1970.

Dobbs, Betty Jo Teeter. *The Foundations of Newton's Alchemy*. Cambridge: Cambridge University Press, 1975.

———. 'Newton as Final Cause and First Mover'. In *Rethinking the Scientific Revolution*. Ed. M Osler. Cambridge: Cambridge University Press, 2000: 25–39.

Dodds, E R. *The Ancient Concept of Progress and Other Essays on Greek Literature and Belief*. Oxford: Clarendon Press, 1973.

Donahue, William H. *The Dissolution of the Celestial Spheres*. New York: Arno Press, 1981.

Donne, John. *Devotions upon Emergent Occasions*. London: T Jones, 1624.

———. *The Epithalamions, Anniversaries and Epicedes*. Ed. W Milgate. Oxford: Clarendon Press, 1978.

Drabkin, Israel Edward and Stillman Drake (eds.). *Mechanics in Sixteenth-century Italy*. Madison: University of Wisconsin Press, 1969.

Drake, Stillman. *Cause, Experiment and Science: A Galilean Dialogue Incorporating a New English Translation of Galileo's 'Bodies that Stay Atop Water, or Move in It'*. Chicago: University of Chicago Press, 1981.

Drayton, Michael. *Poly-Olbion*. London: M Lownes, 1612.

Dreyer, John Louis Emil. *History of the Planetary Systems from Thales to Kepler*. Cambridge: Cambridge University Press, 1906.

Dryden, John. *Of Dramatic Poesie: An Essay*. London: H Herringman, 1668.

Ducheyne, Steffen. 'The Status of Theory and Hypotheses'. In *The Oxford Handbook of British Philosophy in the Seventeenth Century*. Ed. PR Anstey. Oxford: Oxford University Press, 2013: 169–91.

Ducoux, François Joseph. *Notice sur Denis Papin, inventeur des machines et des bateaux à vapeur.* Blois: H Morard, 1854.

Duhem, Pierre. '*Un précurseur français de Copernic: Nicole Oresme (1377)*'. *Revue générale des sciences pures et appliquées* 20 (1909): 866–73.

———. '*Le Principe de Pascal: Essai historique*'. *Revue générale des sciences pures et appliquées* 16 (1905): 599–610.

———. *Le Système du monde: Histoire des doctrines cosmologiques de Platon à Copernic.* 10 vols. Vol. 9: *La Physique Parisienne au XIVe siècle.* Paris: Hermann, 1958.

———. *Le Système du monde: Histoire des doctrines cosmologiques de Platon à Copernic.* 10 vols. Vol. 10: *La Cosmologie du XVe siècle.* Paris: Hermann, 1959.

———. *To Save the Phenomena: An Essay on the Idea of Physical Theory from Plato to Galileo.* Chicago: University of Chicago Press, 1969.

Dunn, Jane. *Read My Heart: Dorothy Osborne and Sir William Temple.* London: Harper, 2008.

Dunn, John. *Modern Revolutions: An Introduction to the Analysis of a Political Phenomenon.* Cambridge: Cambridge University Press, 1972.

Dupleix, Scipion. *La Physique ou science naturelle, divisée en 8 livres.* Paris: Veuve D Salis, 1603.

Eagleton, Catherine. 'Medieval Sundials and Manuscript Sources: The Transmission of Information about the Navicula and the *Organum Ptolomei* in Fifteenth-century Europe'. In *Transmitting Knowledge: Words, Images and Instruments in Early Modern Europe.* Ed. S Kusukawa and I Maclean. Oxford: Oxford University Press, 2006: 41–71.

Eamon, William. *Science and the Secrets of Nature: Books of Secrets in Medieval and Early Modern Culture.* Princeton: Princeton University Press, 1994.

Eastwood, Bruce S. 'Grosseteste's "Quantitative" Law of Refraction: A Chapter in the History of Non-experimental Science'. *Journal of the History of Ideas* 28 (1967): 403–14.

———. 'Medieval Empiricism: The Case of Grosseteste's Optics'. *Speculum* 43 (1968): 306–21.

———. 'On the Continuity of Western Science from the Middle Ages: A. C. Crombie's Augustine to Galileo'. *Isis* 83 (1992): 84–99.

———. 'Robert Grosseteste's Theory of the Rainbow'. *Archives internationales d'histoire des sciences* 19 (1966): 313–32.

Edgerton, Samuel Y. *The Heritage of Giotto's Geometry: Art and Science on the Eve of the Scientific Revolution.* Ithaca: Cornell University Press, 1991.

———. *The Renaissance Rediscovery of Linear Perspective.* New York: Basic Books, 1975.

Eisenstein, Elizabeth L. *The Printing Press as an Agent of Change.* 2 vols. Cambridge: Cambridge University Press, 1979.

———. *The Printing Revolution in Early Modern Europe.* Cambridge: Cambridge University Press, 1983.

———. 'An Unacknowledged Revolution Revisited'. *The American Historical Review* 107 (2002): 87–105.

Elia, Pasquale M d'. *Galileo in China: Relations through the Roman College between Galileo and the Jesuit Scientist-Missionaries (1610–1640).* Cambridge, Mass.: Harvard University Press, 1960.

Elias, A C. *Swift at Moor Park: Problems in Biography and Criticism.* Philadelphia: University of Pennsylvania Press, 1982.

Elton, Geoffrey Rudolph. 'Herbert Butterfield and the Study of History'. *Historical Journal* 27 (1984): 729–43.

———. 'A High Road to Civil War?' In *Studies in Tudor and Stuart Politics and Government.* 4 vols. Vol. 2: *Parliament and Political Thought.* Cambridge: Cambridge University Press, 1974: 164–82.

Empson, William. *Essays on Renaissance Literature.* Ed. J Haffenden. 2 vols. Vol. 1: *Donne and the New Philosophy.* Cambridge: Cambridge University Press, 1993.

Erasmus, Desiderius. *Ye Dyaloge Called Funus.* London: R Copland, 1534.

Ernouf, Alfred-Auguste. *Denis Papin: Sa vie et son oeuvre (1647–1714).* Paris: Hachette, 1883.

Estienne, Henri. *The Frankfurt Book Fair.* Ed. JW Thompson. Chicago: Caxton Club, 1911.

Evelyn, John. *The Diary.* Ed. ES de Beer. 6 vols. Vol. 1. Oxford: Clarendon Press, 1955.

Farr, James. 'The Way of Hypotheses: Locke on Method'. *Journal of the History of Ideas* (1987) 51–72.

Fattori, Marta. '*La diffusione di Francis Bacon nel libertinismo francese*'. *Rivista di storia della filosofia* 2 (2002): 225–42.

Favaro, Antonio. '*Libreria di Galileo Galilei*'. *Bullettino di bibliografia e di storia delle scienze matematiche e fisiche* 19 (1886): 219–93.

Febvre, Lucien. '*De l'à peu près à la précision en passant par ouï-dire*'. *Annales. Économies, Sociétés, Civilisations* 5 (1950): 25–31.

———. *The Problem of Unbelief in the Sixteenth Century: The Religion of Rabelais.* Cambridge, Mass.: Harvard University Press, 1982.

———. *Le Problème de l'incroyance au XVIe siècle: La Religion de Rabelais.* Paris: A Michel, 1942.

Feingold, Mordechai. 'Giordano Bruno in England, Revisited'. *Huntington Library Quarterly* 67 (2004): 329–46.

———. *Jesuit Science and the Republic of Letters.* Cambridge Mass.: MIT Press, 2002.

———. *The Newtonian Moment: Isaac Newton and the Making of Modern Culture.* New York: Oxford University Press, 2004.

———. 'When Facts Matter'. *Isis* 87 (1996): 131–9.

Fernel, Jean. *On the Hidden Causes of Things: Forms, Souls and Occult Diseases in Renaissance Medicine.* Ed. J Henry and JM Forrester. Leiden: Brill, 2005.

———. *Therapeutice, seu medendi ratio.* Venice: P Bosellus, 1555.

Ferrand, Jacques. *Erotomania, or A Treatise Discoursing of the Essence, Causes, Symptomes, Prognosticks and Cure of Love or Erotic Melancholy.* Oxford: Printed for Edward Forrest, 1645.

Feyerabend, Paul K. 'Against Method'. In *Analyses of Theories and Methods of Physics and Psychology.* Ed. M Radner and S Winokur. Minneapolis: University of Minnesota Press, 1970: 17–130.

———. *Against Method.* New York: Schocken, 1975.

———. 'Classical Empiricism'. In *The Methodological Heritage of Newton.* Ed. RE Butts and JW Davis. Oxford: Blackwell, 1970: 150–70.

———. *Farewell to Reason.* London: Verso, 1987.

———. *Science in a Free Society.* London: NLB, 1978.

Field, Judith Veronica. *The Invention of Infinity: Mathematics and Art in the Renaissance.* Oxford: Oxford University Press, 1997.

Figuier, Louis. *Exposition et histoire des principales découvertes scientifiques modernes.* 3 vols. Vol. 3. Paris: Langlois & Leclerq, 1851–2.

Filarete, Antonio Averlino detto il. *Trattato di architettura.* Milan: Il Polifilo, 1972.

Findlen, Paula. 'A Forgotten Newtonian: Women and Science in the Italian Provinces'. In *The Sciences in Enlightened Europe.* Ed. W Clark, J Golinski and S Schaffer. Chicago: University of Chicago Press, 1999: 313–49.

———. 'Natural History.' In *The Cambridge History of Science.* Vol. 3. Ed. K Park and L Daston. Cambridge: Cambridge University Press, 2008: 435–68.

Finlay, R. 'China, the West and World History in Joseph Needham's *Science and Civilisation in China*'. *Journal of World History* 11 (2000): 265–303.

Finley, Moses I. 'Aristotle and Economic Analysis'. *Past and Present* 47 (1970): 3–25.

Finn, Bernard S. 'Laplace and the Speed of Sound'. *Isis* 55 (1964): 7–19.

Finocchiaro, Maurice A. *The Galileo Affair: A Documentary History.* Berkeley: University of California, 1989.

———. *Retrying Galileo, 1633–1992.* Berkeley: University of California Press, 2007.

Fish, Stanley. 'Professor Sokal's Bad Joke'. Op-ed. *The New York Times*, 1996.

Fleck, Ludwik. *Genesis and Development of a Scientific Fact.* Ed. TJ Trenn and RK Merton. Chicago: University of Chicago Press, 1979.

Fleming, Donald. 'Latent Heat and the Invention of the Watt Engine'. *Isis* 43 (1952): 3–5.

Fleming, Fergus. *Barrow's Boys.* London: Granta Books, 1998.

Fleming, James Dougal (ed.). *The Invention of Discovery, 1500–1700.* Burlington, VT: Ashgate, 2011.

Fletcher, John Edward and Elizabeth Fletcher. *A Study of the Life and Works of Athanasius Kircher.* Leiden: Brill, 2011.

Fontenelle, Bernard le Bovier de. *Entretiens sur la pluralité des mondes. Digression sur les anciens et les modernes.* Ed. R Shackleton. Oxford: Clarendon Press, 1955.

Foucault, Michel. *L'archéologie du savoir.* Paris: Gallimard, 1969.

———. *Dits et écrits.* Ed. D Defert, F Ewald and J Lagrange. 2 vols. Paris: Gallimard, 2001.

Fowles, Grant R. *Introduction to Modern Optics.* New York: Dover Publications, 1989.

Fox, Robert (ed.). *Thomas Harriot: An Elizabethan Man of Science.* Aldershot: Ashgate, 2000.

Fraassen, Bas C van. *The Scientific Image*. Oxford: Clarendon Press, 1980.
Franklin, James. *The Science of Conjecture: Evidence and Probability before Pascal*. Baltimore: Johns Hopkins University Press, 2001.
Freedberg, David. 'Art, Science and the Case of the Urban Bee'. In *Picturing Science, Producing Art*. Ed. CA Jones, P Galison and AE Slaton. New York: Routledge, 1998: 272–96.
Frisch, Andrea. *The Invention of the Eyewitness: Witnessing and Testimony in Early Modern France*. Chapel Hill: University of North Carolina Press, 2004.
Froidmont, Libert. *Meteorologicorum libri sex*. Antwerp: Moretus, 1627.
Funkenstein, Amos. *Theology and the Scientific Imagination from the Middle Ages to the Seventeenth Century*. Princeton: Princeton University Press, 1986.
Galilei, Galileo. *Dialogue Concerning the Two Chief World Systems, Ptolemaic and Copernican*. Trans. S Drake. Berkeley: University of California Press, 1967.
———. *Discorsi e dimostrazioni matematiche intorno a due nuoue scienze attenenti alla mecanica e i mouimenti locali*. Leiden: Elsevier, 1638.
———. *The Essential Galileo*. Ed. MA Finocchiaro. Indianapolis: Hackett, 2008.
———. *Le opere di Galileo Galilei. Edizione Nazionale*. Ed. A Favaro. 20 vols. Florence: Barberà, 1890–1909.
Galilei, Galileo and Christoph Scheiner. *On Sunspots*. Ed. E Reeves and AV van Helden. Chicago: University of Chicago Press, 2008.
Galilei, Vincenzo. *Dialogue on Ancient and Modern Music*. Ed. CV Palisca. New Haven: Yale University Press, 2003.
Galloway, Elijah and Luke Hebert. *History and Progress of the Steam Engine with a Practical Investigation of Its Structure and Application*. London: T Kelly, 1836.
Galloway, Robert L. *The Steam Engine and Its Inventors*. London: Macmillan, 1881.
Galluzzi, Paolo. *The Art of Invention: Leonardo and Renaissance Engineers*. Florence: Giunti, 1999.
Galton, Francis. *English Men of Science, Their Nature and Nurture*. London: Macmillan, 1874.
Garber, Daniel. 'On the Frontlines of the Scientific Revolution: How Mersenne Learned to Love Galileo'. *Perspectives on Science* 12 (2004): 135–63.
Garzoni, Leonardo. *Trattati della calamità*. Ed. M Ugaglia. Milan: FrancoAngeli, 2005.
Gascoigne, John. 'Crossing the Pillars of Hercules: Francis Bacon, the Scientific Revolution and the New World'. In *Science in the Age of Baroque*. Ed. O Gal and R Chen-Morris. Dordrecht: Springer, 2012: 217–37.
———. 'A Reappraisal of the Role of the Universities in the Scientific Revolution'. In *Reappraisals of the Scientific Revolution*. Ed. D Lindberg and R Westman. Cambridge: Cambridge University Press, 1990: 207–60.
Gassendi, Pierre. *Animadversiones in decimum librum Diogenis Laertii*. Lyons: Barbier, 1649.
———. *Opera omnia*. 6 vols. Florence: J Cajetan, 1727.
Gatti, Hilary. 'Bruno and the Gilbert Circle'. In *Giordano Bruno and Renaissance Science*. Ithaca: Cornell University Press, 1999: 86–98.

———. *Essays on Giordano Bruno*. Princeton: Princeton University Press, 2011.

Gaukroger, Stephen. *Descartes: An Intellectual Biography.* Oxford: Clarendon Press, 1995.

———. *The Emergence of a Scientific Culture: Science and the Shaping of Modernity 1210–1685*. Oxford: Clarendon Press, 2006.

Gaulke, Karsten. '*Die Papin–Savery-Kontroverse*'. In *Denis Papin: Erfinder und Naturforscher in Hessen-Kassel.* Ed. F Tönsmann and H Schneider. Kassel: Euregioverlag, 2009: 105–22.

Gaurico, Luca, Prosdocimus and others. *Spherae tractatus.* Venice: Ginuta, 1531.

Geertz, Clifford. *Local Knowledge: Further Essays in Interpretive Anthropology.* New York: Basic Books, 1983.

Geis, Gilbert and Ivan Bunn. *A Trial of Witches: A Seventeenth-century Witchcraft Prosecution.* London: Routledge, 1997.

Gellner, Ernest. 'Concepts and Society'. In *Rationality.* Ed. B Wilson. Oxford: Blackwell, 1970: 18–49.

———. *Relativism and the Social Sciences.* Cambridge: Cambridge University Press, 1985.

Gerbino, Anthony and Stephen Johnston. *Compass and Rule: Architecture as Mathematical Practice in England, 1500–1750.* New Haven: Yale University Press, 2009.

Gerland, Ernst. '*Das sogenannte Dampfschiff Papin's*'. *Zeitschrift des Vereins für Hessische Geschichte und Landeskunde* 18 (1880): 221–7.

Gerson, Jean. *Opera.* Basle: N Kesler, 1489.

———. *Opera omnia.* 5 vols. Antwerp: Societas, 1706.

Gerth, Jerome. '*Der Dampfkochtopf = Digestor – Eine Erzählung*'. In *Denis Papin und die Eisenhütte Veckerhagen.* Reinhardshagen: Gemeindevorstand Reinhardshagen, 1987: 2–14.

Gibbon, Nicholas. *A Summe or Body of Divinitie Real. Stating Ye Fundamentall, in Modell, for Ye Evidencing & Fixing the Dogmaticall Truths after Ye Way of Demonstration.* London: [n.p.], 1651.

Gigerenzer, Gerd, Zeno Swijtink and others. *The Empire of Chance: How Probability Changed Science and Everyday Life.* Cambridge: Cambridge University Press, 1989.

Gilbert, Creighton. 'When Did a Man in the Renaissance Grow Old?' *Studies in the Renaissance* 14 (1967): 7–32.

Gilbert, Felix. *Machiavelli and Guicciardini: Politics and History in Sixteenth-century Florence.* Princeton: Princeton University Press, 1965.

Gilbert, William. *De magnete.* Trans. P Fleury Mottelay. New York: Dover, 1951.

———. *De magnete, magneticisque corporibus, et de magno magnete tellure: Physiologia nova.* London: P Short, 1600.

———. *De mundo nostro sublunari philosophia nova.* Amsterdam: Elzevir, 1651.

———. *On the Magnet, Magnetick Bodies Also, and on the Great Magnet of the Earth: A New Physiology.* Trans. SP Thompson. London: Chiswick Press, 1900.

Gimpel, Jean. *The Medieval Machine: The Industrial Revolution of the Middle Ages.* New York: Holt, Rinehart and Winston, 1976.

Gingerich, Owen. *An Annotated Census of Copernicus' 'De revolutionibus' (Nuremberg, 1543 and Basel, 1566)*. Leiden: Brill, 2002.

———. *The Book Nobody Read: Chasing the Revolutions of Nicolaus Copernicus*. London: Penguin, 2005.

———. 'Circles of the Gods: Copernicus, Kepler and the Ellipse'. *Bulletin of the American Academy of Arts and Sciences* 47 (1994): 15–27.

———. 'Did Copernicus Owe a Debt to Aristarchus?' *Journal for the History of Astronomy* 16 (1985): 37–42.

———. 'From Copernicus to Kepler: Heliocentrism as Model and as Reality'. *Proceedings of the American Philosophical Society* 117 (1973): 513–22.

———. 'Johannes Kepler'. In *The General History of Astronomy*. 4 vols. 2A: *Planetary Astronomy from the Renaissance to the Rise of Astrophysics*. Ed. R Taton and C Wilson. Cambridge: Cambridge University Press, 1989: 54–78.

———. 'Sacrobosco as a Textbook'. *Journal for the History of Astronomy* 19 (1988): 269–73.

———. 'Sacrobosco Illustrated'. In *Between Demonstration and Imagination: Essays in the History of Science and Philosophy Presented to John D. North*. Ed. AJ Vanderjagt and L Nauta. Leiden: Brill, 1999: 211–24.

———. 'Tycho Brahe and the Nova of 1572'. In *1604–2004: Supernovae as Cosmological Lighthouses*. Ed. M Turatto, S Benetti, L Zampieri and W Shea. San Francisco: Astronomical Society of the Pacific, 2005: 3–12.

Gingerich, Owen and Albert van Helden. 'From Occhiale to Printed Page: The Making of Galileo's *Sidereus nuncius*'. *Journal for the History of Astronomy* 34 (2003): 251–67.

Gingerich, Owen and JR Voelkel. 'Tycho Brahe's Copernican Campaign'. *Journal for the History of Astronomy* 29 (1998): 1–34.

Gingerich, Owen and Robert S Westman. 'The Wittich Connection: Conflict and Priority in Late-sixteenth-century Cosmology'. *Transactions of the American Philosophical Society* 78 (1988): 1–148.

Ginsburg, Jekuthiel. 'On the Early History of the Decimal Point'. *American Mathematical Monthly* 35 (1928): 347–9.

Ginzburg, Carlo. *Myths, Emblems, Clues*. London: Hutchinson Radius, 1990.

Glanvill, Joseph. *Plus ultra, or The Progress and Advancement of Knowledge since the Days of Aristotle*. London: J Collins, 1668.

———. *Saducismus triumphatus, or Full and Plain Evidence Concerning Witches and Apparitions*. London: J Collins, 1681.

———. *The Vanity of Dogmatizing*. London: H Eversden, 1661.

Gleeson-White, Jane. *Double Entry: How the Merchants of Venice Shaped the Modern World*. Crows Nest, NSW: Allen & Unwin, 2011.

Goddu, André. 'Reflections on the Origin of Copernicus's Cosmology'. *Journal for the History of Astronomy* 37 (2006): 37–53.

Godwin, Francis. *The Man in the Moone*. Ed. W Poole. Peterborough, Ont.: Broadview Press, 2009.

Goldberg, Jonathan. 'Speculations: Macbeth and Source'. In *Shakespeare Reproduced: The Text in History and Ideology*. London, 1987: 242–64.

Goldie, Mark. 'The Context of the Foundations'. In *Rethinking the Foundations of Modern Political Thought*. Ed. A Brett, J Tully and H Hamilton-Bleakley. Cambridge: Cambridge University Press, 2006: 3–19.

Goldstein, Bernard R. 'Theory and Observation in Medieval Astronomy'. *Isis* 63 (1972): 39–47.

Goldstein, Bernard R and Giora Hon. 'Kepler's Move from Orbs to Orbits: Documenting a Revolutionary Scientific Concept'. *Perspectives on Science* 13 (2005): 74–111.

Goldstein, Thomas. 'The Renaissance Concept of the Earth in Its Influence upon Copernicus'. *Terrae incognitae* 4 (1972): 19–51.

Golinski, Jan. 'New Preface'. In *Making Natural Knowledge: Constructivism and the History of Science*. Chicago: University of Chicago Press, 2005: vii–xv.

Gombrich, Ernst Hans. *Art and Illusion*. London: Phaidon, 1960.

Goulding, Robert. 'Henry Savile and the Tychonic World-system'. *Journal of the Warburg and Courtauld Institutes* 58 (1995): 152–79.

Grafton, Anthony. *The Footnote: A Curious History*. Cambridge, Mass.: Harvard University Press, 1997.

———. 'Review: The Importance of Being Printed'. *Journal of Interdisciplinary History* 11 (1980): 265–86.

Grafton, Anthony, April Shelford and Nancy G Siraisi. *New Worlds, Ancient Texts: The Power of Tradition and the Shock of Discovery*. Cambridge, Mass.: Harvard University Press, 1992.

Granada, Miguel A. 'Aristotle, Copernicus, Bruno: Centrality, the Principle of Movement and the Extension of the Universe'. *Studies in History and Philosophy of Science Part A* 35 (2004): 91–114.

———. '*Bruno, Digges, Palingenio: Omogeneità ed eterogeneità nella concezione dell'universo infinito*'. *Rivista di storia della filosofia* 47 (1992): 47–73.

Granada, Miguel A, Adam Mosley and Nicholas Jardine. *Christoph Rothmann's Discourse on the Comet of 1585: An Edition and Translation with Accompanying Essays*. Leiden: Brill, 2014.

Graney, Christopher M. 'Anatomy of a Fall: Giovanni Battista Riccioli and the Story of G'. *Physics Today* 65 (2012): 36–40.

———. 'Science Rather than God: Riccioli's Review of the Case For and Against the Copernican Hypothesis'. *Journal for the History of Astronomy* 43 (2012): 215–26.

———. *Setting Aside All Authority: Giovanni Battista Riccioli and the Science against Copernicus in the Age of Galileo*. Notre Dame: University of Notre Dame Press, 2015.

———. 'The Work of the Best and Greatest Artist: A Forgotten Story of Religion, Science and Stars in the Copernican Revolution'. *Logos: A Journal of Catholic Thought and Culture* 15 (2012): 97–124.

Grant, Edward. 'In Defense of the Earth's Centrality and Immobility: Scholastic Reaction to Copernicanism in the Seventeenth Century'. *Transactions of the American Philosophical Society* 74 (1984): 1–69.

———. *The Foundations of Modern Science in the Middle Ages*. Cambridge: Cambridge University Press, 1996.

———. 'God and Natural Philosophy: The Late Middle Ages and Sir Isaac Newton'. *Early Science and Medicine* 5 (2000): 279–98.

———. 'God, Science and Natural Philosophy in the Late Middle Ages'. *Studies in Intellectual History* 96 (1999): 243–68.

———. *Planets, Stars and Orbs: The Medieval Cosmos, 1200–1687*. Cambridge: Cambridge University Press, 1994.

——— (ed.). *A Source Book in Medieval Science*. Cambridge, Mass.: Harvard University Press, 1974.

Graunt, John. *Natural and Political Observations ... Made upon the Bills of Mortality*. London: T Roycroft, 1662.

Gray, John. *Heresies*. London: Granta Books, 2004.

Greeley, Horace. 'The Age We Live In'. *Nineteenth Century* 1 (1848): 50–4.

Greenblatt, Stephen. 'Invisible Bullets'. In *Shakespearean Negotiations*. Oxford: Clarendon Press, 1988: 21–65.

———. *The Swerve: How the Renaissance Began*. London: Bodley Head, 2011.

Greenblatt, Stephen and Joseph L Koerner. 'The Glories of Classicism'. *New York Review of Books*, 21 February 2013.

Grendler, Marcella. 'Book Collecting in Counter-Reformation Italy: The Library of Gian Vincenzo Pinelli (1535–1601)'. *Journal of Library History* 16 (1981): 143–51.

Griffith, Alexander. *Mercurius Cambro-Britannicus, or News from Wales*. London: [s.n.], 1652.

Griffiths, Ralph. 'Select Dissertations from the *Amoenitates academicae*'. *Monthly Review* 65 (1781): 296–304.

Grünbaum, Adolf. 'The Duhemian Argument'. *Philosophy of Science* 27 (1960): 75–87.

Grynaeus, Simon. *Novus orbis regionum ac insularum veteribus incognitarum*. Basle: J Hervagius, 1532.

Guerlac, Henry. 'Can We Date Newton's Early Optical Experiments?' *Isis* 74 (1983): 74–80.

Guicciardini, Francesco. *Maxims and Reflections (Ricordi)*. Philadelphia: University of Pennsylvania Press, 1972.

Gulliver, Lemuel. *The Anatomist Dissected, or The Man-Midwife Finely Brought to Bed*. Westminster: A Campbell, 1727.

Haack, Susan. *Manifesto of a Passionate Moderate: Unfashionable Essays*. Chicago: University of Chicago Press, 1998.

Hacking, Ian. *The Emergence of Probability: A Philosophical Study of Early Ideas about Probability, Induction and Statistical Inference*. Cambridge: Cambridge University Press, 2006.

———. 'Five Parables'. In *Philosophy in History*. Ed. R Rorty, JB Schneewind and Q Skinner. Cambridge: Cambridge University Press, 1984: 103–24.

———. *Historical Ontology*. Cambridge, Mass.: Harvard University Press, 2002.

———. 'How Inevitable are the Results of Successful Science?' *Philosophy of Science* 67 Supplement (2000): 58–71.

———. Inaugural Lecture: Chair of Philosophy and History of Scientific Concepts at the Collège de France. *Economy and Society* 31 (2002): 1–14.

———. 'Introductory Essay'. In Thomas S Kuhn, *The Structure of Scientific Revolutions*. Chicago; London: University of Chicago Press, 2012: i–xxxvii.

———. 'Language, Truth and Reason'. In *Rationality and Relativism*. Ed. M Hollis and S Lukes. Cambridge, Mass.: MIT Press, 1982: 48–66.

———. 'The Self-vindication of the Laboratory Sciences'. In *Science as Practice and Culture*. Ed. A Pickering. Chicago: University of Chicago Press, 1992: 29–64.

———. *The Social Construction of What?* Cambridge, Mass.: Harvard University Press, 1999.

———. '"Style" for Historians and Philosophers'. *Studies in History and Philosophy of Science Part A* 23 (1992): 1–20.

———. 'Was There Ever a Radical Mistranslation?' *Analysis* 41 (1981): 171–5.

Hahn, Nan L. 'Medieval Mensuration: *Quadrans vetus* and *Geometrie due sunt partes principales*'. *Transactions of the American Philosophical Society* 72 (1982): lxxxv, 204.

Hale, John Rigby. *The Civilization of Europe in the Renaissance*. London: HarperCollins, 1993.

———. 'The Early Development of the Bastion: An Italian Chronology *c.*1450–*c.*1534'. In *Europe in the Late Middle Ages*. Ed. JR Hale. London: Faber, 1965: 466–94.

———. 'Warfare and Cartography, *c.*1450 to *c.*1640'. In *The History of Cartography*. 6 vols. Vol. 3: *Cartography in the European Renaissance*. Ed. D Woodward. Chicago: University of Chicago Press, 2007: 719–37.

Hale, Matthew. *Difficiles nugae, or Observations Touching the Torricellian Experiment*. London: W Shrowsbury, 1674.

Hall, A Rupert. *All was Light: An Introduction to Newton's Opticks*. Oxford: Clarendon Press, 1993.

———. *Ballistics in the Seventeenth Century: A Study in the Relations of Science and War*. Cambridge: Cambridge University Press, 1952.

———. 'Engineering and the Scientific Revolution'. *Technology and Culture* 2 (1961): 333–41.

———. *Philosophers at War: The Quarrel between Newton and Leibniz*. Cambridge: Cambridge University Press, 1980.

———. 'What Did the Industrial Revolution in Britain Owe to Science?' In *Historical Perspectives: Studies in English Thought and Society, in Honour of J. H. Plumb*. Ed. N McKendrick. London: Europa, 1974: 129–51.

———. 'William Wotton and the History of Science'. *Archives internationales d'histoire des sciences* 9 (1949): 1047–62.

Hamblyn, Richard. *The Invention of Clouds: How an Amateur Meteorologist Forged the Language of the Skies*. London: Picador, 2001.

Hamel, Jürgen. *Studien zur 'Sphaera' des Johannes de Sacrobosco*. Leipzig: Akademische Verlagsanstalt, 2014.

Hannam, James. *God's Philosophers: How the Medieval World Laid the Foundations of Modern Science.* London: Icon Books, 2009.

Hanson, Norwood Russell. 'An Anatomy of Discovery'. *Journal of Philosophy* 64 (1967): 321–52.

———. '*Hypotheses fingo*'. In *The Methodological Heritage of Newton.* Ed. RE Butts and JW Davis. Oxford: Blackwell, 1970: 14–33.

———. *Patterns of Discovery: An Inquiry into the Conceptual Foundations of Science.* Cambridge: Cambridge University Press, 1958.

Harle, Jonathan. *An Historical Essay on the State of Physick in the Old and New Testament.* London: R Ford, 1729.

Harley, John Brian. 'Maps, Knowledge and Power'. In *The New Nature of Maps: Essays in the History of Cartography.* Ed. P Laxton. Baltimore: Johns Hopkins University Press, 2001: 51–82.

Harris, John. *Lexicon technicum, or An Universal English Dictionary of Arts and Sciences Vol. I.* London: D Brown, 1704.

Harrison, Peter. *The Bible, Protestantism and the Rise of Natural Science.* Cambridge: Cambridge University Press, 1998.

———. 'Curiosity, Forbidden Knowledge and the Reformation of Natural Philosophy in Early Modern England'. *Isis* 92 (2001): 265–90.

———. 'The Development of the Concept of Laws of Nature'. In *Creation: Law and Probability.* Ed. FN Watts. Minneapolis: Fortress Press, 2008: 13–35.

———. *The Fall of Man and the Foundations of Science.* Cambridge: Cambridge University Press, 2007.

———. 'Newtonian Science, Miracles and the Laws of Nature'. *Journal of the History of Ideas* 56 (1995): 531–53.

———. 'Reassessing the Butterfield Thesis'. *Historically Speaking* 8 (2006): 7–10.

———. 'Voluntarism and Early Modern Science'. *History of Science* 40 (2002): 63–89.

———. 'Voluntarism and the Origins of Modern Science: A Reply to John Henry'. *History of Science* 47 (2009): 223–31.

Harvey, Gabriel. *Gabriel Harvey's Marginalia.* Ed. GCM Moore Smith. Stratford-upon-Avon: Shakespeare Head Press, 1913.

Harvey, Gideon. *The Vanities of Philosophy and Physick.* London: A Roper, 1699.

Harvey, William. *Anatomical Exercitations, Concerning the Generation of Living Creatures.* London: O Pulleyn, 1653.

Haugen, Kristine Louise. *Richard Bentley: Poetry and Enlightenment.* Cambridge, Mass.: Harvard University Press, 2011.

Hay, Denys. *Polydore Vergil: Renaissance Historian and Man of Letters.* Oxford: Clarendon Press, 1952.

Hayton, Darin. 'Instruments and Demonstrations in the Astrological Curriculum: Evidence from the University of Vienna, 1500–1530'. *Studies in History and Philosophy of Science Part C* 41 (2010): 125–34.

Headley, John M. 'The Sixteenth-century Venetian Celebration of the Earth's Total Habitability: The Issue of the Fully Habitable World for Renaissance Europe'. *Journal of World History* 8 (1997): 1–27.

Hedrick, Elizabeth. 'Romancing the Salve: Sir Kenelm Digby and the Powder of Sympathy'. *British Journal for the History of Science* 41 (2008): 161–85.

Heeffer, Albrecht. 'On the Curious Historical Coincidence of Algebra and Double-entry Bookkeeping'. In *Foundations of the Formal Sciences VII*. Ed. K François, B Löwe and T Müller. London: College Publishers, 2011: 109–30.

Heilbron, John L. *Galileo*. Oxford: Oxford University Press, 2010.

———. *Physics at the Royal Society During Newton's Presidency*. Los Angeles: William Andrews Clark Memorial Library, 1983.

Heisenberg, W. '*Über quantentheoretische Umdeutung kinematischer und mechanischer Beziehungen*'. *Zeitschrift für Physik* 33 (1925): 879–93.

Helas, Philine. '*Die Erfindung des Globus durch die Malerei – zum Wandel des Weltbildes im 15. Jahrhundert*'. In *Die Welt im Bild: Weltentwürfe in Kunst, Literatur und Wissenschaft seit der Frühen Neuzeit*. Ed. U Gehring. Munich: W Fink, 2010: 43–86.

———. '*Mundus in rotundo et pulcherrime depictus: Nunquam sistens sed continuo volvens: Ephemere Globen in den Festinszenierungen des italienischen Quattrocento*'. *Der Globusfreund* 45–6 (1998): 155–75.

Helden, Albert van. 'The Invention of the Telescope'. *Transactions of the American Philosophical Society* 67 (1977): 1–67.

———. *Measuring the Universe*. Chicago: University of Chicago Press, 1985.

———. 'Roemer's Speed of Light'. *Journal for the History of Astronomy* 14 (1983): 137–41.

Helden, Anne C van. 'The Age of the Air-pump'. *Tractrix* 3 (1991): 149–72.

Hellman, C Doris. 'Additional Tracts on the Comet of 1577'. *Isis* 39 (1948): 172–4.

———. 'A Bibliography of Tracts and Treatises on the Comet of 1577'. *Isis* 22 (1934): 41–68.

———. *The Comet of 1577: Its Place in the History of Astronomy*. New York: AMS Press, 1971.

Hellman, Hal. *Great Feuds in Mathematics: Ten of the Liveliest Disputes Ever*. Hoboken, NJ: John Wiley, 2006.

Hellyer, Marcus (ed.). *The Scientific Revolution: The Essential Readings*. Malden, Mass.: Blackwell, 2003.

Helmont, Jan Baptist van. *Deliramenta catarrhi, or The Incongruities, Impossibilities and Absurdities Couched under the Vulgar Opinion of Defluxions*. Ed. W Charleton. London: William Lee, 1650.

———. *Ortus medicinae, id est, initia physicae inaudita*. Amsterdam: Elsevier, 1652.

Helmont, Jan Baptist van and Walter Charleton. *A Ternary of Paradoxes. The Magnetick Cure of Wounds. Nativity of Tartar in Wine. Image of God in Man*. London: W Lee, 1649.

Henninger-Voss, Mary. 'Measures of Success: Military Engineering and the Architectonic Understanding of Design'. In *Picturing Machines*. Ed. W Lefèvre. Cambridge, Mass.: MIT Press, 2004: 143–69.

Henry, John. 'Metaphysics and the Origins of Modern Science: Descartes and the Importance of Laws of Nature'. *Early Science and Medicine* 9 (2004): 73–114.

———. *The Scientific Revolution and the Origins of Modern Science*. Houndmills, Basingstoke: Palgrave, 2008.

———. 'Voluntarist Theology at the Origins of Modern Science: A Response to Peter Harrison'. *History of Science* 47 (2009): 79–113.

Hesse, Mary. 'Comment on Kuhn's "Commensurability, Comparability, Communicability"'. *PSA: Proceedings of the Biennial Meeting of the Philosophy of Science Association* (1982): 704–11.

Hessen, Boris and Henryk Grossman. *The Social and Economic Roots of the Scientific Revolution*. Ed. P McLaughlin and G Freudenthal. Dordrecht: Kluwer Academic Publishers, 2009.

Hessler, John W. *The Naming of America: Martin Waldseemüller's 1507 World Map and the 'Cosmographiae introductio'*. London: Giles, 2008.

Hevelius, Johannes and Jeremiah Horrocks. *Mercurius in Sole visus Gedani: Anno christiano 1661 ... cui annexa est, Venus in Sole visa, Anno 1639*. Gdansk: Reiniger, 1662.

Hexter, Jack H. 'The Historian and His Day'. *Political Science Quarterly* 69 (1954): 219–33.

———. *Reappraisals in History*. Evanston, Ill.: Northwestern University Press, 1961

Hiatt, Alfred. *Terra incognita: Mapping the Antipodes before 1600*. Chicago: University of Chicago Press, 2008.

Hill, Christopher. *Intellectual Origins of the English Revolution*. Oxford: Clarendon Press, 1965.

———. 'The Word "Revolution" in Seventeenth-century England'. In *For Veronica Wedgwood These Studies in Seventeenth-century History*. Ed. R Ollard and P Tudor-Craig. London: William Collins, 1986: 134–51.

Hill, Nicholas. *Philosophia epicuraea democritiana theophrastica*. Ed. S Plastina. Pisa: Fabrizio Serra, 2007.

Hills, Richard Leslie. *Power from Steam: A History of the Stationary Steam Engine*. Cambridge: Cambridge University Press, 1989.

Himmelstein, Franz Xaver. *Synodicon herbipolense: Geschichte und Statuten der im Bisthum Würzburg gehaltenen Concilien und Dioecesansynoden*. Würzburg: Stahel, 1855.

Hine, W L. 'Inertia and Scientific Law in Sixteenth-century Commentaries on Lucretius'. *Renaissance Quarterly* 48 (1995): 728–41.

Hintikka, Jaakko. 'Aristotelian Infinity'. *Philosophical Review* 75 (1966): 197–218.

Hoare, Michael Rand. *The Quest for the True Figure of the Earth: Ideas and Expeditions in Four Centuries of Geodesy*. Burlington, VT: Ashgate, 2004.

Hobbes, Thomas. *Critique du 'De mundo' de Thomas White*. Ed. J Jacquot and HW Jones. Paris: J Vrin, 1973.

———. *Elements of Philosophy, the First Section, Concerning Body*. London: A Crooke, 1656.

———. *Humane Nature, or The Fundamental Elements of Policie*. London: F Bowman, 1650.

———. *Leviathan, or The Matter, Forme and Power of a Common Wealth, Ecclesiasticall and Civil.* London: A Crooke, 1651.

———. *Of Libertie and Necessitie: A Treatise.* London: F Eaglesfield, 1654.

———. *Philosophicall Rudiments Concerning Government and Society.* London: Royston, 1651.

Hobson, Anthony. 'A Sale by Candle in 1608'. *The Library* 5 (1971): 215–33.

Hollis, Martin and Steven Lukes (eds.). *Rationality and Relativism.* Cambridge, Mass.: MIT Press, 1982.

Holmes, Geoffrey S. 'Gregory King and the Social Structure of Pre-Industrial England'. *Transactions of the Royal Historical Society* 27 (1977): 41–68.

Hooke, Robert. *Lectiones Cutlerianæ, or A Collection of Lectures, Physical, Mechanical, Geographical & Astronomical.* London: J Martyn, 1679.

———. *Lectures de potentia restitutiva, or Of Spring, Explaining the Power of Springing Bodies.* London: J Martyn, 1678.

———. *Micrographia, or Some Physiological Descriptions of Minute Bodies.* London: J Martyn, 1665.

———. *The Posthumous Works.* London: S Smith, 1705.

Hooker, Richard. *Of the Lawes of Ecclesiasticall Politie, Eight Bookes.* London: J Windet, 1604.

Hooykaas, Reijer. *G. J. Rheticus's Treatise on Holy Scripture and the Motion of the Earth.* Amsterdam: North-Holland, 1984.

———. *Religion and the Rise of Modern Science.* Grand Rapids, MI.: Eerdmans, 1972.

Horrocks, Jeremiah. *Venus Seen on the Sun: The First Observation of a Transit of Venus.* Ed. W Applebaum. Leiden: Brill, 2012.

Horton, Robin. *Patterns of Thought in Africa and the West: Essays on Magic, Religion and Science.* Cambridge: Cambridge University Press, 1997.

Hoskin, Michael. 'The Discovery of Uranus, the Titius–Bode Law, and the Asteroids'. In *The General History of Astronomy.* 4 vols. Vol. 2B: *Planetary Astronomy from the Renaissance to the Rise of Astrophysics.* Ed. R Taton and C Wilson. 1995: 169–80.

Hoyningen-Huene, Paul. 'Three Biographies: Kuhn, Feyerabend and Incommensurability'. In *Rhetoric and Incommensurability.* Ed. RA Harris. West Lafayette, IN: Parlor Press, 2005: 150–75.

———. 'Two Letters of Paul Feyerabend to Thomas S. Kuhn on a Draft of *The Structure of Scientific Revolutions*'. *Studies in History and Philosophy of Science Part A* 26 (1995): 353–87.

Hues, Robert. *A Learned Treatise of Globes, Both Coelestiall and Terrestriall.* London: A Kemb, 1659.

———. *Tractatus de globis, coelesti et terrestri eorumque usu.* Amsterdam: J Hondius, 1617.

Huff, Toby E. *Intellectual Curiosity and the Scientific Revolution: A Global Perspective.* Cambridge: Cambridge University Press, 2011.

Hull, David L. 'In Defense of Presentism'. *History and Theory* 18 (1979): 1–15.

Hull, Gordon. 'Hobbes and the Premodern Geometry of Modern Political Thought'. In *Arts of Calculation: Quantifying Thought in Early Modern Europe*. Ed. D Glimp and MR Warren. New York: Palgrave Macmillan, 2004: 115–35.

Humboldt, Alexander von. *Examen critique de l'histoire de la géographie du nouveau continent: Et des progrès de l'astronomie nautique aux 15me et 16me siècles*. 3 vols. Paris: Gide, 1836–9.

Hume, David. *Philosophical Essays Concerning Human Understanding*. London: A Millar, 1748.

———. *Political Discourses*. Edinburgh: A Kincaid, 1752.

Hunter, Michael. 'Alchemy, Magic and Moralism in the Thought of Robert Boyle'. *British Journal for the History of Science* 23 (1990): 387–410.

———. *Boyle: Between God and Science*. New Haven: Yale University Press, 2009.

———. 'The Decline of Magic: Challenge and Response in Early Enlightenment England'. *The Historical Journal* 55 (2012): 399–425.

———. *Establishing the New Science: The Experience of the Early Royal Society*. Woodbridge, Suffolk: Boydell Press, 1989.

———. 'New Light on the "Drummer of Tedworth": Conflicting Narratives of Witchcraft in Restoration England'. *Historical Research* 78 (2005): 311–53.

———. *The Occult Laboratory: Magic, Science and Second Sight in Late-seventeenth-century Scotland*. Woodbridge: Boydell Press, 2001.

——— (ed.). *Robert Boyle by Himself and His Friends: With a Fragment of William Wotton's Lost Life of Boyle*. London: W Pickering, 1994.

———. *The Royal Society and Its Fellows, 1660–1700: The Morphology of an Early Scientific Institution*. Chalfont St Giles, Bucks: British Society for the History of Science, 1982.

———. 'The Royal Society and the Decline of Magic'. *Notes and Records of the Royal Society* 65: 103–19 (2011).

———. 'Science and Astrology in Seventeenth-century England: An Unpublished Polemic by John Flamsteed'. [1987] In *Science and the Shape of Orthodoxy: Intellectual Change in Late-seventeenth-century Britain*. Boydell & Brewer, 1995: 245–85.

———. 'Science and Heterodoxy: An Early Modern Problem Reconsidered'. In *Reappraisals of the Scientific Revolution*. Ed. D Lindberg and R Westman. Cambridge: Cambridge University Press, 1990: 437–60.

Hunter, Michael and Lawrence M Principe. 'The Lost Papers of Robert Boyle'. *Annals of Science* 60 (2003): 269–311.

Hunter, Michael and Paul B Wood. 'Towards Solomon's House: Rival Strategies for Reforming the Early Royal Society'. *History of Science* 24 (1986): 49–108.

Huppert, George. 'The Life and Works of Louis Le Roy, by Werner L. Gundersheimer'. *History and Theory* 7 (1968): 151–8.

Ibn Al-Haytham. *Alhacen's Theory of Visual Perception: The First Three Books of Alhacen's 'De aspectibus'*. Ed. AM Smith. Philadelphia: American Philosophical Society, 2001.

———. *The Optics: Books I–III, on Direct Vision*. Ed. AI Sabra. 2 vols. London: Warburg Institute, University of London, 1989.

Ilardi, Vincent. *Renaissance Vision from Spectacles to Telescopes*. Philadelphia: American Philosophical Society, 2007.

Iliffe, R. '"In the Warehouse": Privacy, Property and Priority in the Early Royal Society'. *History of Science* 30 (1992): 29–68.

Isaac, Joel. *Working Knowledge: Making the Human Sciences from Parsons to Kuhn*. Cambridge, Mass.: Harvard University Press, 2012.

Ivins, William Mills. *On the Rationalization of Sight: With . . . Three Renaissance Texts*. New York: Da Capo Press, 1975.

———. *Prints and Visual Communication*. London: Routledge, 1953.

Jackson, Thomas. *Justifying Faith, or The Faith by which the Just Do Live*. London: J Beale, 1615.

Jacob, Margaret C. 'Science Studies after Social Construction: The Turn toward the Comparative and the Global'. In *Beyond the Cultural Turn: New Directions in the Study of Society and Culture*. Ed. VE Bonnell and L Hunt. University of California Press, 1999: 95–120.

———. *Scientific Culture and the Making of the Industrial West*. New York: Oxford University Press, 1997.

Jacob, Margaret C and Larry Stewart. *Practical Matter: Newton's Science in the Service of Industry and Empire, 1687–1851*. Cambridge, Mass.: Harvard University Press, 2004.

Jacquot, Jean. 'Thomas Harriot's Reputation for Impiety'. *Notes and Records of the Royal Society of London* 9 (1952): 164–87.

Jalobeanu, Dana. 'A Natural History of the Heavens: Francis Bacon's Anti-Copernicanism'. In *The Making of Copernicus: Early Modern Transformations of the Scientist and His Science*. Ed. W Neuber, T Rahn and C Zittel. Leiden: Brill, 2015: 64–87.

James, GO. 'The Problem of Mechanical Flight'. *Science* 36 (1912): 336–40.

James, William. 'Humanism and Truth (1904)'. In *Pragmatism: A New Name for Some Old Ways of Thinking: [and] the Meaning of Truth, a Sequel to Pragmatism*. Cambridge, Mass.: Harvard University Press, 1978.

Jansen, Paule. *De Blaise Pascal à Henry Hammond: Les Provinciales en Angleterre*. Paris: J Vrin, 1954.

Jardine, Nicholas. *The Birth of History and Philosophy of Science: Kepler's 'A Defence of Tycho against Ursus'*. Cambridge: Cambridge University Press, 1984.

———. *The Scenes of Inquiry: On the Reality of Questions in the Sciences*. Oxford: Clarendon Press, 2000.

———. 'Uses and Abuses of Anachronism in the History of the Sciences'. *History of Science* 38 (2000): 251–70.

———. 'Whigs and Stories: Herbert Butterfield and the Historiography of Science'. *History of Science* 41 (2003): 125–40.

Jarrige, Pierre. *A Further Discovery of the Mystery of Jesuitisme*. London: R Royston, 1658.

Jervis, Jane L. *Cometary Theory in Fifteenth-century Europe*. Dordrecht: D Reidel, 1985.

Jesseph, Douglas M. 'Galileo, Hobbes and the Book of Nature'. *Perspectives on Science* 12 (2004): 191–211.

Jobe, Thomas Harmon. 'The Devil in Restoration Science: The Glanvill–Webster Witchcraft Debate'. *Isis* 72 (1981): 343–56.

Johns, Adrian. 'How to Acknowledge a Revolution'. *American Historical Review* 107 (2002): 106–25.

———. 'Identity, Practice and Trust in Early Modern Natural Philosophy'. *Historical Journal* 42 (1999): 1125–45.

———. *The Nature of the Book: Print and Knowledge in the Making*. Chicago: University of Chicago Press, 1998.

———. 'Science and the Book in Modern Cultural Historiography'. *Studies in History and Philosophy of Science Part A* 29 (1998): 167–94.

Johnson, Christine R. *The German Discovery of the World: Renaissance Encounters with the Strange and Marvelous*. Charlottesville: University of Virginia Press, 2008.

———. 'Renaissance German Cosmographers and the Naming of America'. *Past and Present* 191 (2006): 3–43.

Johnson, Francis R and Sanford V Larkey. 'Thomas Digges, the Copernican System and the Idea of the Infinity of the Universe in 1576'. *Huntington Library Bulletin* 5 (1934): 69–117.

Johnson, Samuel. 'The Vanity of Authors'. In *The Rambler [No.1 March 20, 1750 – No.208 March 14, 1752]*. 6 vols. Vol. 4 (no. 106). London: J. Payne and J. Bouquet, 1752: 46–54.

Johnston, Stephen. 'Theory, Theoric, Practice: Mathematics and Magnetism in Elizabethan England'. *Journal de la Renaissance* 2 (2004): 53–62.

Jones, Richard Foster. *Ancients and Moderns: A Study of the Background of the Battle of the Books*. St Louis: Washington University Press, 1936.

Jonkers, ART. *Earth's Magnetism in the Age of Sail*. Baltimore: Johns Hopkins University Press, 2003.

Joy, Lynn S. 'Scientific Explanation: From Formal Causes to Laws of Nature'. In *The Cambridge History of Science*. 7 vols. Vol. 3: *Early Modern Science*. Ed. K Park and LJ Daston. Cambridge: Cambridge University Press, 2006: 70–105.

Jurin, James. *A Letter to the Right Reverend the Bishop of Cloyne Occasion'd by His Lordship's Treatise on the Virtues of Tar-water*. London: J Robinson, 1744.

Kant, Immanuel. *Critique of Pure Reason*. Ed. N Kemp Smith. New York: Macmillan, 1949.

Kassell, Lauren. *Medicine and Magic in Elizabethan England: Simon Forman – Astrologer, Alchemist and Physician*. Oxford: Clarendon, 2005.

Kastan, David Scott. *Shakespeare and the Book*. Cambridge: Cambridge University Press, 2001.

Kaye, Joel. *Economy and Nature in the Fourteenth Century: Money, Market Exchange and the Emergence of Scientific Thought*. Cambridge: Cambridge University Press, 1998.

Kemp, Martin. 'Science, Non-science and Nonsense: The Interpretation of Brunelleschi's Perspective'. *Art History* 1 (1978): 134–61.

———. *The Science of Art: Optical Themes in Western Art from Brunelleschi to Seurat*. New Haven: Yale University Press, 1990.

Kepler, Johannes. *Dioptrice, seu demonstratio eorum quae visui et visibilibus propter conspicilla non ita pridem inventa accidunt: Praemissae epistolae Galilaei de ijs quae post editionem nuncij siderij ope perspicilli, nova et admiranda in coelo deprehensa sunt*. Augsburg: Franck, 1611.

———. *Dissertatio cum nuncio sidereo*. Ed. I Pantin. Paris: Les Belles Lettres, 1993.

———. *Epitome astronomiae Copernicanae*. Frankfurt: Schönwetter, 1635.

———. *Epitome of Copernican Astronomy, Books IV and V*. Amherst, NY: Prometheus Books, 1995.

———. *L' Étoile nouvelle dans le serpentaire*. Paris: A Blanchard, 1998.

———. *L'Étrenne, ou La Neige sexangulaire*. Ed. R Halleux. Paris: Vrin, 1975.

———. *Kepler's Conversation with Galileo's Sidereal Messenger*. Ed. E Rosen. New York: Johnson Reprint Corporation, 1965.

———. *Kepler's Dream*. Ed. J Lear. Berkeley: University of California Press, 1965.

———. *Kepler's Somnium: The Dream or Posthumous Work on Lunar Astronomy*. Ed. E Rosen. Madison: University of Wisconsin Press, 1967.

———. *New Astronomy*. Trans. WH Donahue. Cambridge: Cambridge University Press, 1992.

———. *The Six-cornered Snowflake*. Ed. C Hardie. Oxford: Clarendon Press, 1966.

———. *The Six-cornered Snowflake: A New Year's Gift*. Ed. JF Nims. Philadelphia: Paul Dry Books, 2010.

Kerker, Milton. 'Science and the Steam Engine'. *Technology and Culture* 2 (1961): 381–90.

Ketterer, David. '"The Wonderful Effects of Steam": More Percy Shelley Words in Frankenstein?' *Science Fiction Studies* 25 (1998): 566–70.

Keynes, Geoffrey. *John Evelyn, a Study in Bibliophily with a Bibliography of His Writings*. Cambridge: Cambridge University Press, 1937.

King, Henry C and John R Millburn. *Geared to the Stars: The Evolution of Planetariums, Orreries and Astronomical Clocks*. Toronto: University of Toronto Press, 1978.

King, Peter. 'Mediaeval Thought-experiments: The Metamethodology of Mediaeval Science'. In *Thought Experiments in Science and Philosophy*. Ed. T Horowitz. Lanham, MD: Rowman and Littlefield, 1991: 43–64.

Kirk, GS, JE Raven and Malcolm Schofield. *The Presocratic Philosophers: A Critical History with a Selection of Texts*. Cambridge: Cambridge University Press, 1983.

Klein, Judy L. *Statistical Visions in Time: A History of Time Series Analysis, 1662–1938*. Cambridge: Cambridge University Press, 1997.

Klein, Ursula. 'Origin of the Concept of Chemical Compound'. *Science in Context* 7 (1994): 163–204.

Koyré, Alexandre. *The Astronomical Revolution: Copernicus, Kepler, Borelli*. Paris: Hermann, 1973.

———. 'Concept and Experience in Newton's Scientific Thought'. [1956] In *Newtonian Studies*. London: Chapman & Hall, 1965: 25–52.

———. '*Du monde de "l'à-peu-près" à l'univers de la précision*'. In *Études d'histoire de la pensée philosophique*. Paris: Colin, 1971: 311–29.

———. *Études d'histoire de la pensée scientifique*. Paris: Gallimard, 1973.

———. *Études Galiléennes*. Paris: Hermann, 1966.

———. *From the Closed World to the Infinite Universe*. Baltimore: Johns Hopkins University Press, 1957.

———. '*Galilée et l'expérience de Pise: À propos d'une légende*'. In *Études d'histoire de la pensée scientifique*. Paris: Gallimard, 1973: 213–23.

———. 'Galileo and the Scientific Revolution of the Seventeenth Century'. *The Philosophical Review* 52 (1943): 333–48.

———. *Newtonian Studies*. London: Chapman & Hall, 1965.

Kren, Claudia. 'The Rolling Device of Naṣir al-Dīn al-Tūsī in the *De Spera* of Nicole Oresme?' *Isis* 62 (1971): 490–8.

Kristensen, Leif Kahl and Kurt Møller Pedersen. 'Roemer, Jupiter's Satellites and the Velocity of Light'. *Centaurus* 54 (2012): 4–38.

Kubovy, Michael. *The Psychology of Perspective and Renaissance Art*. Cambridge: Cambridge University Press, 1986.

Kuhn, Thomas S. *The Copernican Revolution: Planetary Astronomy in the Development of Western Thought*. Cambridge, Mass.: Harvard University Press, 1957.

———. 'Dubbing and Redubbing: The Vulnerability of Rigid Designation'. *Minnesota Studies in the Philosophy of Science* 14 (1990): 298–318.

———. *The Essential Tension: Selected Studies in Scientific Tradition and Change*. Chicago: University of Chicago Press, 1977.

———. 'Historical Structure of Scientific Discovery'. *Science* 136 (1962): 760–64.

———. 'Mathematical versus Experimental Traditions in the Development of Physical Science'. *The Journal of Interdisciplinary History* 7 (1976): 1–31.

———. 'The Principle of Acceleration: A Non-dialectical Theory of Progress: Comment'. *Comparative Studies in Society and History* 11 (1969): 426–30.

———. *The Road since Structure: Philosophical Essays, 1970–1993, with An Autobiographical Interview*. Ed. J Conant and J Haugeland. Chicago: University of Chicago Press, 2000.

———. *The Structure of Scientific Revolutions*. Chicago: University of Chicago Press, 1962.

———. *The Structure of Scientific Revolutions*. Chicago: University of Chicago Press, 1970.

———. *The Structure of Scientific Revolutions*. Chicago: University of Chicago Press, 1996.

———. *The Trouble with the Historical Philosophy of Science: Robert and Maurine Rothschild Distinguished Lecture, 19 November 1991*. Cambridge, Mass.: Department of the History of Science, Harvard University, 1992.

———. 'What are Scientific Revolutions?' [1987] In *The Road since Structure: Philosophical Essays, 1970–1993, with An Autobiographical Interview*. Ed. J Conant and J Haugeland. Chicago: University of Chicago Press, 2000: 13–32.

Kusch, Martin. 'Annalisa Coliva on Wittgenstein and Epistemic Relativism'. *Philosophia* 41 (2013): 37–49.

———. 'Hacking's Historical Epistemology: A Critique of Styles of Reasoning'. *Studies in History and Philosophy of Science Part A* 41 (2010): 158–73.

Kusukawa, Sachiko. *Picturing the Book of Nature: Image, Text and Argument in Sixteenth-century Human Anatomy and Medical Botany.* Chicago: University of Chicago Press, 2011.

———. 'The Sources of Gessner's Pictures for the *Historia animalium*'. *Annals of Science* 67 (2010): 303–28.

Kwa, Chunglin. *Styles of Knowing.* Pittsburgh: University of Pittsburgh Press, 2011.

Labinger, Jay A. and Harry Collins (eds.). *The One Culture? A Conversation about Science.* Chicago: University of Chicago Press, 2001.

La Boëtie, Étienne de. *De la servitude volontaire, ou Contr'un.* Ed. MC Smith. Geneva: Droz, 1987.

Laird, W R. 'Archimedes among the Humanists'. *Isis* 82 (1991): 629–38.

Lakatos, Imre. *The Methodology of Scientific Research Programmes.* Cambridge: Cambridge University Press, 1978.

Lamb, David and Susan M Easton. *Multiple Discovery.* Amersham: Avebury, 1984.

La Mettrie, Julien Offray de. *La Mettrie's 'L'Homme machine': A Study in the Origins of an Idea.* Ed. A Vartanian. Princeton: Princeton University Press, 1960.

Landes, David S. 'Why Europe and the West? Why Not China?' *Journal of Economic Perspectives* 20 (2006): 3–22.

Langbein, John H. *Torture and the Law of Proof: Europe and England in the Ancien Régime.* Chicago: University of Chicago Press, 1977.

Laqueur, Thomas Walter. *Making Sex: Body and Gender from the Greeks to Freud.* Cambridge, Mass.: Harvard University Press, 1990.

Laski, Harold Joseph. *The Rise of European Liberalism: An Essay in Interpretation.* London: Allen & Unwin, 1936.

Laslett, Peter. 'Commentary'. In *Scientific Change.* Ed. AC Crombie. New York: Basic Books, 1963: 861–5.

Latham, RE (ed.). *Dictionary of Medieval Latin from British Sources.* London: British Academy, 1975– .

Latour, Bruno. 'For David Bloor ... and beyond: A Reply to David Bloor's "Anti-Latour"'. *Studies in History and Philosophy of Science* 30 (1999): 113–30.

———. 'The Force and the Reason of Experiment'. In *Experimental Inquiries.* Ed. HE Legrand. Dordrecht: Kluwer, 1990: 49–80.

———. 'One More Turn after the Social Turn: Easing Science Studies into the Non-modern World'. In *The Social Dimensions of Science.* Ed. E McMullin. Notre Dame: Notre Dame University Press, 1992: 272–92.

———. 'On the Partial Existence of Existing and Non-existing Objects'. In *Biographies of Scientific Objects.* Ed. LJ Daston. Chicago: University of Chicago Press, 2000: 247–69.

———. *Pandora's Hope: Essays on the Reality of Science Studies*. Cambridge, Mass.: Harvard University Press, 1999.

———. 'Visualisation and Cognition: Drawing Things Together'. In *Representation in Scientific Activity*. Ed. M Lynch and S Woolgar. Cambridge, Mass.: MIT Press, 1990: 19–68.

———. *We Have Never been Modern*. Cambridge, Mass.: Harvard University Press, 1993.

Lattis, James M. *Between Copernicus and Galileo: Christoph Clavius and the Collapse of Ptolemaic Cosmology*. Chicago: University of Chicago Press, 1994.

Laudan, Larry. 'The Clock Metaphor and Probabilism: The Impact of Descartes on English Methodological Thought, 1650–65'. *Annals of Science* 22 (1966): 73–104.

———. 'A Confutation of Convergent Realism'. *Philosophy of Science* 48 (1981): 19–49.

———. 'Demystifying Underdetermination'. *Minnesota Studies in the Philosophy of Science* 14 (1990): 267–97.

———. 'The Nature and Sources of Locke's Views on Hypotheses'. *Journal of the History of Ideas* 28 (1967): 211–23.

———. 'The Pseudo-science of Science?' *Philosophy of the Social Sciences* 11 (1981): 173–98.

Law, John. 'Technology and Heterogeneous Engineering: The Case of Portuguese Expansion'. In *The Social Construction of Technological Systems*. Ed. WE Bijker, T Hughes and TJ Pinch. Cambridge, Mass.: MIT Press, 1987: 111–34.

Layton Jr, Edwin T. 'Technology as Knowledge'. *Technology and Culture* 15 (1974): 31–41.

Leavis, FR. *Two Cultures? The Significance of C. P. Snow*. Ed. S Collini. Cambridge: Cambridge University Press, 2013.

Leblanc, Vincent. *The World Surveyed, or The Famous Voyages and Travailes of V. Le Blanc, or White*. London: J Starkey, 1660.

Le Clerc, Daniel. *The History of Physick, or an Account of the Rise and Progress of the Art and the Several Discoveries Therein from Age to Age*. London: D Brown, 1699.

Leeuwen, Henry G van. *The Problem of Certainty in English Thought, 1630–1690*. The Hague: Martinus Nijhoff, 1963.

Lefèvre, Wolfgang. 'The Limits of Pictures: Cognitive Functions of Images in Practical Mechanics, 1400–1600'. In *The Power of Images in Early Modern Science*. Ed. W Lefèvre, J Renn and U Schoepflin. Basle: Birkhäuser, 2003: 69–88.

Lehoux, Daryn. 'Tropes, Facts and Empiricism'. *Perspectives on Science* 11 (2003): 326–45.

———. *What Did the Romans Know? An Inquiry into Science and Worldmaking*. Chicago: University of Chicago Press, 2012.

Leibniz, Gottfried Wilhelm, Christiaan Huygens and Denis Papin. *Leibnizens und Huygens' Briefwechsel mit Papin, nebst der Biographie Papins und einigen zugehörigen Briefen und Actenstücken*. Ed. E Gerland. Berlin: Akademie der Wissenschaften, 1881.

Lennox, James G. 'The Disappearance of Aristotle's Biology: A Hellenistic Mystery'. In *Aristotle's Philosophy of Biology: Studies in the Origins of Life Science.* Cambridge: Cambridge University Press, 2001: 110–25.

———. 'William Harvey: Enigmatic Aristotelian of the Seventeenth Century'. In *Teleology in the Ancient World: The Dispensation of Nature.* Ed. J Rocca. Cambridge: Cambridge University Press, forthcoming.

Leonardo da Vinci. *Trattato della pittura.* Ed. G de Rossi. Rome: Stamperia de Romanis, 1817.

———. *Trattato della pittura (1651) = Traité de la peinture.* Ed. A Sconza. Paris: Les Belles Lettres, 2012.

———. *Treatise on Painting: Codex urbinas latinus 1270.* Ed. AP McMahon. Princeton: Princeton University Press, 1956.

Leplin, Jarrett (ed.). *Scientific Realism.* Berkeley: University of California Press, 1984.

Lerner, Michel-Pierre. *Le Monde des sphères.* 2 vols. Paris: Les Belles Lettres, 1997.

Leroi, Armand Marie. *The Lagoon: How Aristotle Invented Science.* New York: Viking, 2014.

Leroy, Louis. *De la vicissitude ou variété des choses de l'univers.* Paris: P L'Huilier, 1575.

———. *Of the Interchangeable Course or Variety of Things.* London: C Yetsweirt, 1594.

Lessing, Karl G. *Gotthold Ephraim Lessings Leben, nebst seinem noch übrigen litterarischen Nachlasse.* 3 vols. *Berlin: In der Vossischen Buchhandlung,* 1793–5.

Lessius, Leonard. *Rawleigh, His Ghost, or A Feigned Apparition of Syr W. Rawleigh, to a Friend of His, for the Translating into English, the Booke of L. Lessius.* St Omer: [s.n.], 1631.

Lester, Toby. *The Fourth Part of the World.* London: Profile, 2009.

Lestringant, Frank. *L'Atelier du cosmographe, ou L'Image du monde à la Renaissance.* Paris: A Michel, 1991.

Leurechon, Jean. *Selectae propositiones in tota sparsim mathematica pulcherrimae ad usum et exercitationem celebrium academiarum.* Pont-à-Mousson: G Bernardus, 1629.

Levenson, Jay A. 'Jacopo de' Barbari'. *Print Quarterly* 25 (2008): 207–9.

Levine, Joseph M. *The Battle of the Books: History and Literature in the Augustan Age.* Ithaca, NY: Cornell University Press, 1991.

———. *Between the Ancients and the Moderns: Baroque Culture in Restoration England.* New Haven: Yale University Press, 1999.

Lévy-Bruhl, Lucien. *How Natives Think.* New York: AA Knopf, 1925.

Lewis, Eric. 'Walter Charleton and Early Modern Eclecticism'. *Journal of the History of Ideas* 62 (2001): 651–64.

Lindberg, David C. 'Alhazen's Theory of Vision and Its Reception in the West'. *Isis* 58 (1967): 321–41.

———. *The Beginnings of Western Science: The European Scientific Tradition in Philosophical, Religious and Institutional Context, 600 BC to AD 1450.* Chicago: University of Chicago Press, 1992.

Lindberg, David C and Ronald L Numbers (eds.). *God and Nature: Historical Essays on the Encounter between Christianity and Science*. Berkeley: University of California Press, 1986.

Lindberg, David C and Robert S Westman (eds.). *Reappraisals of the Scientific Revolution*. Cambridge: Cambridge University Press, 1990.

Line, Francis. *Tractatus de corporum inseparabilitate; in quo experimenta de vacuo, tam Torricelliana, quam Magdeburgica, & Boyliana, examinantur*. London: T Roycroft, 1661.

Livingstone, David N and Charles WJ Withers (eds.). *Geography and Revolution*. Chicago: University of Chicago Press, 2005.

Locke, John. *An Essay Concerning Humane Understanding*. London: T Basset, 1690.

Lohne, JA. 'Isaac Newton: The Rise of a Scientist 1661–1671'. *Notes and Records of the Royal Society of London* (1965): 125–39.

LoLordo, Antonia. *Pierre Gassendi and the Birth of Early Modern Philosophy*. New York: Cambridge University Press, 2007.

Long, Pamela O. 'Invention, Authorship, "Intellectual Property" and the Origin of Patents – Notes toward a Conceptual History'. *Technology and Culture* 32 (1991): 846–84.

———. *Openness, Secrecy, Authorship: Technical Arts and the Culture of Knowledge from Antiquity to the Renaissance*. Baltimore: Johns Hopkins University Press, 2001.

———. 'Picturing the Machine: Francesco di Giorgio and Leonardo da Vinci in the 1490s'. In *Picturing Machines*. Ed. W Lefèvre. Cambridge, Mass.: MIT Press, 2004: 117–41.

———. 'Power, Patronage and the Authorship of Ars: From Mechanical Know-how to Mechanical Knowledge in the Last Scribal Age'. *Isis* 88 (1997): 1–41.

Lower, Richard. *Richard Lower's Vindicatio: A Defence of the Experimental Method*. Ed. K Dewhurst. Oxford: Sandford, 1983.

Luria, AR. *Cognitive Development, Its Cultural and Social Foundations*. Cambridge, Mass.: Harvard University Press, 1976.

Lüthy, Christoph H. 'Where Logical Necessity Turns into Visual Persuasion: Descartes' Clear and Distinct Illustrations'. In *Transmitting Knowledge: Words, Images and Instruments in Early Modern Europe*. Ed. S Kusukawa and I Maclean. Oxford: Oxford University Press, 2006: 97–133.

Lynall, Gregory. *Swift and Science*. London: Palgrave Macmillan, 2012.

Lynes, John A. 'Brunelleschi's Perspectives Reconsidered'. *Perception* 9 (1980): 87–99.

Lyotard, Jean-François. *La Condition postmoderne: rapport sur le savoir*. Paris: Éditions de Minuit, 1979.

Maas, Harro and Mary S Morgan. 'Timing History: The Introduction of Graphical Analysis in 19th-century British Economics'. *Revue d'histoire des sciences humaines* 7 (2002): 97–127.

McCord, Sheri L. 'Healing by Proxy: The Early-modern Weapon-salve'. *English Language Notes* 47 (2009): 13–24.

McCormick, Ted. *William Petty and the Ambitions of Political Arithmetic*. Oxford: Oxford University Press, 2009.

McDonald, Joseph F. 'Russell, Wittgenstein, and the Problem of the Rhinoceros'. *Southern Journal of Philosophy* 31 (1993): 409–24.

Macfarlane, Alan. 'Civility and the Decline of Magic'. In *Civil Histories: Essays in Honour of Sir Keith Thomas*. Ed. P Slack, P Burke and B Harrison. Oxford: Oxford University Press, 2000: 145–60.

MacGregor, Neil. *Shakespeare's Restless World*. London: Allen Lane, 2012.

McGrew, Timothy J, Marc Alspector-Kelly and Fritz Allhoff (eds.). *The Philosophy of Science: An Historical Anthology*. Chichester: Wiley-Blackwell, 2009.

McGuire, JE and Piyo M Rattansi. 'Newton and the "Pipes of Pan"'. *Notes and Records of the Royal Society of London* 21 (1966): 108–43.

Machiavelli, Niccolò. *Selected Political Writings*. Trans. D Wootton. Indianapolis: Hackett, 1994.

McIntosh, Gregory C. *The Johannes Ruysch and Martin Waldseemüller World Maps: The Interplay and Merging of Early-sixteenth-century New World Cartographies*. Cerritos, Calif.: Plus Ultra Publishing, 2012.

MacIntyre, Alasdair C. *After Virtue: A Study in Moral Theory*. London: Duckworth, 1981.

———. 'Epistemological Crises, Dramatic Narrative and the Philosophy of Science in Historicism and Epistemology'. *Monist* 60 (1977): 453–72.

MacKay, R Jock and R Wayne Oldford. 'Scientific Method, Statistical Method and the Speed of Light'. *Statistical Science* (2000): 254–78.

Mackinnon, Nick. 'The Portrait of Fra Luca Pacioli'. *The Mathematical Gazette* 77 (1993): 130–219.

McLaughlin, Martin L. 'Humanist Concepts of Renaissance and Middle Ages in the Tre- and Quattrocento'. *Renaissance Studies* 2 (1988): 131–42.

Maclean, Ian. 'Foucault's Renaissance Episteme'. *Journal of the History of Ideas* 59 (1998): 149–66.

———. *Logic, Signs and Nature in the Renaissance: The Case of Learned Medicine*. Cambridge: Cambridge University Press, 2002.

McMullin, Ernan. 'Bruno and Copernicus'. *Isis* 78 (1987): 55–74.

———. 'Giordano Bruno at Oxford'. *Isis* 77 (1986): 85–94.

———. 'The Impact of Newton's *Principia* on the Philosophy of Science'. *Philosophy of Science* 68 (2001): 279–310.

McNally, Peter (ed.). *The Advent of Printing*. Montreal: McGill University, 1987.

McNulty, Robert. 'Bruno at Oxford'. *Renaissance News* 13 (1960): 300–5.

Maffioli, Cesare S. *Out of Galileo: The Science of Waters 1628–1718*. Rotterdam: Erasmus, 1994.

———. *La via delle acque, 1500–1700: Appropriazione delle arti e trasformazione delle matematiche*. Florence: LS Olschki, 2010.

Malcolm, Noel. *Aspects of Hobbes*. Oxford: Clarendon Press, 2002.

———. 'Hobbes and Roberval'. In *Aspects of Hobbes*. Oxford: Clarendon Press, 2002: 156–99.

———. 'Hobbes's Science of Politics and His Theory of Science'. In *Aspects of Hobbes*. Oxford: Clarendon Press, 2002: 146–55.

———. 'Robert Boyle, Georges Pierre des Clozets and the Asterism: A New Source'. *Early Science and Medicine* 9 (2004): 293–306.

Manetti, Antonio. *Vita di Filippo Brunelleschi*. Ed. C C Perrone. Rome: Salerno, 1992.

Margolis, Howard. *Patterns, Thinking and Cognition: A Theory of Judgment*. Chicago: University of Chicago Press, 1987.

———. *It Started with Copernicus: How Turning the World inside out Led to the Scientific Revolution*. New York: McGraw-Hill, 2002.

Martens, Rhonda. *Kepler's Philosophy and the New Astronomy*. Princeton: Princeton University Press, 2000.

Martinet, Monique. '*Science et hypothèses chez Descartes*'. *Archives internationales d'histoire des sciences* 24 (1974): 319–39.

Massa, Daniel. 'Giordano Bruno's Ideas in Seventeenth-century England'. *Journal of the History of Ideas* 38 (1977): 227–42.

Massey, Lyle. *Picturing Space, Displacing Bodies*. University Park, PA: Pennsylvania State University Press, 2007.

Mattern, Susan P. *Galen and the Rhetoric of Healing*. Baltimore: Johns Hopkins University Press, 2008.

May, Christopher. 'The Venetian Moment: New Technologies, Legal Innovation and the Institutional Origins of Intellectual Property'. *Prometheus* 20 (2002): 159–79.

Mayer, Anna-K. 'Setting Up a Discipline: Conflicting Agendas of the Cambridge History of Science Committee, 1936–1950'. *Studies in History and Philosophy of Science Part A* 31 (2000): 665–89.

Mayer, Thomas F. *The Roman Inquisition: A Papal Bureaucracy and Its Laws in the Age of Galileo*. Philadelphia: University of Pennsylvania Press, 2013.

———. *The Roman Inquisition: Trying Galileo*. Philadelphia: University of Pennsylvania Press, 2015.

Mayr, Ernst. 'When is Historiography Whiggish?' *Journal of the History of Ideas* 51 (1990): 301–9.

Mayr, Otto. *Authority, Liberty & Automatic Machinery in Early Modern Europe*. Baltimore: Johns Hopkins University Press, 1986.

Mazur, Joseph. *Enlightening Symbols: A Short History of Mathematical Notation and Its Hidden Powers*. Princeton: Princeton University Press, 2014.

Mela, Pomponius. *De orbis situ libri tres. Adiecta sunt praeterea loca aliquot ex Vadiani commentarijs*. Ed. J Vadianus. Paris: C Wechel, 1530.

Melchior-Bonnet, Sabine. *The Mirror: A History*. New York: Routledge, 2002.

Merchant, Carolyn. '"The Violence of Impediments": Francis Bacon and the Origins of Experimentation'. *Isis* 99 (2008): 731–60.

Merton, Robert K. 'The Normative Structure of Science'. In *The Sociology of Science*. Chicago: University of Chicago Press, 1973: 267–78.

———. *On the Shoulders of Giants: A Shandean Postcript*. New York: Free Press, 1965.

———. 'Priorities in Scientific Discovery: A Chapter in the Sociology of Science'. *American Sociological Review* 22 (1957): 635–59.

———. 'Resistance to the Systematic Study of Multiple Discoveries in Science'. *European Journal of Sociology* 4 (1963): 237–82.

———. 'Science and Technology in a Democratic Order'. *Journal of Legal and Political Sociology* 1 (1942): 115–26.

———. 'Science, Technology and Society in Seventeenth-century England'. *Osiris* 4 (1938): 360–63.

———. *Science, Technology and Society in Seventeenth-century England.* New York: Harper & Row, 1970.

———. 'Singletons and Multiples in Scientific Discovery: A Chapter in the Sociology of Science'. *Proceedings of the American Philosophical Society* 105 (1961): 470–86.

———. *The Sociology of Science: Theoretical and Empirical Investigations.* Chicago: University of Chicago Press, 1973.

———. 'The Unanticipated Consequences of Purposive Social Action'. *American Sociological Review* 1 (1936): 894–904.

Merton, Robert K and Elinor G. Barber. *The Travels and Adventures of Serendipity.* Princeton: Princeton University Press, 2006.

Meurer, Peter H. 'Cartography in the German Lands, 1450–1650'. In *The History of Cartography.* 6 vols. Vol. 3: *Cartography in the European Renaissance.* Ed. D Woodward. Chicago: University of Chicago Press, 2007: 1172–245.

Michele, Agostino. *Trattato della grandezza dell'acqva et della terra.* Venice: N Moretti, 1583.

Middleton, WE Knowles. *The History of the Barometer.* Baltimore: Johns Hopkins University Press, 1964.

Midgley, Robert. *A New Treatise of Natural Philosophy.* London: J Hindmarsh, 1687.

Mignolo, Walter D. *The Darker Side of the Renaissance: Literacy, Territoriality and Colonization.* Ann Arbor: University of Michigan Press, 2010.

Mill, John Stuart. *Principles of Political Economy.* London: Longmans, Green & Co., 1909.

Miller, DP. *James Watt, Chemist: Understanding the Origins of the Steam Age.* London: Pickering & Chatto Ltd, 2009.

Milliet de Chales, Claude-François. *Cursus seu mundus mathematicus.* 3 vols. Lyons, 1674.

———. *Cursus seu mundus mathematicus.* 4 vols. Lyons, 1690.

Milton, John R. 'Laws of Nature'. In *The Cambridge History of Seventeenth-century Philosophy.* 2 vols. Vol. 1. Ed. D Garber and M Ayers. Cambridge: Cambridge University Press, 1998: 680–701.

———. 'The Origin and Development of the Concept of the "Laws of Nature"'. *European Journal of Sociology* 22 (1981): 173–95.

Minnis, AJ. *Medieval Theory of Authorship: Scholastic Literary Attitudes in the Later Middle Ages.* Aldershot: Wildwood House, 1988.

Mirowski, Philip. 'A Visible Hand in the Marketplace of Ideas: Precision Measurement as Arbitrage'. *Science in Context* 7 (1994): 563–90.

Mizauld, Antoine. *Cosmologia: Historiam coeli et mundi.* Paris: F Morellus, 1570.

Moffitt, John F. *Painterly Perspective and Piety: Religious Uses of the Vanishing Point, From the 15th to the 18th Century.* Jefferson, NC: McFarland, 2008.

Mokyr, Joel. *The Enlightened Economy: An Economic History of Britain, 1700–1850.* New Haven: Yale University Press, 2009.

———. *The Gifts of Athena: Historical Origins of the Knowledge Economy.* Princeton: Princeton University Press, 2004.

———. 'The Intellectual Origins of Modern Economic Growth'. *Journal of Economic History* 65 (2005): 285–351.

———. *The Lever of Riches: Technological Creativity and Economic Progress.* New York: Oxford University Press, 1990.

Montaigne, Michel de. *The Complete Essays.* Trans. MA Screech. London: Allen Lane, 1991.

———. *Essayes: Written in French.* Trans. J Florio. London: E Blovnt, 1613.

———. *Oeuvres complètes.* Ed. M Rat. Paris: Gallimard, 1962.

Moore, George Edward. *A Defence of Common Sense.* London: Allen & Unwin, 1925.

Morando, Bruno. 'The Golden Age of Celestial Mechanics'. In *The General History of Astronomy.* 4 vols. Vol. 2B: *Planetary Astronomy from the Renaissance to the Rise of Astrophysics.* Ed. R Taton and C Wilson. 1995: 211–39.

More, Henry. *Divine Dialogues, Containing Sundry Disquisitions and Instructions Concerning the Attributes and Providence of God.* London: J. Flesher, 1668.

———. *The Immortality of the Soul, So Farre Forth as It is Demonstrable from the Knowledge of Nature and the Light of Reason.* London: W Morden, 1659.

Morison, Samuel Eliot. *Portuguese Voyages to America in the Fifteenth Century.* Cambridge, Mass.: Harvard University Press, 1940.

Mornet, Daniel. *Les Origines intellectuelles de la Révolution française: 1715–1787.* Paris: Armand Colin, 1933.

Mosley, Adam. *Bearing the Heavens: Tycho Brahe and the Astronomical Community of the Late Sixteenth Century.* Cambridge: Cambridge University Press, 2007.

Muir, Edward. *The Culture Wars of the Late Renaissance.* Boston: Harvard University Press, 2007.

Muraro, Luisa. *Giambattista della Porta, mago e scienziato.* Milan: Feltrinelli, 1978.

Murdoch, John E. 'Philosophy and the Enterprise of Science in the Later Middle Ages'. In *The Interaction between Science and Philosophy.* Ed. Y Elkana. Atlantic Highlands, NJ: Humanities Press, 1974: 51–74.

———. 'Pierre Duhem and the History of Late-Medieval Science and Philosophy in the Latin West'. In *Gli studi di filosofia medievale fra otto e novecento.* Ed. A Maier and R Imbach. Rome: Edizioni di Storia e Letteratura, 1991: 253–302.

Musson, AE and Eric Robinson. *Science and Technology in the Industrial Revolution.* Manchester: Manchester University Press, 1969.

Münster, Sebastian. *A Treatyse of the Newe India with Other New Founde Landes and Islandes.* London: E Sutton, 1553.

Nagel, Thomas. 'What is It Like to be a Bat?' *The Philosophical Review* 83 (1974): 435–50.

Naudé, Gabriel. *Instructions Concerning Erecting of a Library Presented to My Lord, the President de Mesme.* Trans. J Evelyn. London: G. Bedle, 1661.

Needham, Joseph. 'Human Laws and Laws of Nature in China and the West (I)'. *Journal of the History of Ideas* 12 (1951): 3–30.

———. 'Human Laws and Laws of Nature in China and the West (II)'. *Journal of the History of Ideas* 12 (1951): 194–230.

———. *The Sceptical Biologist (Ten Essays).* London: Chatto & Windus, 1929.

———. *The Shorter Science and Civilisation in China: An Abridgement.* Ed. Colin A Rowan. 5 vols. Cambridge: Cambridge University Press, 1978–95.

Newcastle, Margaret Cavendish. *Philosophical Letters, or Modest Reflections upon Some Opinions in Natural Philosophy.* London: [s.n.], 1664.

Newman, William Royall. *Atoms and Alchemy: Chymistry and the Experimental Origins of the Scientific Revolution.* Chicago: University of Chicago Press, 2006.

———. 'Brian Vickers on Alchemy and the Occult: A Response'. *Perspectives on Science* 17 (2009): 482–506.

———. *Gehennical Fire.* Chicago: University of Chicago Press, 2003.

———. 'How Not to Integrate the History and Philosophy of Science: A Reply to Chalmers'. *Studies in History and Philosophy of Science Part A* 41 (2010): 203–13.

———. *Promethean Ambitions: Alchemy and the Quest to Perfect Nature.* Chicago: University of Chicago Press, 2004.

———. 'What Have We Learned from the Recent Historiography of Alchemy?' *Isis* 102 (2011): 313–21.

Newman, William Royall and Lawrence M Principe. *Alchemy Tried in the Fire.* Chicago: University of Chicago Press, 2005.

———. 'Alchemy versus Chemistry: The Etymological Origins of a Historiographic Mistake'. *Early Science and Medicine* 3 (1998): 32–65.

Newton, Isaac. *The Correspondence of Isaac Newton.* Ed. HW Turnbull. 7 vols. Cambridge: Cambridge University Press, 1959–77.

———. *Isaac Newton's Papers & Letters on Natural Philosophy and Related Documents.* Ed. IB Cohen. Cambridge, Mass.: Harvard University Press, 1958.

———. 'A Letter of Mr Isaac Newton, Professor of the Mathematicks in the University of Cambridge; Containing His New Theory about Light and Colors: Sent by the Author to the Publisher From Cambridge, Febr. 6. 1671/72; in Order to be Communicated to the R. Society'. *Philosophical Transactions* 6 (1672): 3075–87.

———. *The Mathematical Principles of Natural Philosophy.* Trans. A Motte. 2 vols. London: B Motte, 1729.

———. *Opticks, or A Treatise of the Reflexions, Refractions, Inflexions and Colours of Light.* London: Samuel Smith, 1704.

———. *Unpublished Scientific Papers of Isaac Newton: A Selection from the Portsmouth Collection in the University Library, Cambridge*. Ed. AR Hall and MB Hall. Cambridge: Cambridge University Press, 1962.

Newton, Isaac and Roger Cotes. *Correspondence of Sir Isaac Newton and Professor Cotes*. Ed. J Edleston. London: JW Parker, 1850.

Newton, Robert R. 'The Authenticity of Ptolemy's Parallax Data – Part 1'. *Quarterly Journal of the Royal Astronomical Society* 14 (1973): 367–88.

Niceron, Jean François. *La Perspective curieuse*. Paris: Veuve F Langlois, 1652.

Nicholl, Charles. *Leonardo da Vinci: The Flights of the Mind*. London: Allen Lane, 2004.

Nield, Ted. *Incoming! Or, Why We Should Stop Worrying and Learn to Love the Meteorite*. London: Granta, 2011.

Norman, Robert. *The New Attractive: Containing a Short Discourse of the Magnes or Lodestone*. London: R Ballard, 1581.

North, John David. *God's Clockmaker: Richard of Wallingford and the Invention of Time*. London: Hambledon and London, 2005.

Nummedal, Tara. *Alchemy and Authority in the Holy Roman Empire*. Chicago: University of Chicago Press, 2007.

———. 'On the Utility of Alchemical Fraud'. In *Chymists and Chymistry: Studies in the History of Alchemy and Early Modern Chemistry*. Ed. L Principe. Sagamore Beach, Mass.: Science History Publications, 2007: 173–80.

Nye, Mary Jo. *Michael Polanyi and His Generation: Origins of the Social Construction of Science*. Chicago: University of Chicago Press, 2011.

Oakley, Francis. 'Christian Theology and the Newtonian Science: The Rise of the Concept of the Laws of Nature'. *Church History* 30 (1961): 433–57.

———. *Natural Law, Laws of Nature, Natural Rights: Continuity and Discontinuity in the History of Ideas*. New York: Continuum, 2005.

Oberman, Heiko A. 'Reformation and Revolution: Copernicus's Discovery in an Era of Change'. In *The Cultural Context of Medieval Learning*. Ed. JE Murdoch and ED Sylla. Springer, 1975: 397–435.

Ogborn, Miles and Charles WJ Withers. 'Introduction: Book Geography, Book History'. In *Geographies of the Book*. Ed. M Ogborn and CWJ Withers. Farnham: Ashgate, 2010: 1–25.

Ogilvie, Brian W. *The Science of Describing: Natural History in Renaissance Europe*. Chicago: University of Chicago Press, 2008.

O'Gorman, Edmundo. *The Invention of America: An Inquiry into the Historical Nature of the New World and the Meaning of Its History*. Bloomington: Indiana University Press, 1961.

O'Grady, Paul. 'Wittgenstein and Relativism'. *International Journal of Philosophical Studies* 12 (2004): 315–37.

Ong, Walter Jackson. *Orality and Literacy: The Technologizing of the World*. London: Routledge, 1982.

———. *Ramus, Method and the Decay of Dialogue: From the Art of Discourse to the Art of Reason*. Cambridge, Mass.: Harvard University Press, 1958.

Ophir, Adi and Steven Shapin. 'The Place of Knowledge: A Methodological Survey'. *Science in Context* 4 (1991): 3–21.

Oresme, Nicholas. *Le Livre du ciel et du monde*. Ed. AD Menut. Madison: University of Wisconsin Press, 1968.

———. *'The Questiones de spera' of Nicole Oresme: Latin Text with English Translation, Commentary and Variants*. Ed. G Droppers. Milwaukee, MI: University of Wisconsin, 1966.

———. *Traité de l'espère*. Ed. L McCarthy. Toronto: University of Toronto, 1943.

Orgel, Stephen. *Impersonations: The Performance of Gender in Shakespeare's England*. Cambridge: Cambridge University Press, 1996.

Osler, Margaret J. 'John Locke and the Changing Ideal of Scientific Knowledge'. *Journal of the History of Ideas* 31 (1970): 3–16.

——— (ed.). *Rethinking the Scientific Revolution*. Cambridge: Cambridge University Press, 2000.

Owen, GEL. '*Tithenai ta phainomena*'. [1967] In *Articles on Aristotle*. 4 vols. Vol. 1: *Science*. Ed. J Barnes, M Schofield and R Sorabji. London: Duckworth, 1975: 113–26.

Padoa, Alessandro. *La Logique déductive dans sa dernière phase de développement*. Paris: Gauthier-Villars, 1912.

Palingenius, Marcellus. *The Zodiake of Life*. London: R Newberye, 1565.

———. *The Zodiake of Life*. Ed. R Tuve and B Googe. New York: Scholars' Facsimiles & Reprints, 1947.

Palisca, Claude V. '*Vincenzo Galileo, scienziato sperimentale, mentore del figlio Galileo*'. *Nuncius* 15 (2000): 497–514.

Palmerino, Carla Rita. 'Experiments, Mathematics, Physical Causes: How Mersenne Came to Doubt the Validity of Galileo's Law of Free Fall'. *Perspectives on Science* 18 (2010): 50–76.

Palmieri, Paolo. 'The Cognitive Development of Galileo's Theory of Buoyancy'. *Archive for History of Exact Sciences* 59 (2005): 189–222.

———. 'Galileo and the Discovery of the Phases of Venus'. *Journal for the History of Astronomy* 32 (2001): 109–29.

———. 'Re-examining Galileo's Theory of Tides'. *Archive for History of Exact Sciences* 53 (1998): 223–375.

Panofsky, Erwin. *Perspective as Symbolic Form*. New York: Zone Books , 1991.

———. *Renaissance and Renascences in Western Art*. London: Paladin, 1970.

Pantin, Isabel. 'New Philosophy and Old Prejudices: Aspects of the Reception of Copernicanism in a Divided Europe'. *Studies in History and Philosophy of Science Part A* 30 (1999): 237–62.

Papin, Denis. 'An Account of an Experiment Shewn before the Royal Society, of Shooting by the Rarefaction of the Air'. *Philosophical Transactions (1683–1775)* 16 (1686): 21–2.

———. *A Continuation of the New Digester of Bones, Its Improvements, and New Uses It Hath Been Applyed to, Both for Sea and Land: Together with Some Improvements and New Uses of the Air-pump, Tryed Both in England and in Italy*. London: J Streater, 1687.

———. 'A Demonstration of the Velocity wherewith the Air Rushes into an Exhausted Receiver, Lately Produced before the Royal Society'. *Philosophical Transactions (1683–1775)* 16 (1686): 193–5.
———. *La Manière d'amolir les os*. Amsterdam: Desbordes, 1688.
———. *Nouvelle Manière pour élever l'eau par la force du feu mise en lumière*. Cassell: J Estienne, 1707.
———. *Recueil de diverses pièces touchant quelques nouvelles machines*. Kassel: JE Marchand, 1695.
———. *La Vie et les ouvrages de Denis Papin*. Ed. A Péan, LD Belenet and L de La Saussaye. 8 vols. Blois: C. Migault, 1894.
Park, Katharine. 'The Rediscovery of the Clitoris'. In *The Body in Parts: Fantasies of Corporeality in Early Modern Europe*. Ed. D Hillman and C Mazzio. New York: Routledge, 1997: 171–93.
———. 'Response to Brian Vickers, "Francis Bacon, Feminist Historiography and the Dominion of Nature"'. *Journal of the History of Ideas* 69 (2008): 143–6.
Parker, Geoffrey. *The Army of Flanders and the Spanish Road, 1567–1659*. Cambridge: Cambridge University Press, 1972.
Parker, Samuel. *Disputationes de Deo et providentia divina*. London: J Martyn, 1678.
———. *A Free and Impartial Censure of the Platonick Philosophie*. Oxford: R Davis, 1666.
Parronchi, Alessandro. '*Un tabernacolo brunelleschiano*'. In *Filippo Brunelleschi: La sua opera e il suo tempo*. Ed. G Soadolini. Florence: Centro Di, 1980: 239–55.
Parsons, Robert. *The Seconde Parte of the Booke of Christian Exercise*. London: S Waterson, 1590.
Pascal, Blaise. *Les Provinciales*. Cologne: Pierre de la Vallée, 1657.
———. *Les Provinciales, or The Mysterie of Jesuitisme*. London: R Royston, 1657.
———. *Les Provinciales, or The Mystery of Jesuitisme*. London: R Royston, 1658.
———. *Oeuvres*. Ed. P Boutroux and L Brunschvicg. 14 vols. Vol. 2. Paris: Hachette, 1923–5.
———. *Oeuvres complètes*. Ed. J Mesnard. 4 vols. Vol. 2. Paris: Desclée de Brouwer, 1964–1992.
———. *Pensées*. Trans. WF Trotter. New York: EP Dutton, 1958.
———. *The Physical Treatises of Pascal: The Equilibrium of Liquids and the Weight of the Mass of the Air*. Ed. IHB Spiers, AGH Spiers and F Barry. New York: Columbia University Press, 1937.
Passannante, Gerard Paul. *The Lucretian Renaissance: Philology and the Afterlife of Tradition*. Chicago: University of Chicago Press, 2011.
Patrick, Symon. *A Brief Account of the New Sect of Latitude-men*. London: [n.p.], 1662.
Pecquet, Jean. *New Anatomical Experiments*. London: O Pulleyn, 1653.
Peregrinus, Petrus. *Opera*. Ed. RB Thomson and L Sturlese. Pisa: Scuola Normale Superiore, 1995.

Pesic, Peter. 'Proteus Rebound – Reconsidering the "Torture of Nature"'. *Isis* 99 (2008): 304–17.

Peterson, Mark A. *Galileo's Muse*. Cambridge, Mass.: Harvard University Press, 2011.

Petty, William. *A Treatise of Taxes and Contributions*. London: N Brooke, 1662.

Péan, Alonso and Louis de La Saussaye. *La Vie et les ouvrages de Denis Papin vol I*. Paris: Franck, 1869.

Pérez-Ramos, Antonio. *Francis Bacon's Idea of Science and the Maker's Knowledge Tradition*. Oxford: Clarendon Press, 1988.

Phillips, Derek L. *Wittgenstein and Scientific Knowledge: A Sociological Perspective*. London: Macmillan, 1977.

Phillips, Jeremy. 'The English Patent as a Reward for Invention: The Importation of an Idea'. *Journal of Legal History* 3 (1982): 71–9.

Picciotto, Joanna. *Labors of Innocence in Early Modern England*. Cambridge, Mass.: Harvard University Press, 2010.

Piccolomini, Alessandro. *Della grandezza della terra et dell'acqua*. Venice, 1558.

———. *De la sfera del mondo*. Venice: Al Segno del Pozzo, 1540.

———. *La prima parte delle theoriche: overo speculationi de i pianeti*. Venice: Varisco, 1558.

Pickering, Andrew. *The Mangle of Practice: Time, Agency and Science*. Chicago: University of Chicago Press, 1995.

Pinch, Trevor J. *Confronting Nature: The Sociology of Solar-neutrino Detection*. Dordrecht: D Reidel, 1986.

———. 'Kuhn – The Conservative and Radical Interpretations: Are Some Mertonians "Kuhnians" and Some Kuhnians "Mertonians"?' *Social Studies of Science* 27 (1997): 465–82.

———. 'Opening Black Boxes: Science, Technology and Society'. *Social Studies of Science* 22 (1992): 487–510.

Pinch, Trevor J and Wiebe E Bijker. 'The Social Construction of Facts and Artefacts'. In *The Social Construction of Technological Systems*. Ed. WE Bijker, TP Hughes and TJ Pinch. MIT Press, 1987: 17–50.

Pinto-Correia, Clara. *The Ovary of Eve: Egg and Sperm and Preformation*. Chicago: University of Chicago Press, 1997.

Pliny the Elder. *L'Histoire du monde*. Trans. A du Pinet. Lyons: C Senneton, 1562.

———. *Natural History*. Trans. H Rackham. 10 vols. Cambridge, Mass.: Harvard University Press, 1938–63.

Plutarch. 'The Face of the Moon'. In *Moralia*. Vol. 11. Trans. H Cherniss and WC Helmbold. Cambridge, Mass.: Harvard University Press, 1957: 1–223.

Polanyi, Michael. *Personal Knowledge: Towands a Post-critical Philosophy*. Chicago: University of Chicago Press, 1958.

Pomata, Gianna. 'Observation Rising: Birth of an Epistemic Genre, 1500–1650'. In *Histories of Scientific Observation*. Ed. E Lunbeck and LJ Daston. Chicago: University of Chicago Press, 2011: 44–80.

Poovey, Mary. *A History of the Modern Fact: Problems of Knowledge in the Sciences of Wealth and Society.* Chicago: University of Chicago Press, 1998.

Popkin, Richard H. *The History of Scepticism from Erasmus to Spinoza.* Berkeley: University of California Press, 1979.

Popper, Karl Raimund. *The Logic of Scientific Discovery.* London: Hutchinson, 1959.

———. *Objective Knowledge: An Evolutionary Approach.* Oxford: Clarendon Press, 1972.

———. *The Open Society and Its Enemies.* London: Routledge, 1945.

Popplow, Marcus. 'Setting the World Machine in Motion: The Meaning of *Machina mundi* in the Middle Ages and the Early Modern Period'. In *Mechanics and Cosmology in the Medieval and Early Modern Period.* Ed. M Bucciantini, M Camerota and S Roux. Florence: LS Olschki, 2007: 45–70.

Porter, Roy. 'The Scientific Revolution: A Spoke in the Wheel?' In *Revolution in History.* Cambridge: Cambridge University Press, 1986: 290–316.

———. 'The Scientific Revolution and Universities'. In *A History of the University in Europe.* 4 vols. Vol. 2. Ed. W Rüegg. Cambridge: Cambridge University Press, 1996: 531–62.

Post, Heinz R. 'Correspondence, Invariance and Heuristics: In Praise of Conservative Induction'. *Studies in History and Philosophy of Science Part A* 2 (1971): 213–55.

Powell, Thomas. *The Passionate Poet. With a Description of the Thracian Ismarus. By T. P.* London: Valentine Simmes, 1601.

Power, Henry. *Experimental Philosophy, in Three Books Containing New Experiments Microscopical, Mercurial, Magnetical.* London: J Martin, 1664.

Powers, John C. '*Ars sine arte*: Nicholas Lemery and the End of Alchemy in Eighteenth-century France'. *Ambix* 45 (1998): 163–89.

Principe, Lawrence M. 'Alchemy Restored'. *Isis* 102 (2011): 305–12.

———. *The Aspiring Adept: Robert Boyle and His Alchemical Quest.* Princeton: Princeton University Press, 1998.

———. 'Georges Pierre des Clozets, Robert Boyle, the Alchemical Patriarch of Antioch, and the Reunion of Christendom: Further New Sources'. *Early Science and Medicine* 9 (2004): 307–20.

———. *The Scientific Revolution: A Very Short Introduction.* Oxford: Oxford University Press, 2011.

Principe, Lawrence M and Lloyd DeWitt. *Transmutations: Alchemy in Art.* Philadelphia: Chemical Heritage Foundation, 2002.

Pritchard, Duncan. 'Epistemic Relativism, Epistemic Incommensurability and Wittgensteinian Epistemology'. In *Blackwell Companion to Relativism.* Ed. S Hales. Oxford: Blackwell, 2010: 266–85.

Proclus and Euclid. *In primum Euclidis elementorum librum commentariorum.* Ed. F Barozzi. Padua: G Perchacinus, 1560.

Psillos, Stathis. *Scientific Realism: How Science Tracks Truth.* London: Routledge, 1999.

Pugliese PJ. 'The Scientific Achievement of Robert Hooke: Method and Mechanics'. Cambridge, Mass.: Harvard University Press, 1982.

Pumfrey, Stephen. 'Harriot's Maps of the Moon: New Interpretations'. *Notes and Records of the Royal Society* 63 (2009): 163–8.

———. *Latitude: The Magnetic Earth*. Cambridge: Icon, 2001.

———. '"*O tempora, O magnes!*" A Sociological Analysis of the Discovery of Secular Magnetic Variation in 1634'. *British Journal for the History of Science* 22 (1989): 181–214.

———. 'The Selenographia of William Gilbert: His Pre-telescopic Map of the Moon and His Discovery of Lunar Libration'. *Journal for the History of Astronomy* 42 (2011): 193–203.

———. '"Your Astronomers and Ours Differ Exceedingly": The Controversy over the "New Star" of 1572 in the Light of a Newly Discovered Text by Thomas Digges'. *British Journal for the History of Science* 44 (2011): 29–60.

Pumfrey, Stephen, Paul Rayson and John Mariani. 'Experiments in 17th-century English: Manual versus Automatic Conceptual History'. *Literary and Linguistic Computing* 27 (2012): 395–408.

Purs, Ivo. '*Anselmus Boëtius de Boodt, Pansophie und Alchemie*'. *Acta Comeniana* 18 (2004): 43–90.

Putnam, Hilary. *Meaning and the Moral Sciences*. London: Routledge & Kegan Paul, 1978.

———. *Mind, Language and Reality*. Cambridge: Cambridge University Press, 1975.

Quine, Willard Van Orman. 'A Comment on Grünbaum's Claim'. In *Can Theories be Refuted?* Ed. SG Harding. Dordrecht: D Reidel, 1976: 132.

———. 'Main Trends in Recent Philosophy: Two Dogmas of Empiricism'. *Philosophical Review* 60 (1951): 20–43.

Quintilian, Marcus Fabius. *The Orator's Education*. Ed. DA Russell. 5 vols. Vol. 2. Cambridge, Mass.: Harvard University Press, 2001.

Rabb, Theodore K. 'Religion and the Rise of Modern Science'. *Past & Present* 31 (1965): 111–26.

Radelet de Grave, Patricia and D Speiser. '*Le "De magnete" de Pierre de Maricourt. Traduction et commentaire*'. *Revue d'histoire des sciences* 28 (1975): 193–234.

Ragep, F Jamil. 'Copernicus and His Islamic Predecessors: Some Historical Remarks'. *History of Science* 45 (2007): 65–81.

Ramazzini, Bernardino and Robert St Clair. *The Abyssinian Philosophy Confuted, or Telluris theoria Neither Sacred, nor Agreeable to Reason*. London: W Newton, 1697.

Randall, John H. 'The School of Padua and the Emergence of Modern Science'. *Journal of the History of Ideas* 1 (1940): 177–206.

Randles, William Graham Lister. 'The Atlantic in European Cartography and Culture from the Middle Ages to the Renaissance [1992]'. In *Geography, Cartography and Nautical Science in the Renaissance*. Aldershot: Ashgate, 2000: No. 2, 1–28.

———. 'Classical Models of World Geography and Their Transformation Following the Discovery of America'. In *The Classical Tradition and the Americas, Vol. 1: European Images of the Americas and the Classical Tradition*. Ed. W Haase and M Reinhold. Berlin: Walter de Gruyter, 1994: 5–76.

———. 'The Evaluation of Columbus' "India" Project by Portuguese and Spanish Cosmographers in the Light of the Geographical Science of the Period'. *Imago mundi* 42 (1990): 50–64.

———. *Geography, Cartography and Nautical Science in the Renaissance*. Aldershot: Ashgate, 2000.

———. '*Le Nouveau Monde, l'autre monde et la pluralité des mondes*' [1961]. In *Geography, Cartography and Nautical Science in the Renaissance*. Aldershot: Ashgate, 2000: No. 15, 1–39.

———. *De la Terre plate au globe terrestre: Une mutation épistémologique rapide (1480–1520)*. Paris: A Colin, 1980.

———. *The Unmaking of the Medieval Christian Cosmos, 1500–1760: From Solid Heavens to Boundless Æther*. Aldershot: Ashgate, 1999.

Ranea, Alberto Guillermo. 'Theories, Rules and Calculations: Denis Papin Before and After the Controversy with G. W. Leibniz'. In *Der Philosoph im U-Boot*. Ed. M. Kempe. Hanover: Gottfried Willhelm Leibniz Bibliothek, 2015: 59–83.

Rapin, René. *Reflexions upon Ancient and Modern Philosophy*. London: W Cademan, 1678.

Ravetz, Jerry and Richard S Westfall. 'Marxism and the History of Science'. *Isis* 72 (1981): 393–405.

Rawson, Michael. 'Discovering the Final Frontier: The Seventeenth-century Encounter with the Lunar Environment'. *Environmental History* 20 (2015): 194–216.

Ray, Meredith K. *Daughters of Alchemy: Women and Scientific Culture in Early Modern Italy*. Cambridge, Mass., Harvard University Press, 2015.

Raynaud, Dominique. *L'Hypothèse d'Oxford: Essai sur les origines de la perspective*. Paris: Presses Universitaires de France, 1998.

Redondi, Pietro. '*La nave di Bruno e la pallottola di Galileo: Uno studio di iconografia della fisica*'. In *Il piacere del testo: saggi e studi per Albano Biondi*, Vol. 2. Ed. A Prosperi. Rome: Bulzoni, 2001: 285–363.

Reiss, Timothy J and Roger H Hinderliter. 'Money and Value in the Sixteenth Century: The *Monete cudende ratio* of Nicholas Copernicus'. *Journal of the History of Ideas* 40 (1979): 293–313.

Rey, Abel, Lucien Febvre and others (eds.). *L'Outillage mental: Pensée, langage, mathématiques*. Paris: Société de gestion de l'Encyclopédie française, 1937.

Rey, Anne-Lise. 'The Controversy between Leibniz and Papin'. In *The Practice of Reason: Leibniz and His Controversies*. Ed. M Dascal. Amsterdam: John Benjamins, 2010: 75–100.

Reynolds, John. *Death's Vision Represented in a Philosophical, Sacred Poem*. London: J Osborn, 1713.

Reynolds, Terry S. *Stronger than a Hundred Men: A History of the Vertical Water Wheel*. Baltimore: Johns Hopkins University Press, 1983.

Rheticus, Georg Joachimus. *De libris revolutionum ... Nicolai Copernici ... Narratio Prima*. Gdansk: F Rhodus, 1540.

Righter, Anne. *Shakespeare and the Idea of the Play*. London: Chatto & Windus, 1962.

Riskin, Jessica. 'The Defecating Duck, or The Ambiguous Origins of Artificial Life'. *Critical Inquiry* 29 (2003): 599–633.

Roche, John J. 'Harriot, Galileo and Jupiter's Satellites'. *Archives internationales d'histoire des sciences* 32 (1982): 9–51.

Rohault, Jacques. *Traité de physique*. Paris: C Savreux, 1671.

Rolt, L Tom C and JS Allen. *The Steam Engine of Thomas Newcomen*. Hartington: Moorland, 1977.

Rorty, Richard (ed.). *The Linguistic Turn: Recent Essays in Philosophical Method*. Chicago: University of Chicago Press, 1967.

———. 'Science as Solidarity'. In *Objectivity, Relativism and Truth*. Cambridge: Cambridge University Press, 1991: 35–45.

———. 'Thomas Kuhn, Rocks and the Laws of Physics'. In *Philosophy and Social Hope*. New York: Penguin Books, 1999: 175–89.

Rose, Paul Lawrence. 'Copernicus and Urbino: Remarks on Bernardino Baldi's *Vita di Niccolò Copernico* (1588)'. *Isis* 65 (1974): 387–89.

Rosen, Edward. 'Copernicus and the Discovery of America'. *The Hispanic American Historical Review* 23 (1943): 367–71.

———. *Copernicus and His Successors*. London: Hambledon Press, 1995.

——— (ed.). *Three Copernican Treatises*. New York: Dover Publications, 1959.

———. 'Was Copernicus a Neoplatonist?' *Journal of the History of Ideas* 44 (1983): 667–9.

Rosen, William. *The Most Powerful Idea in the World: A Story of Steam, Industry and Invention*. New York: Random House, 2010.

Rosenfeld, Sophia A. *Common Sense: A Political History*. Cambridge, Mass.: Harvard University Press, 2011.

Rosenthal, Earl E. 'The Invention of the Columnar Device of Emperor Charles V at the Court of Burgundy in Flanders in 1516'. *Journal of the Warburg and Courtauld Institutes* 36 (1973): 198–230.

———. '*Plus ultra, non plus ultra*, and the Columnar Device of Emperor Charles V.' *Journal of the Warburg and Courtauld Institutes* 34 (1971): 204–28.

Röslin, Helisaeus. *De opere Dei creationis, seu De mundo hypotheses*. Frankfurt: A Wechel, 1597.

Ross, Alexander. *Arcana microcosmi, or The Hid Secrets of Man's Body Discovered*. London: T Newcomb, 1652.

Ross, Sydney. 'Scientist: The Story of a Word'. *Annals of Science* 18 (1962): 65–85.

Rossi, Paolo. *The Birth of Modern Science*. Oxford: Blackwell, 2001.

———. *Philosophy, Technology and the Arts in the Early Modern Era*. Trans. B Nelson. New York: Harper & Row, 1970.

Rotman, Brian. *Signifying Nothing: The Semiotics of Zero*. Stanford: Stanford University Press, 1993.

Roux, Sophie. '*Le Scepticisme et les hypothèses de la physique*'. *Revue de synthèse* 119 (1998): 211–55.

Rowland, Ingrid D. *Giordano Bruno: Philosopher/Heretic.* New York: Farrar, Straus and Giroux, 2008.

Ruby, Jane E. 'The Origins of Scientific "Law"'. *Journal of the History of Ideas* 47 (1986): 341–59.

Ruestow, Edward G. *The Microscope in the Dutch Republic: The Shaping of Discovery.* Cambridge: Cambridge University Press, 1996.

Russell, Bertrand. 'Obituary: Ludwig Wittgenstein'. *Mind* 60 (1951): 297–8.

Russell, Jeffrey Burton. *Inventing the Flat Earth: Columbus and Modern Historians.* New York: Praeger, 1991.

Russell, JL. 'Kepler's Laws of Planetary Motion: 1609–1666'. *British Journal for the History of Science* 2 (1964): 1–24.

Russo, Lucio. *The Forgotten Revolution: How Science was Born in 300* BC *and Why It Had to be Reborn.* Berlin: Springer, 2004.

Rybczynski, Witold. *One Good Turn: A Natural History of the Screwdriver and the Screw.* London: Scribner, 2000.

Ryle, Gilbert. *The Concept of Mind.* London: Hutchinson University Library, 1949.

Sabra, AI. 'The Commentary that Saved the Text'. *Early Science and Medicine* 12 (2007): 117–33.

———. *Theories of Light from Descartes to Newton.* London: Oldbourne, 1967.

Sacrobosco, Johannes de. *Sphaera . . . in usum scholarum.* Leiden: Elzevir, 1647.

———. *Sphaera J. de Sacro Bosco typis auctior quam antehac.* Paris: G Cavellat, 1552.

Sacrobosco, Johannes de, Georg von Peuerbach, and others. *Textus sphaerae Joannis de Sacro Busto.* Venice: J Rubeus, 1508.

Saliba, George. *Islamic Science and the Making of the European Renaissance.* Cambridge, Mass.: MIT Press, 2007.

Salusbury, Thomas (ed.). *Mathematical Collections and Translations.* London: W Leybourn, 1661.

Sankey, Howard. 'Kuhn's Changing Concept of Incommensurability'. *British Journal for the Philosophy of Science* 44 (1993): 759–74.

———. 'Taxonomic Incommensurability'. *International Studies in the Philosophy of Science* 12 (1998): 7–16.

Sarasohn, Lisa T. 'Nicolas-Claude Fabri de Peiresc and the Patronage of the New Science in the Seventeenth Century'. *Isis* 84 (1993): 70–90.

Sargent, Rose-Mary. *The Diffident Naturalist: Robert Boyle and the Philosophy of Experiment.* Chicago: University of Chicago Press, 1995.

Sarnowsky, Jürgen. 'Concepts of Impetus and the History of Mechanics'. In *Mechanics and Natural Philosophy before the Scientific Revolution.* Ed. WR Laird and S Roux. Dordrecht: Springer, 2008: 121–45.

———. 'The Defence of the Ptolemaic System in Late-Medieval Commentaries on Johannes de Sacrobosco's *De sphaera*'. In *Mechanics and Cosmology in the Medieval and Early Modern Period.* Ed. M Bucciantini, M Camerota and S Roux. Florence: LS Olschki, 2007: 29–44.

Sarpi, Paolo. *Pensieri naturali, metafisici e matematici*. Ed. L Cozzi and L Sosio. Milan: R Ricciardi, 1996.

Sarton, George. *The Study of the History of Science*. Cambridge, Mass.: Harvard University Press, 1936.

Savery, Thomas. *Navigation Improv'd, or The Art of Rowing Ships of All Rates, in Calms, with a More Easy, Swift, and Steady Motion, Than Oars Can*. London: J Moxon, 1698.

Sawday, Jonathan. *Engines of the Imagination: Renaissance Culture and the Rise of the Machine*. London: Routledge, 2007.

Scaliger, Joseph Justus. *Opuscula varia ante hac non edita*. Paris: H Beys, 1610.

Scarpa, Antonio. *Réflexions et observations anatomico-chirurgicales sur l'anéurisme*. Paris: Méquignon-Marvis, 1809.

Schaffer, Simon. 'Enlightened Automata'. In *The Sciences in Enlightened Europe*. Ed. W Clark, J Golinski and S Schaffer. Chicago: University of Chicago Press, 1999: 126–65.

———. 'Glass Works: Newton's Prisms and the Uses of Experiment'. In *The Uses of Experiment: Studies in the Natural Sciences*. Ed. D Gooding, TJ Pinch and S Schaffer. Cambridge: Cambridge University Press, 1989: 67–104.

———. 'Godly Men and Mechanical Philosophers: Souls and Spirits in Restoration Natural Philosophy'. *Science in Context* 1 (1987): 53–85.

———. 'Halley's Atheism and the End of the World'. *Notes and Records of the Royal Society of London* 32 (1977): 17–40.

———. 'Machine Philosophy: Demonstration Devices in Georgian Mechanics'. *Osiris* 9 (1994): 157–82.

———. 'Making Up Discovery'. In *Dimensions of Creativity*. Ed. MA Boden. Cambridge, Mass.: MIT Press, 1994: 13–51.

———. 'Scientific Discoveries and the End of Natural Philosophy'. *Social Studies of Science* 16 (1986): 387–420.

———. 'The Show that Never Ends: Perpetual Motion in the Early Eighteenth Century'. *British Journal for the History of Science* 28 (1995): 157–89.

Schechner, Sara J. 'Between Knowing and Doing: Mirrors and Their Imperfections in the Renaissance'. *Early Science and Medicine* 10 (2005): 137–62.

Schemmel, Matthias. *The English Galileo: Thomas Harriot's Work on Motion*. 2 vols. Dordrecht: Springer, 2008.

Schiebinger, Londa L. *The Mind Has No Sex? Women in the Origins of Modern Science*. Cambridge, Mass.: Harvard University Press, 1989.

Schimkat, Peter. '*Denis Papin und die Luftpumpe*'. In *Denis Papin: Erfinder und Naturforscher in Hessen-Kassel*. Ed. F Tönsmann and H Schneider. Kassel: Euregioverlag, 2009: 50–67.

Schmitt, Charles B. 'Experience and Experiment: A Comparison of Zabarella's View with Galileo's in *De Motu*'. *Studies in the Renaissance* 16 (1969): 80–138.

Schneider, Christoph. *Disputatio physica de terrae motu*. Wittenberg: J Gorman, 1608.

Schott, Gaspar. *Anatomia physico-hydrostatica fontium ac fluminum libris VI*. Würzburg: JG Schönwetteri, 1663.

———. *Mechanica hydraulico-pneumatica . . . acc. experimentum novum Magdeburgicum, quo vacuum alij stabilire, alij evertere conantur . . .* Frankfurt: JG Schönwetteri, 1657.
Schüssler, Rudolf. 'Jean Gerson, Moral Certainty and the Renaissance of Ancient Scepticism'. *Renaissance Studies* 23 (2009): 445–62.
Schuster, John A. 'Cartesian Physics'. In *Oxford Handbook of the History of Physics*. Ed. JZ Buchwald and R Fox. Oxford: Oxford University Press, 2013: 56–95.
———. *Descartes-agonistes: Physico-mathematics, Method and Corpuscular-mechanism, 1618–33*. Dordrecht: Springer, 2013.
———. '"Waterworld": Descartes' Vortical Celestial Mechanics'. In *The Science of Nature in the Seventeenth Century*. Ed. PR Anstey and JA Schuster. Dordrecht: Springer, 2005: 35–79.
Schuster, John A and Judit Brody. 'Descartes and Sunspots: Matters of Fact and Systematizing Strategies in the *Principia philosophiae*'. *Annals of Science* 70 (2013): 1–45.
Schuster, John A and Alan BH Taylor. 'Blind Trust: The Gentlemanly Origins of Experimental Science'. *Social Studies of Science* 27 (1997): 503–36.
Screech, Michael Andrew (ed.). *Montaigne's Annotated Copy of Lucretius: A Transcription and Study of the Manuscript, Notes and Pen-marks*. Geneva: Droz, 1998.
Searle, John R. *The Construction of Social Reality*. New York: Free Press, 1995.
Secord, James A. 'Knowledge in Transit'. *Isis* 95 (2004): 654–72.
———. *Visions of Science: Books and Readers at the Dawn of the Victorian Age*. Oxford: Oxford University Press, 2014.
Segre, Michael. 'Torricelli's Correspondence on Ballistics'. *Annals of Science* 40 (1983): 489–99.
Sen, SN. 'Al-Biruni on the Determination of Latitudes and Longitudes in India'. *Indian Journal of History of Science* 10 (1975): 185–97.
Seneca. *Seneca's Morals Abstracted*. Ed. R L'Estrange. London: T Newcomb, 1679.
Serene, Eileen F. 'Robert Grosseteste on Induction and Demonstrative Science'. *Synthèse* 40 (1979): 97–115.
Serjeantson, Richard. 'Francis Bacon and the "Interpretation of Nature" in the Late Renaissance'. *Isis* 105 (2014): 681–705.
———. 'Testimony and Proof in Early-modern England'. *Studies in History and Philosophy of Science* 30 (1999): 195–236.
Serlio, Sebastiano. *Libro primo [-quinto] d'architettura*. Venice: Sessa Fratelli, 1559.
Serrano, Juan D. 'Trying Ursus: A Reappraisal of the Tycho–Ursus Priority Dispute'. *Journal for the History of Astronomy* 44 (2013): 17–46.
Severinus, Petrus. *Idea medicinae philosophicae, fundamenta continens totius doctrinae Paracelsicae, Hippocraticae, & Galenicae*. Basle: S Henricpetrus, 1571.
Sewell, Keith C. 'The "Herbert Butterfield Problem" and its Resolution'. *Journal of the History of Ideas* 64 (2003): 599–618.

Shank, John Bennett. *The Newton Wars and the Beginning of the French Enlightenment.* Chicago: University of Chicago Press, 2008.
———. 'What Exactly was Torricelli's Barometer?' In *Science in the Age of Baroque.* Ed. O Gal and R Chen-Morriz. Dordrecht: Springer, 2012: 161–95.
Shank, Michael H. 'Mechanical Thinking in European Astronomy (13th–15th Centuries)'. In *Mechanics and Cosmology in the Medieval and Early Modern Period.* Ed. M Bucciantini, M Camerota and S Roux. Florence: LS Olschki, 2007: 3–27.
———. 'Setting Up Copernicus? Astronomy and Natural Philosophy in Giambattista Capuano da Manfredonia's *Expositio* on the Sphere'. *Early Science and Medicine* 14 (2009): 290–315.
Shapere, Dudley. 'The Structure of Scientific Revolutions'. *Philosophical Review* 73 (1964): 383–94.
Shapin, Steven. 'Cordelia's Love: Credibility and the Social Studies of Science'. *Perspectives on Science* 3 (1995): 255–75.
———. 'History of Science and Its Sociological Reconstructions'. *History of Science* 20 (1982): 157–211.
———. 'How to be Antiscientific'. In *Never Pure: Historical Studies of Science.* Baltimore: Johns Hopkins University Press, 2010: 32–46.
———. 'The Invisible Technician'. *American Scientist* 77 (1989): 554–63.
———. 'Possessed by the Idols'. *London Review of Books*, 30 November 2006.
———. 'Pump and Circumstance: Robert Boyle's Literary Technology'. *Social Studies of Science* 14 (1984): 481–520.
———. 'Robert Boyle and Mathematics: Reality, Representation and Experimental Practice'. *Science in Context* 2 (1988): 23–58.
———. *The Scientific Revolution.* Chicago: University of Chicago Press, 1996.
———. *A Social History of Truth: Civility and Science in Seventeenth-century England.* Chicago: University of Chicago Press, 1994.
———. 'Understanding the Merton Thesis'. *Isis* 79 (1988): 594–605.
———. 'A View of Scientific Thought'. *Science* 207 (1980): 1065–6.
Shapin, Steven and Simon Schaffer. *Leviathan and the Air-pump: Hobbes, Boyle, and the Experimental Life.* Princeton: Princeton University Press, 1985.
Shapiro, Alan E. 'The Gradual Acceptance of Newton's Theory of Light and Color, 1672–1727'. *Perspectives on Science* 4 (1996): 59–140.
———. 'Introduction'. In *The Optical Papers of Isaac Newton: The Optical Lectures 1670–1672.* Cambridge: Cambridge University Press, 1984: 1–25.
Shapiro, Barbara J. 'The Concept "Fact": Legal Origins and Cultural Diffusion'. *Albion* 26 (1994): 1–25.
———. *A Culture of Fact: England, 1550–1720.* Ithaca: Cornell University Press, 2000.
———. *John Wilkins, 1614–1672: An Intellectual Biography.* Berkeley: University of California Press, 1969.
Sharratt, Michael. *Galileo: Decisive Innovator.* Oxford: Blackwell, 1994.
Shaw, Peter. *A Treatise of Incurable Diseases.* London: J Roberts, 1723.

Shea, James H. 'Ole Rømer, the Speed of Light, the Apparent Period of Io, the Doppler Effect and the Dynamics of Earth and Jupiter'. *American Journal of Physics* 66 (1998): 561–9.

Shea, William R. *Designing Experiments and Games of Chance: The Unconventional Science of Blaise Pascal.* Canton, MA: Science History Publications, 2003.

———. *Galileo's Intellectual Revolution: Middle Period, 1610–1632.* New York: Science History Publications, 1972.

Sheppard, Samuel. *The Honest Lawyer.* London: Woodruffe, 1616.

Shirley, John William. *Thomas Harriot, a Biography.* Oxford: Clarendon Press, 1983.

Sills, David L and Robert K Merton. *International Encyclopedia of the Social Sciences: Social Science Quotations.* New York: Macmillan, 1991.

Simek, Rudolf. *Heaven and Earth in the Middle Ages: The Physical World before Columbus.* Woodbridge: Boydell Press, 1996.

Singer, Charles Joseph, A Rupert Hall and others. *A History of Technology.* 8 vols. Oxford: Clarendon Press, 1954–84.

Singer, Dorothea Waley and Giordano Bruno. *Giordano Bruno, His Life and Thought. With Annotated Translation of His Work on the Infinite Universe and Worlds.* New York: Schuman, 1950.

Siraisi, Nancy G. *Communities of Learned Experience: Epistolary Medicine in the Renaissance.* Baltimore: Johns Hopkins University Press, 2013.

———. *Taddeo Alderotti and His Pupils: Two Generations of Italian Medical Learning.* Princeton: Princeton University Press, 1981.

Skinner, Quentin. 'Classical Liberty and the Coming of the English Civil War'. In *Republicanism: A Shared European Heritage.* 2 vols. Vol. 2. Ed. M van Gelderen and Q Skinner. Cambridge: Cambridge University Press, 2002: 9–28.

———. 'Meaning and Understanding in the History of Ideas'. *History and Theory* 8 (1969): 3–53.

———. *Reason and Rhetoric in the Philosophy of Hobbes.* Cambridge: Cambridge University Press, 1996.

———. *Visions of Politics.* 3 vols. Vol. 1: *Regarding Method.* Cambridge: Cambridge University Press, 2002.

Slack, Paul. 'Government and Information in Seventeenth-century England'. *Past and Present* 184 (2004): 33–68.

———. 'Measuring the National Wealth in Seventeenth-century England'. *Economic History Review* 57 (2004): 607–35.

Slezak, Peter. 'A Second Look at David Bloor's *Knowledge and Social Imagery*'. *Philosophy of the Social Sciences* 24 (1994): 336–61.

Smeaton, John. *An Experimental Enquiry Concerning the Natural Powers of Water and Wind to Turn Mills.* London: [n.p.], 1760.

Smith, Alan. 'A New Way of Raising Water by Fire: Denis Papin's Treatise of 1707 and Its Reception by Contemporaries'. *History of Technology* 20 (1998): 139–81.

Smith, AM. 'Knowing Things Inside Out: The Scientific Revolution from a Medieval Perspective'. *American Historical Review* 95 (1990): 726–44.

Smith, Margaret M. 'Printed Foliation: Forerunner to Printed Page-numbers?' *Gutenberg Jahrbuch* 63 (1988): 54–70.

Smith, Pamela H. 'Art, Science and Visual Culture in Early Modern Europe'. *Isis* 97: 83–100 (2006).

———. *The Body of the Artisan: Art and Experience in the Scientific Revolution.* Chicago: University of Chicago Press, 2006.

———. *The Business of Alchemy: Science and Culture in the Holy Roman Empire.* Princeton: Princeton University Press, 1994.

———. 'Science on the Move: Recent Trends in the History of Early Modern Science'. *Renaissance Quarterly* 62 (2009): 345–75.

Smith, Robert W. 'The Cambridge Network in Action: The Discovery of Neptune'. *Isis* 80 (1989): 395–422.

Snell, Bruno. 'The Forging of a Language for Science in Ancient Greece'. *Classical Journal* 56 (1960): 50–60.

———. 'The Origin of Scientific Thought'. In *The Discovery of the Mind: The Greek Origins of European Thought.* Trans. T Rosenmeyer. Cambridge, Mass.: Harvard University Press, 1953: 227–45.

Snobelen, Stephen D. '"God of Gods, and Lord of Lords": The Theology of Isaac Newton's General *Scholium* to the *Principia*'. *Osiris* 16 (2001): 169–208.

———. 'Isaac Newton, Heretic: The Strategies of a Nicodemite'. *British Journal for the History of Science* 32 (1999): 381–419.

———. 'The Myth of the Clockwork Universe'. In *The Persistence of the Sacred in Modern Thought.* Ed. CL Firestone and N Jacobs. Notre Dame: University of Notre Dame Press, 2012: 49–184.

———. 'William Whiston, Isaac Newton and the Crisis of Publicity'. *Studies in History and Philosophy of Science Part A* 35 (2004): 573–603.

Snow, Charles Percy. *The Two Cultures and the Scientific Revolution.* Cambridge: Cambridge University Press, 1959.

Snow, Vernon F. 'The Concept of Revolution in Seventeenth-century England'. *Historical Journal* 5 (1962): 167–74.

Sobel, Dava. *Longitude: The True Story of a Lone Genius Who Solved the Greatest Scientific Problem of His Time.* New York: Walker, 1995.

Sokal, Alan D. *Beyond the Hoax: Science, Philosophy and Culture.* Oxford: Oxford University Press, 2008.

Soll, Jacob. *The Reckoning: Financial Accountability and the Making and Breaking of Nations.* London: Allen Lane, 2014.

Spencer, John. *A Discourse Concerning Prodigies.* Cambridge: W Graves, 1663.

Sprat, Thomas. *The History of the Royal-Society of London.* London: J Martyn, 1667.

Stabile, Giorgio. '*Il concetto di esperienza in Galilei e nella scuola galileiana*'. In *Experientia.* Ed. M Veneziani. Florence: LS Olschki, 2002: 217–41.

Standage, Tom. *The Turk: The Life and Times of the Famous Eighteenth-century Chess-playing Machine.* New York: Walker, 2002.

Stanford, P Kyle. *Exceeding Our Grasp: Science, History and the Problem of Unconceived Alternatives.* Oxford: Oxford University Press, 2010.

Starkey, George. *Alchemical Laboratory Notebooks and Correspondence.* Ed. WR Newman and L Principe. Chicago: University of Chicago Press, 2004.

———. *Nature's Explication and Helmont's Vindication.* London: T Alsop, 1657.

Steele, Brett D. 'Muskets and Pendulums: Benjamin Robins, Leonhard Euler and the Ballistics Revolution'. *Technology and Culture* 35 (1994): 348–82.

Stein, Gertrude. *Everybody's Autobiography.* New York: Random House, 1937.

Steinle, F. 'Negotiating Experiment, Reason and Theology: The Concept of Laws of Nature in the Early Royal Society'. In *Ideals and Cultures of Knowledge in Early Modern Europe.* Ed. W Detel and K Zittel. Berlin: Akademie Verlag, 2002: 197–212.

Steinle, F and Friedel Weinert. 'The Amalgamation of a Concept: Laws of Nature in the New Sciences'. In *Laws of Nature: Essays on the Philosophical, Scientific and Historical Dimensions.* Berlin: Walter de Gruyter, 1995: 316–68.

Stewart, Larry. 'A Meaning for Machines: Modernity, Utility and the Eighteenth-century British Public'. *Journal of Modern History* 70 (1998): 259–94.

———. *The Rise of Public Science: Rhetoric, Technology and Natural Philosophy in Newtonian Britain, 1660–1750.* Cambridge: Cambridge University Press, 1992.

Stigler, Stephen M. 'John Craig and the Probability of History: From the Death of Christ to the Birth of Laplace'. *Journal of the American Statistical Association* 81 (1986): 879–87.

———. 'Stigler's Law of Eponymy'. *Transactions of the New York Academy of Sciences* 39 (1980): 147–57.

Stone, Lawrence. *The Causes of the English Revolution, 1529–1642.* New York: Harper & Row, 1972.

Stubbe, Henry. *An Epistolary Discourse Concerning Phlebotomy.* London: [s.n.], 1671.

Stubbes, John. *The Discoverie of a Gaping Gulf.* London: W Page, 1579.

Sutton, Clive. '"*Nullius in verba*" and "*nihil in verbis*": Public Understanding of the Role of Language in Science'. *British Journal for the History of Science* 27 (1994): 55–64.

Sutton, Robert B. 'The Phrase *Libertas philosophandi*'. *Journal of the History of Ideas* 14 (1953): 310–16.

Swerdlow, Noel M. 'Copernicus and Astrology, with an Appendix of Translations of Primary Sources'. *Perspectives on Science* 20 (2012): 353–78.

———. 'The Derivation and First Draft of Copernicus's Planetary Theory: A Translation of the *Commentariolus* with Commentary'. *Proceedings of the American Philosophical Society* 117 (1973): 423–512.

———. 'An Essay on Thomas Kuhn's First Scientific Revolution: *The Copernican Revolution*'. *American Philosophical Society Proceedings* 141 (2004): 64–120.

———. 'Montucla's Legacy: The History of the Exact Sciences'. *Journal of the History of Ideas* 54 (1993): 299–328.

———. '*Urania propitia, tabulae rudophinae faciles redditae a Maria Cunitia* [Beneficent Urania, the Adaptation of the Rudolphine Tables by Maria

Cunitz]'. In *A Master of Science History.* Ed. JZ Buchwald. Dordrecht: Springer, 2012: 81–121.

Swift, Jonathan. *Gulliver's Travels.* Ed. D Womersley. Cambridge: Cambridge University Press, 2012.

———. *On Poetry: A Rhapsody.* London: J Huggonson, 1733.

———. *A Tale of a Tub and Other Works.* Ed. M Walsh. Cambridge: Cambridge University Press, 2010.

Tachau, Katherine H. *Vision and Certitude in the Age of Ockham.* Leiden: EJ Brill, 1988.

Taisnier, Jean. *Opusculum perpetua memoria dignissimum: De natura magnetis, et eius effectibus.* Cologne: J Birckmannus, 1562.

Tanturli, Giuliano. '*Rapporti del Brunelleschi con gli ambienti letterari fiorentini*'. In *Filippo Brunelleschi: La sua opera e il suo tempo.* Ed. G Soadolini. Florence: Centro Di, 1980: 125–44.

Tarrant, Neil. 'Giambattista della Porta and the Roman Inquisition'. *British Journal for the History of Science* 46 (2013): 601–25.

Tassoni, Alessandro. *Dieci libri di pensieri diversi.* Venice: MA Brogiollo, 1627.

Taylor, Eva Germaine Rimington. *The Haven-finding Art: A History of Navigation from Odysseus to Captain Cook.* New York: American Elsevier, 1971.

———. *The Mathematical Practitioners of Tudor and Stuart England.* Cambridge: Cambridge University Press, 1954.

Tedeschi, John. 'The Roman Inquisition and Witchcraft: An Early-seventeenth-century "Instruction" on Correct Trial Procedure'. *Revue de l'histoire des religions* 200 (1983): 163–88.

Temple, William. *Miscellanea. The Third Part: Containing: I. An Essay on Popular Discontents. II. A Defense of the Essay upon Antient and Modern Learning: With Some Other Pieces.* Ed. J Swift. London: B Tooke, 1701.

Thomas, Keith. *The Ends of Life: Roads to Fulfilment in Early Modern England.* Oxford: Oxford University Press, 2009.

———. *Religion and the Decline of Magic: Studies in Popular Beliefs in Sixteenth- and Seventeenth-century England.* London: Weidenfeld & Nicolson, 1997.

Thoren, Victor E. *Lord of Uraniborg: A Biography of Tycho Brahe.* Cambridge: Cambridge University Press, 2007.

Thorndike, Lynn. *A History of Magic and Experimental Science.* 8 vols. New York: Columbia University Press, 1923–58.

———. 'Newness and Craving for Novelty in Seventeenth-century Science and Medicine'. *Journal of the History of Ideas* 12 (1951): 584–58.

———. *Science and Thought in the Fifteenth Century.* New York: Columbia University Press, 1929.

———. *The Sphere of Sacrobosco and Its Commentators.* Chicago: University of Chicago Press, 1949.

Tilling, Laura. 'Early Experimental Graphs'. *British Journal for the History of Science* 8 (1975): 193–213.

de Tocqueville, Alexis. *The Old Regime and the Revolution.* Trans. J Bonner. New York: Harper & Brothers, 1856.

Tolomei, Claudio, Lodovico Guicciardini and Giovanni Botero. *Tre discorsi appartenenti alla grandezza delle citta*. Rome: G Maratinelli, 1588.

Tönsmann, Frank. '*Wasserbauten und Schifffahrt in Hessen um 1700 und die Forschungen von Papin*'. In *Denis Papin: Erfinder und Naturforscher in Hessen-Kassel*. Ed. F Tönsmann and H Schneider. Kassel: Euregioverlag, 2009: 89–103.

Tönsmann, Frank and Helmuth Schneider (eds.). *Denis Papin: Erfinder und Naturforscher in Hessen-Kassel*. Kassel: Euregioverlag, 2009.

Topdemir, Hüseyin Gazi. 'Kamal al-Din al-Farisi's Explanation of the Rainbow'. *Humanity and Social Sciences Journal* 2 (2007): 75–85.

Toscano, Fabio. *La formula segreta: Tartaglia, Cardano e il duello matematico che infiammò l'Italia del Rinascimento*. Milan: Sironi, 2009.

Tosh, Nick. 'Anachronism and Retrospective Explanation: In Defence of a Present-centred History of Science'. *Studies in History and Philosophy of Science Part A* 34 (2003): 647–59.

Trenchard, John and Thomas Gordon. *Cato's Letters, or Essays on Liberty, Civil and Religious, and Other Important Subjects*. Ed. R Hamowy. 4 in 2 vols. Vol. 3. Indianapolis: Liberty Fund, 1995.

Trevor-Roper, Hugh R. 'Nicholas Hill, the English Atomist'. In *Catholics, Anglicans and Puritans: Seventeenth-century Essays*. London: Secker & Warburg, 1987: 1–39.

———. 'The Religious Origins of the Enlightenment'. In *Religion, the Reformation and Social Change*. London: Macmillan, 1967: 193–236.

Trompf, Garry Winston. *The Idea of Historical Recurrence in Western Thought from Antiquity to the Reformation*. Berkeley: University of California Press, 1979.

Trutfetter, Jodocus. *Summa in tota[m] physicen: Hoc est philosophiam naturalem conformiter siquidem ver[a]e sophi[a]e: que est theologia*. Erfurt: M Maler, 1514.

———. *Summa philosophiae naturalis contracta*. Erfurt: M Maler, 1517.

Tuck, Richard. *Natural Rights Theories: Their Origin and Development*. Cambridge: Cambridge University Press, 1979.

———. 'Optics and Sceptics: The Philosophical Foundations of Hobbes's Political Thought'. In *Conscience and Casuistry in Early Modern Europe*. Ed. E Leites. Cambridge: Cambridge University Press, 1988: 235–63.

Tunstall, Kate E and Denis Diderot. *Blindness and Enlightenment: An Essay*. New York: Continuum, 2011.

Turgot, Anne-Robert-Jacques. *Turgot on Progress, Sociology and Economics: A Philosophical Review of the Successive Advances of the Human Mind on Universal History [and] Reflections on the Formation and the Distribution of Wealth*. Ed. RL Meek. Cambridge: Cambridge University Press, 1973.

Ugaglia, M. 'The Science of Magnetism before Gilbert: Leonardo Garzoni's Treatise on the Loadstone'. *Annals of Science* 63 (2006): 59–84.

Valente, Michaela. '*Della Porta e l'Inquisizione: Nuove documenti dell'archivo del Sant' Uffizio*'. *Bruniana e Campanelliana* 5 (1999): 415–34.

Valenza, Robin. *Literature, Language and the Rise of the Intellectual Disciplines in Britain, 1680–1820*. Cambridge: Cambridge University Press, 2009.

Vallisneri, Antonio. '*Lezione accademica intorno all'origine delle fontane*'. In *Opere diverse*. Venice: Ertz, 1715.

Vanini, Giulio Cesare. *De admirandis naturae reginae deaeque mortalium arcanis*. Paris: A Perier, 1616.

Vasari, Giorgio. *The Lives of the Artists. A Selection*. Trans. G Bull. Harmondsworth: Penguin Books, 1965.

Vaughan, MF. 'An Unnoted Translation of Erasmus in Ascham's *Schoolmaster*'. *Modern Philology* 75 (1977): 184–6.

Vergil, Polydore. *An Abridgeme[n]t of the Notable Worke of Polidore Virgile: Conteignyng the Devisers and Fyrst Fynders Out*. Trans. T Langley. London: R Grafton, 1546.

———. *On Discovery*. Ed. BP Copenhaver. Cambridge, Mass.: Harvard University Press, 2002.

———. *A Pleasant and Compendious History of the First Inventers and Instituters of the Most Famous Arts, Misteries, Laws, Customs and Manners in the Whole World*. Trans. T Langley. London: J Harris, 1686.

———. *The Works of the Famous Antiquary, Polidore Vergil*. London: S Miller, 1663.

Verlinden, Charles. '*Lanzarotto Malocello et la découverte portugaise des Canaries*'. *Revue belge de philologie et d'histoire* 36 (1958): 1173–209.

Vickers, Brian. 'Francis Bacon, Feminist Historiography and the Dominion of Nature'. *Journal of the History of Ideas* 69 (2008): 117–41.

———. 'The "New Historiography" and the Limits of Alchemy'. *Annals of Science* 65 (2008): 127–56.

Vitruvius Pollio, Marcus. *De architectura: libri dece*. Como: G da Ponte, 1521.

———. *Zehen Bücher von der Architectur und Künstlichem Bawen*. Trans. GGH Rivius. Nuremberg: Petreius, 1548.

Vlastos, Gregory. '*Wege und Formen frühgriechischen Denkens* by Hermann Fränkel'. *Gnomon* 31 (1959): 193–204.

Vogel, Klaus A. '*America: Begriff, geographische Konzeption und frühe Entdeckungsgeschichte in der Perspektive der deutschen Humanisten*'. In *Von der Weltkarte zum Kuriositatenkabinett: Amerika im deutschen Humanismus und Barock*. Ed. K Kohut. Frankfurt: Vervuert, 1995: 11–43.

———. 'Cosmography'. In *The Cambridge History of Science*. 7 vols. Vol. 3: *Early Modern Science*. Ed. K Park and LJ Daston. Cambridge: Cambridge University Press, 2006: 469–96.

———. '*Das Problem der relativen Lage von Erd- und Wassersphäre im Mittelalter und die kosmographische Revolution*'. *Mitteilungen der österreichischen Gesellschaft für Wissenschaftsgeschichte* 13 (1993): 103–43.

———. *Sphaera terrae – das mittelalterliche Bild der Erde und die kosmographische Revolution*. Göttingen: University of Göttingen,1995.

Voltaire. *Letters Concerning the English Nation*. London: C Davis, 1733.

———. '*Micromégas': A Study in the Fusion of Science, Myth, and Art.* Ed. I Wade. Princeton: Princeton University Press, 1950.

W., G. *The Modern States-man.* London: H Hill, 1653.

Waard, Cornelis de. *L'Expérience barométrique, ses antécédents et ses explications, étude historique* . Thouars: Impr. nouvelle, 1936.

Wagner, David Leslie. *The Seven Liberal Arts in the Middle Ages.* Bloomington: Indiana University Press, 1983.

Waldseemüller, Martin. *The Cosmographiæ introductio of Martin Waldseemüller in Facsimile Followed by the Four Voyages of Amerigo Vespucci, with Their Translation into English.* Ed. CG Herbermann. New York: United States Catholic Historical Society, 1907.

Wallace, Anthony FC. *The Social Context of Innovation: Bureaucrats, Families and Heroes in the Early Industrial Revolution.* Princeton: Princeton University Press, 1982.

Wallis, Helen. 'What Columbus Knew'. *History Today* 42 (1992): 17–23.

Wallis, John. 'An Essay of Dr John Wallis, Exhibiting His Hypothesis about the Flux and Reflux of the Sea'. *Philosophical Transactions* 1 (1666): 263–81.

Walsham, Alexandra. 'The Reformation and "The Disenchantment of the World" Reassessed'. *Historical Journal* 51 (2008): 497–528.

Walton, Steven A. *Wind and Water in the Middle Ages: Fluid Technologies from Antiquity to the Renaissance.* Tempe, AZ: ACMRS, 2006.

Washburn, Wilcomb E. 'The Meaning of "Discovery" in the Fifteenth and Sixteenth Centuries'. *American Historical Review* 68 (1962): 1–21.

Waters, David W. 'Nautical Astronomy and the Problem of Longitude'. In *The Uses of Science in the Age of Newton.* Ed. JG Burke. Berkeley: University of California Press, 1983: 143–69.

Watson, James D. *The Double Helix: A Personal Account of the Discovery of the Structure of DNA.* London: Weidenfeld & Nicolson, 1968.

Weber, Eugen. *Peasants into Frenchmen: The Modernization of Rural France 1870–1914.* Stanford: Stanford University Press, 1976.

Weber, Max. *The Vocation Lectures.* Ed. TB Strong and DS Owen. Trans. R Livingstone. Indianapolis: Hackett, 2004.

Webster, Charles. 'The Discovery of Boyle's Law, and the Concept of the Elasticity of Air in the Seventeenth Century'. *Archive for History of Exact Sciences* 2 (1965): 441–502.

———. *The Great Instauration: Science, Medicine and Reform, 1626–1660.* London: Duckworth, 1975.

———. 'Henry More and Descartes, Some New Sources'. *British Journal for the History of Science* 4 (1969): 359–77.

———. 'Henry Power's Experimental Philosophy'. *Ambix* 14 (1967): 150–78.

——— (ed.). *The Intellectual Revolution of the Seventeenth Century.* London: Routledge & Kegan Paul, 1974.

———. 'New Light on the Invisible College: The Social Relations of English Science in the Mid-seventeenth Century'. *Transactions of the Royal Historical Society (Fifth Series)* 24 (1974): 19–42.

———. 'William Harvey's Conception of the Heart as a Pump'. *Bulletin of the History of Medicine* 39 (1965): 508–17.

Webster, John. *The Displaying of Supposed Witchcraft*. London: JM, 1677.

Weeks, Sophie. 'Francis Bacon and the Art–Nature Distinction'. *Ambix* 54 (2007): 117–45.

———. 'The Role of Mechanics in Francis Bacon's *Great Instauration*'. In *Philosophies of Technology: Francis Bacon and His Contemporaries*. Ed. C Zittel, G Engel, R Nanni and N Karafyllis. Leiden: Brill, 2008: 133–97.

Weinberg, Steven. *To Explain the World: The Discovery of Modern Science*. 2015.

———. 'Sokal's Hoax'. *New York Review of Books*, 8 August 1996.

Weiner, Stephen A. 'The Civil Jury Trial and the Law–Fact Distinction'. *California Law Review* 54 (1966): 1867–938.

Weld, Charles Richard. *A History of the Royal Society, with Memories of the Presidents*. 2 vols. London: JW Parker, 1848.

Wengenroth, Ulrich. 'Science, Technology and Industry'. In *From Natural Philosophy to the Sciences: Writing the History of Nineteenth-century Science*. Ed. D Cahan. Chicago: University of Chicago Press, 2003: 221–53.

Wesley, Walter G. 'The Accuracy of Tycho Brahe's Instruments'. *Journal for the History of Astronomy* 9 (1978): 42–53.

Westfall, Richard S. 'The Development of Newton's Theory of Color'. *Isis* (1962): 339–58.

———. *Never at Rest: A Biography of Isaac Newton*. Cambridge: Cambridge University Press, 1980.

———. 'Newton and the Fudge Factor'. *Science* 179 (1973): 751–8.

———. 'Science and Technology during the Scientific Revolution: An Empirical Approach'. In *Renaissance and Revolution. Humanists, Scholars, Craftsmen and Natural Philosophers in Early Modern Europe*. Ed. JV Field and FA James. Cambridge: Cambridge University Press, 1997: 63–72.

———. 'The Scientific Revolution Reasserted'. In *Rethinking the Scientific Revolution*. Ed. M Osler. Cambridge: Cambridge University Press, 2000: 41–55.

———. 'Unpublished Boyle Papers Relating to Scientific Method: I'. *Annals of Science* 12 (1956): 63–73.

Westman, Robert S. *The Copernican Question: Prognostication, Skepticism and Celestial Order*. Berkeley: University of California Press, 2011.

———. 'The Copernican Question Revisited: A Reply to Noel Swerdlow and John Heilbron'. *Perspectives on Science* 21 (2013): 100–36.

Westman, Robert S and JE McGuire. *Hermeticism and the Scientific Revolution*. Los Angeles: William Andrews Clark Memorial Library, 1977.

Westrum, Ron. 'Science and Social Intelligence about Anomalies: The Case of Meteorites'. *Social Studies of Science* 8 (1978): 461–93.

Whewell, William. 'On the Connexion of the Physical Sciences'. *Quarterly Review* 51 (1834): 54–68.

———. *The Philosophy of the Inductive Sciences, Founded upon Their History*. 2 vols. London: John W Parker, 1840.

White, Gilbert. *The Natural History and Antiquities of Selborne, in the County of Southampton*. London: B White, 1789.

White, John. *The Birth and Rebirth of Pictorial Space*. Cambridge, Mass.: Belknap Press, 1987.

White, Lynn Townsend. 'The Medieval Roots of Modern Technology and Science' [1963]. In *Medieval Religion and Technology: Collected Essays*. Berkeley: University of California Press, 1978: 75–91.

Whitley, Richard. 'Black Boxism and the Sociology of Science: A Discussion of the Major Developments in the Field'. *Sociological Review* 18 (1970): 61–92.

Wierzbicka, Anna. *Experience, Evidence and Sense: The Hidden Cultural Legacy of English*. Oxford: Oxford University Press, 2010.

Wigelsworth, Jeffrey R. *Selling Science in the Age of Newton: Advertising and the Commoditization of Knowledge*. Farnham: Ashgate, 2011.

Wilding, Nick. *Galileo's Idol: Gianfrancesco Sagredo and the Politics of Knowledge*. Chicago: University of Chicago Press, 2014.

———. 'The Return of Thomas Salusbury's Life of Galileo (1664)'. *British Journal for the History of Science* 41 (2008): 241–65.

Wilkins, John. *A Discourse Concerning a New World and Another Planet*. London: J Maynard, 1640.

———. *An Essay towards a Real Character, and a Philosophical Language*. London: S Gellibrand, 1668.

———. *Mathematicall Magick*. London: S Gellibrand, 1648.

———. *Of the Principles and Duties of Natural Religion*. London: T Basset, 1675.

Williams, Bernard. *Essays and Reviews, 1959–2002*. Princeton: Princeton University Press, 2014.

———. 'Wittgenstein and Idealism'. *Royal Institute of Philosophy Lectures* 7 (1973): 76–95.

Williams, Glyndwr. *Voyages of Delusion: The Quest for the Northwest Passage*. New Haven: Yale University Press, 2002.

Willmoth, Frances. 'Römer, Flamsteed, Cassini and the Speed of Light'. *Centaurus* 54 (2012): 39–57.

Wilson, Adrian and Timothy G Ashplant. 'Whig History and Present-centred History'. *Historical Journal* 31 (1988): 1–16.

Wilson, Bryan R (ed.). *Rationality*. Oxford: Blackwell, 1970.

Wilson, Catherine. *The Invisible World: Early Modern Philosophy and the Invention of the Microscope*. Princeton: Princeton University Press, 1995.

———. 'From Limits to Laws: The Construction of the Nomological Image of Nature in Early Modern Philosophy'. In *Natural Law and Laws of Nature in Early Modern Europe*. Ed. LJ Daston and M Stolleis. Farnham: Ashgate, 2008: 13–28.

Wilson, Curtis A. 'From Kepler's Laws, So-called, to Universal Gravitation: Empirical Factors'. *Archive for History of Exact Sciences* 6 (1970): 89–170.

Wilson, G. 'On the Early History of the Air-pump in England'. *Edinburgh New Philosophy Journal* 46 (1849): 330–54.

Winch, Peter. *The Idea of a Social Science and Its Relation to Philosophy*. London: Routledge & Kegan Paul, 1958.

Wintzer, E. *Denis Papins Erlebnisse in Marburg, 1688–1695*. Marburg: N Elwert, 1898.

Withington, Phil. *Society in Early Modern England*. Cambridge: Polity, 2010.

Wittgenstein, Ludwig. *On Certainty*. Ed. GEM Anscombe and GHV Wright. Oxford: Blackwell, 1969.

———. *Philosophical Investigations*. Oxford: Blackwell, 1953.

———. 'Remarks on Frazer's Golden Bough'. In *Philosophical Occasions, 1912–1951*. Ed. JC Klagge and A Nordmann. Indianapolis: Hackett, 1993: 115–55.

———. *Tractatus Logico-Philosophicus*. London: Kegan Paul, Trench, Trubner, 1933.

Wolper, Roy S. 'The Rhetoric of Gunpowder and the Idea of Progress'. *Journal of the History of Ideas* 31 (1970): 589–98.

Womersley, David. 'Dean Swift Hears a Sermon: Robert Howard's Ash Wednesday Sermon of 1725 and *Gulliver's Travels*'. *Review of English Studies* 60 (2009): 744–62.

Wood, Paul B. 'Methodology and Apologetics: Thomas Sprat's History of the Royal Society'. *British Journal for the History of Science* 13 (1980): 1–26.

Woodward, David (ed.). *The History of Cartography*. 6 vols. Vol. 3: *Cartography in the European Renaissance*. Chicago: University of Chicago Press, 2007.

———. 'The Image of the Spherical Earth'. *Perspecta* 25 (1989): 2–15.

Woodward, John. *Dr Friend's Epistle to Dr Mead*. London: J Roberts, 1719.

Wootton, David. 'Accuracy and Galileo: A Case Study in Quantification and the Scientific Revolution'. *Journal of The Historical Society* 10 (2010): 43–55.

———. *Bad Medicine: Doctors Doing Harm Since Hippocrates*. Oxford: Oxford University Press, 2006.

———. 'Galileo: Reflections on Failure'. In *Causation and Modern Philosophy*. Ed. K Allen and T Stoneham. Routledge, 2011: 13–30.

———. *Galileo: Watcher of the Skies*. New Haven: Yale University Press, 2010.

———. 'The Hard Look Back'. *Times Literary Supplement* 14 (2003): 8–10.

———. 'Hume's "Of Miracles": Probability and Irreligion'. In *Studies in the Philosophy of the Scottish Enlightenment*. Ed. MA Stewart. Oxford: Oxford University Press, 1990: 191–229.

———. 'Hutchinson, Francis'. In *Encyclopedia of Witchcraft: The Western Tradition*. 4 vols. Vol. 2. Ed. RM Golden. Santa Barbara: ABC-CLIO, 2006: 531–2.

———. 'Liberty, Metaphor and Mechanism: "Checks and Balances" and the Origins of Modern Constitutionalism'. In *Liberty and American Experience in the Eighteenth Century*. Ed. D Womersley. Indianapolis: Liberty Fund, 2006: 209–74.

———. 'Lucien Febvre and the Problem of Unbelief'. *Journal of Modern History* 60 (1988): 695–730.

Wotton, William. *A Defense of the Reflections upon Ancient and Modern Learning*. London: Goodwin, 1705.

———. *Reflections upon Ancient and Modern Learning*. London: P Buck, 1694.

Wotton, William and Richard Bentley. *Reflections upon Ancient and Modern Learning. The Second Part, with a Dissertation upon the Epistles of Phalaris.* London: PB, 1698.

Wright, John Kirtland. *The Geographical Lore of the Time of the Crusades.* New York: American Geographical Society, 1925.

Wussing, Hans. *Die grosse Erneuerung: Zur Geschichte der wissenschaftlichen Revolution.* Basle: Birkhäuser, 2002.

Yates, Frances Amelia. *Giordano Bruno and the Hermetic Tradition.* Chicago: University of Chicago Press, 1991.

Yeomans, Donald K, Juergen Rahe and Ruth S Freitag. 'The History of Comet Halley'. *Journal of the Royal Astronomical Society of Canada* 80 (1986): 62–86.

Yiu, Yvonne. 'The Mirror and Painting in Early Renaissance Texts'. *Early Science and Medicine* 10 (2005): 187–210.

Yolton, John W. *Thinking Matter: Materialism in Eighteenth-century Britain.* Minneapolis: University of Minnesota Press, 1983.

Zambelli, Paola. '*Introduzione*'. In Alexandre Koyré, *Dal mondo del pressappoco all'universo della precisione.* Turin: Einaudi, 1967: 7–46.

Zammito, John H. *A Nice Derangement of Epistemes: Post-positivism in the Study of Science from Quine to Latour.* Chicago: University of Chicago Press, 2004.

Zanden, Jan Luiten van. *The Long Road to the Industrial Revolution.* Leiden: Brill, 2009.

Zarlino, Gioseffo. *Dimostrationi harmoniche.* Venice: Francesco de i Franceschi, 1571.

Zhmud, Leonid. *The Origin of the History of Science in Classical Antiquity.* Trans. A Chernoglazov. Berlin: Walter de Gruyter, 2006.

Zilsel, Edgar. 'The Genesis of the Concept of Scientific Progress'. *Journal of the History of Ideas* 6 (1945): 325–49.

———. 'The Origin of William Gilbert's Scientific Method'. *Journal of the History of Ideas* 2 (1941): 1–32.

———. 'The Sociological Roots of Science'. *American Journal of Sociology* 47 (1942): 544–62.

Index

by Ian Craine

Locations for illustrations within the text are entered in *italics*.